ENERGY CONVERSION SYSTEMS

Harry A. Sorensen
Washington State University

John Wiley & Sons

New York Chichester Brisbane Toronto Singapore

Library of Congress Cataloging in Publication Data

Sorensen, Harry A.
Energy conversion systems.

Includes indexes.
1. Power (Mechanics) 2. Power-plants.
I. Title.
TJ163.9.S67 1983 621.4 82-20076
ISBN 0-471-08872-2

Printed in the United States of America

12 13 14 15 16 17 18 19 20

PREFACE

From the study of thermodynamics, fluid mechanics, and heat transfer, the engineering student acquires the basic concepts of energy transfer and energy conversion. At this point however, he usually has only limited comprehension of the physical systems in which the conversion of energy is accomplished. The need to enhance the student's understanding of energy conversion systems becomes clear in the light of the concentrated effort currently directed to conservation of energy and development of nonconventional energy sources.

This text was prepared with the expectation of achieving a comprehensive treatment of the systems, both conventional and nonconventional, that convert primary forms of energy into usable mechanical and electrical energy. The operating principles of the energy conversion system are established, and an examination of the design and construction of the equipment serves as a means to relating the theoretical concepts to the physical system. Finally, the utility of the energy conversion system is investigated in the light of the performance, operating difficulties, and particularly the anticipated operating economy.

The introductory chapter on power generation precedes the chapter that deals with commercial fuels. Power generation systems are presently highly dependent upon the consumption of fossil and nuclear fuels. An understanding of the availability and the physical and chemical characteristics of fuels is important to the selection of the most appropriate fuel for a given system.

A comprehensive treatment of energy conversion must necessarily include an examination of the machines that consume high-level energy. The text thus includes sections that deal with cooling systems and with various machines used for moving and pumping gases and liquids. Large amounts of energy, usually electrical energy, are consumed by these machines, hence their proper selection and operation must necessarily be observed.

Steam engines on a few occasions are used for industrial power applications. Within the past few years, a modest investigation has been directed to the application of the steam engine to highway vehicles. While the steam engine is not likely to regain its former position of importance, it is nevertheless a basic energy converter, and a limited examination of this machine is in order.

A substantial portion of the book relates to internal combustion and gas turbine engines and to fossil fuel-fired and nuclear power plants. These power plants, together with the hydro power plants, produce virtually all the mechanical and electrical energy consumed by the domestic and commercial sectors of the economy. Because these power plants will for many years remain the major power producers, they are examined in considerable detail. Reliability, high performance, and good economy are emphasized, while attention is directed to the restrictions imposed by environmental control regulations.

In the final two chapters, nonconventional energy conversion systems are examined, and potential commercial applications are investigated. The principal problems of low-level performance, low power-producing capability, and poor economy are noted. It is particularly significant to observe whether any of these energy converters might in time serve to alleviate the heavy dependence upon petroleum-based fuels or contribute to an improvement in the environment.

The subject matter included in the text is arranged to allow considerable flexibility in developing a particular course sequence. Various sections and chapters can be omitted without incurring a lack of continuity.

The text material was prepared with the anticipation that the reader is grounded in thermodynamics and fluid mechanics and has some understanding of the basic concepts of heat transfer.

The International System of Units (SI) is employed throughout this text. English units appear occasionally in order to introduce a familiar or commonly used quantity. Some energy resources are reported in English units in order to avoid altering the original reference.

Engineering and scientific publications were important sources of information that contributed to the development of this textbook. Numerous industrial companies and various organizations generously provided illustrations and technical data for which I am most appreciative. I should like to thank the members of the Mechanical Engineering Department at Washington State University for their many helpful suggestions. I am particularly indebted to Richard W. Crain, Jr., who read portions of the manuscript, and to James S. Englund for his contributions to the section on solar energy collection.

The manuscript for this text was reviewed by Thomas L. Eddy, Georgia Institute of Technology; James W. Leach, North Carolina State University; Michael Moran, Ohio State University; Jerald D. Parker, Oklahoma State University; C. M. Simmang, Texas A&M University; and Wayne C. Turner, Oklahoma State University. The suggestions and comments that arose from the review were of substantial assistance in the development of the manuscript. I am indeed pleased to take this opportunity to thank the reviewers for their contributions to my effort to treat the subject of energy conversion engineering.

Pullman, Washington *H.A.S.*

CONTENTS

ENERGY CONVERSION SYSTEMS

CHAPTER ONE

INTRODUCTION TO POWER GENERATION

For many years, the inhabitants of the world have had at their disposal abundant sources of energy that could be utilized at low cost in many ways beneficial to mankind. In some respects, energy has been used wastefully, and in the more affluent nations high energy consumption is now judged to be somewhat irresponsible. It has become painfully apparent in the past few years that nonrenewable energy sources are finite and are in danger of depletion, to various degrees, in the not too distant future. Conservation and efficient use of energy must be observed in order to ensure a strong and stable world economy.

Solar and geothermal energy can be used directly for heating. Other energy sources are not directly usable, hence some kind of conversion process must be employed to change the energy to a different form, that is, to one of direct utility. These highly important energy conversion processes produce thermal energy and generate power. An examination of the various energy conversion systems will disclose the economic and practical limitations imposed on their use and indicate the effectiveness that may be anticipated in achieving the energy conversion.

1.1 THE NEED FOR POWER

Industrial development and improvement in our way of life are highly dependent upon an abundant supply of inexpensive energy. A human being is a comparatively feeble power plant. The muscular power of man is about 70 W when working at a normal rate, with higher power ratings possible over short periods of time. Dependence upon labor force alone would place any nation in the class of a most primitive society. Energy may be regarded as a multiplying factor that greatly enhances man's ability to fashion resource materials into useful products and provide a wide variety of essential services.

The total energy consumed annually in a country is a measure of the level of the national economy. Figure 1.1 shows the increase in the total energy consumption that has occurred over the past five to six decades in the United States. The general upward trend in the consumption of energy was interrupted during the depression years of the 1930 decade and in the early years of the 1970 decade, when energy conservation measures were adopted.

The sustained increase in the production of electrical energy in the United States is shown in Fig. 1.2. For several decades, the production of electrical energy has essentially doubled every 10 years. Substantial increases in the consumption of electrical energy have occurred in all sectors of the economy—domestic, commercial, and industrial—and indicate a continuous improvement in the productivity of the nation.

A brief survey is made in this chapter of the various types of power plants as to their development, applications, and probable future status. In later chapters, a detailed examination and analysis is presented for each type of power plant. Today, environmental restrictions influence to a substantial degree the design, location, and operation of virtually all power plants.

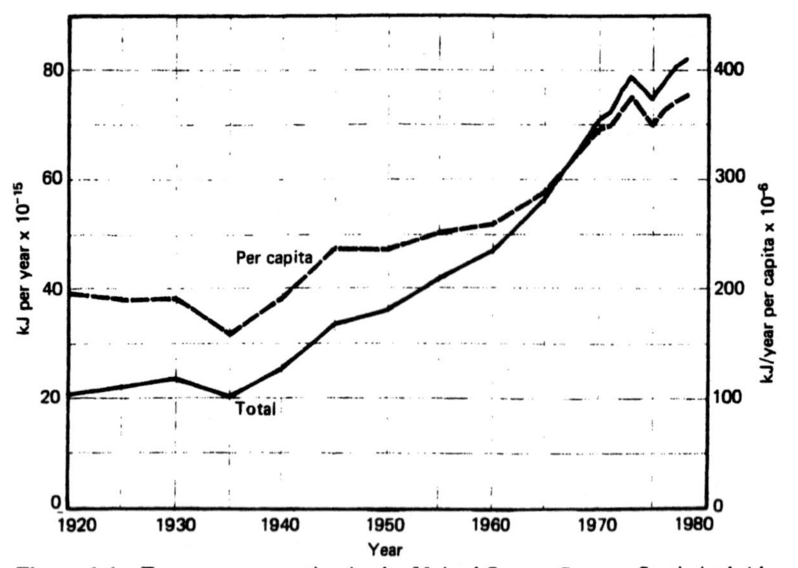

Figure 1.1 Energy consumption in the United States. ***Source:*** Statistical Abstract of the United States, U.S. Dept. of Commerce, Bureau of the Census; and U.S. Dept. of Energy, Electric Power Statistics.

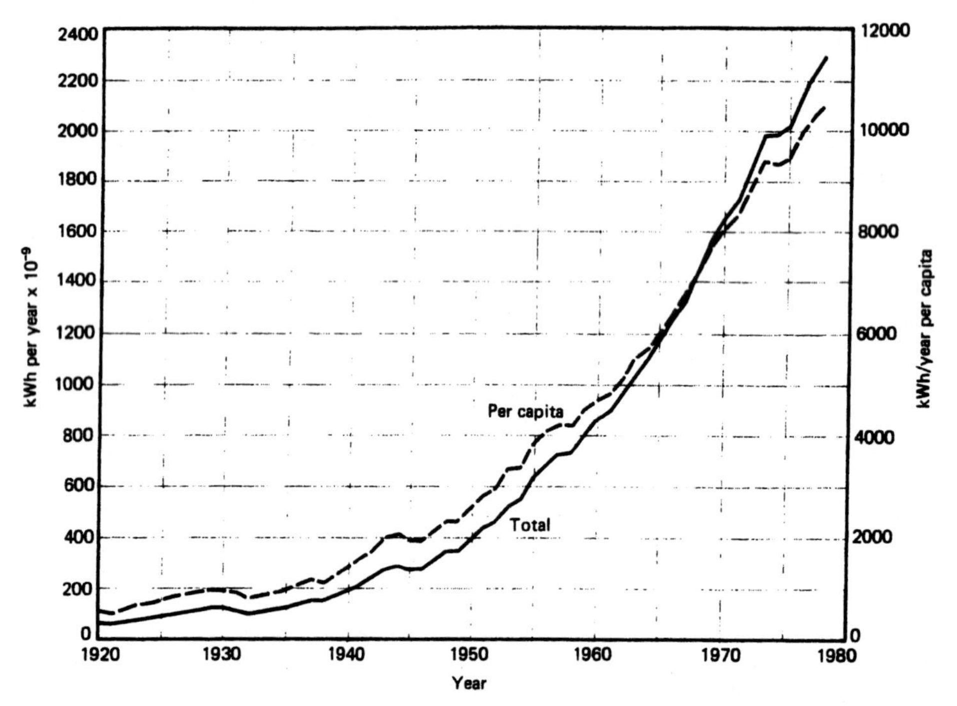

Figure 1.2 Net production of electrical energy in the United States (utility and industrial generating plants). Alaska and Hawaii included in 1959 and subsequent years. ***Source:*** Historical Statistics of the United States, Colonial Times to 1970; U.S. Dept. of Commerce, Bureau of the Census; and the U.S. Dept. of Energy, Electrical Power Statistics.

The survey will cover mainly the power systems that have acquired commercial status. Later sections of the text will explore possible applications of other energy conversion devices presently on test or under development.

Table 1.1 presents a classification of energy conversion systems used in power production. A small number of systems of minor importance are not included in Table 1.1. For some of the systems, a subdivision would be employed to designate a fossil or nuclear fuel or other source of energy.

TABLE 1.1 Power Generation Systems

- I Thermal
  - (a) Steam (vapor) power plants
    - 1 Reciprocating engine
    - 2 Turbine
  - (b) Internal combustion engine—reciprocating
  - (c) External combustion engine—reciprocating
  - (d) Gas turbine engine
- II Hydroelectric
  - (a) Water source: rivers
  - (b) Water source: tides
- III Wind
- IV Direct energy conversion
  - (a) Fuel cell (chemical)
  - (b) Photovoltaic
  - (c) Thermal
    - 1 Magnetohydrodynamic
    - 2 Thermionic
    - 3 Thermoelectric

In general, commercial- and industrial-type power plants are designed and operated so as to conform to the economic principle of achieving an acceptable return on the investment. Stated somewhat differently, the total cost, that is, the capital and operating costs, for producing power should be in a competitive range.

Economy however may not be the primary consideration for power plants constructed for special applications. For example, a power plant installed in a space vehicle is required to meet certain criteria that are paramount to economy.

1.2 WIND POWER

Currently, the amount of power produced by wind is negligible. The first power plants developed by man were relatively feeble water wheels and windmills that utilized the naturally occurring sources of energy available in moving masses of water and wind. Early applications of these energy conversion machines were for grinding grain and pumping water. As might be expected, there are only incomplete records of these early power developments, since they were constructed so many years ago. Despite worldwide use of wind power plants throughout many centuries, the development of these machines was generally of limited scope and confined to low-output installations.

The operating principle of the windmill is comparatively simple, and workable designs have resulted from trial-and-error procedures. Application of engineering concepts, including aerodynamic theory, to the design of wind power plants will however improve the performance of the machines. Air undergoes a change in momentum as it flows across the rotor vanes. The resulting force acting on the vanes causes the rotor to turn against an external, resisting torque. Work is performed on the rotor at the expense of the kinetic energy in the moving air.

In 1941, a wind turbine-generator rated at 1250 kW was installed on a peak in the Green Mountains of Vermont. The power plant was operated intermittently until 1945, when it was removed from service. The rated output of this turbine-generator was developed at wind velocities of 48 km/h or higher. Meteorologic records indicated that velocities of this magnitude could be expected to prevail for about 50 percent of the time during the year. Mechanically, a wind turbine-generator of this kind can be expected to perform satisfactorily. The feasibility of utilizing the energy in wind for power generation centers principally on the economy of the operation.

The interest in wind power generation has been revived, and numerous wind turbine-generators are now operating in the United States. While many of the units are small, the rated capability of the experimental machines extends over a wide range, up to 3 MW.

1.3 WATER POWER

Since the early colonial days, water has been used in this country, first for the production of mechanical energy and later for the generation of electric power. For many years, numerous water power plants, scattered across the nation, supplied power for the operation of small mills and factories. Eventually, a number of these installations became uneconomical to operate in competition with central stations producing electrical energy on a large scale.

Today, utility and industrial companies and governmental agencies operate water power plants that range from a low to a very high generating capability. Within the past 40 years, the federal government has undertaken the construction of several hydroelectric power plants of exceptionally high generating capability. Very often, these water power projects have multipurpose objectives, namely, power development, flood control, recreation, improvement of navigation, and irrigation of adjacent farm land.

The first water power machines were simple paddle wheels. Vanes which project into the flowing water are fastened to the periphery of a wheel that is supported on a rotating, horizontal shaft. The paddle wheel is not particularly effective because only the kinetic energy of the flowing water can be utilized in the production of power.

Erection of dams provides a means to utilizing the potential energy of stored water. Water wheels of the undershot, breast, and overshot types, all horizontal-axis machines, were installed in the low-head plants that were constructed prior to the introduction of the hydraulic turbine. These three types of water wheels differ principally in the way in which the water is admitted to the wheel. For the undershot and breast types, water is directed to the buckets on the lower portion of the wheel, while for the overshot type, water is conveyed in a flume to the top of the wheel. The conversion efficiency for a water wheel is 0.70 for an undershot-type machine and about 0.85 for the breast and overshot types.

In the United States, development of the hydraulic turbine started about 1820. The modern hydraulic turbine is a closed structure consisting of a casing in which the impeller rotates under the control of the flow regulating mechanism (Fig. 1.3). The high-capability hydraulic turbines are vertical-axis machines. Design of the turbine impeller is influenced by the operating head. Propeller-type and runners of the mixed-flow design are used in turbines installed in low-head plants. The mixed-flow design is also characteristic of the medium-head plant. Turbines installed in moderately high-head plants are equipped with runners designed for inward flow through the reaction wheel. In all of these machines, the impeller runs submerged in the casing. Conversion efficiencies are 0.90 to 0.94 for reaction-type and 0.85 to 0.93 for propeller-type turbines.[1]

The bulb-type water turbine, developed in Europe on a large scale over the past decade, is particularly effective in producing power at river sites with a high rate of flow and a head ranging up to 18 m. The bulb turbine, a horizontal-axis machine, achieves a high operating efficiency because of the straight-through flow characteristics (Fig. 1.4). In comparison with conventional machines, bulb turbines are physically smaller, a factor that causes a reduction on the order of 10 percent in the overall capital cost of a water power project.

The Pelton or impulse turbine is installed in high-head water power plants. In this design, the water is accelerated in a nozzle and directed against the buckets fastened to the periphery of a wheel that rotates on a horizontal shaft (Fig. 1.5). The pressure in the casing is essentially atmospheric; and in the vicinity of the nozzle, the casing is constructed so as to allow free discharge of the water from the buckets. The conversion efficiency for an impulse hydraulic turbine is 0.85 to 0.93.[1]

1.4 TIDAL POWER

The proposal to utilize tidal currents for large-scale power generation has, for several decades, been supported by a number of individuals who visualize an inexpensive source of electrical energy. Unlike the conventional hydroelectric power plant, constructed above sea level on a body of

Figure 1.3 Hydraulic turbine-generator, 1000-MW class, Francis-type runner. (Photo courtesy of Allis-Chalmers Corporation.)

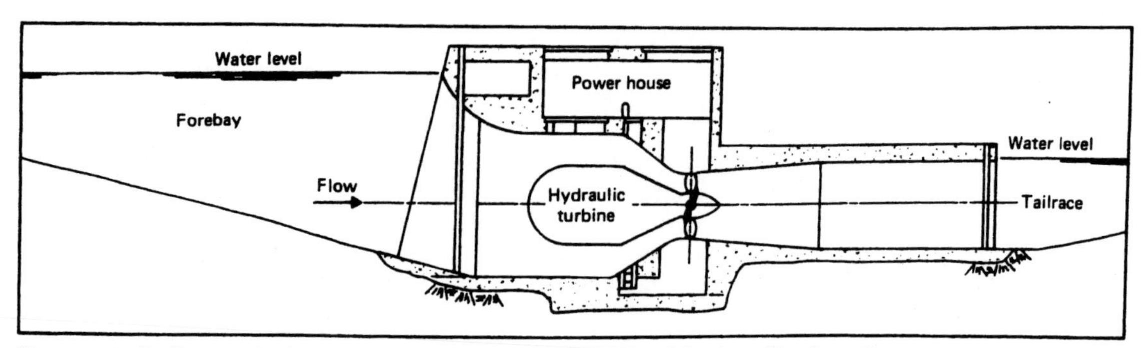

Figure 1.4 Bulb-type hydraulic turbine-generator. (Courtesy Morrison-Knudsen Co.)

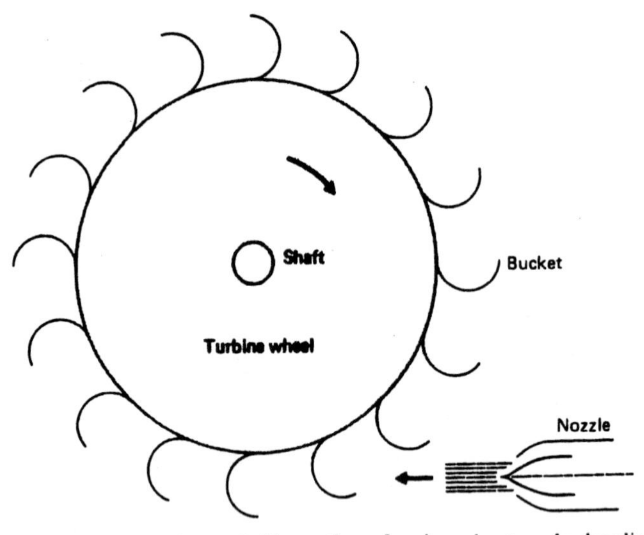

Figure 1.5 Schematic illustration of an impulse-type hydraulic turbine.

water or river, where the flow can be regulated, the tidal power plant is subjected to periods of slack water, reversed currents, and high and low water levels. Continuous output from a tidal power plant can only be achieved by providing within the system some kind of facility for storing energy.

In comparison with conventional thermal and hydro power plants, the capital cost for a tidal power plant is likely to be high, and usually most designs prove to be uneconomical. During the 1930 decade, limited backing was obtained for construction of a tidal power plant on Passamaquoddy Bay, located between Maine and New Brunswick. Some preliminary work on the project was undertaken, but the project was ultimately abandoned because of a number of objections that were raised, principally that of poor economy. In addition, the location of the power plant was remote from the probable market for the power to be produced by the plant.

The Passamaquoddy Bay power project may be revived in the light of the present endeavor to develop alternative power sources. Throughout the world, there are at least 50 other developed or potential tidal power plant sites, notably a tidal power installation in France and another in Holland that generate electrical energy on a commercial scale. Under favorable economic and technologic conditions, development of additional sites may be anticipated within the next two to three decades.

1.5 STEAM POWER

Steam, as a working substance, was introduced probably about the year 200 B.C. Several devices operated by steam were constructed, one of them the well-known reaction turbine attributed to Hero. Some of these early machines performed no useful work, while others operated simple mechanical devices that were scarcely more than novelties.

After many years of relatively unsuccessful development, the steam engine became an important source of power in the latter part of the eighteenth century. Subsequently, the steam power industry has been highly successful in maintaining a continuous program of development and expansion. The steam engine was extensively improved but was later replaced by other power machines, principally the steam turbine.

Steam-producing equipment has advanced from the simple boiler to the highly complex and very large steam generator.

The first practical "steam engine" was developed by Thomas Newcomen in 1705. Several "engines" had been built prior to the construction of Newcomen's machine, but operating difficulties precluded industrial applications. The machine developed by Newcomen was actually a steam-operated pump used for mine drainage. Flooding of the lower levels of the deep coal mines in England had by this time become a serious problem, one that required an early, economical solution. Newcomen's pump was fairly successful with respect to mechanical performance, but the operating economy was poor. The use of water jets to achieve condensation of the steam inside the cylinder caused the steam consumption to be prohibitive for many applications.

James Watt acquired a Newcomen machine in the year 1764 and began a series of experiments that 20 years later produced a steam engine capable of economical power generation. Watt and his business partner, Matthew Boulton, made many improvements on the engine, including double-acting operation, expansion of the steam in the cylinder, external condensation of the exhaust steam, and transformation of the reciprocating motion of the piston to the rotary motion of the crankshaft.

Watt and his contemporary workers, and later other investigators, further modified and improved the steam engine until it reached a high degree of mechanical perfection. The steam engine was widely used for the generation of electrical energy and as a power source for pumping water, driving machinery, propelling ships, and moving trains. Some of these engines were very large and consisted of several stages of expansion, though all turned at comparatively slow speeds. Currently, a number of investigators are engaged in the development of a steam engine suitable for application to highway vehicles.

The higher-speed steam turbine was introduced toward the end of the nineteenth century

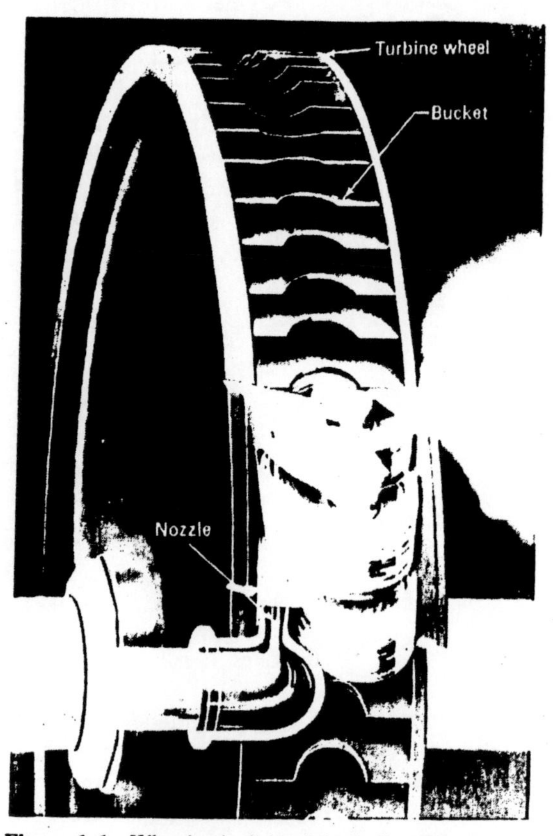

Figure 1.6 Wheel principle of the helical-flow steam turbine. (Courtesy Terry Corporation—Steam Turbine Division.)

(Fig. 1.6). The turbine, with its generally superior characteristics, soon began to replace the steam engine, particularly for driving electric generators. The turbine, similarly to the steam engine, owes its existence to the efforts of a large number of engineers. Early experimental work was started about the year 1830, but it was not until the period extending from 1880 to 1900 that a vigorous program of development was instituted.

The electric power industry originated in the small generating companies that produced electrical energy for street, home, and commercial lighting. The Pearl Street station of the New York Edison Company was the first significant central

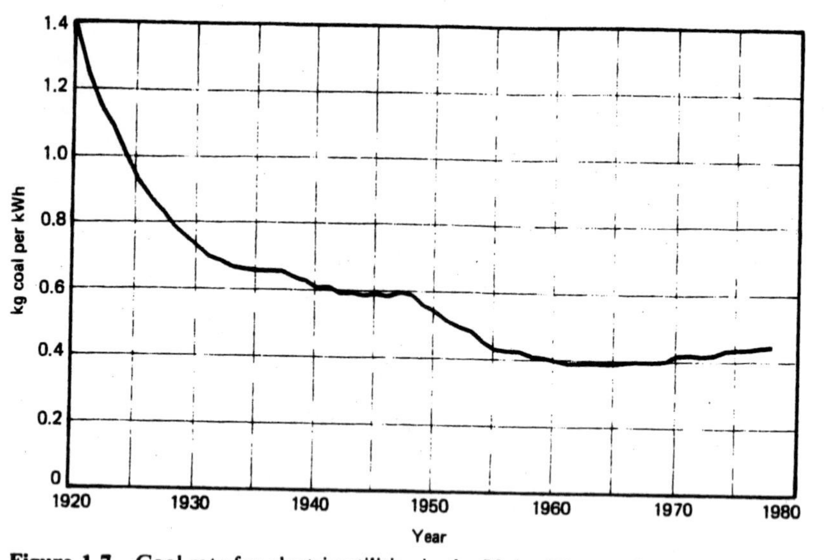

Figure 1.7 Coal rate for electric utilities in the United States. ***Source:*** Statistical Abstract of the United States, U.S. Dept. of Commerce, Bureau of the Census; and U.S. Dept. of Energy, Electric Power Statistics.

electric power plant erected in the United States. The station was constructed under the direction of Thomas Edison and began operation in the year 1882. Many new types of equipment were installed in this early power station. Subsequently, over the past 100 years, improvements in design and operating procedures have contributed to essentially a continuous increase in the thermal efficiency of the steam-electric power plant. The improved thermal performance is observed in the downward trend of the annual average coal rate for the steam-electric generating stations (Fig. 1.7).

1.6 INTERNAL COMBUSTION ENGINES

Following a number of unsuccessful experiments, the first internal combustion engine of commercial significance was constructed circa 1876. The early engines were of the flame-ignition or spark-ignition type. Development of this power plant has produced the modern lightweight, efficient "gasoline" engine that has extensive applications in the transportation field. Principal installations are in motor vehicles, aircraft, boats, light construction machinery, and miscellaneous small power plants.

Toward the end of the nineteenth century, Rudolf Diesel introduced into the power field the compression ignition internal combustion engine that subsequently became a highly important source of power. It is of interest to note that Dr. Diesel developed this engine following an unsuccessful attempt to construct an engine capable of operating on powdered coal. The modern, high-compression Diesel engine operates on a petroleum-based fuel of about the same specific gravity as kerosene. In general, the heavy, slow-speed Diesel engines are used for stationary power generation and marine propulsion. Medium- and high-speed engines are employed in motor vehicles and locomotives and in industrial machines of a great variety.

1.7 GAS TURBINE ENGINES

The first gas turbine power plant was constructed and operated in Paris in the year 1903. Owing principally to limitations imposed by the low ef-

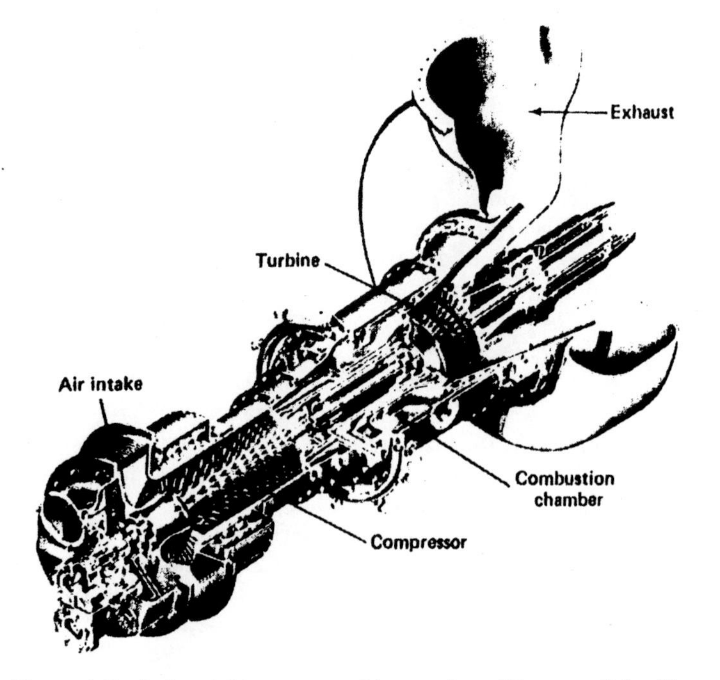

Figure 1.8 Industrial-type gas turbine engine. (Courtesy Solar Turbines Incorporated.)

ficiency of the compressor and the inability of available materials to withstand the effects of high temperature, this machine was unable to develop a useful output. As a result of technologic advances, these two principal limitations were in time largely eliminated, and the development of the gas turbine engine moved forward at a rapid pace.

The inherent characteristics of high speed and high power output promote the use of the gas turbine engine in a wide variety of applications (Fig. 1.8). Initially, the most rapid and intensive development of the gas turbine engine was in the turbojet category. The turbojet engine has replaced the reciprocating internal combustion engine for propelling large- and medium-sized aircraft. Currently, the relatively new gas turbine engine is used, in addition to propelling aircraft, to generate electrical energy, propel ships, and drive industrial machines. Contrary to the early optimistic predictions, the gas turbine engine, for a number of reasons, has not penetrated the automotive field, but investigative work in this area will continue.

1.8 NUCLEAR POWER

After fission of the uranium atom by Otto Hahn et al. (circa 1939) power generation through a controlled fission reaction was recognized as a distinct possibility. Many different investigations were required prior to the design of a nuclear power plant. Some of these studies involved the effect of radiation on the properties of structural materials, heat transfer from the reactor core, fluid flow within the core, and protection against harmful radiation.

Present-day nuclear power plants are thermal engines. Energy released in the reactor core is transferred as heat directly to the working fluid, or to an intermediate fluid and then to the working fluid. The working fluid is usually water, while liquid sodium, pressurized water, or helium gas is employed as the intermediate fluid.

Several basic designs have been proposed for the nuclear power plant. In the United States, almost all the commercial nuclear power plants are equipped with pressurized water or boiling water reactors. Water, the working fluid for both types of plants, is expanded as steam in a conventional turbine-generator.

The breeder reactor, which produces as much, or more, fuel than it consumes, has received considerable attention because of its superior fuel-burning characteristics. Adoption of the breeder reactor as a principal source of power in the United States will be dependent, in part, upon the experience derived from the operation of a demonstration plant currently in the design and early construction stage. Strong opposition to the construction of breeder-type nuclear power plants has however developed because of the high capital cost and the potential hazards that exist in the use and handling of plutonium.

1.9 ELECTRIC POWER STATIONS

Many factors are considered in planning an expansion of the generating capability of a power system. The type of power plant selected for construction should, in general, be one that produces electrical energy at a minimum cost. There are however often sound reasons for departing from this principle. Environmental constraints, an uncertain fuel supply, and probable changes occurring in the system power demand are some of the reasons that can be cited. Base load power generation is usually provided by high-capability hydroelectric and fossil-fuel and nuclear steam-electric power plants.

In the United States, most of the electrical energy is generated in steam and hydro power stations. Currently, the energy output ratio is approximately 5 to 1 for thermal and hydro power plants, respectively. As the demand for power increases, it will be necessary to construct many additional steam stations and to develop some of the remaining available water power sites. Certain of these undeveloped sites are not particularly appropriate for construction of a power plant because of the high capital cost that will result from the erection of long transmission lines, environmental restrictions, acquisition of land, and extensive construction involving the dam and the power house.

A steam power plant requires an ample supply of cooling water and adequate transportation facilities for supplying fuel. Transportation of fuel is usually achieved without major difficulty, but unrestricted use of water for thermal power plant cooling is in most locations no longer possible, hence air cooling may be required. The problems that develop with air cooling are discussed in a later chapter.

A combination of steam and hydro power plants provides an opportunity to arrange a flexible operating schedule that can improve the system performance. When the stream flow is high, more of the system load can be shifted to the hydro stations; and during periods of low runoff, the steam stations can carry a greater part of the load. In addition, the water turbine can be operated as a peak load machine, because it is readily started and stopped and there are no standby losses.

The gas turbine power plant is not, as yet, a primary power-generating unit for central station operation. In the United States, gas turbine generating sets have been installed for peaking or auxiliary service in main power plants, or enclosed in small shelters strategically located at various points in the system. The gas turbine power plant is particularly well adapted to this type of service. Delivery and erection time is comparatively short, and the machine can be started "cold" and quickly loaded. To a limited extent, the gas turbine engine has been integrated in the combined cycle with the steam power plant.

The continued increase in the demand for electric power must be met largely through expansion of steam power-generating facilities. Some improvement in station thermal efficiencies can be anticipated in the event that steam pressures and temperatures are increased and larger machines are constructed, but such gains are likely to be modest. A significant increase in the thermal efficiency of the electric generating station is expected with the introduction of an advanced-type combined-cycle power plant. Improved thermal performance of nuclear power plants will be achieved with the development of plant designs that effect an increase in the operating pressure and temperature of the steam cycle.

The direct conversion of solar energy into electrical energy has been accomplished with good results for small-scale power units used for special applications. Commercial-scale photovoltaic power production is currently not feasible because of the high capital cost. The prospects for utilizing this power source will improve with a substantial reduction in the manufacturing cost of the solar cell.

During the next two decades, the number of experimental and demonstration solar-thermal, wind, and geothermal power plants will be constructed. Subsequently, construction of some commercial-scale plants of the three types can be anticipated, provided that these plants are capable of producing electrical energy at a competitive cost. The contribution by these nonconventional power sources to the production of electrical energy subsequent to the year 2000 is a matter of conjecture.

Although there are a few exceptions, power plants are subject to environmental constraints that influence, to a marked degree, both the construction and the operation of the plant. Because of air, water, and land-use control regulations, the acquisition of a site for an electric power plant has become, in most cases, a difficult undertaking. Permission to construct and operate a power plant requires review of the proposal by numerous regulatory commissions; and as a result, construction schedules have, on the average, lengthened substantially. Power plant capital costs have increased materially because of extended construction schedules, installation of pollution control equipment, and the restrictions on the use of once-through cooling.

1.10 FORECAST OF THE ENERGY DEMAND FOR THE UNITED STATES

The present rate of energy consumption in the United States is about 78 q per year; and various studies estimate that in the year 2000, the rate will be about 200 q per year. (***Note:***[1] q = 10^{15} Btu.) An increase in the energy supply over a period of about 20 years from 78 q to 200 q per year represents a formidable undertaking; thus, it is important to identify the energy sources that may be utilized to achieve an increase of this magnitude.

G. L. Decker[2] estimates for the year 2000 an energy demand in the United States of 195 q. This figure corresponds to an average growth rate from 1972 to 2000 of about 3.5 percent per year, that is lower than the recent historical rate of better than 4 percent per year. A business-as-usual climate is assumed, but some reduction is allowed because of higher energy costs. Table 1.2 indicates the technologies that may provide the energy the United States will probably require in the year 2000. The new technologies expected to be in effect in the year 2000 are examined in Ref. 2.

Prediction of the energy demand for the year 2000 or later is based essentially on the projection of the energy consumption for the past several decades. Unforeseen circumstances or events could however alter to a substantial degree the established trends in the consumption of the several forms of primary energy, hence the predicted demands for energy may not materialize.

Estimates of the contributions from new sources of energy are often made unrealistically because of the failure to recognize fully the length of time required to complete the development of an energy conversion process. After the successful demonstration of a process in the laboratory, as many as 25 years or perhaps more can pass before there is implementation of the process on a commercial scale.

TABLE 1.2 Energy Supplied or Saved by Various Technologies (United States)

	Energy (q/year)[a]		
	1972	*2000*	
Domestic oil or gas	45	28	↑
Coal	13	14	Present
Nuclear	1	13	technology[b]
Hydro	3	5	↓
Imports	12	0	
Improved recovery (oil and gas fields)		6	
New coal technology		40	
Improved nuclear technology		31	
Other probable and possible sources		27	
Conservation and higher efficiencies		31	
Total	74	195	

Source: G. L. Decker.[2]

[a] $q = 10^{15}$ Btu.

[b] 60 q in 2000.

1.11 THE COST OF CONVERTING ENERGY

In general, there are two different kinds of costs incurred in the conversion of energy, namely, the dollar cost and the energy cost. The monetary cost consists of the fixed charges that, in part, account for the system capital cost and the operating costs, principally for fuel, labor, maintenance, and supplies. The energy costs are placed in three categories, namely, the conversion loss, the energy expended in processing or manufacturing a fuel, and the energy consumed in the construction of the physical plant. Each of the energy costs will influence the overall conversion efficiency of the system.

Losses are encountered in converting energy from one form to another, for example, the conversion of chemical energy in coal to electrical energy in a steam power plant. Because of the second law limitation, irreversible processes, and combustion and heat losses, only 35 to 40 percent of the energy in the coal is converted to electrical energy, while the remainder represents the conversion loss.

In the conversion of coal to a liquid fuel, the chemical energy of the product fuel will perhaps equal one-half of the chemical energy of the coal. The conversion loss would be acceptable, because the liquid fuel has far greater utility in transportation service than coal has. Similarly, electrical energy has utility in many applications that coal alone cannot fulfill.

The energy conversion effectiveness of a thermal power system is ordinarily evaluated on the basis of the energy in the fuel as supplied to the energy converter. It is important to note however that there has been a prior expenditure of energy, or an energy cost, in extracting, processing, and transporting the fuel. For example, the energy expended in mining coal and transporting the coal to the power plant is typically equal to about 1.4 percent of the energy originally in the coal.[3] Comparable figures for a light-water nuclear power plant and an oil-fired power plant are, respectively, 5.0 and 21 percent of the energy originally in the fuel. The high figure for oil firing is attributed mainly to the energy consumed in the refining process.

In the development of alternative energy sources, the input energy is a factor that must be realistically observed. This is particularly true for ethanol, which apparently will have expanded use as an engine fuel. The energy expended in the manufacture of ethanol through advanced technology is predicted to be equal to about 50 percent of the energy value of the final product. A figure of 100 percent or higher is indicated for current practice.

The energy consumed in the construction of a power plant or an energy conversion device is a third energy cost. For a coal-fired steam power plant, the energy consumed in constructing the power plant and the electrical transmission system is equivalent to about 0.7 percent of the original energy in the coal burned in the plant over a 30-year period.[3] A comparable figure of

TABLE 1.3 Conversion Efficiency for Selected Energy Cycles

Cycle	*Gross Energy Input [kWh(t)/yr ÷ 10⁹]*	*Construction Energy [kWh(e) ÷ 10⁹]*	*Net Cycle Efficiency (Percent)*
Oil energy	26.1	1.4	26.6
Coal energy	21.4	1.4	32.5
Gas from coal	39.3	3.1	17.5
Liquid fuel from coal[a]	35.5	3.2	19.4
Light-water reactor	27.8	2.5	24.9
Breeder reactor	20.0	2.6	34.6
Fusion reactor	27.0	3.1	25.5
Solar collector[b]	209	58	2.4
Solar cell[b]	70.0	335	Negative

Source: S. Baron.[3]

[a] *Mine mouth plant (solvent-refined coal).*

[b] *Includes pumped water storage and electric generation.*

1.2 percent is given for a light-water nuclear power plant. The total energy required for processing the raw material and manufacturing silicon solar cells is high, roughly equivalent to the electrical energy produced by the cell over a period of 40 to 45 years.

Table 1.3 shows pertinent data for several energy cycles. The electrical energy output for each cycle is 7.0×10^9 kWh/year. The difference between the original energy in the fuel (the gross energy input) and the energy in the fuel, as supplied to the plant, is a comparatively small amount. Conversion losses thus account for most of the difference between the gross energy input and the electrical energy output. The electrical energy consumed in fabricating and constructing the plants is charged against the electrical energy output over a 30-year period.

The net cycle efficiency is given by

$$\eta = \frac{7.0 \times 10^9 - \{\text{construction energy [kWh(e)]}/30\}}{\text{gross energy input [kWh(t)/year]}}$$

Fossil fuel-fired and nuclear power plants have the highest net efficiency for the energy cycle, while solar power plants show poor performance with respect to overall energy utilization. Pumped water storage is prescribed for the solar power plants in order to avoid intermittent operation. Elimination of energy storage would increase the net cycle efficiency of the solar collector system to 3.2 percent but would not eliminate the negative value for the solar cell power plant. A substantial energy penalty is incurred through the use of substitute gaseous and liquid fuels.

Dollar costs for power production are difficult to predict because of the sharply escalating capital cost for power plants and the continuous increase in the cost of fuel. For some of the power generation systems, improved technology may however reduce the severity of the anticipated increase in the costs for fuel and plant construction.

1.12 POWER PLANT LOAD

The power demand on an electric generating station may vary widely during a continuous 24-h time interval, usually prescribed from midnight to midnight. The output of the station must meet this demand, and a continuous record is maintained of the power produced during the day. Figure 1.9 shows a typical daily load curve for a utility electric generating station.

The shape of the daily load curve for a particular utility power plant or system will depend

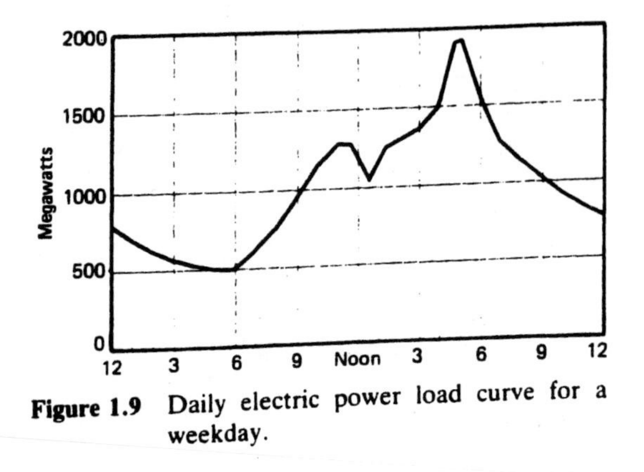

Figure 1.9 Daily electric power load curve for a weekday.

upon the character of the load, the time of the week, and the time of the year. The load components are classified as industrial, commercial, residential, street lighting, and traction. Each of these components consumes power in a different manner with respect to the end use and the time of day. The traction load for operation of electrified rail lines is limited principally to a few large cities and their surroundings and to the New York-to-Washington corridor.

Street lighting produces a block-type load with all of the lights turned on and off at about the same time. In the winter, a relatively heavy lighting load is extended into the late afternoon hours. In some sections of the country, the summertime air conditioning load is very heavy, particularly during the afternoon and early evening hours. Generally, the industrial load exists to some degree over the 24-h period, but the heaviest industrial demand is likely to occur during the daytime working hours.

Industrial-type power plants are operated differently from the utility power plants, and the form of the load curve will be influenced by the electrical and thermal energy requirements for the particular industry. Considerable variation can be observed in these requirements because of the diverse nature of the industrial sector.

The load on an electrical power system is apportioned among the power plants that comprise the system. The generating capability of the system must be sufficient to meet the peak or maximum demand of the year, with some reserve generation to provide for possible failure of equipment and outage for regular maintenance. The spinning reserve consists of those machines synchronized with the electrical network and ready for loading. Other machines held in reserve are not ready immediately to take load but can be made ready for loading following a proper start-up procedure. The flexibility in operating a system will improve with an increase in the number of machines installed in the power plants.

An examination of the daily load curve, Figure 1.9, will show that there can be considerable unused generating capability during a substantial portion of the day. Thus, a wide difference between the average and peak loads is usually observed. It is considered to be good practice to attempt to level off the load curve, that is, reduce the peak load, if possible, and increase the off-peak power consumption. Load adjustment is facilitated by breaking down the total load into the various classes noted above. A study of each class is made in order to achieve an improved load distribution with respect to time. Then, by offering reduced rates for certain off-peak power consumption, or through other inducements, it may be possible to reduce the peak load, build up the load during the low periods, and perhaps increase the total energy consumed per day to improve the operating economy of the system.

The load curve shown in Fig. 1.9 is intended to represent the weekday power demand. Reduced demand, particularly the peak demand, would be expected on Saturday and Sunday when the industrial and commercial loads drop. The area under the daily load curve is proportional to the energy produced during the 24-h period. The load curve may also be constructed for the average day of the month, or the average day of the year.

The load factor is defined as the ratio of the average load on the power plant during a prescribed period of time to the peak load occurring within that period. A high load factor represents good power plant operating practice. High peak loads over short periods of time are particularly undesirable because of the adverse effect on the load factor. From another point of view, a low load factor is indicative of excessive unused generating capability.

The most efficient machines in an electric power system are usually operated continuously at high loads. In Fig. 1.9, for example, the base load machines might carry the load up to about 750 MW. Higher loads in this case would be carried by the less efficient machines and machines installed for peak load operation. Loading schedules are intended in this scheme to achieve the most economical operation of the system.

The relationship between the connected load

and the maximum demand on the power plant or system is useful in determining the required generating capability of the power plant or system. The demand factor is defined as the ratio of the maximum demand to the total connected load. The total connected load would occur if all the connected apparatus in the system were to operate simultaneously at the "name plate" rating. Because such operation does not occur, the magnitude of the demand factor is less than 1, that is, ordinarily in the range of 0.4 to 0.6.

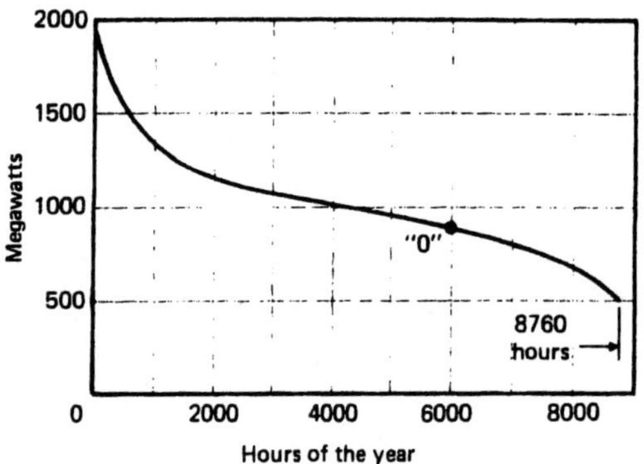

Figure 1.10 Electric power load duration curve.

Example 1.1

A power plant has an installed generating capability of 750 MW and a connected load of 1000 MW. On a particular day, the peak load is 620 MW, and the electrical energy produced during the 24-h period is 6240 MWh. Determine the load factor and the demand factor.

1 Load factor.

$$\text{Average load} = \frac{6240}{24} = 260 \text{ MW}$$

$$\text{Load factor} = \frac{\text{average load}}{\text{peak load}} = \frac{260}{620} = 0.419$$

2 Demand factor.

$$\text{Demand factor} = \frac{\text{maximum load}}{\text{connected load}}$$

$$= \frac{620}{1000} = 0.620$$

The load duration curve is a useful record of the rate at which electrical energy is produced during the year. Figure 1.10 is a typical power plant load duration curve. For this particular curve, the load does not drop below 500 MW at any time during the year, while the highest load will correspond with the maximum peak load of the year. Point 0 in this illustration indicates that the load is equal to or exceeds 900 MW during 6000 h of the year. The area under the load duration curve is proportional to the electrical energy generated during the year. The load duration curve indicates graphically the extent to which use is made of the power plant generating capability.

The annual plant factor is defined as the ratio of the electrical energy produced during the year to the product of the number of hours per year (usually 8760) and the total rated generating capability of all of the machines in the plant. A high annual plant factor is an important objective in power plant operation, because it indicates good use of the generating equipment. A daily plant factor may also be used.

Example 1.2

For the data given in Example 1.1, determine the daily plant factor.

Daily plant factor

$$= \frac{\text{electrical energy produced per day}}{(\text{hours per day}) \times \left(\begin{matrix}\text{rated generating} \\ \text{capability}\end{matrix}\right)}$$

$$= \frac{6240}{24 \times 750} = 0.347$$

The power output from a fuel-burning power plant is seldom restricted because of a short supply of fuel, the primary energy source. Hydro, solar, and wind power plants, on the other hand, may on occasion not attain the expected power level owing to subnormal river flow, sunshine, and air movement, respectively. Consequently, for such power plants a deficiency in the normal primary energy supply can cause a drop in the power output and thus effect a reduction in the annual plant factor.

1.13 INTRODUCTION TO POWER PLANT ECONOMICS

Selection of the type of power plant, location, fuel, and the principal station components is based in part on an economic study to determine the minimum cost of producing a unit of electrical energy. Studies of this kind are not confined to the power industry; however, applications of economic principles are particularly pertinent to power plant planning. The subsequent discussion demonstrates why economy is of paramount importance and why thermal efficiency can be less significant in the design and operation of a power plant.

The cost to generate electrical energy is divided into two main categories, namely, the fixed charges and the operating charges or production cost. The fixed charges are independent of the amount of electrical energy generated and must be carried whether the plant is operated. The production cost is somewhat proportional to the quantity of energy generated. Each of the two main charges is in turn subdivided as follows:

1 Fixed charges.
  - **(a)** Interest on the investment.
  - **(b)** Taxes.
  - **(c)** Insurance.
  - **(d)** Depreciation and obsolescence.
  - **(e)** Salaries (general).
  - **(f)** General maintenance.

2 Operating costs.
  - **(a)** Fuel.
  - **(b)** Supervision and labor.
  - **(c)** Maintenance (equipment and buildings).
  - **(d)** Supplies.

A brief discussion that omits accounting procedures will be presented for each of the items in the two main categories.

Sale of stock is a common procedure used to raise money for plant construction. The purchaser of the stock becomes a co-owner of the corporation and expects to be paid for the use of the invested capital. This payment is made through a dividend declared by the corporate directors, and under normal business conditions such dividends appear at regular intervals and constitute a fair return on the investment. This interest on the investment, or on the capital expenditure, will continue indefinitely, or until the business is liquidated.

Plant financing may also be achieved through the sale of bonds. Bonds are certificates of indebtedness and promise repayment of the loan on a specified date, with provision for payment of interest until maturity. Plant construction may also be financed through reinvestment of company earnings.

In common with other business establishments, power companies pay taxes to various governments—federal, state, and local. Insurance may be carried with an industrial insurance company and written to cover losses owing to fire, theft, power outage, equipment failure, and natural events of unusual severity.

Plant equipment may be made to operate indefinitely by proper maintenance, but eventually it becomes more economical to replace the unit in question when the cost of maintenance exceeds the interest on the capital expenditure for a new unit. The replacement is usually made at the end of an estimated useful life. A useful life of 30 years for power plant equipment and about 50 years for structures is commonly assumed. Some equipment may become obsolete before judged to be worn out and will be replaced by units that are, because of technologic development, more economical to operate. Whether replacements are made for one reason or the other, a portion of the revenue obtained from the sale of power must be set aside to cover this expenditure. In short, provision must be made for replacing equipment that ultimately deteriorates in order that management can show something of value for the capital expenditure for which it is responsible.

Certain salaries are paid whether a plant is in operation or not, and a minimum amount of general maintenance must likewise be provided.

These two items are small in comparison with the first four items. For utility power plants the annual fixed charges amount to approximately 15 to 20 percent of the investment. Higher percentages are often encountered for industrial plants where special problems peculiar to the industry must be considered.

For steam stations, the cost of fuel is a substantial portion of the operating costs. Salaries and wages for supervision and labor and expenditures for maintenance of equipment, buildings, and so on, constitute the remainder of the operating costs, except for an additional small amount to cover supplies.

In comparison with a utility steam station, a hydroelectric power plant of the same generating capability shows substantially lower operating costs, because no fuel is required and less labor is needed for operation of the plant. The total cost for generating a unit of electrical energy is dependent upon a large number of factors, hence no attempt will be made to compare power generation costs for the two types of plants.

An increase in the power plant thermal efficiency can be achieved through various improvements in the plant design that, however, raise the installed cost of the plant. An increase in the thermal performance of the plant follows the principle of diminishing return, requiring an ever-increasing capital expenditure for each unit gain in thermal efficiency (see Fig. 1.11).

Because the most efficient power plant is not necessarily the most economical generating station, the optimum design for any particular installation is the one that produces a unit of electrical energy at the lowest total cost, regardless of the thermal efficiency. The application of economic principles to the design and operation of a power plant is demonstrated by use of a simplified illustration. Figure 1.12 shows, for a specified generating capability, the fixed charges and the production cost plotted against the investment. The fixed charges are taken as a constant percentage of the capital expenditure; thus, a straight, sloping line represents the variation in this portion of the total cost. Because of improved thermal performance, the production cost decreases with increase in the investment, but such decrease occurs at a diminishing rate (Fig. 1.12). The combination of these two curves, fixed charges and production cost, produces the total cost curve.

Attention is first directed to full-load operation. Over the range of the investment cost, the income remains constant while the total cost of

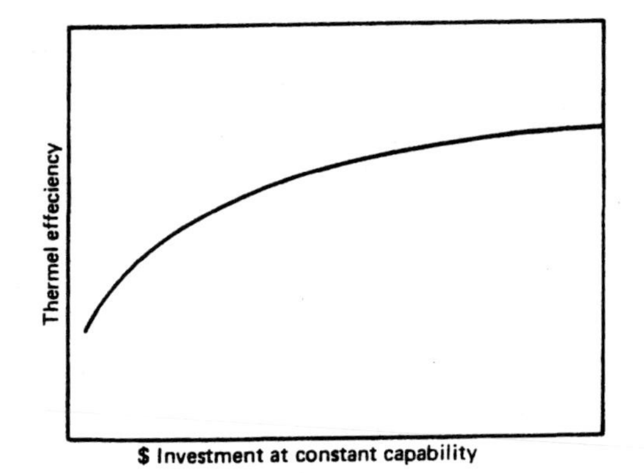

Figure 1.11 Cost of improved thermal performance.

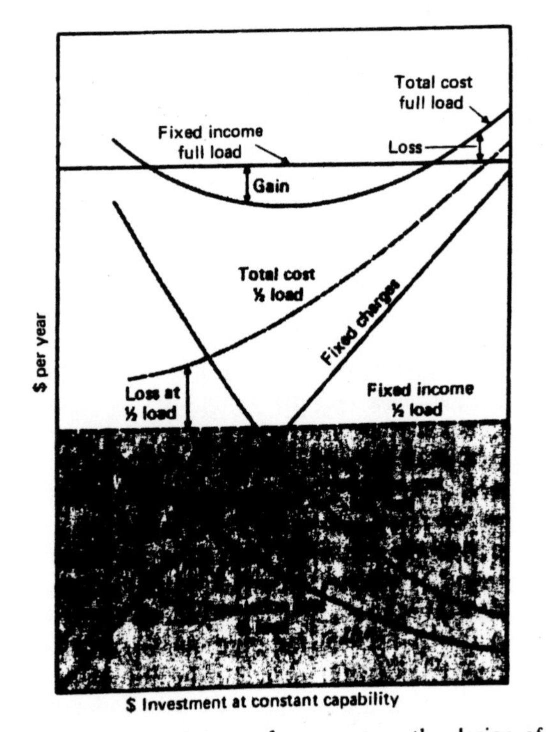

Figure 1.12 Influence of economy on the design of a power plant.

generating electrical energy initially decreases to a low point and then increases. It is thus possible to operate at a deficit, as shown, if the plant efficiency is too low or too high. The low point on the total cost curve indicates the optimum capital cost.

The effect of a decrease in the power plant load is also shown in Fig. 1.12. If the load drops to one-half the design load, the income is decreased by 50 percent, and the production cost is reduced by approximately the same amount. The result is to place the total cost of power generation well above the reduced income. A considerable reduction in load is thus accompanied by an operating deficit, because of the fixed charges that must be carried regardless of the magnitude of the load on the power plant.

1.14 SOME FINANCIAL CONSIDERATIONS

Depreciation, a particularly important part of power plant accounting, is defined as the annual charge, against the revenue, that is used to replace the plant equipment no longer useful, or for return of the original capital to the investors. The depreciation charge made over the useful life of a power plant unit should be equal to the total amount expended in acquiring the unit. Depreciation is employed principally for tax accounting. Tax laws may define the depreciation methods that are acceptable and prescribe the number of years for the useful life of the item.

The commonly used straight-line depreciation method, applied over the useful life of the plant item, provides for an equal annual reduction of the unrecovered investment in the item. There are other depreciation methods that may be employed. An accelerated form of depreciation allows the taxpayer to deduct, during the early years of the life of a plant item, depreciation expense at a higher rate than would be permissible using the straight-line method. In the later years of the life of the item, the reverse is observed, that is, the annual reduction under the straight-line method exceeds that under the accelerated form of depreciation. The total deduction is the same for all depreciation methods.

Money for replacement of equipment is made available usually by creating a sinking or amortization fund. A portion of the revenue is placed in the fund at a prescribed number of times during the year over the useful life of the plant item. The money transferred to the fund is invested in some manner in order to earn interest that contributes substantially to building up the fund.

The annual contribution to the sinking fund is given by

$$A = \frac{Si}{[(1 + i/n)^{ny} - 1]} \qquad (1.1)$$

where

A = the annual contribution to the depreciation reserve
S = the sinking fund required for replacement of equipment
n = the number of contributions, per year, of equal amount to the depreciation reserve
i = the annual rate of interest earned on the accumulated depreciation reserve, compounded n times per year
y = the useful life of the equipment, years

Example 1.3

A power plant machine has a useful life of 30 years and an estimated salvage value of $2500. The probable replacement cost of the machine is $100,000. An annual interest rate of 8 percent is earned on the reserve. Contributions to a sinking fund are made semiannually. Determine the annual payment to the sinking fund.

$$\text{Sinking fund} = (\text{cost of item}) - (\text{salvage value})$$
$$= \$100{,}000 - \$2500 = \$97{,}500$$

Applying Eq. 1.1, with $y = 30$, $n = 2$, $i = 0.08$,

$$A = \frac{Si}{[(1 + i/n)^{ny} - 1]}$$
$$= \frac{\$97500 \times 0.08}{[(1 + 0.08/2)^{60} - 1]} = \$819.36$$

Each semiannual payment = $409.68

Expansion of the generating capability of a power system requires major financing. Money is obtained from some source and is expended during construction. While construction is under way, interest must be paid on the borrowed capital despite the lack of any income from the sale of power, that is, from the plant being built. A protracted construction schedule can, because of this interest payment, place a serious financial burden on the power company.

Attention is occasionally directed to the ease of "paying back" the capital cost of installing power plants of various kinds and sizes. Very often, the cost of borrowing or using money has been ignored. Regardless of the source of the capital required for construction of the power unit, interest charges are very real and must be properly charged to the project.

1.15 AVAILABLE ENERGY

Various examples can be cited to show that equal quantities of two different kinds of energy do not have the same worth. The distinguishing characteristic in fixing the worth is the availability of the energy. Availability is a thermodynamic property that is indicative of the potential of a system to perform work under certain prescribed conditions.

The measure of availability is taken as the reversible work performed by a system that interacts with the infinite atmosphere as the system moves to the dead state at pressure P_0 and temperature T_0, the pressure and temperature of the infinite atmosphere, respectively. The reversible work W_{rev} for a steady-flow transformation from state 1 to state 2 is given by

$$W_{rev} = h_1 - h_2 - T_0(s_1 - s_2) \tag{1.2}$$

When the system moves to the dead state 0, the reversible work is designated as the availability ϕ. Thus,

$$\phi = h - h_0 - T_0(s - s_0) \tag{1.3}$$

High-level energy can be entirely available, such as electrical energy and the potential energy in stored water. Theoretically, electrical energy can be completely converted to work in a motor, although practically there are motor losses that reduce the degree of conversion. Combustion and heat transfer processes, on the other hand, occur with a loss in availability.

In an energy conversion process, the system availability is altered by a decrease in the availability $E_{av,1}$ of the source and by an increase in the availability $E_{av,2}$ of the prescribed output. The effectiveness ε of the conversion is given by

$$\varepsilon = \frac{\Delta E_{av,2}}{-\Delta E_{av,1}} \tag{1.4}$$

The following example illustrates the need to evaluate the effectiveness of an energy conversion process.

Example 1.4

The exhaust gas discharged at 425 C from an internal combustion engine is used to increase the temperature of water from 30 to 80 C. The gas leaves the heat exchanger at 100 C. For the gas, c_p = 1.088 kJ/kg·K. T_0 = 25 C. Determine the effectiveness of the heat exchange process.

1 Gas/water mass ratio. For water at 30 C, h = 125.79 kJ/kg, s = 0.4369 kJ/kg·K; at 80 C, h = 334.91 kJ/kg, s = 1.0753 kJ/kg·K.*

$$m_g c_{p,g}(T_{g,i} - T_{g,o}) = m_w(h_{w,o} - h_{w,i})$$
$$m_g\, 1.088(425 - 100) = 1(334.91 - 125.79)$$
$$m_g = 0.591 \text{ kg gas/kg water}$$

2 Availability change for the gas.

$$\begin{aligned}\phi_2 - \phi_1 &= h_2 - h_1 - T_0(s_2 - s_1)\\ &= c_{p,g}(T_{g,2} - T_{g,1}) - T_0 c_{p,g} \ln \frac{T_{g,2}}{T_{g,1}}\\ &= 1.088(100 - 425) - 298.2 \times 1.088 \ln \frac{373.2}{698.2}\\ &= -353.60 + 203.22\\ &= -150.38 \text{ kJ/kg gas}\end{aligned}$$

* Property values for steam used in this text were selected in some instances from G. J. Van Wylen and R. E. Sonntag, *Fundamentals of Classical Thermodynamics*, SI Version, John Wiley and Sons, Inc., New York, 1976.

3 Availability change for the water.

$$\begin{aligned}\phi_2 - \phi_1 &= h_2 - h_1 - T_0(s_2 - s_1) \\ &= (334.91 - 125.79) \\ &\quad - 298.2(1.0753 - 0.4369) \\ &= 209.12 - 190.37 \\ &= 18.75 \text{ kJ/kg water}\end{aligned}$$

4 Conversion effectiveness.

$$\varepsilon = \frac{\Delta E_{av,2}}{-\Delta E_{av,1}} = \frac{m_w(\phi_2 - \phi_1)_w}{-m_g(\phi_2 - \phi_1)_g}$$

$$= \frac{1 \times 18.75}{-0.591\,(-150.38)} = 0.211$$

In the absence of a heat loss to the surroundings, the heat exchange process described in Example 1.4 will have a conversion efficiency equal to 1.0. The conversion effectiveness however is considerably lower, that is, $\varepsilon = 0.211$. As the cost of energy increases, there is greater need to examine energy conversion processes to determine whether some improvement in the conversion effectiveness can be achieved.

Despite a low conversion effectiveness, a particular energy conversion process may be judged acceptable because of the simplicity of the equipment and the absence of unduly high economic restrictions. For example, a gas or electric water heater has a low conversion effectiveness, but the heater is a comparatively simple device; and at the present time, most users accept the operating charge despite rising costs for energy.

PROBLEMS

1.1 The hourly power demand on an electric generating station is:

Hour	*MW*	*Hour*	*MW*
12 Midnight	160	12:30 P.M.	223
1	139	1	300
2	118	2	319
3	113	3	332
4	113	4	365
5	113	5	420
6	134	6	395
7	210	7	336
8	315	8	323
9	328	9	281
10	323	10	239
11	319	11	206
12 Noon	305	12 Midnight	160

The installed generating capability is 500 MW, and the total connected load is 700 MW.
(a) Construct the daily load curve from the above data.
(b) Determine the average load and the load factor.
(c) Calculate the daily plant factor and the demand factor.

1.2 Plot the load duration curve from the following data:

Hours of the year	*Load (MW)*
1	480
1000	350
2000	296
3000	272
4000	250
5000	210
6000	150
7000	95
8000	65
8760	30

Determine:
(a) The energy output, MWh/year.
(b) The yearly average load, MW.
(c) The annual plant factor for an installed generating capability of 650 MW.

1.3 The power plant load on a certain day is distributed as follows:

Hour	*Load (MW)*			
	Commercial	*Industrial*	*Residential*	*Street Lighting*
12 Midnight	25	50	98	42
2 A.M.	22	41	75	42
4	20	36	65	42
6	26	51	77	42
8	53	155	124	0
10	105	150	129	0
12 Noon	101	148	136	0
12:30 P.M.	42	74	147	0
1	112	152	131	0
2	107	161	124	0
4	98	169	115	0
6	51	151	123	0
8	36	75	242	42
10	30	59	148	42
12	24	50	102	42

Plot the subdivision and total load curves.

1.4 A power plant is to be constructed for 100-MW generating capability. For design A, the estimated capital and operating costs are $625/kW and 27.5 mills/kWh. Similarly for design B, the costs are $875 per kilowatt and 17.5 mills/kWh. The fixed charges are 15 percent of the installed cost. Select the most economical design for operation with an annual plant factor of 0.33. Repeat the analysis for an annual plant factor of 0.65. Base the comparison on the cost per kilowatt hour for generating electrical energy.

1.5 A study is made of the type of power plant to be selected for a generating capability of 250 MW. A steam plant will have a capital cost of $750/kW and an operating cost of 30 mills/kWh. For a hydroelectric plant, the corresponding costs are $1650/kW and 8 mills/kWh. The fixed charges are 16 percent for the steam plant and 11.5 percent for the hydroelectric plant. The annual plant factor is 0.45. For each type of power plant, determine the total cost per kilowatt hour for generating electrical energy.

1.6 A power plant provides service to a connected load of 350 MW. On a particular day, the maximum load carried by the station is 185 MW and the average load is 105 MW.

(a) Determine the demand factor and the load factor.
(b) All of the machines have a 50-MW generating capability. Allow for sufficient reserve and determine the number of machines that should be installed in the power plant.
(c) Calculate the daily plant factor.

1.7 The installed cost of a 300-MW steam power plant is $800/kW. The fixed charges are 15 percent of the installed cost. The annual plant factor is 0.52. One hundred men are employed in the plant at an average individual cost of $1800 per month. Maintenance, excluding labor, costs $450,000 per year. Supplies cost $140,000 per year. The average coal rate is 0.48 kg/kWh, and the cost of coal is $30 per metric ton. Determine the total cost to produce 1 kWh of electrical energy.

1.8 A manufacturing company is investigating the expansion of its hydroelectric generating capability at a cost of $10 million. All of the power generated by the new unit will be sold locally to yield an income of $1 million per year. The useful life of the new machine is taken as 40 years. The required capital is to be acquired by the sale of bonds that mature in 10 years and pay 9 percent interest. The initial bond sale is for $10 million.

Subsequent bond sales are planned for $7.5, $5.0, and $2.5 million at the end of 10, 20, and 30 years, respectively, in order to reduce the bonded indebtedness by $2.5 million at the end of each decade. A sinking fund will be created to retire the bonds. Money placed quarterly in the reserve is expected to earn interest at an annual rate of 10 percent, compounded quarterly. The fixed charges for taxes, insurance, general salaries, and maintenance are 2.5 percent of the capital cost. The operating costs are $40,000 per year. Evaluate the economic feasibility of the proposal. *Hint:* Examine the financial return for each decade.

1.9 A 500-kW steam turbine-generator is to be installed in an industrial power plant. The existing steam generation capability is sufficient for operating either a condensing or noncondensing turbine. For a preliminary study, capital costs for the two machines are assumed to be about equal. The fuel consumption for steam generation is expected to be 327 kg/h for noncondensing operation and 226 kg/h for condensing operation of the turbine. Fuel cost is 28 ¢/kg. The power input to the condenser circulating pump motor is about 2 kW, and the cost of electrical energy is 5.5 ¢/kWh. The capital cost of the condenser is $90,000, and the fixed charges are 15 percent of the installed cost. Condenser maintenance is taken as $1500 per year. The turbine-generator will operate 2000 h per year. Determine the economic advantage, if any, for condensing operation.

1.10 A power plant presently uses city water for boiler make-up. The cost of city water is 5.3 ¢/m^3. The make-up water requirement is 22,500 kg/h. A proposal is made to draw the make-up water from an underground source. Power required for pumping this water is 7 kW, and the cost of electrical energy is 5.2 ¢/kWh. The pump costs $7150, and the fixed charges are 20 percent of the installed cost. Pump maintenance is $200 per year. A shift from city water to the independent source will require some additional treatment of the make-up water at a cost of $1000 per year. Assume continuous operation of the power plant during the year and evaluate the feasibility of the proposal.

REFERENCES

1 *Kent's Mechanical Engineers' Handbook*, Power Volume, 12th ed., John Wiley & Sons, Inc., New York, 1957.

2 Decker, G. L. "The Impact of the New Energy Technologies," *Mechanical Engineering*, May 1977.

3 Baron, S. "Energy Cycles: Their Cost Interrelationship for Power Generation," *Mechanical Engineering*, June 1976.

BIBLIOGRAPHY

DeGarmo, E. P. and Canada, J. R., *Engineering Economy*, The Macmillan Co., New York, 1973.

Grant, E. L. and Ireson, W. G. *Principles of Engineering Economy*, The Ronald Press Co., New York, 1970.

Reistad, G. M. "Available Energy Conversion and Utilization in the United States," ASME Paper No. 74-WA/Pwr-1, New York, 1974.

Riggs, J. L. *Engineering Economics*, McGraw-Hill Book Co., New York, 1977.

Skrotski, B. G. A. and Vopat, W. A. *Power Station Engineering and Economy*, McGraw-Hill Book Co., New York, 1960.

Van Wylen, G. J. and Sonntag, R. E. *Fundamentals of Classical Thermodynamics*, SI Version, John Wiley & Sons, Inc., New York, 1976.

CHAPTER TWO

FUELS

A fuel is virtually any substance that will participate in an exothermic reaction with relative ease. For a specified application, however, the substance must be readily available in an adequate amount and perform satisfactorily. Furthermore, in most instances its use must be justified economically.

2.1 SELECTION OF FUELS

The principal fuels used today are either fossil or nuclear. A general classification of fuels, as shown in Table 2.1, would include other substances that are of minor utility or exhibit potential use.

Shale oil and oil obtained from tar sands may in time provide major fuels and supplement the dwindling supply of fuels manufactured from petroleum. Except for alcohol, the fuels in categories 3 to 6 are not likely to become principal sources of energy because of inherently limited quantities or light demand.

Usually, the fuel that is selected for a particular application will yield the lowest overall cost for acceptable operation of the power-producing system. The criteria for acceptable operation, however, vary widely depending upon the location, type, and function of the system and ordinarily upon the environmental impact.

2.2 FOSSIL FUELS

The term fossil refers to an earlier geologic age. Fossil fuels were formed a great many years ago and are not renewable. Once consumed, fossil fuels cannot be replaced, and consequently these fuels exist in deposits of finite extent.

TABLE 2.1 Types of Fuel

1 Fossil.
- **(a)** Coal, petroleum, natural gas (including derived fuels).
- **(b)** Shale oil and oil extracted from tar sands.

2 Nuclear.
- **(a)** Uranium.
- **(b)** Thorium, plutonium (potential use).
- **(c)** Deuterium.

3 Manufactured.
- **(a)** Coke.
- **(b)** Alcohol.
- **(c)** Various gases.

4 Wood.
- **(a)** Industrial wastes (logging and manufacturing).
- **(b)** Logs (for household use).

5 Waste material.
- **(a)** Municipal and industrial solids.
- **(b)** Organic.

6 Miscellaneous, including chemical fuels for special applications.

Fossil fuels are solar derived and are found in the three phases of matter. The principal fossil fuels commercially available today are designated as coal, petroleum, and natural gas.

Fossil fuels have truly served as the world's "work horse" at an ever-increasing rate. These fuels exist in abundant quantities and may be burned without exceptional difficulty, actually in many instances quite easily. Fossil fuels provide energy for heating, generating electricity, operating all modes of transportation, and for processing a great variety of materials. The demand for these fuels has become heavy, particularly for petroleum. Coal supplies are ample, but owing to distribution problems, natural gas shortages may occur on occasion in certain regions of the United States.

While petroleum is a major source of energy, a substantial portion of the crude oil production is consumed by the petrochemical industry for the manufacture of a wide variety of products. Currently, an ample supply of crude oil is available throughout the world, but inevitably the petroleum resources will be depleted. Whether the supply of petroleum will, in the next one to two decades, continue to meet the demand depends upon a number of uncertain factors, including the increase in the rate of consumption of crude oil, improved production in the existing oil fields, and the discovery of new deposits of petroleum, in particular at offshore locations. Crude oil shortages can of course be arbitrarily created by reducing production for political or economic reasons.

2.3 COAL

Millions of years ago, vegetation grew luxuriously in the hot, moist climate of the earth. With the passage of time, huge trees and ferns fell successively to the ground to form layers of partially decomposed vegetation. Some of these heavy deposits of vegetable material, impervious to the free flow of air, were eventually covered with earth to varying depths. In a subsequent period of time, extending over millions of years, the vegetable material was subjected to heat and pressure and transformed from carbohydrates to carbon and hydrocarbons.

The transformation of the original vegetation and wood to coal occurred progressively. Initially, the deposited material was converted to peat, a partly decayed plant material; subsequently, under the influence of heat and pressure, peat was converted to lignite. In time, the lignite was successively changed to subbituminous, bituminous, and anthracite coal. Not all of the deposits were subjected to the same degree of conversion. Throughout various regions of the world, deposits are found of peat, lignite, and the several ranks of coal that differ in composition and combustion characteristics. Coal deposits vary considerably in the thickness and inclination of the coal seam and in the depth of the overburden, conditions that influence the type of mining to be employed, that is, strip mining or underground mining.

Most coal is mined mechanically and, following cleaning and classification, burned in power plants located close to the mine or transported to other points of consumption. Two other possibilities for utilizing coal are being explored. One method provides for underground gasification of the coal, while the second method effects conversion of coal to a liquid or gaseous fuel. Commercial underground gasification of coal depends, in part, upon the solution to a number of technical problems. A fuel obtained by the conversion of coal will be expected generally to compete economically with the currently available sources of energy.

The continued increase in the demand for energy in the United States will place greater dependence upon coal as an industrial fuel, particularly for the generation of electrical energy. Oil and natural gas, because of limited supply and high cost, will not compete strongly with coal in the planning and design of future fossil fuel-fired power plants. Conversion to coal firing is proposed for some of the power plants that are currently burning oil.

For 1981, coal production in the United States is expected to be about 770 million metric tons per year.

2.4 COAL ANALYSIS

Two analyses are used in reporting the composition of coal. The proximate analysis reports moisture, volatile matter, fixed carbon, and ash. The moisture in the coal consists of (1) extraneous moisture that comes from external sources, and (2) inherent moisture that has its origin in the vegetable matter that formed the coal. Mine samples, unless taken from a wet face, contain little extraneous moisture, while varying amounts of such moisture are found in exposed coal.

Ash represents the residue of the mineral matter following combustion. The combustible matter in coal consists of the gaseous and solid portions. The composition of the volatile matter varies widely with different coals and consists of both combustible gases and noncombustible gases. Included in the combustible classification are hydrogen, carbon monoxide, methane, and other hydrocarbons. The noncombustible gases are principally carbon dioxide and water vapor. The solid portion, or fixed carbon, remains together with the mineral matter after the moisture and volatile matter have been expelled.

The ultimate analysis reports the composition of coal in percent by mass of carbon, hydrogen, nitrogen, sulfur, oxygen, ash, and moisture. The ultimate analysis is employed principally for combustion calculations.

The proximate analysis is particularly useful for classifying coals and providing subsequently an indication of the performance of the fuel in boiler furnaces. A separate determination of the sulfur content of the coal is often made in conjunction with the proximate analysis. In addition, the softening and fusing temperatures of the coal ash are considered to be essential to predicting the effect of the ash on the boiler performance.

2.5 CHARACTERISTICS OF COAL

Anthracite is in general a hard coal with a glossy, black appearance. Because of a low volatile matter content, these coals do not ignite readily; but during combustion, they burn freely, on grates, with a short flame. Since the supply is small, anthracite coal has limited industrial use.

Bituminous coal, mined generally in the eastern section of the United States, is used extensively as an industrial fuel. Bituminous coal, a soft coal, contains a relatively high percentage of volatile matter, hence ignition is readily established. The low-volatile bituminous coals burn with a clear flame, while medium- and high-volatile bituminous coals tend to smoke if not properly burned. In general, bituminous coals tend to cake when burned on a stoker. Most of the bituminous coals have a laminated structure and thus possess considerable resistance to weathering and crumbling during handling.

Subbituminous coals and lignites, found principally in the western region of the United States, have been comparatively unimportant fuels, despite the vast supply. These coals burn without caking but crumble in handling and possess rather poor weathering qualities. In comparison with bituminous coals, the western subbituminous coals and lignite have relatively low heating values and high moisture content. (See Table A.4 for heating values of different coals.)

Table 2.2 shows the proximate analysis and sulfur content for several classes of coal. The data are mean values derived from analyses of a representative number of as-received coal samples.

The content and the form of the sulfur in a fuel are important characteristics. Initially, during combustion, sulfur is oxidized to sulfur dioxide, and later some of the SO_2 is converted to sulfur trioxide. The latter compound will combine with condensed water vapor in the stack gas to form low-concentration sulfuric acid. Corrosion of heat transfer surfaces is avoided by maintaining the surface temperature above the dew point temperature of the gas. The dew point temperature of the gas however increases with an increase in the sulfur content of the fuel. Thus, the sulfur content of the fuel will influence the minimum temperature at which the heat transfer surfaces may be maintained and consequently regulate the temperature of the discharged flue gas.

The discharge of oxides of sulfur into the atmosphere is limited by air pollution control measures, a factor that must be considered in the selection of fuels. Firing a low-sulfur coal is

TABLE 2.2 Proximate Analysis and Sulfur Content of Some Coals

	Percent by Mass				
Class	*Fixed Carbon*	*Volatile Matter*	*Ash*	*Moisture*	*Sulfur*
Anthracite (Pennsylvania)	72.5	8.9	15.5	3.1	(0.5)
Low- and medium-volatile bituminous (Appalachian field)	71.7	20.0	6.2	2.1	(1.1)
Bituminous (Appalachian field)	53.8	35.6	7.8	2.8	(1.2)
Bituminous (Illinois field)	44.6	38.8	10.2	6.4	(3.0)
Bituminous (Rocky Mountain field)	47.9	36.1	10.5	5.5	(0.9)
Subbituminous (Rocky Mountain field)	38.6	31.4	6.4	23.6	(0.3)
Lignite	27.9	27.0	6.8	38.3	(0.6)

Source: For the proximate analysis, U.S. Bureau of Mines, Technical Paper 586, 1941. Values for sulfur, not included in the proximate analysis, were taken from various reports.

thus likely to be advantageous from the standpoint of minimizing air pollution, although other characteristics of the fuel might create difficult operating problems.

Low-sulfur content is an important characteristic of the western coals. Consequently, the demand for the western subbituminous coal is very likely to increase despite a long haul, in some instances, to the point of consumption.

Proposals are currently under consideration for local consumption of western coals, including lignite. Coal would be burned directly or gasified in large plants constructed adjacent to the coal fields. The end product, electrical energy or high-energy gas, will be consumed mainly in other regions of the country. Economical transportation of either product by electric transmission lines or gas pipelines is anticipated.

An extensive shift of coal mining operations from the eastern to the western region of the United States will require substantial expenditures. Opening new mines, providing mining and coal-handling equipment, and building large gasification plants will entail a heavy capital outlay. Of particular significance is the possibility of curtailing operation of the eastern mines, which represent earlier investments of substantial magnitude.

2.6 COMBUSTION CALCULATIONS

A number of examples are introduced in this chapter to illustrate the application of stoichiometry to the analysis of combustion systems. In a combustion reaction, the combustible elements of the commercial fossil fuels combine with oxygen. The several combinations are expressed by basic chemical equations that represent the participation of very large numbers of molecules in the reactions. The coefficients in the combining equations define the molal quantities of the elements.

The oxygen content of the atmosphere is 23.19 percent by mass, and the nitrogen/oxygen ratio is taken as 3.76 mol N_2/mol O_2. Molecular weights of various gases are given in Table A.6.

Example 2.1

The ultimate analysis of a coal, in percent by mass, is carbon 73.99, hydrogen 5.39, oxygen 10.02, nitrogen 1.38, sulfur 1.79, and ash 7.43.

Determine the quantity of air required for theoretically complete combustion.

Note: The total quantities of hydrogen and oxygen are reported in this analysis, hence moisture does not appear as a separate item.

For carbon, $\overline{m}$ = 12.012. The combining equations are:

$$C + O_2 \rightarrow CO_2$$

$$1 \text{ mol C} + 1 \text{ mol } O_2 \rightarrow 1 \text{ mol } CO_2$$

$$12.012 \text{ kg C} + 31.999 \text{ kg } O_2 \rightarrow 44.011 \text{ kg } CO_2$$

$$\frac{O_2}{C} = 2.664 \text{ kg } O_2/\text{kg C}$$

$$\frac{CO_2}{C} = 3.664 \text{ kg } CO_2/\text{kg C}$$

$$H_2 + {}^1\!/_2 O_2 \rightarrow H_2O$$

$$1 \text{ mol } H_2 + {}^1\!/_2 \text{ mol } O_2 \rightarrow 1 \text{ mol } H_2O$$

$$2.016 \text{ kg } H_2 + 16.000 \text{ kg } O_2 \rightarrow 18.016 \text{ kg } H_2O$$

$$\frac{O_2}{H_2} = 7.937 \text{ kg } O_2/\text{kg } H_2$$

$$\frac{H_2O}{H_2} = 8.937 \text{ kg } H_2O/\text{kg } H_2$$

$$S + O_2 \rightarrow SO_2$$

$$1 \text{ mol S} + 1 \text{ mol } O_2 \rightarrow 1 \text{ mol } SO_2$$

$$32.064 \text{ kg S} + 31.999 \text{ kg } O_2 \rightarrow 64.063 \text{ kg } SO_2$$

$$\frac{O_2}{S} = 0.998 \text{ kg } O_2/\text{kg S}$$

$$\frac{SO_2}{S} = 1.998 \text{ kg } SO_2/\text{kg S}$$

Oxygen required to burn the combustible elements:

Carbon $\frac{O_2}{C} \times C = 2.664 \times 0.7399$

$= 1.9711$

Hydrogen $\frac{O_2}{H_2} \times H_2 = 7.937 \times 0.0539$

$= 0.4278$

Sulfur $\frac{O_2}{S} \times S = 0.998 \times 0.0179$

$= \underline{0.0179}$

Total $= 2.4168$ kg O_2/kg fuel

Net oxygen = gross oxygen less oxygen in fuel

$= 2.4168 - 0.1002$

$= 2.3166$ kg O_2/kg fuel

Required air = (net O_2) ÷ (oxygen/air mass ratio)

$= 2.3166 \div 0.2319$

$= 9.990$ kg air/kg fuel

Example 2.2

The coal in Example 2.1 is burned completely with 25 percent excess air. The ambient conditions are P_{bar} = 101.00 kPa, abs, T = 25 C, and the relative humidity is 60 percent. Determine the probable flue gas analysis. Calculate the dew point temperature and the specific volume of the wet flue gas at 420 K.

Air supplied $= 1.25 \times 9.9897$

$= 12.4871$ kg air/kg fuel

Oxygen $= 0.2319 \times 12.4871$

$= 2.8958$ kg O_2/kg fuel

Nitrogen $= 0.7681 \times 12.4871$

$= 9.5913$ kg N_2/kg fuel

Excess oxygen = oxygen supplied less net oxygen required

$= 2.8958 - 2.3166$

$= 0.5792$ kg/kg fuel

Carbon dioxide: $\frac{CO_2}{C} \times C = 3.664 \times 0.7399$

$= 2.7110$ kg CO_2/kg fuel

Water vapor from the fuel:

$\frac{H_2O}{H_2} \times H_2 = 8.937 \times 0.0539$

$= 0.4817$ kg water vapor/kg fuel

Water vapor from the combustion air:

$$P_v = \phi(P_v)_{sat,\ 25\ C}$$
$$= 0.60 \times 3.169 = 1.901 \text{ kPa, abs}$$
$$\omega = 0.622 \frac{P_v}{P_{bar} - P_v}$$
$$= 0.622 \frac{1.901}{101.00 - 1.901}$$
$$= 0.01193 \text{ kg water vapor/kg dry air}$$
$$m_w = m_a\omega = 12.4871 \times 0.01193$$
$$= 0.1490 \text{ kg/kg fuel}$$

$$\text{Total water vapor} = 0.4871 + 0.1490 = 0.6307 \text{ kg/kg fuel}$$

The contribution of the nitrogen in the fuel to the flue gas is disregarded. Sulfur dioxide will not be included in the flue gas analysis because the disposition of the sulfur is uncertain.

Flue Gas Analysis

				Volume Fraction (%)	
Gas	$\bar{m}$	*Mass (kg)*	*Mole*	*Wet gas*	*Dry Gas*
CO_2	44.011	2.7110	0.06160	13.48	14.59
O_2	31.999	0.5792	0.01810	3.96	4.29
N_2	28.013	9.5913	0.34239	74.90	81.12
H_2O	18.015	0.6307	0.03501	7.66	—
		13.5122	0.45710	100.00	100.00

Dew Point Temperature of the Wet Gas

Partial pressure of the water vapor:

$$P_v = \frac{N_{H_2O}}{N_t} P_{bar}$$
$$= \frac{0.03501}{0.45710} 101.00 = 7.736 \text{ kPa, abs}$$

The corresponding saturation temperature is obtained from the *Steam Tables*:

$$T_{dp} = 40.8 \text{ C}$$

The dew point temperature of 40.8 C is determined by ignoring the presence of sulfur oxides in the flue gas; thus, only a single condensable gas, that is, the water vapor, is considered to be present. In a real system, however, the dew point temperature of the gas would be elevated above 40.8 C because of the presence of SO_2 and SO_3 in addition to the water vapor.

Specific Volume of the Wet Gas

From the flue gas analysis,

$$m_g = 13.5122 \text{ kg gas/kg fuel}$$
$$N_g = 0.45710 \text{ mol gas/kg fuel}$$
$$\bar{m} = \frac{m_g}{N_g} = \frac{13.5122}{0.45710} = 29.561$$
$$R = \frac{\bar{R}}{\bar{m}} = \frac{8.3143}{29.561} = 0.2813 \text{ kJ/kg·K}$$
$$v = \frac{RT}{P} = \frac{0.2813 \times 420}{101.00} = 1.170 \text{ m}^3\text{/kg}$$

2.7 PETROLEUM

Commercial liquid fuels are produced mainly from petroleum, a naturally occurring liquid that consists principally of a mixture of various hydrocarbons, with small amounts of sulfur, nitrogen, oxygen, and possibly ash usually present. The analysis of crude oil varies somewhat among the different oil fields scattered across the United States. Approximate ranges of the component elements, in percent by mass, are carbon 82 to 85, hydrogen 10 to 15, oxygen and nitrogen 1 to 3, and sulfur 0.5 to 1.5. The ash content is 0 to 0.10 percent. The specific gravity ranges from 0.85 to 0.95 for most crudes.

Petroleum was formed during an earlier geologic age from prolific marine growths. The cover that formed over the deposits of decayed marine life and vegetable matter was eventually compressed into limestone and shale. Subsequently, because of the action of heat and pressure, this organic material decomposed under the rock cover into hydrocarbons of various combinations. The petroleum trapped under the cap rock was forced by gas pressure and the bouyant action of lower-level water into cavities and porous sand. The rock formation was not completely impervious to flow, hence gases and the lighter liquid components escaped to various degrees from the porous strata. Crude oils of different densities and compositions were thus formed. Oil found adjacent to coal fields may possibly have been formed by decomposition of vegetable matter.

Crude oil is commonly classified in accordance with the type of residue that remains from nondestructive distillation. Three main classes are used: paraffin base, asphalt base, and mixed base. The base is identified by the respective separation from the distillation residue of a paraffin wax, asphalt, or a mixture of wax and asphalt. In each class, the more volatile fractions of the crude are expected to have characteristics similar to those of the residue. These more volatile fractions represent the important fuel components and thus would consist mainly of paraffinic or naphthenic hydrocarbons or a mixture of the two groups.

Fractional distillation of the crude oil produces, in the order of increasing boiling points, gasoline, kerosene, fuel oil, and lubricating oil.* The fuel oil cut provides most Diesel engine fuels, furnace oils, and the cracking stock for the manufacture of gasoline. The heavy fuel oils used for steam generation are normally the residual oils from the cracking process. Following chemical treatment, the various products produced at the refinery are blended in prescribed proportions to obtain a final product.

Example 2.3

The composition of an oil fuel, in percent by mass, is carbon 84.2, hydrogen 14.7, and inert material 1.1. The analysis of the dry products of combustion reports carbon dioxide 12.79, carbon monoxide 0.31, oxygen 2.54, and nitrogen 84.36 percent by volume. Calculate the percent excess air supplied for combustion of the fuel. Assume 100 mol dry gas.

Quantity of Fuel Burned

One mole of carbon is burned for each mole of CO_2 and CO produced:

$$m_C = \overline{m}_C (N_{CO_2} + N_{CO})$$
$$= 12.012\,(12.79 + 0.31)$$
$$= 157.36 \text{ kg C}$$
$$m_f = \frac{m_C}{f_C} = \frac{157.36}{0.842}$$
$$= 186.89 \text{ kg fuel/100 mol dry gas}$$

Quantity of Air Supplied

The nitrogen in the gas was admitted as a component of the air supplied for combustion:

$$m_{N_2} = N_{N_2} \times \overline{m}_{N_2} = 84.36 \times 28.013$$
$$= 2363.2 \text{ kg nitrogen/100 mol dry gas}$$
$$m_{air} = \frac{m_{N_2}}{1 - 0.2319} = \frac{2363.2}{1 - 0.2319}$$
$$= 3076.7 \text{ kg air/100 mol dry gas}$$

Note: The quantity 1 − 0.2319 is the fraction of nitrogen in the atmosphere.

Air/Fuel Ratio (Actual)

$$\frac{m_a}{m_f} = \frac{3076.7}{186.89} = 16.463 \text{ kg air/kg fuel}$$

Theoretical Air/Fuel Ratio (see Example 2.1 for the procedure)

Oxygen required for the combustion of the carbon and the hydrogen:

$$\frac{O_2}{C} \times C = 2.664 \times 0.842 = 2.243$$
$$\frac{O_2}{H_2} \times H_2 = 7.937 \times 0.147 = \underline{1.167}$$
$$3.410 \text{ kg } O_2\text{/kg fuel}$$

* Somewhat different terms are used in industry to describe the fractions discharged from the still. They are light gases, naphthas, middle distillates, light gas oil, heavy gas oil, and residues. Naphtha is the basic ingredient of gasoline.

Air required:

$$m_{air} = \frac{m_{O_2}}{0.2319} = \frac{3.410}{0.2319}$$
$$= 14.705 \text{ kg air/kg fuel}$$

Excess Air

$$\text{Percent excess air} = \frac{(a/f)_{actual} - (a/f)_{theo.}}{(a/f)_{theo.}} \times 100$$
$$= \frac{16.463 - 14.705}{14.705} \times 100 = 11.96$$

Hydrocarbons and alcohols are represented generally by chemical formulas, namely, C_mH_n and $C_mH_nO_z$, respectively. For complete combustion with air, the corresponding equations are written as

$$C_mH_n + xO_2 + 3.76xN_2 \rightarrow mCO_2 + \frac{n}{2}H_2O + \left(x - m - \frac{n}{4}\right)O_2 + 3.76xN_2 \quad (2.1)$$

$$C_mH_nO_z + xO_2 + 3.76xN_2 \rightarrow mCO_2 + \frac{n}{2}H_2O + \left(x - m - \frac{n}{4} + \frac{z}{2}\right)O_2 + 3.76xN_2 \quad (2.2)$$

Example 2.4

Liquid $C_{12}H_{26}$ is completely burned in a continuous-flow system with 320 percent theoretical air. Fuel is supplied at 25 C, and the air enters at 500 K. Assume zero heat loss and determine the temperature of the gas at the outlet section of the combustion chamber.

1 Combustion equation

(a) With theoretical air.

$$C_{12}H_{26} + 18.5O_2 + (3.76 \times 18.5)N_2 \rightarrow 12CO_2 + 13H_2O + (3.76 \times 18.5)N_2$$

Moles of air, $N_a = 88.06$ mol air

Moles of products, $N_p = 94.56$ mol products

$N_p - N_a = 6.5$ mol

Note: For complete combustion, $N_p - N_a = 6.5$ regardless of the quantity of excess air.

(b) With 320 percent theoretical air.

$$C_{12}H_{26} + (3.2 \times 18.5)O_2 + 3.76(3.2 \times 18.5)N_2 \rightarrow 12CO_2 + 13H_2O + \left[(3.2 \times 18.5) - 12 - \frac{26}{4}\right]O_2 + 3.76(3.2 \times 18.5)N_2$$

$$C_{12}H_{26} + 59.20O_2 + 222.59N_2 \rightarrow 12CO_2 + 13H_2O + 40.70O_2 + 222.59N_2$$

When combustion is incomplete, the combustible components in the products are determined, or prescribed, and inserted in the right-hand side of the equation. Common combustible components, present usually in small or trace amounts, are carbon monoxide, hydrogen, hydrocarbons, and carbon.

2 Energy equation.

The energy equation for a continuous-flow combustion system is written generally as

$$-H_{RP}^{\circ} + \Delta H_R^{\circ} - \Delta H_P^{\circ} + Q = 0 \quad (2.3)$$

when the change in kinetic energy is a negligible quantity. In this equation, H_{RP}° represents the enthalpy of combustion at the standard state, taken at 1 atm, 25 C, and ΔH_R° and ΔH_P° are, respectively, the enthalpy values of the reactants and the products of combustion measured with respect to the standard state. The heat transfer Q is commonly a comparatively small quantity.

The energy equation for the complete combustion of $C_{12}H_{26}$, as described above, is

$$[12(\Delta\bar{h}^{\circ})_{CO_2} + 13(\Delta\bar{h}^{\circ})_{H_2O} + 40.70(\Delta\bar{h}^{\circ})_{O_2} + 222.59(\Delta\bar{h}^{\circ})_{N_2}]_T = -(\bar{h}_{RP}^{\circ})_{C_{12}H_{26}} + [59.20(\Delta\bar{h}^{\circ})_{O_2} + 222.59(\Delta\bar{h}^{\circ})_{N_2}]_{500\text{ K}}$$

For the fuel at 25 C, $(\Delta\bar{h}^{\circ})_{C_{12}H_{26}} = 0$.

For liquid $C_{12}H_{26}$ and water vapor in the products of combustion,

$h_{RP}^{\circ} = -44085$ kJ/kg (from Table A.3)

$\bar{m} = 170.34$

At 500 K,

$$(\Delta\bar{h}^\circ)_{O_2} = 6088 \text{ kJ/kgmol} \qquad \text{(from Table A.5)}$$

$$(\Delta\bar{h}^\circ)_{N_2} = 5912 \text{ kJ/kgmol}$$

Substitution of the above quantities into the right-hand side of the energy equation yields

$$-[170.34(-44085)] + (59.20 \times 6088) + (222.59 \times 5912) = 9{,}185{,}801 \text{ kJ}$$

The temperature T_p of the products of combustion is determined by a trial-and-error procedure. At this temperature, the energy of the products will be equal to 9,185,801 kJ. Enthalpy values for the combustion gases are selected from Table A.5.

		T = 1200 K		*T = 1300 K*	
Gas	*N (mol)*	*$\Delta\bar{h}^\circ$ (kJ/kgmol)*	*$N \times \Delta\bar{h}^\circ$ (kJ)*	*$\Delta\bar{h}^\circ$ (kJ/kgmol)*	*$N \times \Delta\bar{h}^\circ$ (kJ)*
CO_2	12	44,484	533,808	50,158	601,896
H_2O	13	34,476	448,188	38,903	505,739
O_2	40.70	29,765	1,211,436	33,351	1,357,386
N_2	222.59	28,108	6,256,560	31,501	7,011,808
			8,449,992		9,476,829

By interpolation, $T_p = 1272$ K.

Dissociation of the products of combustion has not been observed in Example 2.4. At a temperature of about 1300 K, dissociation of the component gases is of minor consequence.

2.8 GASEOUS FUELS

Various gaseous fuels are available for industrial and residential use. In comparison with solid and liquid fuels, a gaseous fuel is less difficult to handle, and the combustion process is relatively free of operating problems. The characteristics of the gaseous-phase place limitations on marketing some of these fuels, that is, the economical limits on transportation of the gas, are dependent upon a number of factors, including the energy value per unit volume of the gas. Storage of a gaseous fuel in very large quantities is ordinarily not feasible, either physically or economically.

Natural gas is by far the principal gaseous fuel. The origin of natural gas is generally believed to be similar to that of petroleum. The location of the gas-producing fields in the United States coincides to some extent with the oil fields. Natural gas lines extend for hundreds of miles from the producing fields into the industrial and domestic market areas of the country. Compressors located in pumping stations along the pipeline provide the increase in gas pressure necessary to overcome the frictional losses in the line. Storage of natural gas in caverns and depleted wells is employed to some extent.

Natural gas is also transported in liquefied form (LNG) in specially constructed tankers and barges. Distribution systems for LNG are in the developmental stage. Liquefied petroleum gas (LPG) is a product of the refining process and is also obtained by "drying" natural gas. These gases perform well in internal combustion engines because of good combustion characteristics and minor contribution to air pollution. Other uses for these liquefied gases are found in domestic and industrial applications and for augmenting gas supplies in distribution systems.

Natural gas burns readily with a blue flame. It is an ideal fuel for industrial and residential heating and for firing internal combustion engines. Large quantities of natural gas are burned in boiler furnaces for the production of steam in electric power plants. Abrasive material is removed from the gas at the wellhead in order to avoid wear of pumping, control, and firing equipment. Additional preparation prior to firing natural gas is limited to removal of liquid components and sulfur compounds.

Methane is the principal component of natural gas. Smaller quantities of ethane and a few heavier hydrocarbons, together with carbon dioxide and nitrogen, complete the composition. The heavier hydrocarbons are often included under the ethane heading. The noncombustible components are usually in the range of 1 to 5 percent by volume.

The higher heating value of natural gas in the United States ordinarily ranges between 37 and 41 MJ/m^3, measured at 1 atm, 15 C. The specific gravity of the gas relative to air varies from 0.60 to 0.65.

The gases discharged from blast furnaces contain sufficient amounts of carbon monoxide and hydrogen to justify firing these gases in boiler furnaces or burning them in the cylinders of large gas engines. The available gas depends upon the output of the blast furnaces, and the excess gas that cannot be used at the mill may be economically transported in large mains to local steam power plants where the gas is usually fired simultaneously with oil or coal.

The heating value of blast furnace gas is low, approximately 3.7 MJ/m^3, because of the high content of inert gas. Considerable variation is observed in the composition of blast furnace gas. A somewhat typical analysis, in percent by volume, may be taken as carbon dioxide 8.7, nitrogen 56.5, carbon monoxide 32.8, hydrogen 1.8, and methane 0.2.

Refinery gas, coke-oven gas, and gas produced in sewage treatment are by-product fuels that are used in the plant for heating stills, firing boilers, and operating gas engines. Excess gas may be sold locally. Producer gas, coal gas, and water gas are fuels manufactured for on-site industrial use. These fuels have relatively limited use in gas engine operation and virtually no use in steam production.

Example 2.5

The analysis of a gaseous fuel, in percent by volume, is CO 10, CH_4 28, H_2 22, N_2 17, CO_2 13, and O_2 10. Determine the volumetric analysis of the products of combustion when the fuel is burned with 110 percent of theoretical air.

Combining Equations

$$CO + \tfrac{1}{2}O_2 \rightarrow CO_2$$
$$CH_4 + 2O_2 \rightarrow CO_2 + 2H_2O$$
$$H_2 + \tfrac{1}{2}O_2 \rightarrow H_2O$$

Required Air

COMPONENT GAS (mol)		REQUIRED OXYGEN (mol)
CO	10	5
CH_4	28	56
H_2	22	11
		72 (gross)
		− 10 mol (supplied in the fuel)
		62 mol O_2/100 mol fuel (net)

Air supplied for combustion

Oxygen: 1.10 × 62
= 68.2 mol O_2/100 mol fuel

Nitrogen: 3.76 × 68.2
= 256.4 mol N_2/100 mol fuel

Products of Combustion

1 Nitrogen.

Supplied in the air	256.4 mol
Contained in the fuel	17
	273.4 mol N_2/100 mol fuel

2 Oxygen.

Supplied in the air	68.2 mol
Consumed (net)	− 62
Excess	6.2 mol O_2/100 mol fuel

3 Carbon dioxide.

COMPONENT GAS (mol)		CO_2 PRODUCED (mol)
CO	10	10
CH_4	28	28
		38 mol CO_2/100 mol fuel

From combustion	38 mol
Contained in the fuel	13
Total	51 mol CO_2/100 mol fuel

4 Water vapor.

COMPONENT GAS (mol)		H_2O PRODUCED (mol)
CH_4	28	56
H_2	22	22
		78 mol water vapor/100 mol fuel

Analysis of the Products of Combustion

		Percent by Volume	
Gas	*(mol)*	*Wet Gas*	*Dry Gas*
CO_2	51	12.5	15.4
O_2	6.2	1.5	1.9
N_2	273.4	66.9	82.7
H_2O	78	19.1	—
	408.6	100.0	100.0

Example 2.6

Determine the actual air/fuel ratio and the percent of theoretical air for a combustion process for which the fuel and combustion gas analyses are given.

FUEL	(PERCENT BY VOLUME)	COMBUSTION GAS	(PERCENT BY VOLUME)
CO_2	12.4	CO_2	24.6
CO	27.0	O_2	1.0
H_2	2.2	N_2	74.4
N_2	58.4		100.0
	100.0		

1 Ratio of fuel to the dry gas.

For the gas: 24.6 mol C/100 mol dry gas

For the fuel: 39.4 mol C/100 mol fuel

$24.6 \div 39.4 = 0.624$ mol fuel/mol dry gas

2 Air supplied.

$$N_2 \text{ in air} = \begin{pmatrix} N_2 \text{ in the} \\ \text{dry gas} \end{pmatrix} - \begin{pmatrix} N_2 \text{ in the} \\ \text{fuel} \end{pmatrix}$$

$$= 74.4 - (0.624 \times 58.4)$$

$$= 38.0 \text{ mol } N_2/100 \text{ mol dry gas}$$

$$38.0 \div 0.790 = 48.1 \text{ mol air}/100 \text{ mol dry gas}$$

(Atmosphere: 79.0 mol N_2/100 mol air)

3 Actual air/fuel ratio.

$0.481 \div 0.624 = 0.771$ mol air/mol fuel

4 Theoretical air/fuel ratio.

COMPONENT GAS (mol)		REQUIRED OXYGEN (mol)
CO	27.0	13.5
H_2	2.2	1.1
		14.6 mol O_2/100 mol fuel

$4.76 \times 14.6 = 69.5$ mol air/100 mol fuel

(Atmosphere: 4.76 mol air/mol oxygen)

5 Percent theoretical air.

$$\text{Percent theoretical air} = \frac{(a/f)_{actual}}{(a/f)_{theo}} \times 100$$

$$= \frac{0.771}{0.695} \times 100 = 110.9$$

2.9 MISCELLANEOUS FUELS

In addition to coal, petroleum, and natural gas, many substances are less extensively employed as fuels in certain unique applications. In some instances the substance may be burned both to recover energy and for disposal as a waste material.

Refuse wood, including bark, and other waste vegetable material, for example, dried sugar cane, are burned locally, usually at the mill. Wood and waste materials often contain a good deal of water and are burned with some difficulty, although combustion may be improved by firing in conjunction with coal or oil. Combustion of refuse wood has diminished in recent years, because of the adoption of conservation measures and the restrictions that have been placed on air pollution.

The spent liquors from the wood pulping processes are burned, usually without auxiliary fuel firing, following suitable concentration of the dilute liquor. Disposal of these liquors by burning, rather than by discharge into a waterway, is an essential step in combating pollution. In some of these combustion processes economical recovery may be achieved of chemicals used for pulping.

Coke, produced by the distillation of bituminous coal, is used almost exclusively for metallurgical processing. Distillation removes the moisture and most of the volatile matter from the coal. The residue is a fused, porous structure consisting essentially of carbon and ash. The gas, distilled from the coal and known as coke-oven gas, is used locally at the mill as an energy source for heating and power generation.

The feasibility of burning ammonia in internal combustion and gas turbine engines has been explored. While test results were generally satisfactory, ammonia does not appear to be a particularly appropriate engine fuel. Certain operating problems can be anticipated, and the engine exhaust gas contains a relatively large amount of water, that is, in comparison with conventional operation on hydrocarbon fuels. Moreover, discharge of unburned ammonia in the engine exhaust could not be tolerated.

Ethanol (C_2H_5OH), produced by the fermentation of grain, has long been proposed as a principal fuel for the internal combustion engine. When used as an internal combustion engine fuel, ethanol is blended with gasoline. The antiknock characteristics of a low-octane-number gasoline can be improved by the addition of alcohol. Because of the availability of high-quality gasolines today, the antiknock property of alcohol is not currently an important factor.

Methanol (CH_3OH) can be burned satisfactorily in internal combustion engines and in gas turbine engines, particularly the latter. Extensive use of this fuel will be influenced by the world supply and the price of crude oil, and will depend, in part, on the ability to derive methanol economically from coal, or biomass, in sufficiently large quantities.

Commercial-scale manufacture of methanol is achieved by the reaction between hydrogen and carbon monoxide under high pressure, $CO + 2H_2 \rightarrow CH_3OH$. Re-formed natural gas is a common source of the methane that in a reaction with carbon dioxide and steam produces the synthesis gas, hydrogen and carbon monoxide. Raw materials other than natural gas, for example, the heavy residues from oil refineries, are also used to obtain the synthesis gas. A very substantial increase in the production of methanol is not likely to be achieved by greater use of natural gas and crude oil, because of the limited supplies of these raw materials.

Incidentally, a simple process can be employed to convert coal-derived methanol into high-grade gasoline.[1] Currently the proprietary process is used to a limited extent for commercial-scale manufacture of gasoline from methanol produced from natural gas.

In comparison with gasoline, methanol is a superior engine fuel. Thermal efficiency values for the engine are higher, and there are no emission problems. Because of a high octane number, namely, 106, methanol is an excellent fuel for high-compression engines.

On the basis of a unit mass, methanol has a lower energy value than gasoline. The lower heating value of the liquid fuel is 19,953 kJ/kg for methanol and 44,397 kJ/kg for C_8H_{18}. The stoichiometric air/fuel ratios are, however, substantially different, 6.43 for methanol and 15.08 for octane. There is consequently only a small difference in the energy values of the air–fuel mixtures, 2685 kJ/kg for methanol and 2760 kJ/kg for octane. Thus, with respect to the energy value of the charge, the engine power is not reduced appreciably when operated on methanol rather than gasoline. A shift from one fuel to the other, however, requires some adjustment of the engine.

The energy of vaporization for methanol, 1170 kJ/kg, is substantially higher than that for octane, 300 kJ/kg. A somewhat lower cylinder temperature is thus observed, at the start of compression, when the engine operates on methanol. A reduction in the cylinder temperature causes an increase in the density of the charge, and consequently the engine power increases with the increase in the flow rate of the inducted methanol–air mixture.

Example 2.7

A gasoline is represented by $C_{8.5}H_{18.4}$ ($\overline{m}$ = 120.65). Calculate the theoretical air/fuel ratio for this fuel and for methanol (CH_3OH). Determine for each fuel the energy of the reactants per kilogram of the air-fuel mixture, based on the lower heating value of the liquid fuel. The lower heating value of the gasoline is taken as 44,315 kJ/kg.

1 Air/fuel ratio for $C_{8.5}H_{18.4}$.

$$C_mH_n + xO_2 + 3.76xN_2 \rightarrow mCO_2 + \frac{n}{2}H_2O + \left(x - m - \frac{n}{4}\right)O_2 + 3.76xN_2$$

$$x - m - \frac{n}{4} = 0$$

$$x = m + \frac{n}{4} = 8.5 + \frac{18.4}{4}$$

$$= 13.1 \text{ mol } O_2/\text{mol fuel}$$

Oxygen

$13.1 \times 31.999 = 419.19$

Nitrogen

$3.76 \times 13.1 \times 28.013 = 1379.81$

Air $\quad$ 1799.00 kg/mol fuel

$$\frac{a}{f} = \frac{m_{air}}{m_{fuel}} = \frac{1799.0}{120.65}$$

$$= 14.91 \text{ kg air/kg gasoline}$$

2 Air/fuel ratio for methanol (CH_4O) ($\overline{m}$ = 32.04).

$$C_mH_nO_z + xO_2 + 3.76xN_2 \rightarrow mCO_2 + \frac{n}{2}H_2O + \left(x - m - \frac{n}{4} + \frac{z}{2}\right)O_2 + 3.76xN_2$$

$$x - m - \frac{n}{4} + \frac{z}{2} = 0$$

$$x = m + \frac{n}{4} - \frac{z}{2}$$

$$= 1 + \frac{4}{4} - \frac{1}{2}$$

$$= 1.5 \text{ mol } O_2/\text{mol fuel}$$

Oxygen

$= 1.5 \times 31.999 = 48.00$

Nitrogen

$= 3.76 \times 1.5 \times 28.013 = 157.99$

Air $\quad = 205.99$ kg air/mol fuel

$$\frac{a}{f} = \frac{m_{air}}{m_{fuel}} = \frac{205.99}{32.04}$$

$$= 6.43 \text{ kg air/kg methanol}$$

3 Energy—air/gasoline.

$$\frac{\text{LHV}}{a/f + 1} = \frac{44{,}315}{14.91 + 1}$$

$$= 2785 \text{ kJ/kg air–gasoline mixture}$$

4 Energy—air/methanol.

$$\frac{\text{LHV}}{a/f + 1} = \frac{19{,}953}{6.43 + 1}$$

$$= 2685 \text{ kJ/kg air–methanol mixture}$$

(See Table A.3 for the lower heating value of liquid methanol.)

A discussion of the lower heating value (LHV) is presented in Section 8.18.

2.10 COAL CONVERSION

The dwindling supply of petroleum and natural gas in the United States has promoted a number of studies directed to the conversion of coal into substitute gaseous and liquid fuels. Some of these studies are in the research stage, while others involve construction of coal conversion pilot plants.

Underground coal gasification has for some time been under investigation by actually firing coal seams. Incomplete combustion of the coal produces a gas mixture that consists of various combustible components. While this method of coal gasification eliminates the mining operation, it has not proved, as yet, to be commercially feasible. Underground coal gasification is particularly appropriate for recovering coal from thin or sharply inclined seams or from deposits that are otherwise difficult to mine.

At the present time, the most promising approach to coal conversion is the production of a substitute for natural gas. This substitute gas (SNG) would be sulfur free and clean burning. In comparison with the direct combustion of coal, conversion of coal to gas and the subsequent combustion of the gas will result in a reduction in the overall emission of nitrogen oxides. These oxides of nitrogen, NO and NO_2, are known as NO_x, a major air pollutant.

The Lurgi process, developed by Lurgi Gesellschaften GmbH of West Germany, is a commercially available high-pressure process for coal gasification. The process is complex and involves the reaction, under pressure, of coal with oxygen and steam. The reaction is accomplished in a gasifier that operates at a pressure of about 2750 kPa. The crude gas discharged from the gasifier has a higher heating value, on a dry basis, of 11 to 18.5 MJ/m^3, measured at 1 atm, 15 C. The crude gas mixture contains methane, carbon monoxide, hydrogen, carbon dioxide, hydrogen sulfide, ammonia, and unconverted steam.

Subsequent to discharge from the gasifier, the crude gas, which is low in methane, is subjected to a number of processes for removal of coal byproducts and for improving the quality of the gas. A shift conversion process and the catalytic methanation reaction are employed to generate additional methane from the crude gas. The end product of the Lurgi process is SNG that has a higher heating value of 33.5 to 37 MJ/m^3. This high-quality gas is then compressed to a pressure of 7 MPa and discharged as pipeline gas.

Because of temperature and mechanical limitations, the diameter of the Lurgi gasifier is restricted to about 4 m, hence many parallel units are normally required. Thirty gasifiers would be installed for a plant output of 7 million m^3 per day of SNG, measured at 1 atm, 15 C. Each kilogram of coal yields 0.25 kg SNG.

In addition to the Lurgi process, other coal gasification processes are being investigated and developed. Some of these processes will yield a product gas that has a heating value considerably below the energy level of SNG. Two energy ranges of potential interest are 11 to 18.5 and 5.5 to 9.5 MJ/m^3; the product gases are designated, respectively, as intermediate-energy gas and low-energy gas. The intermediate-energy gas can be produced by scrubbing the crude gas in the SNG process for removal of H_2S.[4] On a heating value basis, this gas can be produced at a relatively lower cost than SNG. Low-energy gas can be produced by injecting air directly into the gasifier and scrubbing to remove the H_2S. Low-energy gas contains 40 to 50 percent nitrogen, and production costs are below the cost for producing intermediate-energy gas.

The economics of coal gasification and consumption of the product gas are complex and involve many factors. In general, these gases must be competitive with other fuels, and the competitive position of any fuel is susceptible to change with time. SNG, like natural gas, can be economically transported for a considerable distance by pipeline. Transportation of intermediate-energy gas and low-energy gas is economically limited to comparatively short distances. A low-energy gas should be consumed in the vicinity of the production plant, most likely in a gas turbine engine or in a steam generator.

Extensive gasification of coal will require large amounts of capital for construction of the conversion facilities and for establishing the supporting mining operations. Existing pipelines can be used for transporting SNG, although construction of some additional pipelines may be required. Coal gasification plants for the most part are likely to be built in certain of the western states in order to tap the vast coal deposits known to exist in this region of the United States.

Liquid fuels derived from coal could become important supplements to petroleum-based fuels, provided that these converted fuels perform satisfactorily when burned in conventional engines and power plants. In the Fischer–Tropsch process, which was developed in Germany, the coal reacts with hydrogen under the influence of heat and pressure.[5] The first by-product of liquefaction is a thick, low-sulfur oil inherently suitable for burning in electric generating plants located in densely populated areas. Evidently, lighter fuels, for example, gasoline, can be produced by extending the conversion process. Certain coals in the United States are judged to possess characteristics essential for liquefaction.

Today in the United States, coal liquefaction is a highly important part of the overall research and development program directed to coal conversion. The various coal liquefaction processes under investigation yield generally solvent-refined coal, synthetic crude oil, and substitutes for petrochemical feedstocks.

Solvent-refined coal (SRC) is a low-sulfur, very low-ash solid fuel derived from high-sulfur coals. If SRC can be successfully handled and fired in large boiler furnaces, the installation of complex and costly flue gas cleaning equipment can be largely eliminated. Firing SRC in boiler furnaces however, in contrast to firing coal directly, may not prove to be an economical method for burning coal, particularly for electric power generation.

Bituminous coal contains about 5 percent hydrogen, while crude oil pumped in the United States has a hydrogen content of about 13 percent. Apparently, a synthetic crude oil derived from coal could be obtained by hydrogenation of the coal. Briefly, in the basic process, pulverized coal reacts with hydrogen under the influence of heat and pressure to form hydrocarbon liquids suitable for conversion to fuels, or for use as chemical feedstocks. Unreacted carbon would be discharged as char, a material that can be fired as a solid fuel.

Liquid fuels derived from coal would in today's market be priced substantially above the current price for comparable conventional fuels. The price differential may to some extent diminish with the anticipated increase in the cost of crude oil. In time, say, within the next one to two decades, the coal-derived liquid fuels may attain a competitive position with the petroleum-based fuels.

The conversion energy losses associated with the processes developed for coal liquefaction cannot be ignored because such losses are likely to be comparatively large. The ratio of the energy originally in the coal to the energy in the final product could be as high as 2 to 1, depending upon the nature of the particular conversion process. Large-scale production of liquid fuels derived from coal should be achieved with the lowest possible conversion energy loss, otherwise a severe drain on the country's coal reserves must be anticipated.

2.11 HYDROGEN

Extensive use of hydrogen, virtually as an all-purpose fuel, is currently receiving serious attention. Hydrogen would be burned in power plants and engines and used in a variety of heating and chemical processes, and perhaps for space heating.

There are several methods available for producing hydrogen, an element that is not present in the earth's atmosphere except in a minute quantity. Steam re-forming with hydrocarbons obtained from fossil fuels is presently the most economical method for commercial production of hydrogen. Water may be converted to hydrogen and oxygen by electrolysis, thermal dissociation at high temperatures, and thermochemical splitting. The high temperatures required for sustaining some of the processes under investigation can be achieved in solar furnaces and nuclear reactors.

An economy based on extensive use of hydrogen would require the construction of vast facilities for the production, transportation, and storage of this fuel. The most economical method for large-scale production of hydrogen will most likely be achieved by electrolysis of water through use of electrical energy generated at low cost in high-capability, breeder-type nuclear power plants. These plants could be located offshore in order to avoid objectionable thermal pollution of the surroundings and eliminate the opposition that often develops in the selection of a site for a power plant.

The hydrogen would be transported from the production plant either as a gas or as liquefied hydrogen. Significant differences appear in these two modes of transportation, and each mode will require detailed examination prior to possible adoption. Liquefied hydrogen, in comparison with gaseous hydrogen, offers the advantage of low pressure and low bulk, but containment difficulties arise from the cryogenic nature of the liquefied gas. The use of liquefied hydrogen as a fuel in rocket engines has demonstrated that hydrogen can be safely handled when reasonable precautions are observed.

Hydrogen, as a substitute for fossil fuels, would be burned in gas turbine and internal combustion engines for the production of mechanical energy and the propulsion of aircraft. The production of electrical energy in fuel cells and in steam power plants fueled with hydrogen requires a careful examination of the conversion losses. The energy conversion chain is now somewhat extended: nuclear energy → electrical energy → chemical energy (hydrogen) → electrical energy. In this energy conversion chain, the hydrogen serves as an energy carrier, and each individual conversion involves some loss. Production of electrical energy in this manner may not prove to be universally practicable, thus the decision to burn hydrogen in a steam-electric power plant will be strongly influenced by the overall operating economy.

Unlike electrical energy, hydrogen can be economically stored, particularly during the off-peak periods that occur in the operation of electric power systems. Further, the cost of transporting a gas by pipeline is relatively low in comparison with conventional transmission costs for electrical energy.

Hydrogen is a clean-burning fuel. The products of combustion, when hydrogen is burned completely with air, consist of water, oxygen, and nitrogen. Oxides of nitrogen will form during the combustion reaction, but the products will be free of the pollutants that are ordinarily present in the gases discharged from fossil fuel-fired power plants and engines. Laboratory tests conducted on internal combustion engines burning hydrogen demonstrate good performance. In comparison with an engine burning gasoline, the emission of nitrogen oxides is far less for the engine fueled with hydrogen.

Petroleum-based fuels possess a characteristic important to the various types of vehicles used in transportation. A comparatively large quantity of energy is associated with the volume occupied by the liquid fuel. The heating value per unit mass of hydrogen is relatively high, but the low density of liquid hydrogen signifies a low volumetric concentration of energy. For a prescribed quantity of energy, liquid hydrogen would weigh less than gasoline; but considerably larger tankage, about four times larger, would be required. This increase in tank size would present a problem in vehicle design, particularly in the light of a trend toward the manufacture of smaller automobiles.

It is not feasible to charge an automobile fuel tank with gaseous hydrogen. The volumetric energy content of the fuel would be very low despite high tank pressures, which in turn would necessitate a heavy structure.

A metal hydride storage system for installation in automobiles has been developed.[6] The fuel tank contains a bundle of tubes manufactured from an iron–titanium alloy that has a hydrogen absorption capability. Hydrogen is held inside the metal powder by chemical bonds which can be broken by heat supplied from the engine cooling system. Should the fuel tank be accidently ruptured, the fuel would escape so slowly that the possibility of a fire casualty would be remote.

Of particular significance is the recycling characteristic of hydrogen when produced by the dissociation of water. The hydrogen however is a secondary fuel, and the primary source of energy, for example, a nuclear fuel, is consumable. Incidentally, the electrolytic process for producing hydrogen would also yield oxygen, a useful by-product that has numerous industrial applications.

Example 2.8

Hydrogen is completely burned in a continuous-flow system with 320 percent theoretical air. The hydrogen is supplied at 25 C, and the air enters at 500 K. Assume zero heat loss and determine the temperature of the gas at the outlet section of the combustion chamber.

1 Combustion equation.

(a) With theoretical air.

$$H_2 + 0.5O_2 + (3.76 \times 0.5)N_2 \rightarrow H_2O + (3.76 \times 0.5)N_2$$

(b) With 320% theoretical air.

$$H_2 + 1.6O_2 + 6.016N_2 \rightarrow H_2O + 1.1O_2 + 6.016N_2$$

2 Energy equation.

$$[(\Delta\bar{h}^\circ)_{H_2O} + 1.1(\Delta\bar{h}^\circ)_{O_2} + 6.016(\Delta\bar{h}^\circ)_{N_2}]_T = -(\bar{h}^\circ_{RP})_{H_2} + [1.6(\Delta\bar{h}^\circ)_{O_2} + 6.016(\Delta\bar{h}^\circ)_{N_2}]_{500\text{ K}}$$

For hydrogen—water vapor in the products of combustion,

$\bar{h}^\circ_{RP} = -119{,}874$ kJ/kg (from Table A.3)

$\bar{m} = 2.016$

At 500 K,

$(\Delta\bar{h}^\circ)_{O_2} = 6088$ kJ/kgmol (from Table A.5)

$(\Delta\bar{h}^\circ)_{N_2} = 5912$

Substitution into the right-hand side of the energy equation yields

$$-[2.016(-119874)] + (1.6 \times 6088) + (6.016 \times 5912) = 286{,}973 \text{ kJ}$$

At temperature T_p, determined by trial and error, the energy of the products equals 286,973 kJ. Energy values of the product gases are selected from Table A.5.

		T = 1300 K		T = 1400 K	
Gas	*N (mol)*	$\Delta\bar{h}^\circ$ *(kJ/kgmol)*	$N \times \Delta\bar{h}^\circ$ *(kJ)*	$\Delta\bar{h}^\circ$ *(kJ/kgmol)*	$N \times \Delta\bar{h}^\circ$ *(kJ)*
H_2O	1	38903	38903	43447	43447
O_2	1.1	33351	36685	36966	40663
N_2	6.016	31501	189510	34936	210175
			265099		294285

By interpolation $T_p = 1375$ *K.*

The temperature $T_p = 1375$ K that results from combustion of hydrogen is compared with the temperature $T_p = 1272$ K achieved by burning dodecane under similar conditions, as shown in Example 2.4. The higher temperature T_p for combustion of hydrogen can be expected when a comparison is made of the energy values per kilogram of air for the two fuels. Thus,

(a) Hydrogen.

7.616 mol air/mol fuel

$\overline{m}_{air} = 28.970$

LHV of hydrogen = 119,874 kJ/kg

$\overline{m}_f = 2.016$

$$\frac{\overline{m}_f \times \text{LHV}}{N_{air}\overline{m}_{air}} = \frac{2.016 \times 119874}{7.616 \times 28.970}$$

$$= 1095.3 \text{ kJ/kg air}$$

(b) Dodecane.

281.79 mol air/mol fuel

LHV of liquid dodecane = 44,085 kJ/kg

$\overline{m} = 170.34$

$$\frac{\overline{m}_f \times \text{LHV}}{N_{air}\overline{m}_{air}} = \frac{170.34 \times 44085}{281.79 \times 28.970}$$

$$= 919.9 \text{ kJ/kg air}$$

Combustion of gaseous dodecane, $(\text{LHV})_g$ = 44443 kJ/kg, instead of liquid dodecane, $(\text{LHV})_\ell$ = 44085 kJ/kg, under the same conditions would produce a slightly higher value of T_p, namely, T_p = 1278 K.

2.12 WASTE MATERIAL

The quantity of waste material collected each year in the United States is estimated at 180 million metric tons. This refuse material is dumped into a land fill or burned, usually in an incinerator where recovery of energy can be achieved. More waste material will be burned as the practice of dumping diminishes in response to a move to conserve energy. Further, dumping areas are unsightly, and land for this purpose is becoming less available. During most of the year, the refuse collected in this country has a heating value of 11,500 to 14,000 kJ/kg, not an insignificant quantity.

Municipal solid waste material can be converted by pyrolysis to a combustible gas.[7] Pyrolysis is the thermal decomposition of a substance in the absence of oxygen. The product, in this case, is an intermediate-energy gas that can be used for power generation or as a source of heat for a variety of processes. The energy ratio, heating value of the product gas to the heating value of the waste material, is somewhat below 0.7.

Mixed household and commercial wastes can be burned as a supplementary fuel in boiler furnaces.[8] The steam generator firing equipment is modified in order to allow firing the waste material simultaneously with a conventional fuel. Tests conducted on a pulverized coal-fired boiler demonstrated that the waste material should be fired in relatively small percentages, on the order of 10 to 20 percent of the full-load fuel requirement.

While the combustion of refuse in boiler furnaces can be achieved, certain operating problems will develop, particularly with respect to the variation in the homogeneity of this material. The refuse has a high and variable moisture content, and the material is more corrosive than most coal and oil fuels. Storage facilities for the refuse must be provided because the refuse collection schedule differs from the operating schedule for the power plant. Test results indicate however that supplementary firing of municipal wastes in steam-electric power plants may under certain circumstances prove to be economically feasible.

The gases produced from solid waste material can, in addition to being burned directly as fuel, be converted to useful chemicals, such as methanol and ammonia. The feasibility of achieving this conversion has been demonstrated in a large-scale system in which a product gas is initially produced by partial oxidation (pyrolysis) of the waste material. The product yield for the conversion process is estimated to be 185 kg of methanol or 225 kg of ammonia per metric ton of solid waste.[9] The primary purpose of installing a chemical conversion system of this kind is, at present, waste disposal. The overall operation however, starting with the collection of the waste material and terminating with the sale of the end product, must be economically justified. In this respect, the market price of the end product is an important factor.

Despite an earlier optimistic outlook, solid waste resource recovery is presently regarded as an unsuccessful approach to the disposal of municipal waste. A number of resource recovery plants have been constructed, but the operation of most of these plants has proved to be uneconomical. Further, various technical difficulties have developed, and economic markets for recovered resources have not materialized. Because of firing problems, power plant operators are reluctant to accept refuse-derived fuels. Difficulties are experienced in shredding the waste, usually the first step in the recovery process. Severe shredder explosions have occurred, and often a substantial fraction of the collected waste is unsuitable for shredding and must be diverted directly to a land fill.

2.13 SHALE OIL

Deposits of oil shale occur in many areas of the world, including the extensive Green River Formation located in the states of Colorado, Utah, and Wyoming. Oil found in shale is actually a solid, high-molecular-weight organic compound called kerogen that has the following ultimate analysis in percent by mass: carbon 80.3, hydrogen 10.4, nitrogen 2.3, sulfur 1.1, and oxygen 5.9.[10]

The origin of kerogen, a fossil substance, is attributed to plankton that for millions of years existed in prehistoric lakes. The remains of the plankton sank to the bottom of the lake in layers, embedded with the silt washed into the water from the surrounding ranges. The streaked oil shale was thus formed to appear today as deposits in the dry basins of the early lakes. Actually, the rock formation is more appropriately described as a marlstone, rather than shale.

Briefly, in order to extract the oil, the shale is mined and subsequently heated in a retort to about 550 C. Approximately 10 percent by mass of the shale is recovered as oil or gas, while the remaining substance is waste material. The raw shale oil is refined to produce a commercial product that is likely to have a very low sulfur content. This method of recovering shale oil has some disadvantages. The disposal of large quantities of waste material could create an unacceptable impact on the environment. Processing the shale requires large amounts of water that may not be available in sufficient quantity to support the operation.

The possibility of retorting oil shale *in situ* has been investigated. The rock is first fractured using either explosive or mining techniques, and then heated to release the oil. Air is introduced to effect *in situ* combustion of part of the organic material in the oil shale. The hot combustion gases move in advance of the combustion zone and cause the bulk of the oil to be released from the shale. Provision is made for collecting the oil and pumping it to the surface. In comparison with surface retorting of the mined shale, *in situ* processing offers the advantages of lower cost for extracting the oil and a less severe impact on the environment.

The shale oil content of the Green River Formation is estimated to be 2 trillion barrels. Not all of this oil is economically recoverable. If the lower limit for economic recovery is set at about 100 liters of oil per metric ton of shale, the recoverable amount of oil is perhaps 500 billion barrels, an amount some 15 times the known reserves of crude oil in the United States.[11] The 1974 estimate of recoverable oil from the Green River Formation is 600 billion barrels.

Although oil shale is a very large fossil fuel reserve, to date no attempt has been made to produce shale oil on a commercial scale. Various investigations however are currently directed to exploring the means to extract effectively the oil from the rock. Preliminary reports indicate that a high percentage, about 95 percent, of the oil present in the rock can be extracted, along with large quantities of useful gas. Conversion energy losses are comparatively low, on the order of 10 to 15 percent. A quantity of refined fuel derived from shale oil will be tested in selected commercial vehicles in order to evaluate the performance of the fuel for use in automotive engines.

The significance of the oil shale as an energy source was demonstrated by Hubbert[12] in estimating fossil fuel reserves within the United States, as follows:

Coal	78 percent	(73)
Oil shale	16 percent	(17)
Natural gas	3 percent	(9)
Petroleum	3 percent	

These figures represent Hubbert's estimate in 1962 of production plus known reserves, and reserves yet to be discovered. The parenthetical figures represent other estimates.

Oil and gas shale deposits considerably more extensive than the deposits in the Green River Formation are believed to exist in the eastern and midwestern regions of the United States. The eastern shale reserve is estimated to represent more than 60 times the current crude oil reserves (30 billion barrels) of the United States.

2.14 TAR SANDS

Tar sands are another possible source of vast quantities of oil. In the United States, tar sands can be expected to yield an estimated 10 billion barrels of oil. A much larger quantity of oil may be recovered from the Athabasca Tar Sands located in northern Alberta. Estimates of the amount of oil contained in these sands range from 100 to 600 billion barrels. Initially, oil will be recovered in an area where the tar sands are saturated with up to 25 percent by mass of oil.

Tar sand oil is a bitumen or naturally occurring asphalt. Some difficulty is experienced in processing tar sands, because of the abrasive and sticky characteristics of these sands.

Commercial production of fuel derived from tar sands is presently somewhat questionable, principally because of the absence of strong economic incentives. For the next several years, the extraction of oil from tar sands is likely to remain in the pilot plant stage. High prices and short supplies of crude oil will however improve the competitive position of fuels manufactured from oil shale and tar sands.

2.15 FUELS FOR SPECIAL APPLICATIONS

The space exploration program that was initiated during the early years of the 1960 decade promoted the utilization of many nonconventional fuels and systems in which these fuels are burned. Because the requirements for these fuels and fuel-burning systems are high performance and maximum reliability, cost is of secondary importance. In other words, economic competition with conventional fuels and combustion systems is not a factor that requires consideration.

Combustion of fossil fuels is normally achieved through a reaction between the combustible components of the fuel and the oxygen in the air supplied to the combustion chamber. For some of the nonconventional systems, oxygen is not a reacting substance, although one of the substances participating in the reaction is designated as the oxidizer. For example, in the hydrogen–fluorine reaction, fluorine is designated as the oxidizer and hydrogen as the fuel. Most of the high-performance fuels are used in rocket engines.

2.16 NUCLEAR FUELS

Power reactors are fueled with uranium, the heaviest element that occurs in nature in more than trace quantities. Uranium ores are found in various regions throughout the world. The most abundant uranium resources are located in the United States, Canada, Australia, South Africa, and South-West Africa (also known as Namibia). In the United States, most of the commercial-grade uranium ore is found in the western region of the country. The major uranium-producing areas are central Wyoming and the Colorado Plateau (northwestern New Mexico and adjacent portions of Arizona, Utah, and Colorado).

Underground mines supply most of the uranium ore, although some shallow deposits are mined by open-pit techniques. Two principal uranium minerals are pitchblende and carnotite, which contain, respectively, 50 to 80 and 50 to 55 percent uranium oxide (U_3O_8). Because of the presence of other materials, the uranium ore as mined contains only a small amount of U_3O_8, usually on the order of 2.5 to 3 kg per metric ton of ore.

The U_3O_8 is separated from the ore and subsequently, in a series of processes, refined and converted to uranium dioxide (UO_2). The refined product contains three isotopes of uranium, namely, ^{234}U, ^{235}U, and ^{238}U. The ^{234}U isotope exists in minute traces and may consequently be disregarded. The ^{235}U content is 0.72 percent of the total quantity. Because the ^{235}U isotope fissions with comparative ease in light-water reactors, it is presently of greater significance than the ^{238}U isotope, which occurs much more abundantly.

Nuclear power reactors, constructed in accordance with currently established light-water designs, are fueled with enriched uranium (3 to 5% ^{235}U). Breeder reactors, now under development, will convert, as a secondary effect, an unfissionable isotope into a fissionable isotope of another element. The two predominant isotopes that exhibit significant possibility of conversion are uranium 238 and thorium 232. The source of uranium is noted above.

In the breeder reactor the normally unfissionable ^{238}U is converted into plutonium 239, a fissionable isotope. Similarly, the fertile ^{232}Th is converted into fissionable ^{233}U. The fissionable ^{239}Pu or ^{233}U is subsequently used as a primary nuclear fuel.

Monazite sands are the only known source of thorium. These sands result from weathering of granite, and the alluvial deposits contain a great deal of silica. Concentration of the monazites is achieved by ore dressing. Monazites, a mixture of rare earth phosphates, usually contain 1 to 5 percent thorium dioxide (ThO_2). The largest deposits of monazite sands are found in Brazil, India, and Canada, and to a lesser extent in the United States. The monazites found in Brazil and India have a higher-than-average (5 to 10 percent) concentration of ThO_2.

2.17 SOLAR-DERIVED FUELS

The geologic formation of the nonrenewable fossil fuels, which occurred millions of years ago, started with organic material produced by photosynthesis. Commercial-scale application of the photosynthetic process to present-day production of fuels is viewed as a possible approach to augmenting the energy resources of the world. A considerable amount of the basic vegetable material is available in the form of agricultural wastes, forest biomass, and industrial residues. Large-scale production of solar-derived fuels would however require cultivation of energy crops. Large tracts of land in the humid tropics and in the temperate zones are not suitable for growing foodstuffs but can be used to grow a variety of plants that would become a renewable resource for the production of chemicals and fuels.

Conversion of animal and vegetable matter into various chemicals and fuels has been demonstrated in a number of processes.[13] Low-temperature pyrolysis of vegetable material, achieved by reduced air flow, will yield solid char, combustible gases, and volatile organic vapors. A gaseous product is produced by high-temperature pyrolysis accomplished by increasing the air flow to the converter. The pyrolytic products are subsequently used as fuels or in the manufacture of chemicals, such as methanol. Vegetable material can be converted into ethanol by fermentation and distillation, while methane can be produced by anaerobic digestion of vegetable and animal matter.

Evidently, land is available in the United States for extensive cultivation of fuel crops, but any such activity must necessarily be dependent to a considerable degree upon economic factors, and presently this approach to augmenting fuel resources is viewed only with passing interest. Land requirements for crop production should be noted. An arable land area on the order of 1000 km^2 would be required for growing sufficient combustible material for fueling a 1000-MW power plant.

2.18 FUEL RESOURCES

An evaluation of the world's nonrenewable energy resources usually considers the proved and currently recoverable reserves and the estimated total remaining recoverable amount of each different kind of fuel. Because estimates of fuel deposits are made with some, or perhaps considerable, uncertainty, the various available fuel resource evaluations can vary appreciably in the total amount of fuel, in any category, that may ultimately be recovered. At any particular time, the quantity of fuel considered to be recoverable will be dependent in part on economic factors, for example, the current price of uranium oxide (U_3O_8).

Proved reserves do not represent exact evaluations of the extent of subterranean fuel or mineral deposits. In order to determine a proved reserve, the field's strata are studied by drilling for rock and fluid samples and subsequently predicting the probable output of the field by comparison with the production from similar fields already in operation. The actual output of a field seldom corresponds with the estimated yield. The possible or total resources of a deposit are considerably more uncertain than the proved reserves.

Tables 2.3 and 2.4 provide estimates of the world's nonrenewable energy resources. The total remaining fossil energy that may ultimately be recovered is estimated in these two tables at 208 and about 150 Q, respectively (1 Q = 10^{18} Btu). These two estimates differ principally in the quantities of coal and in the quantities of oil that may be extracted from shale and tar sands. Coal is by far the predominant fossil fuel resource.

The world total energy consumption is currently about 0.25 Q, a comparatively small quantity in the light of the total remaining recoverable fossil energy, say, 150 to 200 Q. A direct comparison of 0.25 Q per year with a total 175 Q is however misleading. The world consumption of crude oil is equivalent to about 0.12 Q per year, or one-half of the total energy consumption of 0.25 Q per year.

The heavy dependence upon petroleum for transportation, heating, power generation, and production of petrochemical feedstocks is currently causing throughout the world a critical situation with respect to the utilization of fossil fuels. Severe economic dislocations have been brought about by the high price, rather than the supply, of crude oil. A shift from oil to coal could alleviate the problem to some degree, but any substantial decrease in the demand for crude oil over the next several years cannot be easily achieved, or is it believed to be likely.

TABLE 2.3 Energy Content of the World's Supply of Recoverable Fossil Fuels

	Initially Available[a]		*Available at the end of 1975 (Q units)*
	Quantity	*Q Units*	
Coal and lignite	7.6×10^{12} metric tons	191	187
Petroleum liquids	2000×10^9 bbl	11.1	9.2
Natural gas	$10{,}000 \times 10^{12}$ ft^3	10.0	8.6
Tar sand oil	300×10^9 bbl	1.74	1.74
Shale oil	190×10^9 bbl	1.09	1.09
Total		215	208

[a] *From M. K. Hubbert.*[14] *1Q = 10^{18} Btu; bbl = barrels.*

TABLE 2.4 Total Nonrenewable World Energy Resources* (As of December 31, 1975)

	Proved and Currently Recoverable	*Estimated Total Remaining Recoverable*
Coal		
10^9 Short tons	668	5374–6126
10^{18} Btu (= 1Q)	13.79	107.48–122.52
Natural gas		
10^{12} ft^3	2050–2507	9150–9550
10^{18} Btu	2.11–2.58	9.43–9.85
Natural gas liquid		
10^9 bbl	54–66	243–253
10^{18} Btu	0.22–0.27	0.99–1.04
Crude oil		
10^9 bbl	547–644	1520–1860
10^{18} Btu	3.17–3.74	8.82–10.79
Syncrude from oil shale and bitumen		
10^9 bbl	270	2415 (17,500)
10^{18} btu	1.57	14.01 (101.5)
Uranium oxide at < \$30/lb_m		
10^3 Short tons	2403	6614
10^{18} Btu in burner reactors	0.961	2.646
10^{18} Btu in breeder reactors	72.08	198.4
Total fossil energy, 10^{18} Btu	20.86–21.95	140.73–158.21
Total fossil and fissile energy, burner reactors, 10^{18} Btu	21.8–22.9	143.4–160.8
Total fossil and fissile energy, breeder reactors, 10^{18} Btu	92.9–94.0	339.2–356.6

Source: Compiled by the Institute of Gas Technology.[15]

A reasonably good supply of crude oil will be available for perhaps another decade, unless production is arbitrarily curtailed. The price of oil however is expected to increase continually. The data presented in Table 2.4 indicate that 1000 to 1200 billion barrels (5.80 to 6.96 Q) of oil remain to be discovered. Other estimates fix the amount of oil at 925 to 1275 billion barrels.[16]

Locating "new" oil is becoming more difficult, and the cost of drilling wells is sharply increasing. Toward the end of the century, or perhaps much sooner, the world demand for crude oil is expected to exceed the production.

The world consumption of energy is expected to increase substantially during the next century. A possible evaluation of the energy increase is given by W. Häfele[17]:

1970	0.24 Q/year	4×10^9 people	2 kW(t)/capita
2000	2.1	7×10^9	10
2050	6	10×10^9	20

2.19 ENERGY STATUS OF THE UNITED STATES

During the year 1974, the United States, with 5.5 percent of the world's population, consumed 30.5 percent of the world energy consumed during the year. The following tabulation is a comparison for the year 1974 of the energy consumption for certain selected industrial countries.

	Per Capita Energy Consumption (kg Coal Equivalent)	*Percent of World Population*	*Percent of Total World Energy Consumption*
United States[a]	11,485	5	30
U.S.S.R.	5252	7	17
Western Europe	4243	9	19
Japan	3839	3	5
World average	2059		

Source: U.N. Statistical Papers.[18]

[a] *In 1979, the United States consumed 28.8 percent of the worldwide consumption of energy. The respective energy quantities were, in tons of coal equivalent, 2506 and 8706 million metric tons. (Statistical Abstract of the United States 1981.* U.S. Dept. of Commerce Bureau of Census U.S. Government Printing Office, Washington, D.C. 1981.)

The heavy consumption of energy in the United States raises the question whether self-sufficiency in the utilization of energy can be achieved. In past years, imports of natural gas, crude oil, and petroleum products supplemented domestic supplies. While importation of these fuels could be economically justified, the dependence to a substantial degree upon foreign sources of fuel proved to have unfortunate consequences with respect to price escalation and supply. An examination of the remaining energy sources of the United States will provide an insight into the possibility of achieving self-sufficiency in the use of energy.

The United States has vast coal resources and a relatively good uranium resource. The supply of natural gas is presently adequate, but the domestic production of petroleum falls well below the demand. The total remaining and recoverable energy resources of the United States are shown in Table 2.5

TABLE 2.5 United States Energy Resources

Resource	*Quantity*	*Energy (Q units)*
Coal	600×10^9 tons	14
Petroleum	160×10^9 bbl	0.928
Natural gas	750×10^{12} ft^3	0.787
Uranium, proved	700,000 tons	0.610[a] 61[b]
Uranium, speculated (total in ore ≥ 200 ppm)	3×10^6 tons?	2.6[a] 260[b]
Lithium 6 for fusion	10^6 tons?	3×10^5

Source: D. J. Rose, P. W. Walsh, and L. L. Leskovjan.[19]

[a] *If used in present light-water reactors only.*

[b] *If used in breeder reactors.*

For the year 1979, the energy consumed in the United States was derived from the following resources:

Resource	*q*[a]	*Fraction*
Coal	15.1	0.191
Petroleum	37.1	0.470
Natural gas	20.7	0.262
Hydroelectric	3.2	0.041
Nuclear	2.7	0.034
Geothermal, other	0.1	0.001
Total	(79.0)	(1.000)

Source: *Statistical Abstract of United States 1981* U.S. Government Printing Office, Washington, D.C., 1981.
[a] *1q = 10^{15} Btu.*

The petroleum resources of the United States are 160×10^9 bbl, or an equivalent 928 q units of energy (See Table 2.5). The proved petroleum reserves are however considerably lower, about 30 billion barrels (174 q). The current domestic yearly production of crude oil in the United States is equivalent to about 20 q, a large quantity in comparison with the proved reserves of approximately 200 q. The petroleum resources will ultimately yield more oil than can be extracted solely from the proved reserves, but "new" oil is difficult and costly to find. The probability of drilling a "dry" hole has increased, and much of the exploration for oil is undertaken at offshore sites. Currently, of the domestic oil resources, three barrels of oil are consumed for each two barrels that are found. In 1980, the consumption of crude oil and refined petroleum products in the United States was somewhat above 16 million barrels per day, of which nearly 45 percent was imported.

The present rate of consumption of natural gas in the United States is 20×10^{12} ft^3 (21 q) per year. The proved reserves of natural gas are about 200×10^{12} ft^3 (210 q). A low reserves-to-consumption ratio of about ten years indicates the need for conservation and improved production. The total natural gas resources, equivalent to roughly 800 q, can be expected to yield additional gas. Presently in the United States, natural gas is consumed at a rate twice that at which gas is discovered. The untapped proved gas reserves in Alaska are (22 to 26) $\times 10^{12}$ ft^3, but the total may reach 360×10^{12} ft^3.

The preceding statistical information indicates that, except for petroleum and natural gas, the United States has ample energy resources. The continual increase in the consumption of petroleum products and the growing and heavy dependence upon imported oil are cause for grave concern. There are however several steps that can be taken to offset the deficiencies in the domestic oil and gas supplies.

The vast coal deposits in the United States will support conversion from gas and oil firing to coal firing, particularly in power plants and industrial boilers. Substitute liquid and gaseous fuels can be manufactured from coal and oil shale. Accelerated construction of nuclear power plants can be undertaken because the supply of uranium is relatively abundant. After the year 2000, contributions may be anticipated from other sources of energy, such as solar, geothermal, and wind.

Some difficulty is likely to be experienced in

any attempt to increase substantially the use of coal. Conversion from oil and gas firing to coal firing will in many cases be costly and will require extensive changes in power plant equipment. New fossil fuel-fired power plants will however for the most part burn coal.

The outlook for a rapid increase in coal production is somewhat pessimistic because of the numerous constraints on the production and consumption of coal. These constraints arise from environmental and safety regulations, equipment and skilled workforce limitations, inadequate transportation, productivity decline, and the long lead time and high capital cost inherent in starting a new mining operation. Improved rail transportation, and possible construction of coal pipelines, could alleviate any difficulty in transporting coal economically in sufficient quantity.

If today coal were to replace imported oil solely on an equivalent energy basis, the production of coal would be essentially doubled. The mining industry is judged to have the capability of increasing coal production from 750 million metric tons in 1980 to about a billion metric tons per year in 1985, provided a realistic national energy plan can be adopted.

An estimate of the future rate of energy consumption is usually based on the projection of past demand for energy, with adjustments made to allow for probable changes in economic trends. Electrical energy generation in the United States has for the past several decades increased at an average yearly rate of 7.4 percent, that is, demand for electrical energy doubled every 10 years. This rate of increase is in a general sense projected forward to the year 2000. However, in the actual planning for increased generating capability, a somewhat more conservative estimate is likely to be used (see Fig. 2.1). Production of electrical energy in the United States for the year 1979 was 2319 billion kWh, and the estimated production for the year 2000 is 9036 billion kWh.[21] The increase in the consumption of electrical energy for the decade 1970–1980 when projected to the year 2000 indicates the demand for electrical energy will be substantially below the estimated 9036×10^9 kWh/year, a figure based on an optimistic growth in the national economy.

A number of estimates have from time to time been made of the total energy consumption over the next several decades. For the period 1965–1975, the total energy consumption in the United States increased at a rate of about 6 percent per year. For future years, extending to the year 2000, a somewhat lower rate is anticipated, probably on the order of 3.5 percent per year.

A forecast of the energy requirements of the United States to the year 2000 was prepared by the Bureau of Mines. Certain data abstracted from the report are presented in Table 2.6. The several assumptions on which the forecast of energy consumption is based relate to the gross national product, population, industrial production, fuel availability, technology, and life style.

Table 2.6 shows the distribution of the gross energy consumption to the major consuming sectors of the energy market. The consumption of raw fuel for the production of synthetic gas is expected to become a significant quantity sometime prior to the year 2000.

The electric generation and the synthetic gas sectors are not the final consumers of the energy input. When the electrical energy and the syn-

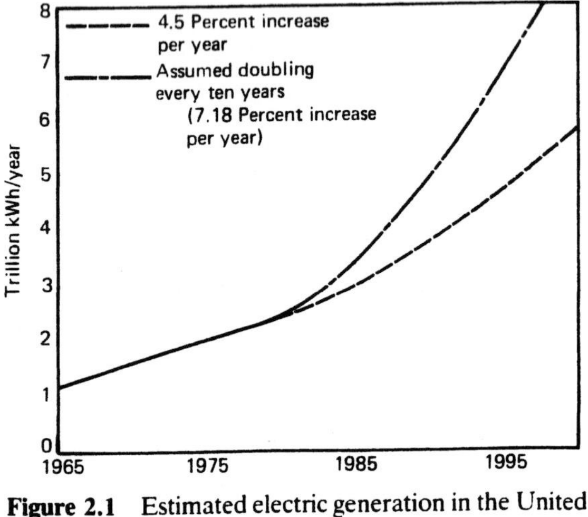

Figure 2.1 Estimated electric generation in the United States from 1980 to 2000.

TABLE 2.6 United States Energy Consumption (in q Units)

	1971[a]		*2000*	
	Gross	*Net*	*Gross*	*Net*
Household and commercial	14.281	(17.441)	21.920	(39.630)
Industrial	20.294	(22.623)	39.300	(57.780)
Transportation	16.971	(16.989)	42.610	(42.660)
Electric generation	17.443		80.380	
Synthetic gas	—		7.690	
(Conversion losses)		(11.936)		(51.830)
Total	68.989	68.989	191.900	191.900

Source: W. G. Dupree, Jr.[22]
[a] *Actual*

thetic gas produced are distributed to the final consuming sectors, the resulting figures give the net energy input to the economy as shown parenthetically in Table 2.6.

Significant differences appear between the total net energy consumed and the total energy consumed in the form of the raw fuel. These differences represent the conversion losses experienced in converting primary energy into secondary sources, in the electrical and synthetic gas sectors. The conversion losses are expected to increase from 11.936 q in 1971 to 51.830 q in the year 2000. The corresponding percentages of the total gross energy input are 17.3 and 27.0.

The forecast of an energy consumption of 191.9 q for the year 2000 is in line with similar estimates reported by other investigators. Today, it is difficult to indicate with any assurance the energy sources that will meet the anticipated energy demand of about 190 q in the year 2000. Energy studies conducted circa 1970 allocated substantial increases through the years 1970 to 2000 to the energy supplied by petroleum, from 30 q to 71 q, and by nuclear fuels, from 0.4 q to 49 q.[22] It is doubtful that energy increases of such magnitudes can be achieved in these two categories because of the unstable world supply of petroleum and the lag in the construction of nuclear power plants. In the same light, an increase from 12.5 q to 31 q over a 30-year period in the energy supplied by coal would represent a formidable undertaking.

PROBLEMS

2.1 A power plant rated at 500 MW can burn a bituminous coal from either Kentucky or Wyoming. The power plant thermal efficiency is 0.38 based on the higher heating value of the fuel. The Kentucky coal has a higher heating value (HHV) of 28,079 kJ/kg and a sulfur content of 3.5 percent. For the Wyoming coal, HHV = 30,125 kJ/kg and S = 0.50 percent. For continuous full-load operation, determine the annual SO_2 production that can be anticipated when burning these two coals.

2.2 A coal-fired power plant rated at 600 MW operates with a thermal efficiency of 0.40 based on the higher heating value of the fuel. For the coal fired in the plant, the HHV is 30,915 kJ/kg, and the theoretical air requirement is 0.325 kg per 1000 kJ HHV. The coal is fired with 20 percent excess air, and the air is measured at 105 kPa, abs, and 32 C. Estimate the combustion air flow, in m^3/min, at the rated output.

2.3 A power plant burns a fuel oil that has a specific gravity of 0.904 and a higher heating value of 44,746 kJ/kg. The thermal efficiency achieved at the rated output of 200 MW is 0.35. Determine the size of the tank, in m^3, required for storing a 30-day supply of fuel. Assume continuous operation at full load.

2.4 An internal combustion engine develops an output of 100 kW with a thermal efficiency of

0.28 based on the lower heating value of the fuel (43,900 kJ/kg). The fuel contains 85 percent carbon by mass. The engine exhaust gas contains, on a volumetric basis, 12.5 percent CO_2 and 2.1 percent CO. Calculate the discharge of CO from the engine in grams per hour.

2.5 A 350-MW steam power plant is fired with natural gas that has a higher heating value of 36,694 kJ/m^3 measured at 1 atm, 15.5 C. The thermal efficiency for operation at the rated load is 0.37 based on the higher heating value of the fuel. The gas in the pipeline that supplies the plant is at a pressure of 800 kPa, gage, and a temperature of 25 C. Estimate the diameter of the pipeline for a prescribed gas velocity of 20 m/s.

2.6 Assume a coal that contains 79.2 percent carbon and 5.4 percent hydrogen by mass can be converted to gasoline by hydrogenation. Determine the required amount of hydrogen that is added and the gasoline yield per kilogram of coal.

2.7 A fuel tank installed in an automotive vehicle has a capacity of 100 liters. The fuel is assumed to be either liquid propane or liquid hydrogen. The density and lower heating value are, respectively, 579.9 kg/m^3 and 45,960 kJ/kg for liquid propane and 70.96 kg/m^3 and 119,420 kJ/kg for liquid hydrogen. Assume that the same thermal efficiency is achieved with either fuel. When operating on propane, the fuel consumption is 0.12 liter/km. Calculate for each fuel the distance the vehicle can travel on a full tank of fuel.

2.8 A power plant burns a Wyoming bituminous coal that has a higher heating value of 30,940 kJ/kg. The power plant has a rating of 2000 MW, and the average operating thermal efficiency is 0.35. The coal is transported from the mine to the plant by unit trains that consist of 100 cars, each of which has a capacity of 90 metric tons. Assume an annual plant factor of 0.60. How many trains will arrive at the plant during a period of one year?

2.9 Estimate the energy conversion ratio for the Lurgi process.

2.10 Transportation of hydrogen gas by pipeline requires pressure "boosting" in a pumping station. At the station, the pressure of the gas is increased from 700 kPa, abs, to 3000 kPa, abs, by a centrifugal compressor that is driven by a gas turbine engine. The efficiency of the compressor is 0.78 and the thermal efficiency of the gas turbine engine is 0.32, based on the lower heating value of the hydrogen fuel. The hydrogen gas enters the compressor at 310 K. For hydrogen, $k = 1.399$ and $c_p = 14.46$ kJ/kg · K. How much hydrogen is required to fuel the gas turbine engine for each kilogram of gas that enters the pumping station?

2.11 Methanol and octane are assumed to burn with the theoretical amount of air. Determine the energy value of a cubic meter of methanol–air and octane–air mixture at 1 atm, 25 C. The lower heating values for methanol vapor and octane vapor are, respectively, 21,106 and 44,759 kJ/kg.

2.12 The present crude oil production in the United States may be taken as equivalent to 21 q. If the production rate is continued unchanged, when will the recoverable petroleum resources be depleted? In how many years will the resources be depleted if the production of crude oil increases at a yearly rate of 5 percent? Determine the annual production, in q units, at the time of depletion.

2.13 The heating system for a small factory is designed for a building heat loss of 500,000 kJ/h. The cost of fuel oil is 15¢ per liter, while coal costs $30 per metric ton. The heating system efficiency is 0.70 for oil firing and 0.60 for coal firing. Calculate for the design conditions the fuel cost, in cents per hour, for burning oil and for burning coal. Which fuel would you select? Why?

2.14 A coal-fired power plant generates 200 MW with a thermal efficiency of 0.35. The higher heating value of the coal is 31,736 kJ/kg, and the coal analysis, in percent by mass, is C 0.7544, H 0.0535, O 0.1096, N 0.0199, S 0.0139, and ash 0.0487. The coal is burned with 25 percent excess air. The flue gas is discharged to the stack at 100.0 kPa, abs, and 150 C. The gas velocity in the stack is 10 m/s. Calculate the diameter of the stack.

2.15 The composition of a natural gas in percent by volume, is methane 78.1, ethane 19.9, and nitrogen 2.0. Verify the reported higher heating value of 41,700 kJ/m^3 for the gas, measured at 1 atm, 20 C.

2.16 $C_{10}H_{20}$ is burned with 108 percent theoretical air. Calculate the probable volumetric analysis of the dry products of combustion.

2.17 Repeat Problem 2.16 for combustion of the fuel with 95 percent theoretical air. The hydrogen is assumed to burn completely. Determine the dew point temperature of the products of combustion. The barometric pressure is 100.0 kPa, abs. Neglect the water vapor in the atmosphere.

2.18 The composition of a gaseous fuel is CH_4 60, C_2H_6 30, and N_2 10 percent by volume. Calculate the air/fuel ratio and the analysis of the wet and the dry products of combustion. The fuel is burned with 110 percent theoretical air. The combustion air is supplied at 100.0 kPa, abs, 30 C, and 55 percent relative humidity.

2.19 The analysis for a bituminous coal reports carbon 76.6, hydrogen, 4.9, sulfur 1.3, oxygen 3.9, nitrogen 1.6, ash 9.1, and moisture 2.6 percent by mass. Calculate the required theoretical air.

2.20 The analysis for a lignite fuel is carbon 42.4, hydrogen 2.8, sulfur 0.7, oxygen 12.4, nitrogen 0.7, ash 6.2, and moisture 34.8 percent by mass. Calculate the required theoretical air.

2.21 A bituminous coal has the following composition: carbon 62.8, hydrogen, 4.6, sulfur 4.3, oxygen 6.6, nitrogen 1.0, ash 8.6, and moisture 12.1 percent by mass. The coal is burned completely with 25 percent excess air. Calculate the probable analysis of the dry products of combustion.

2.22 Repeat Problem 2.21 for incomplete combustion of the carbon, that is, 0.008 kg carbon/kg coal burns to produce carbon monoxide.

2.23 An oil fuel has the following composition: carbon 85.5, hydrogen 14.1, and inert material 0.4 percent by mass. The fuel is burned with 12 percent excess air. Calculate the actual air/fuel ratio and the probable volumetric analysis of the products of combustion.

2.24 The composition of a gaseous fuel is CO_2 10, CO 40, H_2 15, and N_2 35 percent by volume. The volumetric analysis of the combustion gas reports CO_2 24.9, O_2 1.3, and N_2 73.8 percent. Determine the percent theoretical air used in firing the fuel.

2.25 Determine the actual air/fuel ratio for the following firing conditions. Coal analysis: carbon 83.84, hydrogen 4.65, sulfur 0.47, oxygen 5.36, nitrogen 1.17, and ash 4.51 percent by mass. Stack gas analysis: CO_2 13.74, CO 0.25, O_2 5.39, and N_2 80.62 percent by volume.

2.26 Liquid $C_{10}H_{22}$ is burned completely with air. The fuel and air are supplied, respectively, at 25 C and 600 K. The gas temperature at the outlet of the combustion chamber is 1600 K. Determine the air/fuel ratio.

2.27 Methane is burned completely with 250 percent theoretical air. The fuel and air are supplied at 400 K. Determine the temperature of the products of combustion.

2.28 A combustion system is supplied with carbon monoxide at 500 K. The inlet air temperature is 300 K, and the gas leaves at 2000 K. Calculate the air/fuel ratio.

2.29 Ethanol is completely burned in air. The temperature of the products of combustion is 2000 K. The fuel and air are supplied at 25 C. Combustion is achieved in a continuous-flow system with virtually zero heat loss. Calculate the air/fuel ratio.

2.30 Hydrogen is completely burned in a continuous-flow system with 150 percent theoretical air. The air is supplied at 400 K, and the hydrogen is supplied at 600 K. Calculate the temperature of the products of combustion. Assume zero heat loss.

REFERENCES

1 *Mechanical Engineering,* June 1976, page 53.

2 Papamarcos, J. "Gas from Coal," *Power Engineering,* Feb. 1973.

3 Goodholm, P. R. "Coal Gasification—An Alternative in Clean Energy Production," ASME Paper No. 73-Pet-1, New York, Sept. 1973.

4 Siegel, H. M. and Kalina, T. "Technology and Cost of Coal Gasification," *Mechanical Engineering,* May 1973.

5 *Mechanical Engineering,* Dec. 1973, page 83.

6 *Mechanical Engineering,* Mar. 1974, pages 40, 41.

7 Burton, R. S. and Baile, R. C. "Fluid Bed Pyrolysis of Solid Waste Materials," *Combustion,* Feb. 1974.

8 Singer, J. G. and Mullen, J. F. "Closing the Refuse Power Cycle," *Combustion,* Feb. 1974.

9 Corlett, R. F. "Conversion of Seattle's Solid Waste to Methanol or Ammonia" *The Trend in Engineering,* University of Washington, April 1975.

10 Prien, C. H. "The Current Status of Oil Shale Technology," Annual One-Day Meeting, Am. Inst. of Chem. Engrs., Los Angeles, Calif., April 24, 1964.

11 Savage, H. K. and Holt, B. "Shale Oil as a Future Energy Source," ASME Paper No. 68-PWR-2, New York, Sept. 1968.

12 Hubbert, M. K. "Energy Resources," Publication 1000-D, National Academy of Sciences—National Research Council, Washington, D.C., 1962.

13 Pollard, W. G. "The Long-Range Prospects for Solar-Derived Fuels," *American Scientist,* Sept.–Oct. 1976.

14 Hubbert, M. K. "The Energy Resources of the Earth," *Scientific American,* Sept. 1971.

15 *Mechanical Engineering,* April 1977, page 21.

16 Meyerhoff, A. A. "Economic Impact and Geopolitical Implications of Giant Petroleum Fields," *American Scientist,* Sept.–Oct. 1976.

17 Häfele, W. "A Systems Approach to Energy," *American Scientist,* July–August 1974.

18 *World Energy Supplies 1950–1974,* U.N. Statistical Papers, Series J, No. 19.

19 Rose, D. J., Walsh, P. W., and Leskovjan, L. L. "Nuclear Power—Compared to What?" *American Scientist,* May–June 1976. Reprinted by permission of *American Scientist,* The Scientific Research Society.

20 Bennett, R. R. "Energy for the Future," *Combustion,* April 1970 (Reprinted from *Ebasco News,* Vol. XXII, No. 8).

21 West, J. A. "Outlook for Fossil Fuels in Power Generation," ASME Paper No. 70-Pwr-8, New York, Sept. 1970.

22 Dupree, W. G., Jr. "United States Energy Requirements to the Year 2000," ASME Paper No. 73-ICT-105, New York, Sept. 1973.

23 "Energy Forecast: Coal and Sunny," *Mechanical Engineering,* Dec. 1976, page 66.

BIBLIOGRAPHY

Chigier, N. *Energy, Combustion, and Environment,* McGraw-Hill Book Co., New York, 1981.

Combustion Engineering, Combustion Engineering, Inc., New York, 1966.

Glassman, I. *Combustion,* Academic Press, New York, 1977.

Steam/Its Generation and Use, Babcock & Wilcox Co., New York, 1978.

CHAPTER THREE

RECIPROCATING MACHINES

Reciprocating machines may be classified as (1) vapor and gas expanders, (2) compressors (gases) and pumps (liquids), (3) internal combustion engines, and (4) external combustion engines. All of these machines effect some kind of energy conversion. The internal combustion engine, a versatile and widely used power plant, is examined in some detail together with the external combustion engine in the following chapter.

3.1 STEAM ENGINES

A vapor or gas expander is one component of a complete power system the operation of which is governed by an appropriate thermodynamic cycle. The expander, for example, a steam engine, in following a mechanical cycle participates only in a portion of the cyclic change that occurs in the working fluid.

While various pneumatic devices are in general use, most of the power-producing, reciprocating expanders have been steam engines. The steam engine is a comparatively simple machine and inherently free of many of the flow problems associated with rotating machines. The engine consists mainly of a cylinder and piston assembly and of a valve mechanism for control of the steam flow. Machining operations of reasonably good quality are adequate for producing an engine capable of functioning with minimum difficulty.

Engine speed is sufficiently high to establish virtually continuous flow of the steam to and from the cylinder. Actually, steam is admitted in discrete quantities, and each portion of the fluid is allowed to expand within the cylinder as the piston moves to the end of the stroke. The characteristic handling of a succession of small quantities of fluid ensures relatively successful operation of reciprocating-type machines. It is not particularly difficult to construct a basic engine that will operate with relatively low mechanical losses and with very little leakage of steam. Although the energy conversion efficiency of this basic machine is likely to be low, the engine will function. Numerous improvements can be made in order to increase the machine efficiency of the simple steam engine.

Shortly after World War II, the motive power of the railroads in the United States was converted, within a few years, from the steam locomotive to the Diesel-electric locomotive. In comparison with the steam locomotive, the Diesel-electric locomotive is more economical to operate, principally because of decreased fuel costs, a result of improved thermal efficiency, and reduced maintenance, both of the rolling equipment and the track. In addition, this conversion eliminated the necessity of supplying at frequent "stops" boiler water of good quality.

When the steam locomotive was removed from service, the steam engine virtually ceased to exist as a prime mover of any consequence. In recent years, interest in the steam engine has been revived because of possible application to automobiles and commercial highway vehicles. Since the steam engine may not be wholly extinct, a brief discussion will be presented on the construction and operation of this machine that approximately 200 years ago contributed strongly to the start of the industrial age.

The basic steam engine consists of a cylinder and piston assembly, connecting rod and crank, a single slide valve, and a governor. The flywheel counteracts the effect of the pulsating steam flow, thus a more uniform torque is developed. In designing the engine, attention is directed to the effect of the heavy inertia forces caused by the change in the direction of travel of the reciprocating parts. The governor controls the speed of the engine, within narrow limits, as the load varies. Almost all of the steam engines have been double acting, with the steam admitted alternately to each end of the cylinder.

A single slide valve can control the flow of steam to and from each end of the cylinder. Improved operation, however, can be achieved by installing at each end of the cylinder an inlet valve and an exhaust valve. This four-valve design, in comparison with the single slide-valve construction, provides for improved control of the steam flow through independent timing of the inlet and exhaust valves.

In either the single-valve or the four-valve engine, the steam enters and leaves the cylinder at the same end. These engines are designated as counterflow engines because of the two-directional steam flow within the cylinder.

In the simple steam engine, steam expands between the throttle and exhaust pressures within a single cylinder. Several such cylinders operating in parallel constitute a multicylinder, simple engine. Improved operating economy can be achieved by expanding the steam in series flow through two, three, or four cylinders. The thermodynamic gain derived from multistaging is discussed in a later section.

In 1857, a patent was granted in the United States for a uniflow-type steam engine. An improved design of the uniflow engine was developed in 1908 by Professor Johann Stumpf of Charlottenburg, Germany. The uniflow engine is equipped with a relatively long piston, and a centrally located ring of exhaust ports in the cylinder wall allows single direction of steam flow, that is, from either end toward the center of the cylinder. The piston serves as the exhaust valve by uncovering the ports at the end of the stroke. In particular, Stumpf's design introduced steam-heated cylinder heads, the long piston that uncovers the exhaust ports, and poppet-type steam inlet valves. A modern uniflow engine is shown in Fig. 3.1.

3.2 ENGINE INDICATORS

Reciprocating engines and compressors are indicated by mechanical or electronic instruments that provide a record of the variation of the cylinder pressure with the piston movement. Figures 3.2 and 3.3 respectively show typical indicator cards for a counterflow and a uniflow steam engine. The card is produced graphically by the indicator or constructed from instrument readings, depending upon the type of instrument employed. The indicator card is used in valve timing and for calculating the engine internal work and hence the internal power.

The mechanical cycle, shown by the indicator card, consists of four events. The inlet valve opens at point 1 and closes at point 2. The exhaust port is opened at point 3 and closed at point 4. These points are designated as admission (1), cutoff (2), release (3), and compression (4). The corresponding events are admission (1–2), expansion (2–3), exhaust (3–4), and compression (4–1). Note the high degree of compression achieved in the uniflow engine. The engine valves are adjusted or timed to achieve optimum performance.

The base length of the indicator card must be a true reduction of the piston travel. The mean ordinate of the enclosed area of the card is then proportional to the mean effective pressure acting on the piston face. The net work of the mechanical cycle is given by $P_m A l$, where P_m is the mean effective pressure for a double stroke, A is the transverse area of the cylinder, and l is the length of the stroke. On the forward stroke, work is transferred from the steam to the piston; while on the return stroke, work is transferred from the piston to the steam.

At one time, extensive use was made of direct-acting steam pumps. The power and pump cylinders of these machines are arranged in line for

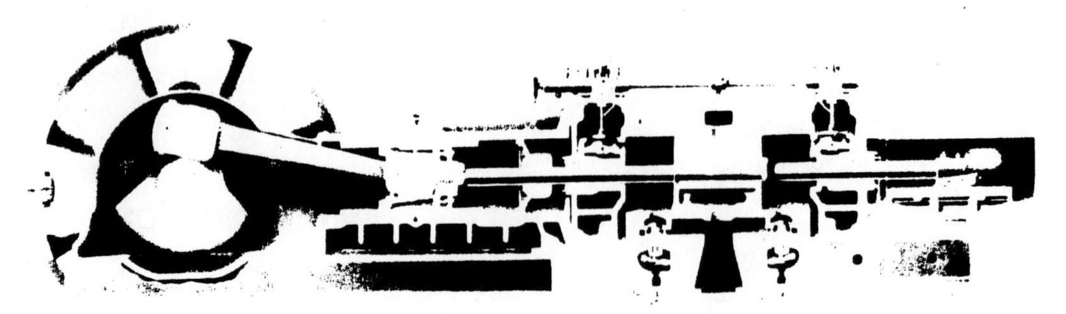

Figure 3.1 The "Universal Unaflow" steam engine. (Courtesy Skinner Engine Co.)

tandem operation, and the two pistons are attached to a common piston rod. Steam is admitted to the power cylinder at line pressure for the entire length of the stroke. In the absence of expansion of the steam, the theoretical indicator card takes a rectangular shape. The failure to expand the steam in the power cylinder contributes to the comparatively poor thermal performance of the direct-acting steam pump.

Example 3.1

The following data relate to a double-acting steam engine: Bore, 200 mm; stroke, 300 mm; piston

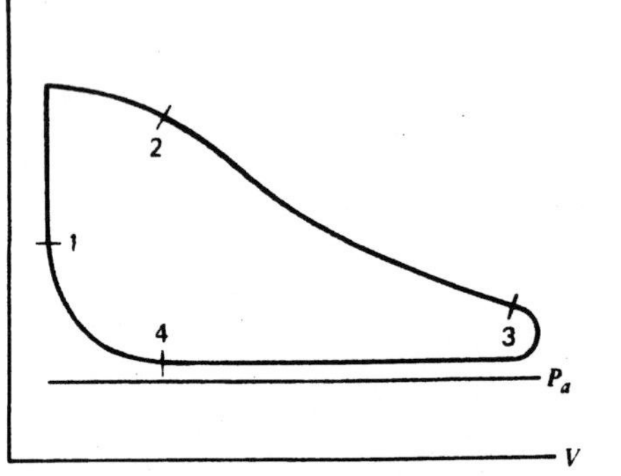

Figure 3.2 Indicator card for counterflow steam engine—noncondensing operation.

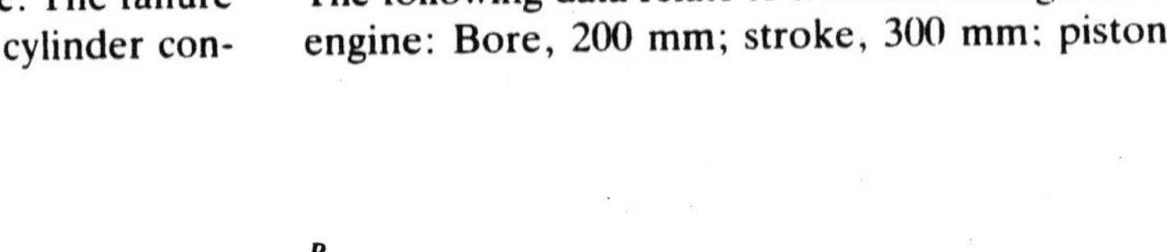

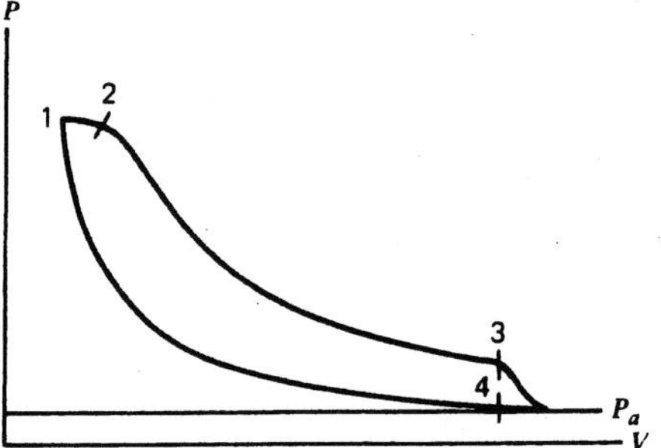

Figure 3.3 Indicator card for uniflow steam engine—noncondensing operation.

rod diameter, 38 mm; speed, 200 rpm. The indicator has head-end card area of 9.70 cm^2, crank-end card area of 10.45 cm^2, card length of 6.95 cm, and a spring scale of 300 kPa/cm deflection of the tracing point. Calculate the engine internal power.

(a) Head-end power.
Mean effective pressure

$$P_m = \frac{a}{l}s = \frac{9.70}{6.95}300 = 419 \text{ kPa}$$

Cylinder transverse area

$$A = 0.0314 \text{ m}^2$$

Work (for one double stroke)

$$W_i = P_m A l = 419 \times 0.0314 \times 0.300 = 3.944 \text{ kJ}$$

Power

$$\dot{W}_i = Wn = 3.944\frac{200}{60} = 13.15 \text{ kW}$$

(b) Crank-end power.

$$P_m = \frac{10.45}{6.95}300 = 451 \text{ kPa}$$

Piston rod cross-sectional area
= 0.0011 m^2

Cylinder net transverse area
= 0.0314 − 0.0011 = 0.0303 m^2

$$W_i = 451 \times 0.0303 \times 0.300$$
$$= 4.100 \text{ kJ per double stroke}$$

$$\dot{W}_i = 4.100\frac{200}{60} = 13.67 \text{ kW}$$

(c) Total internal power.

$$\dot{W}_i = 13.15 + 13.67 = 26.82 \text{ kW}$$

The term W_i is used generally in this chapter to designate the net indicated or internal work performed in the cylinder of a reciprocating machine. The net transfer of work for an engine is from the working fluid to the piston face, while the reverse net transfer of work occurs in a pump or compressor. Because of the mechanical losses W_f in the machine parts, the brake or shaft work W_b for the engine is less than the internal work. Thus,

$$W_b = W_i - W_f \qquad \text{(engine)}$$

On the other hand, the shaft work for a pump or compressor is greater than the internal work because of the mechanical losses or friction. Thus,

$$W_b = W_i + W_f \qquad \text{(compressor or pump)}$$

Note: W_b is used generally to designate the shaft work for any type of machine, engine or compressor.

3.3 ENGINE EFFICIENCIES

The machine or relative efficiency η_r is defined as the ratio of the engine internal work W_i to the reversible adiabatic work W_s based on expansion of the steam from the inlet conditions to the exhaust pressure. Thus,

$$\eta_r = \frac{W_i}{W_s} \tag{3.1}$$

The machine efficiency is also designated as the Rankine cycle ratio (RCR).

The mechanical efficiency is defined as the ratio of the shaft work W_b to the internal work W_i of the engine. Thus,

$$\eta_m = \frac{W_b}{W_i}$$

Because $W_b = W_i - W_f$, $\eta_m = (W_i - W_f)/W_i$, or

$$\eta_m = 1 - \frac{W_f}{W_i} \tag{3.2}$$

where W_f is the work equivalent of the engine friction; W_f is essentially a function of the engine speed and varies inappreciably with the load on the engine. Thus, for an engine operating at constant speed, the fraction W_f/W_i continues to decrease, or η_m increases, with an increase in the load.

The shaft power of the engine may be measured by a friction brake, fluid brake, or electric dynamometer. The engine tends to rotate the friction brake or the casing of the fluid brake, or dynamometer. Motion is prevented by applying a restraining torque through a force-measuring device. Shaft work is given by

$$W_b = 2\pi LF \qquad \text{per revolution} \tag{3.3}$$

where

L = the length of the moment arm
F = the restraining force

Example 3.2

Calculate the shaft power and the mechanical efficiency for the steam engine described in Example 3.1. The length of the brake moment arm is 1.5 m, and the restraining force is 0.770 kN.

$$W_b = 2\pi LF = 2\pi 1.5 \times 0.770$$

$$= 7.257 \text{ kJ} \qquad \text{per revolution}$$

$$\dot{W}_b = W_b n = 7.257 \frac{200}{60} = 24.19 \text{ kW}$$

$$\eta_m = \frac{\dot{W}_b}{\dot{W}_i} = \frac{24.19}{26.82} = 0.902$$

3.4 STEAM ENGINE LOSSES

The machine efficiency for the steam engine is comparatively low, particularly for high-vacuum exhaust conditions. A simple slide-valve engine, operating noncondensing, will have a machine efficiency of about 0.4. Machine efficiencies for the uniflow engine are approximately 0.75 for noncondensing operation and 0.65 for condensing operation. Machine efficiencies of about the same magnitude are achieved by four-valve, multistage engines. Machine efficiency values are normally given for the optimum operating conditions for the engine.

Exhaust steam from an engine or turbine is commonly used for industrial heating and processing. The pressure of the exhaust steam often exceeds the barometric pressure. The machine efficiency for a uniflow engine can approach 0.90 when operating noncondensing at a back pressure of about 35 kPa, gage. Although a machine efficiency of this magnitude compares favorably with that achieved by a steam turbine under similar operating conditions, the turbine will prove to be more economical for most applications.

The structure and the operating characteristics of the steam engine promote high thermal losses. In the light of the renewed interest in the application of the steam engine to automotive propulsion, these losses are examined and evaluated below. Procedures employed for reducing these losses are also investigated.

The thermodynamic and mechanical losses inherent in the operation of the counterflow slide-valve engine are:

1 Condensation and reevaporation.
2 Throttling.
3 Incomplete expansion.
4 Leakage.
5 Heat transfer.
6 Mechanical friction.

The equilibrium temperature of the engine cylinder wall varies somewhat, but generally this temperature may be taken roughly as the mean of the inlet and exhaust steam temperatures. At the end of compression, prior to admission, the clearance space contains relatively cool steam. Incoming steam comes in contact with this clearance steam and with the relatively cool metallic cylinder wall. As a result, a portion of the incoming steam is condensed before it can expand and perform useful work. This loss is especially severe at light loads, because a high percentage of the steam admitted is thus initially condensed.

During the first portion of the forward stroke of the piston, the temperature of the steam in the cylinder is higher than the wall temperature. This temperature difference causes transfer of heat from the steam to the cylinder wall. As expansion continues, the decreasing steam temperature drops below the wall temperature. The direction of heat flow is now reversed. Water droplets formed on the cylinder wall during condensation are reevaporated. In addition, heat flow from the cylinder wall to the steam continues during the exhaust stroke.

The condensation and reevaporation process does not represent a heat loss from the engine to the surroundings. Energy transferred from the steam is subsequently regained during the single cycle of operation. There is however a loss in availability of the energy so transferred, because the temperature at which the energy is transferred from the steam is higher than the temperature at which the reverse flow of energy occurs. During the exhaust stroke, energy transferred from the cylinder wall to the steam becomes completely unavailable.

Throttling of the steam during flow through the inlet and exhaust ports reduces the internal operating pressure ratio of the engine. As a consequence, a loss in availability is incurred.

The loss that results from incomplete expansion of the working fluid is particularly significant, because the conditions that promote this loss are present in any reciprocating expander where the working fluid is not expanded to the cycle minimum pressure (see Fig. 3.2). At point 3, when the piston is near the end of the stroke and the exhaust port is opened, the pressure P_3 in the cylinder is still above the internal exhaust pressure P_4. The degree of incomplete expansion varies with the point of cutoff or with the load on the engine. It is possible to design the engine to achieve complete expansion of the steam when operating on the design point. However, the modest increase in work achieved by complete expansion does not represent an economical gain. A longer, and heavier, cylinder would be required, and frictional losses will increase as a result of the lengthened piston stroke.

Losses because of mechanical friction result from the relative motion of the various parts of the engine. A full-load mechanical efficiency of 90 percent or somewhat higher can be achieved with good machine construction and adequate lubrication.

Leakage of steam from the engine can be held to a negligible amount through use of properly designed seals that are maintained in good operating order. The heat loss from the engine, because of radiation and convection, will not be appreciable when the cylinder is properly insulated.

The losses that result from condensation and reevaporation and from throttling are relatively high in the simple counterflow steam engine equipped with a slide valve. These losses can be reduced by utilizing improved designs for the steam engine.

Appreciable throttling is caused by the ordinary slide valve as it moves to cover and uncover the steam ports. Throttling losses are effectively reduced by employing separate inlet and exhaust valves. These valves are quick acting, and large-area ports and short passages are incorporated into the design. The use of separate inlet and exhaust valves allows the point of cutoff to be controlled independently of the other three points, namely, release, compression, and admission.

Superheated steam supplied to the engine reduces initial condensation because of the poorer characteristics for heat transfer from superheated steam than from saturated steam. For most engines however, only a moderate amount of superheat can be tolerated because of the necessity for maintaining adequate internal lubrication.

A reduction in the time required for completion of a single mechanical cycle will decrease the heat flow associated with condensation and reevaporation. With respect to this particular loss, the high-speed engine shows superior performance in comparison with the engines that operate at medium and low speeds.

A compound steam engine consists of a high-pressure cylinder and a low-pressure cylinder for two-stage expansion of the steam between the throttle and the exhaust pressures. An intermediate receiver may be installed in order to reduce pressure fluctuations at the inlet to the low-pressure cylinder.

Because of compound expansion, the temperature range in either the high-pressure or low-pressure cylinder is considerably less than the corresponding temperature range for single-stage expansion. The decrease in the difference between the temperature of the steam and the temperature of the cylinder wall causes a reduction in the heat flow that occurs during the condensation and reevaporation process, and consequently a smaller loss in the available energy is observed. Further, the portion of the available energy that is not converted to work in the first stage increases the available energy of the steam for expansion in the second stage.

Each additional stage of expansion shows a gain in thermal performance in accordance with the preceding analysis. A very large number of stages of expansion would virtually eliminate the condensation and reevaporation loss. In practice, the maximum number of stages of expansion is limited to three or four. The relatively large triple- and quadruple-expansion steam engines were used for marine propulsion and were installed in stationary pumping and power plants.

The uniflow steam engine is equipped with quick-acting, large-area poppet-type valves that provide for steam admission with minimum throttling. The large-area exhaust ports, covered and uncovered by the piston, allow the exhaust steam to flow from the cylinder with a minimum pressure drop.

The uniflow steam engine inherently has a low condensation and reevaporation loss. Because the point of compression occurs early in the return stroke of the piston, the expanded steam, trapped in the cylinder, is compressed essentially to the line pressure (see Fig. 3.3). Condensation at the ends of the cylinder is largely eliminated by the use of steam-heated cylinder heads. The high degree of compression of the residual expanded steam and the single-direction flow of the steam jointly establish in the cylinder wall a temperature gradient that extends from the inlet section to the central ring of exhaust ports. The absence of low-temperature steam in the clearance space and the small temperature difference that exists between the steam and the cylinder wall combine effectively to reduce the condensation and reevaporation loss.

3.5 STEAM ENGINE MEAN EFFECTIVE PRESSURE

The mean effective pressure used in the design of a steam engine is preferably obtained from test data taken on a similar engine. Lacking such test data, the next best procedure is the construction of an indicator card. The method used for construction of the card may be found in handbooks that cover the steam engine in detail.

The probable mean effective pressure may be determined from the prescribed operating pressures, the cutoff ratio, and an experimentally determined diagram factor. Calculation of the theoretical mean effective pressure is based on the theoretical indicator card which has zero clearance and 100 percent release (see Fig. 3.4). The broken line shows the corresponding actual indicator card. The expansion curve 1–2 is assumed to conform to a pressure–volume variation described by PV = constant. This relationship is based on observation of steam engine performance and does not represent isothermal expansion of the steam.

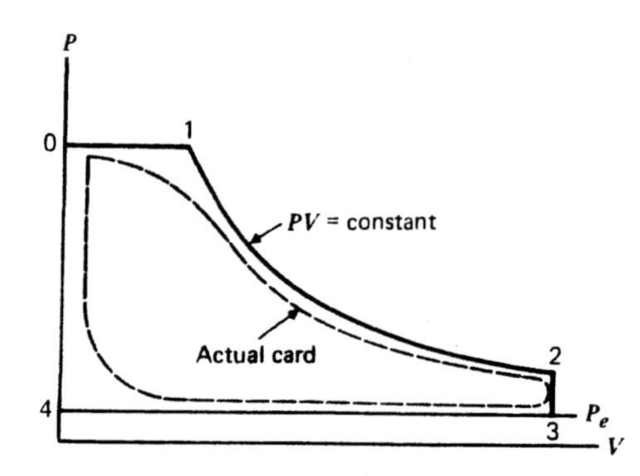

Figure 3.4 Steam engine theoretical and actual indicator cards.

The work for the theoretical cycle is given by

$$W_c = P_1V_1 + P_1V_1 \ln \frac{V_2}{V_1} - P_eV_2$$

The theoretical mean effective pressure is determined by dividing the cycle work by the displacement V_2. Thus,

$$\text{Theoretical mep} = \frac{P_1}{r}(1 + \ln r) - P_e \quad (3.4)$$

where P_1 and P_e are, respectively, the inlet and exhaust absolute pressures; and r is the expansion ratio V_2/V_1 (the reciprocal of the cutoff ratio).

The probable mep is determined by multiplying the theoretical mep by the diagram factor that has a value of less than 1. The diagram factor allows for the difference between the actual and the theoretical indicator cards (see Fig. 3.4). Tabulated values of the diagram factor are typical or average, and thus are not sufficiently precise for use in engine design.

Example 3.3

Determine the probable internal power for a double-acting steam engine that operates with a throttle pressure of 900 kPa, abs. and an exhaust pressure of 110 kPa, abs. The bore and stroke are, respectively, 200 mm and 250 mm. The speed is 210 rpm. The cutoff is 20 percent. The diagram factor is 0.76.

$$\text{Theoretical mep} = \frac{P_1}{r}(1 + \ln r) - P_e$$

$$= \frac{900}{5}(1 + \ln 5) - 110$$

$$= 360 \text{ kPa}$$

$$\text{Probable mep} = 0.76 \times 360 = 274 \text{ kPa}$$

$$\text{Internal power } \dot{W}_i = 2PLAn$$

$$= 2(274 \times 0.250 \times 0.0314 \times 3.5)$$

$$= 15.06 \text{ kW}$$

3.6 STEAM ENGINE PERFORMANCE

The various performance factors for the steam engine are customarily plotted against the internal power of the engine (see Fig. 3.5).

The variation in speed between the operation of the engine at no load and at the rated load is likely to range from 1 to 4 percent. The decrease in speed is set by the governor requirements for stable operation.

The mechanical efficiency continues to increase with increase in the load on the engine. The point of zero mechanical efficiency corresponds to zero shaft output, that is, the idling operation of the engine.

As the engine load is increased, the steam consumption (kg/h) increases at an increased rate. The steam rate (kg/kWh) is a minimum when the engine operates at approximately two-thirds of the rated load. These trends are typical of the cutoff-governed engine.

Throttle-governed engines show steam consumption curves that vary linearly with the change in the load. In this case, the steam rate continues to decrease to the point of maximum load.

The machine efficiency, or Rankine cycle ratio (RCR), of the steam engine may be defined as the ratio of the theoretical steam rate to the ac-

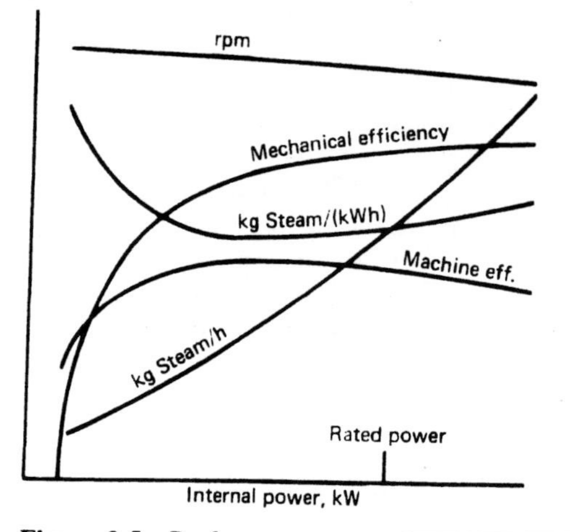

Figure 3.5 Performance curves for uniflow steam engine.

tual steam rate. The theoretical steam rate is a function of the inlet steam conditions and the exhaust pressure. The theoretical steam rate thus remains constant as the load changes. It follows that the machine efficiency curve shown in Fig. 3.5 varies inversely with respect to the steam rate curve.

A change from noncondensing operation to condensing operation for the steam engine causes a reduction in the machine efficiency, for example, approximately 10 points for a uniflow engine that operates with a throttle pressure of 1.5 MPa. Condensing operation increases the temperature range in the engine, thus increasing the condensation loss. In addition, the reduced exhaust pressure generally causes an increase in the loss owing to incomplete expansion.

All engines, whether reciprocating or rotating, require some kind of control. Some machines, particularly those used for vehicular propulsion, operate over a wide speed range, while for power generation the engine is required to operate at constant speed or with a comparatively slight variation in speed. For steady-state operation of the engine, the control system provides a balance between the energy input and the energy output.

The function of the steam engine governing system is to regulate the engine speed in relation to the applied load by controlling the flow of steam to the cylinder. The amount of steam admitted to the engine cylinder is regulated by throttling the inlet flow or by varying the point of cutoff. Either function is accomplished by a mechanical type of governor that operates with a decrease in speed over the range from no load to full load. The method of cutoff governing is economically superior to the control achieved by throttling the steam, because cutoff variation does not impose a reduction in the internal pressure ratio and, as a consequence, the associated loss in availability is not incurred.

3.7 POSSIBLE STEAM ENGINE APPLICATIONS

In considering possible applications of the steam engine today, attention is directed to the inherent disadvantages of the machine, namely:

1 Low machine efficiency, particularly for high-vacuum exhaust conditions.
2 Bulky and heavy.
3 Low-speed limitation, except for very small engines.
4 Limited to moderate steam pressures and low superheat.
5 Internal lubrication causes contamination of the exhaust condensate.
6 Low power capability.

Because of these negative characteristics, the steam engine is generally inferior to other power-producing machines, both economically and technologically, and as a consequence the steam engine has very limited use.

Early in the 1970 decade, renewed interest in the steam engine developed for possible application to motor vehicles. Numerous experimental engines were constructed to determine whether such application is feasible. The results of these investigations are examined briefly here in the light of predicting the future status of the steam engine. Support for the development of the automotive steam engine was derived, in part, from the overall program to investigate energy sources and energy conversion systems of all types.

Continuous-flow combustion in the steam engine power plant is expected to yield, in contrast to the uncontrolled automotive engine, an exhaust gas that contains low concentrations of hydrocarbons and carbon monoxide. The development of the catalytic converter however reduces somewhat the significance of this advantage. As in most combustion systems, NO_x emission is likely to exceed the permissible limit and thus require some effort to reduce the concentration of this pollutant.

In addition to the steam engine, there must also be installed in the vehicle a boiler, a combustion chamber, and a condensing system. The fuel, probably a petroleum derivative, would not require the high refining quality of gasoline.

Test installations of the automotive steam engine were for the most part made in large vehicles, such as buses and trucks. Some passenger automobiles have been equipped with steam engines, but space requirements for the entire power plant remain a problem in certain designs.

There is today only a flagging interest in the automotive steam engine. Experience acquired from the operation of a few experimental steam power plants is not particularly encouraging with respect to future commercial applications. The low air pollution characteristic of the steam power plant is an asset, but there are several problem areas that could impede further development and subsequent large-scale production of steam-operated vehicles.

The automotive steam power plant is large and heavy in comparison with the conventional automobile engine of the same power capability. Because of space limitations in the engine compartment, the size of the condenser is severely restricted. Under adverse operating conditions, such as full load or hot weather, complete condensation of the exhaust steam is consequently not achieved, and some loss of water from the system is experienced.

Water, because of its collective properties, is the most appropriate fluid capable of serving as the working medium in a vapor cycle power plant. The comparatively high freezing temperature of water is a significant disadvantage when water is used in a power plant designed for automotive service. Standby heating is required during the nonoperating periods when subfreezing temperatures are experienced. Such protective heating adds to the complexity and cost of the power plant, and further, the auxiliary heater must possess a high degree of reliability. Other fluids, for example, freon, may possibly prove to perform satisfactorily in the vapor cycle automotive power plant.

Adequate lubrication of the expander is difficult to achieve. The lubricant, which is injected into the steam at a point between the steam generator outlet and the expander inlet, must be removed from the condensate prior to return to the boiler. Failure to remove the lubricant from the boiler feedwater will cause formation of deposits on the heat transfer surface and a resulting decline in the steam-generating capability.

In contemplating a substitute for the spark-ignition automotive engine, the comparative economy must be investigated. Fuel rates for the spark-ignition engine and the automotive steam power plant are approximately in the ratio of 1 to 1.5..The capital cost of the steam power plant may be about two times the capital cost of the spark-ignition engine. As noted in an earlier section, the reciprocating steam expander is a comparatively simple machine, but experience indicates that the other components, namely, the steam generator, condenser, burner, and water controls, are complex and will require extensive development leading to cost reduction before the "steam automobile" can become a reality.

Reduction in the fuel rate for the automotive steam power plant is a matter of considerable interest and speculation. The fuel rates for the spark-ignition engine and the automotive steam power plant appear to correlate with their respective Carnot cycle thermal efficiencies despite the very simple relationship. The dependence, in part, of the fuel rate on the cycle maximum temperature is indicated by this analysis. Although the Carnot cycle thermal efficiency is only a rough performance indicator, the fuel rate for the steam power plant in automotive service may remain high unless the maximum operating temperature can be increased. Higher operating temperatures however often cause metallurgical problems and in other ways adversely affect the operation of the power plant.

3.8 GAS COMPRESSORS

The reciprocating gas compressor is one of an array of "pumping" machines that serve generally to move various gases and liquids. Gas compression is an integral part of many industrial processes, and compressed air is used widely in industry for operation of power tools and as the working medium in pneumatic devices and controls. Because significant quantities of energy are required for pumping, energy conversion in pumping machines is a topic of some importance.

The gas compressor operates with a noticeable increase in pressure and decrease in the specific volume of the working fluid. Within the machine, the mechanical energy, supplied by a driving motor or engine, is converted principally into enthalpy of the gas, with a smaller portion of this energy transferred from the machine as heat.

Usually, compressor operation does not effect an appreciable change in the kinetic energy of the gas.

Single-stage reciprocating gas compressors are operated effectively with discharge pressures that range up to 550 to 700 kPa, gage. When higher discharge pressures are required, service is provided by multistage compressors equipped with intercoolers.

Reciprocating compressors are usually double acting (see Fig. 3.6). An intake valve and a discharge valve are located at each end of the cylinder. Throttling through the ports is minimized by providing large flow areas. The valves, which operate automatically, are unseated by a small pressure differential, particularly the intake valve. Cylinder walls and heads are water cooled in order to maintain an effective lubricating oil film between the wall and the piston rings and to provide for some cooling of the gas during compression. The cylinders of small-capacity air compressors are usually air cooled.

A typical indicator card for a compressor is shown in Fig. 3.7. Several deviations from the theoretical mechanical cycle are noted. Because of the mechanical requirements for clearance, the piston in its travel must stop short of contacting the cylinder head. Consequently, on the discharge stroke, a portion of the compressed gas in the cylinder is not delivered to the line. The compressed gas in the clearance space expands, during the first portion of the charging stroke, to the intake pressure in the cylinder. Fluctuating pressures that occur during the discharge and intake strokes result from valve flutter and surging flow.

The clearance volume is equivalent to the volume of the gas contained in the cylinder with the piston at the corresponding extreme position, and with both the intake and discharge valves closed. Clearance volume is expressed as a percentage of the piston displacement.

The system operating characteristics influence the selection of an appropriate type of control for the compressor. It is essential however in making this selection to avoid a type of control that will cause the system to operate with poor economy.

Reciprocating compressors are ordinarily operated at a single prescribed speed, with the discharge from the compressor regulated in response to changes in the downstream conditions. Usually, the receiver pressure is maintained within prescribed limits, and the flow is varied by intermittent operation or by continuous operation with intermittent delivery. Delivery can be interrupted by means of an automatically operated unloading device that holds the intake valve in

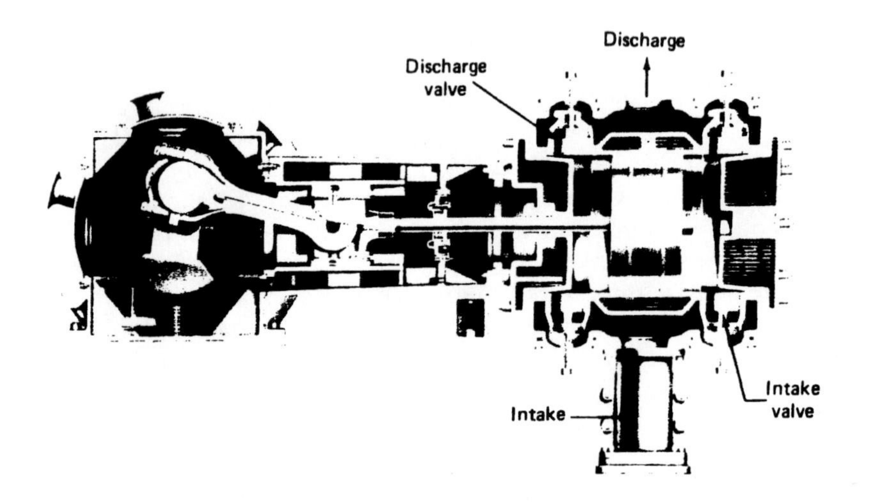

Figure 3.6 Reciprocating compressor. (Courtesy Ingersoll-Rand Co.)

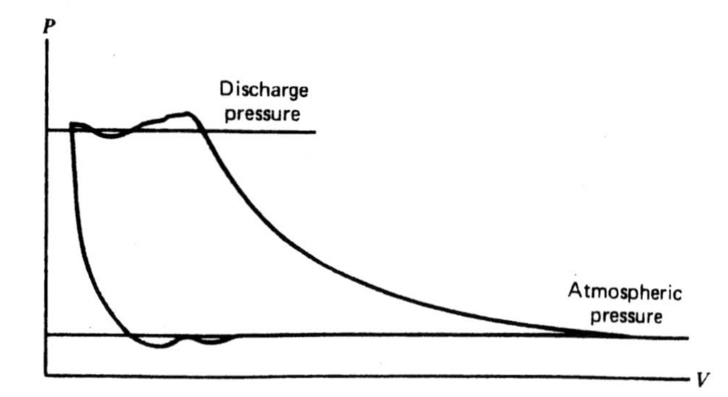

Figure 3.7 Indicator card for a reciprocating gas compressor.

either a closed or open position. A second method uses clearance pockets, located in the cylinder head, to effect changes in the clearance volume and thereby vary the compressor intake.

3.9 COMPRESSOR VOLUMETRIC EFFICIENCY

The compressor displacement, expressed in m^3/min, is equal to the volume swept by the piston face on the intake stroke during a time interval of 1 min. Most gas compressors are double acting, and for these machines the displacement of both piston faces must be considered.

The actual volume of air handled by the compressor, based on free air conditions, is less than its displacement. Free air conditions refer to the ambient pressure and temperature. The volumetric efficiency is defined as the ratio of the volume of free air handled by the compressor to the displacement. The definition implies that both quantities in this ratio are evaluated for the same time interval.

The free air volume is less than the compressor displacement because of the presence of the clearance air, throttling of the air in passing through the intake and discharge valves, heating of the incoming air, and leakage. The most effective procedure for the determination of the volumetric efficiency is to measure the actual flow of air to the compressor. Various types of meters and devices are available for measuring gas flow.

The volumetric efficiency of the compressor may be determined approximately from the indicator card by comparing the length of the effective intake stroke with the base length of the diagram. In the absence of an indicator card, a somewhat similar procedure can be used to calculate the volumetric efficiency. The volume of the expanded clearance air is determined from the relationship PV^n = constant, where n may be taken as 1.34. Proper allowance is made for pressure drops through the valves and for leakage of gas around the valves and the piston rings (see Fig. 3.8).

Example 3.4

A reciprocating air compressor operates with a discharge (line) pressure of 700 kPa, gage. The atmospheric pressure is 100 kPa, abs. The pressure drops through the intake and discharge valves

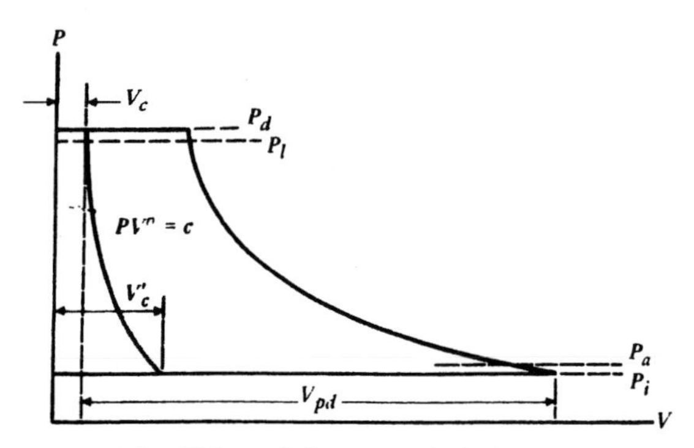

Figure 3.8 Effect of the expanded clearance gas on the volumetric efficiency of compressor.

are, respectively, 4 and 15 kPa. The clearance volume is equal to 5 percent of the piston displacement. Estimate the volumetric efficiency of the compressor (see Fig. 3.8).

Internal Pressures

$$P_i = 800 \text{ kPa, abs}$$

$$P_d = P_l + \Delta P_v = 800 + 15$$
$$= 815 \text{ kPa, abs}$$

$$P_i = P_a - \Delta P_v = 100 - 4$$
$$= 96 \text{ kPa, abs}$$

Volume of the Expanded Clearance Air

$$V_c' = V_c\left(\frac{P_d}{P_i}\right)^{1/n}$$
$$= 0.05\left(\frac{815}{96}\right)^{1/1.34}$$
$$= 0.247$$

Intake Volume at Pressure P_i

$$V_i = V_{pd} + V_c - V_c'$$
$$= 1.000 + 0.050 - 0.247$$
$$= 0.803$$

Intake Volume at Pressure P_a

$$V_a = V_i \frac{P_i}{P_a} = 0.803 \frac{96}{100}$$
$$= 0.771$$

Volumetric Efficiency

$$\eta_v = \frac{V_a}{V_{pd}} = \frac{0.771}{1.000} = 0.771$$

To allow for heating and leakage effects, the calculated value of 0.771 for the volumetric efficiency is reduced to 0.75.

Two conclusions may be drawn from an examination of Fig. 3.8. For a fixed clearance volume, the volumetric efficiency of a compressor decreases with an increase in the operating pressure ratio. For a fixed discharge pressure, an increase in the clearance volume causes a decrease in the effective intake volume, and as a consequence the volumetric efficiency of the compressor is reduced.

Compressed gas trapped in the clearance space reexpands to the intake pressure. Because of irreversibility, the mechanical energy returned to the piston is less than the work required for compression of the gas. This difference in the work transfer however is small, hence a relatively inappreciable amount of energy is expended in the compression and reexpansion of the clearance gas.

A reduction in the volumetric efficiency of the compressor increases the number of piston strokes required for delivery of a specified quantity of gas. The increased travel of the piston raises the mechanical losses and consequently causes an increase in the compressor shaft work.

Compressed gas that leaks by the piston rings and valves is throttled to the intake pressure. The mechanical energy that is expended for compression of this gas is consequently dissipated through throttling and represents a loss in the work input to the compressor.

3.10 COMPRESSOR POWER

Because reciprocating gas compressors are cooled through use of water jackets, the isothermal compression cycle is taken as the theoretical mechanical cycle for the compressor. Isothermal compression represents the limiting degree of cooling that may be achieved in compressor operation.

The work of the actual mechanical compression cycle is not readily evaluated by application of thermodynamic relationships. Intake and discharge strokes do not occur at constant pressure.

The processes that comprise the cycle are irreversible and difficult to define.

Compressor work is estimated by applying an experimentally determined efficiency factor to the calculated work required for theoretical isothermal compression. Calculation of the isothermal work is based on the delivered free air volume and the external operating pressures. The compressor efficiency η_C, a ratio of the theoretical isothermal work to the compressor internal work, is likely to range from 0.70 to 0.75.

The compressor internal power can be measured by installing an indicator on the machine. The shaft input may be measured by driving the compressor through a torque meter or by an electric dynamometer. The difference between the shaft power and the internal power represents the compressor frictional power.

Mechanical losses for reciprocating machines vary approximately with the square of the speed and remain relatively constant with changes in the load. The mechanical efficiency for the compressor is taken as the ratio of the internal work to the shaft work, or $\eta_m = W_i/W_b$. The internal work is equal to the shaft work minus the friction work W_f. Thus, $\eta_m = (W_b - W_f)/W_b$, or

$$\eta_m = 1 - \frac{W_f}{W_b} \qquad (3.5)$$

For constant-speed operation, the friction work W_f may be assumed to remain essentially constant, hence the mechanical efficiency of the compressor continues to increase with an increase in the load represented by the shaft work W_b. At full-load operation, the mechanical efficiency of the reciprocating compressor may be taken as about 0.90.

Example 3.5

A reciprocating compressor delivers 8.50 m^3/min of free air. The barometric pressure is 100.0 kPa, abs. The discharge pressure is 415.0 kPa, gage. The isothermal compressor efficiency is 0.72, and the mechanical efficiency is 0.88. Estimate the shaft power.

Theoretical Isothermal Power

$$\dot{W}_T = P_1\dot{V}_1 \ln\frac{P_1}{P_2}$$

$$= 100.0 \times 8.50 \ln\frac{100.0}{515.0}$$

$$= -1393 \text{ kJ/min} \qquad -23.22 \text{ kW}$$

Internal Power

$$\dot{W}_i = \frac{\dot{W}_T}{\eta_C} = \frac{-23.22}{0.72} = -32.25 \text{ kW}$$

Shaft Power

$$\dot{W}_b = \frac{\dot{W}_i}{\eta_m} = \frac{-32.25}{0.88} = -36.65 \text{ kW}$$

The negative sign indicates a power input to the compressor.

The compressor shaft work may also be determined by applying the first law equation to steady-state, steady-flow operation of the machine. When changes in the kinetic energy of the gas are neglected, $W = h_1 - h_2 + Q$. In this equation, Q is numerically negative and consists of the energy transferred to the cooling water and the heat that is otherwise lost to the surroundings. The latter quantity is comparatively small and can be estimated in establishing the energy balance.

Example 3.6

The following data were recorded during a test made on a reciprocating air compressor: temperature increase of the air, 110 C; air flow, 9.10 kg/min; temperature increase of the jacket water, 13.9 C; jacket water flow, 4.27 kg/min.

The heat lost to the surroundings, Q_l, is assumed to be equivalent to 10 percent of the energy removed by the jacket water, Q_j. For water, c_w = 4.18 kJ/kg·K; and for air, $c_{p,a}$ = 1.00 kJ/kg·K.

Estimate the power input to the compressor.

Total Heat Transfer

$$\dot{Q} = \dot{Q}_j + \dot{Q}_l = 1.10\dot{Q}_j$$

$$= 1.10\dot{m}_w c_w \Delta T$$

$$= 1.10 \times 4.27 \times 4.18 \times 13.9$$

$$= 272.9 \text{ kJ/min}$$

Shaft Power

$$\dot{W} = \dot{m}_a(h_1 - h_2) + \dot{Q}$$

$$= \dot{m}_a c_{p,a}(T_1 - T_2) + \dot{Q}$$

$$= [9.10 \times 1.00(-110)] - 273$$

$$= -1001 - 273$$

$$= -1274 \text{ kJ/min} \qquad -21.23 \text{ kW}$$

3.11 MULTISTAGE COMPRESSORS

The degree of compression, that is, the pressure ratio developed in a single compressor cylinder, is limited by the temperature increase of the gas, the clearance, and the need to reduce shaft work. Excessively high gas temperatures may cause ignition of the lubricating oil vapor or decomposition of the lubricant. As previously noted, an increase in the pressure ratio causes greater expansion of the clearance gas and a consequent decrease in the volumetric efficiency of the compressor.

The discharge pressure for single-stage compressors is usually limited to approximately 700 kPa, gage. For higher discharge pressures, compression is achieved in two or more stages, with intercooling employed between the stages. Gas cooling between the stages provides the means to achieving an appreciable reduction in the compressor work and maintaining gas temperatures within safe operating limits. In multistage operation, a moderate pressure ratio is developed in each cylinder, hence excessive expansion of the clearance gas is avoided. Consequently, despite the high overall pressure ratio for the compressor, the volumetric efficiency for the individual stages, and for the entire machine, can be held to an acceptable level.

Maximum discharge pressures for multistage compressors will vary somewhat with the manufacturer. Two-stage compressors are limited to approximately 2000 kPa, gage. For three-, four-, and five-stage compressors, maximum discharge pressures are not specifically listed. Five-stage machines however often develop discharge pressures in excess of 35 MPa. Pressure limits for compressors are based on an atmospheric intake pressure.

A shell-and-tube design is usually employed for the construction of compressor intercoolers. The gas ordinarily flows over the outside surface of the tubes while the water flows through the tubes. Energy is transferred from the gas, the source fluid, through the metallic wall of the tube to the cooling medium or receiver fluid.

The principles of multistage compression are illustrated through use of theoretical mechanical cycles that describe compressor operation (see Fig. 3.9). The compression processes are internally reversible and are defined by PV^n = con-

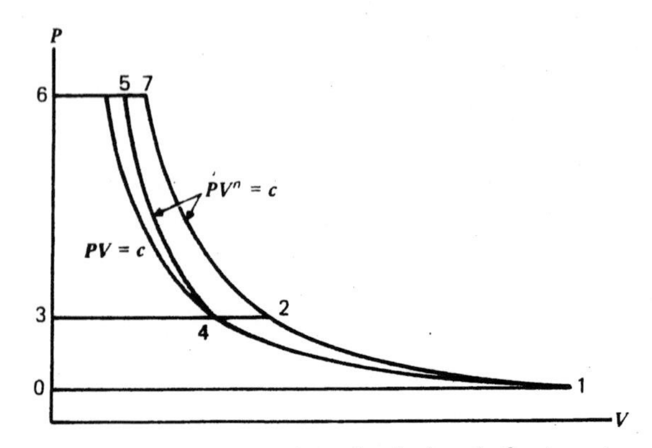

Figure 3.9 Theoretical mechanical cycle for two-stage compression.

stant, where $n < k$. The reversible isothermal process, defined by PV = constant, is taken as the theoretical limit for compressor cooling.

The displacement processes 2–3 and 3–4 represent, respectively, the discharge stroke for the low-pressure compressor and the intake stroke for the high-pressure compressor. The gas is returned from the intercooler at temperature T_4, which is equal to the initial temperature T_1. Pressure losses in the valves, lines, and intercooler are neglected in the theoretical analysis. The volume reduction effected by intercooling is equal to $V_2 - V_4$.

The work required for single-stage compression is equivalent to the area 012760. The work expended for compression in the first stage is equivalent to the area 01230, and in the second stage, to the area 34563. The area 0124560 represents the total work input for two-stage compression, while the area 27542 is equivalent to the reduction in work achieved by employing two-stage rather than single-stage compression. This reduction in work is the result of interstage gas cooling.

Some reduction in work is accomplished by cylinder cooling, although this quantity of energy is comparatively small. Theoretically, the work required to compress the gas is less for reversible polytropic compression, $PV^n = c$, than for reversible adiabatic compression, $PV^k = c$, where $n < k$.

A theoretical analysis shows that the minimum work for compound compression is achieved when the work input is divided equally between the two stages. In order to demonstrate this conclusion, several assumptions are made. Theoretical polytropic compression, PV^n = constant, is prescribed for both stages (see Fig. 3.9). The temperature at the start of compression is the same for both stages, that is, $T_4 = T_1$. Incidentally, on passing through the intercooler, the temperature of the gas is reduced from T_2 to T_4.

The compressor intake and discharge pressures are designated, respectively, P_i and P_d. The constant intercooler pressure is designated P_{ic}.

The first-stage work is given by

$$W_1 = \frac{n}{1-n} R(T_2 - T_1)$$

$$= \frac{n}{1-n} RT_1\left(\frac{T_2}{T_1} - 1\right)$$

$$\frac{T_2}{T_1} = \left(\frac{P_{ic}}{P_i}\right)^x \qquad \text{where } x = \frac{n-1}{n}$$

$$W_1 = \frac{n}{1-n} RT_1\left[\left(\frac{P_{ic}}{P_i}\right)^x - 1\right]$$

In a similar manner, the second-stage work is given by

$$W_2 = \frac{n}{1-n} R(T_5 - T_4)$$

$$= \frac{n}{1-n} RT_4\left(\frac{T_5}{T_4} - 1\right)$$

$$\frac{T_5}{T_4} = \left(\frac{P_d}{P_{ic}}\right)^x$$

$$W_2 = \frac{n}{1-n} RT_4\left[\left(\frac{P_d}{P_{ic}}\right)^x - 1\right]$$

Since $T_4 = T_1$, the total work for the two stages is expressed by

$$W_C = W_1 + W_2$$

$$= \frac{n}{1-n} RT_1$$

$$\left[\left(\frac{P_{ic}}{P_i}\right)^x + \left(\frac{P_d}{P_{ic}}\right)^x - 2\right]$$

W_C is a minimum when the quantity within the brackets is a minimum. W_C is differentiated with respect to P_{ic}, and the result is set equal to zero. The resulting equation is readily reduced to

$$P_{ic} = \sqrt{P_i P_d} \qquad (3.6)$$

and

$$\frac{P_{ic}}{P_i} = \frac{P_d}{P_{ic}}$$

Because the two compressor stages have like pressure ratios, the discharge temperatures are equal, or $T_5 = T_2$ (Fig. 3.9).

The theoretical analysis can be extended for determination of the intermediate, or intercooler, pressures for gas compression in three or more stages. In all cases, the minimum total work is achieved when the work quantities and pressure ratios for the several stages are equal.

For three-stage compression, the theoretical intercooler pressures are

$$P_{ic,1} = \sqrt[3]{P_i^2 P_d} \quad (3.7a)$$

$$P_{ic,2} = \sqrt[3]{P_i P_d^2} \quad (3.7b)$$

For four-stage compression, Eq. 3.6 is first used to determine the second intercooler pressure,

$$P_{ic,2} = \sqrt{P_i P_d} \quad (3.8a)$$

and subsequently to determine the first and third intercooler pressures,

$$P_{ic,1} = \sqrt{P_i P_{ic,2}} \quad (3.8b)$$

$$P_{ic,3} = \sqrt{P_{ic,2} P_d} \quad (3.8c)$$

Example 3.7

A gas is compressed in a two-stage reciprocating machine from 100 kPa, abs, to 600 kPa, abs. The inlet gas temperature is 300 K. For the gas, $\overline{m} = 16.04$ and $k = 1.299$. Calculate for theoretical polytropic compression, $n = 1.22$, the compressor work and the heat transfer (refer to Fig. 3.9).

$$R = \frac{\overline{R}}{\overline{m}} = \frac{8.3143}{16.04} = 0.518 \text{ kJ/kg·K}$$

$$P_{ic} = \sqrt{P_i P_d} = \sqrt{100 \times 600}$$
$$= 244.9 \text{ kPa, abs}$$

$$W_1 = \frac{n}{1-n} RT_1 \left[\left(\frac{P_{ic}}{P_i} \right)^x - 1 \right]$$

$$x = \frac{n-1}{n} = \frac{1.22 - 1}{1.22} = 0.180$$

$$W_1 = \frac{1.22}{1 - 1.22} 0.518 \times 300 \cdot \left[\left(\frac{244.9}{100} \right)^{0.180} - 1 \right]$$
$$= -150.76 \text{ kJ/kg gas}$$

$$W_C = 2W_1 = 2(-150.76)$$
$$= -301.52 \text{ kJ/kg gas}$$

Heat Transfer

The steady-flow energy equation is applicable to the intercooler and the compressor stages:

$$T_2 = T_1 \left(\frac{P_{ic}}{P_i} \right)^x$$
$$= 300(244.9/100)^{0.180} = 352.5 \text{ K}$$

$$c_p = \frac{kR}{k-1}$$
$$= \frac{1.299 \times 0.518}{1.299 - 1} = 2.250 \text{ kJ/kg·K}$$

Intercooler

$$Q_{ic} = c_p(T_4 - T_2)$$
$$= 2.250(300 - 352.5)$$
$$= -118.12 \text{ kJ/kg gas}$$

First-Stage Compressor Cylinder

$$Q_1 = (h_2 - h_1) + W = c_p(T_2 - T_1) + W$$
$$= 2.250(352.5 - 300) - 150.76$$
$$= -32.64 \text{ kJ/kg gas}$$

Second-Stage Compressor Cylinder

$$Q_2 = Q_1 = -32.64 \text{ kJ/kg gas}$$

Total Amount of Heat Transferred from the Compressor

$$Q_{total} = Q_{ic} + Q_1 + Q_2$$
$$= -118.12 + (-32.64) + (-32.64)$$
$$= -183.40 \text{ kJ/kg gas}$$

Note: For theoretical single-stage compression, $n = 1.22$, the compressor work is -327.99 kJ/kg gas. The reduction in the compressor work, 26.47 kJ/kg gas, is attributed to intercooling.

3.12 RECIPROCATING PUMPS

The operating principle of the reciprocating pump used to move liquids is relatively simple. On the intake stroke, the spring-loaded inlet valves are unseated and the supply liquid flows into the cylinder. On the return stroke, the pressure exerted on the essentially incompressible fluid causes the spring-loaded discharge valve to open, and the liquid is delivered to the high-pressure line. Single-cylinder or multicylinder pumps are constructed for either single- or double-acting operation.

The reciprocating pump is widely used for industrial applications. Included among the advantages of this machine are the ability to develop high delivery pressures, operate with superior economy in certain installations, and operate with good reduced-load characteristics. Through speed control, a constant discharge pressure can be maintained over a wide range of delivery, because for normal operation the discharge pressure is not a function of speed.

The pump internal power is determined from the energy equation that is generally applicable to continuous-flow systems. Thus,

$$\dot{W}_i = \dot{m}(h_1 - h_2) + \frac{\dot{m}}{2}(C_1^2 - C_2^2) + \dot{Q} \quad (3.9)$$

Because of mechanical friction in the external bearings, the pump shaft power is somewhat higher than the internal power given by Eq. 3.9.

Equation 3.9 is especially applicable to high-pressure pumping where the fluid cannot be considered as incompressible. $\dot{Q}$ is numerically negative when heat is transferred from the pump casing.

The internal machine efficiency of the pump is evaluated from

$$(\eta_P)_i = \frac{h_1 - h_{2,s}}{W_i} \quad (3.10)$$

when the change in the kinetic energy of the fluid is comparatively small and can be neglected. The numerator in this ratio is the reversible adiabatic work for pumping the fluid between the prescribed pressure limits. The denominator W_i is the corresponding actual internal work of the pump.

Ordinarily, the shaft work of the pump is related to the theoretical work through use of the overall pump efficiency $(\eta_P)_b$ that accounts for all of the fluid and mechanical losses. Then, for pumping a compressible fluid,

$$(\eta_P)_b = \frac{h_1 - h_{2,s}}{W_b} \quad (3.11)$$

When liquids are pumped through a moderate pressure range, the fluid may be taken as incompressible, or v = constant. The quantity $u + Pv$ is substituted for h in Eq. 3.9, and the resulting equation is written in differential form. Thus,

$$d\dot{W}_i = \dot{m}(-du - v\,dP - P\,dv) - \dot{m}C\,dC + d\dot{Q}$$

Then, for v = constant, $dv = 0$, and the above equation is reduced to

$$d\dot{W}_i = \dot{m}(-du - v\,dP) - \dot{m}C\,dC + d\dot{Q}$$

and subsequent integration of the equation yields

$$(\dot{W}_i)_v = \dot{m}(u_1 - u_2) + \dot{m}v(P_1 - P_2) + \frac{\dot{m}}{2}(C_1^2 - C_2^2) + \dot{Q} \quad (3.12)$$

Because of fluid friction and turbulence, the internal energy u will increase slightly, but the difference $u_2 - u_1$ cannot be evaluated precisely. Further, the heat transfer $\dot{Q}$ is also a small quantity that is not readily measured. The terms u_2

$-u_1$ and $\dot{Q}$ are consequently set equal to zero, thus reducing Eq. 3.12 to

$$(\dot{W}_i)_{th} = \dot{m}v(P_1 - P_2) + \frac{\dot{m}}{2}(C_1^2 - C_2^2) \quad (3.13)$$

where $(\dot{W}_i)_{th}$ represents the theoretical internal power for pumping an incompressible fluid in the absence of fluid friction and transfer of heat to the surroundings.* In order to determine the pump shaft power $(\dot{W})_b$, the quantity $(\dot{W}_i)_{th}$ is increased by introduction of an efficiency term to account for the fluid and mechanical losses. Thus,

$$(\eta_P)_b = \frac{(\dot{W}_i)_{th}}{\dot{W}_b} \quad (3.14)$$

where $(\eta_P)_b$ is the overall pump efficiency applicable to pumping an incompressible fluid.

Pump inlet and discharge pressures are often measured at points that are above or below the center line of the pump. It is then essential to correct the measured pressures to the equivalent center line readings or include in the energy equation a term for the change in the potential energy of the fluid.

* An alternative development for Eq. 3.13 is achieved by observing that

$$dW = -dh - C\,dC + dQ$$

and for a thermodynamic process, $dh = T\,ds + v\,dP$. Then

$$dW = -T\,ds - v\,dP - C\,dC + dQ$$

Optimum pumping is accomplished in a reversible adiabatic process, hence

$$T\,ds = 0 \quad \text{and} \quad dQ = 0$$

Thus

$$dW = -v\,dP - C\,dC$$

When pumping liquids v can ordinarily be assumed to remain constant. Then, by integration

$$W_{v=c} = -v(P_2 - P_1) - \frac{C_2^2 - C_1^2}{2}$$

$W_{v=c}$ is taken as the theoretical work for pumping an incompressible fluid and provides a close approximation to the reversible adiabatic work.

Example 3.8

The inlet and discharge pressure gages installed on a pump read, respectively, −35 and 830 kPa. The inlet pressure gage is installed on the center line of the pump. The discharge pressure gage is located at a point 2 m above the center line. The specific gravity of the liquid handled by the pump is 0.92. The fluid velocity in the inlet line is 1.5 m/s, and in the outlet line, 3 m/s. The barometric pressure is 96.0 kPa, abs. The pump overall efficiency, which allows for the internal and external losses, is 0.70. The pump discharge rate is 15 kg/s. Calculate the pump shaft power.

Discharge pressure P_2 = 926 kPa, abs

Inlet pressure P_1 = 61 kPa, abs

$$v = \frac{1}{\rho} = \frac{1}{0.92 \times 1000} = 0.001087\ \text{m}^3/\text{kg}$$

(for water ρ = 1000 kg/m³)

Equation 3.13 is modified to include a term that represents the change in the potential energy of the fluid:

$$\begin{aligned}(\dot{W}_i)_{th} &= \dot{m}v(P_1 - P_2) + \frac{\dot{m}}{2}(C_1^2 - C_2^2) + \dot{m}g(z_1 - z_2) \\ &= 15\{0.001087(61 - 926)10^3 + \tfrac{1}{2}[(1.5)^2 - (3.0)^2] + 9.807(0 - 2)\} \\ &= 15\{-940.3 - 3.4 - 19.6\} \\ &= -14{,}450\ \text{W} = -14.45\ \text{kW}\end{aligned}$$

$$\dot{W}_b = \frac{(\dot{W}_i)_{th}}{(\eta_P)_b} = \frac{-14.45}{0.70} = -20.64\ \text{kW}$$

PROBLEMS

3.1 A steam engine designed for automotive service consists of six single-acting cylinders. Bore and stroke are, respectively, 65 mm and 82 mm. Steam is supplied to the engine dry, saturated at 3500 kPa, abs. The exhaust pressure is 50 kPa, abs. The cutoff ratio is 15 percent. The engine design speed is 3000 rpm. The diagram factor is taken as 0.82. The mechanical efficiency is 0.90. Estimate the shaft output of the engine.

3.2 For the engine described in Problem 3.1, determine the steam rate, in kg/kWh, based on the shaft output. The machine efficiency is 0.60. Calculate the steam flow, in kg/h.

3.3 A small factory is equipped with a 100-kW steam-electric power plant that has not operated for several years. A proposal has been made to operate the plant, at full load, for a short period of time each day to meet the peak electrical demand and thus avoid the high demand charge for purchased electrical energy. The boiler is fired with an oil fuel that has a higher heating value of 45,560 kJ/kg and a specific gravity of 0.85. The boiler efficiency is 0.82, based on the higher heating value of the fuel. The steam engine machine and mechanical efficiencies are, respectively, 0.60 and 0.88, and the generator efficiency is 0.95. The boiler supplies dry, saturated steam at 1400 kPa, abs. The condenser pressure is 30 kPa, abs. The fuel cost is 16¢/liter, and the cost of the electrical energy, including the demand charge, is 15¢/kWh. Labor cost for operating the power plant is $15/hr. Determine the economic feasibility of operating the power plant. The capital cost of the power plant is no longer a significant factor.

3.4 Determine for the steam engine theoretical mechanical cycle the loss because of incomplete expansion for two different exhaust pressures, namely, 40 and 125 kPa, abs. The steam is initially dry and saturated at 1200 kPa, abs. The cutoff is 18 percent of the stroke.

3.5 A double-acting reciprocating compressor is required to deliver air at the minimum rate of 3.20 m^3/min, based on free air conditions (1 atm, 25 C). The compressor speed is 250 rpm. The piston rod diameter is 35 mm. The discharge (line) pressure is 750 kPa, gage. The pressure drop through the intake valve is 3.5 kPa, and through the discharge valve, 15.2 kPa. The compressor clearance is 4 percent of the displacement. For expansion of the clearance air, n = 1.34. The compressor stroke-to-bore ratio is 1.20. The isothermal compressor efficiency is 0.70, and the mechanical efficiency is 0.87. Determine the compressor bore and stroke, and the size of the motor used to drive the compressor.

3.6 A reciprocating gas compressor operates with a pressure ratio of 10 to 1. For the compression process, n = 1.32. Estimate the reduction in work achieved by employing two-stage compression rather than single-stage compression.

3.7 A double-acting reciprocating pump handles a liquid that has a specific gravity of 0.85. The bore and stroke are, respectively, 150 mm and 175 mm. The piston rod diameter is 30 mm. The inlet pressure is 50 kPa, gage. The pump volumetric efficiency is 0.95. The pump speed is 180 rpm. The overall pump efficiency is 0.82. The inlet and outlet fluid velocities are essentially equal. The maximum output of the driving motor is 30 kW. Determine the maximum discharge pressure that can be developed by the pump.

3.8 The bore and stroke for a single-acting air compressor are, respectively, 180 mm and 200 mm. The compressor speed is 240 rpm. The internal discharge and intake pressures are, respectively, 800 and 95 kPa, abs. The compressor clearance is increased from 3 to 8 percent by activating a clearance pocket. The atmospheric pressure is 101 kPa, abs. For expansion of the clearance air, n = 1.32. Calculate the change in the volume of air delivered, in m^3/min, because of the increase in the compressor clearance.

3.9 A double-acting steam engine is designed to deliver 50 kW. The stroke-to-bore ratio is 1.25. The steam pressure at the throttle is 1000 kPa, abs, and the prescribed exhaust pressure is 45 kPa, abs. The engine speed is 225 rpm, and the mechanical efficiency is taken as 0.91. The design cutoff ratio is 0.35, and the probable diagram factor is 0.80. Determine the bore and stroke for the engine.

3.10 Saturated water at a pressure of 2 MPa, abs, is compressed in a reciprocating pump to 20 MPa, abs. The overall pump efficiency is 0.60. Determine the pump shaft work for pumping (1) a compressible fluid and (2) and assumed incompressible fluid.

3.11 The compressor described in Problem 3.8 is redesigned for compound compression. The intercooler pressure drop is 20 kPa. The clearance for both the high- and low-pressure cylinders is 3 percent. The piston strokes for the two cylinders are equal. Determine the air flow, in m^3/min, for the compressor and the bore for the high-pressure cylinder.

3.12 The following data are reported for a test conducted on a double-acting steam engine: speed, 102 rpm; area of the head-end indicator card, 16.85 cm^2; area of the crank-end indicator card, 17.25 cm^2; length of the indicator card, 7.20 cm; indicator spring scale, 175 kPa/cm; length of the brake arm, 1.50 m; brake load (net), 1700 N. The engine dimensions are: bore, 250 mm; stroke, 450 mm; piston rod diameter, 45 mm. Determine the engine internal power, shaft power, and mechanical efficiency.

3.13 The bore and stroke for the steam cylinder of a direct-acting pump are, respectively, 180 mm and 200 mm. The internal steam pressures, measured by an indicator, are 750 kPa, gage, and 50 kPa, gage. The diagram factor is 0.95. The pump speed is 30 double strokes per minute. The overall pump mechanical efficiency is 0.90. The diameter of the water cylinder is 120 mm. The water enters the pump cylinder at 70 kPa, gage, and 50 C. Neglect the effect of the piston rod. Calculate the pump delivery, in kg/min, and the discharge pressure.

3.14 A gas is compressed (n = 1.32) from 98 kPa, abs, to 3500 kPa, abs, in a three-stage reciprocating machine equipped with intercoolers. The gas inlet temperature is 280 K. For the gas, R = 0.297 kJ/kg·K, k = 1.400, and c_p = 1.042 kJ/kg·K. Calculate for theoretical compression the compressor work and the heat transfer.

3.15 A gas is compressed in a four-stage reciprocating machine from 120 kPa, abs, to 5000 kPa, abs. The gas enters the compressor at 290 K. For the gas, k = 1.34. The compression process is described by n = 1.25. Calculate the intercooler pressures and the temperature of the gas leaving the compressor.

REFERENCES

1 Joseph, J. "Inside Bill Lear's 'Steam Motor,' " *Diesel and Gas Turbine Progress,* June 1969.

2 "GM Progress of Power," A General Motors Report on Vehicular Power Systems, May 1969.

BIBLIOGRAPHY

Kent's Mechanical Engineers' Handbook, Power Volume, J. K. Salisbury, ed., John Wiley & Sons, Inc., New York, 1957.

Marks' Standard Handbook for Mechanical Engineers, T. Baumeister, E. A. Avallone, and T. Baumeister III, eds., McGraw-Hill Book Co., New York, 1978.

CHAPTER FOUR

INTERNAL COMBUSTION ENGINES

The internal combustion engine unquestionably occupies an important position in the field of power production. Internal combustion engines are dependable, compact, self-contained power plants that characteristically show good operating thermal efficiencies. Unlike other thermal engines, the entire thermodynamic cycle for the internal combustion engine is completed within a single element, commonly a cylinder-and-piston machine. An equivalent thermodynamic cycle is followed in the Wankel engine, a rotary machine.

4.1 GENERAL

Maximum temperatures developed within the cylinder of the internal combustion engine are exceptionally high, approaching 2200 C. These high temperatures can be tolerated in part because of the charging process during which cool atmospheric air, or a fuel–air mixture, enters the cylinder subsequently to the discharge of the hot exhaust gas. The principal effect of the air or mixture cooling is to hold piston temperatures within safe operating limits. Incidentally, this cooling action is a good deal more effective in the four-stroke-cycle engine than in comparable two-stroke-cycle machines, unless provision is made for a high flow of scavenging air.

Because the entire cycle of operation for the internal combustion engine occurs within the single element, that is, the cylinder-and-piston machine, there is no transfer of energy between system primary fluids. The absence of this kind of energy transfer avoids a degradation of temperature and the resultant loss in availability. High cycle maximum temperatures, combined with a freedom from cascading energy processes, promote high thermal efficiencies for the internal combustion engine.

Principally because of a compact structure and a good operating thermal efficiency, the internal combustion engine is admirably suited to service in the transportation field. The internal combustion engine is used to power automotive vehicles, locomotives, small- and medium-sized vessels, and light aircraft. Extensive use is made of the internal combustion engine for operating a wide variety of industrial machines. Internal combustion engines of various sizes are installed in small central stations for generation of electrical energy, but the total amount of energy produced in these power plants is comparatively small.

Because of certain inherent disadvantages, the internal combustion engine is not well suited to large-scale power production. A high-output internal combustion engine is a bulky machine. Many moving parts are used in the construction of the engine, and the reciprocating motion of some of these parts causes the mechanical efficiency of the engine to be somewhat low. Except for low-output machines, rotating speeds must remain low in order to avoid excessively high inertia forces in the reciprocating mechanism. In comparison with other prime movers, consumption of lubricating oil is high, and a higher-quality fuel must be used. For electric power generation, the internal combustion engine, in capabilities exceeding roughly 10,000 kW, usually cannot compete economically with the steam turbine generating unit. Higher capital, maintenance, and fuel costs are contributing factors to this limitation in the area of electric power generation.

The automotive engine is a major contributor to air pollution in metropolitan areas. Reduction in the emission of pollutants has been accomplished by the development of the catalytic converter and by introducing certain design changes.

Except for experimental engines, the single-cylinder construction is used only for machines of very low output. Basically, multicylinder engines conform either to an in-line or radial arrangement. Two banks of in-line cylinders are commonly combined on a single frame to form a V-engine. Radial banks may be arranged in parallel, for example, two banks each consisting of nine cylinders. A few engines have incorporated the opposed piston design, an arrangement in which two mechanically synchronized pistons move in opposite directions in a common cylinder.

The cylinders of air-cooled engines are equipped with external fins to provide an extended surface and thereby increase the rate of heat transfer to the surrounding air. If not air cooled, the engine is constructed with jacketed cylinders and hollow heads for liquid cooling. Fans or pumps are provided for circulation of the cooling medium, which is air or an appropriate liquid.

For discussion purposes, the following categories are used to classify the internal combustion engine.

1 Spark ignition (S-I) or compression ignition (C-I).
2 Two-stroke or four-stroke cycle.
3 Supercharged or nonsupercharged.
4 Single cylinder or multicylinder.

4.2 SPARK-IGNITION ENGINES

Figure 4.1 illustrates the operation of a spark-ignition engine operating on the four-stroke cycle. On the intake stroke, the piston moves downward causing a reduction in the cylinder pressure. A mixture of air and fuel flows through the open intake port into the cylinder. On the subsequent upward stroke of the piston, the mixture is compressed to a pressure of approximately 1100 kPa, gage, in this case for a compression ratio of 7 to 1. Both the intake and exhaust valves remain closed during compression.

Shortly before the piston reaches the top-center position, a spark is discharged between the electrodes of the plug, igniting the mixture. The combustion reaction normally continues for a short increment of piston travel from the top-center position. Again for this case, a peak pressure of about 2800 kPa, gage, will be attained. During

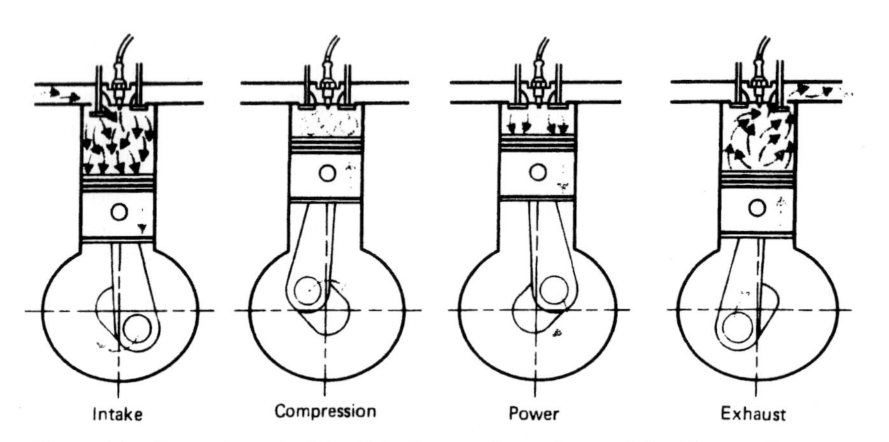

Figure 4.1 Operating principle of the four-stroke-cycle spark-ignition engine. (Courtesy Texaco's magazine *Lubrication*.)

the downward movement of the piston, the combustion gases expand and transfer energy to the mechanical system. On the following stroke, the upward movement of the piston creates a pressure differential between the interior of the cylinder and the atmosphere and thereby causes the gases to flow through the open exhaust port.

A brief description of the operation of the two-stroke-cycle S-I engine will be presented. These engines are used mainly in low-power installations, for example, as marine outboard-engine drives. The charging operation is effected by crankcase induction (see Fig 4.2). On the compression stroke, the upward movement of the piston causes the pressure in the crankcase to drop below the atmospheric pressure. Because of this pressure differential, the fuel–air mixture flows into the crankcase through an intake valve. On the subsequent power stroke, the mixture is compressed in the crankcase. As the piston uncovers the exhaust ports, the combustion gases, under a relatively high pressure, are discharged from the cylinder. As the downward movement of the piston continues, the inlet ports in the cylinder wall are uncovered and the compressed fuel–air mixture flows from the crankcase through an interconnecting passage into the cylinder. In addition to charging the cylinder, the mixture

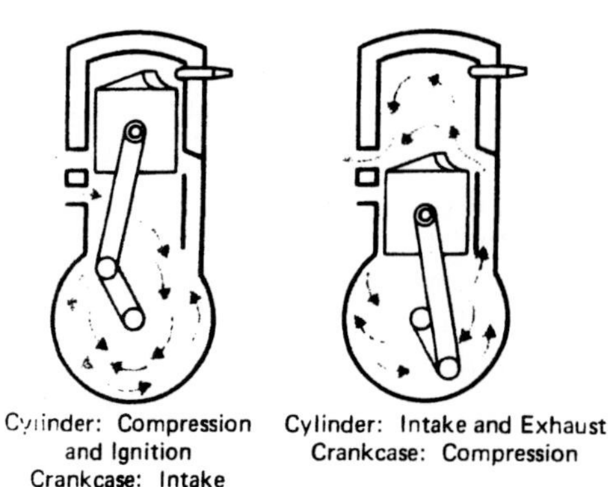

Figure 4.2 Operating principle of the two-stroke-cycle spark-ignition engine. (Courtesy Texaco's magazine *Lubrication*.)

flow assists in the removal of the residual exhaus gases, a scavenging operation.

Theoretically, for the same bore, stroke, an speed, the two-stroke-cycle engine should de velop twice the power output of the four-stroke cycle engine. Actually, the ratio is less than 2 For the two-stroke-cycle engine described above the effective power stroke is shortened by earl opening of the exhaust ports to increase the tim available for scavenging; consequently, the los owing to incomplete expansion is comparativel high. Cylinder charging is not particularly effec tive because of incomplete scavenging of the cy inder and poor volumetric efficiency for crank case induction. The thermal efficiency of th engine is reduced by loss of fuel that occurs whe some of the fuel–air mixture escapes from th cylinder with the exhaust gas during the scav enging operation.

4.3 THE ENGINE INDICATOR CARD

The exhaust and intake strokes for the four-stroke cycle engine constitute the pumping process i which the exhaust gases are expelled from th cylinder and a fresh charge is inducted. Me chanical energy must be expended for this pump ing operation.

A typical indicator card for full-load operatio of the four-stroke-cycle S-I engine is shown i Fig. 4.3a. The upper loop of the diagram rep resents the gross power developed in the engin cylinder, while the area of the lower loop i equivalent to the pumping power.

For reduced-load operation of the engine, th throttle valve in the intake system is partiall closed in order to restrict the flow of the charg to the cylinder, thus causing a decrease in th area of the upper loop of the diagram (se Fig. 4.3b). In addition, throttling causes a re duction in the intake pressure and a resultan increase in the area of the lower loop of th indicator card. The energy expended in pumpin will thus increase with increased throttling o reduced engine load.

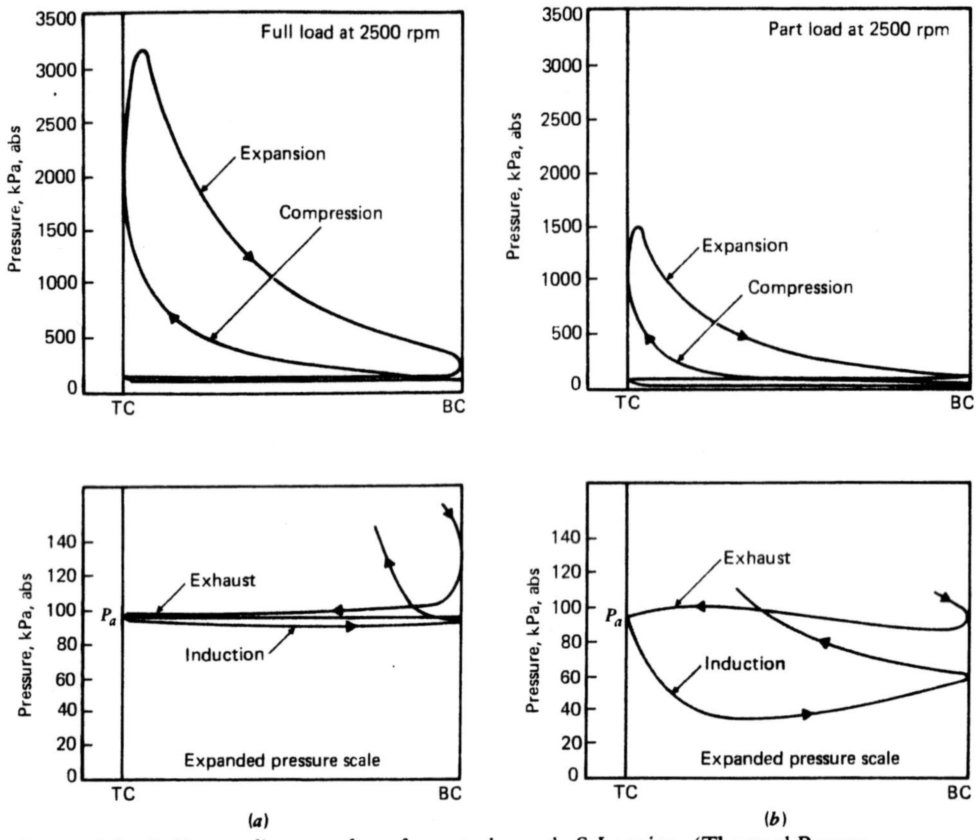

Figure 4.3 Indicator diagrams for a four-stroke-cycle S-I engine. (Thermal Power Laboratory, Washington State University.)

The internal power $\dot{W}_i$ of the engine is given by

$$\dot{W}_i = NP_i lAn \tag{4.1}$$

where

N = number of cylinders
P_i = indicated mean effective pressure
l = length of the piston stroke
A = cylinder transverse area
n = number of power strokes per unit time, per cylinder

The indicated mean effective pressure for the four-stroke-cycle engine is calculated from the net area of the indicator card. The net area, in turn, is determined by subtracting the area of the lower, or negative, loop from the area of the upper loop of the card.

When used to measure the internal power of the two-stroke-cycle S-I engine, indicators are installed on the power cylinder and on the crankcase. Separate determinations are thus made for the gross internal power and the pumping power expended for crankcase induction of the charge. The net internal power of the engine is equal to the difference between the gross internal power and the pumping power.

Example 4.1

A two-cylinder, two-stroke-cycle S-I engine operates at 600 rpm. The engine bore and stroke are, respectively, 110 mm and 140 mm. The engine operates with crankcase induction. The area and length of the card produced by the indicator installed on the crankcase are, respectively, 3.85 cm^2 and 7.50 cm. The indicator scale is 25 kPa/cm. For the indicator installed on the power cylinder, the card area is 5.15 cm^2, length 7.50 cm, and scale 700 kPa/cm. The mechanical efficiency of the engine is 0.82. Determine the engine shaft power.

Power Cylinder

$$P_i = \frac{a}{l}s = \frac{5.15}{7.50}700$$
$$= 480.7 \text{ kPa}$$
$$A = 0.009503 \text{ m}^2$$

$$(\dot{W}_i)_p = NP_i lAn$$
$$= 2 \times 480.7 \times 0.140 \times 0.009503 \times 600$$
$$= 767.4 \text{ kJ/min} = 12.79 \text{ kW}$$

Crankcase

$$P_i = \frac{a}{l}s = \frac{3.85}{7.50}25$$
$$= 12.83 \text{ kPa}$$

$$(\dot{W}_i)_c = NP_i lAn$$
$$= 2 \times 12.83 \times 0.140 \times 0.009503 \times 600$$
$$= 20.48 \text{ kJ/min} = 0.34 \text{ kW}$$

Net Internal Power

$$\dot{W}_i = (\dot{W}_i)_p - (\dot{W}_i)_c$$
$$= 12.79 - 0.34 = 12.45 \text{ kW}$$

Shaft Power

$$\dot{W}_b = \eta_m \dot{W}_i = 0.82 \times 12.45$$
$$= 10.21 \text{ kW}$$

Mechanical and electronic devices are both used to generate or construct the indicator card for the internal combustion engine. In addition to being employed for power measurements, the indicator card is useful for examining the operating characteristics of the engine as influenced by the combustion process and valve timing.

Because the indicated mean effective pressure of the engine cannot in every case be readily determined by direct measurement, use is made of an equally important performance factor, the brake mean effective pressure. The brake or shaft power $\dot{W}_b$ of the engine, measured by a dynamometer or brake, is introduced into Eq. 4.1 in place of the internal power $\dot{W}_i$. Then the brake mean effective pressure bmep is given by

$$\text{bmep} = \frac{\dot{W}_b}{NlAn} \tag{4.2}$$

Example 4.2

Estimate the probable shaft power for an eight-cylinder, four-stroke-cycle S-I engine operating at 2400 rpm. Bore and stroke are, respectively, 120 mm and 125 mm. The anticipated brake mean effective pressure is 820 kPa.

$$A = 0.01131 \text{ m}^2$$
$$n = 20 \text{ power strokes per second, per cylinder}$$

$$\dot{W}_b = NP_b lAn$$
$$= 8 \times 820 \times 0.125 \times 0.01131 \times 20$$
$$= 185.5 \text{ kW}$$

4.4 COMBUSTION IN THE SPARK-IGNITION ENGINE

Although the fuel may not be completely vaporized when the mixture enters the cylinder, complete vaporization will be achieved before the start of the combustion reaction. Energy is transferred to the charge from the warm cylinder walls and, in the mixing process, from the residual exhaust gas. In additon, the mixture temperature is increased as a result of compression. During the intake and compression strokes, there is consequently sufficient energy and ample time available for complete evaporation of the fuel droplets.

Example 4.3

Fuel and air enter the carburetor of a S-I engine at a temperature of 36.8 C (310.0 K). The air/fuel ratio is 15.09 to 1, and the fuel is octane (C_8H_{18}). Determine the mixture temperature that results from the evaporation of 35 percent of the fuel in the absence of heat transfer. The mixture pressure is 92.50 kPa, abs.

The energy equation is

$$m_a c_{p,a} (T_1 - T_2)_{air} + [h_{f,1} - (1 - x)h_{f,2} - x h_{v,2}]_{fuel} = 0$$

Only the final trial is shown for the trial-and-error solution. Assume $T_2 = 302.7$ K:

$$15.09 \times 1.004\,(310.0 - 302.7) + [22.72 - (1 - 0.35)6.13 - (0.35 \times 370.78)] = R$$

$$R = 110.598 - 111.038 = -0.440$$

Enthalpy values for octane are selected from *Thermodynamic Properties in SI* by W. C. Reynolds, Stanford University. The equation of state for the ideal gas is assumed to be valid for the octane vapor. Then,

$$P_v = (P_t - P_v)\frac{m_v R_v}{m_a R_a}$$

$$= (92.50 - P_v)\frac{0.35 \times 0.0728}{15.09 \times 0.287}$$

$$= 0.5410 \text{ kPa, abs}$$

The saturation pressure for octane at 302.7 K is 2.451 kPa, abs.

Note: The octane vapor is not in equilibrium with the liquid.

At a downstream point in the engine, the fuel is completely vaporized, and the temperature and pressure of the mixture are 51.8 C (325.0 K) and 89.0 kPa, abs. Determine the quantity of heat transferred to the mixture.

$$T_1 = 310.0 \text{ K} \qquad T_2 = 325.0 \text{ K}$$

$$m_a c_{p,a}(T_1 - T_2)_{air} + (h_{f,1} - h_{v,2})_{octane} + Q = 0$$

$$15.09 \times 1.004(310.0 - 325.0) + (22.72 - 408.67) + Q = 0$$

$$-227.26 - 385.95 + Q = 0$$

$$Q = 613.21 \text{ kJ} \qquad 38.11 \text{ kJ/kg mixture}$$

$$P_v = (89.0 - P_v)\frac{1 \times 0.0728}{15.09 \times 0.287}$$

$$P_v = 1.4713 \text{ kPa, abs}$$

$$P_a = 87.5287 \text{ kPa, abs}$$

Combustion in the S-I engine is normally a progressive process that extends throughout the mixture. Verification of this progressive reaction has been made through high-speed photography. When the spark plug fires, the temperature of the fuel and air molecules, in the path of the electrical discharge, is raised to the ignition point, and the combustion reaction starts. The flame front that is established moves outward through the fuel–air mixture. As each portion of the charge in a thin layer burns and releases energy, the temperature of this small mass increases at constant pressure, thus producing a tendency to expand. Expansion of the burning charge causes compression of both the burned and the unburned portions, with a consequent increase in the pressure of the entire mass. It follows that the first portion of the charge burns at the lowest cylinder pressure and the final portion at the highest pressure. Except for piston movement, the combustion reaction occurs at constant volume, yet because of progressive burning there is a tendency for each element of mass to burn at constant pressure. In addition to the contribution from the expansion of the burning portion of the charge, the flame front is also propagated by turbulence within the entire mass and by the reaction velocities acquired by the molecules.

It is highly desirable to have the combustion process occur at the top-center position of the piston in order to achieve high cycle temperatures and maximum availability. In comparison with the time required for completion of one cycle, the combustion time interval is appreciable, extending over several degrees of crank rotation. Engine design should include consideration of those factors that can increase the velocity at which the flame front travels and reduce the length of travel of the flame front. Turbulence within the mass is a primary factor that influences the velocity at which the flame front travels, hence attention is directed to the design of the cylinder head and piston face with the objective of promoting such turbulence during the combustion reaction. A central location of the spark plug will tend to reduce the length of travel of the flame front and thus shorten the time required for completion of the combustion process.

Carbon deposits, overheated spark plugs, and faulty engine design have a tendency to cause hot spots to develop inside the cylinder. As previously noted, the temperature of the mixture increases during compression; but in normal operation of the engine, the ignition temperature is not attained prior to firing the plug. With hot spots present, local temperature of the mixture may however reach the ignition temperature and start the combustion reaction prior to the spark discharge. Early combustion is undesirable because of the increased heat loss from the gas, and the consequent reduction in the cycle maximum temperature and decrease in the thermal efficiency.

The point at which the spark discharge occurs in normal engine operation is designated in degrees of crank rotation in advance of the top-center position of the piston. Spark timing, controlled automatically by the distributor, is varied in accordance with changes in the engine speed and load. An increase in the engine speed promotes increased gas turbulence, and consequently the flame front will advance more rapidly and reduce the time required for combustion. The increase in the combustion rate however is not as great as the increase in the engine speed, hence the spark is advanced in order to burn the fuel in the vicinity of the top-center position of the piston.

Example 4.4

When a S-I engine operates at 600 rpm, the average flame speed is 8.53 m/s, and 95 percent of the flame travel occurs during 55° rotation of the crankshaft. At an operating speed of 1800 rpm, the crankshaft rotation is 73° for the corresponding flame travel. Calculate the average flame speed for operation at 1800 rpm and the time interval for 95 percent of the flame travel at 600 rpm and 1800 rpm.

At 600 rpm operating speed,

$$\tau = \frac{55}{10 \times 360} = 0.01528 \text{ s}$$

$$l = \tau \times C_{\text{flame}} = 0.01528 \times 8.53$$

$$= 0.1303 \text{ m}$$

At 1800 rpm operating speed,

$$\tau = \frac{73}{30 \times 360} = 0.006759 \text{ s}$$

$$C_{\text{flame}} = \frac{l}{\tau} = \frac{0.1303}{0.006759}$$

$$= 19.28 \text{ m/s}$$

A 3-to-1 increase in the piston speed causes the average flame speed to increase in a ratio of 2.26 to 1.

The increase in the number of degrees of crankshaft rotation, from 55° to 73°, for the flame to travel 95 percent of the length of the combustion chamber is caused principally by the low flame speed in the vicinity of the spark plug. In this zone, the high heat transfer to the relatively cool cylinder wall causes a combustion lag. The effect of this lag is to increase the number of degrees of crankshaft rotation for the flame to travel 10 percent of the chamber length. This increase in degrees of crankshaft rotation is significantly higher than the increase in degrees of crankshaft rotation observed for the flame to travel from 10 percent to 95 percent of the length of the combustion chamber.

The spark is also advanced to compensate for the decrease in the velocity of flame propagation caused at part load by the reduced pressure of the fuel–air mixture and by the increased exhaust gas dilution. In addition, the spark is advanced to compensate for an increase in the humidity of the combustion air. The water molecules have the same diluting effect as the inert exhaust gas molecules.

4.5 DETONATION IN THE SPARK-IGNITION ENGINE

Detonation is a highly significant phenomenon associated with combustion in the S-I engine. The detonation process is complex, but a general discussion of the subject is within the scope of this text. Incidentally, detonation in the S-I engine has been studied in great detail by many investigators.

As the flame front advances, the temperature of the most remote portion of the unburned charge continues to increase. Under certain conditions, present within the cylinder, the temperature of the unburned charge reaches the spontaneous ignition temperature of the mixture. At several points in this unburned portion, ignition occurs simultaneously, and the entire mass burns very rapidly and in effect at constant volume. The rate of pressure rise is high, and the resulting pressure waves travel through the gas and strike the cylinder wall with an audible ping. In addition to the objectionable noise that is caused by vibration of the metal wall, detonation results in loss of power and reduced thermal efficiency. Severe detonation, if continued, will cause damage to the engine, particularly the piston and the exhaust valves. Destruction of the engine parts is apparently a result of high metal temperatures that are caused by an increased rate of heat transfer associated with the detonation phenomenon.

For a specified mixture ratio, that is, fuel to air, the tendency for detonation to occur can be reduced by increasing the velocity of propagation of the flame front and by decreasing the temperature of the most remote portion of the unburned charge. If the flame front moves with sufficient speed, normal progressive combustion of the entire charge can be achieved before the final portion reaches the self-ignition temperature. High temperatures in the unburned portion of the charge can be avoided principally by reducing the compression ratio. A decrease in the engine compression ratio causes a decrease in the temperature of the mixture at the start of combustion, and hence lower temperatures will occur in both the burned and unburned portions of the charge.

Modifications in engine design may also effect a decrease in the combustion gas temperature and thus contribute to a reduced tendency for detonation to occur. These changes may incorporate improved engine cooling and a decrease in the length of travel of the flame front. The latter change would reduce the length of time during which the temperature of the end gas is increasing.

Detonation is the principal limiting factor in the effort to increase the compression ratio and thereby improve the thermal efficiency of the spark-ignition engine. Certain substances, for example, benzene, alcohol, and isooctane, possess inherent antiknock characteristics; but because of high cost or other reasons, they are not used commercially as engine fuels. Gasoline, the common fuel for the S-I engine, varies somewhat in the ability to resist detonation. The resistance to detonation can be improved by blending a gasoline stock with another fuel component, such as isooctane. The octane number of a gasoline fuel is ordinarily increased by the addition of a small quantity of tetraethyl lead (TEL). This compound, when present in gasoline, has a marked ability to suppress detonation.

Tetraethyl lead and other antiknock additives are believed to cause a break in a sufficient number of chain reactions in which high-energy particles and radicals are produced. The net effect of the break in the chain reactions is to delay autoignition in the end gas and thereby allowing normal progression of the flame front through the gas without the occurrence of a combustion knock. Other compounds are blended with TEL to avoid deposits of lead compounds in the engine.

4.6 CARBURETION

The function of the carburetor is to inject liquid fuel, usually gasoline, into the stream of air inducted into the cylinders of the spark-ignition engine. More specifically, the carburetor provides for atomization of the fuel and metering the correct amount of fuel into the air stream. Atomization consists of finely dividing the fuel to form a mist composed of very small droplets. The proper quantity of fuel is dependent upon the engine operating conditions, which may vary widely and often rapidly. Vaporization of the liquid droplets occurs in the distributing manifold and within the engine cylinder.

The effect of the variation in the mixture ratio on the engine performance is shown in Fig. 4.4. Data on the power output and the thermal efficiency are obtained for constant-speed operation at a fixed throttle setting. Maximum power is achieved with an air/fuel ratio of about 12 to 1. This ratio corresponds approximately to the mixture ratio that supports maximum flame speed. In leaner or richer mixtures, within the limits of inflammability, the velocity of flame propagation is lower and the power output is reduced. The limits of inflammability for a gasoline–air mixture are approximately $a/f = 7{:}1$ (rich) and $a/f = 20{:}1$ (lean). Maximum thermal efficiency is achieved with an air/fuel ratio of about 15 to 1. This mixture is a trifle lean in comparison with the stoichiometric ratio, which is slightly under 15 to 1 for the average gasoline.

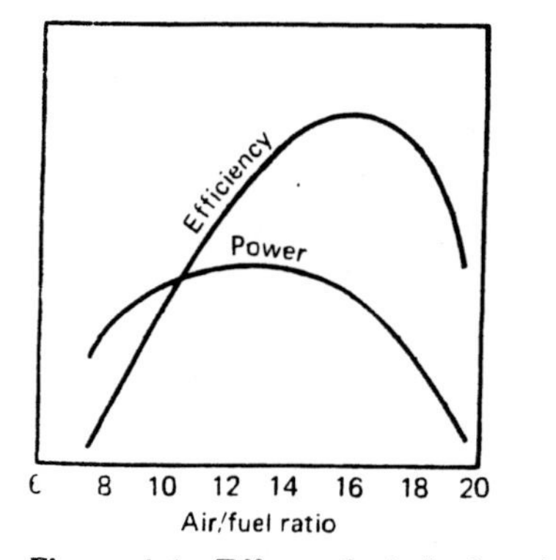

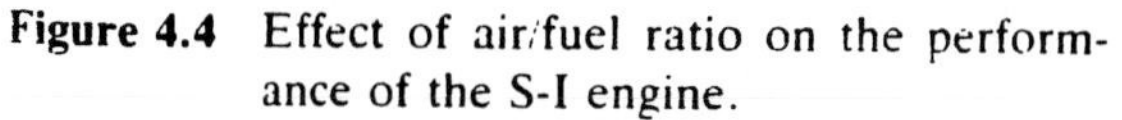
Figure 4.4 Effect of air/fuel ratio on the performance of the S-I engine.

The simple carburetor cannot supply the correct fuel–air mixture required under all load conditions imposed on the engine. Typical automotive carburetors are relatively complex devices. The variation in the mixture ratio required for the different load conditions can be achieved through use of a combination of fuel jets and air bleeds, an accelerating pump, and a choke valve. Multicylinder automotive engines are commonly equipped with two- or four-barrel carburetors to facilitate distribution of a more uniform mixture to the individual cylinders.

Example 4.5

A simple carburetor is equipped with a Venturi tube, through which the air flows. A fuel nozzle is inserted in the throat of the Venturi tube. The carburetor operating data are

Atmospheric pressure P_{01} = 101.000 kPa, abs
Atmospheric temperature T_{01} = 20 C (293.15 K)
Venturi throat diameter = 35 mm
Venturi throat pressure = −90 cm water (with respect to atmospheric pressure)
Discharge coefficient of Venturi tube = 0.82
Discharge coefficient of fuel nozzle orifice = 0.75
Gasoline specific gravity = 0.743
Pressure in the float chamber = atmospheric

Determine for an air/fuel ratio of 12 to 1 the diameter of the fuel nozzle orifice.

Note: The actual flow of the fluid, air or fuel, is determined by applying a coefficient of discharge to the theoretical flow.

Air Flow

$$1 \text{ cm of water} = 97.969 \text{ Pa}$$
$$\Delta P = 97.969(-90) = -8817 \text{ Pa}$$
$$P_2 = P_{01} + \Delta P = 101.000 - 8.817 = 92.183 \text{ kPa, abs}$$

$$T_{2s} = T_{01}\left(\frac{P_2}{P_{01}}\right)^{(k-1)/k} = 293.15\left(\frac{92.183}{101.000}\right)^{0.2857} = 285.60 \text{ K}$$

$$\frac{C_2^2}{2} = c_p(T_{01} - T_2) = (1.004 \times 1000)(293.15 - 285.60)$$

$$C_2 = 123.128 \text{ m/s}$$

$$\rho_2 = \frac{P_2}{RT_2} = \frac{92.183}{0.287 \times 285.60} = 1.1246 \text{ kg/m}^3$$

$$A_2 = 0.96211 \times 10^{-3} \text{ m}^2$$

$$\dot{m}_a = C_d\rho_2A_2C_2 = 0.82 \times 1.1246(0.96211 \times 10^{-3})123.128 = 0.10924 \text{ kg/s}$$

Fuel Flow

$$\dot{m}_f = \dot{m}_a \times f/a = 0.10924 \times \frac{1}{12} = 0.0091033 \text{ kg/s}$$

$$\rho_2 = 0.743 \times 998.2 = 741.7 \text{ kg/m}^3$$

$$\frac{C_2^2}{2} = v(P_1 - P_2) = \frac{1}{741.7}8817$$

$$C_2 = 4.8760 \text{ m/s}$$

$$A_2 = \frac{\dot{m}_f}{C_d\rho_2C_2} = \frac{0.0091033}{0.75 \times 741.7 \times 4.8760} = 3.3562 \times 10^{-6} \text{ m}^2$$

$$d = 2.0672 \text{ mm}$$

For a fixed diameter of the fuel nozzle orifice, the carburetor performance is examined for a Venturi tube throat pressure of (a) −15 cm water and (b) −160 cm water. The calculated air and fuel flow rates are reported below.

Engine Load	*P (cm water)*	*Air Flow (kg/s)*	*Fuel Flow (kg/s)*	*Air/Fuel Ratio*
Light	−15	0.046575	0.0037170	12.53
Medium	−90	0.10924	0.0091033	12.00
Heavy	−160	0.13971	0.012138	11.51

As seen in the above tabulation, the simple carburetor characteristically delivers a progressively richer mixture as the engine output is increased.

A spark-ignition engine operating at constant speed typically achieves maximum thermal efficiency over a range extending from no load to about 70 percent of full load. The air/fuel ratio increases from 12 to 1 at no load (idling) to about 15 to 1 and 16 to 1, respectively, at 20 and 70 percent of full-load operation. Further increase in the load is accomplished with a reduction in the air/fuel ratio to about 13 to 1 when maximum power is developed at full load. Some reduction in the thermal efficiency is experienced in achieving full power. Because of exhaust gas dilution, a comparatively rich mixture is required for idling.

Limited use has been made of solid fuel injection for the S-I engine, with early applications made to aircraft engines. Today, some automotive engines are equipped with solid fuel injection systems. Basically, a control element, acting in response to the load conditions, meters the correct quantity of fuel to the injection pump for subsequent distribution to each of the several cylinders. The fuel is injected through nozzles either into the intake manifold or directly into the cylinders. When placed in the manifold, the fuel nozzles are located directly above each intake valve, and on the intake stroke the inducted air carries the fuel into the cylinder. In comparison with conventional carburetion, solid fuel injection should show improved fuel economy because of more precise distribution of the fuel and improved control of the mixture ratio. Some other operational gains are possible, namely, less dilution of the lubricating oil, increased peak power, and improved acceleration.

4.7 SPARK-IGNITION ENGINE DEVELOPMENT

A great deal of developmental work, including a vast amount of testing, has been conducted in an effort to improve the operation of the spark-ignition (S-I) internal combustion engine. Much of this work has been directed toward improving the combustion process, because the activity that occurs during this portion of the cycle has the most pronounced influence on the effectiveness of the energy conversion process. An important objective is the complete combustion of the fuel while the piston is close to the top-center position as a means to achieving a high maximum temperature and a resulting high thermal efficiency. Incomplete combustion is however characteristic of the S-I engine, demonstrated by the emission of hydrocarbons and carbon monoxide in the exhaust gas.

The thin quench zone adjacent to the combustion chamber wall is generally believed to be a major source of the hydrocarbons emitted by the engine. The thin layer of combustion mixture in this zone is cooled by convective and radiative heat transfer to the combustion chamber wall, which is maintained at a comparatively low temperature. The propagating flame is consequently extinguished when it arrives at the quench zone, thereby causing formation of unburned hydrocarbons. Recent investigations however disclose that the quench zone may only be a minor source of the hydrocarbons emitted by the engine.

There are several other in-cylinder hydrocarbon (HC) mechanisms that appear to influence the hydrocarbon emission to a greater degree than wall quenching. These mechanisms are ring-crevice quenching and hydrocarbon storage in the oil film, the surface deposits, and the ring crevices. The unburned hydrocarbons from the various sources intermingle with the main body of exhaust gas and are discharged from the engine cylinder.

Further investigation is needed for constructing a model that has a predictive capability with respect to the emission of hydrocarbons by the various HC mechanisms in the engine cylinder.

For many years, exhaust gases from automotive vehicles were discharged into the atmosphere with virtually no restriction. Pollution resulting from the ever-increasing number of these vehicles, and from other sources, ultimately reached unbearable concentrations in centers of heavy population. The air quality control measures that were subsequently enacted required a stepwise abatement in the emission of pollutants from all sources.

The exhaust emissions from light-duty automotive vehicles are required to comply with the provisions of the Clean Air Act Amendments of 1977, as shown in Table 4.1. Control of the exhaust with respect to hydrocarbon and carbon monoxide emissions was initially accomplished by engine modification, that is, changes were made in carburetion, spark timing, and combustion chamber surface-to-volume ratio. A shift from a rich to a lean mixture ratio tends to effect a reduction in the concentration of hydrocarbons (HC) and CO in the engine exhaust, because complete combustion of a fuel is normally enhanced by the presence of excess oxygen (see Fig. 4.5). Lower peak combustion temperatures and higher exhaust gas temperatures, which result from retarded spark timing, cause a reduction, respectively, in the concentration of NO and HC in the engine exhaust.

TABLE 4.1 Clean Air Act Amendments of 1977

Beginning in Model Year	*Maximum Permissible Exhaust Emissions (g/mile)*		
	Hydro-carbons	*Carbon Monoxide*	*Nitrogen Oxides*
1978–1979	1.5	15.0	2.0
1980	0.41	7.0	2.0
1981	0.41	3.4	1.0

While the corrective measures described above, applied to the automotive engines for the model years 1968 through 1974, achieved a substantial reduction in the emission of CO and HC, the engine performance deteriorated somewhat.

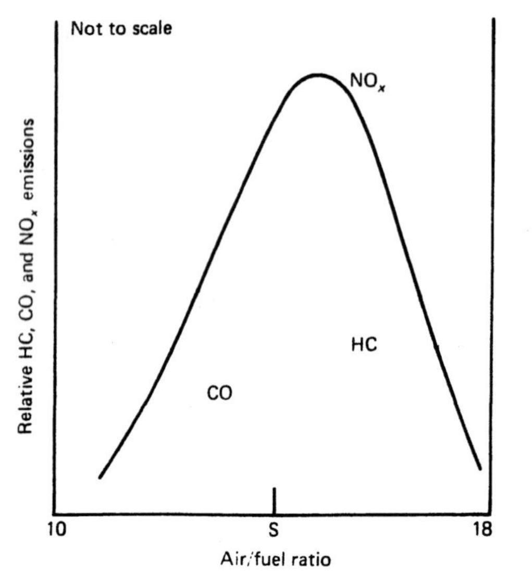

Figure 4.5 Effect of air/fuel ratio on the exhaust gas emissions from the S-I engine.

Various operating problems and increased maintenance developed with many of these engines, and generally a decline in the fuel economy was experienced.

The catalytic converter, first installed in a number of the 1975-model-year automobiles, offered an improved method for control of the CO and HC emissions. The converter, packed with platinum–palladium elements, for example, beads, is installed in the automobile exhaust system. Carbon monoxide and HC are catalytically oxidized to CO_2 and H_2O prior to discharge into the atmosphere. The catalytic converter does not interfere with the operation of the engine, hence the earlier engine modifications were eliminated and improved performance and fuel economy were achieved for the 1975-model-year automobiles and for automobiles produced in subsequent years.

While the platinum–palladium catalytic converter provides an effective and apparently satisfactory solution to the CO and HC emission problem, it does possess one rather significant flaw. Lead in the engine exhaust gas will quickly render the catalytic converter inoperative, thus requiring automobile engines equipped with these converters to burn only lead-free gasoline. Considerable success has been reported in the development of a catalytic converter that will not be deactivated by the presence of lead in the exhaust gas. This improved converter is comprised of a combination of nonnoble metals on a ceramic base.

4.8 NITROGEN OXIDE EMISSION

Emission of HC and CO from the spark-ignition internal combustion engine is a result of incomplete combustion. The production of nitrogen oxides is incidental to the combustion process and represents unwanted and undesirable chemical combinations. While control of HC and CO emissions has proved to be feasible, particularly with installation of the catalytic converter, acceptable control of the emission of nitrogen oxides is not readily achieved.

Several different oxides of nitrogen are formed within the engine cylinder. The principal and most significant oxide of nitrogen is NO, formed from atmospheric nitrogen during combustion. Subsequently, a portion of the nitric oxide NO is converted to nitrogen dioxide NO_2. Of these two oxides of nitrogen, NO_2 is chemically more reactive. The combined emission of NO and NO_2 is often reported as a single quantity, NO_x. The actual NO concentration in the automotive engine exhaust is much higher than the amount calculated for equilibrium at exhaust gas temperatures. The NO formed at high temperatures, up to 2500 K, during combustion is "frozen" when the gas temperature quickly drops on the expansion stroke.

In a typical combustion process involving fuel and air, the quantity of NO formed increases with an increase in the reaction temperature. Consequently, adjustment of the engine operating variables to effect a decrease in the peak combustion temperature will in turn decrease the quantity of NO formed in the engine cylinder. Peak combustion temperatures can be reduced by retarding the spark and by decreasing the compression ratio. Both of these adjustments however lower the thermal efficiency of the engine.

Prior to the development of the catalytic converter, emission of HC and CO was controlled, in part, by operating the engine with a relatively lean mixture ratio. The oxidizing atmosphere created by the lean mixture is however conducive to the formation of NO (see Fig. 4.5).

Concentration of NO_x in the engine exhaust gases initially increases with an increase in the air/fuel ratio. A peak concentration is reached at an air/fuel ratio of approximately 16–17 to 1, depending upon the compression ratio of the engine. This range of 16–17 to 1 places the air/fuel ratio in the lean region, that is, above the stoichiometric ratio of about 15 to 1. Beyond the peak, the NO_x concentration decreases sharply with an increase in the air/fuel ratio.

Actually, the formation of NO at peak combustion temperatures, the subsequent decomposition of NO on the expansion stroke, and the formation of NO_2 are complex mechanisms that are dependent upon numerous variables. Further, the presence of NO is known to affect the combustion reaction in a number of ways.

For a prescribed pressure and air/fuel ratio, the equilibrium concentration of NO varies almost linearly with the equilibrium temperature.[4] Any engine operating variable that effects a reduction in the peak combustion temperature should also promote a decrease in the NO concentration. Investigations have been conducted on the feasibility of controlling the NO_x emission by recirculation of exhaust gas. A small portion, on the order of 2 percent, of the engine exhaust gas is recirculated to the engine intake as a means to diluting the air–fuel mixture in order to lower the peak combustion temperature. While control of NO_x emission by exhaust gas recirculation offers some possibility, the system imposes certain penalties on the operation of the engine.

The ultimate NO_x emission limit is 0.4 g per mile for light-duty automotive vehicles. Experimental investigations conducted to date indicate that this requirement will be difficult to meet. The control of NO_x emission may eventually be achieved by the development of a converter that would be placed in the exhaust system for the purpose of decomposing NO_x into O_2 and N_2.

The concentration of NO_x in the exhaust of the S-I engine is not particularly high, typically 0.1 to 0.2 percent. NO_x is however a principal ingredient in the formation of photochemical smog. Concentrations of smog may occasionally reach dangerous levels in heavily populated areas at such times when poor atmospheric ventilation prevails. Smog is considered to be biologically harmful as a source of eye irritation and as a contributor to respiratory problems. Further, smog reduces visibility, produces disagreeable odors, and may cause certain materials to deteriorate.

Smog is formed by a complex mechanism. Hydrocarbons and NO react, in the presence of oxygen and strong sunlight, to form a variety of smog products. The conversion occurs in a number of simultaneous reactions. Apparently, smog formation could be eliminated if either or both of the primary reactants, HC and NO, were not discharged into the atmosphere. The automotive engine however is not the only source of these two reactants. Hydrocarbons are emitted from various nonautomotive combustion processes and from natural sources. NO_x emissions are attributed to thermal power plants and engines, industrial plants, and a variety of combustion processes.

4.9 INTERNAL COMBUSTION ENGINE FUELS

Petroleum, or crude oil, is the predominant source of liquid hydrocarbon fuels. A commercial hydrocarbon fuel is a mixture of a large number of hydrocarbon compounds. Gasoline, for example, consists principally of liquid hydrocarbons in the C_4-to-C_{10} range and small amounts of lighter and heavier hydrocarbons. Additives and crude oil impurities are present in minute quantities.

Hydrocarbon compounds are grouped in accordance with the hydrogen-carbon ratio and the structure of the molecule. The predominant hydrocarbon groups that compose petroleum are listed in Table 4.2.

TABLE 4.2 Characteristics of Selected Hydrocarbon Families

Name	*Chemical Formula*	*Type of Structure*	*Carbon Bonds*
Paraffin	C_nH_{2n+2}	Chain	Saturated
Olefin	C_nH_{2n}	Chain	Unsaturated
Diolefin	C_nH_{2n-2}	Chain	Unsaturated
Naphthene	C_nH_{2n}	Ring	Saturated
Aromatic			
Benzene	C_nH_{2n-6}	Ring	Unsaturated
Naphthalene	C_nH_{2n-12}	Ring	Unsaturated

The members of the paraffin family are saturated compounds, that is, no double bonds exist between adjacent carbon atoms. The name of each member of this family includes the suffix ane, which is added to a term identifying the number of carbon atoms in the molecule. A normal paraffin hydrocarbon has a straight-chain structure with respect to the arrangement of the carbon atoms. Thus, for *n*-octane,

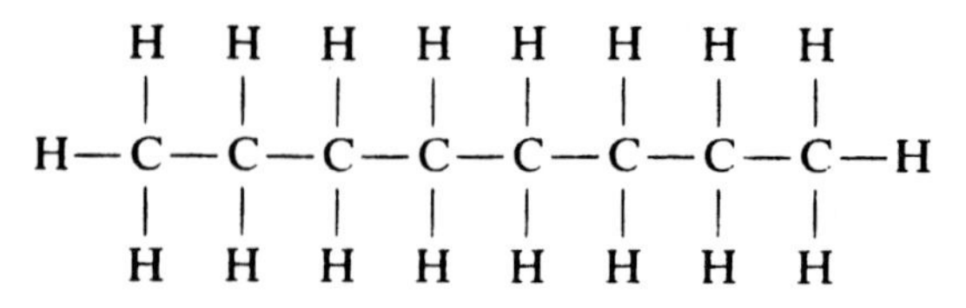

Characteristically, the normal paraffin hydrocarbons show a progressive decrease in the self-ignition temperature with increase in the length of the chain. The tendency for these normal paraffin hydrocarbons to detonate correlates roughly with the self-ignition temperature. The heavier normal paraffin hydrocarbons consequently show poorer resistance to detonation than do the lighter normal members.

An isomer has the same chemical formula as the corresponding normal compound but different chemical and physical properties. For example, many different molecular arrangements other than the normal structure are possible for octane. The molecular arrangement for 2,2,4-trimethylpentane, an isooctane, is a branched structure shown by

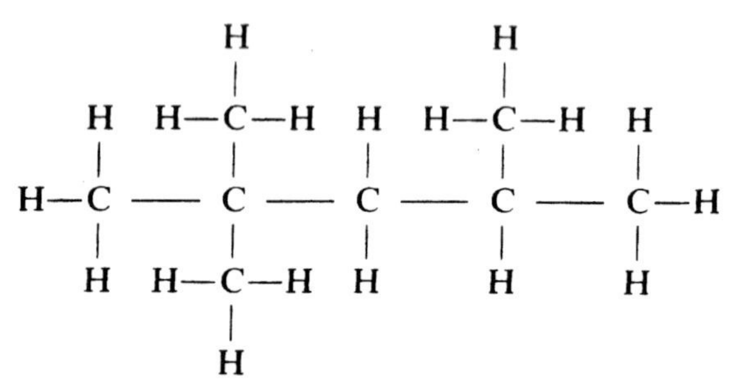

Because this compound (2,2,4-trimethylpentane) is used in the knock rating of gasoline, it is often designated simply as isooctane.

Isomers are of particular significance in the manufacture of gasoline because some isomers resist detonation to a considerably greater degree than does the normal compound. Engine operation at a substantially higher compression ratio is possible if the fuel, for example, is isooctane rather than *n*-octane.

Olefins, identified by the suffix ene, are unsaturated, open-chain compounds. The straight carbon chain has a single double bond. The olefins and the corresponding paraffin compounds display closely similar physical properties, although the olefins are superior with respect to antiknock characteristics. Some of the more common olefins are given two names, for example, propylene for propene and butylene for butene. Because of the free carbon bond, the olefins are chemically reactive. Reaction with oxygen produces a gum, an undesirable residue. Straight-run products contain few olefins. Substantial quantities of olefins are found however in some of the highly cracked fuels.

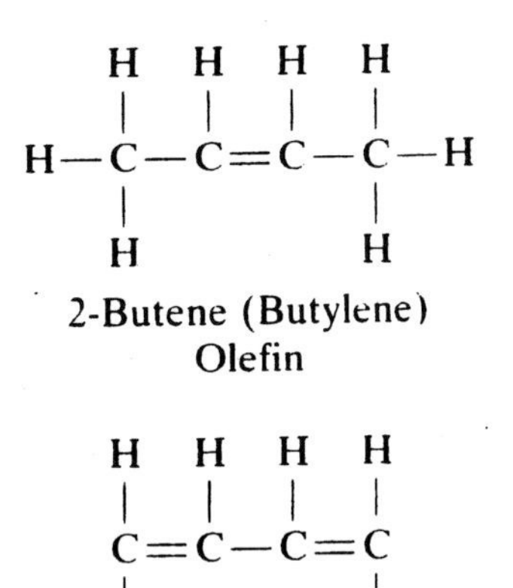

2-Butene (Butylene)
Olefin

1,3-Butadiene
Diolefin

The diolefins, like the olefins, are unsaturated, open-chain compounds but have two double bonds in the carbon link. The diolefin suffix is diene. Because of chemical reactivity, diolefins are not desirable fuel components, particularly with respect to formation of gum. Operational problems develop in the engine from the deposits that are formed by the gum present in the fuel. Some of the effects are plugged carburetor jets, sticking carburetor parts, and sticking intake valves.

Naphthenes, as a group, are principal components of crude oil and are saturated, ring-structured compounds. Naphthenes and olefins have the same chemical formula but dissimilar structures, saturated ring as opposed to unsaturated chain. The naphthenes are named by adding the prefix cyclo to the corresponding paraffin name, for example, cyclobutane. Because of favorable chemical and physical properties, the naphthenes are desirable components of automotive engine fuels.

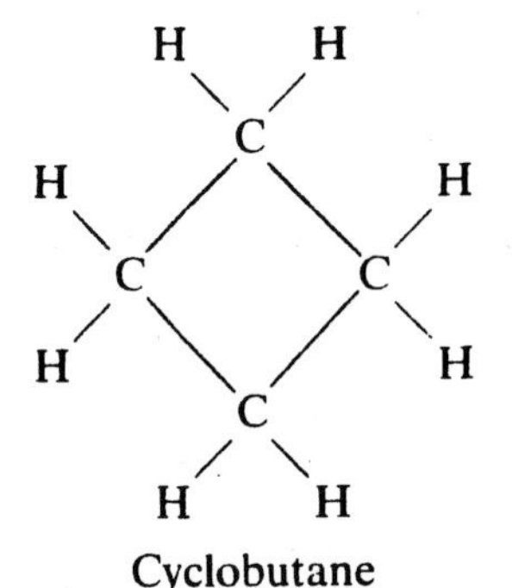

Cyclobutane

The benzene and naphthalene families, included in the aromatic grouping, are unsaturated, ring-structured compounds. The benzene molecule (C_6H_6) is the central structure of the aromatic families.

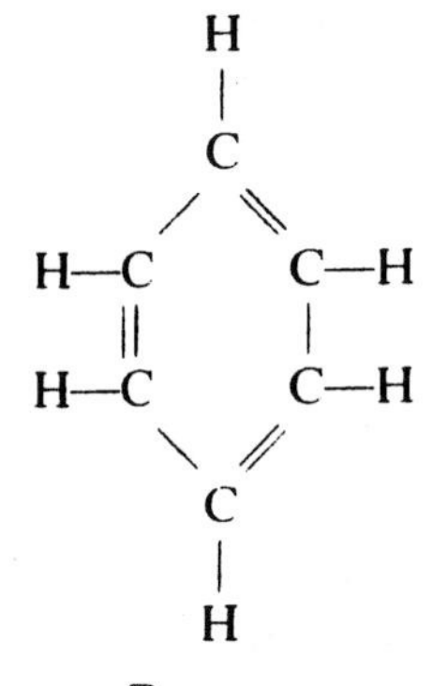

Benzene

Although the benzene molecule is unsaturated, the chemical properties of benzene suggest a saturated hydrocarbon. The stability of the benzene molecule is attributed to the oscillation or resonance of the single and double carbon bonds, that is, the single and double bonds alternate positions. Other aromatic compounds are formed by replacing one or more of the benzene hydrogen atoms with an organic radical. For example, toluene ($C_6H_5CH_3$) is formed by replacing one hydrogen atom with a methyl group (CH_3). The aromatics have particularly good autoignition characteristics and hence display high octane ratings.

The naphthalenes are double-ring-structured compounds formed by the combination of two benzene rings. The naphthalene double ring consists of 10 carbon atoms, two of which are shared by each of the basic benzene rings. The aromatics are found in many of the crudes and in some of the refined products.

The feasibility of burning alcohol, ammonia, and hydrogen in the S-I engine was explored in Chapter 2. Of these three substances, only alcohol has been used commercially as a fuel. The light alcohols—methanol, ethanol, and butanol—are usually blended with other fuels. Other alcohols, such as propanol and amyl alcohol, apparently are not used to any extent as engine fuels. Ethanol, in particular, has for some time been blended with gasoline in order to improve the antiknock characteristics of low-octane gasoline.

Alcohols are saturated, straight-chain structures somewhat similar to the paraffin hydrocarbons. See, for example, ethanol (C_2H_5OH). Alcohol is not a significant component of crude oil.

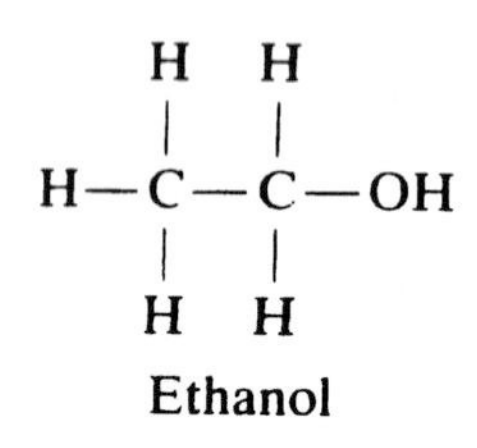

Ethanol

The octane numbers of several alcohols, namely, methanol, ethanol, and butanol, range from 98 to 100+.* Alcohol heating values are lower than the heating values for the hydrocarbons. Because of a lower stoichiometric air/fuel ratio, the energy value of the charge, on a volumetric basis, for an alcohol is only slightly below that for a hydrocarbon. The energy of vaporization for an alcohol is high in comparison with gasoline. A high energy of vaporization produces a low mixture temperature which in turn causes an increase in the density of the charge and a consequent increase in the power output of the engine.

* The octane rating of a fuel is an indication of the detonating characteristics of the fuel when burned in a spark-ignition engine. The detonating quality of the fuel is measured in a standard test engine and reported as an octane number. Untreated gasolines may have a rating below 50 octane, while certain aviation gasolines are rated above 100 octane.

4.10 GASOLINE

With minor exceptions, gasoline is the principal fuel for the spark-ignition engine. The composition of gasoline, excluding possible impurities, may be taken as 85.5 percent carbon and 14.5 percent hydrogen, for an average specific gravity of about 0.74. The higher heating value of gasoline is about 47,600 kJ/kg.

A gasoline should have proper volatility characteristics in order to facilitate ease in starting the engine, and avoid formation of vapor locks and any tendency to cause dilution of the lubricating oil. Gum formation in gasoline should be avoided because of the deleterious effect of gum deposits on the operation of the engine. The total sulfur content of gasoline is preferably held to less than 0.1 percent in order to minimize formation of corrosive acids in the engine. The octane number of a gasoline should be proper for the engine in which it is burned.

Primary distillation of crude oil produces, among other products, straight-run gasoline. Distillation separates the crude-oil hydrocarbons into various groups or fractions that are identified by boiling point ranges. The overall yield of gasoline is increased by blending straight-run gasoline with gasoline obtained from natural gas and from various conversion processes. Prior to marketing, the blended product is chemically treated for removal of sulfur and those compounds that contribute to the formation of gum and sludge. The finished product has a light color and is also free from an objectionable odor.

Conversion processes employed for increasing the yield of gasoline or raising the octane number of the finished product are generally complex and are costly operations. The yield of gasoline obtained from crude oil can be increased substantially by cracking those fractions that are heavier than the compounds included in the gasoline range. Both thermal and catalytic processes are used to split the heavier molecules into smaller molecules of lower boiling points. The cracking operation is accompanied by other simultaneous reactions, such as polymerization, alkylation, and dehydrogenation.

Polymerization is a process used for increasing the gasoline yield by combining the light, unsaturated molecules in refinery gas to form a larger molecule. Both the feed hydrocarbons and the product hydrocarbons are olefins. Hydrogenation involves the addition of hydrogen to an unsaturated hydrocarbon. A paraffin, for example, is formed by the addition of hydrogen to an olefin. Other conversion processes are described below.

An increase in the thermal efficiency of the spark-ignition engine can generally be achieved by an increase in the engine compression ratio. Higher compression ratios however will raise combustion pressures and temperatures and thus promote a greater tendency for detonation to occur. A high-performance engine with a compression ratio of about 10.5 to 1 is required, for satisfactory operation, to burn a premium fuel of about 100 RON (research octane number). Engines designed for a lower compression ratio of about 8.5 to 1 are able to burn, without knocking, "regular" gasoline, of 92 to 95 RON.

The research octane number of a gasoline is measured in the lower operating range and with fuel mixture temperatures described as "cool." When tests are conducted on the same gasoline under relatively severe conditions of higher speeds and higher fuel mixture temperatures, the knock potential of the fuel is more pronounced and the result is designated as the motor octane number (MON). For proper performance in an engine, the RON and the MON of a fuel should be in close agreement.

The desired octane numbers of the different grades of gasoline have been achieved usually by adding a small quantity of tetraethyl lead $[(C_2H_5)_4Pb]$ (TEL) to the particular hydrocarbon blend. Although other additives may be effectively employed, TEL is the most economical additive currently available. The increase in the research octane number in response to the quantity of TEL added to the fuel is nonlinear. An increase of three numbers in the octane rating is achieved by an initial TEL addition of 0.5 g/gallon, while a second addition of 1.5 g/gallon is required to raise the RON the next three numbers.

The presence of TEL in gasoline is responsible for certain undesirable secondary effects. The lead oxide (PbO_2) formed during the combustion reaction, together with other lead compounds and scavengers, causes corrosion of certain parts of the engine, namely, spark plugs, exhaust pipes, and mufflers, and plugging of exhaust recirculation systems. Under similar operating conditions, the hydrocarbon emission is somewhat higher when a leaded fuel is burned in comparison with burning a lead-free fuel of the same RON, say, 91. The catalytic converter employed to control HC and CO exhaust emissions is very quickly deactivated when the engine is operated on fuels that contain TEL. Discharge of lead compounds into the atmosphere is considered to have a deleterious environmental effect.

The clear pool is the total stock of raw gasoline obtained from refining, prior to mixing or blending. The United States average clear-pool RON is 88.5.

Lead-free gasoline could be produced by separation of the clear-pool components into high-octane (91 RON) and low-octane (RON < 88.5) stocks. The available quantity of lead-free, high-octane gasoline produced in this manner is not likely to be sufficient for meeting the expected demand. Further, a relatively large addition of TEL to the low-octane stock would be required to produce a marketable product, that is, a leaded fuel of sufficiently high RON.

Because total dependence upon the separation process is not feasible, the alternative is the conversion of some of the clear-pool components into hydrocarbons of higher octane number. Hydrocarbon conversion for the purpose of upgrading the octane number cannot be achieved without incurring certain penalties. The conversion equipment is costly, hence high capital charges are reflected in the ultimate price of the fuel. Further, the gasoline yield is reduced because a portion of the raw feed is discharged as a product of low market value or is used as a refinery fuel. As a consequence of these conversion losses, an increased quantity of crude oil must be processed for the production of a prescribed amount of higher-octane gasoline. Development of more efficient engines will however offset this increased consumption of crude oil.

A project has been initiated to develop an additive for improving the octane rating of lead-free gasoline. The compound, methyl tertiary butyl ether (MTBE), is approved by the Environmental Protection Agency for use in lead-free gasoline in concentrations up to 7 percent. Production of this compound is important to the effort to reduce the quantity of crude oil required for the manufacture of lead-free gasoline.

Three conversion processes—catalytic reforming, alkylation, and isomerization—currently employed for raising the octane number of gasoline components are examined subsequently. Lead-free gasoline with a RON of 91, or perhaps slightly higher, can be produced by blending low-octane fuel with these "high-octane" conversion products. The manufacture of a lead-free, 100-RON gasoline, which would be burned in high-compression engines, would encounter formidable problems.

Reformate is produced by catalytic re-forming, at elevated temperatures and moderate pressures, heavy, low-octane hydrocarbons. The predominant chemical reaction that occurs during re-forming is the dehydrogenation of naphthenes. Straight-chain hydrocarbons are converted to naphthenes in a preliminary reaction. Aromatics, such as benzene, toluene, and xylene, are formed by the dehydrogenation of the saturated, ring-structured naphthenes.

The aromatics, which are centered on the benzene ring, have high octane ratings. The octane rating of xylene, for example, is 110. Clear reformate has a RON of 100. Because of the high octane number, reformate is an important blending component for the manufacture of gasoline. The corresponding MON of 88 is, in comparison, low because the RON and MON separation should be small for good engine performance.

Alkylation, like polymerization, effects a combination of light, gaseous hydrocarbons to form larger molecules. The feed however is different and consists of olefins and isoparaffins. Usually, butene (butylene) (C_4H_8) is united with isobutane (C_4H_{10}), although the olefin charge may include propylene or pentene. The long-chain hydrocarbons produced by alkylation have high octane numbers. The end product, called alkylate, is an important and valuable blending stock.

Of the available nonleaded stocks, alkylate has the second-highest octane rating, particularly with respect to a MON of 91. While the RON (93) for alkylate is considerable below that for reformate, the MON is three numbers higher. Further, the difference between the RON and the MON for alkylate is small. In comparison with reformate, alkylate has certain superior blending characteristics.

Ten percent of the clear pool consists of pentanes and hexanes—light, straight-chain C_5 and C_6 hydrocarbons. These components have very low octane ratings. The octane rating can be substantially improved by isomerization, a process that changes the straight-chain molecules into the branched versions of the same molecule. The octane rating of normal pentane is 61 while isopentane has an octane rating of 92.

Isomerization of the pentane and hexane feedstocks, or C_5/C_6 mixtures, is achieved in a reactor over a platinum-containing catalyst in a hydrogen atmosphere. Reactor pressures and temperatures are moderate. Isomerizate, the end product, does not have a high RON, but the process contributes, through blending, to the increase in the octane rating of lead-free gasoline. In addition, because of the small difference between the RON (85) and the MON (82), isomerizate is an important fuel component when octane sensitivity is considered.

The use of alcohol for fueling the spark-ignition engine is promoted today principally as a means to supplementing the gasoline supply. The gasoline–alcohol blend, known as gasohol, would contain about 10 percent alcohol. The use of gasohol as an engine fuel is feasible, but for a number of reasons gasohol production is likely to remain low for some time. Supplies of raw materials and manufacturing facilities are limited, and production costs are relatively high.

Separation of the two components will occur in a gasoline–alcohol blend if the water content exceeds the equilibrium amount for the particular temperature of the mixture. The tendency for separation to occur could be troublesome in the event the water content of the mixture increases during storage. It is important to note that separation in a blend of 10 percent methanol in gasoline can be caused by a water content of 0.05 percent. A similar ethanol blend will separate if the water content reaches 0.5 percent.

Methanol is cheaper than ethanol; but on an equal-energy basis, the price of methanol is at least twice that of gasoline. While the price of gasoline is certain to increase, so will the cost of producing methanol, pehaps by coal conversion. Thus, from an economic standpoint alone, methanol is not likely to be strongly competitive with gasoline. The production of ethanol from grain and other substances has some backing, but the economic feasibility of a large-scale operation cannot be demonstrated.

Example 4.6

A S-I engine operates on a gasoline fuel that is represented by C_8H_{18}. The exhaust gas analysis, on a dry basis, in percent by volume, is CO_2 10.5, CO 5.8, CH_4 0.9, H_2 2.6, and O_2 0.3. Determine the air/fuel ratio and the energy loss owing to incomplete combustion.

For 100 mol dry gas, $C = 17.2$ mol, a quantity that is supplied by 2.15 mol fuel. The reaction is then expressed by

$$2.15C_8H_{18} + xO_2 + 3.76xN_2 \rightarrow \text{(combustion products)}$$

Hydrogen Balance

4.4 mol in the dry products
19.35 mol in the fuel
$19.35 - 4.40 = 14.95$ mol H_2 in the water

Oxygen Balance

CO_2	10.5 mol O_2
CO	2.9
H_2O	7.5
O_2	0.3
Total O_2	21.2 mol

The equation for the reaction is

$$2.15C_8H_{18} + 21.2O_2 + 79.7N_2 \rightarrow 10.5CO_2 + 5.8CO + 0.9CH_4 + 2.6H_2 + 0.3O_2 + 15.0H_2O + 79.7N_2$$

The energy loss owing to incomplete combustion is taken equal to the enthalpy of combustion of the combustible components in the exhaust gas. Values of the enthalpy of combustion are selected from Table A.3.

CO	$5.8 \times 282817 =$	1,640,339 kJ
CH_4	$0.9 \times 801679 =$	721,511 kJ
H_2	$2.6 \times 242145 =$	629,577 kJ
		2,991,427 kJ/2.15 mol fuel
		1,391,361 kJ/mol fuel

The energy loss is expressed as a fraction of the enthalpy of combustion of the fuel, or lower heating value of the liquid C_8H_{18}.

$$\frac{1391361}{5071469} = 0.274$$

Fuel/Air Ratio

$$f/a = \frac{N_f \bar{m}_f}{N_a \bar{m}_a} = \frac{2.15 \times 114.23}{100.9 \times 28.97}$$
$$= 0.08402$$

The stoichiometric fuel/air ratio for C_8H_{18} is 0.06627. Thus,

$$\Phi = \frac{(f/a)_{\text{actual}}}{(f/a)_{\text{stoichiometric}}} = \frac{0.08402}{0.06627}$$
$$= 1.27$$

A compression ratio of 8.5 to 1 is prescribed for the engine, and for $\Phi = 1.27$, the thermal efficiency of the theoretical cycle is 0.34, Fig. 4.8.

The available portion of the energy loss owing to incomplete combustion is

$$0.34 \times 1391361 = 473{,}063 \text{ kJ/mol of fuel}$$

or 9.33 percent of the lower heating value of the fuel.*

4.11 STRATIFIED-CHARGE ENGINE

An increase in the air/fuel ratio above the stoichiometric value causes generally a decrease in the mean effective pressure and consequently reduced power output for the spark-ignition engine. An adjustment of the mixture ratio could conceivably be employed for control of part-load operation. Above an air/fuel ratio of 16 to 1, for a homogeneous mixture, the engine tends to misfire because flame propagation is not supported in the lean mixture. A stratified charge however will place in the vicinity of the spark plug a mixture sufficiently rich to ensure ignition. The high-temperature gases created by the first-stage combustion ignite the remainder of the charge, a lean mixture, and second-stage combustion is completed. Most S-I engines, however, throttle the intake for load control and thereby regulate the quantity of fuel and air admitted to the cylinders.

In recent years, the stratified-charge concept has been employed as a means to reducing the emission of CO, HC, and NO_x from the S-I engine. Figure 4.6 represents schematically the construction of a stratified-charge engine that utilizes two combustion chambers. The main body of the engine follows conventional design practices. The spark plug is placed in the auxiliary prechamber that is supplied, from a second carburetor, with a rich mixture. The main combustion chamber is supplied with a lean mixture from the main carburetor. Ignition is ensured in the fuel-rich prechamber, and the burning gases expelled into the main combustion chamber promote good flame propagation through the lean mixture.

Because of the excess oxygen in the main combustion chamber, the HC and CO gases are highly oxidized prior to discharge from the cylinder. The lean mixture, in addition, causes a relatively low combustion temperature that is conducive

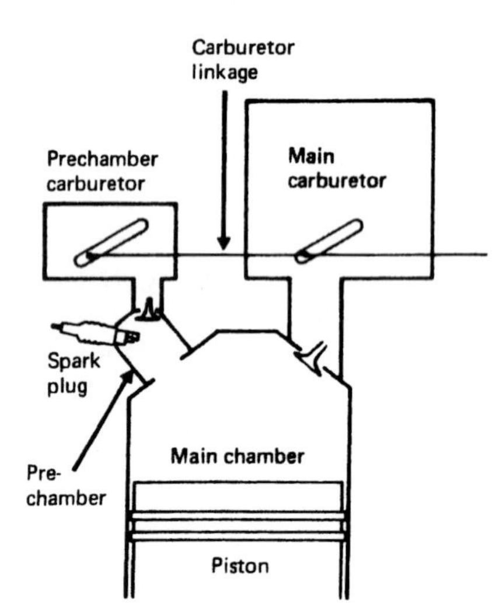

Figure 4.6 Schematic representation of stratified-charge engine. (R. K. Prud'homme, *American Scientist*, March 1974. Reprinted by permission of *American Scientist*, Journal of Sigma Xi, The Scientific Research Society.)

* See Section 8.18 for the note on the use of the lower heating value.

to reduced formation of NO. The engine operates on regular-leaded gasoline with good economy, particularly if only a reduction in the emission of HC and CO is intended. The effectiveness of the stratified-charge engine has been demonstrated in limited application to automotive vehicles, and expanded use of the engine is anticipated.

Stratified-charge combustion can be accomplished in a S-I engine equipped for solid-fuel injection. The fuel is injected directly into the combustion chamber in the vicinity of the spark plug. The charge is ignited by the spark plug, and the flame travels in the swirling gas to any region of the combustion chamber where a combustible mixture exists. The air/fuel ratio varies throughout the combustion chamber from the relatively rich mixture at the spark plug to lean mixtures, or perhaps air, in the outlying zones.

4.12 THE OTTO CYCLE

Although earlier attempts to construct an internal combustion engine are on record, the Lenoir cycle is the principal precursor to the cycle on which the present-day engine operates. The Lenoir engine is arranged for induction of the charge for a portion of the forward stroke. Charging terminates at some prescribed point in the forward stroke, at which time ignition occurs. The cylinder pressure and temperature rapidly increase, and the gases transfer work to the piston while expanding through the remainder of the forward stroke. During the return stroke, the exhaust gases are expelled from the cylinder. The Lenoir engine operates with a low thermal efficiency because the cycle of events does not include compression of the charge prior to combustion and the subsequent expansion of the gases.

In 1862, Beau de Rochas proposed a four-stroke operating sequence for a reciprocating-type internal combustion engine. Today, most of the S-I engines operate in accordance with this sequence of events. N. A. Otto, a German engineer, constructed in 1876 a four-stroke-cycle engine based on the principles presented by Beau de Rochas. Because of Otto's success in this endeavor, the operating cycle for this type of engine has been named the Otto cycle.

The term Otto cycle is often applied to the sequence of events that describes both the theoretical and the actual operation of the engine. In order to avoid possible confusion, the term Otto cycle, in this text, is employed only for the theoretical cycle. Consideration is first given to the theoretical cycle and later to the deviations from the theoretical cycle that are encountered in practice.

The Otto cycle is customarily depicted on the P–v and T–s planes (see Fig. 4.7). The four processes that comprise the cycle are

Reversible adiabatic compression. (1–2)
Constant-volume heat addition. (2–3)

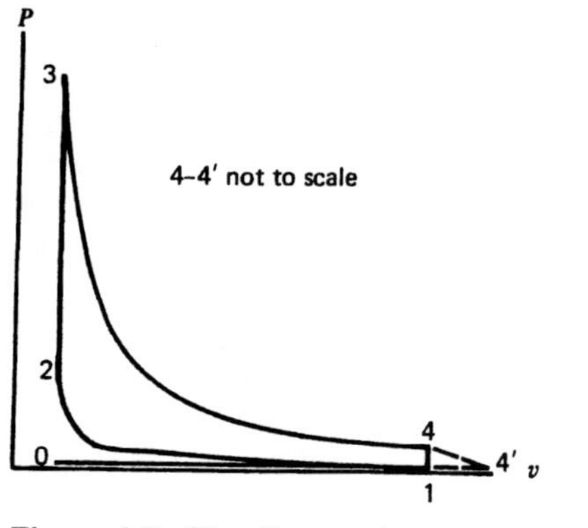

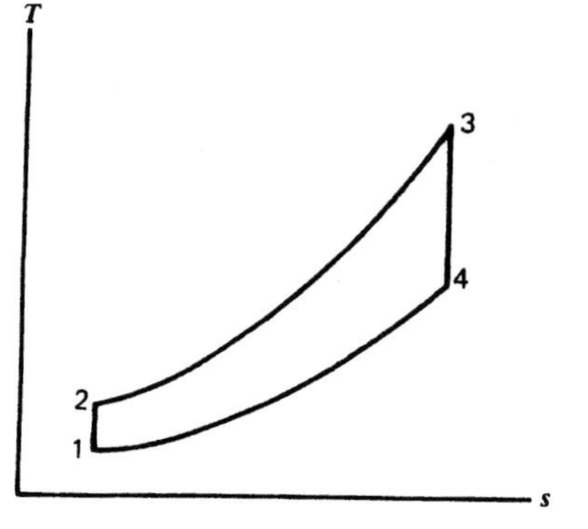

Figure 4.7 The Otto cycle.

Reversible adiabatic expansion. (3–4)
Constant-volume heat rejection. (4–1)

This sequence of events does not describe the operation of the actual engine.

Usually, an analysis of the Otto cycle disregards the events that effect the charge and discharge of the working fluid. In Fig. 4.7, the intake and exhaust strokes are represented, respectively, by the displacement processes 0–1 and 1–0, which combined involve no net work.

A simplified analysis of the Otto cycle can be made, provided that the air standard principles are observed. The air standard analysis is described in thermodynamics textbooks and need not be repeated at this point. The analysis demonstrates that for the air standard Otto cycle the thermal efficiency is a function solely of the compression ratio. Thus,

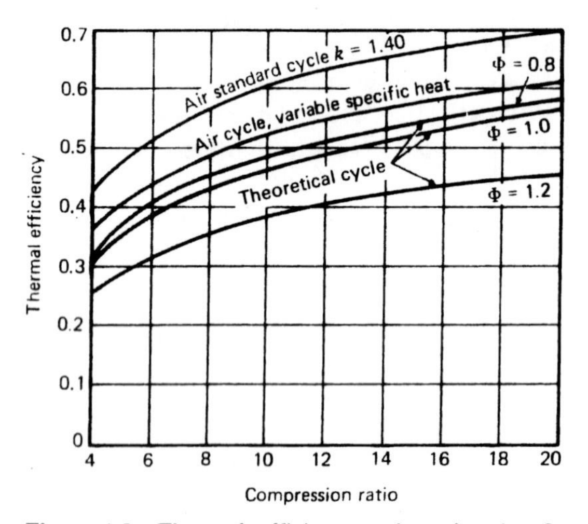

Figure 4.8 Thermal efficiency values for the Otto cycle, based on the lower heating value of liquid isooctane. Air and liquid fuel are supplied at 1 atm, 15 C. (For the air standard and air cycles, the energy supplied is 2935 kJ/kg air contained in the displaced volume. Air cycle calculations are based on the Keenan, Chao, and Kaye Gas Tables. Theoretical cycle calculations are based on the charts provided in Ref. 6.)

$$\eta_t = 1 - \frac{1}{r^{k-1}} \tag{4.3}$$

where r is the compression ratio v_1/v_2 for the cycle.

Equation 4.3 indicates that the thermal efficiency of the air standard Otto cycle continues to increase with an increase in the compression ratio (see Fig. 4.8).

Example 4.7

Calculate the thermal efficiency and the mean effective pressure for an air standard Otto cycle that has a compression ratio of 6 to 1. At the start of compression, P = 100 kPa, abs., and T = 290 K. Q = 2800 kJ/kg air (refer to Fig. 4.7).

$$\eta_t = 1 - \frac{1}{r^{k-1}} = 1 - \frac{1}{(6)^{0.4}} = 0.512$$

$$v_1 = \frac{RT_1}{P_1} = \frac{0.287 \times 290}{100}$$
$$= 0.832 \text{ m}^3/\text{kg}$$

$$v_2 = \frac{v_1}{r} = \frac{0.832}{6} = 0.139 \text{ m}^3/\text{kg}$$

$$v_1 - v_2 = 0.693 \text{ m}^3/\text{kg}$$

$$W = \eta_t Q_{2-3} = 0.512 \times 2800$$
$$= 1434 \text{ kJ/kg air}$$

$$\text{mep} = \frac{W}{v_1 - v_2} = \frac{1434}{0.693} = 2069 \text{ kPa}$$

The simplifying assumptions on which the analysis of the air-standard Otto cycle is based lead to high values of the thermal efficiency and the mean effective pressure. The mep, incidentally, is dependent upon the thermal efficiency or compression ratio, and the quantity of energy, per unit mass of air in the displacement volume, transferred to the cycle working fluid. While the results of an air standard analysis are not realistic, the analysis does provide a means to achieving a qualitative evaluation of the cycle. In particular, the analysis shows that a comparatively high mean effective pressure is a characteristic of the Otto cycle. A high mean effective pressure is desirable for a reciprocating engine because of the low ensuing displacement.

The unrealistically high maximum temperatures predicted by the air standard Otto cycle are caused, in part, by assuming constant values for the specific heats and by neglecting the effect of dissociation of the combustion products. Values of the specific heats of gases increase appreciably with an increase in temperature. Calculated gas temperatures are thus lower when real values of the specific heats are used rather than the "room-temperature" values assumed for the air standard analysis. Allowance for dissociation of the product gases would entail complex calculations that can be facilitated by employing the digital computer.

The "air" analysis applied to the Otto cycle observes the specific heat variation but assumes that the equation of state $Pv = RT$ is valid. Specific heat variation is conveniently introduced into the calculations through use of the Keenan, Chao, and Kaye Gas Tables. The effect of specific heat variation is a general decrease in the cycle pressures and temperatures, thermal efficiency, and mean effective pressure.

For the cycle described in Example 4.7, the effect of specific heat variation can be seen in Table 4.3.

TABLE 4.3 Otto Cycle, $r = 6$ to 1

	Air Standard Analysis	*Air Analysis*
η_t	0.512	0.433
mep	2070 kPa	1750 kPa
P_3	9322 kPa, abs	7337 kPa, abs
T_3	4515 K	3546 K

The theoretical Otto cycle analysis is based on the following assumptions:

1 The fuel–air charge, inducted on the intake stroke, mixes prior to compression with the residual exhaust gases that remain, from the preceding cycle, in the combustion chamber (0–1, Fig. 4.7).
2 Compression of the mixture is reversible and adiabatic (1–2).
3 Combustion occurs at constant volume (2–3).
4 Chemical equilibrium is maintained in the combustion gas mixture as the temperature drops during the expansion stroke.
5 Expansion of the gases is reversible and adiabatic (3–4).
6 At the end of the expansion stroke, with the piston at the bottom-center position, the cylinder pressure drops to the exhaust pressure as the major portion of the exhaust gas is discharged from the cylinder (4–4′). Simultaneously, the gases that remain in the cylinder expand reversibly and adiabatically to the exhaust pressure (4–1).
7 On the subsequent exhaust stroke, most of the remaining gas is expelled from the cylinder (1–0).

Because of specific heat variation, dissociation, and real-gas mixtures of different composition, the theoretical cycle analysis is complex. Theoretical cycles can however be conveniently evaluated by use of combustion charts.[5,6] These charts are devised for various fuel/air ratios and, in comparison with the air cycle analysis, provide the means to achieving a more realistic evaluation of the engine cycle.

Figure 4.8 shows for three different cycles the influence of the compression ratio on the cycle thermal efficiency. The several curves for the theoretical cycle indicate the effect of the variation in the equivalence ratio Φ, a term that corresponds to the reciprocal of the percent theoretical air. For example, an equivalence ratio of 0.8 signifies 125 percent theoretical air. As the value of Φ decreases toward zero, the theoretical cycle curve approaches the air cycle curve as an upper limit with respect to the cycle thermal efficiency.

Because of air dilution, combustion temperatures decrease as Φ drops below a value of 1.0, which corresponds to the stoichiometric ratio for the mixture. As a consequence of lower temperatures, specific heat values are lower, but in turn higher pressures and temperatures per unit of energy supplied result from the decrease in the specific heat values. The overall effect of a decrease in the value of Φ is an increase in the thermal efficiency and a reduction in the mep of the cycle (see Figs. 4.8 and 4.9).

4.13 REAL ENGINES

The actual S-I engine, for a number of reasons, does not perform as efficiently as indicated by the corresponding theoretical cycle that serves as the thermodynamic model. The several deviations from the model will be examined briefly.

Combustion occurs over a finite time interval and not at the instant when the piston is at the top-center position. Further, combustion timing may not be correct. The net result of these deviations from the theoretical combustion process is a decrease in the maximum gas temperature and a loss in availability.

Example 4.8

Combustion of the fuel in a S-I engine is assumed to occur instantaneously 30° before the top-center position of the piston. The engine bore and stroke are 125 mm and 150 mm, respectively.

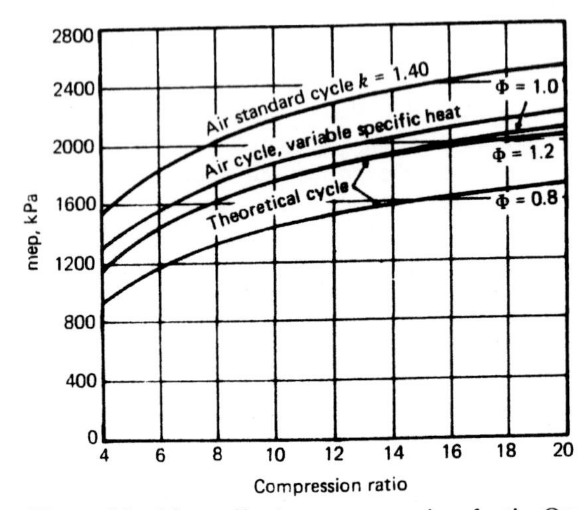

Figure 4.9 Mean effective pressure values for the Otto cycle. (See explanatory notes for Figure 4.8.)

The length of the connecting rod is 300 mm. The compression ratio is 8 to 1. A stoichiometric air/fuel ratio is prescribed. Determine the effect of "early" ignition on the engine performance.

For a connecting rod/crank ratio of 4 to 1, the piston position before top center is equal to 8.3 percent of the stroke. At this position of the piston, the effective compression ratio is 5.06.

The thermal efficiency of the theoretical cycle, for $\Phi = 1.0$, is taken from Fig. 4.8. Thus,

$$\eta_t = 0.435 \text{ for a compression ratio of 8 to 1}$$

$$\eta_t = 0.355 \text{ for a compression ratio of 5 to 1}$$

For 100 kJ of energy supplied, the available energy is

43.5 kJ	for r = 8 to 1
35.5 kJ	for r = 5 to 1

a reduction of 18.4 percent. The loss in available energy is 8.0 kJ per 100 kJ energy supplied.

The above analysis is indicative of the influence of combustion timing on the thermal performance of the engine. In the actual engine, combustion of the fuel extends over a finite interval of time, but the reaction should occur in the vicinity of the top-center position of the piston.

A variation in the mixture ratio of the charge supplied to the several cylinders disturbs the conditions that promote optimum combustion. Incomplete combustion results from imperfect mixing, flame quenching, and misfiring because of an excessively lean mixture.

While dissociation of the product gases occurs, chemical equilibrium is not truly established. Because of "freezing," complete recombination of the dissociated products is not achieved as the temperature of the combustion gases decreases during the expansion stroke.

Irreversibility, particularly on compression and expansion, causes a reduction in the net work output of the engine.

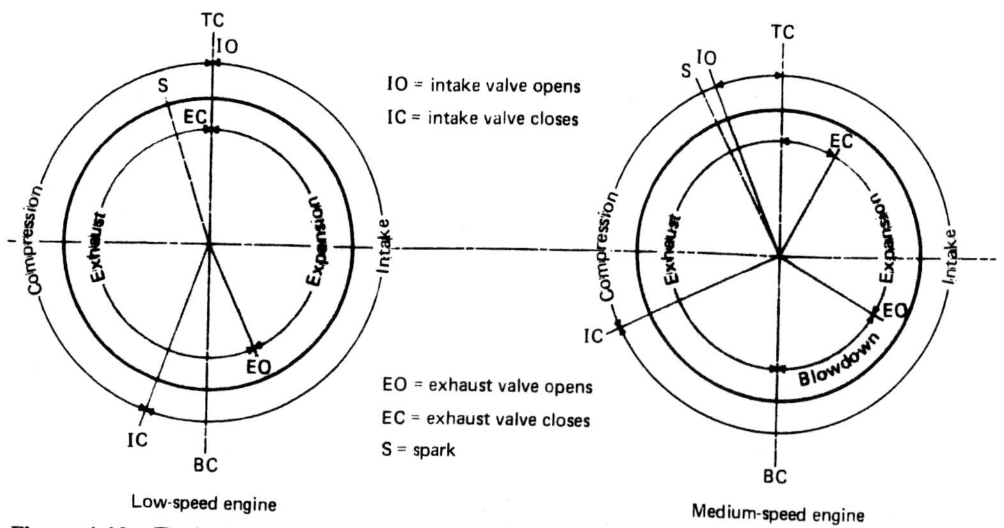

Figure 4.10 Typical valve timing diagrams for the four-stroke-cycle S-I engine.

The intake and the exhaust valves do not open and close at the dead-center positions of the piston. The extent of the deviation from the theoretical point of operation is dependent upon the speed range for which the engine is designed (see Fig. 4.10). Closure of the intake valve is delayed in order to allow the rapidly moving charge to continue to flow into the cylinder despite the opposing motion of the piston. The delayed closure of the intake valve promotes increased volumetric efficiency, although improper valve timing may cause reversed flow of the charge, particularly at low speeds. Early opening of the exhaust valve expedites the flow of exhaust gas from the cylinder by effecting self-scavenging and thereby avoiding a high back pressure on the following exhaust stroke. The effecive length of the expansion stroke is however reduced to some degree, and a balance between these two effects should be observed. The reduction in pumping work thus compensates for the decrease in the work performed on the expansion stroke. Because the valve timing is fixed by design, appreciable speed variation will affect to some extent the charging and exhaust processes.

The intake and exhaust pressures are assumed to be equal for the theoretical cycle. For the nonsupercharged engine, the exhaust pressure exceeds the intake pressure because of fluid pressure losses incurred in the manifolds, valve passages, and other elements in the intake and exhaust systems. Pumping work, required to move the fluids into and from the engine, is proportional to this pressure difference.

The compression and expansion processes for the actual engine are nonadiabatic. A relatively large quantity of the energy in the fuel is transferred as heat to the cooling system during combustion and expansion, and during the exhaust stroke. The heat so transferred does not represent an equivalent loss in potential work because much of this energy is unavailable. The heat transferred from the gas during the exhaust stroke is entirely unavailable because expansion within the engine cylinder has been terminated. Availability is a maximum when the gases are at the peak combustion temperature and drops to zero at the end of the expansion stroke.*

* The term availability is used at this point to define the capability of the combustion gas to perform work in the engine. At the end of the exhaust stroke, this capability is zero. With respect to the infinite atmosphere however, the availability of the high-temperature gas discharged from the engine is in excess of zero.

The losses examined in the preceding discussion constitute thermodynamic losses that are responsible for causing the internal thermal efficiency of the actual engine to be lower than the thermal efficiency of the corresponding theoretical cycle. A secondary loss, superimposed on the thermodynamic losses, is the result of mechanical friction and the work required to drive the auxiliary equipment. These mechanical losses represent the difference between the internal work and the shaft work, or output, of the engine.

Comparison of the internal performance of the actual engine with the corresponding theoretical cycle reveals the margin of possible improvement in engine design. Engine development could conceivably reduce this difference in performance. Similarly, the second group of losses, the mechanical losses, can be examined for possible reduction.

The internal machine efficiency, or internal relative efficiency $\eta_{r,i}$, is defined as the ratio of the internal thermal efficiency of the engine to the thermal efficiency of the corresponding theoretical cycle. This term should not be confused with the mechanical efficiency. Values of $\eta_{r,i}$ are typically within the range of 0.60 to 0.75 for nonthrottled operation. The value of $\eta_{r,i}$ for a high-performance engine may approach 0.80.

An approximate value of 0.85 may be assumed for the mechanical efficiency η_m of the S-I engine for nonthrottled operation.

The performance of the actual engine is commonly reported as the shaft, or brake, thermal efficiency $\eta_{t,b}$. Shaft power and the corresponding fuel consumption are measured quantities. Then,

$$\eta_{t,b} = \frac{3600\dot{W}_b}{\dot{m}_f(\text{LHV})} \tag{4.4}$$

where

$\dot{W}_b$ = shaft power, kW
$\dot{m}_f$ = fuel consumption, kg/h
LHV = lower heating value of the fuel, kJ/kg

Another common performance factor is the brake specific fuel consumption bsfc, expressed in kg/(kWh)_b.

If the engine mechanical losses are measured or the engine indicated, the internal thermal efficiency $\eta_{t,i}$ is determined by replacing the shaft power $\dot{W}_b$ in Eq. 4.4 with the internal power $\dot{W}_i$.

The approximate distribution of the energy in the fuel may, for the S-I engine, be taken as

1 Shaft power, 25 to 30 percent.
2 Exhaust gas (sensible and chemical energy), 40 to 35 percent.
3 Cooling system, including heat dissipation resulting from mechanical losses, 35 percent.

4.14 VOLUMETRIC EFFICIENCY

The amount of work transferred to the piston is dependent basically upon the quantity of fuel burned in the engine cylinder and upon the efficiency of energy conversion. The energy delivered to the piston is thus governed by

1 Quantity (mass) of the charge in the cylinder.
2 Energy of the charge per unit mass.
3 Effectiveness of the energy conversion process, that is, the internal thermal efficiency.

The quantity of the charge in the cylinder is determined from the volumetric efficiency, the engine displacement, and the air/fuel ratio.

The volumetric efficiency is actually a mass ratio. Thus,

$$\eta_v = \frac{m_i}{m_{pd}} \tag{4.5}$$

where

m_i = the mass of air inducted by the engine on the intake stroke
m_{pd} = the mass of air that would occupy, at atmospheric conditions, the volume displaced by the piston

Calculation of the volumetric efficiency of the engine produces only a rough approximation of this performance factor because of the dependence, in part, upon internal pressures and temperatures, which are difficult to measure precisely. Direct measurement of the air flow to the engine by an appropriate flow meter provides for a considerably more precise determination of the volumetric efficiency.

The volumetric efficiency is influenced inappreciably by the quantity and temperature of the residual gas that fills the combustion space at the start of the intake stroke. The principal factors that control the volumetric efficiency are the position of the piston and the pressure and temperature of the charge at the time the intake valve closes. Because of throttling through the intake system, the cylinder pressure at this point is likely to be lower than the atmospheric pressure. Heat transfer from the cylinder walls raises the temperature of the charge above that of the atmospheric air. The combined effect of throttling and heating is to increase the specific volume of the charge and thereby reduce the quantity of free air inducted by the engine.

Figure 4.11 illustrates the variation in the volumetric efficiency with a change in the engine speed. The intake valve has been timed to develop maximum volumetric efficiency at medium-speed operation. In the low-speed range, the ramming effect of the incoming charge is slight, and a portion of the charge is likely to be pushed back into the intake manifold. At high speeds, above the optimum point, the increase in fluid pressure losses through the intake passage dominates the inertia effect and hence causes a decrease in the volumetric efficiency. The intake valve timing may be altered in order to shift the point of optimum volumetric efficiency in either direction, depending upon the probable operating requirements. The maximum value of the volumetric efficiency for a variable-speed S-I engine is in the range of 0.80 to 0.85 for nonthrottled, nonsupercharged operation.

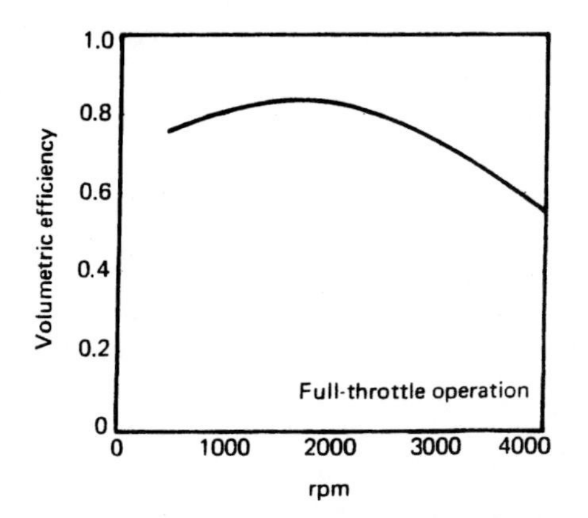

Figure 4.11 Variation in the volumetric efficiency with the change in the engine speed for the S-I engine.

Example 4.9

An eight-cylinder, four-stroke-cycle S-I engine operates at 2400 rpm with an air/fuel ratio of 15.08 to 1 (the stoichiometric ratio). Engine bore and stroke are, respectively, 100 mm and 120 mm. The compression ratio is 10 to 1. The volumetric efficiency is 0.82. The lower heating value of the liquid fuel is 44,400 kJ/kg. The internal relative efficiency is 0.70 and the mechanical efficiency is 0.85. Atmospheric conditions are 100 kPa, abs., and 290 K. Calculate the engine power output for nonthrottled operation.

Piston displacement P.D.

$$\left[\frac{\pi}{4}(0.10)^2 0.12\right] 8 \times 1200 = 9.048 \text{ m}^3/\text{min}$$

Air

$$v = 0.832 \text{ m}^3/\text{kg} \qquad \text{at ambient conditions}$$

Inducted air $\dot{m}_a$

$$\frac{\eta_v(\text{P.D.})}{v} = \frac{0.82 \times 9.048}{0.832} = 8.918 \text{ kg/min}$$

Fuel consumption $\dot{m}_f$

$$\frac{\dot{m}_a}{a/f} = \frac{8.918}{15.08} = 0.5913 \text{ kg/min}$$

Thermal efficiency for the theoretical cycle

$$(\eta_t)_{th} = 0.47 \quad \text{(from Fig. 4.8, } \Phi = 1.0 \text{ and } r = 10 \text{ to } 1\text{)}$$

Brake thermal efficiency

$$\eta_{t,b} = \eta_m \eta_{r,i} (\eta_t)_{th}$$
$$= 0.85 \times 0.70 \times 0.47 = 0.28$$

Power output

$$\dot{W}_b = \eta_{t,b} \dot{m}_f (\text{LHV})_f$$
$$= 0.28 \times 0.591 \times 44400$$
$$= 7347 \text{ kJ/min} = 122.5 \text{ kW}$$

Brake mean effective pressure

$$\text{bmep} = \frac{\dot{W}_b}{\text{P.D.}} = \frac{7347}{9.048} = 812 \text{ kPa}$$

The mep for the theoretical cycle is 1710 kPa (Fig. 4.9):

$$\text{bmep} = \eta_m \eta_{r,i} \eta_v (\text{mep})_{th}$$
$$= 0.85 \times 0.70 \times 0.82 \times 1750$$
$$= 854 \text{ kPa (approx.)}$$

The rated maximum bmep for a nonsupercharged, automotive S-I engine is likely to be in the range of 830 to 900 kPa. Within this range, the higher values of the bmep are observed for passenger automobile engines and the lower values, for truck and bus engines.

Consideration has been given in the foregoing presentation only to nonthrottled cycles. Theoretically, for a nonthrottled cycle, the intake and exhaust pressure are constant and equal. In the actual, self-charging engine, the exhaust pressure is higher than the intake pressure. As a consequence of the pressure difference, fluid displacement, or pumping, requires a transfer of work from the piston.

Pumping work is usually determined from an indicator card (see Section 4.3). Although pumping work represents a reduction in the engine power, this decrease is not particularly severe for operation of the engine with the throttle fully opened. At part-load operation of the engine however, the pressure drop across the partially closed throttle causes a substantial increase in the pumping power while at the same time the gross power is reduced. Throttle control thus contributes to decreased thermal efficiency that is observed for part-load operation of the engine.

4.15 ALTERNATIVE ENGINE DESIGNS

A number of nonconventional designs that have been proposed for reciprocating internal combustion engines are based on the concept of compressing and expanding, respectively, the charge and the combustion gases in separate cylinders. One design provides for compression of the charge in a cooled cylinder and subsequent transfer of the charge to a larger, hot, insulated cylinder, where combustion and expansion of the gases occur.[7] Another design arranges for intake of air to the compressor cylinder and subsequent flow of the air to an intercylinder combustion chamber, where fuel is injected. Following combustion, the gases are directed to the expander cylinder.[8] Expected advantages to be derived from employing these designs are higher thermal efficiencies because of reduced heat loss, improved combustion, and decreased emission of pollutants.

The concept of employing separate compressor and expander cylinders is not new. Brayton's patent for an engine of this type was granted in 1874, and numerous variations of this design have since been proposed. Provision is made for effecting combustion in the expander or in an intercylinder combustion chamber. Although several advantages can be cited for the compressor–expander designs, the end product is likely to be a comparatively heavy and bulky machine.

Many attempts have been made to develop and construct rotary-type engines that would have the capability of reproducing the cycle of events followed by the conventional reciprocating S-I engine. Despite the construction of numerous rotary-type engines and the expenditure of large sums of money, no significant success was achieved until the Wankel engine was developed circa 1960.

The construction and operation of the Wankel engine is shown in Fig. 4.12. The three chambers that contain the working fluid are formed between the sides of the rotor and the casing and are separated from each other by the seals at the rotor tips. As the rotor turns, each chamber moves around the inner wall of the casing. The geometric configuration of the casing and the rotor planetary drive cause the volume of the chamber to change as required for four-stroke-cycle operation. For each revolution of the rotor, a complete cycle is performed in each chamber. Consequently, for each revolution of the rotor, three expansion events will occur.

The Wankel engine is basically a compact machine of comparatively low weight and volume, perhaps its most significant characteristic. The engine has fewer parts and operates with less vibration and a smoother output torque in comparison with the conventional S-I engine. Multi-rotor construction can be employed to increase the power output of the engine. Because of the absence of reciprocating parts, the rotary engine possesses an inherent high-speed capability, a factor that contributes to the comparatively high power output. The fuel rate, in kg/kWh, for the Wankel engine is somewhat higher than the fuel rate for the conventional engine.

Sealing against gas leakage is particularly difficult to achieve in the rotary engine. The rotor apex seals are subjected to severe operating conditions. Unlike the piston rings of the reciprocating S-I engine, the contact between the apex seals and the casing approaches "line contact," and the openings in the casing produce surface discontinuities. The piston speed of the reciprocating engine is low near the top-center position during the time when the gas pressure is at a maximum. In the Wankel engine, on the other hand, the rotor speed remains high at the time when peak gas pressures are developed in the rotating chamber.

Oil consumption is higher in the Wankel engine than in the piston-type internal combustion engine. In addition to the customary "oil change,"

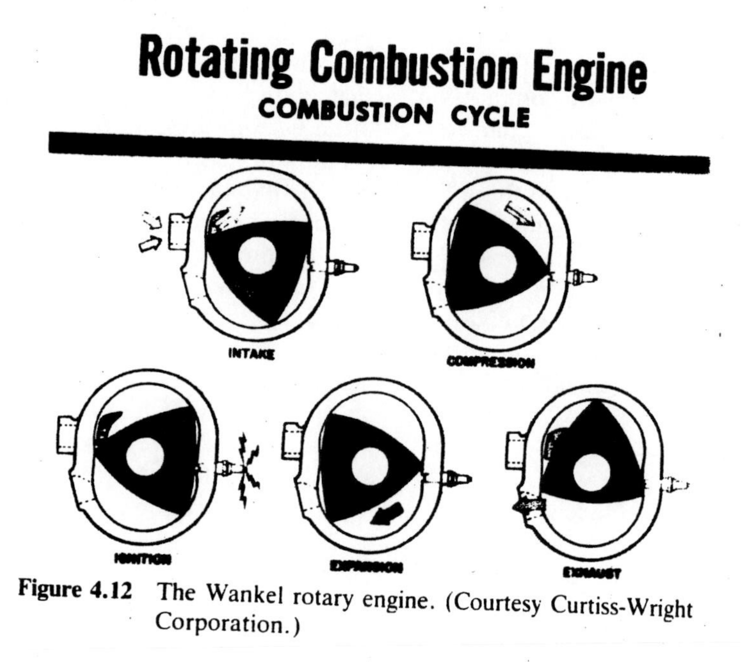

Figure 4.12 The Wankel rotary engine. (Courtesy Curtiss-Wright Corporation.)

oil must be injected into the fuel for lubrication of the rotor seals.

A limited number of Wankel engines have been installed in automobiles and in various experimental and test vehicles, boats, and engine–generator sets. Extensive application of the rotary engine to automotive service is presently doubtful because the earlier interest in this engine has largely diminished.

The compact construction of the rotary engine is cited as one reason for installing this engine in an automobile. In effect, more room would be available in the engine compartment for installation of additional emission control devices and safety equipment. The development of the catalytic converter has however lessened to a considerable degree the significance of this high power-to-bulk ratio. Emission of pollutants from the Wankel engine is inherently high for uncontrolled operation, hence emission control equipment must be installed when the engine serves as an automotive power plant.

4.16 COMPRESSION-IGNITION ENGINES

Internal combustion engines in the compression-ignition (C-I) classification, known as Diesel engines, are usually constructed with compression ratios in the range of 14:1 to 22:1. Characteristically, only air is compressed in the C-I engine. On the compression stroke, the pressure in the cylinder is raised to 3500 kPa or higher, and a corresponding air temperature of about 540 C is attained. Near the top-center position of the piston, the fuel is injected into the combustion chamber, and ignition and combustion quickly follow. Fuel injection pressures usually range from 7 to 70 MPa.

Compression ratios on the order of 15 to 1 are required for starting the engine without the use of auxiliary ignition systems. Once started however, an engine that has a lower compression ratio would continue to operate on compression ignition alone.

Compression-ignition engines are classified in accordance with their operating speeds, that is, low speed below 500 rpm, medium speed 500 to 1000 rpm, and high speed above 1000 rpm. Large, slow-speed engines, constructed with cylinder diameters ranging from 300 to about 900 mm, are used for stationary and marine power plants. Cylinder diameters of the medium-speed engines range from approximately 150 to 300 mm. These engines are installed in railroad locomotives and in the smaller stationary and marine power plants. High-speed engines are used extensively for powering automotive and construction equipment and for miscellaneous small power plants.

Diesel engine designs are based on either the four-stroke or the two-stroke cycle. Both of these cycles are employed for engines constructed with cylinder diameters that range up to about 600 mm. For larger cylinder diameters, the engines operate exclusively on the two-stroke cycle. Turbocharging is widely used for both types of engines in order to increase the power output. A comparison on the basis of equal power rating discloses that the weight of the two-stroke-cycle engine is 25 to 35 percent below the weight of the four-stroke-cycle engine.

Construction of the four-stroke-cycle engine places the intake and exhaust valves and the fuel injection nozzle in the cylinder head. Dual valves, that is, two intake and two exhaust valves, are often used in these engines.

Four-stroke-cycle C-I and S-I engines operate in a similar manner in regard to the induction process. There is however one important exception. In the S-I engine, unlike the C-I engine, the working medium at the start of compression includes the fuel component in addition to the air and residual exhaust gas. The volumetric efficiency of the S-I engine will thus be influenced by the quantity of fuel that is vaporized at the time the intake valve closes. The procedure used in estimating the volumetric efficiency is illustrated by application to the C-I engine.

Example 4.10

The compression ratio for a 250 mm × 300 mm C-I engine is 15 to 1. Air enters the cylinder at

32 C, and the temperature of the exhaust gas is 480 C. At the end of the exhaust stroke, the pressure in the cylinder is +14 kPa. The intake valve closes when the piston has traveled 6 percent of the stroke from the bottom-center position. At this time, the pressure in the cylinder is −4 kPa. Heat transfer from the cylinder wall to the working fluid is 0.25 kJ. The atmospheric pressure is 100 kPa, abs. For the exhaust gas, c_p = 1.173 kJ/kg·K and R = 0.291 kJ/kg·K. Estimate the volumetric efficiency for the engine.

$$\text{Piston displacement} = 0.014726\ \text{m}^3/\text{stroke}$$

$$\text{Combustion chamber volume } V_{cc} = 0.001052\ \text{m}^3$$

Exhaust Gas

$$m_g = \frac{P_e V_{cc}}{R_g T_g} = \frac{114 \times 0.001052}{0.291 \times 753} = 0.5473 \times 10^{-3}\ \text{kg}$$

Inducted Air

Cylinder volume at the time the intake valve closes:

$$V_c = V_{cc} + fV_{pd} = 0.001052 + (0.94 \times 0.014726) = 0.014894\ \text{m}^3$$

Assume T = 340 K at the time the intake valve closes. Then,

$$P_g = \frac{m_g R_g T_g}{V_c} = \frac{(0.5473 \times 10^{-3})0.291 \times 340}{0.014894} = 3.6357\ \text{kPa, abs.}$$

$$m_a = m_g \frac{P_a R_g}{P_g R_a} = m_g \times 1.0139 \frac{P_t - P_g}{P_g} = (0.5473 \times 10^{-3})1.0139 \frac{96.0 - 3.6357}{3.6357} = 0.014097\ \text{kg}$$

The energy equation applicable to the charging process is

$$[(m_g u_g) + (m_a u_a)]_f - (m_g u_g)_i = [m_a(u + Pv)_a]_{\text{in}} + Q - W$$

The subscripts i and f designate, respectively, the state of the fluid in the cylinder at the start and at the end of the induction process. The subscript in refers to the incoming charge of air.

The energy equation is used to determine whether the assumed temperature T_f is correct. The equation is modified to read

$$m_g c_{v,g}(T_f - T_i)_{\text{gas}} + m_a[c_{v,a}(T_f - T_{\text{in}}) - (Pv)_{\text{in}}]_{\text{air}} - Q + W = 0$$

$$(0.5473 \times 10^{-3})0.882(340 - 753.2) + 0.014097[0.717(340 - 305.2) - (0.287 \times 305.2)] - 0.25 + (96 \times 0.013842) = -0.00368$$

Only the final trial is shown for the trial-and-error solution.

Volumetric Efficiency

$(m_a)_{pd}$ = mass of air, at atmospheric conditions, contained in a volume equivalent to the piston displacement

$$(m_a)_{pd} = \frac{P_a V_{pd}}{R_{\text{air}} T_a} = \frac{100 \times 0.014726}{0.287 \times 305.2} = 0.016812\ \text{kg}$$

$$\eta_v = \frac{m_a}{(m_a)_{pd}} = \frac{0.014097}{0.016812} = 0.839$$

The two-stroke-cycle C-I engine is commonly designed for uniflow of the exhaust gases within the cylinder. Scavenging air for the engine enters the cylinder through ports placed in the cylinder liner. The swirling, upward flow of air sweeps the combustion gases through the exhaust valve located in the cylinder head.

Both the inlet and the exhaust ports may be located in the cylinder liner (see Fig. 4.13). Port construction provides for turbulent, upward flow of air in order to achieve effective scavenging of the cylinder through the exhaust ports placed in the opposite side of the liner. In comparison with the uniflow of the gases, this arrangement of the ports may lead to some difficulty in thoroughly scavenging the cylinder. The absence of exhaust valves in the cylinder head however simplifies the engine construction.

The exhaust valve is also eliminated in the two-stroke-cycle opposed-piston engine. Upper and lower ports, placed in the cylinder wall, provide for admission of air and expulsion of the combustion gases. Two fuel injection nozzles are ordinarily installed in opposed locations, one on each side of the cylinder. In the opposed-piston design, the cylinder head is eliminated, and the two cranks, upper and lower, are linked mechanically through a vertical drive. The exhaust piston leads the intake piston by about 10° of crank rotation, an arrangement that provides for opening the exhaust port prior to uncovering the intake port. Cylinder pressures are thus quickly reduced shortly before admission of the air. Some supercharging is achieved by closing the exhaust port in advance of the intake port closure. Among other advantages, the opposed-piston engine weighs less and requires less floor space than the more conventional two-stroke-cycle C-I engines of comparable power output. Opposed-piston engines operate with excellent fuel economy.

Two-stroke-cycle C-I engines are equipped with blowers that supply scavenging air at a pressure of approximately 35 kPa, gage. The volume of free air supplied to the engine is equivalent to about 140 percent of the piston displacement. Some of this air passes through the exhaust port during the scavenging operation. The major por-

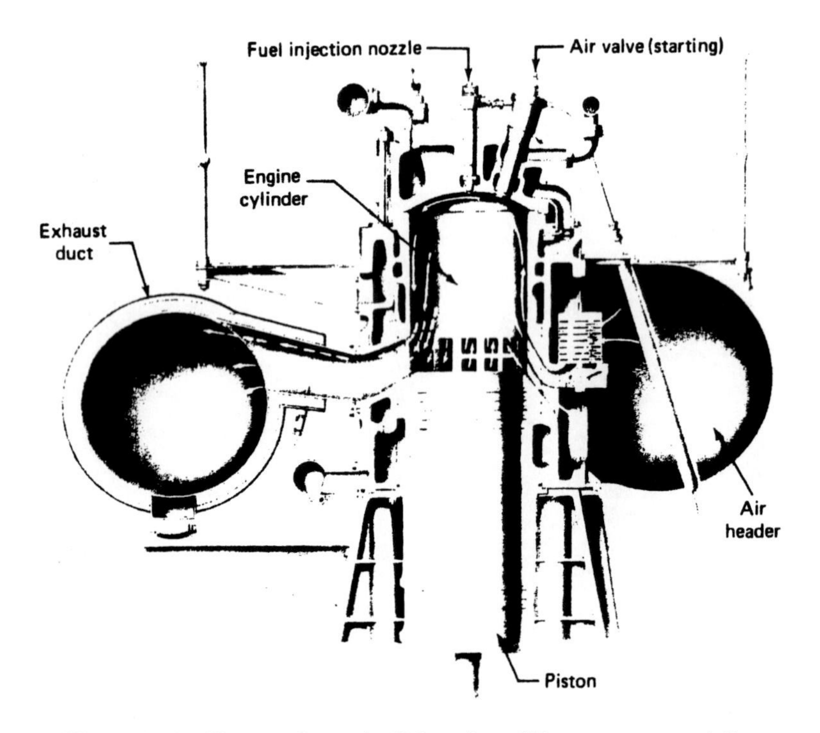

Figure 4.13 Two-stroke-cycle C-I engine. (Photo courtesy of Cooper Energy Services.)

tion of the air however remains in the cylinder and is subsequently compressed. The blower power consumption ranges from 10 to 15 percent of the engine output. Rotary lobe-type blowers are commonly installed on the engine.

Two-stroke-cycle operation requires timing that differs substantially from the valve timing employed for the four-stroke-cycle engines (see Fig. 4.14). The intake ports are uncovered during the expansion stroke, but some time after the opening of the exhaust valve. Because of the short time available for scavenging, the exhaust valve opens very early in the expansion stroke and closes after the intake ports are closed. A substantial portion of the gas is discharged through the exhaust ports, prior to admission of the air for completion of the scavenging process. Scavenging continues after closure of the intake ports.

The effectiveness of the expansion stroke is reduced by opening the exhaust valve, or ports, before the piston reaches the bottom-center position. The extent of the reduction in the expansion work is illustrated in the following example.

Example 4.11

The exhaust valve in a C-I engine opens 45° before the bottom-center position of the crank. The compression ratio is 16 to 1, $\Phi = 0.8$. The connecting rod/crank ratio is 4 to 1. Evaluate the effect of the early opening of the exhaust valve (refer to Figure 4.22).

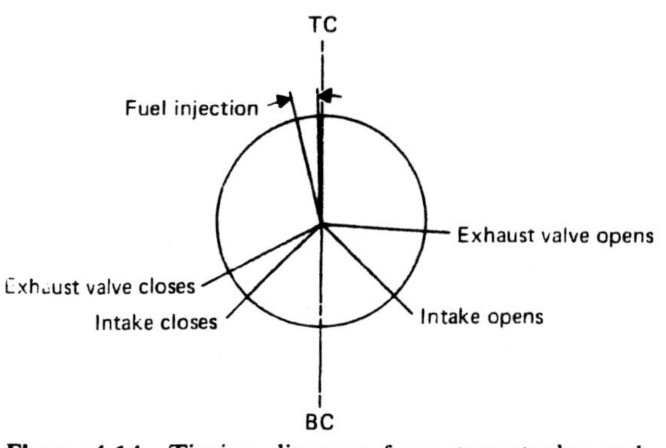

Figure 4.14 Timing diagram for a two-stroke-cycle C-I engine.

At a crank angle of 135°, the availability of the energy is about 4 percent.

For complete expansion, 56 percent of the energy supplied in the theoretical cycle is converted to work. When the piston reaches a position corresponding to a crank angle of 135°, the availability is low, hence only a minor reduction in the expansion work is experienced, because of early opening of the exhaust valve.

4.17 COMBUSTION IN THE COMPRESSION-IGNITION ENGINE

Fuel oils used in C-I engines vary approximately in specific gravity from 0.85 to 0.93. A typical higher heating value may be taken as 45,300 kJ/kg. A representative fuel analysis, in percent by mass, is carbon 85+, hydrogen 14+, and inert material < 1.

Significant characteristics of C-I engine fuels are volatility, viscosity, pour point, sulfur content, and ignition quality. The sulfur content is important with respect to possible corrosion of the engine parts. The pour point indicates the temperature at which the fuel tends to congeal and lose its freedom of flow. Other fuel properties will be examined subsequently.

The C-I engine injector delivers the liquid fuel to the cylinder in the form of a spray that varies in density from the dense liquid core to the surrounding thin envelope of fog. The liquid fuel, commonly under an injection pressure of 14 to 35 MPa, is forced through very small orifices in the nozzle tip. The injector must achieve adequate penetration of the spray into the dense, high-temperature air and cause the fuel to break down into the form of minute globules, sufficiently small for ignition to occur. Viscosity has a direct influence on the size of the fuel spray droplets, which in turn governs the atomization and penetrating qualities of the fuel spray.

Unlike the homogeneous mixture that is characteristic of the S-I engine, a stratified fuel–air mixture exists in the C-I engine cylinder. Because of stratification, it is essential to develop a turbulent combustion process and to provide an excess of oxygen for completion of the combustion reaction. The air/fuel ratio for the C-I engine ranges from 20 to 1 at full-load operation to approximately 100 to 1 when the engine is idling.

The combustion process for the C-I engine may be divided, for full-load operation, into four stages. The first stage, the ignition lag, extends from the start of injection to the start of ignition. During this interval of time, the fuel is accumulating in the cylinder. Simultaneously, because of heat transfer from the high-temperature air, the fuel droplets are evaporating and the fuel temperature is increasing. Soon after the fuel temperature reaches the ignition temperature, ignition takes place at several points in the cylinder. This is the start of the second stage of combustion. Because of the release of energy by the burning fuel, evaporation and combustion are accelerated. During this second stage, the fuel in the cylinder burns very rapidly with a high rate of pressure rise (see Fig. 4.15). The ignition lag period and much of the second stage occur during the upward movement of the piston, with the second-stage combustion terminating at the point where the peak combustion pressure is reached.

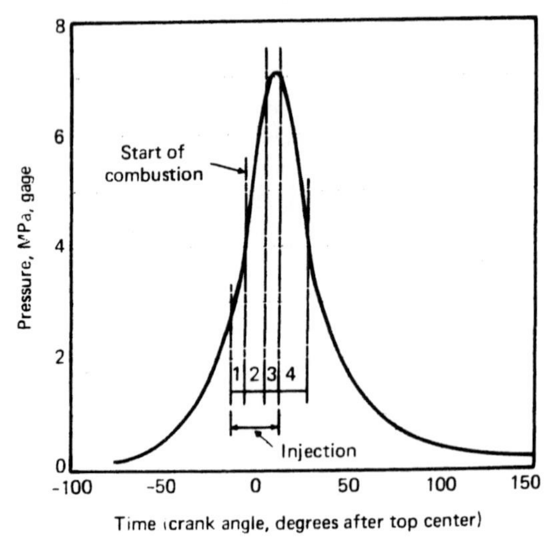

Figure 4.15 Pressure–time diagram for a C-I engine.

The unburned fuel remaining from the second stage, together with any fuel subsequently injected, burns during the third stage, which occurs with the downward movement of the piston. Because this injected fuel is introduced into a very high-temperature atmosphere, combustion starts immediately. During the third stage, combustion is controlled, to some degree, by the fuel injection rate, and the temperature increase occurs with little change in pressure. The third stage terminates at the point where combustion is measurably complete, virtually at the end of fuel injection.

The combustion process is however not homogeneous, hence afterburning occurs in the fourth stage. In order to minimize afterburning, the velocity of flame propagation should be sufficiently high to ensure complete combustion of the fuel within a short interval of time.

From a microscopic point of view, the ignition and combustion processes are exceedingly complex and occur during a very short interval of time. As heat is transferred to a fuel droplet, vaporization causes a layer of vapor to form around the remaining globule of liquid. Ignition and combustion of the vapor will surround the sphere with a layer of combustion gases. Oxygen from the surrounding atmosphere will diffuse through these gases into the inner layer of vapor. In order to speed the process however, turbulence is required for sweeping these gases away and allowing earlier contact between the vapor and the oxygen. In addition, because of stratification, it is essential to promote mixing, through turbulence, within the entire mass. In particular, the liquid core of the spray must be broken up and mixed with the combustion air. In the absence of thorough mixing, the engine exhaust will appear smoky because of the presence of unburned carbon.

A high rate of pressure rise during the second stage of the combustion reaction will result in a knock that is generally characteristic of the C-I engine. Some knocking can be tolerated, but severe knocking is undesirable both from the standpoint of noise and possible damage to the engine. This knock is quite similar to the detonation knock of the S-I engine, because both phenomena result from autoignition of the fuel. In the C-I engine, knocking occurs in the early stage of the combustion reaction and results from the desired self-ignition characteristic of the fuel. In the S-I engine, detonation is associated with the final stage of the reaction as a result of an undesirable self-ignition property of the fuel.

The intensity of the C-I engine knock is related principally to the quantity of fuel that accumulates in the cylinder during the ignition lag period. Thus, early ignition will reduce the quantity of fuel available for combustion during the second stage and consequently diminish the rate of pressure rise and the severity of the knock. In general, early ignition is achieved by a fuel that has a high volatility and a low autoignition temperature, properties that are also important to cold starting of the engine.

The ignition quality of a fuel for the C-I engine is defined by the cetane number, a property that is influenced by the volatility and the self-ignition temperature of the fuel. The cetane number of a fuel is measured in a special test engine by comparing, under prescribed operating conditions, the ignition characteristics of the test fuel with the ignition characteristics of standard reference fuels. A fuel with a medium cetane number has a tendency to burn at constant volume, without knocking, by imposing the necessary limitation on the rate of pressure rise. The approach to constant-volume combustion in the second stage is essential to achieving high combustion temperatures and hence a high thermal efficiency.

Generally, a fuel with a cetane number of 50 to 60 shows the best performance in a high-speed engine. Because of a greater length of time available for the combustion reaction, a fuel that has a lower cetane number will perform satisfactorily in medium- and low-speed engines.

It is of interest to note that, in general, those qualities that produce a superior fuel for the C-I engine are in turn indicative of an inferior gasoline for fueling the S-I engine, and vice versa. For example, a fuel with a medium cetane number will burn in a C-I engine with minimum knocking, but the same fuel would cause severe detonation if burned in a S-I engine.

Certain gases, for example, natural gas or sewage gas, are burned successfully in the internal combustion engine. In the S-I engine, the gaseous fuel is injected into the air passage in advance of the intake valve by a mechanically operated gas inlet valve that times the admission of the fuel to each cylinder. Dual-fuel engines are of the compression-ignition type. Low-pressure admission of the gas, a preferable method for admitting the gaseous fuel component, provides for introducing the gas into the cylinder at the end of the scavenging event, or into the air passage in advance of the intake valve. Because of the higher self-ignition temperature of gas in comparison with oil, compression of the mixture of gas and air is not limited by preignition. The dual-fuel engines operate on a combination of liquid and gaseous fuels over a range extending from 100 percent fuel oil to a small percentage of oil required for piloting the combustion reaction.

Reasonably good test results have been achieved in burning a coal–oil fuel in slow-speed Diesel engines. Finely ground, untreated coal, 32 percent by mass, mixed with Diesel oil was handled by an injection system developed for high-viscosity fuels. No functional problems with the engine were encountered during the test. Injection and combustion proved to be satisfactory. Fuel consumption, based on the same energy value, was slightly higher over the entire load range, compared with conventional operation on an oil fuel. Erosion of various parts of the engine by the ash in the fuel demonstrated the necessity of reducing the ash content of the coal to an acceptable level.

Spark-ignition engine control is accomplished by throttling the incoming charge, while in the C-I engine regulation is achieved by varying the quantity of fuel injected into the combustion air. For constant-speed operation, the air flow to the C-I engine remains essentially constant as the load changes, and as a consequence a wide variation in the air/fuel ratio is observed. No difficulty is experienced in establishing combustion in very lean air–fuel mixtures because normal operation of the engine is achieved with a stratified mixture; that is, in the combustion zone the mixture is sufficiently rich to support the combustion reaction.

4.18 COMBUSTION SYSTEMS FOR THE COMPRESSION-IGNITION ENGINE

The function of the fuel injection system is to provide for (1) pressure buildup, (2) injection timing, (3) metering, and (4) atomization and dispersion. As noted earlier, fuel injection pressures normally range from 7 to 70 MPa, although higher pressures may be required for some systems. Fuel injection must start at the proper point in the cycle, and the correct amount of fuel must be metered to the cylinder. These functions are controlled by the governing system. For a given fuel, dispersion and the degree of atomization are dependent upon the nozzle design and the delivery pressure.

Several mechanical systems are used for solid injection of the fuel to the C-I engine cylinder. In this text, only the widely used individual-pump system is examined. This system uses a single pump, in addition to the fuel transfer pump, to deliver the high-pressure fuel to the injector installed in the engine cylinder. The several fuel pumps, one for each cylinder, required for a multicylinder engine are usually combined in a single housing. These cam-operated pumps effect the necessary pressure buildup and control the injection timing and metering. The high-pressure fuel is transported in a small-bore steel tube from the pump to the injector nozzle. A spring-loaded delivery valve in the nozzle is actuated by the change in the fuel pressure.

In order to achieve a high combustion efficiency and a proper balance of the load among the several cylinders of the engine, small quantities of fuel must be precisely metered and delivered uniformly to each cylinder. The clearance between the plunger and the cylinder wall of the fuel injection pump must necessarily be very small. Lubrication of the pump is ensured by using a fuel of proper viscosity. The fuel injectors, as well as the fuel pumps, are manufactured with very close tolerances.

Nozzle design has a direct influence on fuel atomization and penetration of the fuel into the combustion chamber. Dispersion or spread of the fuel within the cylinder is dependent upon both nozzle and cylinder design. Injection from the nozzle must start and stop sharply, with no dribbling between successive injections.

Pintle-type nozzles are customarily used in engine cylinders that incorporate energy or air cells, turbulent chambers, or precombustion chambers (see Fig. 4.16). The pintle-type orifice delivers, through a single annular opening, a conical spray for which the included angle may range from 0 to 60°, depending upon the combustion system requirements.

The hole nozzle is constructed with one or more orifices in the form of straight, circular holes that are located in the nozzle tip, below the valve seat. The spray from each orifice is relatively dense and compact, and the general

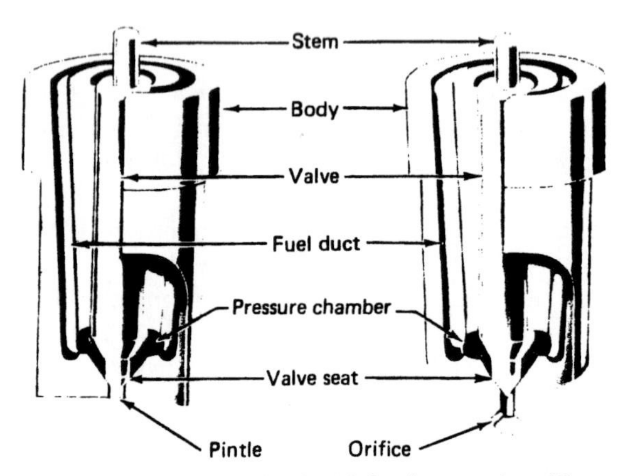

Figure 4.16 C-I engine fuel injection nozzles. (Courtesy United Technologies—American Bosch.)

spray pattern is determined by the number and arrangement of the orifices. Symmetrical or non-symmetrical spray patterns can be generated, depending upon the combustion chamber design and fuel distribution requirements. Hole-type nozzles are customarily used with combustion chambers of the open design.

Turbulence is the most significant factor to be observed in the design of the combustion chamber. Adequate turbulence is highly essential in achieving complete combustion of the fuel at the proper time in the cycle. Design factors that have a strong influence on engine performance are the shape of the cylinder head, piston head, and piston face; location of the fuel nozzle; and the use of auxiliary chambers.

The general trend is to employ open combustion chambers for most C-I engine applications, with some exceptions observed for engines in the high-speed range. The cylinder head and piston in the open-chamber design are usually "shaped" in order to exercise improved control over the combustion reaction, including promotion of adequate turbulence. The fuel injection spray pattern is designed to fit the combustion chamber in order to achieve the proper dispersion of the fuel into the air mass. The open-type combustion chamber has the highest thermal efficiency, lowest heat loss, and the easiest cold-starting capability for a comparatively low compression ratio. Figure 4.17 shows an engine constructed with an open combustion chamber.

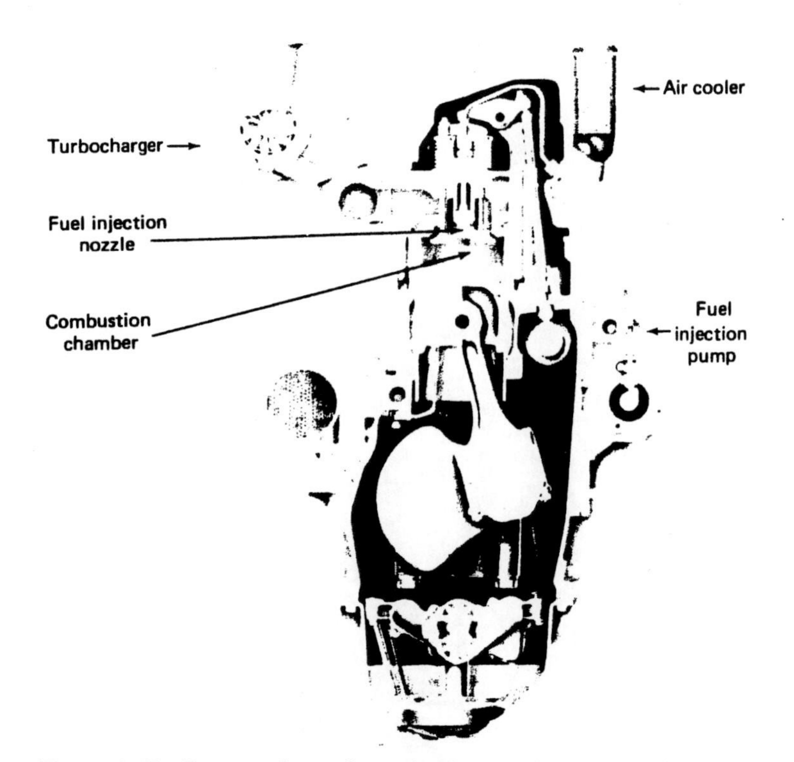

Figure 4.17 Four-stroke-cycle, turbocharged C-I engine—direct injection fuel system. Engine specifications: 6-cylinder, in-line; bore, 137 mm; stroke, 165 mm; 291 kW at 2100 rpm.

The design of the shape of the open-type combustion chamber is directed, in part, to creating a swirl in the compressed air or in the combustion gases. The primary swirl assists in mixing the fuel and air prior to combustion. Secondary swirl and secondary turbulence arise from the combustion process and contribute to mixing the reactants and combustion products. Swirl is a rotary motion of the air or gas, generally about the combustion chamber axis, and differs from turbulence in that the latter is a random-type motion.

Combustion chambers of the divided type have been developed, mainly for small, high-speed engines, in an attempt to surmount certain shortcomings of the open chamber. The divided-chamber designs, which promote highly turbulent combustion for operation in the high-speed range, employ auxiliary devices classified as (1) precombustion chamber, (2) turbulent chamber, (3) air cell, and (4) energy cell. The use of these auxiliary devices causes some increase in the fuel rate because of the cooling effect attributed to the increased surface of the combustion chamber. Somewhat higher compression ratios are required for starting in order to effect ignition. In comparison with the open chamber, a more complex construction is required for a divided combustion chamber.

The precombustion chamber, into which the fuel is injected, is an antechamber incorporated into the main engine cylinder (see Fig. 4.18). Because the fuel is not injected directly into the main chamber, engine performance, in comparison with that of the open-chamber design, is less dependent upon the nozzle and spray characteristics. Unlike the walls of the main chamber, the precombustion chamber is not well cooled, hence the ignition delay period is shortened. This an-

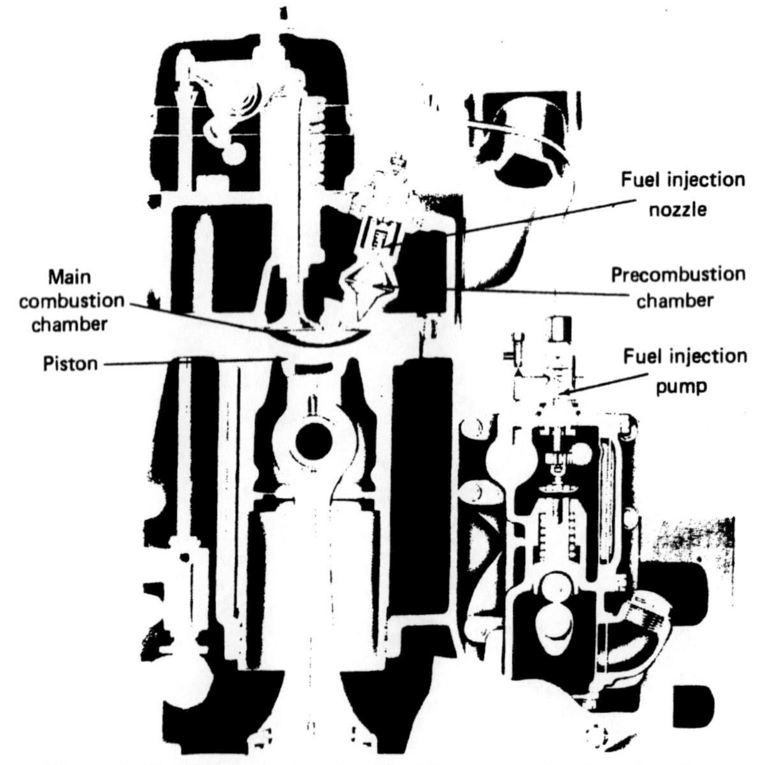

Figure 4.18 C-I engine equipped with a precombustion chamber.

techamber contains about one-third of the full air charge; consequently, combustion in this section of the divided chamber is only partially completed when the engine operates at full load. Turbulence is effected within the precombustion chamber by forcing the air from the main cylinder through the restricted interconnecting passage or throat. As a result of rapid combustion, a high rate of pressure rise is developed in the antechamber. Gases and unburned fuel are discharged and dispersed at a high velocity into the main chamber where combustion is completed in the turbulent atmosphere.

Precombustion chambers are commonly used for automotive engines and for some industrial engines. Engines equipped with precombustion chambers have the capability to operate on fuels of poorer ignition quality, and combustion is less sensitive to the nozzle and spray characteristics. Cylinder construction however is more complex and expensive, and poorer fuel economy results from increased pumping and higher heat loss.

The turbulent chamber is also an antechamber, but it is larger than the precombustion chamber and hence contains at the end of compression most of the air charge in the cylinder. Considerable turbulence is imparted to the air on passing from the main chamber through the interconnecting throat. The fuel is injected into the turbulent chamber where high turbulence ensures good combustion.

Air and energy cells are small antechambers employed as a means to increasing turbulence and improving combustion. Compressed air is trapped in these cells, and at the time of injection, a substantial portion of the fuel is blown into the energy cell. Turbulence in the cylinder is augmented by the reverse flow of air from the air cell and by the back flow of gas and fuel from the energy cell.

The C-I engine operates normally with an air fuel ratio that exceeds the stoichiometric value. The excess air promotes complete combustion, hence the exhaust from the C-I engine contains a low concentration of HC and CO. The concentration of NO_x in the exhaust gases is relatively high and must be reduced in order to meet the requirements specified by the air quality control standards. For normal, uncontrolled operation of the engine, the C-I to S-I engine concentration ratios of the pollutants in the engine exhaust may be taken approximately as CO 1 to 10, HC 1 to 2, and NO_x 1 to 1. A smoky exhaust from the C-I engine is an indication of possible overloading. Particulates in the exhaust gas of the C-I engine are believed to cause health problems, hence the discharge of particulates from the engine is subject to regulation.

An increase in the compression ratio from 14 to 1 to 22 to 1, the typical range for the C-I engine, does not yield a significant gain in the thermal efficiency of the engine. High compression ratios tend to cause increased engine friction, excessively high peak cylinder pressures, and increased starting torque. The C-I engine requires a heavy structure in order to withstand the high gas pressures developed in the cylinders. Structural limitations are hence observed in the selection of the compression ratio for the C-I engine.

Selection of the compression ratio for a C-I engine is usually governed largely by the starting conditions. Large engines installed in heated enclosures are started with relative ease and for starting require a compression ratio on the order of 14 or 15 to 1. Starting conditions are particularly severe for the automotive C-I engines when low ambient temperatures prevail. These engines usually have high compression ratios, 20 to 1 or higher, in order to achieve the high compression temperatures essential for causing ignition of the fuel. Starting may be assisted by glow plugs or other devices that are deactivated when the engine is "running."

4.19 THE DIESEL CYCLE

The engine that Rudolf Diesel constructed was designed to burn fuel essentially at constant pressure. Some of the earlier, slow-speed engines virtually achieved constant-pressure combustion by regulating the rate of fuel injection in relation

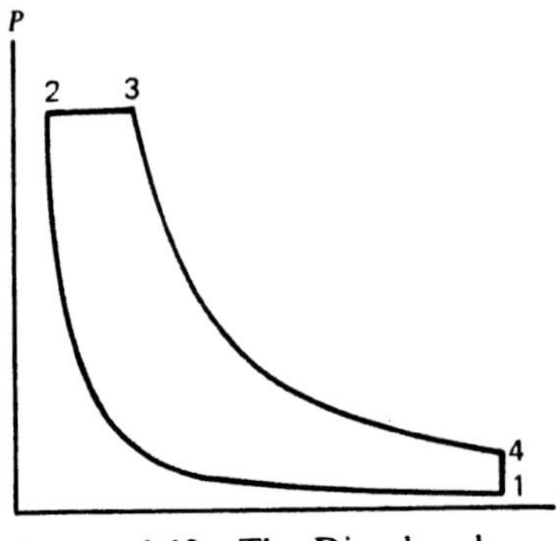

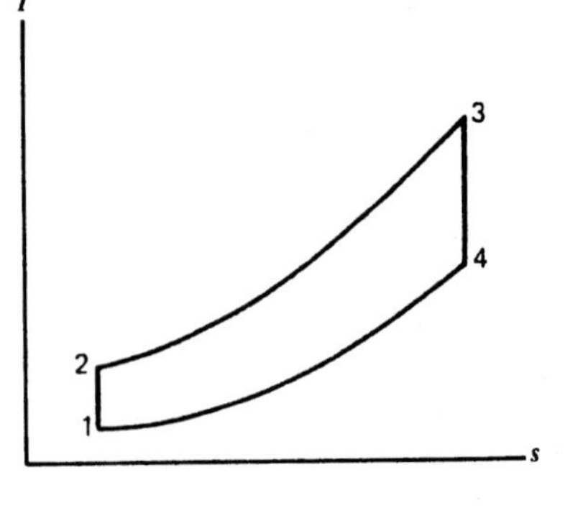

Figure 4.19 The Diesel cycle.

to the movement of the piston. The Diesel cycle then differs from the Otto cycle only with respect to the process that governs the transfer of heat to the working fluid.

An examination of the C-I engine cycle includes the air standard analysis, the theoretical cycle, and the actual engine cycle. The air standard Diesel cycle is shown in Fig. 4.19. Because this cycle is not symmetrical, the thermal efficiency of the cycle is not a function solely of the compression ratio.

The thermal efficiency for the air standard Diesel cycle is expressed by

$$\eta_t = 1 - \frac{1}{r^{k-1}}\left[\frac{1}{k}\frac{\omega^k - 1}{(\omega - 1)}\right] \tag{4.6}$$

where

r = the compression ratio v_1/v_2
ω = the load ratio v_3/v_2

The air standard analysis is used to compare the Otto and Diesel cycles. Equations 4.3 and 4.6 differ in the term included within the brackets. Because $\omega > 1$, the quantity within the brackets is greater than 1. Thus, for a prescribed compression ratio, the thermal efficiency of the Diesel cycle will be less than the thermal efficiency of the Otto cycle.

For a given compression ratio, the thermal efficiency of the Diesel cycle decreases with an increase in the maximum temperature T_3 of the cycle. An examination of Eq. 4.6 confirms this observation. The increase in the quantity of heat transferred to the working fluid will effect an increase in T_3 and in ω, and thus the quantity within the brackets will increase and cause η_t to decrease.

The constant-pressure heat addition for the air standard Diesel cycle is not indicative of the actual combustion process for the C-I engine. In practice, the fuel is burned primarily in two stages that approach constant-volume and constant-pressure combustion. Theoretically, the engine operation can be described by the limited-pressure cycle. This cycle however has little utility as a thermodynamic model, in part because of the indefinite split of the heat addition between the constant-volume and the constant-pressure processes.

4.20 THE COMPRESSION-IGNITION ENGINE THEORETICAL CYCLE

The theoretical cycle for the C-I engine rests upon the same assumptions that are made for the theoretical cycle established for the S-I engine, that is, the working fluid is a real and representative gas mixture. Actual fluid properties are used, and chemical equilibrium is observed in the evaluation of the cycle. The selection of a representative combustion process for the C-I engine theoretical cycle is however not explicitly indicated.

A constant-pressure combustion process is not

a realistic representation of the actual process because much of the fuel burns essentially at constant volume. The theoretical Diesel cycle consequently will yield comparatively low values of the thermal efficiency against which the performance of the actual engine would be measured. Higher thermal efficiency values however will be indicated by the theoretical Otto cycle. Whether this cycle devised for the S-I engine is also an appropriate theoretical cycle for the C-I engine will be contingent upon the deviation observed in the calculated values of the respective thermal efficiencies of the two cycles.

The thermal efficiency values reported in Fig. 4.8 are based on isooctane fuel (C_8H_{18}) and compression of a mixture of air and a gaseous fuel. The comparable theoretical Otto cycle for the C-I engine would most likely be based on dodecane fuel ($C_{12}H_{26}$) and compression of the fuel in the liquid phase, separately from the compression of the air. Calculated thermal efficiency values are reduced slightly by the substitution of dodecane for isooctane, and are increased virtually by the same amount by compression of the fuel in the liquid phase rather than in the gaseous phase. The theoretical Otto cycle, for which thermal efficiency values are reported in Fig. 4.8, may be employed as a standard in measuring the performance of either the S-I or the C-I engine.

The operation of the actual C-I engine deviates from the theoretical cycle in essentially the same manner as the S-I engine. Relative efficiency values for the C-I and S-I engines are generally similar (see Section 4.13). Principally because of operation with higher compression ratios and leaner air/fuel ratios, the brake thermal efficiency for the C-I engine falls within a range above that for the S-I engine.

4.21 ENGINE PERFORMANCE

The brake mean effective pressure bmep and the brake fuel rate bfr are the two most significant and useful factors employed for reporting the performance of the internal combustion engine. The bmep is an indication of the engine loading, while the bfr is a measure of the thermal performance.

Figure 4.20 is a typical plot of the operating characteristics of a variable-speed S-I engine for full-throttle setting. Similar curves are plotted for part-throttle operation. Performance curves for the C-I engine are generally similar to those observed for the S-I engine. The characteristics of an individual engine will influence the trend of a performance curve and the relative position of the point of maximum or minimum value.

Curves showing the variation in the mechan-

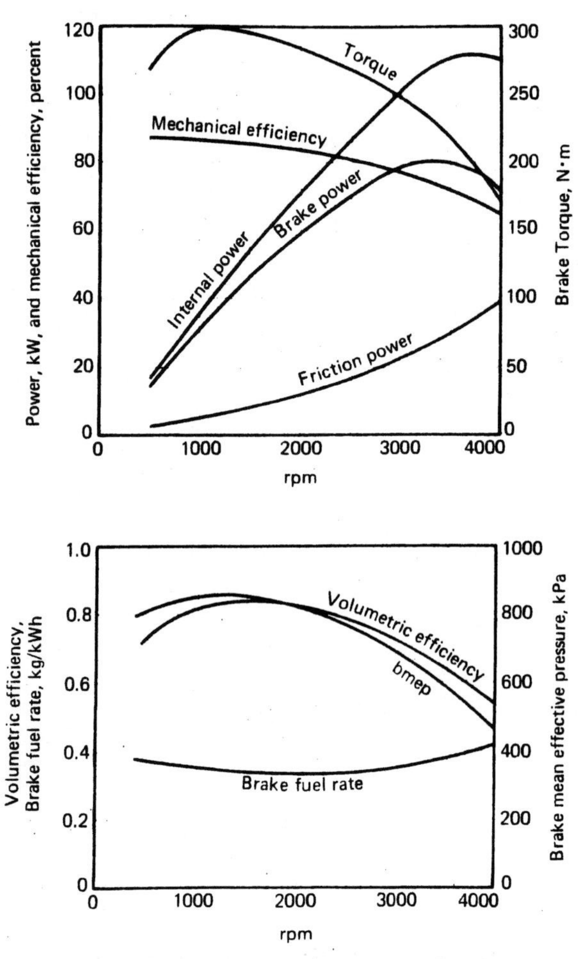

Figure 4.20 Typical performance curves for the S-I engine, full-throttle setting.

ical efficiency and in the volumetric efficiency are particularly significant because these two factors have a direct bearing on other characteristics of the engine. In the high-speed range, the decrease in the volumetric efficiency causes a decline in the internal power of the engine. Friction power $\dot{W}_f$ increases at a faster rate than does the speed. The mechanical efficiency is equal to $1 - (\dot{W}_f/\dot{W}_i)$. Because the friction power increases at a faster rate than the internal power $\dot{W}_i$, the mechanical efficiency decreases with an increase in speed. In the high-speed range, the friction power is particularly high, consequently the brake power falls appreciably below the internal power of the engine.

Torque is proportional to $\dot{W}_b/N$. For most of the operating range, $\dot{W}_b$ does not increase as rapidly as the speed N, hence the engine torque essentially decreases with an increase in speed.

Thermal performance of the internal combustion engine is usually reported by the brake fuel rate, in kg/kWh, which varies inversely with the brake thermal efficiency. The bfr is normally a minimum within the medium-speed range. At low speeds, the increase in the bfr is a result of a reduction in the internal thermal efficiency. At high speeds, the increase in the bfr is caused by the increase in mechanical friction. Minimum fuel rates are about 0.35 kg/kWh for the S-I engine and 0.25 kg/kWh for the C-I engine.

Figure 4.21 shows, for constant-speed operation, typical brake fuel rate curves for S-I and C-I engines.

Typical values of the rated maximum bmep for the S-I and C-I engines are

Nonsupercharged engines	850 to 900 kPa
Supercharged engines	1250 to 1300 kPa

The sensible energy and the latent energy in the exhaust gas, and the energy transferred as heat to the cooling system represent substantial fractions of the energy in the fuel. The energy in the exhaust gas discharged from large industrial engines is often partially recovered in heating and processing applications. The feasibility

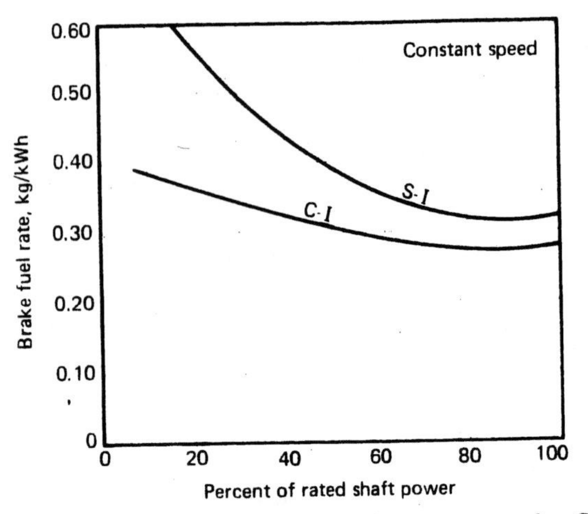

Figure 4.21 Typical brake fuel rate curves for S-I and C-I engines.

of recovering exhaust gas energy in a secondary power plant is explored in a later section.

Engine cooling is required for maintaining the structure at safe operating temperatures. As a consequence of decreasing availability incurred on the expansion and exhaust strokes, much of the heat lost to the cooling system has low availability (see Section 4.13). Complete elimination of the engine cooling would not produce a large increase in the thermal efficiency of the engine.

Example 4.12

The heat loss from the combustion gases in a C-I engine is distributed as follows: 6 kJ for the combustion process, 8 kJ for the expansion process, and 16 kJ for the exhaust process, per 100 kJ energy supplied. The compression ratio is 16 to 1; $\Phi = 0.8$. The connecting rod/crank ratio is 4 to 1. Determine the availability of the energy lost as heat.

The effective compression ratio is calculated at each of several prescribed positions of the crank. At each of these positions, the availability ratio is equivalent to the thermal efficiency of the theoretical cycle for the particular compression ratio (see table below). The availability ratio–crank angle relationship is shown in Fig. 4.22.

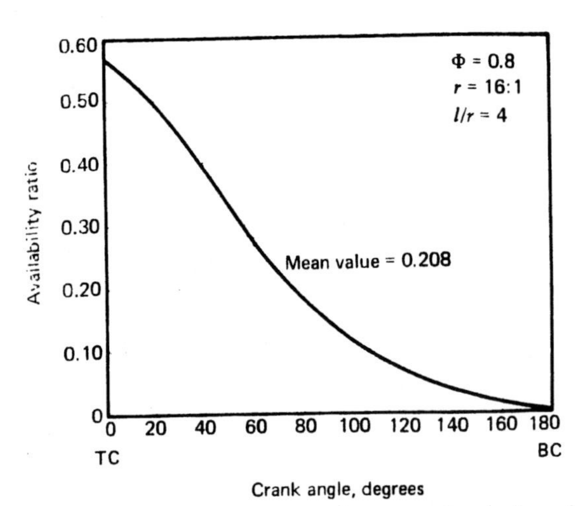

Figure 4.22 Availability ratio-to-crank angle relationship.

The heat transfer rate is assumed to remain constant during the expansion process.

Process	*Heat Loss (kJ)*	*Availability Ratio*	*Availability Loss (kJ)*
Combustion	6	0.56	3.36
Expansion	8	0.208	1.66
Exhaust	16	0	0

The total loss in availability is equal to 5.02 kJ per 100 kJ energy supplied. The heat loss during compression is a minor quantity, hence the availability loss for this process is not particularly significant.

Combustible components, mainly hydrocarbons (HC) and carbon monoxide (CO), in the engine exhaust gas are indicative of the loss charged to incomplete combustion. Excessive discharge of HC and CO from the engine cylinder can be avoided by maintaining the engine in good operating order.

Depending upon the power output, an appreciable fraction of the heat rejected by the engine may be transferred to the surroundings by radiation from the engine structure. The radiant energy transfer can be estimated from test data.

Example 4.13

Test data for a C-I engine are as follows:

Shaft output = 520 kW
Fuel consumption = 120 kg/h
Fuel, liquid $C_{12}H_{26}$, LHV = 44,085 kJ/kg

Cooling water $\dot{m}$ = 17,790 kg/h
ΔT = 25 C
c = 4.18 kJ/kg·K
Exhaust gas temperature = 650 C
Temperature of the fuel and air = 25 C
Exhaust gas analysis (dry), in percent by volume: CO_2 12.1, CO 3.4, O_2 0.5, H_2 1.5, N_2 82.5

Determine the radiant energy transfer from the engine.

The exhaust gas analysis provides the data used to write the combustion equation. Thus,

$$1.292C_{12}H_{26} + 21.94O_2 + 82.5N_2 \rightarrow 12.1CO_2 + 3.4CO + 0.5O_2 + 1.5H_2 + 15.3H_2O + 82.5N_2$$

The corresponding energy equation is given by

$$N_f(-\bar{h}^\circ_{RP})_f - [(N\,\Delta\bar{h}^\circ)_{CO_2} + (N\,\Delta\bar{h}^\circ)_{CO} + (N\,\Delta\bar{h}^\circ)_{O_2} + (N\,\Delta\bar{h}^\circ)_{H_2} + (N\,\Delta\bar{h}^\circ)_{H_2O} + (N\,\Delta\bar{h}^\circ)_{N_2}]_{923.2\text{ K}} + Q - W = 0$$

$$W = \frac{W}{\dot{m}_f} = \frac{520}{120/3600} = 15600 \text{ kJ/kg fuel}$$

Values of $\Delta\bar{h}^\circ$, that is, $\bar{h}^\circ_T - \bar{h}^\circ_{298\text{ K}}$, are selected from the *JANAF Thermochemical Tables* (Appendix A.5). Because the reactants are at a temperature of 25 C (298.2 K), $\Delta\bar{h}^\circ$ values for the reactants are equal to zero. (Note: $-h^\circ_{RP}$ = LHV) Then,

$$(1.292 \times 170.34 \times 44085) - [(12.1 \times 29285) + (3.4 \times 19160) + (0.5 \times 20049) + (1.5 \times 18379) + (15.3 \times 22865) + (82.5 \times 18972)] + Q - (1.292 \times 170.34 \times 15600) = 0$$

$$Q = -3,896,848 \text{ kJ}$$
$$= -17,706.6 \text{ kJ/kg fuel}$$

$$\dot{Q} = \dot{m}_f Q = 120(-17706.6)$$
$$= -2,124,792 \text{ kJ/h}$$

$$\dot{Q}_{cw} = \dot{m}_w c_w \Delta T$$
$$= 17790 \times 4.18 \times 25$$
$$= 1,859,055 \text{ kJ/h}$$

$$\dot{Q}_{total} = \dot{Q}_{cw} + \dot{Q}_{rad}$$

$$-2124792 = -1859055 + \dot{Q}_{rad}$$

$$\dot{Q}_{rad} = -265,737 \text{ kJ/h}$$

Effective cylinder scavenging of the two-stroke-cycle C-I engine is important with respect to achieving a high power output and an acceptable fuel rate. The air delivered by the blower has two functions, namely, removal of the residual exhaust gas and support of the subsequent combustion reaction. Delivery of the proper quantity of air is an important requirement. An excessive amount of air increases the blower power, and too little air impairs the removal of the residual exhaust gas and reduces the air charge.

Example 4.14

The bore and stroke for a six-cylinder, two-stroke-cycle C-I engine are, respectively, 250 mm and 300 mm. The compression ratio is 15 to 1. The gross internal relative efficiency is 0.77, and the mechanical losses are 50 kW for the operating speed of 400 rpm. The free air volume delivered by a rotary lobe blower is equivalent to 140 percent of the maximum cylinder volume of the engine. The blower discharge pressure is 35 kPa, gage. The blower overall efficiency, the ratio of the theoretical power to the shaft power, is 0.65. The fuel composition is 85.5 percent carbon and 13.5 percent hydrogen, and LHV = 44,165 kJ/kg. The air/fuel ratio is 21 to 1. The atmospheric pressure and temperature are 100 kPa, abs, and 22 C, respectively. Calculate the shaft power, brake thermal efficiency, and the brake fuel rate for the engine.

Piston Displacement

$$\dot{V}_{pd} = NlAn$$
$$= 6 \times 0.300 \times 0.04909 \times 400$$
$$= 35.345 \text{ m}^3\text{/min}$$

Maximum Cylinder Volume

$$\dot{V}_c = \frac{r}{r-1}\dot{V}_{pd}$$
$$= \frac{15}{14} 35.345 = 37.870 \text{ m}^3\text{/min}$$

Volume of the Charge

$$\text{Scavenging effectiveness } \eta_s = 1 - e^{-x}$$

where

x = the scavenging ratio $\dot{V}_b/\dot{V}_c = 1.40$

$\dot{V}_b$ = the volume of air delivered by the blower

$$\eta_s = 1 - \frac{1}{e^{1.40}} = 0.753$$

$$\dot{V}_a = \eta_s \dot{V}_c = 0.753 \times 37.870$$
$$= 28.516 \text{ m}^3\text{/min}$$

$$\dot{m}_a = \frac{P_a \dot{V}_a}{RT_a} = \frac{100 \times 28.516}{0.287 \times 295.2}$$
$$= 33.658 \text{ kg/min}$$

Gross Internal Power

Theoretical air = 14.53 kg/kg of fuel (from the fuel analysis). For $a/f = 21$, $\Phi \cong 0.7$

$$r = 15{:}1 \qquad \eta_{th} = 0.547 \qquad \text{(from Fig. 4.8)}$$

$$\dot{m}_f = \frac{\dot{m}_a}{a/f} = \frac{33.658}{21} = 1.6028 \text{ kg/min}$$

$$\dot{W}_i = \eta_{r,i}\eta_{th}\dot{m}_f(\text{LHV})$$
$$= 0.77 \times 0.547 \times 1.6028 \times 44165$$
$$= 29{,}815 \text{ kJ/min} \qquad 496.9 \text{ kW}$$

Blower Power

$$\dot{V}_b = 1.40\dot{V}_c = 1.40 \times 37.870$$
$$= 53.018 \text{ m}^3\text{/min}$$

$$\dot{W}_{bl} = \frac{\Delta P\,\dot{V}_b}{\eta_o} = \frac{35 \times 53.018}{0.65}$$
$$= 2854.8 \text{ kJ/min} \qquad 47.58 \text{ kW}$$

Shaft Power

$$\dot{W}_b = \dot{W}_i - \dot{W}_{bl} - \dot{W}_f$$
$$= 496.9 - 47.6 - 50 = 399.3 \text{ kW}$$

Brake Thermal Efficiency

$$\dot{m}_f = 0.026713 \text{ kg/s}$$

$$\eta_{t,b} = \frac{\dot{W}_b}{\dot{m}_f(\text{LHV})}$$
$$= \frac{399.3}{0.026713 \times 44165} = 0.338$$

Brake Fuel Rate

$$\text{bfr} = \frac{\dot{m}_f}{\dot{W}_b}$$
$$= \frac{3600 \times 0.026713}{399.3} = 0.241 \text{ kg/kWh}$$

4.22 ENGINE SUPERCHARGING

The power output of the internal combustion engine is closely related to the air flow rate because the quantity of fuel that may be burned in the cylinders is limited by the amount of air inducted during the charging process. The volumetric efficiency of a self-charging engine normally falls off at high speeds. When climbing into high altitudes, the power output of the aircraft engine tends to decrease sharply as a result of reduced air density. In both of these cases, the mass flow of air may be increased by supercharging, a process that is also used for increasing the output of engines that normally have a relatively high volumetric efficiency.

Supercharging can be applied to both S-I and C-I engines, either two- or four-stroke-cycle machines. A low-pressure process effecting a pressure increase of 20 to 40 kPa is commonly used for this purpose. The supercharging process increases the cylinder pressure at the start of compression, hence there is a resulting increase in the maximum pressure and temperature of the cycle. For the S-I engine, detonation places a limit on the extent of supercharging. Supercharging applications to the C-I engine must conform to the thermal and structural limitations of the machine.

Centrifugal compressors and either the rotary lobe-type or vane-type positive displacement superchargers are driven from the engine shaft and thus absorb a portion of the gross power output. The turbocharger (turbine-compressor), a high-speed machine, utilizes the energy in the exhaust gases to power the turbine, which in turn drives directly the centrifugal compressor. During compression, the air temperature is increased about 30 C, thus requiring the installation of an intercooler to reduce the air temperature to approximately the level at the compressor inlet. Such cooling is essential because the higher air temperature causes a reduction in the air mass flow and an increase in the cylinder maximum temperature.

A substantial increase in the bmep may be achieved by highly supercharging the C-I engine. The compressor pressure ratio for industrial supercharged C-I engines is usually limited to about 2.5 to 1. The increase in the bmep, because of supercharging, effects a reduction in the specific mass, in kg/kW, of the engine. Supercharging causes a modest increase in the cylinder maximum pressure, a significant factor pertinent to the structural design of the engine. In general, supercharging tends to effect a reduction in the engine brake fuel rate.

Supercharging systems provide for scavenging in addition to supercharging the two-stroke-cycle C-I engine. Supercharging however will not be accomplished unless the exhaust valves are timed to retain within the cylinders, prior to compression, a portion of the relatively high-density air.

Example 4.15

The gross internal power for a nonsupercharged, four-stroke-cycle C-I engine is 500 kW. The compression ratio is 16 to 1. The mechanical friction is equivalent to 75 kW. The gross internal relative efficiency is 0.75. The air/fuel ratio is 22 to 1. The stoichiometric air/fuel ratio is 14.5 to 1, and the lower heating value of the fuel is 44,100 kJ/kg. The volumetric efficiency is 0.82. The average pressures during the intake and exhaust strokes are, respectively, − 10 and + 12 kPa, gage. The atmospheric pressure and temperature are 100 kPa, abs, and 25 C, respectively. Determine (a) the fuel and air flow rates, (b) the piston displacement, (c) the pumping power, (d) the shaft power, and (e) the brake fuel rate.

Fuel and Air Flow Rates

$$\Phi = 14.5 \div 22 = 0.66 \qquad r = 16{:}1$$

$$\eta_{th} = 0.56 \qquad \text{(from Fig. 4.8)}$$

$$\eta_{t,i} = \eta_{r,i} \times \eta_{th} = 0.75 \times 0.56 = 0.42$$

$$\dot{m}_f = \frac{\dot{W}_i}{\eta_{t,i} \times \text{LHV}} = \frac{3600 \times 500}{0.42 \times 44100} = 97.182 \text{ kg/h}$$

$$\dot{m}_a = \frac{a}{f}\dot{m}_f = 22\,\frac{97.182}{60} = 35.633 \text{ kg/min}$$

Piston Displacement

$$\dot{V}_a = \frac{\dot{m}_a R T_a}{P_a} = \frac{35.633 \times 0.287 \times 298.2}{100} = 30.496 \text{ m}^3\text{/min}$$

$$\dot{V}_{pd} = \frac{\dot{V}_a}{\eta_v} = \frac{30.496}{0.82} = 37.190 \text{ m}^3\text{/min}$$

Pumping Power

$$\dot{W}_p = \Delta P \dot{V}_{pd} = 22 \times 37.190 = 818.2 \text{ kJ/min} \qquad 13.64 \text{ kW}$$

Shaft Power

$$\dot{W}_b = \dot{W}_i - \dot{W}_p - \dot{W}_f = 500 - 14 - 75 = 411 \text{ kW}$$

Brake Fuel Rate

$$\text{bfr} = \frac{\dot{m}_f}{\dot{W}_b} = \frac{97.182}{411} = 0.236 \text{ kg/kWh}$$

The engine is now supercharged by means of a rotary lobe blower. The blower intake and delivery pressures are, respectively, −4 and +40 kPa, gage. The blower overall efficiency is 0.66. The power equivalent of the mechanical friction is increased to 80 kW because of higher cylinder pressures. The average exhaust pressure is + 15 kPa, gage, and the average pressure during intake is +30 kPa, gage. The volumetric efficiency of the engine is 1.06. Other operating data remain unchanged. Calculate (a) the air and fuel flow rates, (b) gross internal power, (c) pumping power, (d) blower power, (e) shaft power, and (f) the brake fuel rate.

Fuel and Air Flow Rates

$$\rho_a = \frac{P_a}{RT_a} = \frac{100}{0.287 \times 298.2}$$

$$= 1.1685 \text{ kg/m}^3 \quad \text{(at atmospheric conditions)}$$

$$\dot{m}_a = \eta_v \rho_a \dot{V}_{pd}$$

$$= 1.06 \times 1.1685 \times 37.190$$

$$= 46.064 \text{ kg/min}$$

$$\dot{m}_f = \frac{\dot{m}_a}{a/f} = \frac{46.064}{22}$$

$$= 2.0938 \text{ kg/min} \qquad 125.63 \text{ kg/h}$$

Gross Internal Power

$$\dot{W}_i = \eta_{t,i} \dot{m}_f(\text{LHV})$$

$$= 0.42 \times 2.0938 \times 44100$$

$$= 38{,}781 \text{ kJ/min} \qquad 646.4 \text{ kW}$$

Pumping Power

$$\dot{W}_p = \Delta P \dot{V}_{pd} = 15 \times 30.496$$

$$= 457.4 \text{ kJ/min} \qquad 7.6 \text{ kW}$$

Pumping power is delivered to the piston because the intake pressure exceeds the exhaust pressure.

Blower Power

At intake, $P = -4$ kPa, gage, or 96 kPa, abs., $T = 298.2$ K.

$$v_a = \frac{RT}{P} = \frac{0.287 \times 298.2}{96}$$

$$= 0.8915 \text{ m}^3/\text{kg} \quad \text{(at the blower inlet)}$$

$$\dot{W}_{bl} = \frac{\Delta P \dot{m}_a v_a}{\eta_o}$$

$$= \frac{44 \times 46.064 \times 0.8915}{0.66}$$

$$= 2737.7 \text{ kJ/min} \qquad 45.6 \text{ kW}$$

Shaft Power

$$\dot{W}_b = \dot{W}_i + \dot{W}_p - \dot{W}_{bl} - \dot{W}_f$$

$$= 646 + 8 - 46 - 80 = 528 \text{ kW}$$

Brake Fuel Rate

$$\text{bfr} = \frac{\dot{m}_f}{\dot{W}_b} = \frac{125.63}{528} = 0.238 \text{ kg/kWh}$$

The preceding analysis indicates that the shaft power of the internal combustion engine can be increased substantially by supercharging. In this case, supercharging causes a slight increase in the brake fuel rate.

4.23 THE FREE PISTON ENGINE

The turbocharged C-I engine consists of the reciprocating elements and a turbine-driven compressor. In this installation, the turbine-compressor delivers no shaft work.

The C-I engine could conceivably be supercharged to a pressure of 6 to 7 atm. abs., achieved with a compression pressure ratio of 6 to 1 to 7 to 1. The supercharging compressor is mechanically driven from the engine shaft, and the power required to drive the supercharger is now equivalent to the entire power output of the engine. The engine then functions as a gas generator, and the power plant output is derived solely from the exhaust gas-driven turbine. There is, in this arrangement, no purpose in gearing the turbine to the engine shaft. Because of the problems associated with starting and the necessity of avoiding excessively high peak combustion pressures, this combination of the power elements is not considered to be practicable.

The free piston engine operating in a combination with a gas turbine is a more feasible arrangement for a reciprocating gas generator and a power turbine. The above-mentioned starting and combustion-pressure problems are avoided in this type of power plant. Elimination of the crankshaft and connecting rods simplifies to some degree the mechanical construction of the gas generator.

Figure 4.23 illustrates the construction of the free piston engine or gas generator. The central portion of the element is a two-stroke-cycle, opposed piston Diesel cylinder with the pistons mechanically synchronized through a linkage or pair of racks meshing with a common pinion. The increase in the cylinder pressure following the combustion reaction causes the pistons to move apart. Air for the combustion process is simultaneously compressed in the two air cylinders. In addition, the cushion air is compressed by the two bounce pistons that are located at the outer end of each reciprocating assembly. This compressed cushion air reexpands and forces the two power pistons to move toward each other while effecting compression of the combustion air. The combustion gases discharged from the Diesel cylinder are directed to the power turbine.

The free piston engine–gas turbine power plant can be operated at any rating from full load to turbine idling conditions with good part-load thermal efficiency. An overall thermal efficiency of about 35 percent can be anticipated for full-load operation. The controls for the free piston engine are not especially complex and consist of conventional pneumatic and hydraulic devices.

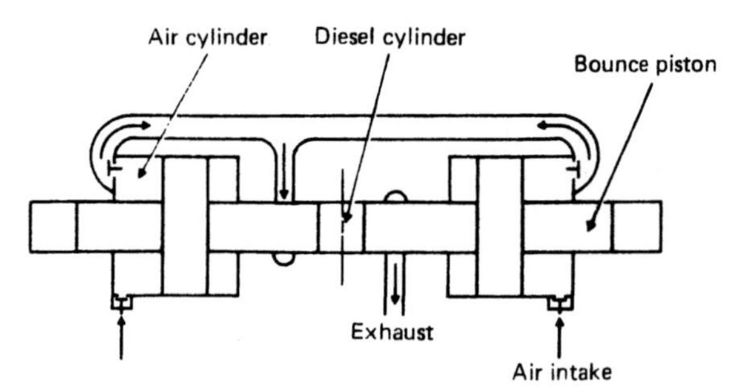

Figure 4.23 Free piston gas generator. (A. J. Ehrat, *Mechanical Engineering*, March 1955.)

The first free piston engine was constructed in France by A. J. Pascara circa 1925. Subsequently, most of the developmental work on this engine has been accomplished in France. Numerous applications of the free piston engine have been proposed, but the interest in the engine that was demonstrated circa 1950 has not been sustained, and presently the early interest appears to have largely waned.

4.24 THE STIRLING ENGINE

Although the Stirling engine is not classified as an internal combustion engine, the construction and application of the engine may logically be examined at this point, particularly in comparison with the internal combustion engine. The elementary concept of the Stirling engine is a cylinder-and-piston mechanical arrangement and a heat source that is external to the cylinder. In several respects, the engine is more complex than the conventional S-I or C-I engine.

Robert Stirling developed the original engine in 1816. A large number of these engines were subsequently constructed and demonstrated good performance with modest power output. Eventually, other types of engines largely replaced the Stirling engine. However, in some parts of the world, limited use of the engine continued, primarily for pumping water. In recent years, interest in the Stirling engine has been revived, and the developmental work, carried out principally in Europe, is directed to the elimination of the operating difficulties that restrict extensive use of the engine.

The theoretical Stirling cycle is shown on the P–v and T–s planes in Fig. 4.24. All of the heat obtained from an external source is supplied isothermally at temperature T_H. Heat is rejected isothermally at temperature T_L to an external receiver or sink. The two interconnecting constant-volume processes, 2–3 and 4–1, are achieved

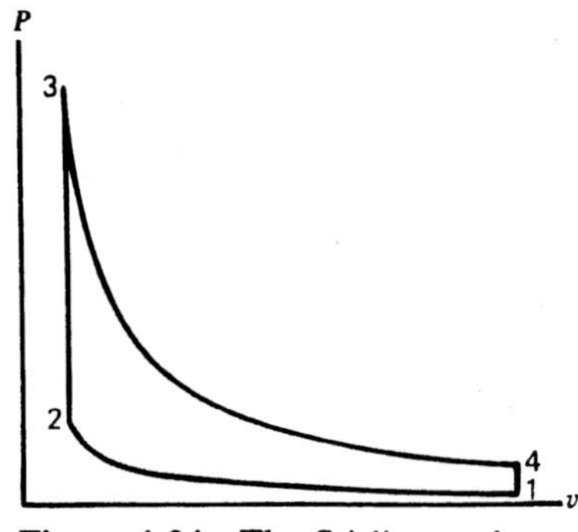

Figure 4.24 The Stirling cycle.

by internal transfer of heat, $Q_{2-3} = -Q_{4-1}$. This internal transfer of heat is particularly significant because external heat transfer is effected solely in the two isothermal processes at the high and low temperatures of the cycle. The thermal efficiency of the theoretical Stirling cycle is consequently equal to the thermal efficiency of the Carnot cycle for prescribed values of temperatures T_H and T_L.

In comparison with the internal combustion engine, the actual Stirling engine is considerably more complex. A simple cylinder-and-piston machine cannot be employed, primarily because of the prescribed regenerative heat transfer and the difficulties encountered in attempting to achieve isothermal compression and expansion of the working fluid.

Stirling's original design was based on the concept of utilizing a single cylinder equipped with two pistons. A modern version of this design is shown in Fig. 4.25. Two pistons, a displacer and a power piston, operate from a rhombic drive. Usually, the displacer leads the power piston by 90°. The closed system is charged with a permanent gas, usually hydrogen or helium. Hydrogen however is considered to be the most suitable working fluid for the Stirling engine. A combustion chamber, located at the top of the cylinder, provides for indirect heating of the gas. High gas pressures, on the order of 100 to 200 atm, are observed in the Stirling engine.

The total enclosed volume of gas varies with the movement of the tightly fitted power piston.

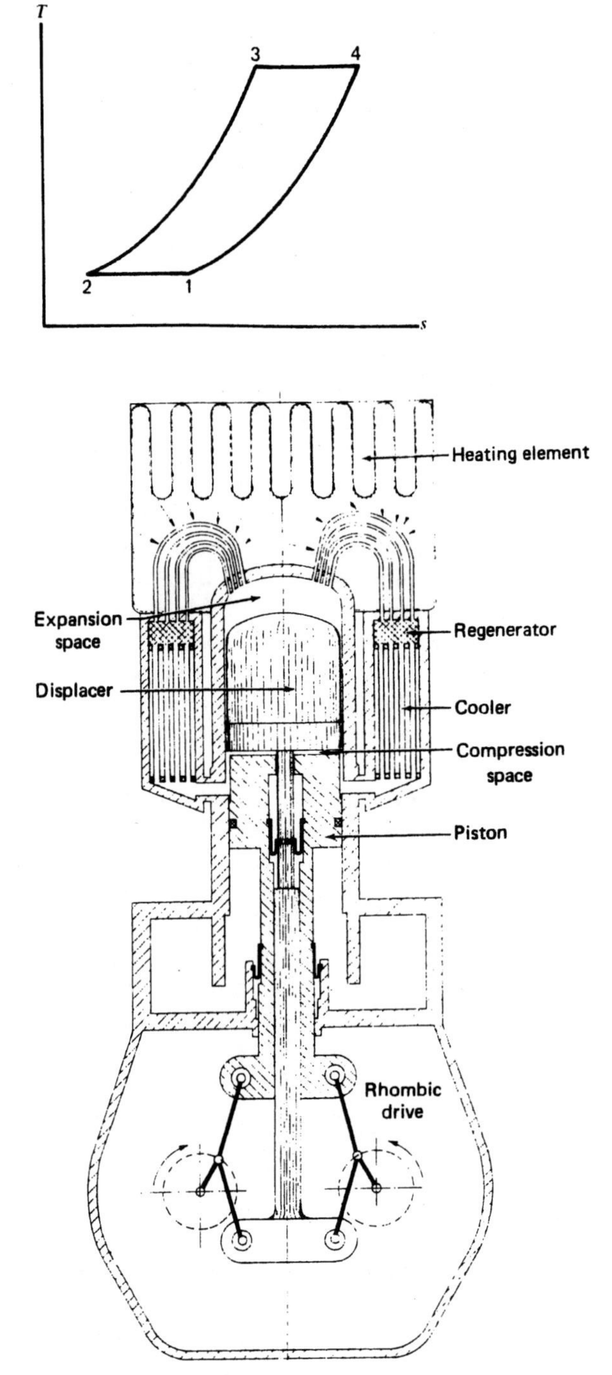

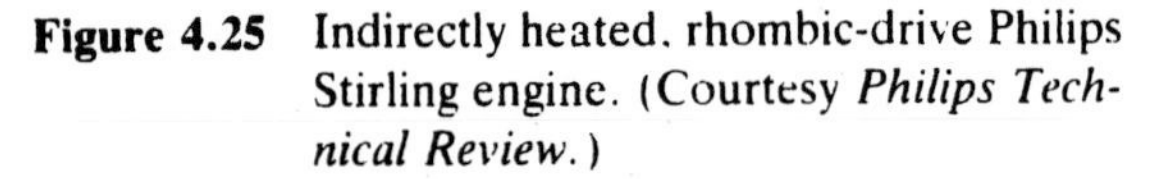

Figure 4.25 Indirectly heated, rhombic-drive Philips Stirling engine. (Courtesy *Philips Technical Review*.)

The loosely fitted displacer divides the enclosed volume into two main regions, the expansion space and the compression space, which are maintained at different temperatures. Movement of the displacer changes the proportion of gas in the two spaces while maintaining essentially the same gas pressure on the opposing faces of the displacer. Because the displacer operates with virtually balanced pressures, a thin steel-shell construction is adequate. Compression and expansion of the total volume of gas are effected by the movement of the power piston.

When the displacer is at the top position in the cylinder, most of the gas is in the cold space. Downward movement of the displacer causes the gas to move from the cold space through the heat exchanger circuit, namely, the cooler, regenerator, and heater, and into the hot space. The net effect of heating the gas is rapid development of a high gas pressure in the expansion space. Because of the tendency to equalize pressures throughout the entire gas volume, a high pressure is exerted on the face of the power piston.

Upward movement of the displacer causes most of the gas to move from the expansion space through the heat exchanger circuit and into the cold space. The major portion of the gas is now at a low temperature and low pressure in the compression space above the face of the power piston. The alternate cooling and heating of the gas produces a variation in the pressure acting on the face of the power piston, and consequently a net transfer of work to the piston is effected.

Various kinds of fuel may be burned in the external combustion chamber of the Stirling engine. In particular, a high-octane-number gasoline is not required. An acceptable fuel however must meet the requirements for low emission of pollutants, combustion with minimum difficulty, and ready response to the engine control system. A low-sulfur, liquid fuel derived from coal is a possibility. The quality of this fuel would not be as high as that of gasoline, a contributing factor in reducing the manufacturing cost of the fuel.

Most of the heat rejected from the Stirling engine is transferred through the cooler to the atmosphere, with a minor quantity of heat rejected in the exhaust gases that are discharged from the combustion chamber. The radiator for the Stirling engine is likely to be about 2.5 times the size of the radiator required for an internal combustion engine of the same power output.

The Stirling engine operates quietly with low emission of pollutants. The thermal efficiency, about 0.35 to 0.40, is in the range achieved by high-economy C-I engines and superior to the thermal efficiency reported for the S-I engine. Thermal efficiency values that exceed 0.4 are predicted for some of the smaller Stirling engines currently under development.

High manufacturing costs, which result principally from the complexity of both the rhombic drive and the heat exchanger elements, tend to restrict widespread use of the Stirling engine. The necessity for construction of a gas-tight system presents additional manufacturing problems. Stirling engine applications are limited to some extent because of the high mass-to-power ratio.

Stirling engines constructed for applications such as vehicular propulsion and electric power generation are likely to burn Diesel engine fuel oil. Engines of low-power capability have been developed for artificial-heart power sources and for operating equipment installed in space ships. For these two applications, in particular, a radioisotope fuel provides the required long-lived primary energy source for the Stirling engine. Small solar-powered Stirling engines are a possibility.

Practical construction of the Stirling engine is dependent upon reciprocating motions. Two different basic machine arrangements have been employed to achieve the same thermodynamic cycle. The earlier arrangement, credited to Robert Stirling, utilizes two pistons moving in a single cylinder, as shown in Fig. 4.25. An alternative mechanical arrangement, invented by A. K. Rider in 1876, employs two working cylinders, each equipped with a piston, as shown in Fig. 4.26. Heat is supplied from an external source at the hot cylinder and removed at the cold cylinder

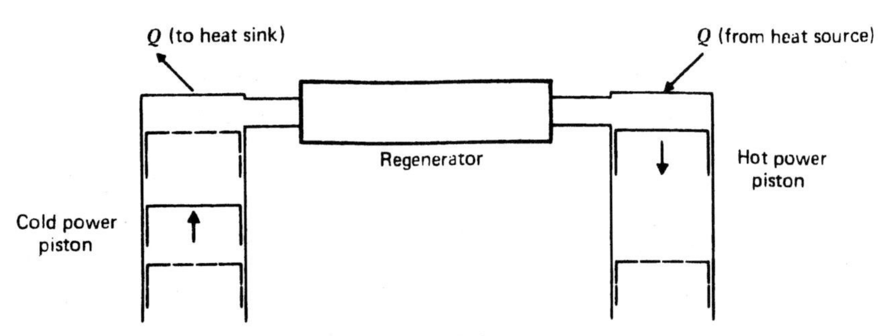

Figure 4.26 Schematic illustration of the two-piston Stirling engine.

for transfer to the surroundings. Two cranks are used, with the hot power piston leading the cold power piston by 90°. The two pistons mutually effect alternate compression and expansion of the working fluid and displacement of the fluid, through the regenerator, from one cylinder to the other. Many of the early engines were constructed in accordance with the operating principle established by Rider, although more often, over the years, engine construction followed the displacer design developed by Stirling in 1816.

Operation of the actual Stirling engine departs significantly from the sequence of events prescribed by the theoretical Stirling cycle. The different events, compression and expansion and heating and cooling, show considerable overlap in time, hence the indicator card for the actual engine will not duplicate the P–v plot shown in Fig. 4.24. The actual work diagram is called the Schmidt cycle, after the first investigator to provide a mathematical description of the engine. The mathematical analysis is complex and provides only an approximate prediction of the power output of the actual Stirling engine.

4.25 COMBINED-CYCLE POWER PLANTS

The large fraction of the energy in the fuel that leaves the internal combustion engine in the exhaust gases raises the question whether a portion of this energy could be effectively recovered in a secondary thermal engine. The possibility of achieving this recovery will be explored first in a theoretical system.

Because the engine exhaust gas is a finite source of energy, the energy in the gas is transferred to a series of elemental reversible engines (see Fig. 4.27). Each elemental engine receives reversibly a quantity of heat, dQ, at temperature T and rejects reversibly a quantity of heat, dQ_{un}, at temperature T_0 to the atmosphere. The first engine in the series receives heat at temperature T_1, and the last engine in the series receives heat at temperature $T_0 + dT$.

The work for each elemental reversible engine is given by

$$dW = \left(1 - \frac{T_0}{T}\right) dQ$$

where $dQ = c_p \, dT$. Then,

$$dW = c_p\left(dT - T_0 \frac{dT}{T}\right)$$

or

$$W = c_p\left[(T_2 - T_1) - T_0 \ln \frac{T_2}{T_1}\right] \quad (4.7)$$

The work W of the series of elemental reversible engines is the equivalent of the available energy E_{av}, represented by the area 1231 in Fig. 4.27. The unavailable energy E_{un} is equivalent to the area 23452.

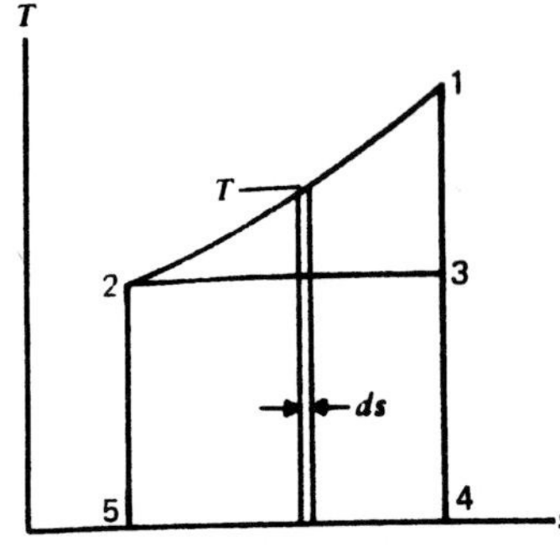

Figure 4.27 Hypothetical recovery of the energy in the engine exhaust gas.

Example 4.16

An internal combustion engine discharges 1165 kg/h gas for a power output of 200 kW. The exhaust gas temperature is 540 C; T_0 = 20 C. For the combustion gas mixture, c_p = 1.09 kJ/kg·K. Determine the theoretical power recovery from the engine exhaust gas.

$$T_1 = 813 \text{ K} \qquad T_2 = 293 \text{ K}$$

Applying Eq. 4.7,

$$W = c_p\left[(T_2 - T_1) - T_0 \ln \frac{T_2}{T_1}\right]$$

$$= 1.09\left[(293 - 813) - 293 \ln \frac{293}{813}\right]$$

$$= -240.86 \text{ kJ/kg gas}$$

The negative sign is attributed to the direction of integration and does not imply a work input.

$$\dot{W} = \dot{m}_g W = 1165 \times 240.86$$

$$= 280{,}602 \text{ kJ/h} \qquad 77.94 \text{ kW}$$

The theoretical engine supplied with energy transferred from the internal combustion engine exhaust gases will produce 77.9 kW. This figure is high when compared with the 200-kW power output of the primary engine. The series of elemental reversible engines is a highly theoretical scheme used for secondary power generation, and any actual recovery of energy from the exhaust gases would be much less promising. This possibility will now be explored.

Example 4.17

Energy in the exhaust gas discharged from the engine in Example 4.16 will be transferred to a steam power plant. The exhaust gas is cooled from 540 C to 205 C. Dry saturated steam is generated at 140 C (361 kPa, abs.). Expansion of the steam terminates at 20 C (see Fig. 4.28). Determine the theoretical output of the steam power plant.

$$h_3 \text{ at } 140 \text{ C, } x = 1.0 = 2733.9 \text{ kJ/kg}$$

$$h_5 \text{ at } 20 \text{ C} = 84.0 \text{ kJ/kg}$$

$$Q_{steam} = h_3 - h_5 = 2733.9 - 84.0$$

$$= 2649.9 \text{ kJ/kg}$$

$$Q_{gas} = c_p(T_2 - T_1)$$

$$= 1.09(205 - 540)$$

$$= -365.15 \text{ kJ/kg}$$

$$\dot{m}_{steam} = \frac{Q_{gas}}{Q_{steam}} \dot{m}_{gas} = \frac{365.15}{2649.9} 1165$$

$$= 160.53 \text{ kg/h}$$

The steam cycle theoretical work is achieved by reversible adiabatic expansion from state 3 to state 4.

$$h_4 = 2028.8 \text{ kJ/kg}$$

$$W_{steam} = h_3 - h_4 = 2733.9 - 2028.8$$

$$= 705.1 \text{ kJ/kg}$$

$$\dot{W}_{steam} = \dot{m}_{steam} W_{steam}$$

$$= 160.53 \times 705.1$$

$$= 113{,}190 \text{ kJ/h} \qquad 31.44 \text{ kW}$$

The theoretical power (31.4 kW) developed by the secondary steam power plant would in an actual system be reduced by miscellaneous heat losses and by irreversible expansion of the steam. Thus, the actual recovery of energy may be taken as equivalent to approximately 20 kW. A gain

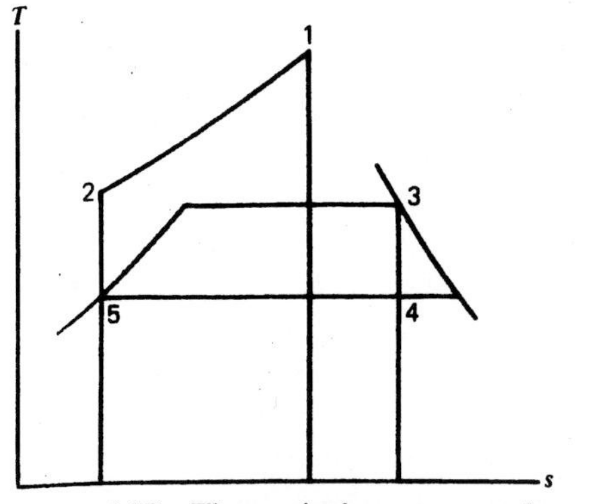

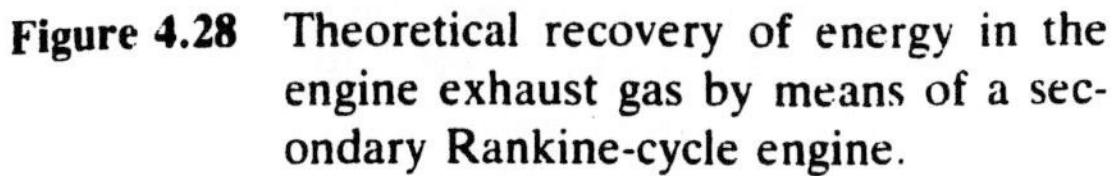

Figure 4.28 Theoretical recovery of energy in the engine exhaust gas by means of a secondary Rankine-cycle engine.

of 10 percent in the power output, achieved by partial recovery of energy in the engine exhaust gases, is not insignificant. Two-stage power generation is however not economically feasible for a low-capability power plant because increased capital and maintenance costs are not compensated by a sufficiently large reduction in the cost of the fuel. Further, the mechanical complexity of a system of this size would not be acceptable.

Example 4.18

Determine the effectiveness for converting the energy in the exhaust gas to work produced in the steam power plant of Example 4.17. The temperature of the infinite atmosphere is 20 C.

1 Gas/steam ratio.

$$\frac{\dot{m}_{gas}}{\dot{m}_{steam}} = \frac{1165}{160.53}$$
$$= 7.257 \text{ kg gas/kg steam}$$

2 Change in the availability of the gas. Applying Eq. 1.3,

$$\phi_2 - \phi_1 = h_2 - h_1 - T_0(s_2 - s_1)$$
$$= c_{p,g}(T_2 - T_1) - T_0 c_{p,g} \ln \frac{T_2}{T_1}$$
$$= 1.09(205 - 540) - 293.2 \times 1.09 \ln \frac{478.2}{813.2}$$
$$= -365.15 + 169.68$$
$$= -195.47 \text{ kJ/kg}$$

3 Change in the availability of the steam.

State 3: $h = 2733.9$ kJ/kg

$s = 6.9299$ kJ/kg·K

State 5: $h = 84.0$ kJ/kg

$s = 0.2966$ kJ/kg·K

$$\phi_3 - \phi_5 = h_3 - h_5 - T_0(s_3 - s_5)$$
$$= 2733.9 - 84.0 - 293.2(6.9299 - 0.2966)$$
$$= 2649.9 - 1944.9 = 705.0 \text{ kJ/kg}$$

4 Conversion effectiveness for the heat transfer process. Applying Eq. 1.4,

$$\varepsilon = \frac{\Delta E_{av,2}}{-\Delta E_{av,1}} = \frac{m_s(\phi_3 - \phi_5)}{m_g(\phi_2 - \phi_1)}$$
$$= \frac{1 \times 705.0}{-7.257(-195.47)} = 0.497$$

5 Conversion effectiveness for the steam plant.

$$\phi_4 - \phi_3 = h_4 - h_3 - T_0(s_4 - s_3)$$

where $s_4 - s_3 = 0$

Thus,

$$\phi_4 - \phi_3 = h_4 - h_3 = 2028.8 - 2733.9$$
$$= -705.1 \text{ kJ/kg}$$

The output for the expansion process 3–4 is the reversible work $W_{3-4} = 705.1$ kJ/kg. Then,

$$\varepsilon = \frac{\Delta E_{av,2}}{-\Delta E_{av,1}} = \frac{705.1}{-(-705.1)} = 1.0$$

The thermal efficiency for the steam plant is determined by

$$\eta_t = \frac{W}{Q_s} = \frac{h_3 - h_4}{h_3 - h_5} = \frac{705.1}{2649.9} = 0.266$$

Because of the second law limitation, only a portion of the supplied heat is converted to work in the steam plant. Complete conversion of the available energy to work is however achieved. In an actual expansion, $h_4 > 2028.8$ kJ/kg and $W_{irrev} < 705.1$ kJ/kg, hence $\Delta E_{av,1} > \Delta E_{av,2}$ and $\varepsilon < 1.0$.

Recovery of energy from the jacket water and the exhaust gas discharged from large C-I engines is a fairly common practice.* The gas is diverted to a boiler for the production of hot water and steam, which are ordinarily used for heating and processing applications. In some installations however, steam is generated in the exhaust gas boiler for operating a noncondensing steam turbine.

A combination of the C-I engine with two Rankine-cycle engines has been proposed.[10] One Rankine-cycle engine utilizes energy transferred from the exhaust gases discharged from the C-I engine, and the other Rankine-cycle engine utilizes energy transferred from the C-I engine cooling water. The two turbines, namely, the vapor-cycle power elements, are coupled directly to the C-I engine shaft, and the three power units drive a single electric generator. In comparison with the C-I engine operating alone, the combined-cycle power plant is expected to show a 25 percent reduction in the brake fuel rate.

While a number of advantages are cited for the C-I engine/dual-Rankine-cycle power plant, the overall economy of the power system can be questioned. For commercial electric power generation, Diesel engine fuel is comparatively high-priced, and the generating capability of the power plant is low. The respective fractions of the total electrical output of the combined-cycle power plant are: the C-I engine 0.806, the exhaust gas-heated power plant 0.174, and the cooling water-heated power plant 0.020. Because only 2 percent of the total output of the power plant is derived from energy recovered from the C-I engine cooling water, a more economical design probably would be achieved by omitting the low-pressure Rankine-cycle power plant, which includes primarily the vapor-generating equipment and the low-pressure turbine.

The continual increase in the cost of fuel will enhance the opportunity to utilize for electric power generation the energy in the exhaust gas discharged from certain existing industrial C-I engine or gas turbine engine generating sets. Energy in the exhaust gas would be recovered in a secondary steam power plant that could, for superior performance, include a condensing steam turbine. The economy of such two-stage power generation would depend upon a number of factors, some local, and would be influenced significantly by the size of the primary power-generating unit.

A number of combinations of the internal combustion engine with the gas turbine engine have been examined, and some power plant development has taken place. A power plant of this type is usually constructed for some unique application. A C-I engine/gas turbine engine power plant, for example, has been developed for driving certain special-duty motor vehicles.

4.26 SUMMARY

Although alternative power plants have been investigated, the probability of replacing the automotive spark-ignition and compression-ignition engines in the foreseeable future appears to be remote. Each potential replacement that has been proposed or investigated to date fails in some respect to meet the requirements for automotive service. Any acceptable replacement

* Cogeneration, a term of somewhat recent origin, may be defined as the sequential use of energy from a single primary source to produce both shaft power and useful thermal energy. Power production may be either at the top or at the bottom of the energy sequence. Cogeneration should not be used to define two-stage power generation.

must compete strongly with the S-I and C-I engines relative to first cost, reliability, size, weight, economy of operation, maintenance costs, and control.

The gas turbine engine and the Stirling engine are currently the foremost thermal power plants proposed as replacements for the automotive S-I and C-I engines. Many difficulties however lie ahead in developing and producing these engines in quantity for automotive service. Further, the use of gas turbine engines and Stirling engines should contribute to a reduction in air pollution and effect conservation of petroleum-based fuels. Fuel conservation may possibly be accomplished by improved thermal efficiency and the use of lower-grade or substitute engine fuels.

Extensive use of the C-I engine in passenger automobiles is viewed by some investigators as a means to reducing air pollution and conserving fuel. In comparison with the S-I engine, the manufacturing cost of the C-I engine is higher; and because of a greater mass-to-power ratio, a somewhat heavier automobile structure is required.

The electrically operated automobile is not a new concept. At the turn of the century, battery-powered automobiles and light-duty trucks were operated in the United States. Because of limited capacity for energy storage, the "electric automobile" is not likely to become a major mode of transportation for another decade or more.

PROBLEMS

4.1 A six-cylinder, four-stroke-cycle internal combustion engine operates at 2760 rpm. Bore and stroke are, respectively, 115 mm and 128 mm. The average indicator card has an upper loop area of 6.95 cm^2 and a lower loop area of 0.42 cm^2. The card length is 7.60 cm. The spring scale is 1000 kPa/cm. The mechanical efficiency is 0.84. Calculate the shaft power.

4.2 Determine the pumping power for the engine described in Problem 4.1.

4.3 Calculate for an air standard Otto cycle the thermal efficiency, work, and mep. At the start of compression P = 105 kPa, abs, and T = 295 K. At the start of expansion P = 10,678 kPa, abs, and T = 3000 K.

4.4 The compression ratio for an air standard Otto cycle is 8 to 1. At the start of compression, P = 100 kPa, abs, and T = 300 K. The heat supplied to the cycle working fluid is 1850 kJ/kg air. (a) Determine for the cycle the thermal efficiency, work, and mep. (b) Repeat the calculations for the air cycle using the actual properties of air (refer to *Gas Tables* in the Appendix).

4.5 A six-cylinder, four-stroke-cycle C-I engine burns $C_{12}H_{26}$ with an air/fuel ratio of 20 to 1. The compression ratio is 15 to 1, and the bore and stroke are, respectively, 200 mm and 250 mm. The following efficiency values are applicable to the engine: internal relative 0.80, mechanical 0.83, and volumetric 0.85. The engine speed is 1200 rpm. Ambient conditions are 1 atm and 15 C. Calculate the shaft power.

4.6 Determine the thermal efficiency, mep, and maximum pressure for an Otto cycle and a Diesel cycle that have similar compression processes; r = 10. At the start of compression, P = 100 kPa, abs, and T = 300 K. The heat added is 2000 kJ/kg air. Base the calculations on the air standard analysis.

4.7 The compression ratio for an air standard Stirling cycle is 10 to 1. At the start of compression, P = 100 kPa, abs, and T = 300 K. The maximum temperature is 2000 K. Calculate the thermal efficiency, mep, and maximum pressure for the cycle.

4.8 A C-I engine operates with a brake thermal efficiency of 0.35 when producing a shaft output of 200 kW. The lower heating value of the fuel is 44,000 kJ/kg, and the air/fuel ratio is 22 to 1. The blower used for supercharging the engine is driven from the crankshaft and delivers 140 percent of the air that is supplied for the combustion reaction. The blower develops a pressure differential of 35 kPa with a machine efficiency of 0.72 and a mechanical efficiency of 0.98. The ambient conditions are 100 kPa, abs, and 20 C. Calculate the blower shaft power.

4.9 An eight-cylinder, four-stroke-cycle S-I engine operates at a speed of 2400 rpm. The bore and stroke are, respectively, 100 mm and 110 mm, and the compression ratio is 10 to 1. A stoichiometric air/fuel ratio (15.12 to 1) is used. The following efficiency values are applicable to the engine: internal relative 0.78, mechanical 0.83, and volumetric 0.80. The lower heating value of the fuel is 44,400 kJ/kg. The ambient conditions are 1 atm and 17 C. Determine the shaft power and the bmep for the engine.

4.10 A test conducted on a four-stroke-cycle internal combustion engine shows a minimum bsfc of 0.322 kg/kWh when operating at 80 percent of the rated speed. The corresponding shaft output, volumetric efficiency, and air/fuel ratio are, respectively, 123 kW, 0.95, and 15 to 1. When the engine operates at 120 percent of the rated speed, the engine develops 162 kW with a bsfc of 0.394 kg/kWh. The volumetric efficiency at the higher speed is 0.85. What is the probable air/fuel ratio at this operating point?

4.11 At the start of compression, the pressure and temperature for an air standard limited-pressure cycle are 100 kPa, abs, and 300 K. The compression ratio is 16 to 1. The total heat supplied is 2800 kJ/kg air. The Q_v/Q_p ratio is 1.5 to 1. Calculate the pressure and temperature at each point of the cycle, the cycle thermal efficiency, and the mep.

4.12 A single-cylinder, two-stroke-cycle S-I test engine operates at 900 rpm. The cylinder indicator cards are 7.65 cm long and have an average area of 3.35 cm^2. The indicator spring scale is 2000 kPa/cm. The engine employs crankcase induction. A light-spring indicator installed on the crankcase has a spring scale of 25 kPa/cm. The length of the crankcase indicator cards is 7.65 cm, and the average area of the cards is 3.45 cm^2. The engine bore and stroke are, respectively, 250 mm and 300 mm. The engine mechanical efficiency is 0.86. Calculate the net mean effective pressure and the shaft power for the engine.

4.13 The exhaust gas from a C-I engine is used to drive a gas turbine. The analysis of the system is based on the properties of air obtained from the Air Table in the Appendix. At the start of compression in the Diesel cycle, the pressure and temperature are 100 kPa, abs, and 290 K. The compression ratio is 20 to 1, and the heat supplied is 2500 kJ/kg air. At the end of the expansion stroke, the gas is discharged quickly from the cylinder. Determine the kinetic energy of the "exhaust gas," the total work of the system, and the system thermal efficiency.

4.14 A low-compression C-I engine develops a shaft output of 260 kW when running at 1200 rpm. The engine, a four-stroke-cycle machine, consists of six cylinders for which the bore and stroke are, respectively, 180 mm and 200 mm. The compression ratio is 10 to 1. The volumetric efficiency is 0.87. The engine operates with 25 percent excess air. Calculate the brake relative efficiency for the engine.

4.15 A two-cylinder, two-stroke-cycle S-I engine employs crankcase induction of the charge. The engine bore and stroke are, respectively, 160 mm and 190 mm. The volumetric efficiency for crankcase induction is 0.85 at the operating speed of 900 rpm. The charge (a/f = 15:1) is delivered by the carburetor to the crankcase. The fuel is octane. Five percent of the charge that enters the cylinders escapes unburned during cylinder scavenging. The engine compression ratio is 8 to 1. The internal relative efficiency is 0.72. The ambient conditions are 1 atm and 16 C. Determine the engine internal power.

4.16 The bore and stroke for an eight-cylinder, four-stroke cycle S-I engine are, respectively, 200 mm and 225 mm. The volumetric efficiency is 0.82 at an operating speed of 2100 rpm. At this operating point the air/fuel ratio is 11.75 to 1. The specific gravity of the gasoline is 0.745. Calculate for a simple carburetor the diameter of the Venturi tube throat and the diameter of the fuel nozzle orifice. The discharge coefficients for the carburetor are: Venturi tube 0.80 and fuel nozzle orifice 0.75. The prescribed pressure at the Venturi tube throat is − 120 cm water. The atmospheric pressure and temperature are 100 kPa, abs, and 18 C.

4.17 The exhaust gas from a C-I engine is directed to a steam plant for secondary power generation. The gas enters the boiler at 282 C and leaves at 175 C. The mean specific heat of the gas is 1.098 kJ/kg·K. The shaft output of the engine is 5000 kW. The fuel consumption is 1260 kg/h. The exhaust gas flow is 64,300 kg/h. The lower heating value of the liquid fuel is 44,100 kJ/kg. The boiler discharges dry, saturated steam at a pressure of 450 kPa, abs. The turbine operates condensing at an exhaust pressure of 10 kPa, abs. The turbine overall efficiency is 0.88. Determine the power output and thermal efficiency of the combined C-I engine/steam power plant.

4.18 A C-I engine test reports the following dry gas analysis, in percent by volume, of the exhaust gas: CO_2 12.7, CO 3.5, H_2 1.4, CH_4 0.4, O_2 0.3, N_2 81.7. The fuel is represented by $C_{13}H_{26}$. The enthalpy of combustion for the fuel is 43,888 kJ/kg. The compression ratio for the engine is 14 to 1. Determine the energy loss and the loss in available energy because of incomplete combustion.

4.19 Investigate the effect of early ignition and combustion, at 25° before the top-center position of the piston, for a S-I engine that has a 150-mm bore and 180-mm stroke. The length of the connecting rod is 378 mm. The compression ratio is 10 to 1. The equivalence ratio for the fuel–air mixture is 0.9.

4.20 The bore and stroke for a four-stroke-cycle C-I engine are, respectively, 300 mm and 365 mm. The compression ratio is 16 to 1. Air enters the cylinder at 27 C, and the exhaust gas temperature is 500 C. At the end of the exhaust stroke, the pressure in the cylinder is + 10 kPa, gage. The intake valve closes when the piston travels 8 percent of the stroke from the bottom-center position. The pressure and temperature are − 5 kPa, gage, and 62 C at the time the intake valve closes. The atmospheric pressure is 98 kPa, abs. For the exhaust gas, c_p = 1.175 kJ/kg·K and R = 0.292 kJ/kg·K. Estimate the volumetric efficiency for the engine and the heat transfer during intake.

4.21 A C-I engine operates with a compression ratio of 20 to 1. The equivalence ratio is 1.0. The heat loss associated with the combustion process is 8 kJ per 100 kJ energy supplied to the engine. Determine the corresponding loss in available energy.

4.22 The shaft output for a C-I engine is 1500 kW. The lower heating value of the fuel ($C_{13}H_{26}$) is 43,888 kJ/kg. The brake thermal efficiency of the engine is 0.365. The fuel and air are supplied to the engine at 25 C. The exhaust gas analysis is given in Problem 4.18. The exhaust gas temperature is 727 C. Calculate the heat transfer from the engine, in kJ/kg fuel, and the percentage distribution of the energy in the fuel. (*Note:* For methane at 1000 K, $\Delta \bar{h}^\circ$ = 38,179 kJ/kgmol.)

4.23 An eight-cylinder, two-stroke-cycle C-I engine operates at 750 rpm. The bore and stroke are, respectively, 350 mm and 425 mm. The compression ratio is 14 to 1. The air/fuel ratio is 18 to 1, and the brake thermal efficiency is 0.35. The lower heating value of the fuel is 43,500 kJ/kg. The blower discharge pressure is 36 kPa, gage. The overall efficiency of the rotary lobe blower is 0.68. The free air delivered by the blower is 142 percent of the maximum cylinder volume. The atmospheric pressure and temperature are 98 psia and 20 C. Calculate the shaft power of the engine and the blower power.

4.24 The compression ratio for a six-cylinder, four-stroke-cycle supercharged C-I engine is 15 to 1. The bore and stroke are, respectively, 130 mm and 150 mm. The internal thermal efficiency is 0.41, and the air/fuel ratio is 18.5 to 1. The lower heating value of the fuel is 44,000 kJ/kg. The intake and delivery pressures for the rotary lobe blower are, respectively, −5 and +42 kPa, gage. The overall efficiency for the blower is 0.65. The average exhaust pressure is +14 kPa, gage, and the average pressure during intake is +32 kPa, gage. The volumetric efficiency for the engine is 1.08. The mechanical friction for the engine is equivalent to 12.5 percent of the gross internal power. The engine operates at 900 rpm. The atmospheric pressure and temperature are 100 kPa, abs, and 25 C. Determine the shaft output for the engine and the brake fuel rate and thermal efficiency.

4.25 An experimental S-I engine operates on propyl alcohol. The air and fuel enter the engine at 295 K. A stoichiometric air/fuel ratio is maintained. During passage through a section of the intake system, the heat transfer to the charge is 1 kJ/kg mixture, and 17 percent of the fuel is vaporized. Determine the final temperature of the mixture.

REFERENCES

1 Prud'homme, R. K. "Automobile Emissions Abatement and Fuels Policy" *American Scientist*, March–April 1974.

2 *Heat Engineering*, Jan.–Feb. 1971.

3 *Mechanical Engineering*, Feb. 1974, page 51.

4 Wimmer, D. B. and McReynolds, L. A. "Nitrogen Oxides and Engine Combustion," Paper No. 380E, SAE 1961 Summer Meeting, New York.

5 Lichty, L. G. *Combustion Engine Processes*, McGraw-Hill Book Co., New York, 1967.

6 Newhall, H. K. and Starkman, E. S. "Digital Calculations of Engine Cycles," SAE Technical Progress Series No. 7, Paper 6-33G, 1963, pages 38–48.

7 DeVries, G. "Low-Pollution Automobile Engine," ASME Paper No. 73-ICT-46, Sept. 1973.

8 "Another New Engine Nudges Detroit" *Business Week*, Oct. 24, 1970, page 86.

9 Martini, W. R. "Developments in Stirling Engines," Presented at the ASME Annual Winter Meeting, New York, N.Y., Nov. 1972.

10 Morgan, D. T., Davis, J. P., Newton, C. L., and Stonehocker, V. T. "High Efficiency Diesel/Organic Rankine Cycle Combined Power Plant," ASME Paper No. 75-DGP-13, April 1975.

BIBLIOGRAPHY

Lewis, A. D. *Gas Power Dynamics*, D. Van Nostrand Co., Princeton, N.J., 1962.

Lichty, L. C. *Combustion Engine Processes*, McGraw-Hill Book Co., New York, 1967.

Obert, E. F. *Internal Combustion Engines and Air Pollution*, Harper & Row, Publishers, New York, 1973.

Taylor, C. F. *The Internal-Combustion Engine in Theory and Practice*, Volume 1, The M.I.T. Press, Cambridge, Mass., and John Wiley & Sons, Inc., New York, 1960.

Taylor, C. F. *The Internal-Combustion Engine in Theory and Practice*, Volume 2, The M.I.T. Press, Cambridge Mass., 1968.

CHAPTER FIVE

ROTATING COMPRESSORS

Virtually every energy conversion system requires the movement of one or more fluids in closed or open loops. All thermal power cycles include an integral compression process in which the pressure of the working medium is increased by fluid pumping. The characteristics of the cycle determine whether liquid or gaseous compression of the fluid is effected.

Large volumes of air and other gases are moved by fans and compressors in a multitude of commercial and industrial applications. Pumps are available in a wide range of designs and sizes and are used extensively for moving many different kinds of liquids in systems of widely divergent characteristics and functions.

The power input for moving a fluid is often high, surprisingly so, hence selection of the motivating machine should be directed to achieving maximum economy of operation as well as effective performance. Selection of the appropriate machine for a particular installation is based primarily on the operating characteristics of the machine; however an understanding of the design principles that influence machine construction will prove to be helpful in this endeavor.

Rotating compressors may be classified generally as radial-flow or axial-flow machines, or as positive displacement machines. The radial-flow types are ordinarily designated as centrifugal machines because of the manner in which the fluid pressure rise is, in part, achieved. In the axial-flow machine, on the other hand, there is little radial displacement of the fluid. Positive displacement machines function by decreasing the volume of discrete quantities of fluid that are essentially in continuous flow.

5.1 CENTRIFUGAL COMPRESSORS

The delivery pressure developed in a single-stage centrifugal compressor is usually in the range of 10 to 150 kPa, gage. An exception is noted in some early turbojet engines where the pressure ratio developed by the single-stage centrifugal compressor is on the order of 4 to 1. Multistage centrifugal compressors are capable of developing discharge pressures as high as 1000 kPa, gage. Tests show that the machine efficiency of the centrifugal compressor tends to decrease with an increase in the pressure ratio for the stage.

Figure 5.1 shows the general construction of

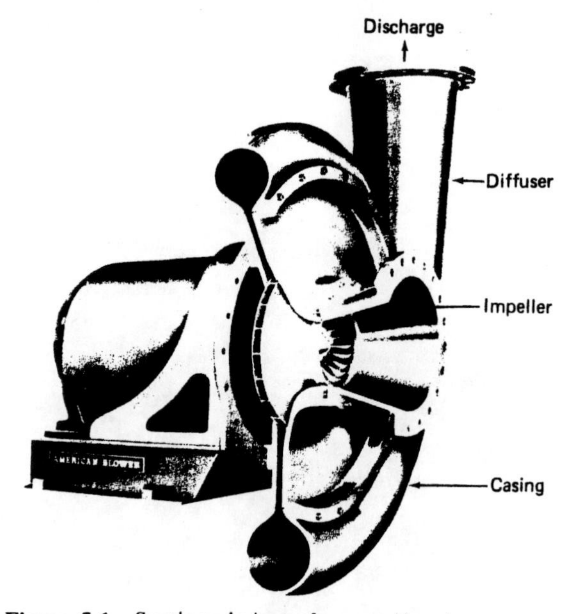

Figure 5.1 Sectioned view of a centrifugal compressor. (Courtesy American-Standard Industrial Products Division.)

a single-stage centrifugal blower. As a result of high-speed rotation of the impeller, air flows to the impeller eye and then through the rotor passages. Because of centrifugal action, compression of the gas occurs in the impeller channels. The working fluid is discharged at a high velocity into the casing that surrounds the impeller. Fluid deceleration within the casing effects an additional increase in pressure.

The machine illustrated in Fig. 5.1 is equipped with a shrouded impeller, while the compressor impeller shown in Fig. 5.2 is nonshrouded. The shroud on the impeller produces a closed rotating channel that may effect an improvement in the compressor efficiency, particularly if the machine operates in the low- or intermediate-efficiency range. The double-inlet design, in which the fluid enters on both sides of the machine, is often used to achieve a reduction in the impeller diameter for a prescribed compressor flow.

Impeller vanes are grouped into forward-curved, radial or straight, and backward-curved types. See Fig. 5.3. The backward-curved vane compressor has generally the most desirable operating characteristics, and the forward-curved vane type, in some respects, the poorest. The characteristics of the different types of impellers are discussed in a later section.

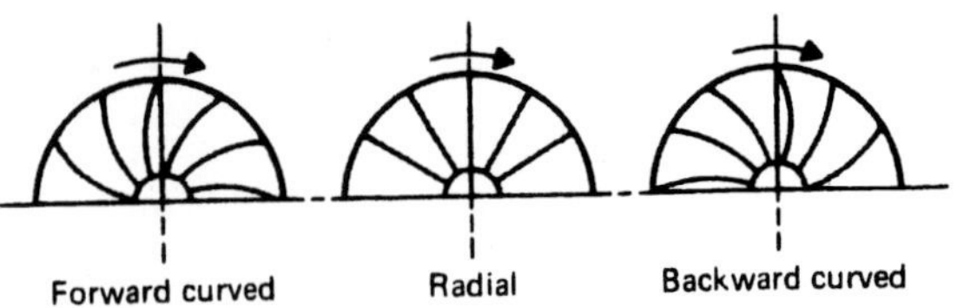

Figure 5.3 Types of centrifugal compressor impeller vanes.

If the pressure rise in a compressor is relatively low, the gas discharged from the impeller may be passed directly into the surrounding casing, or volute, that serves as the collector. When the compressor discharge pressure falls within the high range, diffuser vanes are installed around the impeller in order to achieve a more effective deceleration of the gas prior to delivery to the volute. See Fig. 5.4. The diffuser vanes are a part of the stationary casing.

Fluid losses other than those that occur within the compressor stages are encountered in multistage centrifugal machines because of the turbulent flow of the gas through the interconnecting ductwork installed between stages.

Figure 5.2 Centrifugal compressor impeller. (Courtesy ACE Industries.)

5.2 COMPRESSOR DESIGN AND OPERATION

The design of compressor flow channels and passages is based essentially on the continuity equation, with consideration given to experimentally determined factors and relationships. In this discussion, reference will be made to certain critical

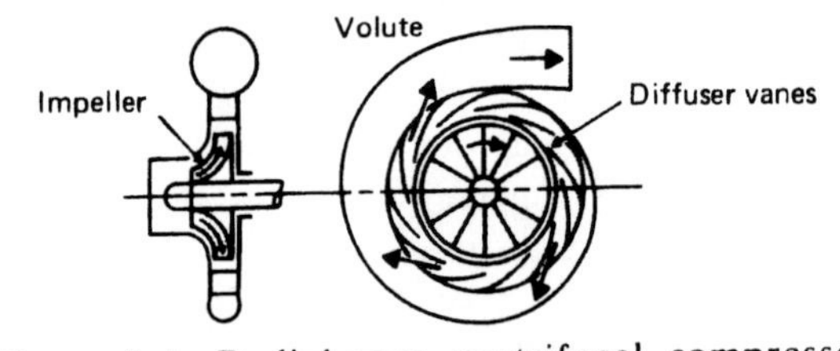

Figure 5.4 Radial-vane centrifugal compressor equipped with diffuser vanes.

concepts that are important to compressor design.

The fluid enters the compressor inlet with an axial velocity C_0 that is also taken as the absolute velocity at the entrance to the impeller vane. See Fig. 5.5. The compressor inlet is relatively large, with maximum diameters approaching values that are approximately two-thirds of the impeller outer diameter. Acceleration of the fluid to the inlet velocity C_0 causes a drop in the static pressure and temperature with respect to the ambient conditions.

The vector diagram shown in Fig. 5.6 demonstrates that for smooth entry of the fluid into the impeller, the lower edge of the vane, rotating in the eye, must be properly curved to meet the inlet relative velocity vector D_0. The curved impeller inlet vanes are shown in Fig. 5.7 for a radial vane impeller. In order to simplify the manufacturing process, these curved vane sections are often formed as a separate element, called the inducer, that is attached to the impeller. The angle β_0 that the inlet vanes make with the impeller face at any radius is determined from the vector diagram. The inlet absolute velocity vector C_0 and the rotor linear velocity vector U_0 are joined to establish the inlet relative velocity vector D_0. U_0 varies directly with the radius, while C_0 is usually assumed to have no radial variation.

In the impeller inlet, the gas is turned for outward flow through the impeller channels. Flow conditions are not uniform across these passages, hence mean conditions are assumed to exist along the channel center line.

The impeller vector diagram is conveniently

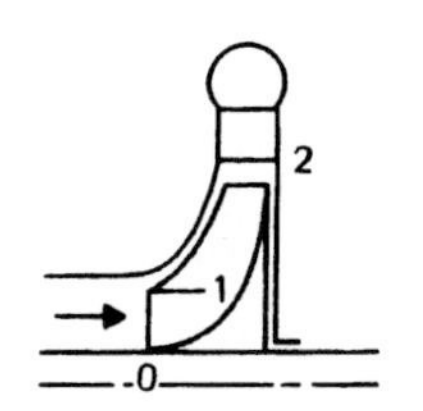

Figure 5.5 Centrifugal compressor gas flow. (0) impeller inlet, (1) entrance to radial flow section, (2) impeller outlet.

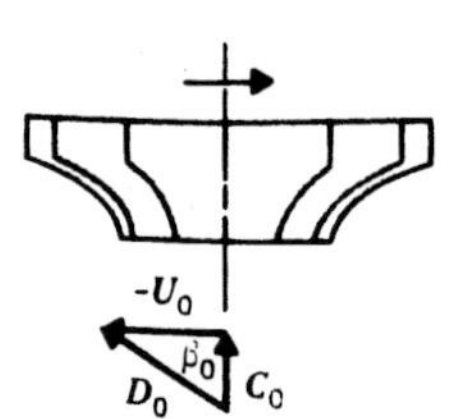

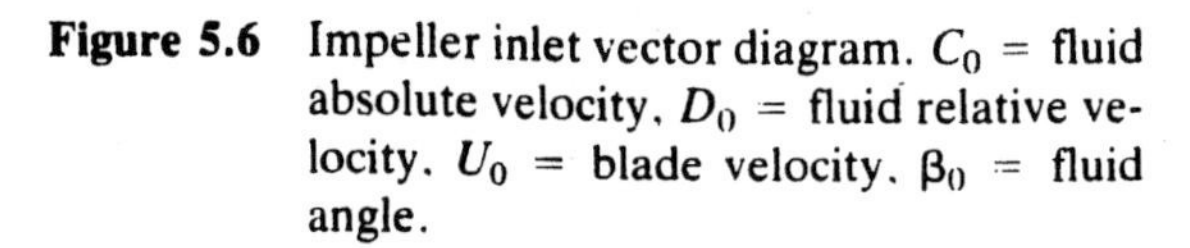

Figure 5.6 Impeller inlet vector diagram. C_0 = fluid absolute velocity, D_0 = fluid relative velocity, U_0 = blade velocity, β_0 = fluid angle.

established on the assumption that the fluid enters the impeller at section 1, Fig. 5.5, with a radial absolute velocity C_1. This assumption does not affect the calculation of the impeller work, because the fluid at both sections 0 and 1 possesses no moment of momentum around the center of rotation.

Figure 5.8 shows the combination of the absolute radial velocity vector C_1 with $-U_1$, the vector that represents the negative linear vane velocity at section 1. The resultant vector represents the fluid inlet relative velocity D_1, while β_1 denotes the inlet fluid angle. The inlet vane angle is normally set equal to the fluid angle β_1 for operation on the design point.

The variation in the relative velocity from D_1 at the inlet section to D_2 at the outlet section is dependent upon channel losses and design criteria. A satisfactory design can be developed by maintaining a constant-flow area through the impeller channel. As the channel width increases from the inlet to the outlet section, the channel

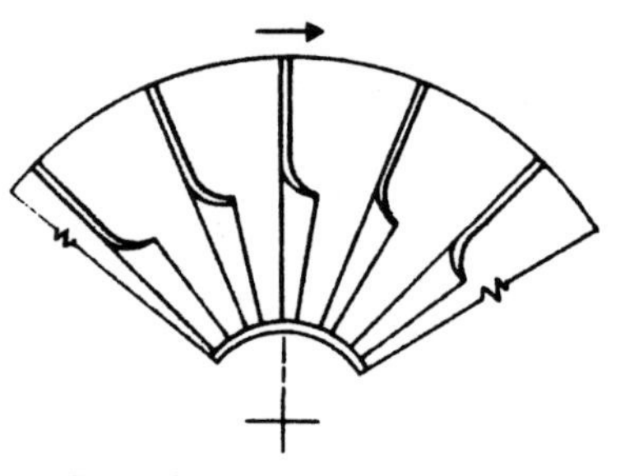

Figure 5.7 Centrifugal compressor impeller, showing curved inlet vanes.

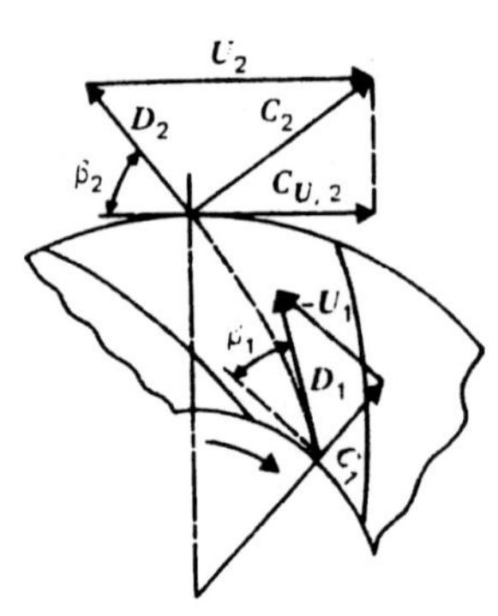

Figure 5.8 Impeller vector diagram. Subscripts 1 and 2 refer, respectively, to the inlet and outlet sections of the flow channel. C = fluid absolute velocity, D = fluid relative velocity, U = blade velocity, β = fluid angle.

depth must decrease in order to maintain a prescribed constant flow area.

The fluid outlet relative velocity vector D_2 is combined with the impeller peripheral velocity vector U_2 to produce the outlet absolute velocity vector C_2. See Fig. 5.8. In passing through the impeller channel, the fluid has accelerated from the absolute velocity C_1 to C_2.

The effect of vane curvature on the fluid absolute velocity at the impeller outlet section is shown in Fig. 5.9. For a prescribed outlet relative velocity and a prescribed impeller tip velocity, the forward-curved vane shows the highest and the backward-curved vane, the lowest outlet absolute velocity C_2.

The fluid leaving the impeller is directed into the stationary diffusing vanes, that is, for those machines equipped with a diffusing ring. The inlet angle of the diffuser vane is determined from the vector C_2 that represents the absolute velocity of the fluid leaving the impeller. See Fig. 5.10. Subsequent deceleration of the fluid in the diffuser vane passages causes an appreciable pressure increase. The diffuser passages discharge into the collector or volute.

Diffuser vanes are designed for smooth entry of the fluid when the compressor is operating under optimum conditions. Off-design operation in certain ranges however can cause sufficient deviation between the inlet fluid angle and the vane angle to affect seriously the diffuser performance. Under these adverse flow conditions, separation of the fluid from the channel wall contributes to a decline in the diffuser efficiency. Some compressors are constructed with movable diffuser vanes that are adjusted to accommodate the off-design flow from the impeller and eliminate, or at least reduce, the deviation angle.

Compressors built without diffuser vanes are capable of operating through wider ranges than those machines that are equipped with a vane-type diffuser. This flexibility achieved in the operation of the compressor is however accompanied by some reduction in the machine efficiency.

Because of the inherent limitations associated with a diffusing flow, the diffuser flow channel must be carefully designed for a very gradual increase in the cross-sectional area. The number of diffuser vanes, 10 to 12, should be held to a minimum that is consistent with meeting the peak efficiency requirement. Impeller designs, on the other hand, usually exhibit a somewhat larger number of vanes, on the order of 12 to 24.

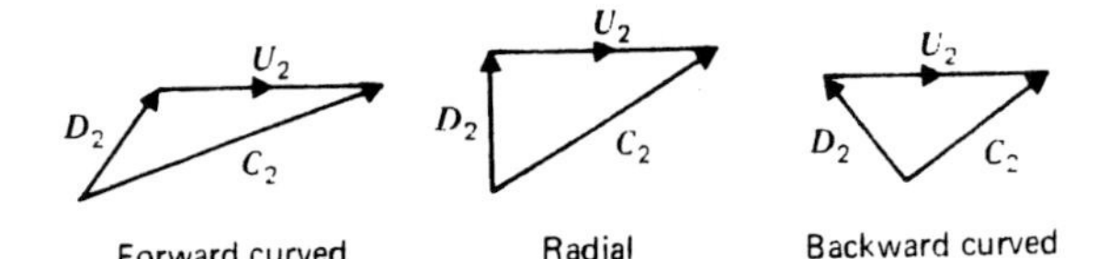

Figure 5.9 Impeller discharge velocity. C = fluid absolute velocity, D = fluid relative velocity, U = blade velocity.

5.3 COMPRESSOR POWER

The work done on the fluid in passing through a compressor is achieved in the rotating impeller. In order to facilitate the analysis of the energy transfer to the fluid, the impeller is assumed to rotate within a control volume that remains fixed in space.

The general case is assumed. The fluid enters the control volume through section 1, at radius

Figure 5.10 Diffuser flow.

r_1, and leaves through section 2, at radius r_2. At the inlet section, the fluid velocity C_1 has three components, namely, the axial component $C_{x,1}$, the radial component $C_{r,1}$, and the rotational component $C_{U,1}$. Similarly, at the outlet section, the three components of C_2 are $C_{x,2}$, $C_{r,2}$, and $C_{U,2}$. In passing through the control volume, the fluid is, for the general case, displaced radially from section 1 to section 2.

Fluid flow rates to and from the control volume are usually determined from the respective axial component C_x of the fluid velocity and the corresponding normal flow area A_x. Thus, $\dot{m}_1 = \rho_1 C_{x,1} A_{x,1}$ and $\dot{m}_2 = \rho_2 C_{x,2} A_{x,2}$. For determination of the compressor work, steady-state, steady-flow conditions are assumed, that is, $\dot{m}_2 = \dot{m}_1 = \dot{m}$, and the rotor turns at constant angular velocity.

The torque T applied to the fluid system by the rotor is equal to the time rate of change of the angular momentum, or of the moment of momentum, the product of the tangential component of momentum mC_U and the radius r. The radial component of momentum has no moment about the center of rotation.

During the time interval $\Delta\tau$, a mass m of the fluid enters and leaves the control volume. The angular momentum of the fluid at the inlet and outlet sections is given, respectively, by $mr_1C_{U,1}$ and $mr_2C_{U,2}$. The change in the angular momentum per unit time is equal to

$$\frac{mr_2C_{U,2} - mr_1C_{U,1}}{\Delta\tau} = \frac{m}{\Delta\tau}(r_2C_{U,2} - r_1C_{U,1})$$

Since $m/\Delta\tau = \dot{m}$, the mass flow rate, the torque is given by

$$T = \dot{m}(r_2C_{U,2} - r_1C_{U,1}) \qquad (5.1)$$

The power, or rate at which work is performed on the fluid, is equal to the product of the angular velocity ω and the torque. Then,

$$\dot{W} = \omega T = 2\pi n T$$

or

$$\dot{W} = \dot{m}2\pi n(r_2C_{U,2} - r_1C_{U,1})$$

But

$$2\pi n r_2 = U_2 \quad \text{and} \quad 2\pi n r_1 = U_1$$

where U_2 and U_1 are the linear velocities of the rotor, respectively, at r_2 and r_1. Then,

$$\dot{W} = \dot{m}(C_{U,2}U_2 - C_{U,1}U_1) \qquad (5.2)$$

In the absence of an inlet whirl component, $C_{U,1} = 0$, and Eq. 5.2 is reduced to

$$\dot{W} = \dot{m}C_{U,2}U_2 \qquad (5.3)$$

Example 5.1

Calculate the power input to the fluid for a centrifugal compressor that discharges 6 kg gas per second. The impeller O.D. is 0.5 m. The speed is 225 rps. The rotational component of the gas leaving the impeller is 300 m/s. The gas has no rotational component at the inlet section.

$$U_2 = \pi(\text{O.D.})n = \pi 0.5 \times 225$$
$$= 353 \text{ m/s}$$

$$\dot{W} = \dot{m}C_{U,2}U_2 = 6 \times 300 \times 353$$
$$= 635{,}000 \text{ W} = 635 \text{ kW}$$

The fluid entering the compressor impeller may be given a prescribed amount of prewhirl by a set of stationary inlet guide vanes. Prewhirl is commonly achieved in the direction of rotation, in part, because of improved flow characteristics at the impeller inlet section. When prerotation is imparted to the fluid, Eq. 5.2 is used for calculating the compressor internal work because the fluid entering the impeller possesses an initial moment of momentum. Prewhirl in the direction of rotation reduces the impeller work, that is, $C_{U,1}U_1$ is a positive quantity.

5.4 COMPRESSOR MACHINE EFFICIENCY

The compressor internal efficiency η_C is defined as the ratio of the reversible adiabatic enthalpy increase to the actual increase in the enthalpy of the fluid for operation between prescribed pressure limits. See Fig. 5.11.

$$\eta_C = \frac{h_{3,s} - h_1}{h_3 - h_1} \tag{5.4}$$

This definition of the compressor machine or internal efficiency is generally acceptable because the heat transfer can ordinarily be neglected and the compressor inlet and outlet fluid velocities do not differ by a significant amount. Stagnation enthalpy values however can be used when such use is deemed appropriate. Section 3 is taken at the compressor outlet, hence $h_3 - h_1$ represents the overall increase in the enthalpy of the fluid achieved in passing through the compressor.

Centrifugal compressors can presently be designed to develop a pressure ratio of about 4 to 1 to 6 to 1, with a machine efficiency of about 0.80. Furthermore, a comparatively wide operating range can be achieved by these machines.

The mechanical losses, caused by bearing friction, are low for a centrifugal compressor; however these losses should be considered in the determination of the shaft power. Mechanical efficiency values for the centrifugal compressor are in the range of 0.98 to 0.99.

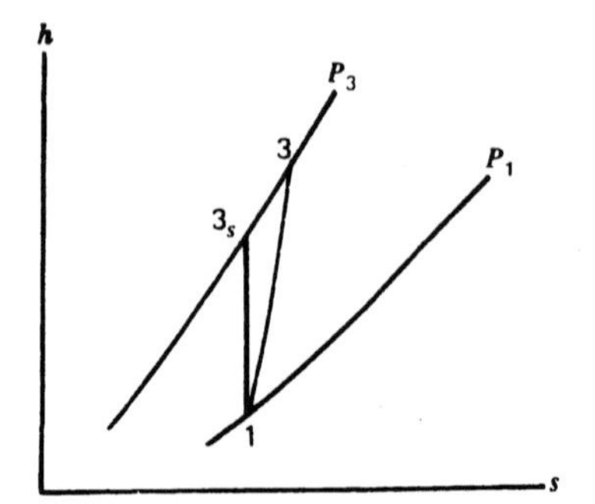

Figure 5.11 Centrifugal compressor overall pressure increase.

Example 5.2

A centrifugal compressor discharges air at 443 K. The inlet air temperature is 300 K. The compressor internal efficiency is 0.72; $k = 1.40$. Calculate the pressure ratio for the machine.

Applying Eq. 5.4,

$$\eta_C = \frac{h_{3,s} - h_1}{h_3 - h_1} = \frac{T_{3,s} - T_1}{T_3 - T_1}$$

$$0.72 = \frac{T_{3,s} - 300}{443 - 300}$$

$$T_{3,s} = 403 \text{ K}$$

$$\frac{P_3}{P_1} = \left(\frac{T_{3,s}}{T_1}\right)^{k/(k-1)}$$

$$= \left(\frac{403}{300}\right)^{3.5} = 2.8$$

5.5 COMPRESSOR PRESSURE RATIO

The pressure of the fluid flowing through the compressor increases both in the rotating impeller and in the stationary diffuser system. The flow processes that cause the pressure to increase in these two parts of the machine are quite different, as noted below. The pressure rise for a radial vane compressor is theoretically about equal for the impeller and the diffuser. For the forward-curved vane compressor, the pressure rise is low in the impeller and high in the diffuser, while the opposite distribution holds true for the backward-curved vane compressor.

The fluid passing through the impeller of the centrifugal compressor is subjected to conditions approaching those of solid-body rotation. Because of the unbalanced centrifugal force acting on an element of mass, a relatively substantial pressure increase is developed in the rotating impeller where little or no provision is made for diffusion.

The theoretical pressure increase in the impeller is based on the assumption that the fluid is subjected to solid-body rotation, $U = r \times$ constant. It can be shown that this pressure increase is given by*

$$P_2 - P_1 = \rho \frac{U_2^2 - U_1^2}{2}$$

Because of turbulence and frictional effects, the actual pressure increase in the impeller is less than the theoretical increase.

The irreversible flow of the fluid through the diffuser is represented by process 2–3 in Fig. 5.12. The pressure increase $(P_3 - P_2)$ in the diffuser is accomplished by decelerating the fluid through an increase in the flow area, thus, $C_3 < C_2$. Diffusers are usually assumed to function in the absence of heat transfer. Then, $h_3 - h_2 = (C_2^2 - C_3^2)/2$.

The diffuser efficiency is defined as the ratio of the reversible adiabatic enthalpy increase to the actual increase in the enthalpy of the fluid for operation between prescribed pressure limits as shown in Fig. 5.12. Then,

$$\eta_D = \frac{h_{3,s} - h_2}{h_3 - h_2} \tag{5.5}$$

The pressure increase is evaluated through use of the reversible adiabatic enthalpy change of the fluid, $h_{3,s} - h_2$.

The pressure increase achieved in the compressor may be estimated from the wheel speed and certain experimentaly determined coefficients. The whirl component $C_{U,2}$ at the impeller outlet section of a radial vane compressor is theoretically equal to the peripheral linear velocity of the impeller. Actually, because of slip, $C_{U,2} < U_2$, hence $C_{U,2} = \sigma U_2$, where σ is the slip factor. As the number of impeller vanes becomes very large, the value of σ approaches 1 as a limit. An increase in the number of impeller vanes however causes an increase in fluid friction and a consequent reduction in the compressor efficiency. The number of impeller vanes, usually

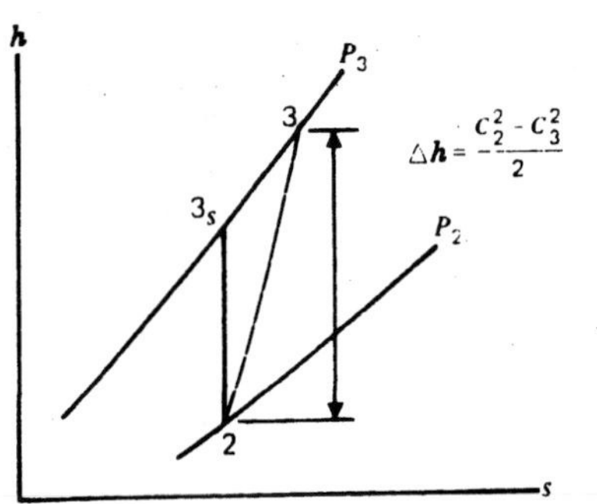

Figure 5.12 Diffuser pressure increase.

* The fluid contained in the compressor impeller is assumed to be in the form of a hollow cylinder that has the radial dimensions r_1 and r_2, respectively, the inner and outer radii. The length of the cylinder in the axial direction is designated by ℓ.

A differential element of fluid at radius r has a radial dimension dr. The mass of the element is

$$dm = \rho\, dV = \rho(2\pi r \ell\, dr)$$

where ρ is the density of the fluid.

The normal component of the resultant of the forces acting on the element, because of rotation, is equal to $mr\omega^2$, where ω is the constant angular velocity of the impeller, or

$$dF_r = dm\, r\omega^2$$

and the radial pressure on the fluid element is

$$dP = \frac{dF_r}{A} = \frac{dm\, r\omega^2}{2\pi r \ell}$$

$$= \frac{\rho(2\pi r \ell\, dr) r\omega^2}{2\pi r \ell} = \rho r\omega^2\, dr$$

$$P_2 - P_1 = \int_1^2 dP = \rho\omega^2 \int_1^2 r\, dr$$

$$= \rho\omega^2 \frac{r_2^2 - r_1^2}{2}$$

Because $\omega^2 r_2^2 = U_2^2$ and $\omega^2 r_1^2 = U_1^2$,

$$P_2 - P_1 = \rho \frac{U_2^2 - U_1^2}{2}$$

12 to 24, is selected from a compromise of the pertinent design factors. For other than radial vane compressors, the theoretical relationship between $C_{U,2}$ and U_2 is established from the probable vector diagram. For most centrifugal compressor impellers, the value of σ is about 0.9.

Because of skin friction, heat losses, and leakage around the vanes, the power input factor z is included in the estimation of the actual compressor work. For a radial vane compressor, Eq. 5.3 can then be modified to read

$$W = z\sigma U^2 \tag{5.6}$$

The power input factor may be considered as a reciprocal of an efficiency term that accounts for the several losses that were noted directly above. The value of z will ordinarily be within the range of 1.03 to 1.04.

Because of the mechanical losses, the compressor shaft work is somewhat higher than the internal work determined from Eq. 5.6. The mechanical efficiency of the centrifugal compressor is relatively high, that is, about 0.98+. See Section 5.4.

The working fluid leaves the impeller at section 2 and enters the diffuser and collector. At the compressor outlet, section 3, the velocity C_3 is substantially below the fluid velocity at section 2. The energy equation written for the overall compressor is

$$h_3 - h_1 = -W$$

when the heat loss can be neglected, and the fluid velocities at the compressor inlet and outlet sections are assumed to be equal. Then,

$$T_3 - T_1 = \frac{-W}{c_p}$$

A mean value of c_p is usually sufficiently precise for this calculation.

Because the work is transferred from the impeller to the fluid, a negative value of $z\sigma U^2$ is substituted into the energy equation. Then,

$$T_3 - T_1 = \frac{z\sigma U^2}{c_p} \tag{5.7}$$

It is essential to recall that in the development of Eq. 5.6, $C_{U,1}$ is assumed to be equal to zero. Actually, when prewhirl is imparted to the fluid, the value of $C_{U,1}U_1$ is ordinarily much smaller than the value of $C_{U,2}U_2$.

Example 5.3

The tip speed for a radial vane compressor impeller is 460 m/s. $T_1 = 290$ K; $z = 1.035$; $\sigma = 0.90$; $\eta_C = 0.75$. For the gas, $k = 1.400$ and $c_p = 1.004$ kJ/kg·K. Calculate the compressor pressure ratio.

$$h_3 - h_1 = z\sigma U^2 = 1.035 \times 0.90(460)^2$$
$$= 197{,}100 \text{ J/kg}$$
$$h_{3,s} - h_1 = \eta_C(h_3 - h_1)$$
$$= 0.75 \times 197.10 = 147.83 \text{ kJ/kg}$$
$$T_{3,s} - T_1 = \frac{h_{3,s} - h_1}{c_p}$$
$$= \frac{147.83}{1.004} = 147.24 \text{ K}$$
$$T_{3,s} = T_1 + (T_{3,s} - T_1)$$
$$= 290.0 + 147.24 = 437.24 \text{ K}$$
$$\frac{P_3}{P_1} = \left(\frac{T_{3,s}}{T_1}\right)^{k/(k-1)}$$
$$= \left(\frac{437.24}{290.0}\right)^{3.5} = 4.22$$

Equation 5.7 is combined with the isentropic pressure–temperature relationship to yield

$$\frac{P_3}{P_1} = \left(\frac{\eta_C z\sigma U^2}{T_1 c_p} + 1\right)^{k/(k-1)} \tag{5.8}$$

Equation 5.8 may be used to estimate the overall pressure ratio developed by a centrifugal compressor. The equation shows the dependence of the compression pressure ratio upon the impeller speed.

Example 5.4

Estimate the impeller speed of a radial vane centrifugal compressor that is required to develop a pressure ratio of 3.0. The impeller O.D. is 0.40 m; $T_1 = 285$ K; $\sigma = 0.90$; $z = 1.032$; $\eta_C = 0.78$. For the gas, $k = 1.667$ and $c_p = 0.520$ kJ/kg·K.

$$\frac{P_3}{P_1} = \left(\frac{\eta_C z \sigma U^2}{T_1 c_p} + 1\right)^{k/(k-1)}$$

$$3.0 = \left(\frac{0.78 \times 1.032 \times 0.90U^2}{285 \times 0.520 \times 10^3} + 1\right)^{2.499}$$

$$U = 336 \text{ m/s}$$

$$\pi(\text{O.D.})n = U \qquad \text{or} \qquad 0.40\pi n = 336$$

$$n = 267 \text{ rps} \qquad N = 16{,}020 \text{ rpm}$$

5.6 COMPRESSOR PERFORMANCE

Figure 5.13 shows a typical centrifugal compressor performance chart. The region to the left of the surge line is an area of unstable operation. The frequency and intensity of the pulsations that develop in a surging flow are dependent upon a number of factors, including the slope of the pressure–flow curve, the quantity of fluid delivered, and the system characteristics. Surging or pulsating flow is undesirable, not only with respect to the resulting unstable system characteristics but also from the standpoint of the severe structural forces that may develop, particularly at high speeds.

The constant-speed pressure–flow curves are in part relatively flat. For a designated speed, acceptable compressor performance can be achieved, provided that the variation in the flow is sufficiently restricted. In each case, constant-speed operation is limited at low flow rates by surging, and at high flow rates by choking.

An examination of Fig. 5.13 discloses the interdependence of the operating speed, flow rate, and pressure ratio of the centrifugal compressor. It is important to note that the machine efficiency of the centrifugal compressor generally declines as the operating point moves to the right of the surge line with an increase in the flow rate.

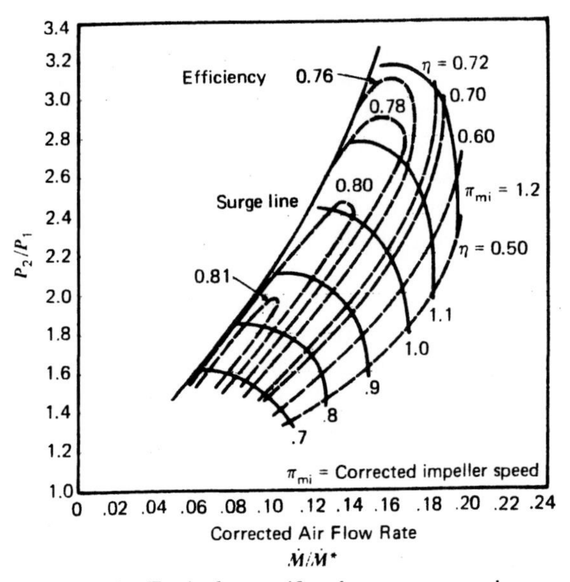

Figure 5.13 Typical centrifugal compressor characteristics. (Courtesy M.I.T. Gas Turbine and Plasma Dynamics Laboratory.)

5.7 POSITIVE DISPLACEMENT COMPRESSORS

The delivery pressure for a rotating positive displacement compressor is independent of speed. Compressor flow is equivalent to the product of the displacement and the volumetric efficiency. The displacement is directly proportional to the speed, while the variation in the volumetric efficiency with the speed is dependent upon the compressor design. Any delivery pressure up to the design limit however may be achieved for a given speed.

The vane-type rotary compressor, a positive displacement machine, is provided with an eccentric impeller equipped with a number of sliding vanes. See Fig. 5.14. Contact with the casing causes the vanes to slide radially in the impeller; or the vanes are operated by an eccentric mechanism, thus contact with the casing is avoided and lubrication is not required. In either case,

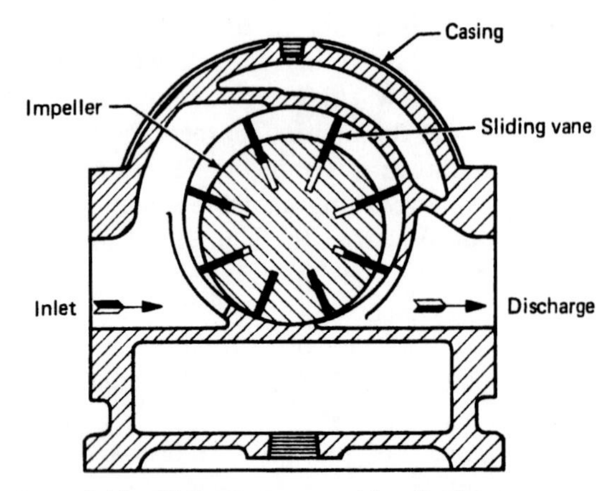

Figure 5.14 Sliding-vane, positive-displacement-type compressor. (Courtesy Allis-Chalmers.)

the inertia of the vanes places an upper limit on the compressor speed. Compression is effected by a reduction in the volume of the gas trapped between adjacent vanes and the casing.

The Lysholm compressor is a lobe-type machine designed to overcome the disadvantages of the conventional rotary compressors. See Fig. 5.15. The advantages of high efficiency and stability of performance normally associated with reciprocating compressors are introduced into the high-speed rotary field by this machine. Two helical rotors transport the fluid from the inlet to the outlet end of the machine, while simultaneously effecting compression of a continuous stream of individual charges. Because the rotors are driven through timing gears, metallic contact is avoided, and the need for lubrication is eliminated, a distinct advantage in many applications.

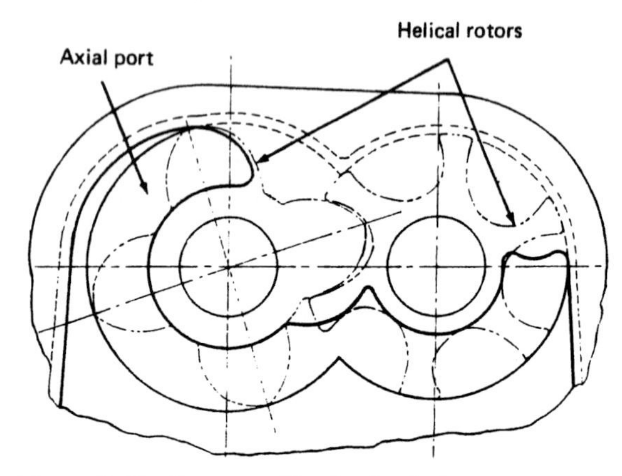

Figure 5.15 Rotor profiles for Lysholm compressor. (W. A. Wilson and J. W. Crocker, *Mechanical Engineering*, June 1946.)

Acceptable machine efficiencies for the Lysholm compressor extend over a wide range of operating conditions defined by the pressure ratio and the flow. Volumetric efficiency curves are flat and indicate that delivery is a function of speed and independent of the pressure ratio. Surging or similar conditions of instability are not present.

Figure 5.16 illustrates a rotary lobe-type blower. This blower, a positive displacement machine, is widely used for industrial applications in moving air and other gases, including the supercharging of internal combustion engines. Single-stage blowers develop delivery pressures ranging from 70 to 100 kPa, gage. For pressures in the range of 100 to 200 kPa, gage, compression is achieved in compound machines.

In Fig. 5.16, the gas enters at the bottom, and discharge takes place at the top of the casing. The two impellers rotate in opposite directions through a gear drive. Clearance between the impellers and the housing is very small. The left-hand impeller rotates in a clockwise direction. As the rotor turns, the inducted fluid is trapped between the impeller and the housing, and further rotation of the impeller causes back flow from the discharge line and compression of the trapped fluid to the discharge pressure.

An analysis of the mechanical cycle for the rotary lobe blower shows that the theoretical work W_C is equal to the difference between the work of the intake and discharge displacement processes, respectively, 0–1 and 2–3. Thus, W_C is equivalent to the area 01230, Fig. 5.17, or

$$W_C = V_1(P_1 - P_2) \tag{5.9}$$

Reversible adiabatic compression of the fluid is described by the cycle 012_s30, Fig. 5.17. The-

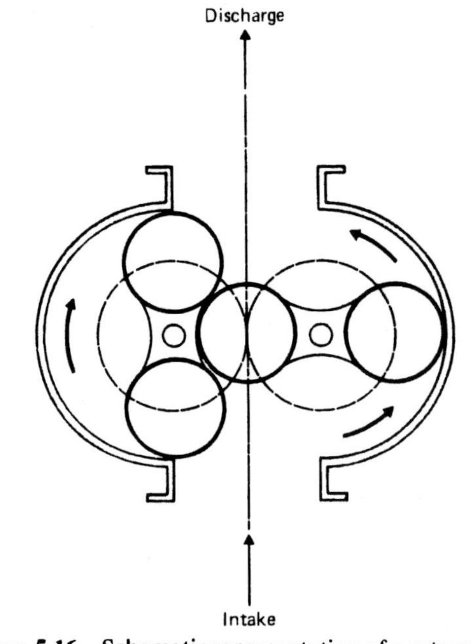

Figure 5.16 Schematic representation of a rotary lobe-type blower.

oretical work of compression for the rotary lobe blower is thus somewhat higher than the work required for the more conventional reversible adiabatic compression. The excess work, equivalent to the area 122_s1, is the result of the reversed flow of a portion of the compressed gas, which causes a substantial increase in the displacement volume at the discharge pressure.

The internal efficiency of the blower is defined as the ratio of the reversible adiabatic work to

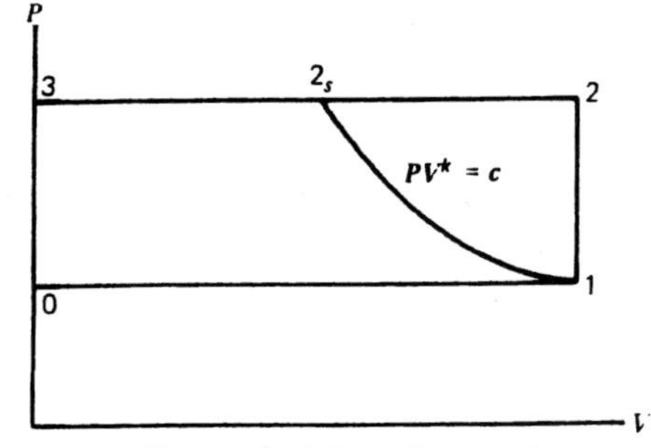

Figure 5.17 Theoretical P–V diagram for rotary lobe blower.

the actual internal work for operation between prescribed pressure limits. In general, the efficiency of the rotary lobe blower is somewhat below the machine efficiency of centrifugal compressors employed for the same class of service. Compressor efficiency values for rotary lobe blowers will fall within the range of 0.5 to 0.7 – .

Because the rotating elements of the rotary lobe blower are not in contact, internal sealing and lubrication are not required. The fluid consequently is not contaminated with a lubricant on passage through the machine. Other advantages of the rotary lobe blower are high volumetric efficiency and low mechanical friction. In comparison with other types of compressors, the rotary lobe blower is somewhat noisy.

5.8 FANS

Extensive use is made of various types of fans for moving air and other gases. In comparison with centrifugal blowers, fans develop relatively low discharge pressures. The range in the static delivery pressure extends from values of less than 1 cm to approximately 90 cm water.

The axial-flow fan, either the propeller or vane type, is used for moving comparatively large volumes of gas at low delivery pressures. The propeller-type fan is simple with respect to the design and construction in comparison with the more complex vane-type fan; however the propeller-type fan operates at a lower efficiency. Static efficiency values for vane-axial fans range from 50 to 65 percent, and for propeller-type fans, from 20 to 40 percent. ***Note:*** Static efficiency is defined under Eq. 5.10.

The principles of the centrifugal fan as related to operation and performance are similar to those previously examined for the centrifugal compressor. Because of the low pressure developed in a fan, light-weight construction is normally adequate for both the rotor and the casing. Rotors are usually "built-up" assemblies, and casings are designed for either a single or a double inlet.

The scroll-shaped housing of the centrifugal fan provides for deceleration, with a simultaneous increase in the static pressure, of the high-velocity fluid leaving the rotor. See Fig. 5.18. Fans commonly operate with an atmospheric inlet or outlet pressure. Various other pressure ranges however are frequently encountered, depending upon the location of the fan in the ductwork.

The shaft power required to drive a fan is determined from

$$\dot{W}_b = \frac{\dot{m} g \, \Delta H_s}{\eta_F} \tag{5.10}$$

where

$\dot{m}$ = mass flow rate of the gas, kg/s
g = acceleration owing to gravity (g_0 = 9.807 m/s^2)
ΔH_s = static head developed by the fan, m fluid
η_F = fan static efficiency

The quantity $\dot{m}g \, \Delta H_s$ is designated as the fluid power.

The term fluid power ($\dot{m}g \, \Delta H_s$) in Eq. 5.10 is determined by employing the static head change ΔH_s developed by the fan. Thus, the fan efficiency that relates the shaft power $\dot{W}_b$ to the fluid power is designated as the static efficiency.

Usually, the static head and the corresponding static efficiency are employed to determine the fan shaft power. Occasionally, in calculating the shaft power, use is made of the total head developed by the fan and the total head efficiency, customarily designated as the mechanical efficiency $(\eta_m)_F$. Then,

$$\dot{W}_b = \frac{\dot{m} g \, \Delta H_t}{(\eta_m)_F} \tag{5.10a}$$

With respect to Eq. 10a, the fluid power is equal to the quantity $\dot{m}g \, \Delta H_t$, where ΔH_t is the total head developed by the fan. The total head H_t at a particular point in the fluid stream is equal to the sum of the pressure and velocity heads.

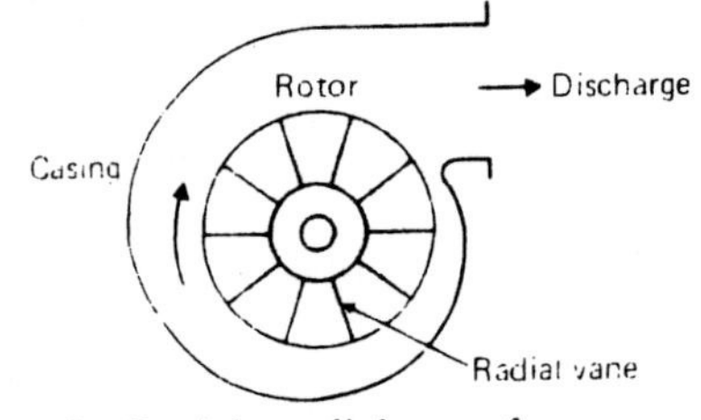

Figure 5.18 Straight radial vane fan.

Example 5.5

A fan discharges air at a static delivery pressure of 5 cm water. The fan inlet is open to the atmosphere. The mean velocity of the air in the outlet duct is 10 m/s. The diameter of the outlet duct is 45 cm. The ambient air pressure and temperature are 100 kPa, abs, and 27 C, respectively. The fan static efficiency is 0.65. Determine the fan shaft power.

$$(\Delta H_s)_w = 5 \text{ cm water} = 0.490 \text{ kPa}$$

$$\text{Discharge pressure} = P_a + \Delta P = 100.0 + 0.49 = 100.49 \text{ kPa, abs}$$

$$\rho_a = \frac{P}{RT} = \frac{100.49}{0.287 \times 300} = 1.167 \text{ kg/m}^3 \quad \text{(at the fan outlet)}$$

$$\text{Duct transverse area} = 0.1590 \text{ m}^2$$

$$\dot{m} = \rho C A = 1.167 \times 10 \times 0.1590 = 1.856 \text{ kg/s}$$

$$(\Delta H_s)_{air} = \frac{\rho_w}{\rho_a}(\Delta H_s)_w = \frac{1000}{1.167} 0.05 = 42.84 \text{ m air}$$

$$\dot{W}_b = \frac{\dot{m} g \, \Delta H_s}{\eta_F} = \frac{1.856 \times 9.807 \times 42.84}{0.65} = 1200 \text{ W} = 1.20 \text{ kW}$$

The performance characteristics of a centrifugal-type fan are dependent upon the shape of the rotor vanes, which in general conform to three basic designs. The rotors are fitted with backward-curved, radial, or forward-curved vanes, although modifications of these basic shapes can be employed to some advantage.

Selection of a fan is usually accomplished by matching the operating characteristics of the fan with the system requirements. The fan must be capable of moving the fluid effectively and efficiently under varying load conditions. In particular, the operating economy of the fan will be adversely affected when the discharge pressure developed by the fan is in excess of the pressure that is required to offset the system resistance.

Figure 5.19 shows typical constant-speed performance curves for a fan equipped with curved blades that are backward inclined. Fans of this type have a steep pressure gradient characteristic and are suitable for high-speed operation at high efficiencies. Because of the backward curve to the vane, the velocity of the fluid leaving the rotor is relatively low, thus permitting high-speed operation. Because the total-pressure curve falls off sharply with increased flow, the power requirements reach a maximum generally at about 65 to 80 percent of the wide-open volume. This characteristic is referred to as nonoverloading, that is, when all resistance to flow is removed, there is no tendency to build up an excessive power demand. The static efficiency for the backward-curved vane fan is in the range of 60 to 80 percent. Efficiency values given for fans are typically the maximum values.

Figure 5.20 illustrates the performance curves for five different types of centrifugal fans. The fan equipped with backward-inclined blades operates with a comparatively high static efficiency. Operation is stable with a nonoverloading power characteristic. The blades have a low resistance to abrasion and are not self-cleaning.

Fans equipped with forward-curved blades operate at moderate speeds and deliver a high output. Static efficiencies are in a medium range, somewhat lower than the efficiencies achieved by fans equipped with backward-inclined vanes. Because the total pressure does not drop rapidly

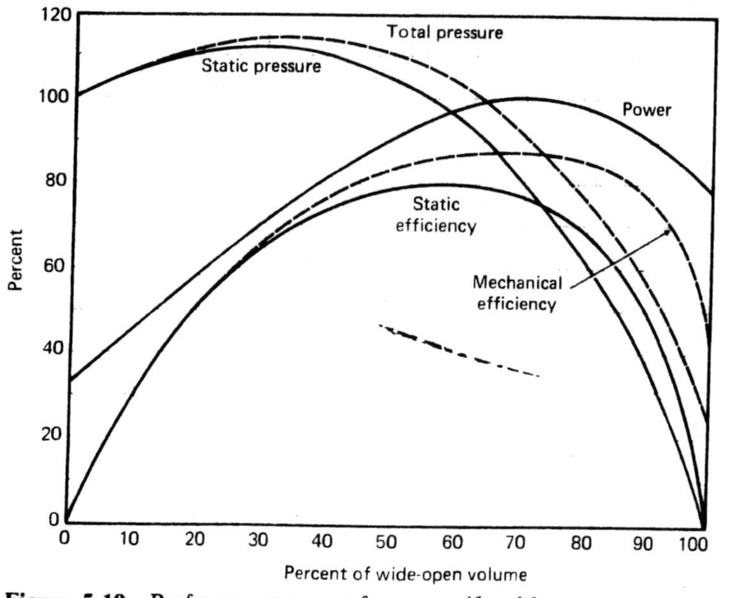

Figure 5.19 Performance curves for a centrifugal fan equipped with backward-curved blades.

with increased flow, the power requirement continues to increase, thus producing an overloading characteristic. The rotor for this type of fan is constructed with a large number of blades. The fan has fair resistance to abrasion. Operation of the fan is unstable if improperly applied. The inherent instability, just below the point of maximum static efficiency, is shown by the dip in the static-pressure curve.

Operating speeds for the straight-radial vane fan are relatively low. The small number of vanes installed on the rotor is typical of this type of fan. Because of good resistance to abrasion, the fan is suitable for severe service, such as handling gases laden with abrasive dust. The power requirement produces an overloading characteristic. Static efficiency values are in a medium range.

The fan equipped with radial-tip blades has excellent abrasion resistance. The fan operates with a comparatively high static efficiency. Operation is very stable, and the rotor is self-cleaning to a high degree. The power requirement tends to "level off" at a high-volume flow.

Airfoil design applied to the blades produces a fan of superior operating characteristics; particularly, a high static efficiency is achieved. Because the airfoil-designed blades are backward curved, the fan has a nonoverloading power characteristic. Operation of the fan is stable, and the noise level is low. The airfoil fan is especially suitable for high-speed applications.

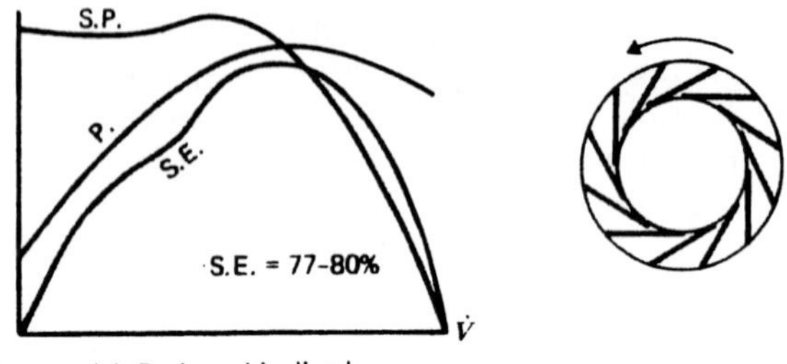

(*a*) Backward inclined

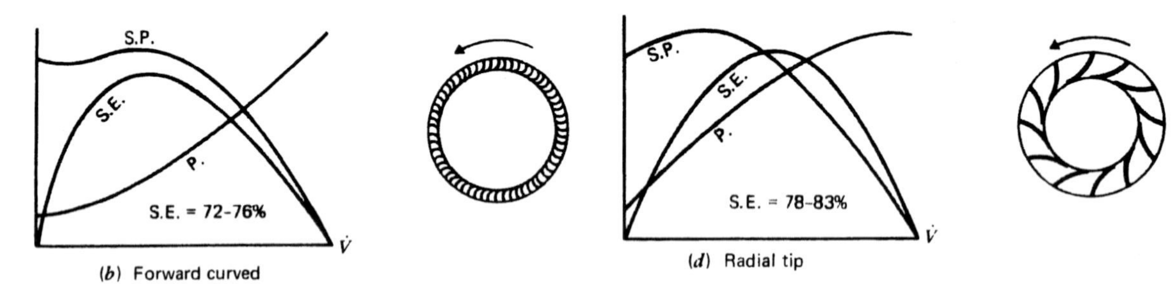

(*b*) Forward curved

(*d*) Radial tip

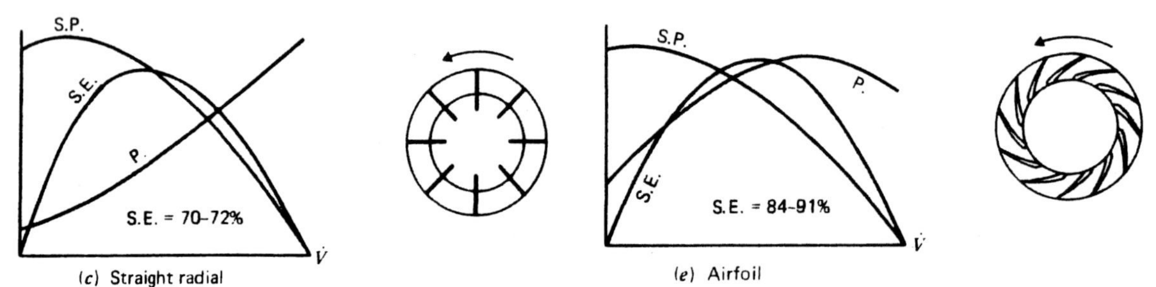

(*c*) Straight radial

(*e*) Airfoil

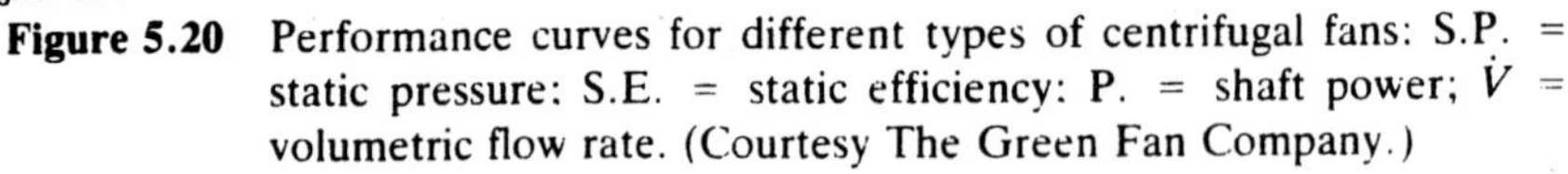

Figure 5.20 Performance curves for different types of centrifugal fans: S.P. = static pressure; S.E. = static efficiency; P. = shaft power; $\dot{V}$ = volumetric flow rate. (Courtesy The Green Fan Company.)

In many installations, the quantity of gas handled by the fan must be varied in accordance with the process or system requirements. Thus, some kind of fan control is essential. The customary types of fan drives are ac and dc electric motors and steam turbines.

The constant-speed ac motor is the most common, and in several respects is the most economical industrial power drive. Fans driven by constant-speed ac motors may be provided with adjustable outlet dampers for control of the gas flow. The static pressure developed by the fan at 100 percent of the rated volume, with the damper wide open, balances exactly the system pressure. See Fig. 5.21. At reduced flow, for example, 70 percent of the rated volume, the damper is adjusted to provide the static pressure that is required to balance the respective system pressure.

The outlet damper is actually a variable resistance to the flow from the fan. An increase in this resistance, at constant fan speed, is accompanied by an increase in the pressure differential across the damper. Thus, at 70 percent of the rated flow, the fan outlet pressure is, for this example, 117 percent of the rated static pressure, while the required pressure on the downstream side of the damper is 49 percent of the rated static pressure. The excess pressure developed by the fan indicates that a considerable amount of power must be supplied above the useful requirement.

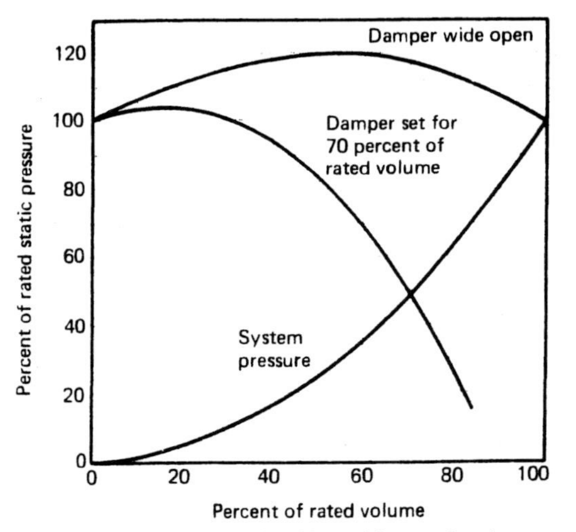

Figure 5.21 Fan control achieved by outlet damper.

Another scheme used for flow control of a constant-speed fan employs vanes installed at the inlet to the machine. Inlet vanes are segment-shaped, pivoted sections that are used to vary the fan inlet flow area and thereby control the flow. However, in a partially closed position, these vanes spin the fluid in the direction of rotation, and as a result the power and pressure characteristics of the fan are altered. The inlet vanes are adjusted to provide a fan delivery pressure that meets the system requirement, and thus the excess pressure encountered in damper control is avoided. On the basis of the power required to drive the fan, inlet vane control is superior to control achieved by use of adjustable outlet dampers.

Minimum power requirements, over a wide range of operation, are achieved generally with complete speed control. The volume flow varies directly with the speed of the fan, and the developed head varies directly with the square of the speed, provided that the system configuration remains unchanged. From these two relationships, it follows that the fluid power varies directly with the cube of the speed.

Variable-speed operation of a fan can be accomplished by changing the speed of the driving motor or engine, or by placing a variable-speed transmission between the fan and the constant-speed driving motor. With respect to the power input, fan control achieved by speed variation is for the most part superior to inlet vane control.

5.9 PUMPS

Several different types of rotating pumps are used in industrial applications for moving liquids under a wide range of delivery pressures. These pumps are classified as centrifugal or rotary machines. The second group includes the positive-displacement types of pumps, namely, the gear, screw, sliding vane, and swinging vane pumps. Rotary pumps will provide satisfactory service, at a low delivery head, in low- and medium-capacity installations. Centrifugal pumps are designed for operation over a wide range of flow and head.

The industrial use of the term "centrifugal pump" includes radial-flow centrifugal pumps, mixed-flow, and axial-flow or propeller pumps. The operating principles and characteristics of the radial-flow pump are similar to those of the centrifugal compressor. The similarity extends to the vector diagram.

The radial-flow type of pump is usually equipped with a backward-curved vane impeller. The casing may be of the volute design (Fig. 5.22) or of the diffuser design (Fig. 5.23). Many single-stage pumps of both types are used for low-pressure pumping. For high-pressure service, either type of pump is constructed as a multistage machine.

Volute pumps are used generally for low- and medium-pressure service where the discharge pressure does not exceed approximately 10 MPa. The discharge pressure for which the diffuser pump is designed can be specified within a wide range, that is, a low- to high-pressure range is covered by these pumps. Pressures well above 10 MPa can be developed in a diffuser pump.

Unlike the volute construction, the diffuser vane construction results in a symmetrical casing, and consequently a uniform radial pressure distribution is developed around the pump impeller. When the diffuser pump operates off the design point, the angle at which the fluid leaves the impeller does not correspond with the diffuser vane angle. Under certain operating conditions, the resulting shock and turbulence may be sufficiently severe to cause an appreciable decline in the pump performance. The operating efficiency and the discharge head will consequently drop sharply.

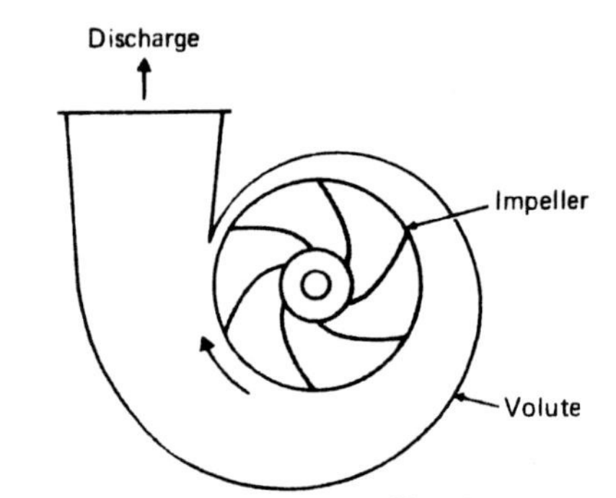

Figure 5.22 Volute-type centrifugal pump.

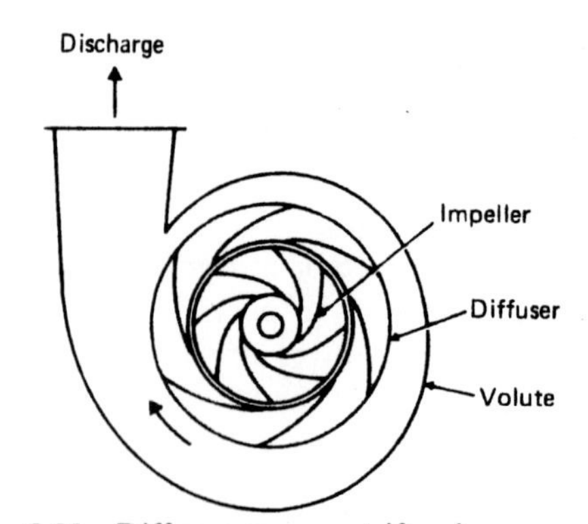

Figure 5.23 Diffuser-type centrifugal pump.

Radial-flow pump impellers are usually shrouded in order to form completely enclosed flow channels. The flow control that is achieved by directing the fluid through the enclosed channels tends in general to raise the efficiency of the pumping operation. Open impellers however are often used in pumps intended for handling liquids that contain solid material. Such pumps normally develop low delivery pressures. Double-inlet, single-stage pumps are relatively common.

The specific speed N_s is a useful pump performance indicator.

$$N_s = \frac{n\sqrt{\dot{V}}}{H_t^{0.75}} \tag{5.11}$$

where

n = pump speed, rpm
$\dot{V}$ = pump delivery, m^3/min
H_t = total developed head, m

Equation 5.11 is developed from the laws of similitude, and the specific speed is a function of the impeller proportions. Thus, two pumps that are exactly geometrically similar have identical specific speeds. The specific speed may be used to classify a pump. The specific speed, which indicates the type of pump, corresponds to the specific speed for the operating point at which the maximum pump efficiency is achieved. Pumps are usually rated at conditions that approximate those of the point of maximum efficiency.

Pumps in any single classification differ somewhat in their proportions, thus each type of pump is classified by a range in the specific speed rather than by a particular number. For various types of impellers, the following ranges in the specific speed are generally applicable.

Radial flow	Usually below 225
Francis	225–675
Mixed flow	675–1350
Axial flow	1200–2100

The total-head pump efficiency is customarily expressed as the ratio of the fluid power to the shaft power. Then,

$$\eta_P = \frac{\dot{m} g \, \Delta H_t}{\dot{W}_b} \qquad (5.12)$$

where

$\dot{m}$ = pump flow, kg/s
g = acceleration owing to gravity, m/s^2
ΔH_t = total developed head, m
$\dot{W}_b$ = pump shaft power, W

The total-head equation, derived from Eq. 3.8, is given by

$$\Delta H_t = \frac{1}{g\rho}(P_1 - P_2) + \frac{1}{2g}(C_1^2 - C_2^2) \qquad (5.13)$$

where

P_1, P_2 = the respective inlet and discharge pressures measured at the center line of the pump
C_1, C_2 = the fluid velocities measured, respectively, at the inlet and outlet sections of the pump
ρ = the density of the fluid

Hydraulic and mechanical losses inherent in the pump account for the difference between the shaft power and the fluid power. Pump efficiencies are related generally to the capacity and specific speed. Low pump efficiencies are associated with low specific speeds because the geometric proportions of pumps in this class are indicative of high fluid losses.

Example 5.6

A radial-flow pump operates with a discharge pressure of 2100 kPa and an inlet pressure of 250 kPa, measured at the center line of the pump. Fluid velocities are 4 m/s at the outlet section and 2 m/s at the inlet section of the pump. The pump flow rate is 0.125 m^3/s. The fluid density is 850 kg/m^3. The total-head pump efficiency is 0.83. Calculate the total head that is developed by the pump and determine the shaft power.

Equation 5.13 is used to calculate the total head.

$$\Delta H_t = \frac{1}{g\rho}(P_1 - P_2) + \frac{1}{2g}(C_1^2 - C_2^2)$$
$$= \frac{(250 - 2100)10^3}{9.807 \times 850} + \frac{2^2 - 4^2}{2 \times 9.807}$$
$$= -221.93 - 0.61 = -222.54 \text{ m}$$

$$\dot{m} = \rho\dot{V} = 850 \times 0.125 = 106.25 \text{ kg/s}$$

$$\dot{W}_b = \frac{\dot{m} g \, \Delta H_t}{\eta_P}$$
$$= \frac{106.25 \times 9.807 \times (-222.54)}{0.83}$$
$$= -279{,}400 \text{ W} = -279.4 \text{ kW}$$

The total head developed by a pump is usually taken as a positive quantity. The negative result obtained in the above example is however consistent with the first law analysis.

Figure 5.24 shows the performance characteristics of a single-stage centrifugal pump for operation at two different constant speeds. The problem of flow control for a constant-speed pump is somewhat similar to that encountered in the

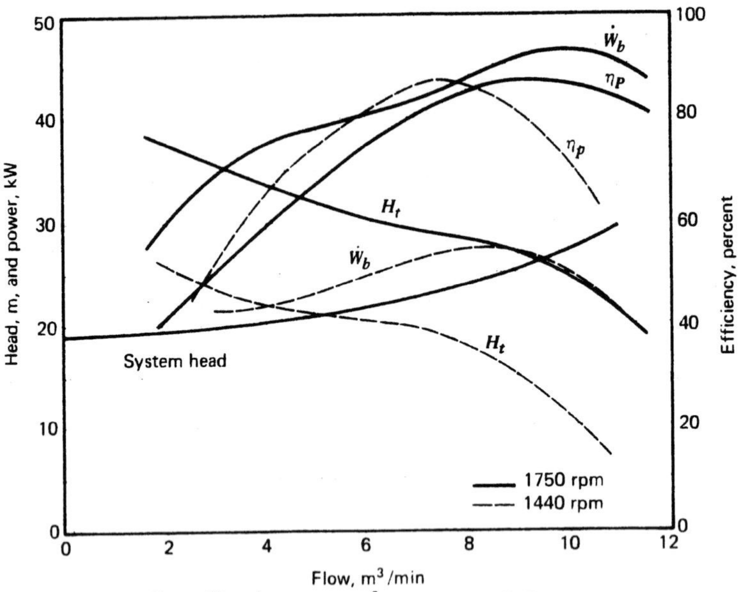

Figure 5.24 Centrifugal pump performance curves.

operation of a fan. However, with respect to the pump, a large part of the system total head is often caused by a relatively constant static pressure. See Fig. 5.24. Thus, the range in the system head from zero to maximum flow may not be especially large. Nevertheless, throttling the pump output in order to control the flow does produce an excessive delivery head over that required to overcome the system resistance.

When the pump operates at 1750 rpm, the maximum flow for the system characteristic shown in Fig. 5.24 is 9.5 m^3/min. If the flow is reduced to 5.1 m^3/min by throttling, the pump will develop a head of 32 m, while the required system head is 21 m. The shaft power is 39 kW. A flow of 5.1 m^3/min can however be established by dropping the pump speed to 1440 rpm and developing a head of 21 m. The shaft power is now 23 kW. Limited pump control achieved by operation at two different constant speeds is a practical arrangement that utilizes a two-speed ac motor.

Variable-speed operation of the pump may be feasible in some installations as a means to improving the overall performance of the pump and, as a consequence, effecting a reduction in the power input. Speed control may be accomplished through use of variable-speed drives, with the control in the driving motor or engine or in the intermediate speed regulator.

The minimum pressure point in a pump normally occurs at the entrance to the impeller vanes. Under certain operating conditions, particularly when pumping a liquid at or near the boiling point, vapor may be released at this point of minimum pressure. Cavitation, the result of vapor formation in the pump, can be a major problem in pumping. Some of the effects of cavitation are noise and vibration, reduced output and efficiency, and erosion of pump parts. Vapor lock will cause the pump to "run hot." Vapor formation in the pump can be avoided by maintaining sufficient pressure on the liquid at the point where the minimum pressure is established.

5.10 AXIAL-FLOW COMPRESSORS

The possibility of designing an axial-flow compressor on the basis of a form of reversed turbine was investigated shortly after the development of the reaction turbine. However, a turbocompressor, which is no more than a reversed turbine, cannot operate with acceptable efficiencies. For low pressure ratios, the efficiency of this machine does not exceed 40 percent and becomes substantially less for higher pressure ratios.

Early investigations, circa 1900, resulted in the construction of axial-flow compressors that were fitted with blades designed on the basis of propeller theory. Although the performance of these machines was not entirely unsatisfactory, the resulting efficiency, approximately 60 percent, was below that of the centrifugal compressor, which is 70 to 80 percent. During the period 1930–1940 the initial development of the gas turbine engine pointed to a definite need for compressor designs that would yield efficiencies well above 80 percent, and consequently attention was again directed to the axial-flow compressor. Subsequent investigations, aided by progress in the field of fluid mechanics, produced satisfactory compressor designs. Multistage axial-flow compressors are capable of operating with peak efficiencies on the order of 90 percent.

Axial-flow blowers, employed for various services, are constructed with one, two, or three stages depending upon the delivery pressure requirements. The compressor element installed in the turbojet engine and in the gas turbine power plant is usually a multistage axial-flow machine. The axial-flow compressor is also an integral unit of numerous industrial processes.

The flow of the working fluid through the blade passages of the axial-flow compressor differs fundamentally from the corresponding turbine flow. Reaction turbine blades are usually formed to produce a convergent nozzle through which stable expansion of the fluid can be achieved. The passage formed by two compressor blades is essentially a diffuser in which the fluid is decelerated and the pressure increased. Unlike the nozzle passage, with its accelerated flow, deceleration of the fluid is subject to restrictions that result in low rates of diffusion and small changes in the direction of the flow. See Fig. 5.25. Any attempt to incorporate high diffusion rates into the blade design will tend to cause increased turbulence and a reduced rate of pressure rise. Excessive changes in the direction of flow will result in flow separation or stalling. The flow over a stalled blade is unstable and often accompanied by a surging discharge.

In the compressor, not only do adverse flow conditions cause a decrease in the efficiency of the compression process, but instability in the fluid stream may seriously affect the operation of the machine. Compressor design procedures are consequently limited to low diffusion rates and to low gas stream deflections; and because of these limitations, low pressure ratios are developed in the individual stages.

The stable, expanding flow in the turbine blade passages may readily be achieved with efficiencies exceeding 90 percent. On the other hand, diffusion or compression through the compressor blade passages, with efficiencies ranging from 85 to 90− percent, can only be achieved through observation of close limits of diffusion rates and changes in the direction of flow.

Figure 5.26 shows the general construction of a multistage, axial-flow compressor that operates with subsonic air velocities. The fluid flows axially from left to right through the bladed annulus composed of alternate rows of rotating blades and stationary vanes. The annular flow area decreases in the direction of flow in order to accommodate the reduced specific volume that results from compression of the fluid. Some com-

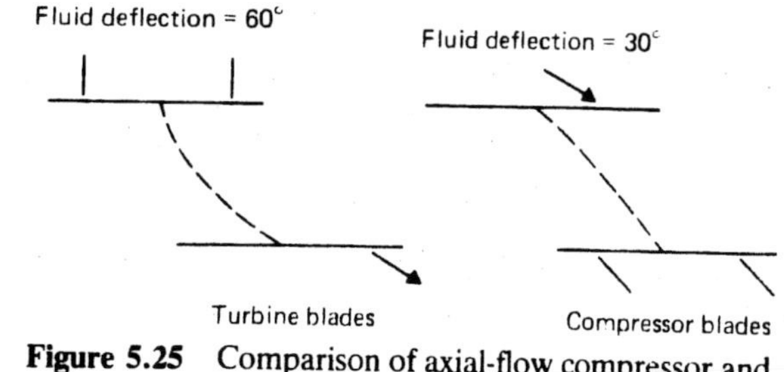

Figure 5.25 Comparison of axial-flow compressor and axial-flow turbine blades.

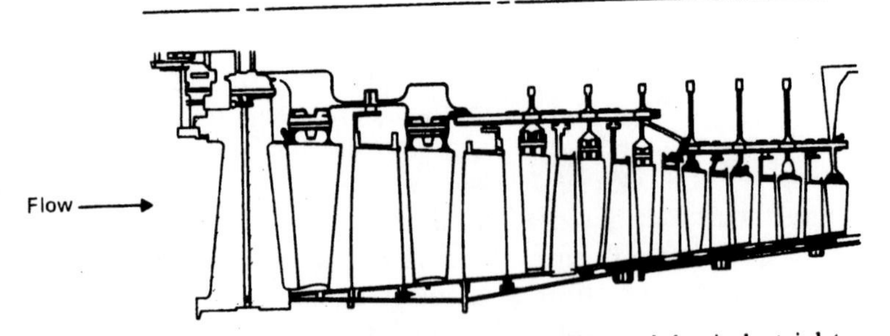

Figure 5.26 Cross-sectional view of the FT4 modular industrial turbine low-pressure compressor. (Courtesy United Technologies Power Systems Division.)

pressor designs prescribe a constant blade root diameter or a constant blade tip diameter. Rotor blades are mounted on disks that, in turn, are secured to one another to form a rigid spindle or mounted on a single drum.

Figure 5.27 shows the pressure increase and the variation in the absolute velocity of the fluid for the flow through several stages of an axial-flow compressor. The absolute velocity of the fluid is alternately increased in the rotor blade row and decreased in the stator vane passages. A continuous pressure rise is normally developed through the several successive stages of the compressor.

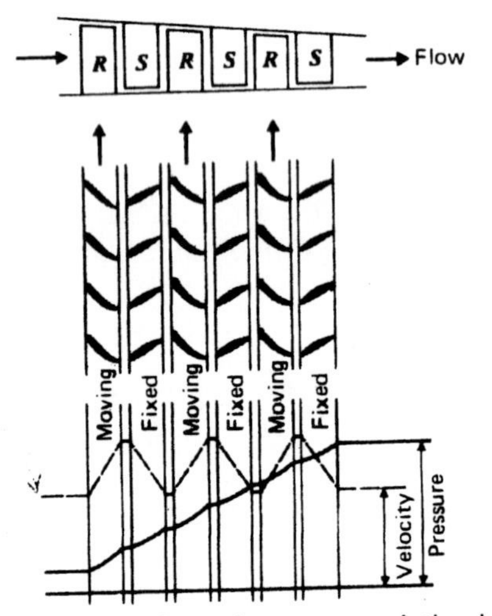

Figure 5.27 Velocity and pressure variation in an axial-flow compressor.

In passing through a rotating blade row, and usually in the stator vane annulus, the fluid is deflected and a reduction in the relative velocity is achieved. Through diffusion, a relatively small pressure rise is accomplished in each row of blades.

5.11 COMPRESSOR BLADE DESIGN

Axial-flow compressor blade design evolved through application of aerodynamic theory in conjunction with experimental design. For a particular blade row, the blades are designed to cause the fluid to flow in a manner described by an appropriate flow pattern. Axial-flow compressors are usually designed for subsonic flow of the fluid in all of the stages. Transonic and supersonic blade designs have been developed for compressors installed in high-performance turbojet engines.

Selection of the compressor blade profile is primarily dependent upon such aerodynamic characteristics as high pressure rise, low fluid losses, and the capability of operating over a wide range without stalling or choking. Figure 5.28 illustrates a compressor blade profile and flow passage. Considerable variation may be observed in the camber angle of the blade profile extending from the root to the tip diameter. The mean camber line, a single or compound curve, can be described mathematically. The blade profile is formed by prescribing an appropriate

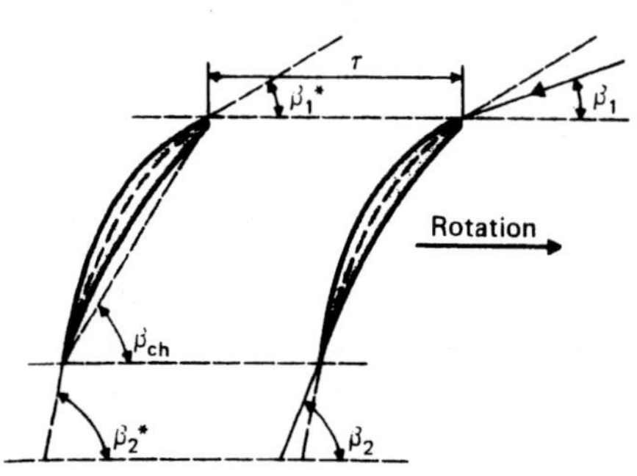

Figure 5.28 Axial-flow compressor fluid and blade angles. β = fluid angle, β^* = blade angle, β_{ch} = chord angle, τ = blade pitch.

thickness distribution in relation to the mean camber line.

Blade angles and the direction of fluid flow are measured with respect to the rotational plane. See Fig. 5.28. The starred and unstarred symbols refer, respectively, to the blade and fluid angles. The blade camber angle θ^*, a function of the maximum camber, is equivalent to $\beta_2^* - \beta_1^*$.

The angle β_1, at which the fluid approaches the blade, usually does not coincide with the blade angle β_1^*. The difference is $\beta_1^* - \beta_1 = i$, the incidence angle, an important performance factor. The average mass of the fluid leaving the blade passage is not turned to the outlet blade angle β_2^*, but to a smaller angle β_2, the outlet fluid angle.

The difference between the outlet and the inlet fluid angles $(\beta_2 - \beta_1)$ is designated the fluid deflection angle θ.

The fluid absolute velocities entering and leaving the blade passage are designated, respectively, C_1 and C_2. The corresponding fluid relative velocities are designated D_1 and D_2. The tangential or linear blade velocity is represented by U.

Aerodynamic design is likely to produce a rotor blade that has a gradual decrease in the cross-sectional area from the root to the tip section. The blade taper is conducive to minimizing the stress at the blade root caused by centrifugal force. Incidentally, rotor blades must be designed to withstand vibrational effects and the bending forces attributed to gas pressures and the momentum change in the fluid. The center of gravity for each rotor blade section is located on a radial line as a means to eliminating bending stresses that would otherwise be caused by centrifugal force. In comparison with the rotor blades, stator vanes are subjected to less severe stresses.

Compressor blades are fabricated from various materials, namely, aluminum, aluminum alloy, magnesium alloy, and titanium, depending upon the structural requirements. Close manufacturing tolerances are required in blade fabrication; otherwise, improperly constructed flow passages will cause an increase in the fluid losses and a reduction in the prescribed flow through the machine.

5.12 THE VECTOR DIAGRAM

Vector diagrams are constructed for the purpose of joining the vectors that represent the linear blade velocity and the fluid absolute and relative velocities. The vector diagram provides data used in the design of compressor blades and in the analysis of the performance of the blade row.

Figure 5.29 shows the vector diagram for a single axial-flow compressor stage. The fluid enters the rotor at a radius r_1 with an axial velocity $C_{x,1}$. The linear or tangential velocity of a point on the blade at radius r_1 is designated U_1. The fluid usually enters the blade row with a rotational, or whirl, component $C_{U,1}$ in the direction of rotation. The inlet velocity vector D_1, relative to the rotor blade, is determined by combining the vectors C_1 and $-U_1$.

A pressure increase is developed in the rotor blade passages as the fluid is turned and decelerated to the outlet relative velocity D_2. The combination of the vectors D_2 and U_2, the blade velocity at the trailing edge at radius r_2, yields the fluid outlet absolute velocity vector C_2. The angles β_1 and β_2 are, respectively, the inlet and outlet fluid angles relative to the rotor blade.

Although the fluid has been decelerated with respect to the rotor blade, there has actually been

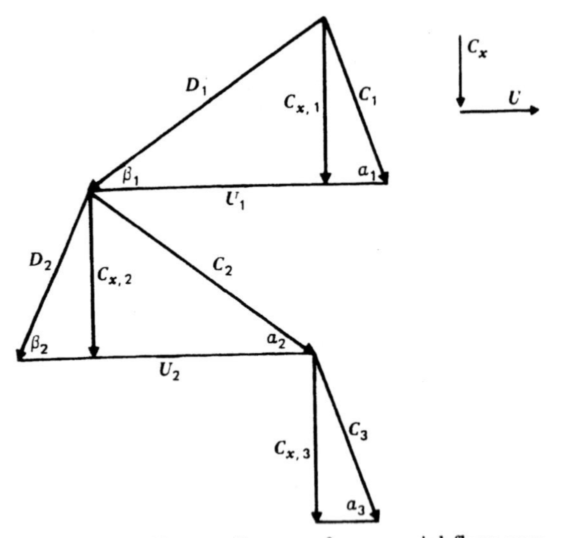

Figure 5.29 Vector diagram for an axial-flow compressor stage. C = fluid absolute velocity, D = fluid relative velocity, C_x = axial velocity component, U = blade velocity, α and β = fluid angles.

an acceleration of the fluid in an absolute sense. The increase in the velocity of the fluid from C_1 to C_2 is a result of rotor action. The rotor consequently causes in the fluid a simultaneous increase in the pressure and in the absolute velocity because of the work done by the rotating blades.

Compression of the fluid in the rotating blade row effects an increase in the density of the fluid; and as a consequence, the flow area and the length of the blades decrease in the direction of flow. Because of a decrease in the blade height from the leading to the trailing edge, a small mass of fluid passing through the blade row is subject to a slight radial displacement, that is, $r_2 < r_1$; and for the corresponding blade velocities, $U_2 < U_1$. In this case, a constant blade root diameter is assumed. The difference in the blade speed $(U_1 - U_2)$ is usually small and may often be neglected.

The fluid absolute velocity C_2 at the rotor outlet is also the stator inlet velocity. Deceleration of the fluid to the outlet absolute velocity C_3 occurs normally in the vane passages. The inlet and outlet fluid angles for the stator are designated, respectively, α_2 and α_3.

In developing a design procedure, it is convenient to assume that the fluid outlet velocity for one blade row is also the inlet velocity for the following blade row, with the magnitude and direction unchanged. Losses because of turbulence between blade rows, and because of blade entrance shock, are accounted for by introducing an appropriate loss factor. The first compressor stage is preceded by a row of inlet guide vanes that impart to the fluid a proper whirl component $C_{U,1}$. Ordinarily, the angles α_1 and α_3, for a particular stage, are of about the same magnitude.

5.13 FREE-VORTEX FLOW

The preceding discussion on the use of the vector diagram was restricted to the flow of a small mass of fluid at a radius r. The vector diagrams that are constructed for the flow at other radii are related to one another in accordance with some prescribed flow pattern. Two basic procedures that can be employed for compressor design will be examined. One of these procedures is based upon the existence of free vortex flow, while the other assumes that a constant-reaction type of flow will be established.

When irrotational, or free vortex, flow is established in a fluid mass, the linear velocity C of an element of mass is described by the relationship Cr = constant, where r is the radius of rotation of the element. Consequently, those elements closest to the center of rotation have the highest linear velocities, while the far-removed elements rotate at the minimum linear velocities.

The free vortex flow pattern applied to the axial-flow compressor stipulates that, in the space between the blade rows, the product of the radius and the rotational component of the fluid absolute velocity is a constant. Thus,

$$r_1 C_{U,1} = (\text{constant})_1$$

and

$$r_2 C_{U,2} = (\text{constant})_2$$

For discussion purposes, it will be assumed

that the radius of rotation r for a small mass of fluid remains unchanged during passage through a blade row, hence $r_2 = r_1$ and $U_2 = U_1$. In an actual design however, a change in r and in U is ordinarily observed. The axial variation in C_x through the compressor is prescribed in the design procedure.

For free vortex flow, equilibrium is established in the fluid in the absence of a radial variation in the axial component of the fluid velocity. Two important conclusions can be drawn: no radial displacement of the fluid occurs in the space between the blade rows, and theoretically there is no radial variation in C_x.

A vector diagram is constructed for each blade section that is designed in order to determine the appropriate fluid velocities and angles. The procedure employed in constructing these diagrams is examined in the following example.

Example 5.7

At the mean, or design, diameter, the following data are applicable.

$$d_m = 45.0 \text{ cm} \qquad U_m = 215 \text{ m/s}$$

$$C_{x,2} = C_{x,1} = 120 \text{ m/s} \qquad \theta = 30.0°$$

$$\alpha_1 = 70° \qquad \alpha_3 = \alpha_1$$

Calculations are made at the mean diameter (45.0 cm) for construction of the vector diagram.

$$C_{U,1} = C_{x,1} \cot \alpha_1 = 120 \times 0.3640 = 43.7 \text{ m/s}$$

$$D_{U,1} = U - C_{U,1} = 215 - 43.7 = 171.3 \text{ m/s}$$

$$\cot \beta_1 = \frac{D_{U,1}}{C_{x,1}} = \frac{171.3}{120} = 1.428 \qquad \beta_1 = 35.0°$$

$$\beta_2 = \beta_1 + \theta = 35.0 + 30.0 = 65.0°$$

$$D_{U,2} = C_{x,2} \cot \beta_2 = 120 \times 0.4663 = 56.0 \text{ m/s}$$

$$C_{U,2} = U - D_{U,2} = 215.0 - 56.0 = 159.0 \text{ m/s}$$

The mean-diameter vector diagram completed from the above data is shown by Fig. 5.29. The vectors D_1 (209.1 m/s) and D_2 (132.4 m/s) represent the fluid relative velocities entering and leaving the rotating blade row. For the stationary blade row, the fluid inlet and outlet velocities are represented by the vectors C_2 (199.2 m/s) and C_3 (127.7 m/s). Because $\alpha_3 = \alpha_1$ and $C_{x,3} = C_{x,1}$, $C_{U,3} = C_{U,1}$.

The flow pattern prescribed for the design of the blade will establish the relationship between the mean-diameter vector diagram and the vector diagram constructed at any other diameter. For free vortex flow, the following relationships are developed.

$$(rC_{U,1})_t = (rC_{U,1})_m \qquad (rC_{U,2})_t = (rC_{U,2})_m$$

$$(C_{x,1})_t = (C_{x,1})_m \qquad (C_{x,2})_t = (C_{x,2})_m$$

$$\frac{U_t}{U_m} = \frac{r_t}{r_m} = \frac{d_t}{d_m}$$

From these relationships, the complete tip diameter vector diagram can be constructed. Similar relationships are developed for the construction of the root diameter vector diagram.

Calculate the data essential for construction of the tip diameter vector diagram. The blade length is 5.0 cm.

$$\text{Tip diameter} = 50.0 \text{ cm}$$

$$U_t = U_m \frac{d_t}{d_m} = 215 \frac{50.0}{45.0} = 238.9 \text{ m/s}$$

$$(C_{U,1})_t = (C_{U,1})_m \frac{d_m}{d_t}$$

$$= 43.7 \frac{45.0}{50.0} = 39.3 \text{ m/s}$$

$$(D_{U,1})_t = U_t - (C_{U,1})_t$$

$$= 238.9 - 39.3 = 199.6 \text{ m/s}$$

$$(C_{U,2})_t = (C_{U,2})_m \frac{d_m}{d_t}$$

$$= 159.0 \frac{45.0}{50.0} = 143.1 \text{ m/s}$$

$$(D_{U,2})_t = U_t - (C_{U,2})_t$$

$$= 238.9 - 143.1 = 95.8 \text{ m/s}$$

$$\beta_{1,t} = 31.0° \qquad \beta_{2,t} = 51.4°$$

$$\theta_t = 20.4°$$

In a similar manner, pertinent data are determined for the root diameter section of the blade.

$d_r = 40.0$ cm $\quad U_r = 191.1$ m/s

$(C_{U,1})_r = 49.2$ m/s $\quad (D_{U,1})_r = 141.9$ m/s

$(C_{U,2})_r = 178.9$ m/s $\quad (D_{U,2})_r = 12.2$ m/s

$\beta_{1,r} = 40.2°$ $\quad \beta_{2,r} = 84.2°$ $\quad \theta_r = 44.0°$

The data derived in Example 5.7 show the radial variation in θ, the fluid deflection angle, from 44.0° at the root diameter section to 20.4° at the tip diameter section. A corresponding variation is observed in the blade camber angle.

From $r_1C_{U,1} = K$ and $U_1/r_1 = K'$, it follows that $C_{U,1}U_1 = K_1$. Similarly, at the trailing edge of the blade, $C_{U,2}U_2 = K_2$. Consequently, $C_{U,2}U_2 - C_{U,1}U_1 = K_3$. An earlier development, presented in Section 5.3, discloses that the difference between the C_UU products is equivalent to the work done on the fluid. Evidently, in free vortex flow, the work done on the fluid is constant along the length of the rotor blade, regardless of the radius at which the fluid rotates.

5.14 CONSTANT-REACTION BLADE DESIGN

The pressure rise in the compressor stage may be divided equally or unequally between the rotor and the stator. The degree of reaction can be defined as the ratio of the enthalpy increase achieved in the rotor to the stage enthalpy increase, that is, $\psi = \Delta h_R/\Delta h_{st}$. A reasonably good approximation for ψ, the degree of reaction, is obtained by assuming that $U_2 = U_1$, $C_3 = C_1$, and $C_{x,2} = C_{x,1}$. Then, at any radius,

$$\psi = 1 - \frac{C_{U,1} + C_{U,2}}{2U} \tag{5.14}$$

As noted in the preceding section, the free vortex and the constant-reaction flow patterns are basic and theoretical flow concepts used in compressor design. The constant-reaction principle requires that the work done on the fluid remain constant with respect to the radius of rotation. Thus, as in free vortex flow, $C_{U,2}U_2 - C_{U,1}U_1$ = constant, along the length of the blade. In a simplified analysis of the constant-reaction flow pattern, the axial velocity along the blade is assumed to be constant, that is, there is no radial variation in C_x. The free vortex and the constant-reaction flow patterns are thus alike with respect to the absence of a radial variation in the axial component of the fluid velocity and in the work done on the fluid.

The difference between the two flow patterns becomes apparent in the third requirement that must be observed in order to define completely the vector diagrams for all of the blade sections extending from the root to the tip diameter. For constant-reaction flow, Eq. 5.14 is applicable, and a unique value of ψ is assigned for the stage. For free vortex flow, the inherent characteristic rC_U = constant is observed. This relationship does not cause constant reaction along the blade, that is, a radial variation in ψ occurs.

A special case of the constant-reaction design is observed when equal enthalpy increases are achieved in the rotor and stator blade rows of a stage. An equal division of the stage enthalpy increase between the rotor and the stator is a reasonably good design assumption, especially because the 50-percent reaction design tends to yield maximum stage efficiency.

Example 5.8

A compressor stage is designed for 50 percent constant reaction; $C_x = 120$ m/s. At the mean diameter (45.0 cm), $U = 210$ m/s and $\Delta C_U = 60$ m/s. The blade length is 5.0 cm. Construct the mean-diameter vector diagram.

$$C_{U,2} - C_{U,1} = 60$$

$$\psi = 1 - \frac{C_{U,1} + C_{U,2}}{2U}$$

$$0.5 = 1 - \frac{C_{U,1} + C_{U,2}}{2 \times 210}$$

$$C_{U,1} = 75 \text{ m/s} \qquad C_{U,2} = 135 \text{ m/s}$$

The vector diagram is shown in Fig. 5.30.

$$\beta_1 = \alpha_2 = 41.6°$$

$$\beta_2 = \alpha_1 = 58.0°$$

$$\theta_m = 16.4°$$

The tip diameter vector diagram is constructed by applying Eq. 5.14 and observing the constant work relationship $(\Delta C_U U)_t = (\Delta C_U U)_m$.

$$U_t = U_m \frac{d_t}{d_m} = 210 \frac{50.0}{45.0} = 233.3 \text{ m/s}$$

$$(C_{U,2} - C_{U,1})_t = (C_{U,2} - C_{U,1})_m \frac{U_m}{U_t}$$

$$= 60 \frac{210}{233.3} = 54.0 \text{ m/s}$$

$$\psi = 1 - \frac{C_{U,1} + C_{U,2}}{2U}$$

$$0.5 = 1 - \frac{C_{U,1} + C_{U,2}}{2 \times 233.3}$$

$$C_{U,1} = 89.6 \text{ m/s} \qquad C_{U,2} = 143.6 \text{ m/s}$$

$$\beta_1 = 39.9° \qquad \beta_2 = 53.2° \qquad \theta_t = 13.3°$$

Similarly, for the root diameter vector diagram,

$$C_{U,1} = 59.6 \text{ m/s} \qquad C_{U,2} = 127.1 \text{ m/s}$$

$$\beta_1 = 43.4° \qquad \beta_2 = 63.6° \qquad \theta_r = 20.2°$$

The free vortex and constant-reaction flow patterns have been employed to illustrate some of the principles of compressor blade design. It is important to note however that the actual flow through the machine will differ for a number of reasons from the theoretical flow.

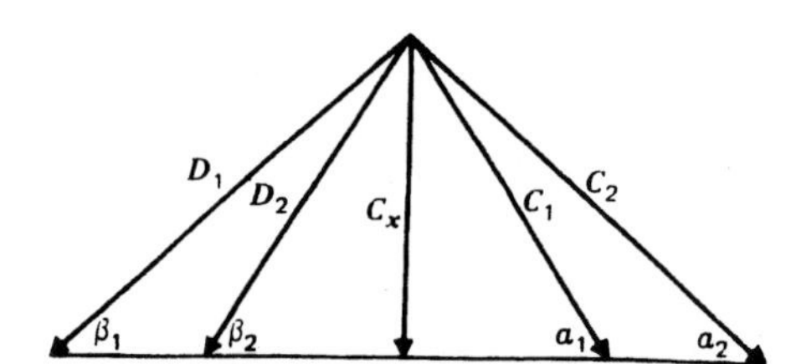

Figure 5.30 Vector diagram for 50 percent reaction in an axial-flow compressor stage. C = fluid absolute velocity, D = fluid relative velocity, C_x = axial velocity component, U = blade velocity, α and β = fluid angles. For the stator vane, $\alpha_3 = \alpha_1$, $C_3 = C_1$.

5.15 STAGE WORK AND PRESSURE RISE

The work done on the fluid during passage through a compressor rotating blade row is determined by the same basic approach presented in Section 5.3. The power input to the fluid is given by Eq. 5.2, that is,

$$\dot{W} = \dot{m}(C_{U,2}U_2 - C_{U,1}U_1)$$

where

$\dot{m}$ = the fluid flow, kg/s
C_U = fluid whirl component, m/s
U = blade linear velocity, m/s

When $U_2 = U_1$, Eq. 5.2 is simplified to read

$$\dot{W} = \dot{m}(C_{U,2} - C_{U,1})U \qquad (5.15)$$

Because $\Delta C_U U$ is constant along the blade for both free vortex and constant-reaction flow, the work per unit mass of fluid is assumed to be constant with respect to the radius for theoretical flow conditions.

The increase in pressure achieved in a compressor stage is determined from the summation of the pressure rise in the individual blade rows.

The enthalpy increase in the fluid, during passage through a blade row, is evaluated by application of the energy equation for adiabatic flow. Then, for the rotor,

$$\Delta h_R = h_2 - h_1$$

$$= \frac{C_1^2 - C_2^2}{2} + (C_{U,2}U_2 - C_{U,1}U_1) \qquad (5.16)$$

The pressure increase across the blade row is

determined from the reversible adiabatic enthalpy increase for the fluid. See Fig. 5.31.

$$h_{2,s} - h_1 = \eta_R(h_2 - h_1) \qquad (5.17)$$

where η_R is the rotor internal efficiency.

In a similar manner, the following equations are developed for the stator.

$$\Delta h_S = h_3 - h_2 = \frac{C_2^2 - C_3^2}{2} \qquad (5.18)$$

$$h_{3,s} - h_2 = \eta_S(h_3 - h_2) \qquad (5.19)$$

where η_S is the stator efficiency.

The compressor discharge pressure is predicted by a row-by-row calculation of the pressure rise. Calculations are based on the conditions at the mean-diameter section of the blade row.

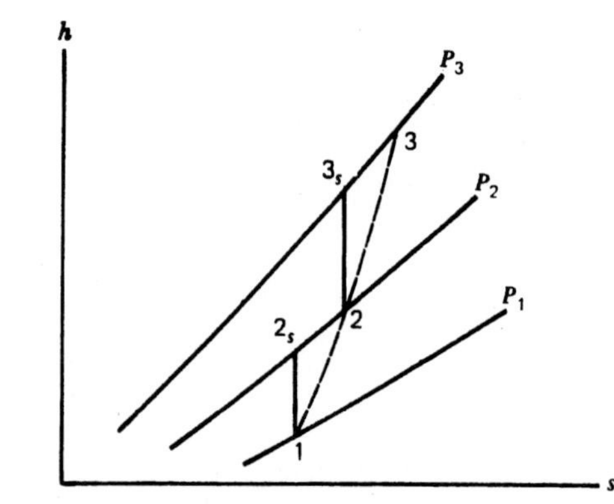

Figure 5.31 Pressure increase in an axial-flow compressor stage.

Example 5.9

Design data for an axial-flow compressor rotor are

$C_x = 165$ m/s $\qquad (C_{U,1})_m = 90$ m/s

$\theta_m = 20.0°$

$\eta_R = 0.835 \qquad \eta_m = 0.98$

fluid : air

At the leading edge of the blade,

$P_1 = 86.0$ kPa, abs $\qquad T_1 = 295$ K

$d_m = 46.0$ cm $\qquad U_m = 310$ m/s

$l_1 = 6.70$ cm $\qquad$ (effective blade length)

Determine the power input to the rotor and the fluid pressure at the rotor outlet section.

From the vector diagram,

$$\beta_1 = 36.9°$$

$$\beta_2 = \beta_1 + \theta$$

$$= 36.9 + 20.0 = 56.9°$$

$$D_{U,2} = C_x \cot \beta_2$$

$$= 165 \times 0.6515 = 107.6 \text{ m/s}$$

Because of compression, the specific volume of the fluid decreases in passing from the leading edge to the trailing edge of the blade. For $C_{x,2} = C_{x,1}$, the trailing edge length is somewhat less than the leading edge length of the blade, hence for a constant root-section diameter, $U_2 < U_1$ at the mean diameter.

Assume $U_{2,m} = 306$ m/s

$$C_{U,2} = U_2 - D_{U,2} = 306.0 - 107.6 = 198.4 \text{ m/s}$$

$$h_2 - h_1 = \frac{C_{U,1}^2 - C_{U,2}^2}{2} + (C_{U,2}U_2 - C_{U,1}U_1)$$

$$= \frac{(90.0)^2 - (198.4)^2}{2} + [(198.4 \times 306) - (90.0 \times 310)]$$

$$= -15{,}631 + 32{,}810 = 17{,}179 \text{ J/kg}$$

$$h_{2,s} - h_1 = \eta_R(h_2 - h_1)$$

$$= 0.835 \times 17.18 = 14.34 \text{ kJ/kg}$$

$$T_2 - T_1 = \frac{h_2 - h_1}{c_p} = \frac{17.18}{1.004}$$

$$= 17.11 \text{ K} \qquad T_2 = 312.11 \text{ K}$$

$$T_{2,s} - T_1 = \frac{h_{2,s} - h_1}{c_p} = \frac{14.34}{1.004}$$

$$= 14.29 \text{ K} \qquad T_{2,s} = 309.29 \text{ K}$$

$$P_2 = P_1\left(\frac{T_{2,s}}{T_1}\right)^{k/(k-1)}$$

$$= 86.0\left(\frac{309.29}{295.0}\right)^{3.5}$$

$$= 101.49 \text{ kPa, abs}$$

At the Leading Edge of the Blade

Blade root diameter $d_r = d_m - l_1$

$$= 46.0 - 6.70 = 39.30 \text{ cm}$$

Effective flow area $A_{x,1} = \pi d_m l_1$

$$= 46.0\pi \times 6.70 = 968.2 \text{ cm}^2$$

$$\rho_1 = \frac{P_1}{RT_1}$$

$$= \frac{86.00}{0.287 \times 295} = 1.016 \text{ kg/m}^3$$

At the Trailing Edge of the Blade

$$\rho_2 = \frac{P_2}{RT_2}$$

$$= \frac{101.49}{0.287 \times 312.11} = 1.133 \text{ kg/m}^3$$

$C_{x,2} = C_{x,1}$, hence

$$A_{x,2} = A_{x,1} \frac{\rho_1}{\rho_2}$$

$$= 968.2 \frac{1.016}{1.133} = 868.2 \text{ cm}^2$$

$d_{r,2} = d_{r,1} = 39.30$ cm (prescribed)

$$A_{x,2} = \pi d_{m,2} l_2 = \pi(d_{r,2} + l_2) l_2$$

$$868.2 = \pi(39.30 + l_2) l_2 \qquad l_2 = 6.09 \text{ cm}$$

$$d_{m,2} = d_{r,2} + l_2 = 39.30 + 6.09 = 45.39 \text{ cm}$$

$$U_2 = \frac{d_{m,2}}{d_{m,1}} U_1 = \frac{45.39}{46.0} 310 = 306 \text{ m/s}$$

Note: Assumed $U_2 = 306$ m/s.

Compressor Shaft Power

$$\dot{m} = \rho_1 C_{x,1} A_{x,1}$$

$$= 1.016 \times 165 \times 0.09682$$

$$= 16.23 \text{ kg/s}$$

$$\dot{W} = \dot{m}(C_{U,2} U_2 - C_{U,1} U_1)$$

$$= 16.23[(198.4 \times 306) - (90 \times 310)]$$

$$= 532{,}513 \text{ W}$$

$$\dot{W}_b = \frac{\dot{W}_i}{\eta_m}$$

$$= \frac{532.5}{0.98} = 543.4 \text{ kW} \qquad \text{(input)}$$

Determine the length of the trailing edge of the stator vane that is matched with the rotor blade.

Assume

$$C_{x,3} = C_{x,1} = 165 \text{ m/s} \qquad \eta_S = 0.830$$

$$C_{U,3} = C_{U,1} = 90 \text{ m/s}$$

From the enthalpy change $h_3 - h_2$, the equivalent of the kinetic energy change $\frac{1}{2}(C_2^2 - C_3^2)$, the following data are derived.

$$T_3 = 327.68 \text{ K} \qquad P_3 = 116.97 \text{ kPa, abs}$$

$$\rho_3 = 1.244 \text{ kg/m}^3 \qquad A_{x,3} = 790.7 \text{ cm}^2$$

$$l_3 = 5.60 \text{ cm} \qquad \theta_S = 21.6°$$

5.16 COMPRESSOR POLYTROPIC EFFICIENCY

A difference is observed between the stage efficiency and the compressor efficiency of multistage compressors. The compressor efficiency is the internal efficiency of the entire machine.

The average stage efficiency and the compressor efficiency are given, respectively, by

$$\eta_{st} = \frac{\Sigma\,(\Delta h_s)_{st}}{W_C}$$

$$\eta_C = \frac{(\Delta h_s)_C}{W_C}$$

where W_C is the compressor internal work.

Because the summation of the isentropic enthalpy increase $\Sigma\,(\Delta h_s)_{st}$ for the several stages of the compressor exceeds the isentropic enthalpy increase $(\Delta h_s)_C$ for the compressor taken as a whole, the average stage efficiency η_{st} exceeds the compressor efficiency η_C. Then, $\eta_{st} = \text{PH}\eta_C$, where PH is the compressor preheat factor given by

$$\text{PH} = \frac{\Sigma\,(\Delta h_s)_{st}}{(\Delta h_s)_C} \tag{5.20}$$

For a very large number of compressor stages, the actual compression path may be represented by the line 1–2, Fig. 5.32. The corresponding theoretical reversible adiabatic compression is indicated by the path 1–2_s.

The actual temperature increase dT is effected by compression from pressure P to pressure $P + dP$. The corresponding isentropic temperature increase is given by

$$T\left[\left(\frac{P + dP}{P}\right)^{(k-1)/k} - 1\right]$$

Therefore,

$$\eta_p = \frac{T\left[\left(\frac{P + dP}{P}\right)^{(k-1)/k} - 1\right]}{dT} = \frac{\left(1 + \frac{dP}{P}\right)^{(k-1)/k} - 1}{dT/T}$$

Integration of the above equation between the prescribed limits yields

$$\eta_p = \frac{k - 1}{k} \frac{\ln(P_2/P_1)}{\ln(T_2/T_1)} \tag{5.21}$$

The polytropic efficiency η_p exceeds the reversible adiabatic or conventional compressor efficiency η_C; and because the compressor usually consists of a relatively large number of stages, η_p closely approximates the individual stage efficiency η_{st}.

Example 5.10

Air enters a 10-stage compressor at 101.35 kPa, abs, and 294.44 K. The energy input to the air is 23.260 kJ/kg per stage. The stage efficiency is 0.85. Determine the outlet conditions and calculate the polytropic or approximate stage efficiency.

Stage-by-stage calculations show P_2 = 575.72 kPa, abs, and T_2 = 523.33 K. The calculated value of the compressor isentropic enthalpy increase between the initial state (101.35 kPa, abs, and 294.44 K) and the final pressure (575.72 kPa, abs) is 189.87 kJ/kg.

Compressor Efficiency

$$\eta_C = \frac{h_{2,s} - h_1}{h_2 - h_1} = \frac{189.87}{10 \times 23.260} = 0.816$$

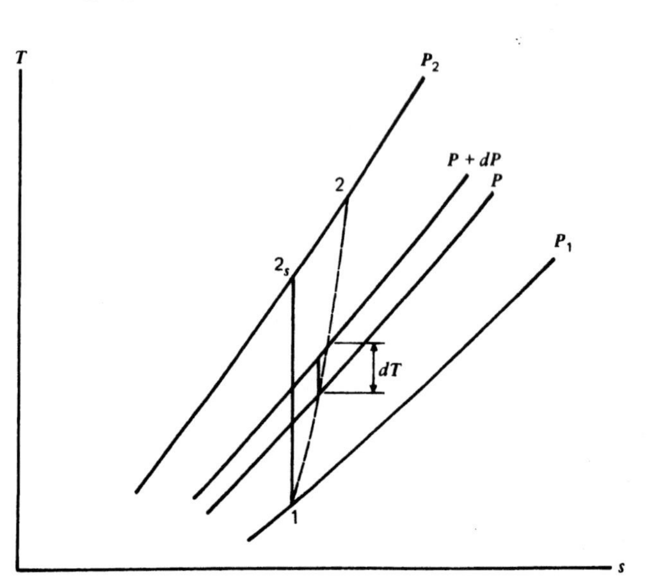

Figure 5.32 Polytropic compression.

Polytropic Efficiency

$$k = 1.395 \qquad \text{at } T_{av} = 409 \text{ K}$$

Substituting the appropriate values of pressure and temperature into Eq. 5.21,

$$\eta_p = \frac{0.395 \ln(575.72/101.35)}{1.395 \ln(523.33/294.44)}$$
$$= 0.855$$

The deviation between the calculated value of the polytropic efficiency and the prescribed value of the stage efficiency in this example is 0.005, a relatively small quantity.

Example 5.11

A multistage axial-flow compressor is to be designed for a pressure ratio of 5 to 1. The inlet conditions are 85.0 kPa, abs, and 280 K. The required stage efficiency is 0.86. Estimate the compressor efficiency. Assume $k = 1.396$ (for air).

Substituting the appropriate quantities into Eq. 5.21,

$$0.86 = \frac{0.396}{1.396} \frac{\ln 5}{\ln (T_2/280)}$$
$$T_2 = 476.4 \text{ K}$$

The isentropic enthalpy increase from 85.0 kPa, abs, and 280 K to 425.0 kPa, abs, is 164.07 kJ/kg. (from Gas Tables). The actual increase in enthalpy from 280 K to 476.4 K is 198.66 kJ/kg.
The compressor efficiency is

$$\eta_C = \frac{h_{2,s} - h_1}{h_2 - h_1} = \frac{164.07}{198.66}$$
$$= 0.826$$

5.17 AXIAL-FLOW COMPRESSOR PERFORMANCE

Performance plots for axial-flow compressors are generally similar to the chart shown in Fig. 5.33. Because the axial-flow compressor is often op-

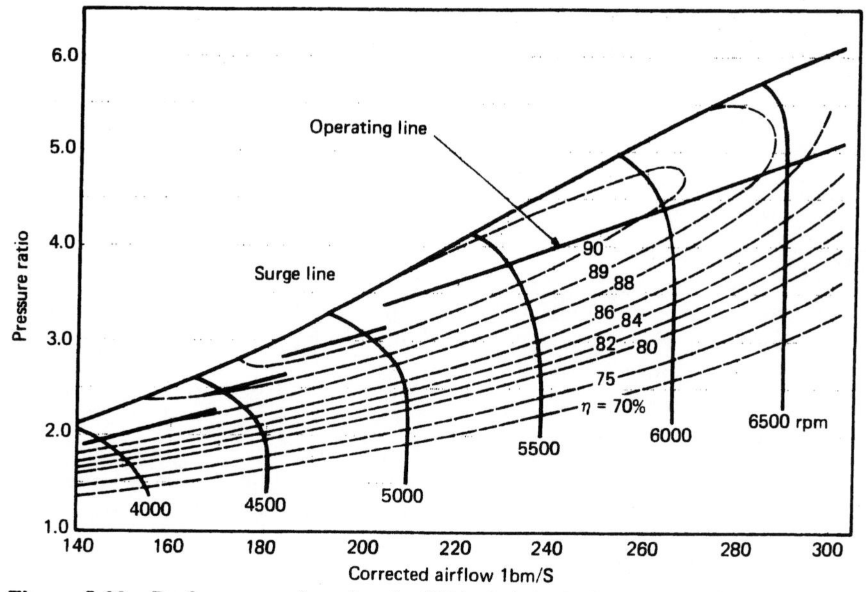

Figure 5.33 Performance chart for the FT4 modular industrial turbine low-pressure compressor. (Courtesy United Technologies Power Systems Division.)

erated as a variable-speed machine, the most significant characteristics, namely, the overall pressure ratio and the efficiency, are plotted for several wheel speeds against the mass flow rate.

For constant-speed operation, the variation in the mass flow rate is limited between surging at one extreme and choking at the other. A relatively small variation in the mass flow rate, at constant-speed operation, causes a significant change in both the pressure ratio and the compressor efficiency.

If the compressor could be operated at constant speed and with essentially no variation in the mass flow rate, operation under optimum conditions would be a possibility. Operation at a fixed point however is usually not observed, hence under certain operating conditions a decline in the compressor performance will be encountered.

The axial-flow compressor is a principal component of the gas turbine power plant. In designing the power plant, the compressor must be matched with other components, principally the turbine, in order to produce a satisfactory operating line. See Fig. 5.33. In this particular performance plot, the operating line for variable-speed operation is located well above the choking limit and crosses into the surging region at low speeds, where the disturbance in the flow is not severe. Except for low-speed operation, the efficiency along the operating line remains relatively high.

5.18 SUMMARY

Because of uncertain supplies and high cost of conventional and nuclear fuels, increasing restraints are placed on the consumption of energy. Conservation of energy is essential with respect to the innumerable secondary energy conversion processes that require an input of mechanical or electrical energy. Pumps, fans, compressors, and many other devices and systems are in this category. Because of improper selection or operation, these machines may function inefficiently and, as a result, consume energy in a wasteful manner. An understanding of the operating principles and characteristics of the various energy-consuming machines will assist in ensuring operation at optimum, or near-optimum, conditions and thereby minimize the energy consumption.

PROBLEMS

5.1 The following dimensions are pertinent to a centrifugal compressor: O.D. 760 mm, inlet diameter 450 mm, and hub diameter 100 mm. Air enters the inlet at P = 95 kPa, abs, and 290 K. The air velocity at the inlet is 10 m/s. The compressor speed is 12,000 rpm, and the slip factor is 0.88. The compressor mechanical efficiency is 0.985. Determine the shaft power.

5.2 Determine for the compressor described in Problem 5.1 the discharge pressure and temperature. The compressor internal efficiency is 0.72. The fluid velocities at the inlet and outlet sections are essentially equal.

5.3 A radial vane centrifugal compressor is capable of developing a pressure ratio of 4.30 to 1 when operating at 15,000 rpm. The inlet temperature of the air is 298 K. The impeller diameter (O.D.) is 600 mm; z = 1.035. The slip factor is 0.91. Estimate the compressor internal efficiency.

5.4 A centrifugal compressor impeller has a 500-mm O.D. and an inlet diameter of 275 mm. For a compressor speed of 13,500 rpm, calculate the theoretical pressure increase developed by the impeller. The inlet conditions are 1 atm, abs, and 300 K; R = 0.275 kJ/kg·K.

5.5 The diffuser vane system in a centrifugal compressor is designed to reduce the air velocity from 400 m/s to 180 m/s. The diffuser efficiency is 0.72. The air enters the diffuser at 160 kPa, abs, and 350 K. Calculate the diffuser outlet-to-inlet area ratio.

5.6 A rotary lobe blower is required to deliver air at the rate of 120 m^3/min, measured at 100 kPa, abs, and 290 K. The delivery pressure is 200 kPa, abs. The blower efficiency based on P ΔV is 0.60. Determine the power input to the blower. Calculate the power input for a centrifugal compressor that provides the same service. The centrifugal compressor efficiency is 0.75.

5.7 A pump is required to deliver water at the rate of 10 m^3/min while developing a total head of 120 m. The pump speed is 1750 rpm. Identify the type of impeller installed in the pump.

5.8 A centrifugal compressor is equipped with a 600-mm-O.D. impeller that turns at 15,000 rpm. The inlet diameter is 320 mm. The inlet air conditions are 96 kPa, abs, and 27 C. The slip factor is 0.89, and the compressor has a prewhirl component of 40 m/s in the direction in which the impeller rotates. The overall machine efficiency is 0.70. Calculate the compressor discharge pressure.

5.9 A centrifugal compressor equipped with a radial vane impeller is required to develop a pressure ratio of 4 to 1 when pumping gas initially at 65 C. The impeller O.D. is 525 mm; z = 1.034; σ = 0.91; k = 1.404; c_p = 1.038 kJ/kg·K. The machine efficiency is 0.72. Calculate the compressor speed.

5.10 A centrifugal fan handles 6000 m^3/min of gas measured at the inlet conditions of -27 cm water and 150 C. The discharge pressure is $+3$ cm water; R = 0.278 kJ/kg·K. The atmospheric pressure is 100 kPa, abs. The fan static efficiency is 0.65. Calculate the fan shaft power.

5.11 A surface condenser requires a cooling water flow of 2×10^6 kg/h. The pump is required to develop a total head of 6.5 m. An axial-flow pump with a midrange specific speed is to be used. Determine the speed of the ac motor used to drive the pump directly and estimate the power output of the motor. A pump efficiency of 0.85 may be assumed.

5.12 A centrifugal pump driven by a constant-speed motor develops a total head of 20 m water. The corresponding delivery is 3.0 m^3 water per minute. The system resistance at this operating point is 20 m water. When the flow is reduced to 1.5 m^3/min, the pump develops a total head of 24 m, while the system resistance drops to 17 m water. The total-head pump efficiency is 0.69 and 0.52, respectively, for the discharge rates of 3.0 and 1.5 m^3/min. Determine the pump power at each discharge rate and the excess power at the reduced load.

5.13 A centrifugal fan delivers 375 m^3 air per minute when operating at a point that is 50 percent of the wide-open volume. The total pressure at this point is 24.2 cm water, and the mechanical efficiency is 0.79. When operating at 100 percent of the wide-open volume, the total pressure and mechanical efficiency are, respectively, 18 cm water and 0.60. Similarly, when operating at 20 percent of the wide-open volume, the total pressure and mechanical efficiency are, respectively, 24 cm water and 0.63. Determine the fan shaft power at each point and describe the overloading characteristic of the fan.

5.14 The rotor of an axial-flow compressor stage turns at 15,000 rpm. The mean diameter at the inlet section of the rotor is 500 mm. The blade length is 80 mm; C_x = 220 m/s; $C_{U,1}$ = 80 m/s. The inlet air pressure and temperature are 120 kPa, abs, and 320 K. The blade row efficiency is 0.87. The pressure of the air at the blade row outlet section is 175 kPa, abs; $C_{x,2} = C_{x,1}$ and $U_{m,2} = U_{m,1}$. Determine the fluid deflection angle.

5.15 Calculate the power input for the rotor described in Problem 5.14. The rotor mechanical efficiency is 0.99.

5.16 Calculate the degree of reaction at the mean, tip, and root diameter sections for the compressor stage described, in part, in Problem 5.14. Assume free vortex flow.

5.17 An axial-flow compressor stage is designed for 50 percent constant reaction. The rotor turns at 10,800 rpm; $C_{x,1} = C_{x,2}$ = 200 m/s; $C_{U,1}$ = 75 m/s at the mean diameter. The rotor mean diameter is 600 mm, and the blade length is 120 mm. Determine the rotor fluid deflection angle at the mean, tip, and root diameter sections and the work input to the fluid, in kJ/kg air.

5.18 Air enters the stator of an axial-flow compressor stage at 200 kPa, abs, and 330 K. At the inlet section, $C_{x,2}$ = 150 m/s; and at the outlet section, $C_{x,3}$ = 145 m/s. The vane root diameter is 650 mm, and the blade length at the inlet section is 150 mm. At the mean diameter, $C_{U,2}$ = 190 m/s, and the fluid deflection angle is 30°. The efficiency of the blade row is 0.86. Calculate the length of the vane at the stator outlet section.

5.19 A multistage axial-flow compressor is designed to develop a pressure ratio of 8 to 1 with an isentropic efficiency of 0.86. Determine the probable average stage efficiency and the fluid power required for compressing the air. The inlet air temperature is 300 K.

5.20 At the mean diameter of an axial-flow compressor stage, $C_{x,1}$ = $C_{x,2}$ = 165 m/s. The inlet blade angle α_1 = 70°, and θ = 25°. The blade speed is 275 m/s. The rotor efficiency is 0.87. The air inlet conditions are 120 kPa, abs, and 300 K. Determine the pressure of the air leaving the rotor.

5.21 The work input for an axial-flow compressor stage is 35 kJ/kg air. At the inlet section, the velocity ratio $C_{U,1}/C_{X,1}$ = 0.470. The blade speed U_m = 285 m/s, C_x = 170 m/s. Calculate the fluid deflection angle at the mean diameter.

5.22 An axial-flow compressor stage is designed for a constant reaction of 70 percent; C_x = 180 m/s. At the mean diameter, $C_{U,1}$ = 65 m/s and U_m = 325 m/s. The stage mean diameter is 500 mm, and the blade length is 80 mm. Calculate the rotor and stator fluid deflection angles at the mean and tip diameters. Determine the work done on a unit mass of fluid at each diameter.

5.23 At the inlet to an axial-flow compressor stage, the pressure and temperature of the air are 110 kPa, abs, and 290 K; C_x for the stage is 175 m/s. At the mean diameter, the blade velocity is 320 m/s, and the fluid deflection angle is 20°. The rotor efficiency is 0.87. Determine the pressure increase developed in the rotor for (a) no prerotation of the air and (b) for prerotation with α_1 = 65°. Calculate the work input to the rotor for each case.

5.24 A multistage axial-flow compressor operates with a stage efficiency equal to 0.88. At the compressor inlet, the pressure and temperature of the air are 98 kPa, abs, and 305 K. The compressor discharge pressure is 700 kPa, abs. Determine the compressor efficiency.

5.25 An axial-flow compressor stage is designed to develop a pressure ratio of 1.30. A 50-percent reaction design is prescribed. Assume C_3 = C_1; C_x = 155 m/s; α_1 = 70°. The stage efficiency is 0.88. At the inlet section, the pressure and temperature of the air are 120 kPa, abs, and 320 K, respectively. Determine the blade velocity at the mean diameter section.

5.26 A multistage axial-flow compressor develops a pressure ratio of 8 to 1. The air enters the compressor at 100 kPa, abs, and 290 K. The compressor efficiency is 0.87. Determine the average stage efficiency.

5.27 An axial-flow compressor stage has a mean diameter of 500 mm. The blade length is 100 mm; C_x = 210 m/s. The compressor speed is 12,000 rpm. At the tip diameter section, $C_{U,1}$ = 50 m/s, and the fluid deflection angle is 15°. Calculate the fluid deflection angle at the mean diameter. Assume free vortex flow.

5.28 Air enters the rotor of an axial-flow compressor stage at P = 80 kPa, abs, and T = 260 K. The axial velocity of the air is 160 m/s. At the inlet section, the blade length is 150 mm, and the mean diameter is 650 mm. The power input to the rotor is 1775 kW. The inlet relative fluid angle is 35°, and the fluid deflection angle is 20°. Determine the rotor speed.

BIBLIOGRAPHY

Csanady, G. T. *Theory of Turbomachines,* McGraw-Hill Book Co., New York, 1964.

Kent's Mechanical Engineers' Handbook, Power Volume, J. K. Salisbury, ed., John Wiley & Sons, Inc., New York, 1957.

Lichty, L. C. *Combustion Engine Processes,* McGraw-Hill Book Co., New York, 1967.

Marks' Standard Handbook for Mechanical Engineers, T. Baumeister, E. A. Avallone, and T. Baumeister, III eds., McGraw-Hill Book Co., New York, 1978.

CHAPTER SIX

AXIAL-FLOW TURBINES

Steam and gas turbines can be classified in accordance with the general direction of the flow of the working fluid through the machine, namely, axial flow or radial flow. Most turbines are designed for axial flow of the steam or gas. Thus, except for the following brief note, this chapter will cover only the axial-flow turbine.

The Ljungstrom steam turbine, used principally in Europe, is a noteworthy example of the radial-flow turbine. The steam flows outward in a radial direction through several rows of rotating blades. Alternate blade rows rotate in opposite directions, hence high relative steam velocities are generated, an indication of high-capability performance. Numerous small radial-flow turbines have been constructed. Both inward- and outward-flow designs are observed for these machines.

Steam and gas turbines, despite the different characteristics of their working fluids, exhibit many points of similarity in design, construction, and operation. Significant differences are observed however in the operating pressures and temperatures of these machines. For steam turbines, the maximum throttle temperature is presently limited to about 540 C. Gas turbines, on the other hand, generally operate with higher fluid inlet temperatures, about 1000 C for industrial machines and up to about 1300 C for aircraft turbines designed for high-level performance. Maximum pressures for the working fluid are about 35 MPa for steam turbines and 4 MPa and 2 MPa for gas turbines operating, respectively, in closed-cycle and open-cycle power plants. High throttle pressures require heavy construction for certain sections of the steam turbine. The gas turbine, on the other hand, is a relatively light-weight machine.

6.1 TURBINE DEVELOPMENT

The development of the steam turbine was achieved through the efforts of many investigators; however, the names of four men appear prominently in the literature of this field. During the 1880 decade, C. G. P. De Laval produced the first steam turbines of commercial importance. Improvements in the De Laval turbine, a simple, single-stage machine, through velocity compounding and pressure compounding are attributed, respectively, to C. G. Curtis and A. C. E. Rateau. These three types of turbines are classified as impulse machines. C. A. Parsons, working independently with a different approach, constructed a successful reaction turbine.

The possibility of constructing a gas turbine engine was recognized many years ago. At the turn of the century, a machine of this kind was constructed, but the performance achieved in subsequent operation proved to be disappointing. The difficulty experienced with available materials and inadequate means of air compression retarded for several decades intensive development of the gas turbine engine. However, small-machine development was motivated significantly by the success achieved in the application of turbosuperchargers to the internal combustion engine. Current activity in the gas turbine industry can be traced to an extensive program of development that originated during the 1930 decade.

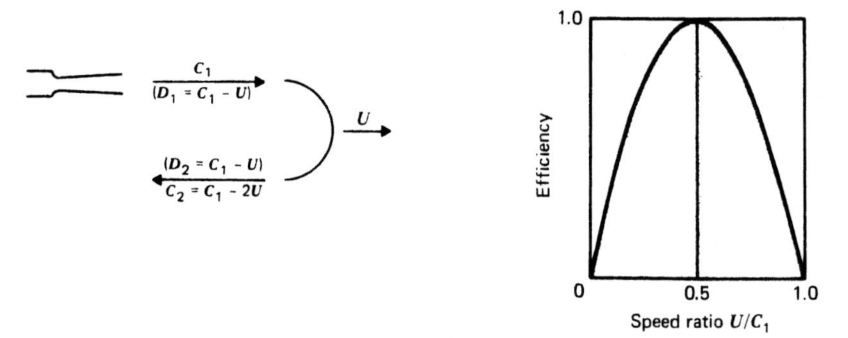

Figure 6.1 Hypothetical single-row impulse turbine stage. D = fluid velocity relative to the blade.

6.2 IMPULSE TURBINES

The impulse turbine is designed on the principle of expanding the working fluid in stationary nozzles and removing the kinetic energy of the high-velocity fluid stream by subsequent impingement on the blades fastened to a rotating wheel. The product of the blade velocity and the momentum change of the fluid in the plane of rotation is equivalent to the work done on the blades.

The operating principle of the De Laval turbine is illustrated by examining a hypothetical turbine equipped with frictionless 180° blades. See Fig. 6.1. (See also Fig. 6.5.) The working fluid is expanded in the nozzle to the outlet velocity C_1. The blade velocity is designated by U. The absolute velocity of the fluid leaving the blade is expressed by $C_2 = C_1 - 2U$. When $U = 0.5C_1$, $C_2 = 0$. All of the kinetic energy in the fluid leaving the nozzle has been converted to work, and consequently a blade efficiency of 100 percent is attained. When $U = C_1$, or $U = 0$, no work is performed, and a blade efficiency of zero is observed. Thus, for the hypothetical De Laval turbine, $U_{opt} = 0.5C_1$.

Expansion of the high-pressure fluid from the inlet to the exhaust pressure in a single set of nozzles produces a very high outlet velocity. For optimum operation, that is, $U = 0.5C_1$, the wheel must have a very large diameter or turn at a very high speed. The small-capability De Laval turbines often operated at speeds well above 30,000 rpm and, for some applications, drove the connected load through a gear speed reducer. For high-capability turbines, neither the use of speed reducers nor use of large wheel diameters would represent a practicable design, hence compound arrangements of the simple turbine have been developed.

The Curtis, or velocity-compounded, turbine utilizes a single expansion of the working fluid in the nozzles, with the resulting kinetic energy of the fluid reduced in two or more rotating blade rows.* The rotating blades are fastened to a common disk. Stationary reversing vanes are installed between the rotating blade rows.

Figure 6.2 shows the operating principle of a hypothetical two-row, velocity-compounded impulse turbine. (See Fig. 6.6.) The blade speed U is assumed to be equal for the two rows. The kinetic energy of the fluid leaving the second blade row is zero when $U = 0.25C_1$. A theoretical blade efficiency of 100 percent is thus achieved at the optimum speed ratio, $\rho_{opt} = U/C_1 = 0.25$. The blade efficiency declines as the speed ratio is increased above the optimum value.

In comparison with the De Laval turbine, the two-row Curtis stage achieves theoretically, for a prescribed nozzle outlet velocity, equivalent performance with the blade velocity reduced by 50 percent. For three and four blade rows, the optimum speed ratios are, respectively, 0.167 and 0.125. Velocity compounding, because of the de-

* See Fig. A.3 for construction details.

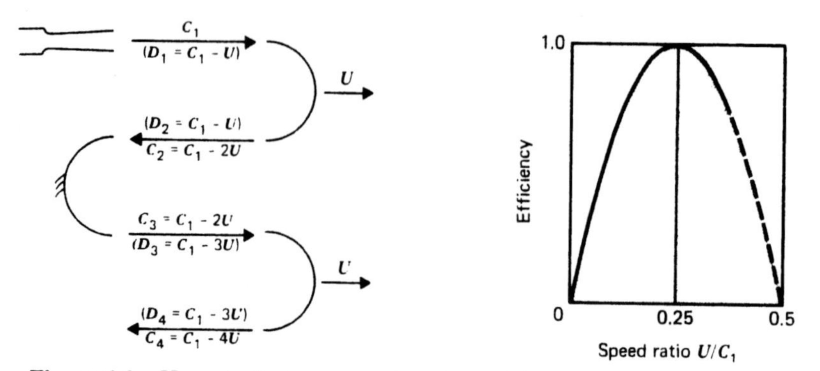

Figure 6.2 Hypothetical two-row impulse turbine stage.

crease in the blade velocity, achieves a reduction in the product of the wheel speed and diameter.

The blade row work is equivalent to the product of the blade velocity and the momentum change of the fluid in the rotational plane; thus, for a two-row Curtis stage, the ratio of the first-row work to the second-row work is theoretically 3 to 1. Similarly, for three- and four-row stages, the theoretical work ratios are, respectively, 5 to 3 to 1 and 7 to 5 to 3 to 1.

From another point of view, for the same blade speed, the pressure drop and nozzle outlet velocity are greater for the Curtis stage than for the comparable De Laval turbine. High fluid velocities however promote high frictional losses, hence Curtis stages generally operate with relatively poor efficiencies. Curtis turbines were originally constructed with several rotating-blade rows, but because of low efficiencies, the number of rows is usually limited to two, or possibly three. Steam turbines consisting of a two-row Curtis stage are used for auxiliary drives and in low-capability power units.

The Rateau turbine, a pressure-compounded machine, consists of a number of single-row stages arranged for series flow of the working fluid. Following expansion in one stage, the working fluid enters the nozzle row of the adjacent downstream stage. Each stage is essentially a De Laval turbine; hence for hypothetical operation, $U_{opt} = 0.5\ C_1$. (Ref.: Fig. 6.1.)

Pressure compounding produces a satisfactory design, although turbine construction costs are relatively high because of the large number of required stages. High turbine efficiencies are achieved, in part, because of the relatively low fluid velocities that are typical of the Rateau turbine.

6.3 REACTION TURBINES

Fluid expansion in both the stationary and the rotating blade passages is the principal distinguishing characteristic of the reaction turbine. Figure 6.3 illustrates the hypothetical blade arrangement for the reaction stage. (Ref.: Fig. 6.8.) For optimum conditions, expansion within the stationary blade row is just sufficient to cause the fluid to enter the moving blades; thus, $U = C_1$ and $D_1 = 0$, and consequently $\rho_{opt} = 1.0$. Subsequent expansion within the moving blade row accelerates the fluid to the point where the outlet relative velocity D_2 is equal to U, hence $C_2 = 0$.

The reaction turbine, because of low fluid velocities, shows a comparatively high level of performance, but the required large number of stages is a disadvantage with respect to the length and the resulting cost of the machine. Although straight pressure-compounded and straight reaction steam turbines were formerly built in large numbers, current practice usually incorporates in the first stage of a high-capability machine a single-row or a two-row impulse-type wheel. This impluse stage achieves a high enthalpy drop at the inlet end of the turbine, with an accompa-

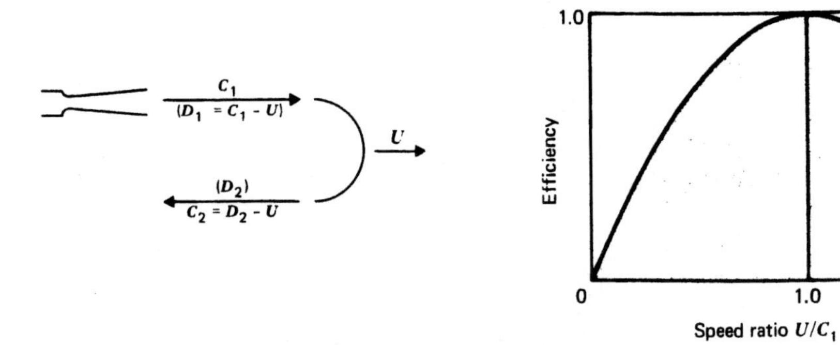

Figure 6.3 Hypothetical reaction turbine stage.

nying rapid decrease in the steam pressure. In addition, there is a significant decrease in the required number of stages for the turbine.

The basic turbine designs are compared theoretically in Table 6.1 for equal values of the overall isentropic energy drop $(\Delta h_s)_T$ and for operation at the optimum speed ratio.

Table 6.1 shows that for a prescribed isentropic energy drop of 125 kJ/kg fluid, theoretical expansion may be achieved in a single De Laval turbine with a blade speed of 250 m/s. Equivalent expansion, with a blade speed of 125 m/s, may be achieved in (a) a single two-row Curtis stage, (b) four Rateau stages, or (c) eight reaction stages. In practice, because of fluid losses and certain design requirements, a somewhat different numerical relationship will be observed.

Gas turbine designs usually incorporate reaction stages. Both the impulse and reaction designs are used in steam turbine practice, although reaction stages are better suited for operation in the low-pressure region where high steam volumes are handled. In commercial practice, the impulse and reaction designs achieve essentially the same level of operating efficiency.

6.4 TURBINE STAGE ARRANGEMENT

The preceding discussion serves as a simple yet effective means for comparing the characteristics of the several types of turbine designs. For the actual machine, the nozzle vane angle, measured from the plane of rotation, usually ranges from 12 to 20 °. A gas turbine nozzle ring is shown in Fig. 6.4. Turbine blades are actually designed to achieve a fluid deflection angle that is substantially less than 180°. These departures from the theoretical concepts are incorporated into the design and construction of the turbine in order to provide an adequate flow area through the nozzle and blade rows.

Figures 6.5 to 6.8 show the nozzle and blade arrangements for the several types of turbines examined in the preceding sections. Velocity and pressure variations through the different stages

TABLE 6.1 Stage Work for Basic Turbine Designs

Turbine Type	*Stage Work*	*Blade Speed (m/s)*	*Stage Work (J/kg)*	*Work Ratio*
De Laval	$2U^2$	250	125,000	1
Curtis (two-row)	$8U^2$	125	125,000	1
Rateau	$2U^2$	125	31,250	0.250
Reaction	U^2	125	15,625	0.125

Figure 6.4 Gas turbine stator ring. (Courtesy Contura Incorporated.)

are also depicted. Figure 6.9 illustrates typical impulse and reaction blade profiles.

6.5 VECTOR DIAGRAMS

Fluid flow through a turbine stage is described graphically by a vector diagram. Figure 6.10 shows the vector diagram for a single-row impulse turbine. The nozzle inlet velocity for a machine of this type is often assumed to be zero; however stagnation properties can be used to account for the inlet kinetic energy. Gas or steam leaves the nozzle with an absolute velocity C_1 at an angle α_1 with the plane of rotation. The combination of the vectors C_1 and $-U_1$, the negative blade velocity, produces the fluid inlet relative velocity vector D_1.

During flow through the blade passage, the relative velocity is reduced by friction and turbulence, hence the fluid outlet relative velocity

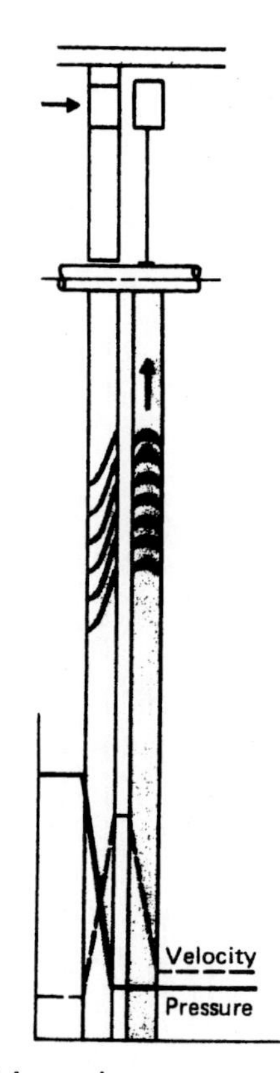

Figure 6.5 Schematic arrangement for a single-row impulse turbine stage.

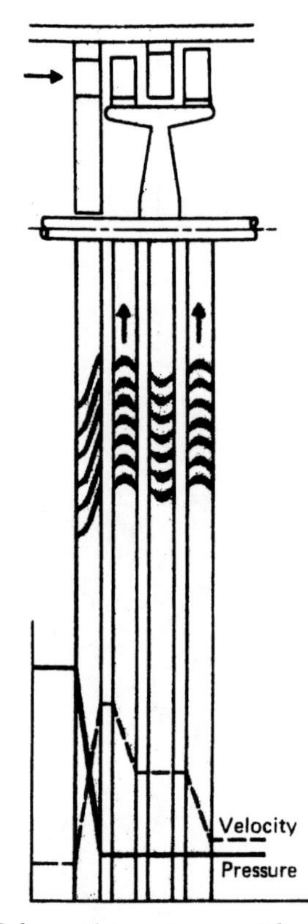

Figure 6.6 Schematic arrangement for a two-row impulse turbine stage.

$D_2 = k_b D_1$, where k_b is the blade velocity coefficient, a factor that has a value of less than 1. The fluid outlet absolute velocity vector C_2 is determined by combining the vectors D_2 and U. The fluid leaves the rotating blade row at an angle α_2, measured from the plane of rotation. Inlet and outlet fluid angles relative to the blade are indicated, respectively, by β_1 and β_2. It was assumed in this discussion that no radial displacement of the fluid occurs during passage through the blade row, that is, $U_2 = U_1$.

The vector diagram for a two-row impulse turbine stage is shown in Fig. 6.11. In the row of fixed blades, the working medium is not expanded but only turned for entry into the second row of moving blades. Because of friction, the fluid velocity is reduced during passage through each blade row, hence $C_3 = k_{b,2}C_2$ and $D_4 = k_{b,3}D_3$. The magnitude of the blade velocity coefficient is dependent upon a number of factors, including the fluid inlet relative velocity; thus, $k_{b,1} < k_{b,2} < k_{b,3}$.

Figure 6.12 shows a typical vector diagram for a reaction turbine stage. Expansion of the fluid in the rotating blades causes an increase in the relative velocity of the fluid from D_1 to D_2. When the reaction turbine stage is constructed with identical blade rows, fixed and moving, the vector diagram may be formed from two congruent triangles.

Considerable difference is observed for some designs in the configuration of the vector diagrams constructed at different radii for a particular blade. This deviation is especially apparent

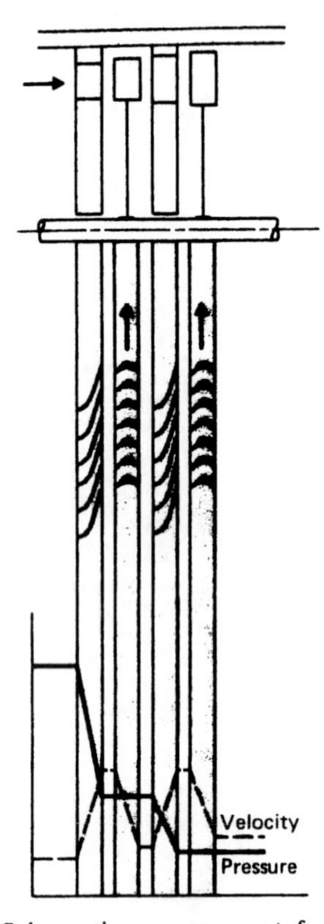

Figure 6.7 Schematic arrangement for a two-stage pressure-compounded turbine.

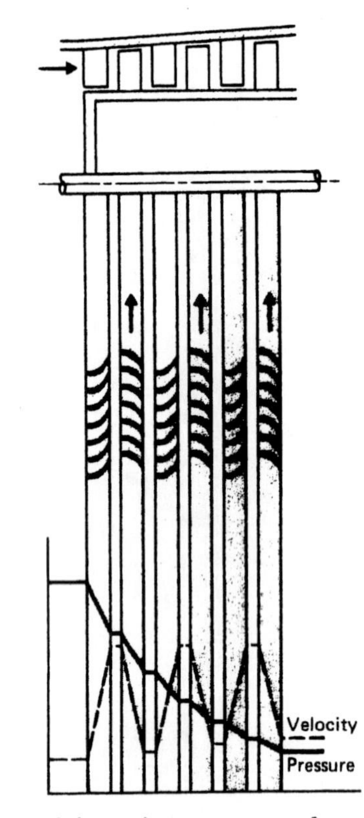

Figure 6.8 Schematic arrangement for a three-stage reaction turbine.

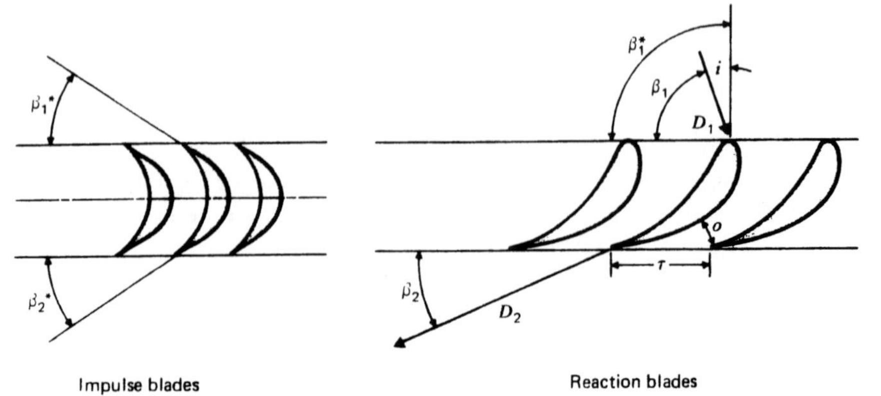

Figure 6.9 Typical turbine blade profiles. D = fluid relative velocity, β = fluid angle, β_1^* = inlet blade angle, i = incidence angle ($\beta_1^* - \beta_1$), o = minimum width of the blade passage, τ = blade pitch.

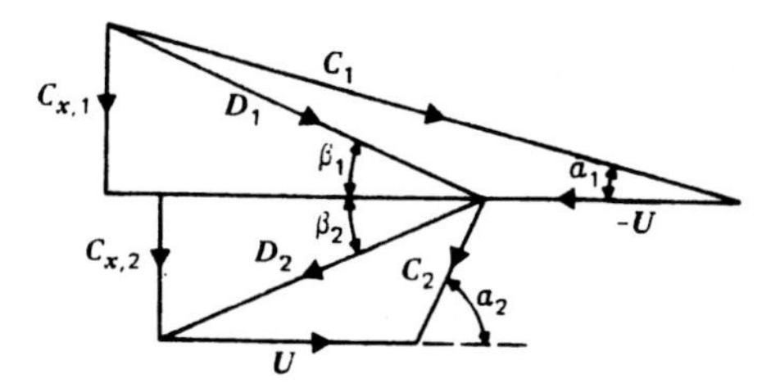

Figure 6.10 Vector diagram for a single-row impulse turbine stage. C = fluid absolute velocity, D = fluid relative velocity, C_x = axial velocity component, U = blade velocity, α and β = fluid angles.

for relatively long blades because of the variation in the linear blade speed with the radius of rotation.

6.6 STAGE WORK

The work performed by the fluid on the rotating blades of a turbine stage is given by

$$W = C_{U,1}U_1 - C_{U,2}U_2 \tag{6.1}$$

This equation is derived from an analysis similar to that presented in Section 5.3 for the compressor. Care must be exercised in the use of Eq. 6.1, because $C_{U,1}$ and $C_{U,2}$ are ordinarily of opposite sign, as shown in Figs. 6.10 and 6.12.

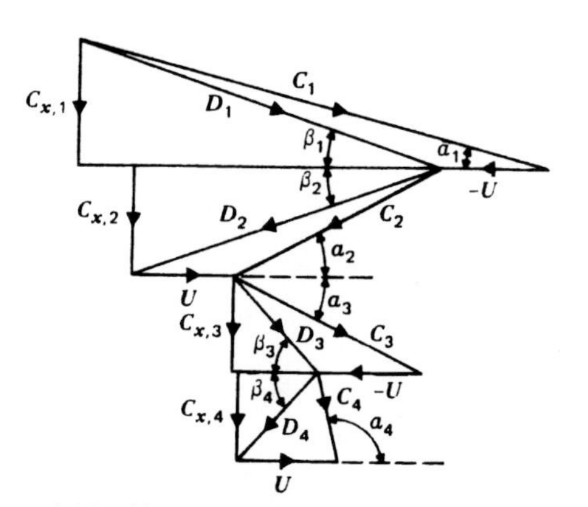

Figure 6.11 Vector diagram for a two-row impulse turbine stage. Symbols are defined under Fig. 6.10.

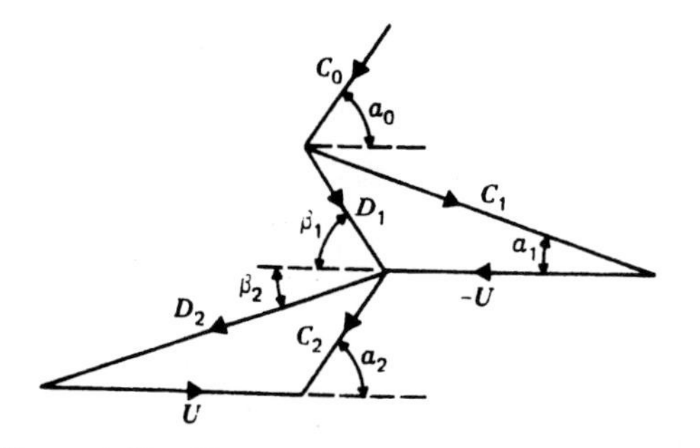

Figure 6.12 Vector diagram for a reaction turbine stage. Symbols are defined under Fig. 6.10.

U_1 and U_2 are often assumed to be of the same magnitude, and Eq. 6.1 can then be reduced to

$$W = (D_{U,1} - D_{U,2})U \tag{6.2}$$

$D_{U,1}$ and $D_{U,2}$ are usually of opposite sign.

Stage work is frequently determined from the first law energy balance made on the rotating blade row. For an impulse turbine stage,

$$W = \frac{C_1^2 - C_2^2}{2} - (h_2 - h_1)$$

The quantity $h_2 - h_1$ is designated as the blade reheat Δh_R. The enthalpy increase, or reheat, in the blade row is caused by fluid friction.

$$\Delta h_R = \frac{D_1^2 - D_2^2}{2} = (1 - k_b^2)\frac{D_1^2}{2}$$

Then,

$$W_{\text{imp}} = \frac{C_1^2 - C_2^2}{2} - (1 - k_b^2)\frac{D_1^2}{2} \tag{6.3}$$

The energy balance made on the complete reaction turbine stage yields

$$W = (h_0 - h_2) + \frac{C_0^2 - C_2^2}{2}$$

$$= (h_0 - h_1) + (h_1 - h_2) + \frac{C_0^2 - C_2^2}{2}$$

An energy balance is made on the stationary blade row and on the rotating blade row. Thus,

$$h_0 - h_1 = \frac{C_1^2 - C_0^2}{2}$$

$$h_1 - h_2 = W - \frac{C_1^2 - C_2^2}{2}$$

$$= (C_{U,1}U_1 - C_{U,2}U_2) - \frac{C_1^2 - C_2^2}{2}$$

$$= \frac{D_2^2 - D_1^2}{2} + \frac{U_1^2 - U_2^2}{2}$$

Therefore,

$$W_{\text{reac}} = \frac{C_1^2 - C_0^2}{2} + \frac{D_2^2 - D_1^2}{2} + \frac{U_1^2 - U_2^2}{2} + \frac{C_0^2 - C_2^2}{2} \tag{6.4}$$

In the analysis of a reaction turbine stage, it is often feasible to assume that the fluid inlet and outlet velocities are equal, that is, $C_2 = C_0$. Further, the radial displacement of the fluid in passing through the stage may on occasion be negligible, or $U_2 = U_1$. Equation 6.4 is then reduced to

$$W_{\text{reac}} = \frac{C_1^2 - C_0^2}{2} + \frac{D_2^2 - D_1^2}{2} \tag{6.5}$$

Stage work, efficiencies, and other performance data are often derived from measurements and calculations made at the mean diameter. Note: Miscellaneous fluid losses have been neglected in the preceding developments.

6.7 AXIAL THRUST

The axial thrust F_x acting on the rotor of an impulse turbine stage is caused by the change in the momentum of the fluid in the axial direction. Thus,

$$F_x = \dot{m}(C_{x,1} - C_{x,2}) \tag{6.6}$$

where $\dot{m}$ is the mass flow rate of the fluid; and $C_{x,1}$ and $C_{x,2}$ are, respectively, the axial velocity components of the fluid entering and leaving the rotating blade row.

The summation of the axial thrust for the several impulse stages that comprise the turbine is not an especially large quantity. The total axial force acting on the rotor can be restrained by a thrust bearing.

In a reaction turbine stage, the momentum change in the fluid causes only a relatively small axial thrust on the rotor. A significant pressure drop exists however across each blade row. The axial thrust applied to a blade ring is equivalent to the product of this pressure difference and the annular flow area. A summation of the axial forces acting on the moving blade rows yields the total axial thrust that tends to move the rotor toward the exhaust end of the turbine. For a multistage reaction turbine, this force is of a substantial magnitude.

Gas turbine engine construction commonly arranges for a tandem installation of the turbine and the compressor elements. For this arrangement, the reaction turbine axial thrust is roughly counterbalanced by a similar but opposing force that is exerted on the rotor of the axial-flow compressor. The residual force is then carried by a thrust bearing of moderate capacity.

Almost all of the stages in a steam turbine are designed for some degree of reaction. The axial force acting on the rotor is partially counterbalanced in tandem compound machines by arranging the several turbine elements for opposed steam flow. (Ref.: Fig. 6.19.) The residual unbalanced force is carried by a thrust bearing.

6.8 TURBINE STAGE EFFICIENCY

The efficiency at the wheel periphery of an impulse turbine stage is defined as the product of the nozzle efficiency and the blade efficiency, $\eta_N \times \eta_b$. The blade efficiency is evaluated from the vector diagram and taken as the ratio of the work done on the rotor blades to the kinetic energy in the fluid leaving the nozzle, that is, $\eta_b = W_b \div (C_1^2/2)$. See Fig. 6.13. The kinetic energy

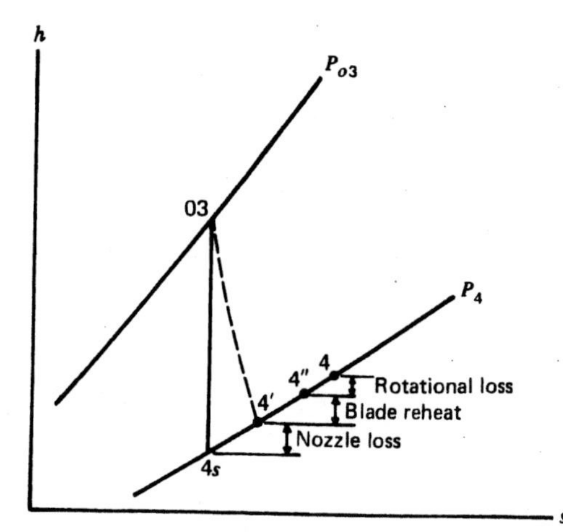

Figure 6.13 Impulse turbine stage h–s diagram.

$C_{4'}^2/2$ of the fluid at the nozzle outlet is equal to $\eta_N(h_{03} - h_{4,s})$, the product of the nozzle efficiency and the isentropic energy drop for the stage.

During passage of the fluid through the blade row, 4′ to 4″, a portion of the inlet kinetic energy is dissipated in friction and turbulence, causing a simultaneous, constant-pressure increase in the enthalpy, $h_{4''} - h_{4'}$, the blade reheat. Because of disk friction and fanning action of the idle blades, a portion of the blade work is returned to the working fluid as an increase in enthalpy Δh_f. Then, $h_4 = h_{4''} + \Delta h_f$, for $C_4 = C_{4''}$.

The efficiency of the impulse turbine stage is equal to the net internal work divided by the isentropic energy drop for the stage. Thus,

$$\eta_{st} = \frac{W_b - \Delta h_f}{h_{03} - h_{4,s}} \tag{6.7}$$

where

W_b = blade work
Δh_f = loss owing to disk friction and fanning action of the idle blades (rotational loss)

The stage work is also determined by applying the energy equation to the fluid flowing through the turbine between the inlet section (03) and the outlet section (4). Thus

$$W_{st} = h_{03} - h_4 - \frac{C_4^2}{2}$$

and the stage efficiency is expressed by

$$\eta_{st} = \frac{h_{03} - h_4 - C_4^2/2}{h_{03} - h_{4,s}} \tag{6.7a}$$

Note: The numerator in Eq. 6.7 and the numerator in Eq. 6.7a are like quantities.

If the residual kinetic energy in the fluid discharged from the last stage of a turbine is not recovered in some manner, this energy is considered to be a leaving loss. The residual kinetic energy for stages other than the last stage may, in some impulse turbines, be dissipated entirely through turbulence. A portion of the residual kinetic energy in the fluid can however be conserved by directing the outlet fluid, with a velocity C_4, into the inlet of the nozzle of the following stage, and thus improve to some degree the performance of the turbine.

For fixed nozzle operating conditions, the kinetic energy of the fluid at the nozzle outlet remains constant, while the stage work is a function of the blade speed. In order to achieve a high level of performance, the speed ratio ρ is selected to yield the maximum stage work that is consistent with maintaining a reasonably high value of the blade efficiency.

The reaction turbine design is based on the carry-over of the kinetic energy of the fluid from one blade row to another. The vector C_1, Fig. 6.12, represents the velocity of the fluid leaving the stationary blade row for the reaction turbine stage; C_0 designates the velocity of the fluid leaving the rotating blade row of the preceding stage. For a prescribed isentropic enthalpy drop Δh_s for the stationary blade row, the kinetic energy of the fluid at the outlet section is determined from

$$\frac{C_1^2}{2} = \phi_E\left(\Delta h_s + \phi_i \phi_v \frac{C_0^2}{2}\right) \tag{6.8}$$

where

$C_1^2/2$ = outlet kinetic energy
ϕ_E = expansion energy coefficient
ϕ_i = incidence coefficient
ϕ_v = kinetic energy coefficient
Δh_s = isentropic enthalpy drop for the blade row
$C_0^2/2$ = kinetic energy at the blade row inlet

The primary loss resulting from turbulence and wall friction within the blade passage is indicated by the expansion energy coefficient ϕ_E. Values of this coefficient for gas turbine blades are shown in Fig. 6.14.

The effective kinetic energy at the outlet section is less when obtained from a specified quantity of inlet kinetic energy than from an equal amount of energy made available by expansion. Because of the higher loss incurred in the carry-over of the inlet kinetic energy, the quantity $C_0^2/2$ is reduced by the inclusion of the coefficient ϕ_v. The value of the kinetic energy coefficient ϕ_v may be taken as approximately equal to the expansion energy coefficient ϕ_E. The incidence coefficient ϕ_i is included in Eq. 6.8 in order to allow for additional losses that result from a non-ideal approach of the fluid at the blade inlet where a positive or negative incidence angle may

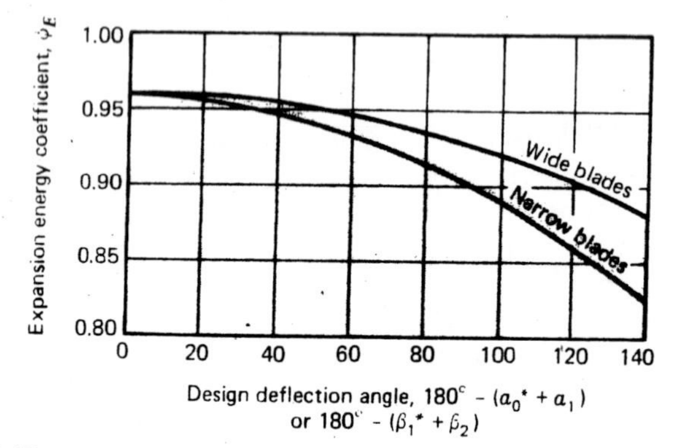

Figure 6.14 Expansion energy coefficient for gas turbine reaction blades. α_0^* and β_1^* = inlet blade angles, α_1 and β_2 = fluid outlet angles. (H. D. Emmert, *Trans. ASME*, Feb. 1950.[1])

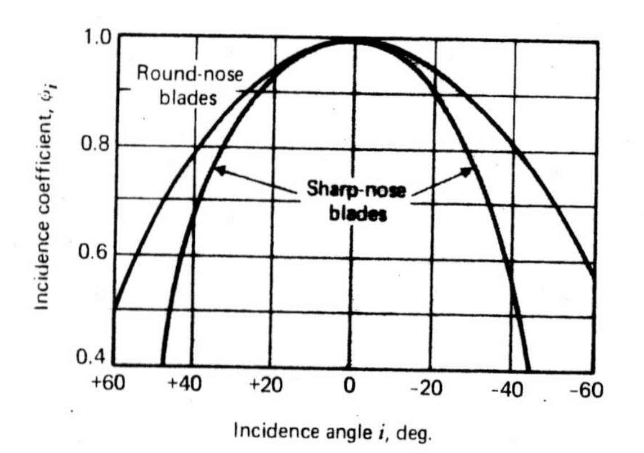

Figure 6.15 Incidence coefficient for gas turbine reaction blades. Ref. Fig. 6.9. i = incidence angle, $(\alpha_0^* - \alpha_0)$ or $(\beta_1^* - \beta_1)$. (H. D. Emmert, *Trans. ASME*, Feb. 1950.[1])

be established. Values of ϕ_i are taken from Fig. 6.15.

Coefficients of approximately the same magnitude as those shown in Figs. 6.14 and 6.15 would be used in the design and in the analysis of the performance of steam turbine reaction blades.

An expression similar to Eq. 6.8 can be written for the rotating blade row of the reaction turbine stage. The vectors D_1 and D_2 represent, respectively, the fluid inlet and outlet relative velocities.

$$\frac{D_2^2}{2} = \phi_E\left(\Delta h_s + \phi_i\phi_v\frac{D_1^2}{2}\right) \qquad (6.9)$$

Equations 6.8 and 6.9 are used to calculate, for a prescribed vector diagram, the isentropic enthalpy drops $(\Delta h_s)_N$ and $(\Delta h_s)_R$, respectively, for the stationary and rotating blade rows of a reaction turbine stage. The isentropic enthalpy drop for the reaction turbine stage is usually taken as the sum of the enthalpy drops for the two blade rows, although this summation is not strictly correct.

Unlike the impulse turbine stage, the efficiency of the reaction turbine stage cannot be subdivided into the nozzle and blade efficiencies because the fluid expansion occurs in both blade rows. Thus, the reaction turbine stage efficiency is given by

$$\eta_{st} = \frac{W_{stage}}{(\Delta h_s)_N + (\Delta h_s)_R} \qquad (6.10)$$

Example 6.1

The following data are applicable to a stationary blade row for a gas turbine reaction stage.

Inlet gas velocity: 125 m/s

Incidence angle: $-12°$

Design deflection angle: 100°

Blade design: average width and average nose construction

Isentropic enthalpy drop for the blade row: 50.0 kJ/kg

Blade row mean diameter: 38 cm

$\sin \alpha_1 = 0.370 \qquad (o/\tau = 0.370)$*

Blade length: 5.0 cm

Number of blades: 77

Blade thickness at the trailing edge: 0.52 mm

Density of the fluid at the outlet of the blade row: 1.169 kg/m³

(a) Determine the gas flow through the blade row.

$$\frac{C_0^2}{2} = \frac{(125)^2}{2} = 7812 \text{ J/kg}$$

$$\phi_E = 0.91 \qquad \text{(Fig. 6.14)}$$

$$\phi_i = 0.95 \qquad \text{(Fig. 6.15)}$$

$$\frac{C_1^2}{2} = \phi_E\left(\Delta h_s + \phi_i \phi_v \frac{C_0^2}{2}\right)$$

$$= 0.91[50.0 + (0.95 \times 0.91 \times 7.81)]10^3$$

$$C_1 = 321.4 \text{ m/s}$$

*The fluid outlet angle from the blade ring is assumed to be approximately equal to the arcsin o/τ, where o is the blade throat opening and τ is the blade circumferential pitch.

The flow area is taken normal to the outlet velocity vector C_1, hence

$$A_1 = (\pi d_m \sin \alpha_1 - nt)l$$

where

n = number of blades
t = blade thickness at the trailing edge
l = blade length

$$A_1 = [(38\pi \times 0.370) - (77 \times 0.052)]5$$

$$= 200.8 \text{ cm}^2$$

$$\dot{m} = \rho_1 C_1 A_1$$

$$= 1.169 \times 321.4 \times 0.02008$$

$$= 7.544 \text{ kg/s}$$

(b) The rotating and stationary blade rows are assumed to be identical for the stage. The vector diagram is comprised of two congruent triangles.

$$180° - (\alpha_0^* + \alpha_1) = \text{design deflection angle}$$

$$180° - (\alpha_0^* + 21.7°) = 100° \qquad \alpha_0^* = 58.3°$$

$$i = \alpha_0^* - \alpha_0$$

$$-12° = 58.3° - \alpha_0$$

$$\alpha_0 = 70.3°$$

Also $\beta_1 = 70.3°$

From the vector diagram, similar to Fig. 6.12,

$$D_1 = 126.3 \text{ m/s} \qquad U = 256.0 \text{ m/s}$$

$$C_{U,1} = 298.6 \text{ m/s} \qquad C_{U,2} = 42.6 \text{ m/s}$$

$$D_2 = C_1 = 321.4 \text{ m/s}$$

$$W = (C_{U,1} - C_{U,2})U$$

$$= (298.6 + 42.6)256.0 = 87{,}347 \text{ J/kg}$$

$$D_1 = 126.3 \text{ m/s} \qquad \frac{D_1^2}{2} = 7.98 \text{ kJ/kg}$$

The isentropic enthalpy drop for the rotor is determined from Eq. 6.9 for $D_2 = 321.4$ m/s, $\phi_E = 0.91$, and $\phi_i = 0.95$.

$$\frac{(321.4)^2}{2} = 0.91[\Delta h_s + (0.95 \times 0.91 \times 7.98)]10^3$$

$$(\Delta h_s)_R = 49.85 \text{ kJ/kg}$$

$$(\Delta h_s)_{st} = (\Delta h_s)_N + (\Delta h_s)_R = 50.0 + 49.85 = 99.85 \text{ kJ/kg}$$

$$\eta_{st} = \frac{W}{(\Delta h_s)_{st}} = \frac{87.35}{99.85} = 0.875$$

$$\dot{W} = \dot{m}W = 7.544 \times 87.35 = 659.0 \text{ kW}$$

6.9 TURBINE OPTIMUM BLADE SPEED

The blade speed U (Fig. 6.11) for a two-row impulse turbine stage is assumed to have the same value for both blade rows. Then,

$$W = [(D_{U,1} + D_{U,2}) + (D_{U,3} + D_{U,4})]U = [(D_1 \cos \beta_1 + D_2 \cos \beta_2) + (D_3 \cos \beta_3 + D_4 \cos \beta_4)]U$$

All of the vectors in the above equations are assumed to be positive quantities.

A number of equivalent terms are now established.

$$D_2 = k_{b,1} D_1 \quad \text{and} \quad D_4 = k_{b,3} D_3$$

$$D_1 \cos \beta_1 = C_1 \cos \alpha_1 - U$$

$$D_3 \cos \beta_3 = C_3 \cos \alpha_3 - U = k_{b,2} C_2 \cos \alpha_3 - U = k_{b,1} k_{b,2} \frac{\cos \beta_2 \cos \alpha_3}{\cos \beta_1 \cos \alpha_2} \cdot (C_1 \cos \alpha_1 - U) + k_{b,2} \frac{\cos \alpha_3}{\cos \alpha_2} U - U$$

Substitution of the equivalent terms for D_2, D_4, $D_1 \cos \beta_1$, and $D_3 \cos \beta_3$ in the equation for stage work yields

$$W = \{K(C_1 \cos \alpha_1 - U) + N[L(C_1 \cos \alpha_1 - U) - U(M + 1)]\}U \tag{6.11}$$

where $K = 1 + k_{b,1} \dfrac{\cos \beta_2}{\cos \beta_1}$

$$L = k_{b,1} k_{b,2} \frac{\cos \beta_2 \cos \alpha_3}{\cos \beta_1 \cos \alpha_2}$$

$$M = k_{b,2} \frac{\cos \alpha_3}{\cos \alpha_2}$$

$$N = 1 + k_{b,3} \frac{\cos \beta_4}{\cos \beta_3}$$

The velocity coefficient and the cosine ratio of the angles for each of the three blade rows are held constant for a specified design. There is then an optimum blade speed U that achieves maximum work and efficiency for each value of $C_1 \cos \alpha_1$.

Equation 6.11 is differentiated with respect to U and the result set equal to zero, $dW/dU = 0$. Then,

$$U_{\text{opt}} = \frac{K + LN}{2[K + N(L + M + 1)]} C_1 \cos \alpha_1 \tag{6.12}$$

For a single-row impulse turbine stage, Eq. 6.11 is reduced to

$$W = K(C_1 \cos \alpha_1 - U)U \tag{6.13}$$

and the corresponding optimum blade speed is given by

$$U_{\text{opt}} = \frac{C_1 \cos \alpha_1}{2} \tag{6.14}$$

In practice, the speed ratio $\rho = U/C_1$ at the design point is likely to be somewhat lower than the optimum value given by Eq. 6.12 or Eq. 6.14. For a specified blade speed, a low value of ρ is indicative of high gas velocities, which in turn effect an increase in the stage work. Although such increased work is accompanied by a lower stage efficiency, the overall design, considering all factors, is usually improved.

Example 6.2

The available energy for a two-row impulse turbine stage is 300 kJ/kg. The nozzle efficiency is 0.90, and the nozzle angle α_1 is 16°. Equiangular blades are prescribed. The blade velocity coefficients are $k_{b,1} = 0.80$, $k_{b,2} = 0.90$, and $k_{b,3} = 0.93$. Determine the optimum blade speed and the corresponding stage work and stage efficiency. Neglect disk friction and the blade fanning loss.

$$\frac{C_1^2}{2} = \eta_N(\Delta h_s)_N = \eta_N(\Delta h_s)_{st}$$

$$C_1^2 = 2 \times 0.90 \times 300{,}000$$

$$C_1 = 734.8 \text{ m/s}$$

Because the blades are equiangular, the cosine ratios are eliminated in the evaluation of U_{opt}.

$$K = 1 + k_{b,1} = 1 + 0.80 = 1.80$$

$$L = k_{b,1}k_{b,2} = 0.80 \times 0.90 = 0.72$$

$$M = k_{b,2} = 0.90$$

$$N = 1 + k_{b,3} = 1 + 0.93 = 1.93$$

Using Eq. 6.12,

$$U_{opt} = \frac{K + LN}{2[K + N(L + M + 1)]} C_1 \cos \alpha_1$$

$$= \frac{1.80 + (0.72 \times 1.93)}{2[1.80 + 1.93(0.72 + 0.90 + 1)]} \cdot 734.8 \times 0.9613$$

$$= 164.3 \text{ m/s}$$

$$\rho = \frac{U}{C_1} = \frac{164.3}{734.8} = 0.224$$

The stage work is determined by using Eq. 6.11.

$$W = \{K(C_1 \cos \alpha_1 - U) + N[L(C_1 \cos \alpha_1 - U) - U(M + 1)]\}U$$

$$= \{1.80(734.8 \times 0.9613 - 164.3) + 1.93[0.72(734.8 \times 0.9613 - 164.3) - 164.3 \times 1.90]\}164.3 = 185{,}065 \text{ J/kg}$$

$$\left(\frac{C^2}{2}\right)_{jet} = \eta_N(\Delta h_s)_N$$

$$= 0.90 \times 300 = 270.0 \text{ kJ/kg}$$

$$\eta_b = \frac{W}{(KE)_{jet}} = \frac{185.06}{270.0}$$

$$= 0.685 \quad \text{(blade efficiency)}$$

$$\eta_{st} = \eta_N\eta_b = 0.90 \times 0.685$$

$$= 0.617 \quad \text{(stage efficiency)}$$

Also,

$$\eta_{st} = \frac{W}{(\Delta h_s)_{st}} = \frac{185.06}{300} = 0.617$$

The comparatively low stage efficiency of 0.617 is somewhat typical of the two-row impulse turbine. Low efficiency values result, in part, from the high fluid losses that are generated by the high fluid velocities relative to the blades.

The outlet blade angles are customarily made smaller than the corresponding inlet blade angles in order to increase the difference between the rotational velocity components, and thereby increase the blade work.

In Example 6.2, the division of work is 86.6 percent for the first row and 13.4 percent for the second row, a ratio of about 6.5 to 1. If the effects of blade friction are neglected, the optimum speed ratio is increased slightly to 0.240, and the corresponding theoretical work is 249 kJ/kg and the division of work is 3 to 1.

Impulse turbine stages are often designed for partial admission, that is, the working fluid is admitted to the rotor through a nozzle arc of less than 360°. The rotating blades periodically pass through the active jet where the above-mentioned frictional losses are incurred. The fanning action of the inactive blades in the idle working fluid causes a frictional loss that reduces the net work transferred to the shaft. Disk friction has a similar effect on the shaft output. These two losses however have little influence on the optimum speed ratio.

Example 6.3

The blade speed for the two-row impulse turbine stage in Example 6.2 remains at 164.3 m/s, but the speed ratio is reduced to 0.18. The blade velocity coefficients are unchanged. Determine the stage work and the blade efficiency for the revised speed ratio.

$$C_1 = \frac{U}{\rho} = \frac{164.3}{0.18} = 912.8 \text{ m/s}$$

$$W = 274.73 \text{ kJ/kg} \qquad \text{(from Eq. 6.11)}$$

$$\left(\frac{C^2}{2}\right)_{jet} = \frac{C_1^2}{2} = \frac{(912.8)^2}{2}$$

$$= 416{,}600 \text{ J/kg}$$

$$\eta_b = \frac{W}{(KE)_{jet}} = \frac{274.73}{416.60} = 0.659$$

The results obtained in Example 6.3 indicate that the reduction in the speed ratio substantially increases the stage work from 185.1 kJ/kg to 274.7 kJ/kg, but a decrease in the blade efficiency from 0.685 to 0.659 is incurred. The division of work between the two rows is altered to some extent, that is, a work ratio of 3.3 to 1 is established. For precise calculations, the values of the blade velocity coefficients used in Examples 6.2 and 6.3 would differ to some degree, hence the increase in the blade work would be somewhat smaller than the quantity indicated above.

A special, but common, reaction turbine stage design prescribes identical stationary blade and rotating blade profiles. The vector diagram is then comprised of two congruent triangles. See Fig. 6.12. For this case, the optimum speed ratio is readily established.

$$\eta_{st} = \frac{W}{(\Delta h_s)_N + (\Delta h_s)_R}$$

$$= \frac{(C_{U,1} - C_{U,2})U}{(\Delta h_s)_{st}}$$

But $C_{U,1} = C_1 \cos \alpha_1$ and $C_{U,2} = -D_{U,1} = -(C_1 \cos \alpha_1 - U)$. Thus,

$$\eta_{st} = \frac{(2\, C_1 \cos \alpha_1 - U)U}{(\Delta h_s)_{st}}$$

For a constant value of $(\Delta h_s)_{st}$, the stage efficiency is a maximum when the numerator, or work, is a maximum. Hence

$$\frac{dW}{dU} = 2\, C_1 \cos \alpha_1 - 2U = 0$$

or

$$U_{opt} = C_1 \cos \alpha_1 \tag{6.15}$$

For actual designs, the reaction turbine stage speed ratio is usually less than the optimum value of $\rho = \cos \alpha_1$. Increased stage work is thus achieved, but some reduction in the stage efficiency is incurred. The variation in the stage efficiency with the speed ratio is shown in Fig. 6.16 for the different types of turbine stages.

Example 6.4

A reaction turbine stage is designed for a mean-diameter blade velocity of 200 m/s and a speed ratio of 0.70. The outlet fluid angle $\alpha_1 = 20°$. Fifty-percent reaction is prescribed; $\phi_E = 0.93$; $\phi_i = 0.94$. Determine the stage work and the stage efficiency.

$$C_1 = \frac{U}{\rho} = \frac{200}{0.70} = 285.7 \text{ m/s}$$

$$D_1 = 119.3 \text{ m/s} \qquad \text{(from the vector diagram; see Fig. 6.12)}$$

$$C_2 = C_0 = D_1$$

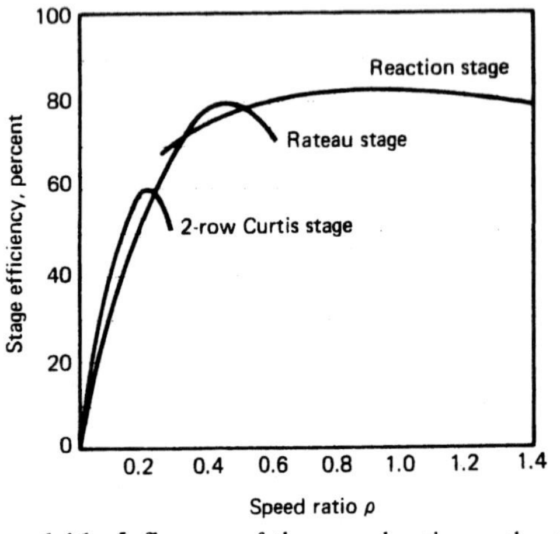

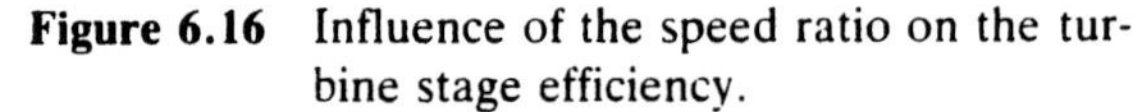

Figure 6.16 Influence of the speed ratio on the turbine stage efficiency.

Stage Work

$$\Delta h_{st} = \Delta h_N + \Delta h_R$$
$$= 2\,\Delta h_N = 2\,\frac{C_1^2 - C_0^2}{2}$$
$$= 2\,\frac{(285.7)^2 - (119.3)^2}{2}$$
$$= 67{,}390 \text{ J/kg}$$

Also,

$$W = (C_{U,1} - C_{U,2})U$$
$$= (2\,C_1 \cos\alpha_1 - U)U$$
$$= (2 \times 285.7 \times 0.9397 - 200)200$$
$$= 67{,}390 \text{ J/kg}$$

Isentropic Enthalpy Drop

Equation 6.8 is applied to the nozzle ring.

$$\frac{C_1^2}{2} = \phi_E\left(\Delta h_s + \phi_i\phi_v\,\frac{C_0^2}{2}\right)$$
$$\frac{(285.7)^2}{2} = 0.93\left[\Delta h_s + 0.94 \times 0.93\,\frac{(119.3)^2}{2}\right]$$
$$(\Delta h_s)_N = 37{,}663 \text{ J/kg}$$

For the rotating blade row,

$$(\Delta h_s)_R = 37{,}663 \text{ J/kg}$$

For the stage,

$$(\Delta h_s)_{st} = 75{,}326 \text{ J/kg}$$

Stage Efficiency

$$\eta_{st} = \frac{W}{(\Delta h_s)_{st}} = \frac{67.39}{75.33} = 0.895$$

Example 6.5

For the turbine stage described in Example 6.4, determine the stage work and stage efficiency achieved at the optimum speed ratio. U is held at 200 m/s; $\phi_E = 0.94$; $\phi_i = 1.0$.

$$\rho_{opt} = \cos\alpha_1 = 0.9397$$
$$C_1 = \frac{U}{\rho} = \frac{200}{0.9397} = 212.8 \text{ m/s}$$
$$D_1 = C_0 = C_2 = 72.79 \text{ m/s}$$

Stage Work

$$\Delta h_{st} = 2\,\frac{C_1^2 - C_0^2}{2}$$
$$= 2\,\frac{(212.8)^2 - (72.79)^2}{2}$$
$$= 40{,}000 \text{ J/kg}$$

Isentropic Enthalpy Drop

$$\frac{(212.8)^2}{2} = 0.94\left[\Delta h_s + 0.94 \times 1.0\,\frac{(72.79)^2}{2}\right]$$
$$(\Delta h_s)_N = 21{,}597 \text{ J/kg}$$
$$(\Delta h_s)_{st} = 2(\Delta h_s)_N = 43{,}194 \text{ J/kg}$$

Stage Efficiency

$$\eta_{st} = \frac{W}{(\Delta h_s)_{st}} = \frac{40.00}{43.19} = 0.926$$

A comparison of Examples 6.4 and 6.5 indicates an expected increase in the stage efficiency, because of the increase in the speed ratio from 0.70 to the optimum value of 0.940. The efficiency gain however is achieved with a substantial reduction in the stage work, caused by the decrease in the fluid velocities.

6.10 SOME TURBINE DESIGN FACTORS

The ratio of the enthalpy drop of the working fluid in the rotor to the corresponding stage enthalpy drop defines the degree of reaction for the stage. The degree of reaction may be held constant with respect to the radius of rotation, or be given a radial variation in accordance with some prescribed flow pattern.

Turbine reheat is a factor that resembles to some degree compressor preheat. See Section 5.16. In a multistage turbine, a portion of the available energy for each stage is converted into work while the remainder of the available energy, because of fluid friction, is the nonutilized energy. In turn, the nonutilized energy increases the available energy of the following stage, the result of an increase in the enthalpy of the fluid at the start of expansion. Thus, the summation of the stage available energy quantities exceeds the available energy for the turbine taken as a whole.

Unlike the efficiency relationship observed for the compressor, the overall efficiency of the multistage turbine exceeds the average individual stage efficiency. The average stage efficiency is given by

$$\eta_{st} = \frac{\Sigma\, W_{st}}{\Sigma\, (\Delta h_s)_{st}}$$

and the overall turbine efficiency is expressed by

$$\eta_T = \frac{W_T}{(\Delta h_s)_T} = \frac{\Sigma\, W_{st}}{(\Delta h_s)_T}$$

But $(\Delta h_s)_T < \Sigma\, (\Delta h_s)_{st}$; hence $\eta_T > \eta_{st}$.

The turbine reheat factor RH is defined as the ratio of the summation of the stage isentropic enthalpy drops to the isentropic enthalpy drop for the turbine. Thus,

$$\text{RH} = \frac{\Sigma\, (\Delta h_s)_{st}}{(\Delta h_s)_T} \tag{6.16}$$

The turbine efficiency and the average stage efficiency are related by

$$\eta_T = (\text{RH})\eta_{st} \tag{6.17}$$

It is possible to determine analytically, from the stage efficiency and the characteristics of the working fluid, a theoretical reheat factor and an approximate value of the turbine efficiency. This analysis, applicable to a gas turbine, is based on a gaseous working medium that can be treated as an ideal gas. Figure 6.17 shows, respectively, the actual and the theoretical expansion lines 3–4 and 3–4_s taken between the pressure limits P_3 and P_4. The actual expansion is assumed to occur in a very large number of stages. Expansion from P to $(P - dP)$ produces an actual temperature difference dT. The corresponding isentropic temperature drop is

$$T\left[1 - \left(\frac{P - dP}{P}\right)^{(k-1)/k}\right]$$

The small-stage or polytropic efficiency is given by

$$\eta_p = \frac{c_p\, dT}{c_p T\left[1 - \left(\frac{P - dP}{P}\right)^{(k-1)/k}\right]} = \frac{dT/T}{1 - \left(1 - \frac{dP}{P}\right)^{(k-1)/k}}$$

Integration between the prescribed limits yields

$$\eta_p = \frac{k}{(k - 1)} \frac{\ln(T_4/T_3)}{\ln(P_4/P_3)} \tag{6.18}$$

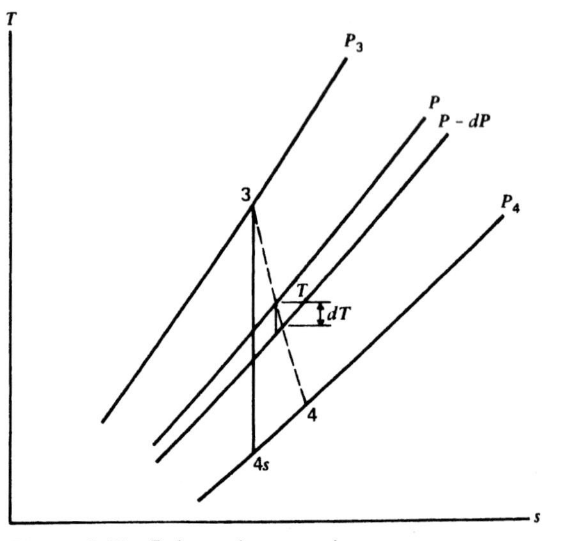

Figure 6.17 Polytropic expansion.

For a very large number of stages, the polytropic efficiency is equal to the average stage efficiency. Because the gas turbine ordinarily consists of a small number of stages, close agreement between the stage efficiency and the polytropic efficiency cannot be expected. The polytropic efficiency is useful for estimating turbine performance.

Example 6.6

A three-stage gas turbine is to operate between the stagnation pressure limits of 415 and 78.85 kPa, abs. The inlet gas temperature is 1030 K. The stage efficiency is 0.85; $k_{av} = 1.337$. Estimate the turbine efficiency and the reheat factor.

The polytropic efficiency is assumed to be equal to the average stage efficiency.

Applying Eq. 6.18,

$$\eta_p = \frac{k}{k-1} \frac{\ln (T_4/T_3)}{\ln (P_4/P_3)}$$

$$0.85 = \frac{1.337}{0.337} \frac{\ln (T_4/1030)}{\ln (78.85/415)}$$

$$T_4 = 721.6 \text{ K}$$

$$T_{4,s} = T_3\left(\frac{P_4}{P_3}\right)^{(k-1)/k}$$

$$= 1030\left(\frac{78.85}{415}\right)^{0.2521} = 677.7 \text{ K}$$

$$\eta_T = \frac{h_3 - h_4}{h_3 - h_{4,s}} = \frac{c_p(T_3 - T_4)}{c_p(T_3 - T_{4,s})}$$

$$= \frac{1030 - 721.6}{1030 - 677.7} = 0.875$$

$$\text{RH} = \frac{\eta_T}{\eta_{st}} = \frac{0.875}{0.85} = 1.03$$

The turbine efficiency and the reheat factor, determined by a stage-by-stage calculation, are, respectively, 0.869 and 1.02. Because of the small number of stages prescribed in this example, deviations of the magnitude indicated can be anticipated.

Bearing friction causes a comparatively small turbine loss, and a relatively small fraction of the shaft output is consumed in driving the auxiliary equipment. Fluid losses constitute the major portion of the overall turbine loss. Losses that result from nozzle, blade, and disk friction and from partial admission of the working fluid have been previously examined. Internal leakage of the working fluid results in a loss of available energy and a consequent reduction in turbine work. Such leakage occurs at the shaft seals in the interstage diaphragms and in reaction turbine stages around the tips of the rotating blades. Leakage at the blade tips is minimized by maintaining close clearances through use of thin metallic strips fastened to the blades or to stationary surfaces. Shaft leakage at the turbine ends is controlled by the installation of appropriate shaft seals.

The residual kinetic energy in the fluid leaving the last stage of the turbine represents a reduction in the turbine work for expansion between a prescribed initial state and final pressure. The turbine exhaust loss is observed in a determination of the turbine internal efficiency, defined as the ratio of the internal work for the several stages to the energy available in a reversible adiabatic expansion from the inlet stagnation state 03 to the exhaust static pressure P_4. At the exhaust pressure, state 4_s is achieved by reversible adiabatic expansion and state 4, by actual expansion of the working fluid.

Then, for adiabatic expansion, the turbine internal work is given by

$$W = h_{03} - h_4 - \frac{C_4^2}{2}$$

But

$$h_{03} - h_4 = \eta_{st}\text{RH}(h_{03} - h_{4,s})$$

Hence

$$\eta_T = \frac{W}{h_{03} - h_{4,s}} = \eta_{st}\text{RH} - \Delta h_e \qquad (6.19)$$

where Δh_e is the exhaust loss, expressed as a fraction of the turbine available energy, or $$\Delta h_e = \frac{\frac{C_4^2}{2}}{h_{03} - h_{4,s}}$$

The magnitude of the exhaust loss is usually controlled, in the design of the turbine, by the size of the last stage annular flow area. Because of the necessity for reducing the nonrecoverable kinetic energy in the exhaust stream, it is essential in the design of certain types of turbines to specify large exhaust annuli, particularly when large volumetric flows are encountered. Aircraft gas turbine engines, on the other hand, are designed to utilize the residual kinetic energy in the propulsion of the airplane, hence high exhaust gas velocities at the turbine outlet will not have an adverse effect on the engine performance.

Example 6.7

The inlet conditions for a steam turbine are 3.00 MPa, abs, and 350 C. The exhaust pressure is 150 kPa, abs. The average stage efficiency is 0.85, and the reheat factor is 1.04. The exhaust velocity is 110 m/s. Determine the turbine efficiency and the internal work.

The theoretical enthalpy drop Δh_s is achieved by isentropic expansion from the inlet conditions to the exhaust pressure.

$$\Delta h_s = 606.5 \text{ kJ/kg}$$

$$\frac{C_e^2}{2} = \frac{(110)^2}{2} = 6050 \text{ J/kg}$$

$$\Delta h_e = \frac{\frac{C_e^2}{2}}{\Delta h_s} = \frac{6.05}{606.5} = 0.00998$$

$$\eta_T = \eta_{st}(\text{RH}) - \Delta h_e$$

$$= (0.85 \times 1.04) - 0.010 = 0.874$$

$$W = \eta_T \, \Delta h_s$$

$$= 0.874 \times 606.5 = 530.1 \text{ kJ/kg steam}$$

Depending upon the prescribed flow pattern, a significant variation, extending from the root diameter to the tip diameter section, can be observed in the rotor blade profiles. For example, a particular design is based on free vortex flow and a prescribed 50-percent reaction at the mean diameter. Some pertinent data for this design are shown in Table 6.2.

TABLE 6.2 Radial Variation in Free Vortex Flow

Section	*Degree of reaction*	*Deflection angles (Degrees)*	
		θ_N	θ_R
Tip diameter	0.633	104.8	64.9
Mean diameter	0.500	111.5	111.5
Root diameter	0.280	118.6	134.8

Considerable variation is observed, from the root to the tip diameter section, in the blade profiles for the rotating blade row. Little radial variation is noted in the blade profiles for the nozzle ring, because the inlet and outlet fluid angles, α_0 and α_1, respectively, do not change appreciably over the length of the blade.

6.11 TURBINE APPLICATIONS

The turbine element of the gas turbine power plant or engine operates with a moderate maximum pressure in an approximate range of 700 to 1500 kPa, gage, for open-cycle machines. Maximum temperatures range from 850 to 1100 C for industrial machines to approximately 1280 C for aircraft engines. The higher turbine inlet temperatures contribute to superior performance, but at the expense of increased maintenance, reduced operating life, and increased initial cost resulting from the use of higher-priced materials.

The power element of a gas turbine engine will consist usually of one to four stages, depending upon the design of the machine. Turbines for aircraft service are designed for high performance in order to achieve low weight and a structure of limited size. In comparison with aircraft turbines, industrial gas turbines can operate at lower speeds and temperatures and may be physically larger for a specified power output.

Larger annular flow areas are prescribed in order to reduce the gas velocities, particularly at the outlet section of the last stage. Figure 6.18 shows the rotor for a typical gas turbine power element.

The central station steam turbine used for electric power generation is classified in accordance with the throttle pressure and temperature, speed, reheat or nonreheat, generating capability, and arrangement of the several elements that comprise the machine. An examination of design trends in steam power engineering shows a general increase, through the years, in the pressure and temperature of the steam supplied to the turbine. At the same time, an increase has been achieved in the power generating capability and in the operating speed of the largest turbine-generators.

Maximum steam temperatures reached a level of 595 C, and steam pressures as high as 34.5 MPa can be observed in some of the supercritical-pressure power plants. Operation with steam temperatures at 595 C has not proved to be economical because of the high cost of austenitic steels and the limitations on the life of the material at these high temperatures. For new construction, current practice limits design steam conditions to 24 MPa and 540 C. A maximum throttle pressure of approximately 17 MPa is observed when the turbine operates with subcritical-pressure steam.

When connected to an electric generator, the operating speed of a steam turbine installed in a central station is either 1800 or 3600 rpm. For other applications, the steam turbine may turn at higher speeds. The operating speed of a 60-Hz turbine-generator may be set at 1800 rpm, in part, because of the structural limitations of the generator. The 3600-rpm turbine-generator is only slightly less efficient than the 1800-rpm machine of comparable generating capability; however, it is more economical to operate, because of reduced size, weight, and cost.

The modern, high-capability steam turbines built for installation in central stations are of the reheat type. Central station steam turbines are designed for condensing operation and for extraction of steam for feedwater heating. The number of points at which steam is extracted from the turbine is dependent upon the prescribed power plant cycle.

Figure 6.18 Gas turbine engine rotor. The turbine blades are shown at the right-hand end of the rotor and the compressor blades, at the left-hand end. (Courtesy Sulzer Brothers Limited.)

Industrial steam turbines operate condensing or noncondensing, depending upon the requirement for low-pressure heating or process steam. These turbines may also be constructed for controlled bleeding of process steam at selected points on the turbine casing.

Figure 6.19 shows a reheat steam turbine that has a rated generating capability within a range of 80 to 125 MW. The several reaction stages of the high-pressure element of the turbine are preceded by a single-row impulse stage. Because of the inherent high-work characteristics of the impulse stage, the pressure and temperature of the steam drop rapidly in this single-row stage, and thus reduce the extent of the turbine structure that is exposed to the most severe operating conditions. The turbine length is also reduced because the impulse stage, in effect, takes the place of a number of reaction stages of comparable work capability. Partial admission of the steam to the impulse stage, as against full admission required for a reaction stage, permits the installation of longer blades in the first stage. An extremely short blade, installed in the first stage, would have an adverse effect on the turbine performance because the loss caused by surface friction is exceptionally high.

Subsequent to expansion of the steam in the high-pressure section of the turbine (Fig. 6.19), the steam is reheated in the boiler and then admitted to the intermediate-pressure section. Expansion of the steam to the exhaust pressure is achieved in the intermediate-pressure section and in the double-flow, low-pressure section of the turbine.

Large, tandem-compound, 3600-rpm reheat steam turbines are built with ratings that are in the range of several hundred megawatts. The machines are comprised ordinarily of a high-pressure, an intermediate-pressure, and three or four low-pressure elements. See Fig. 6.20. Opposed steam flow is employed to balance, in part, the rotor axial thrust. Steam pressures for these large machines range from 12.5 to 16.5 MPa. Turbine design prescribes an inlet steam temperature and a reheat steam temperature of about 540 C.

Steam pressures generated in pressurized-water nuclear power plants are lower than the steam pressures normally observed in the modern fossil fuel-fired power plants. Because of the comparatively high specific volume of the steam, the design for a nuclear power plant is likely to incorporate a turbine-generator that operates at

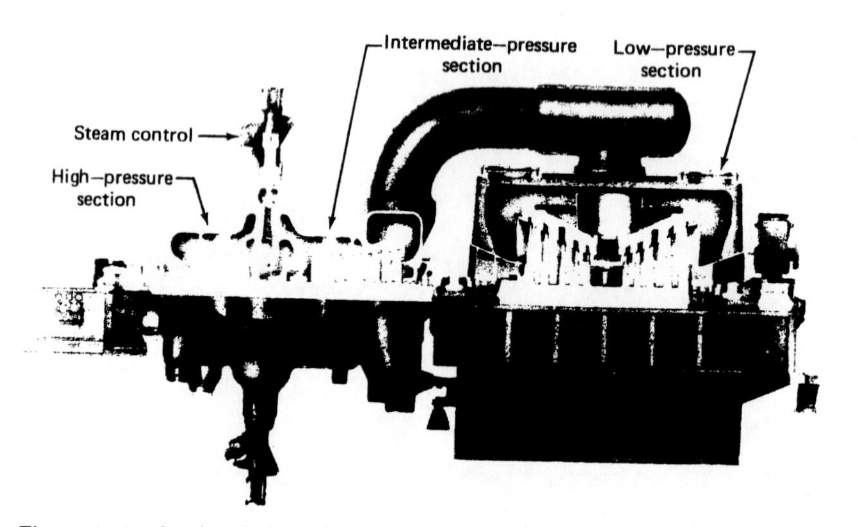

Figure 6.19 Sectional view of a 3600-rpm, tandem-compound, double-flow reheat steam turbine. (Photo courtesy of General Electric Company, Medium Steam Turbine Departments, Lynn, Mass., U.S.A.)

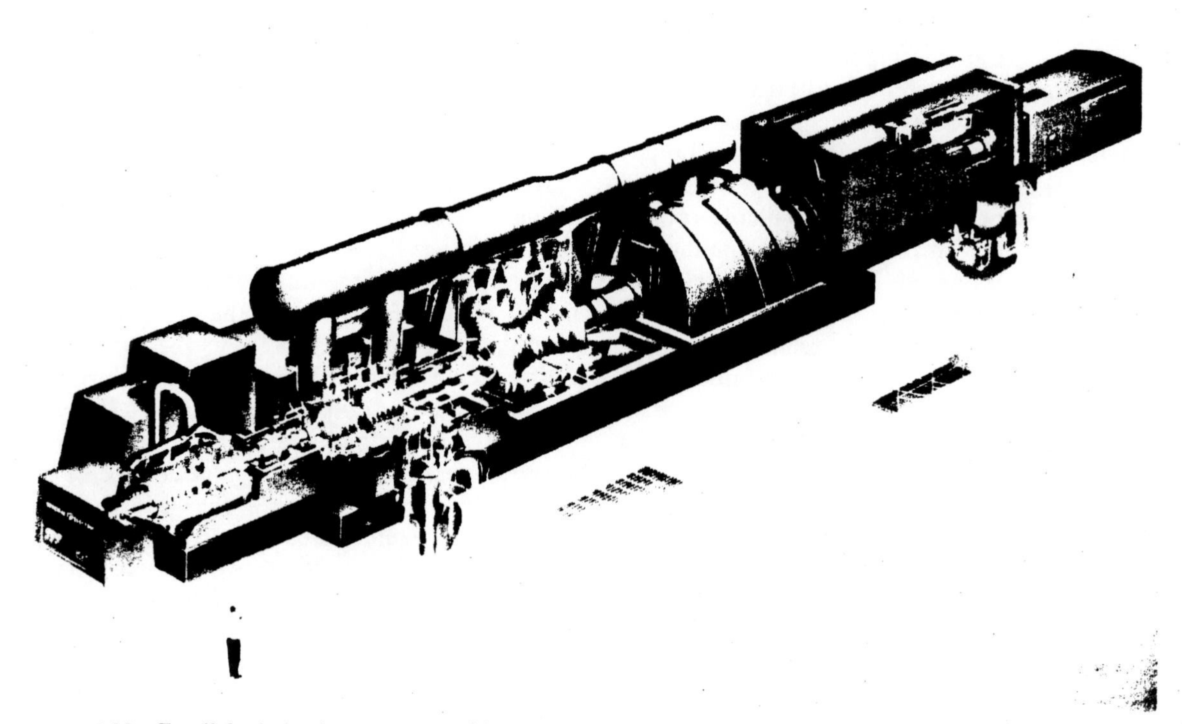

Figure 6.20 Fossil-fueled reheat steam turbine-generator rated at more than 500 MW, 3600 rpm, tandem-compound, four flow, four casing. (Courtesy The General Electric Company.)

1800 rpm. These turbines are commonly tandem-compound machines of high generating capability.

Steam turbines of the highest capability are ordinarily of the cross-compound design. In comparison with the tandem-compound machine, the cross-compound turbine requires more floor area, and the operation is somewhat more complex. Further, the capital cost per kilowatt of generating capability is higher for the cross-compound machine. The efficiency of the cross-compound turbine is slightly higher than that of the tandem-compound machine.

The cross-compound turbine is comprised of two machines arranged in a parallel configuration. Each machine consists of a generator and a number of turbine elements. Steam is admitted to the primary machine, and the subsequent flow of steam between the two machines is through the crossover piping. A balanced design provides for dividing the electrical load equally between the two machines. Other splits however can be prescribed, for example, a 65/35 split. The primary machine operates at 3600 rpm, and the secondary machine operates at either 3600 or 1800 rpm. For a 3600/3600-rpm turbine-generator, four or six low-pressure elements would be provided, while one double-flow, low-pressure element is adequate for a 3600/1800-rpm turbine-generator.

In order to achieve a high turbine efficiency, the exhaust loss must be reduced to the lowest level that can be economically justified. A low exhaust steam velocity is achieved by providing the proper number of low-pressure turbine elements and by establishing a large flow annulus in the last stage of each low-pressure element. The maximum blade length for the turbine is dependent upon the operating speed of the machine. A limit of 660 mm for the length of the blades in the last rotating blade row of a 3600-rpm turbine can be extended to 1168 mm for the comparable blade row in a machine that operates at 1800 rpm. The corresponding area ratio is 1 to 3+ for the same blade stress level. Depending upon the temperature of the available cooling water, a turbine with a large exhaust annulus can advantageously operate with a condenser pressure of about 13 mm mercury, abs.

The supercritical-pressure steam power plant represents one of the most recent significant developments that have been made in the steam power industry. Turbine inlet steam pressures range from 24 to 35 MPa. A double reheat can be economically justified for the steam power plants that operate on the supercritical-pressure steam cycle. Design steam temperatures were comparatively high for the early supercritical-pressure power plants, with turbine inlet steam temperatures ranging as high as 650 C. As noted in an earlier section, power plant designs presently prescribe lower steam temperatures, because of the operating difficulties and cost problems that have developed with steam temperatures in excess of 540 C.

The supercritical-pressure turbine is essentially a topping unit that is incorporated into the design of a modern, subcritical-pressure reheat steam turbine. In the supercritical-pressure turbine, steam is initially expanded into the subcritical-pressure region. Subsequent expansion of the steam, with double reheating, occurs in the high-, intermediate-, and low-pressure elements.

6.12 STEAM TURBINE PERFORMANCE

Steam turbines directly connected to ac generators must necessarily operate at constant speed in order to maintain essentially zero variation in the frequency of the electric current. The turbine speed is controlled by the operating governor. An emergency governor is also provided for protection against overspeed in the event that the control is lost by the operating governor. At a fixed percentage overspeed, the emergency governor operates a trip valve that quickly cuts off the steam flow to the machine.

Reheat turbines are equipped with intercept valves located at the point where the reheated steam enters the turbine. A sudden loss in the load will cause these valves to trip and thus prevent a relatively large quantity of steam stored in the reheater from causing the machine to exceed the overspeed limit.

The steam flow, in kg/h, plotted against the turbine-generator output is essentially a straight line, between no load and the rated load, for a throttle-governed, single-valve machine. See Fig. 6.21. For condensing turbines, the no-load steam flow may be taken approximately as 0.25 kg per kilowatt of rated output. Above the rated load, the form of the Willans line is dependent upon the overload valve arrangement. The derived steam rate curve is also shown in Fig. 6.21. The steam rate, in kg/kWh, is relatively high at light loads, hence the machine is best operated in the vicinity of the rated load.*

Nozzle and bypass governing is employed effectively for the large turbine-generators. Turbines governed in this manner are equipped with several inlet valves. Each primary valve controls the steam flow to a certain number of nozzle passages for the first stage, usually a single-row or two-row impulse stage. These valves are opened successively, with some overlap, as the load on the machine is increased. Throttling losses are reduced by multivalve control of the steam flow, in comparison with single-valve control, and as a consequence improved economy is achieved for operation of the turbine at light loads when multivalve control is employed.

* The term capability is used in steam turbine practice in a special manner to define the maximum output of the turbine-generator. For standard units, the capability is 110 percent of the nominal rating. Unless used in this special sense, the term capability will relate generally to the nominal rating of a power generating unit or system.

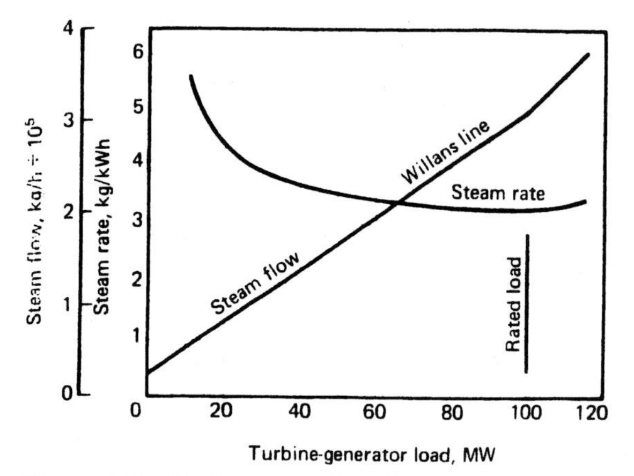

Figure 6.21 Performance chart for throttle-governed steam turbine.

Operation of the turbine with all of the primary control valves open will conform closely to the design conditions for the machine and hence show the lowest heat rate. A greater load may be carried by introducing steam at line conditions to a downstream stage, for example, the second stage. Secondary steam admission, or bypass governing, provides for increased steam flow, but at the expense of decreased turbine efficiency. A still greater load can be carried by tertiary admission, say, to the fourth stage. The reduction in the turbine efficiency observed with bypass governing is caused by off-design operation, particularly in the bypassed stages. Figure 6.22 shows a typical heat rate curve for a turbine designed for nozzle and bypass governing.

The turbine heat rate, in kJ/kWh, is commonly employed for reporting turbine performance, because this factor is influenced by both the turbine efficiency and the operating steam conditions. The turbine heat rate is equal to the product of the turbine steam rate and the energy supplied in the boiler to a unit mass of steam. The turbine steam rate is based on the net power output, or the difference between the gross turbine power and the power required for pumping. The station heat rate is derived from the turbine heat rate by accounting for boiler and miscellaneous losses and the power used by the station auxiliaries.

Selected reat rates for a 500-MW steam turbine-generator are shown in Table 6.3.

TABLE 6.3 Steam Turbine-Generator Heat Rates

Cycle	*Heat Rate (kJ/kWh)* Turbine	Station
(a) Subcritical-pressure single reheat, 17.25 MPa, 540/540 C	8240	9600
(b) Supercritical-pressure single reheat, 34.5 MPa, 540/540 C	8090	9420
(c) Supercritical-pressure double reheat, 34.5 MPa, 565/565/565 C	7780	9060

6.13 ALTERNATING CURRENT GENERATORS

Because of structural limitations, it was formerly necessary to limit rotational speeds of 3600 rpm to the low-capability electric generators. However, because of advanced design concepts and improved methods of construction, the generating capability of 3600-rpm two-pole machines has been extended to approximately 1100 MVA. The maximum generating capability of the 1800-rpm four-pole machine is somewhat higher, that is, about 1300 MVA. The large two-pole and four-pole machines generate three-phase 60-Hz current at 20 to 26 kV.

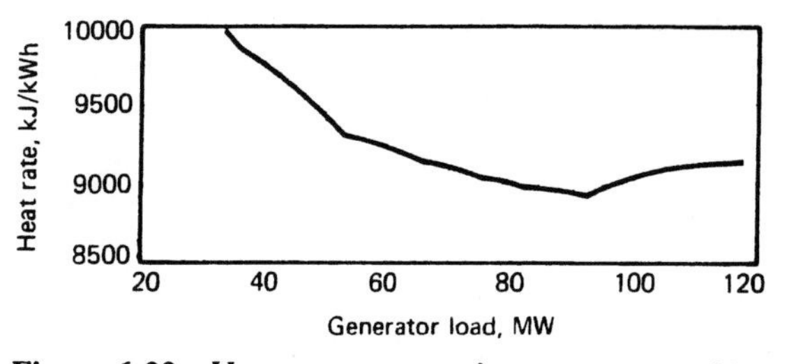

Figure 6.22 Heat rate curve for a steam turbine equipped with a nozzle and bypass control system.

The 3600-rpm generator is not inherently superior to the four-pole 1800-rpm machine. In addition to a number of limitations and problems, the 3600-rpm machine is somewhat less efficient than the slower four-pole generator. The advantages of the 3600-rpm design are related to the turbine portion of the turbine-generator set. Over a period of years, improved thermal performance has been achieved by increasing steam pressures and temperatures. The elevated steam conditions can be most effectively utilized, economically and technically, at high rotating speeds, that is, 3600 rpm for central station machines. Significant structural advantages are achieved by employing the higher operating speed, namely, by a reduction in the diameter of the turbine shell and in the number of turbine stages.[2]

Electrical and frictional losses within the generator produce internal heating that, in the absence of adequate cooling, causes an excessive rise in the temperature of the windings and insulation of both the rotor and the stator. For a designated capability, the size of the generator is dependent, in part, upon the effectiveness of the internal cooling system. Originally, when generators were of low capability, natural cooling at the ends of the machine was adequate. Later units of higher rating were cooled by circulating air through ducts placed in the insulation of both the rotor and the stator. In this arrangement, energy transferred to the air passing through the generator is in turn removed in a water-cooled heat exchanger. The closed system aids in the circulation of dust-free air.

The use of hydrogen, at a pressure of 100 to 200 kPa, gage, in place of air has resulted in improved performance of the generator, principally because of lower fluid losses, cleaner ventilating passages, and superior heat transfer characteristics of the hydrogen gas. There still remains however within the insulation a temperature gradient between the surface of the conductor and the wall of the ventilating duct.

The inner-cooled machine represents an important advancement in generator design. Hydrogen gas, under a pressure of 210 or 315 kPa, gage, is circulated at high velocities through the hollow conductors used in forming the stator coils. Provision is also made for circulating the hydrogen gas through slotted holes in the rotor conductors. Heat is thus removed at the point where it is generated instead of being transferred to the cooling medium through successive layers of coils, insulation, and other structural materials. As a direct consequence of the improved cooling system, the generator, for a specified rating, has been substantially reduced in both size and weight.

A modification of gas inner cooling is accomplished by circulating a liquid through the hollow strands of the generator stator windings.[3] Hydrogen is picked up from the air gap between the stator and the rotor and circulated through the slotted holes in the rotor conductors.

Certain advantages are achieved with liquid cooling in comparison with gas cooling. A higher fluid density and superior heat transfer characteristics contribute to a reduction in the fluid velocity and in the size of the ducts. A structural advantage is derived from the very small temperature difference that, for liquid cooling, exists between the copper and the coolant. Water has proved to be the most effective liquid coolant for generator cooling applications.

Extension of liquid cooling to the generator rotor is feasible; and, if adopted, the use of hydrogen gas as a coolant would be eliminated. Further, liquid cooling of the rotor may effect some reduction in the size of the machine.

PROBLEMS

6.1 Combustion gases enter the nozzle of a single-row impulse turbine at 400 kPa, abs, and 1350 K. The exhaust pressure is 100 kPa, abs. The nozzle efficiency is 0.92, and the blade velocity coefficient is 0.78; c_p = 1.205 kJ/kg·K; k = 1.315. The nozzle angle is 18°. The blades are equiangular. The absolute velocity of the gas leaving the rotating blade row has no rotational component. Construct the vector diagram and calculate the stage work, blade efficiency, and the speed ratio.

6.2 Repeat Problem 6.1 for operation at the optimum speed ratio.

6.3 The turbine in Problem 6.1 operates with a blade velocity of 430 m/s and a speed ratio equal to 0.416. Calculate the stage work, the blade efficiency, and the isentropic energy drop for the stage. Compare these answers with the results obtained for Problems 6.1 and 6.2.

6.4 The nozzle outlet velocity for a single-row impulse turbine stage is 700 m/s. The nozzle angle is 20°. The absolute velocity of the fluid leaving the rotating blade row is 188.7 m/s. The stage work is 199.47 kJ/kg gas; $k_b = 0.85$. Calculate the blade inlet angle and the speed ratio for the stage.

6.5 A two-row impulse turbine is designed to deliver 1500 kW for a steam flow of 10 kg/s. The linear speed of the turbine wheel is 150 m/s. The nozzle angle is 17°, and the nozzle efficiency is 0.92. The blade velocity coefficients are $k_{b,1} = 0.82$, $k_{b,2} = 0.89$, and $k_{b,3} = 0.92$, and the blades are equiangular. Calculate the isentropic energy drop for the nozzle and the stage velocity ratio.

6.6 The mean diameter for a reaction turbine stage is 500 mm. The rotating speed is 5000 rpm. The stage is required to deliver 200 kW. The mechanical efficiency is 0.985. The expansion energy coefficient is 0.92, and the incidence coefficient is 1.0. The speed ratio is 0.85. The nozzle angle is 16°. Equal energy drops occur in the two blade rows. Determine the gas flow rate and the efficiency for the stage.

6.7 Repeat Problem 6.6 for the optimum speed ratio. The expansion energy coefficient is 0.93.

6.8 A steam turbine consists of 15 stages. Each stage has a shaft output of 4000 kW with a stage efficiency of 0.90. The mechanical efficiency is 0.98. The steam flow is 100 kg/s. The turbine reheat factor is 1.03. The exhaust steam leaves the turbine at 10 kPa, abs, and $x = 0.88$. Determine the pressure and temperature of the inlet steam.

6.9 Gas enters a turbine at 1050 kPa, abs, and 1400 K. The exhaust pressure is 100 kPa, abs. The exhaust gas velocity is 160 m/s. The turbine efficiency is 0.90, and the reheat factor is 1.025. Assume that combustion occurs with 400 percent theoretical air. Calculate the turbine internal work, the temperature of the exhaust gas, and the average stage efficiency.

6.10 The stagnation inlet conditions for the nozzle of an impulse gas turbine stage are 380 kPa, abs, and 1350 K. The nozzle outlet pressure is 95 kPa, abs. The nozzle efficiency is 0.91, and the nozzle angle is 18°. The nozzle pitch is 20 mm, and the nozzle height is 30 mm. There are eight passages in the nozzle block, and the vane thickness is 0.80 mm. Calculate the nozzle outlet velocity and the gas flow rate. Assume that combustion occurs with 400 percent theoretical air.

6.11 The turbine stage described in Problem 6.10 is equipped with a single-row rotor. The speed ratio is 0.400, and $\beta_1 - \beta_2 = 4°$. The blade velocity coefficient is 0.78. Calculate the power delivered to the rotor.

6.12 Steam leaves the nozzle of a single-row impulse turbine stage at the velocity of 600 m/s. The nozzle angle is 20°. Calculate the stage work for operation at the optimum speed ratio for (a) equiangular blades and (b) $\beta_1 - \beta_2 = 5°$. The blade velocity coefficient is 0.81 for (a) and 0.80 for (b).

6.13 Steam enters the nozzle of a two-row impulse turbine stage at 1400 kPa, abs, and 300 C. The nozzle outlet pressure is 7.5 kPa, abs, and the nozzle efficiency is 0.91. The nozzle angle is 18°. The blades are equiangular, and the blade velocity coefficients are $k_{b,1} = 0.80$, $k_{b,2} = 0.87$, and $k_{b,3} = 0.90$. Calculate the blade work for operation at the optimum speed ratio, and determine the stage efficiency.

6.14 A reaction turbine stage consists of two identical blade rows. The blade outlet angle is 20°, and the working fluid leaves the blade row with a velocity of 210 m/s. The speed ratio is

0.80. The incidence coefficient is taken as 1.0. Determine the blade work and the stage efficiency.

6.15 The leaving velocity for a single-row impulse turbine is 175 m/s. The inlet and outlet relative gas velocities are, respectively, 400 and 320 m/s. The miscellaneous stage losses are 5.8 kJ/kg gas. The stage available energy is 237 kJ/kg gas. The nozzle efficiency is 0.91, and the mechanical efficiency is 0.98. Determine the shaft output.

6.16 The stationary blade row for a gas turbine stage is designed for an axial gas velocity of 90 m/s. The isentropic energy drop for the blade row is 35 kJ/kg gas. The inlet blade angle is 70°, and the inlet gas angle is 60°. Determine the gas velocity and the fluid angle at the outlet section.

6.17 The moving blade row that follows the stationary blade row described in Problem 6.16 has a blade inlet angle of 70°. The isentropic energy drop for the blade row is 35 kJ/kg gas. The mean-diameter blade velocity is 205 m/s at the leading edge and 210 m/s at the trailing edge of the blade. Determine the relative and absolute gas angles at the outlet section of the blade row. Determine the stage work and the stage efficiency.

6.18 Gas stagnation properties at the nozzle inlet for an aircraft gas turbine engine stage are 380 kPa, abs, and 1375 K. The nozzle efficiency is 0.92, and the nozzle angle is 20°. The Mach number of the gas stream discharged from the nozzle is 1.08. Determine the velocity, the pressure, and the temperature of the gas leaving the nozzle. Assume combustion with 400 percent theoretical air and refer to Tables A.1 and A.2.

6.19 The rotor for the nozzle ring described in Problem 6.18 is equipped with blades that turn the gas through an angle of 100°. The outlet gas angle is 20°. The two blade rows have equal outlet relative gas velocities. The blade row efficiency is 0.92. Calculate the blade velocity, the Mach number of the gas at the inlet section, the stage work, and the gas pressure at the outlet section of the rotor.

6.20 A steam turbine has a steam rate of 3.25 kg/kWh when operating at the most efficient load of 160 MW. The machine is throttle-governed. The no-load steam flow is 8 percent of the steam flow at the most efficient load. When operating at 180 MW, the steam consumption is 608,790 kg/h. Construct the steam flow and the steam rate curves.

REFERENCES

1 Emmert, H. D. "Current Design Practices for Gas Turbine Power Elements," *Trans. ASME*, Feb. 1950.

2 Laffoon, C. M. "Inner-Cooled Generators," *Westinghouse Engineer*, July 1954.

3 Harrington, B. D. and Jenkins, S. C. "Trends and Advancements in the Design of Large Generators," 32nd Annual Meeting of the American Power Conference, Chicago, Ill., April 1970.

BIBLIOGRAPHY

Church, E. F., Jr. *Steam Turbines*, McGraw-Hill Book Co., New York, 1950.

Salisbury, J. K. *Steam Turbines and Their Cycles*, John Wiley & Sons, Inc., New York, 1950.

Sawyer's Gas Turbine Engineering Handbook, J. W. Sawyer, ed., Vol. I, Theory and Design; Vol. II, Applications, 1980. Turbomachinery Publications, Inc., Norwalk, Conn.

Stodola, A. *Steam and Gas Turbines*, Vols. 1 and 2, Translated from the German by L. Lowenstein, Peter Smith Co., New York, 1945.

CHAPTER SEVEN

GAS TURBINE POWER

The first practical gas turbine engine began operation circa 1940, although experimental engines were operated during the 1930 decade. The possibility of constructing a gas turbine engine was however recognized during the early years of steam turbine development. In the preceding two chapters, attention was directed to the principal obstacles that retarded the development of the gas turbine engine, namely, low compressor efficiency and lack of materials capable of withstanding temperatures sufficiently high to ensure an acceptable thermal efficiency for the power plant. In time, heat-resistant alloys and advanced compressor designs were made available, and engine development moved forward at a rapid pace.

7.1 THE THEORETICAL CYCLE FOR THE GAS TURBINE ENGINE

The basic theoretical cycle for the gas turbine engine is comprised of the four processes shown in Fig. 7.1. Reversible adiabatic compression and expansion are achieved, respectively, by processes 1–2 and 3–4. Heat addition occurs at constant pressure from 2 to 3, and heat is rejected at constant pressure from 4 to 1. The cycle is applicable to both open- and closed-cycle engines. Most gas turbine engines operate on the open cycle, and process 4–1 represents direct heat transfer from the exhaust gas to the atmosphere.

The cycle depicted in Fig. 7.1 is attributed to George Brayton, and the initial application of the cycle was to reciprocating machines. The Brayton cycle also describes closely the operating sequence of events for the simple gas turbine engine. The compression and expansion processes for the actual engine are essentially adiabatic but irreversible, and the pressure loss in the combustion chamber is relatively small.

The analysis of the theoretical gas turbine engine cycle is based on steady flow of the working fluid in achieving compression, heat addition,

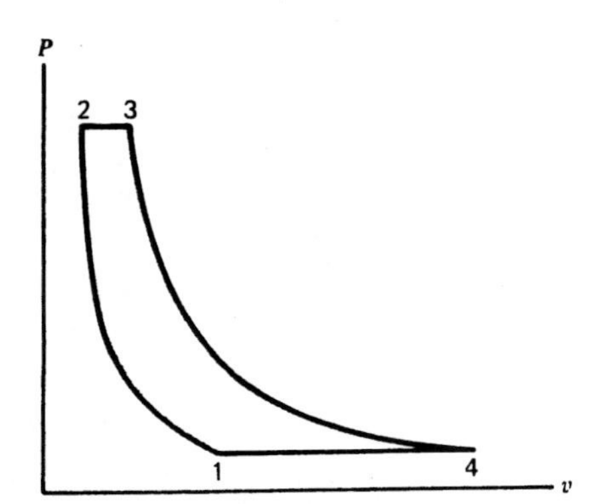

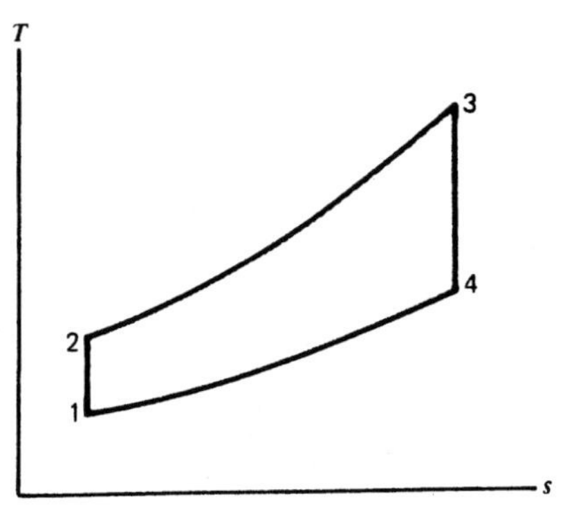

Figure 7.1 The Brayton cycle.

and expansion. The specific heats of the working fluid are assumed to remain constant, and the pressure loss effects are neglected. Energy is supplied to the working fluid by transfer of heat from an external source. Then, applying the first law equation for determination of the cycle thermal efficiency,

$$\text{Compression work } W_c = h_1 - h_2 = c_p(T_1 - T_2)$$

$$\text{Expansion work } W_e = h_3 - h_4 = c_p(T_3 - T_4)$$

$$\text{Heat supplied } Q_s = h_3 - h_2 = c_p(T_3 - T_2)$$

$$\eta_t = \frac{W}{Q_s} = \frac{W_c + W_e}{Q_s} = \frac{(T_1 - T_2) + (T_3 - T_4)}{T_3 - T_2} = 1 - \frac{T_1\,(T_4/T_1 - 1)}{T_2\,(T_3/T_2 - 1)}$$

But

$$\frac{T_4}{T_1} = \frac{T_3}{T_2} \quad \text{and} \quad \frac{T_2}{T_1} = \left(\frac{P_2}{P_1}\right)^{k-1/k}$$

Then,

$$\eta_t = 1 - \frac{1}{\left(\frac{P_2}{P_1}\right)^{k-1/k}} \qquad (7.1)$$

The thermal efficiency of the Brayton cycle is a function solely of the cycle pressure ratio. Further, the magnitude of the thermal efficiency continues to increase with an increase in the pressure ratio of the cycle. See Fig. 7.2.

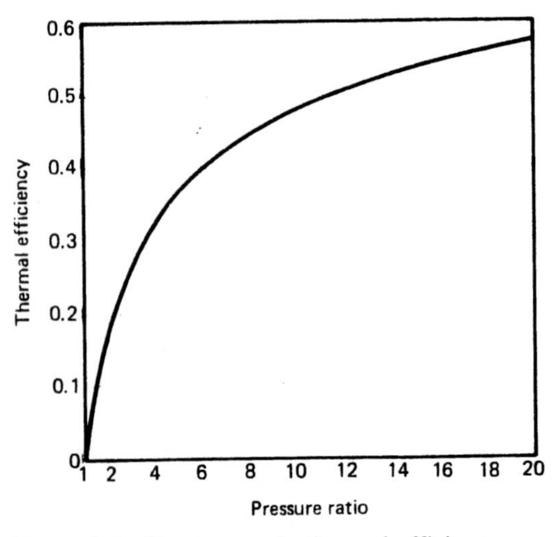

Figure 7.2 Brayton cycle thermal efficiency.

7.2 GAS TURBINE ENGINES

The operating characteristics of the actual gas turbine engine deviate substantially from the sequence of events described by the Brayton or theoretical cycle. These deviations are subsequently examined in an analysis of the several components of the engine.

Gas turbine engines can be classified as (1) dynamic or aircraft engines; and (2) static power plants for stationary, marine, railroad, and automotive applications. The aircraft classification includes turbojet and turboprop engines. The turbojet engine produces a high-velocity stream of gas that propels the airplane, while the other types of gas turbine engines deliver mechanical energy at the shaft.

Aircraft engine designs are based on the simple, open cycle, while some designs for the static-type gas turbine power plants include one or more variations from the basic cycle, thus producing a more complex mechanical system. The flow diagram for the simple, open-cycle gas turbine engine is shown in Fig. 7.3. Air is compressed, usually with a pressure ratio ranging from 4 to 1 to 15 to 1, in the compressor and delivered to the combustion chamber where the fuel is injected. Following combustion, the product gases flow to the turbine for expansion to essentially atmospheric pressure. Gas expansion in the turbojet engine occurs in the turbine and in the exhaust nozzle.

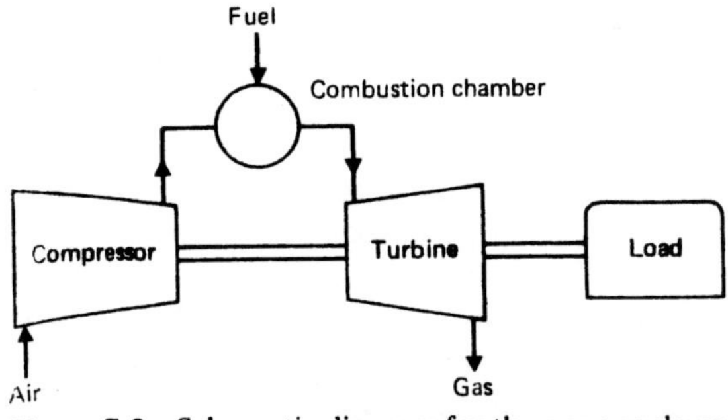

Figure 7.3 Schematic diagram for the open-cycle gas turbine power plant.

The flow diagram for the closed-cycle gas turbine power plant is shown in Fig. 7.4. Provision is made for the recirculation of the working fluid, which is a gas selected for superior operating characteristics. System pressures are independent of the atmospheric pressure and hence may be prescribed to achieve optimum performance. Fuel is externally fired and does not come in contact with the working fluid. Energy is transferred to and from the working fluid through the boundary walls of the heat exchangers that represent the heat source and the heat sink. Although the components installed in the open-cycle and closed-cycle power plants are different in a number of respects, the thermodynamic principles are generally similar for the two types of power plants.

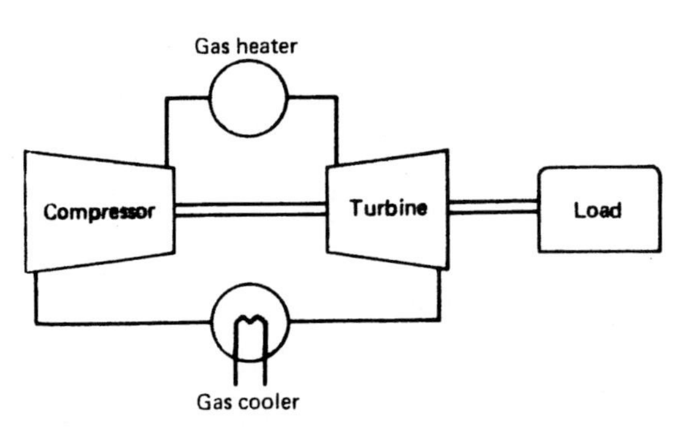

Figure 7.4 Schematic diagram for the closed-cycle gas turbine power plant.

7.3 THE COMPRESSOR INLET

In advance of examining the operation of the compressor, the influence of the air inlet must be investigated in order to identify the fluid properties at the inlet section of the compressor. Air entering the stationary gas turbine engine is accelerated from the quiescent ambient conditions P_0 and T_0 to a velocity C_1 at the compressor inlet section. See Fig. 7.5. At point 1, the compressor inlet section, the static properties of the fluid are P_1, T_1, and h_1, and the air is moving at a velocity C_1.

Because of the frictional losses encountered in the compressor inlet, the actual static pressure P_1 at the compressor inlet section will be lower than the pressure $P_{1,s}$ achieved by isentropic acceleration to the prescribed fluid velocity C_1. A greater pressure drop is required for the actual than for the theoretical acceleration to the same terminal velocity. In turn, the stagnation pressure P_{01} at the compressor inlet section will be lower than the ambient pressure P_0.

The energy equation for the inlet of the stationary engine is given by

$$h_0 = h_1 + \frac{C_1^2}{2} = h_{01}$$

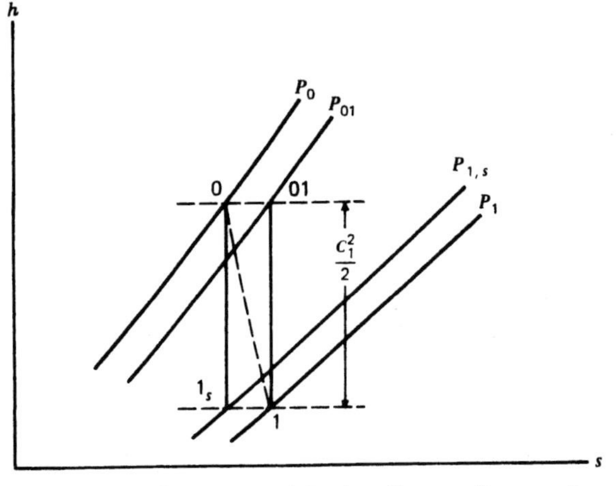

Figure 7.5 Compressor inlet h–s diagram for a static-type gas turbine power plant.

The inlet effectiveness η_I is expressed as the ratio of the actual stagnation pressure, at the compressor inlet section, to the ambient pressure.

$$\eta_I = \frac{P_{01}}{P_0} \tag{7.2}$$

The analysis of the aircraft gas turbine engine is conveniently based on the assumption that the engine remains stationary and the air approaches the inlet at a velocity equal to the forward speed C_p of the airplane through the air. These dynamic-type engines are equipped with forward-facing inlets that are designed to achieve maximum utilization of the ramming effect of the incoming air. Thus, at the compressor inlet section, the stagnation pressure of the air can closely approach the stagnation pressure of the free airstream.

Consideration will first be given to subsonic flight of the airplane. The air approaches the inlet at a subsonic velocity, $M < 1$, and the inlet functions as a subsonic diffuser.

State 0 in Fig. 7.6 represents the air at ambient pressure and temperature and at a velocity C_0 approaching the inlet. Because the inlet is assumed to be stationary, $C_0 = -C_p$. If full pressure recovery were achieved, the air pressure would rise to P_{00}. Because of fluid losses in the inlet, full pressure recovery is not accomplished, and the pressure rise is actually from P_0 to P_{01}. The pressure P_{01} is consequently the stagnation pressure of the air at the compressor inlet section. The static properties of the air at this point are dependent upon the air velocity C_1 at the inlet section of the compressor. Note state 1, Fig. 7.6. The loss in pressure incurred in the inlet is equal to the difference between the two stagnation pressures P_{00} and P_{01}.

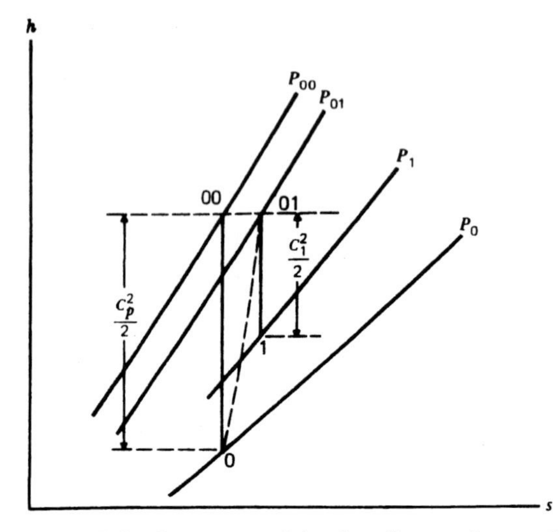

Figure 7.6 Compressor inlet h–s diagram for an aircraft gas turbine engine.

The effectiveness of the inlet is defined as the ratio of the actual pressure increase to the theoretical increase in the pressure of the air. Thus,

$$\eta_R = \frac{P_{01} - P_0}{P_{00} - P_0} \tag{7.3}$$

A high inlet effectiveness is a requirement for achieving an acceptable engine design, because the losses incurred in the inlet have a marked effect on the overall performance of the engine. Acceptable values of the inlet effectiveness exceed 90 percent.

Adiabatic diffusion is usually assumed for the inlet, hence $h_{01} = h_{00}$, and $T_{01} = T_{00}$. The energy equation for the inlet is

$$h_0 + \frac{C_0^2}{2} = h_{00} = h_{01}$$

$$= h_1 + \frac{C_1^2}{2}$$

See Fig. 7.6.

Example 7.1

An airplane travels at a speed of 675 km/h. The ambient air conditions are 70.0 kPa, abs, and 280.0 K; $c_p = 1.004$ kJ/kg·K; $k = 1.401$. The inlet effectiveness is 0.92. Calculate the stagnation pressure at the compressor inlet section.

$$C_p = 187.5 \text{ m/s}$$

$$T_{00} = T_0 + \frac{C_0^2}{2c_p}$$

$$= 280.0 + \frac{(187.5)^2}{2\,(1.004 \times 10^3)}$$

$$= 297.5 \text{ K}$$

$$P_{00} = P_0 \left(\frac{T_{00}}{T_0}\right)^{k/k-1}$$

$$= 70.0 \left(\frac{297.5}{280.0}\right)^{3.494}$$

$$= 86.52 \text{ kPa, abs}$$

$$P_{01} = P_0 + \eta_R(P_{00} - P_0)$$

$$= 70.0 + 0.92\,(86.52 - 70.0)$$

$$= 85.20 \text{ kPa, abs}$$

When the flight speed of the airplane is in the supersonic range, a considerably more complex flow pattern is devised for the compressor inlet. Before the air reaches the compressor inlet section, the velocity of the air must be reduced from a supersonic to a subsonic value. When the airplane is traveling at a relatively low supersonic velocity, an acceptable pressure increase can be achieved in a subsonic diffuser. Deceleration of the air is accomplished in two stages. A normal shock develops at the diffuser inlet section. The velocity of the air changes abruptly from a supersonic to a subsonic value across the normal shock. Subsonic diffusion of the air is subsequently effected within the inlet.

A compressor inlet designed for supersonic flight speeds above approximately Mach 1.5 is necessarily a more complex structure than the simple subsonic diffuser. Briefly, the inlet is designed to develop one or more oblique shocks in advance of the normal shock that occurs near the minimum section of the subsonic diffuser.

7.4 COMPRESSOR PERFORMANCE

The stagnation pressure and temperature of the working fluid entering the compressor are dependent upon the operating conditions of the engine and upon the performance of the inlet. For a prescribed fluid velocity at the compressor inlet section, the corresponding static properties of the fluid are then readily determined.

Because the velocity of the fluid at the inlet section of the compressor does not differ appreciably from the velocity of the fluid at the outlet section, compressor performance can be evaluated by using the fluid stagnation properties. The heat loss from the compressor, per unit mass of fluid, is a small quantity and may ordinarily be neglected.

In Fig. 7.7, the path 01–02_s represents the theoretical reversible adiabatic compression achieved between the stagnation pressures P_{01} and P_{02} taken, respectively, at the compressor inlet and outlet sections. Because of irreversible effects, the actual compression follows the path 01–02 that is typical of gas turbine engine compressors.

The compressor energy equation is given by

$$W_C = h_{01} - h_{02} + Q \tag{7.4}$$

where

W_C = compressor internal work
h_{01} = stagnation enthalpy of the fluid at the inlet section
h_{02} = stagnation enthalpy of the fluid at the outlet section
Q = heat transfer between the compressor and the surroundings, a negative quantity

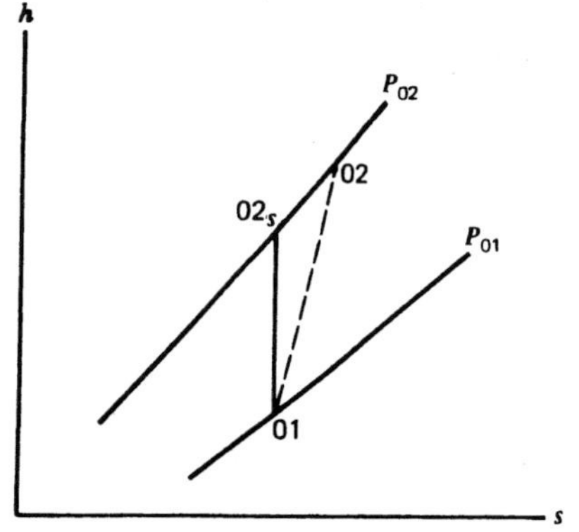

Figure 7.7 Compression process for the gas turbine engine.

The compressor internal isentropic efficiency is the ratio of the theoretical reversible adiabatic work to the actual work required for compression of the fluid between prescribed pressure limits. Then, for $Q = 0$,

$$\eta_C = \frac{h_{01} - h_{02,s}}{h_{01} - h_{02}} \tag{7.5}$$

Typical efficiency values are 0.85 to 0.90 for axial-flow machines and 0.80 for centrifugal-type compressors.

The specific heat values of the fluid are essentially equal for the temperature ranges for reversible adiabatic and actual compression. Equation 7.5 is modified to read

$$\eta_C = \frac{c_p(T_{01} - T_{02,s})}{c_p(T_{01} - T_{02})}$$

The above equation is then reduced to

$$\eta_C = \frac{\left(\frac{P_{02}}{P_{01}}\right)^{k-1/k} - 1}{\frac{T_{02}}{T_{01}} - 1} \tag{7.6}$$

Example 7.2

Air enters a compressor at 95.0 kPa, abs and 300 K (stagnation). The pressure ratio is 6 to 1. The compressor efficiency is 0.86. Calculate the air temperature at the outlet section of the compressor and determine the internal work.

Assume $T_{av} = 415$ K; then, $k_{av} = 1.394$ (Table A.1).

Applying Eq. 7.6,

$$0.86 = \frac{(6)^{0.2826} - 1}{(T_{02}/300) - 1}$$

$$T_{02} = 530.0 \text{ K}$$

At 300.0 K,

$$\bar{h}_{01} = 8696.5 \text{ kJ/kgmol} \qquad \text{(Table A.2)}$$

At 530.0 K,

$$\bar{h}_{02} = 15{,}471.6 \text{ kJ/kgmol}$$

$$W_C = h_{01} - h_{02}$$

$$= 8696.5 - 15471.6$$

$$= -6775.1 \text{ kJ/kgmol}$$

$$- 233.9 \text{ kJ/kg air}$$

Depending upon the design of the gas turbine engine, provision is made for cooling certain parts of the engine by air piped from the compressor. Such cooling may be required for turbine disks, blades, and nozzles, and for bearings, combustion chamber shielding, and portions of the exhaust nozzle casing that are directly or indirectly heated by the high-temperature combustion gases. Cooling air is taken from some point on the compressor casing.

Because the cooling air may represent an appreciable portion of the total air flow, 7 or 8 percent in some designs, a proper allowance for this air must be made in evaluating the performance of the engine. Compressor work is based on the total air flow, with the possibility that some of the air is compressed to a pressure less than that at the outlet section. A portion of the cooling air may expand in the turbine and contribute a small amount of power, while the remainder of the air is throttled through the cooling passages and contributes nothing to the turbine work.

7.5 THE COMBUSTION CHAMBER

The function of the combustion chamber is to introduce the fuel into the compressed air stream and to achieve efficient combustion with minimum pressure and heat losses. Various kinds of fuel may be burned in the gas turbine engine combustion chamber, for example, liquid and gaseous hydrocarbons, methanol, and low-energy gas derived from coal.

The combustion chamber for the gas turbine power plant or engine can be designed to achieve efficient combustion with a moderate amount of excess air, say, 10 to 15 percent, depending upon the fuel fired. However, certain turbine parts, namely, the nozzles, blades, and disks, cannot operate successfully at the elevated temperatures that result from combustion of the fuel with a relatively low percentage of excess air. Consequently, in designing the combustion chamber, provision is made for dilution of the combustion gases with a substantial quantity of air in excess of the theoretical requirement. The air supplied to the combustion chamber is divided into two streams, the primary and secondary flows. Ignition of the fuel is effected in the primary air stream. Secondary air is progressively admitted to the gas stream in order to complete the combustion reaction and lower the gas temperature at the outlet section of the combustion chamber to a value within a range that extends from 850 to 1280 C, depending upon the type of engine or power plant.

Present-day gas turbine engines for the most part burn natural gas or a fuel derived from petroleum. Combustion calculations are facilitated by using, in conjunction with Eq. 7.7, thermodynamic data taken from the Gas Tables.[1]

The following equation is written for complete and adiabatic combustion.

$$-\bar{h}^{\circ}_{RP} = N_p(\bar{h}_p - \bar{h}'_p) - N_a(\bar{h}_a - \bar{h}'_a) - (\bar{h}_f - \bar{h}'_f) \tag{7.7}$$

where

$\bar{h}^{\circ}_{RP}$ = enthalpy of combustion of the fuel at 25 C, for water in the products in the gaseous phase
$\bar{h}_p$ = enthalpy of the products of combustion
$\bar{h}_a$ = enthalpy of the air
$\bar{h}_f$ = enthalpy of the fuel
N_p = mol products per mol fuel
N_a = mol air per mol fuel

The prime superscript designates a property value taken at 25 C. For liquid hydrocarbons, $h_f - h'_f$ may be evaluated with sufficient precision from $2.1(T_f - 25)$ kJ/kg, where T_f is the fuel temperature, in degrees C.

Enthalpy values for air and the products of combustion are available in Table A.2 of the Appendix. Values for the enthalpy of combustion for selected fuels are given in Table A.3.

Equation 7.7 is an extension of the basic energy equation for the combustion process. See Eq. 2.3. The energy values in Eq. 7.7 are for gas mixtures rather than the energy values for the individual gases as demonstrated in Example 2.4.

In an actual combustion process, some heat loss from the combustion chamber will occur, and the reaction will be incomplete to some degree. In order to apply Eq. 7.7 to an actual combustion process, an appropriate efficiency term is introduced.

The combustion efficiency may be defined and used in several different ways depending upon the choice exercised by the investigator. The definition of the combustion efficiency η_B used in this chapter is a comparison of the theoretical and actual fuel/air ratios and accounts for the effects of incomplete combustion and the heat loss from the combustion chamber.

$$\eta_B = \frac{(f/a)_{\text{theo.}}}{(f/a)_{\text{actual}}} \tag{7.7a}$$

The actual fuel/air ratio is determined from the combustor test data. The theoretical fuel/air ratio is calculated for complete and adiabatic combustion that occurs in an ideal combustion chamber operating between the temperature limits observed for the actual combustor.

Example 7.3

Air leaves a compressor at 570.0 kPa, abs, and 530.0 K. See Example 7.2. The fuel, liquid $C_{12}H_{26}$, is supplied to the combustion chamber at 25 C. The stagnation temperature at the combustion chamber outlet section is 1250 K. The combustion efficiency is 0.98. Determine the actual fuel/air ratio.

The theoretical fuel/air ratio is determined from Eq. 7.7.

$$-\bar{h}^{\circ}_{RP} = N_p(\bar{h}_p - \bar{h}'_p) - N_a(\bar{h}_a - \bar{h}'_a) - (\bar{h}_f - \bar{h}'_f)$$

For $C_{12}H_{26}$,

$$N_p - N_a = 6.5 \quad \text{(Ref.: Example 2.4.)}$$

$$h^\circ_{RP} = -44{,}085 \text{ kJ/kg}$$

$$\overline{m} = 170.34 \quad \text{(from Table A.3)}$$

$$\overline{h}^\circ_{RP} = -7{,}509{,}439 \text{ kJ/kgmol}$$

Combustion is assumed to occur with 400 percent of theoretical air. From Table A.2,

$$\overline{h}'_p = 8714.8 \text{ kJ/kgmol}$$

For $T_{03} = 1250$ K,

$$\overline{h}_p = 39{,}533.3 \text{ kJ/kgmol}$$

$$\overline{h}'_a = 8648.5 \text{ kJ/kgmol}$$

For $T_{02} = 530.0$ K,

$$\overline{h}_a = 15{,}482.5 \text{ kJ/kgmol}$$

$$\overline{h}_f - \overline{h}'_f = 0$$

Substitution of the above numerical quantities into Eq. 7.7 yields

$$-(-7509439) = (N_a + 6.5)(39533.3 - 8714.8) - N_a(15482.5 - 8648.5)$$

$$N_a = 304.74 \text{ mol air/mol fuel}$$

Percent Theoretical Air

For stoichiometric combustion of $C_{12}H_{26}$,

$$N_a = 88.06 \text{ mol air/mol fuel}$$

$$\frac{304.74}{88.06} \times 100 = 346 \text{ percent theoretical air}$$

The air supplied, 304.74 mol/mol fuel, is equivalent to 346 percent theoretical air. Because the enthalpy values of the products taken from Table A.2 were based on combustion of the fuel with 400 percent theoretical air, a minor discrepancy will appear in the above calculations. The calculations may be refined by introducing into Eq. 7.7 interpolated values of the enthalpy of the products, say, for 345 percent theoretical air, namely, $\overline{h}_p = 39749.0$ and $\overline{h}'_p =$ 8732.3 kJ/kgmol. Then, $N_a = 302.19$ mol air/mol fuel, equivalent to 343 percent theoretical air.

Theoretical Fuel/Air Ratio

$$(f/a)_{\text{theo.}} = \frac{\overline{m}_f}{N_a \overline{m}_a} = \frac{170.34}{302.19 \times 28.97} = 0.01946 \text{ kg fuel/kg air}$$

Actual Fuel/Air Ratio

Applying Eq. 7.7a,

$$\eta_B = \frac{(f/a)_{\text{theo.}}}{(f/a)_{\text{actual}}}$$

$$0.98 = \frac{0.01946}{(f/a)_{\text{actual}}}$$

$$(f/a)_{\text{actual}} = 0.01986 \text{ kg fuel/kg air}$$

Equation 7.7 does not take account of the effect of dissociation of the products of combustion or of the presence of water vapor in the combustion air on the temperature achieved in the reaction. The effects of dissociation are examined in a later section. Equation 7.7 may be modified to allow for the energy absorbed by the water vapor that enters the combustion chamber with the air.

It is essential to maintain a high combustion efficiency, say, 98 percent, and yet not develop an excessive pressure drop in the combustion chamber. The pressure loss in the combustion chamber has an adverse effect on the turbine output, because of the reduction in the turbine pressure ratio. The pressure loss incurred in the combustion chamber is expressed as a fraction on the inlet stagnation pressure. A pressure loss factor in the range of 0.02 to 0.03 may be anticipated.

7.6 TURBINE PERFORMANCE

The exhaust static pressure for a stationary gas turbine engine is atmospheric or slightly higher when pressure drops are encountered in the exhaust gas ductwork and in the gas side of the regenerator. The turbine exhaust stagnation pressure will be somewhat higher than the corresponding static pressure, depending upon the terminal velocity C_4. See Fig. 7.8. The kinetic energy in the exhaust gases discharged from a static-type power plant represents energy that is not recovered by the turbine. The terminal or leaving velocity of the gas therefore should be reduced to the lowest level that is consistent with the economic objectives of the design.

Unlike in the static-type machines, gas expansion in the aircraft turbojet engine is not completed in the turbine element. Subsequent to expansion in the turbine, the gas is expanded in the exhaust nozzle in order to produce a high-velocity jet for propulsion of the airplane. Because of the pressure drop in the nozzle, the turbine exhaust pressure P_4, Fig. 7.8, will be appreciably above the ambient air pressure.

The energy equation for the turbine is given by

$$\begin{aligned} W_T &= h_{03} - h_{04} + Q \\ &= h_{03} - h_4 - \frac{C_4^2}{2} + Q \end{aligned} \tag{7.8}$$

The heat loss from a well-insulated turbine casing is normally not a significant quantity and may often be neglected.

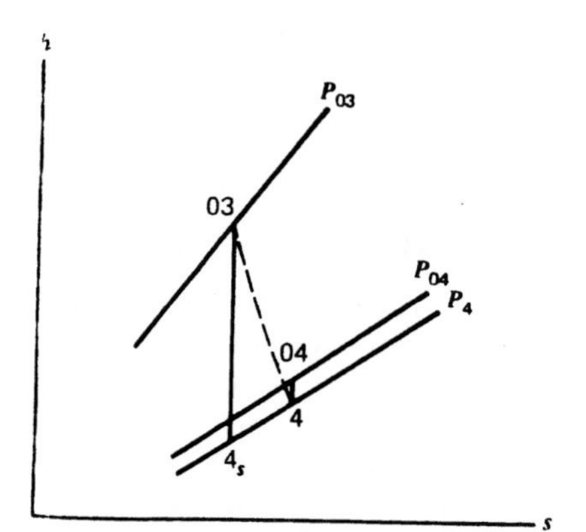

Figure 7.8 Expansion process for the gas turbine engine.

The turbine internal isentropic efficiency is defined as the ratio of the actual internal turbine work to the theoretical work achieved by the corresponding reversible adiabatic expansion. Thus, for $Q = 0$,

$$\eta_T = \frac{h_{03} - h_{04}}{h_{03} - h_{04,s}} \tag{7.9}$$

Because the specific heat of the gas has virtually the same value for the actual process as for the isentropic process, the above equation can be modified to read

$$\eta_T = \frac{1 - \dfrac{T_{04}}{T_{03}}}{1 - \left(\dfrac{P_{04}}{P_{03}}\right)^{k-1/k}} \tag{7.10}$$

The average value of the exponent k for the isentropic process is used in Eq. 7.10.

The machine efficiency for the gas turbine element is approximately 0.90.

Example 7.4

The inlet and outlet stagnation pressures for a gas turbine are, respectively, 550.0 and 110.0 kPa, abs. The inlet stagnation temperature is 1000 K. The turbine internal efficiency is 0.87. Enthalpy values for the gas may be taken for combustion of the fuel with 400 percent theoretical air. Calculate the turbine internal work.

$$\begin{aligned} \text{At } T_{03} &= 1000 \text{ K}, \\ k &= 1.324 \quad \text{(from Table A.1)} \\ \text{Assume } T_{04,s} &= 660 \text{ K}, \\ k &= 1.357 \\ \text{Average value of } k &= 1.340 \end{aligned}$$

$$\begin{aligned} T_{04,s} &= T_{03}\left(\frac{P_{04}}{P_{03}}\right)^{k-1/k} \\ &= 1000\left(\frac{110.0}{550.0}\right)^{0.2537} \\ &= 665 \text{ K} \end{aligned}$$

h_{03} = 30,869.0 kJ/kgmol

$h_{04,s}$ = 19,878.8 kJ/kgmol (Table A.2)

$$W_T = h_{03} - h_{04}$$
$$= \eta_T(h_{03} - h_{04,s})$$
$$= 0.87\ (30869.0 - 19878.8)$$
$$= 9561.5 \text{ kJ/kgmol gas}$$
$$330.3 \text{ kJ/kg gas}$$

7.7 STATIONARY GAS TURBINE POWER PLANTS

The net output of a stationary gas turbine power plant is equivalent to the difference between the turbine shaft output and the compressor shaft input, minus the work required for driving the auxiliary equipment, principally the fuel and lubricating-oil pumps. Additional power would be required in plants that employ compressor intercooling for operation of the cooling-water pumps and in coal-fired plants for operating the fuel pulverizing and handling equipment. Then,

$$W_n = (1 + f/a)\,(W_T)_b - (W_C)_b - W_{aux} \quad (7.11)$$

where

W_n = net power plant output, kJ/kg air
$(W_T)_b$ = turbine shaft work, kJ/kg gas
$(W_C)_b$ = compressor shaft work, kJ/kg air
W_{aux} = work required for operating the auxiliary equipment, kJ/kg air
f/a = fuel/air ratio

Mechanical efficiency values for the turbines and the compressor are comparatively high, that is, 0.98 to 0.99.

Figure 7.9 illustrates the construction of an industrial-type gas turbine engine equipped with a single combustion chamber. In comparison with installations that employ multiple combustion chambers, the single-chamber design achieves generally simplified combustion control, improved reliability, and reduced maintenance.

Several terms are employed for reporting the thermal performance of the power plant, namely, the specific fuel consumption sfc, the heat rate, and the thermal efficiency. Each of these performance factors is based on the net output of the power plant.

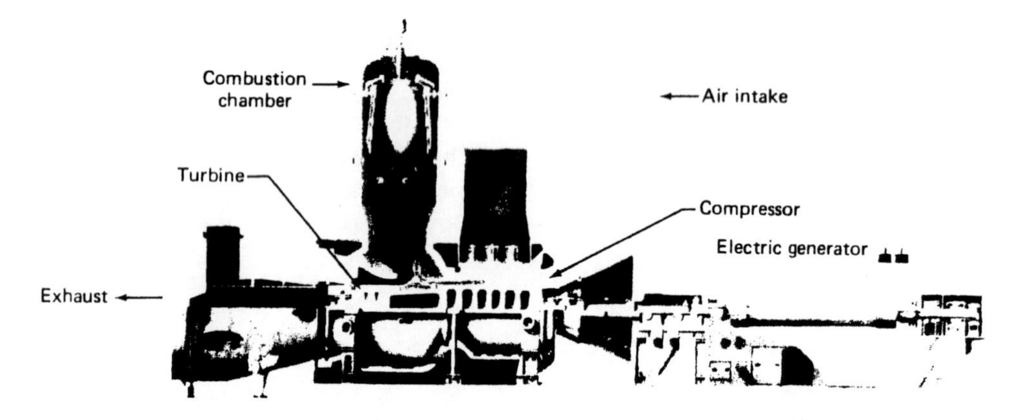

Figure 7.9 Industrial-type gas turbine power plant. The main components of the BBC Type 13 power plant are the electric generator, 17-stage compressor, single combustion chamber, and five-stage turbine. The rated output at the coupling for the three variants of the Type 13 power plant are 60.7, 71.0, and 79.4 MW, respectively. The corresponding mean inlet temperatures to the gas turbine are approximately 820, 880, and 945 C, respectively. (Courtesy Brown Boveri & Cie.)

The specific fuel consumption is defined as the quantity of fuel burned per unit energy output, that is, kg fuel per kWh. The heat rate, in kJ/kWh, is equal to the product of the sfc and the heating value of fuel in kJ/kg. The lower heating value of the fuel is ordinarily used in practice.

The thermal efficiency of the power plant is determined by

$$\eta_t = \frac{3600\dot{W}}{\dot{m}_f(\text{LHV})} \tag{7.12}$$

where

$\dot{W}$ = power output, kW
$\dot{m}_f$ = fuel consumption, kg/h
LHV = lower heating value of the fuel, kJ/kg
3600 = energy equivalent, kJ/kWh

or

$$\eta_t = \frac{3600}{(\text{sfc})(\text{LHV})} \tag{7.12a}$$

Example 7.5

Operating data for a stationary gas turbine power plant are

Compressor internal work: 227.8 kJ/kg air

Turbine internal work: 367.7 kJ/kg gas

Compressor and turbine mechanical efficiencies: 0.98

Fuel/air ratio: 0.0160

Airflow: 40 kg/s

Lower heating value of the fuel: 44,050 kJ/kg

Neglect the power required for driving the auxiliary equipment. Calculate the net output and the thermal efficiency for the power plant.

(a) Net output of the power plant.

$$(W_C)_b = \frac{W_C}{(\eta_m)_C} = \frac{227.8}{0.98} = 232.4 \text{ kJ/kg air}$$

$$\begin{aligned}(W_T)_b &= (1 + f/a)\,[(\eta_m)_T W_T] \\ &= (1 + 0.0160)(0.98 \times 367.7) \\ &= 366.1 \text{ kJ/kg air}\end{aligned}$$

$$\begin{aligned}W_n &= (W_T)_b - (W_C)_b \\ &= 366.1 - 232.4 \\ &= 133.7 \text{ kJ/kg air}\end{aligned}$$

$$\begin{aligned}\dot{W}_n &= \dot{m}_a W_n \\ &= 40 \times 133.7 \\ &= 5348 \text{ kW}\end{aligned}$$

(b) Power plant thermal efficiency.

$$\begin{aligned}\text{sfc} &= \frac{\dot{m}_f}{\dot{W}_n} = \frac{(f/a)\dot{m}_a}{\dot{W}_n} \\ &= \frac{0.0160 \times 40 \times 3600}{5348} \\ &= 0.4308 \text{ kg/kWh}\end{aligned}$$

$$\begin{aligned}\eta_t &= \frac{3600}{(\text{sfc})(\text{LHV})} \\ &= \frac{3600}{0.4308 \times 44050} \\ &= 0.190\end{aligned}$$

$$\begin{aligned}\text{Heat rate} &= \text{sfc} \times \text{LHV} \\ &= 0.4308 \times 44050 \\ &= 18{,}977 \text{ kJ/kWh}\end{aligned}$$

7.8 THE THEORETICAL JET ENGINE

The fundamental principles that apply to a theoretical jet engine are established with reference to Fig. 7.10. The jet engine is moving through the air at the velocity C_p of the airplane, while the exhaust fluid is discharged from the engine at a velocity C_j relative to the casing. Velocity vectors directed to the right are taken as positive quantities.

The engine casing is now assumed to remain stationary; and, as a consequence, the air approaches the inlet at a velocity C_a that is equal to $-C_p$. Both C_a and C_j are relative velocity vectors. It is immaterial in this discussion to consider the means to accelerating the fluid from velocity C_a to C_j. The flow rate $\dot{m}$ of the working fluid through the engine is constant.

The force that tends to move the casing to the left is caused by the momentum change that occurs in the fluid. Then,

$$F_j = \dot{m}(C_j - C_a)$$

The rate of imparting energy to the fluid in order to effect, within the engine, an increase in the fluid velocity from C_a to C_j is equivalent to the rate of change of the fluid kinetic energy. Thus, the fluid power is evaluated, for the theoretical jet engine, by

$$\dot{E}_a = \frac{\dot{m}}{2}(C_j^2 - C_a^2)$$

The jet power, or power input to the airplane, is equal to the product of the jet thrust and the velocity of the airplane. Then,

$$\dot{W}_p = F_j C_p = \dot{m}(C_j - C_a)C_p$$

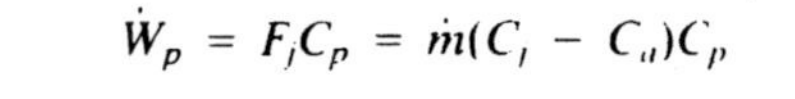
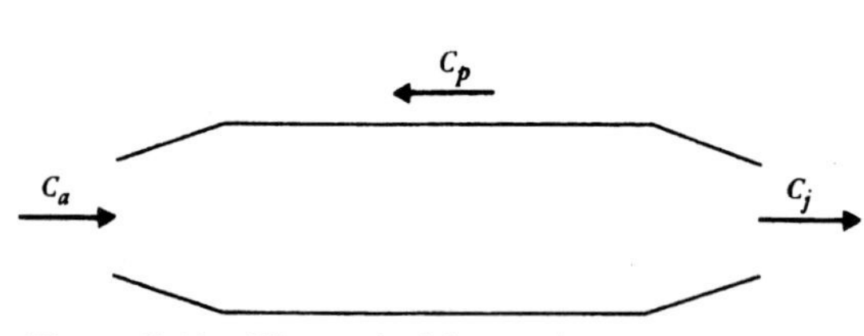

Figure 7.10 Theoretical jet engine.

The propulsive efficiency is defined as the ratio of the jet power to the fluid power. Thus,

$$\eta_p = \frac{\dot{W}_p}{\dot{E}_a} = \frac{\dot{m}(C_j - C_a)C_p}{\frac{\dot{m}}{2}(C_j^2 - C_a^2)}$$

$$= \frac{2(C_j - C_a)C_p}{C_j^2 - C_a^2}$$

Because C_a is numerically equal to C_p,

$$\eta_p = \frac{2}{\frac{C_j}{C_p} + 1} \tag{7.13}$$

Figure 7.11 shows the variation in $\dot{W}_p$, $\dot{E}_a$, and η_p with change in the magnitude of the ratio C_p/C_j. The propulsive power $\dot{W}_p$ is a maximum at $C_p = \frac{1}{2} C_j$, while the fluid power $\dot{E}_a$ decreases from a maximum value at $C_p = 0$ to a zero value at $C_p = C_j$. The combination of these two power terms causes the propulsive efficiency η_p to vary from zero at $C_p = 0$ to the theoretical limit of 100 percent at $C_p = C_j$.

7.9 THE TURBOJET ENGINE NOZZLE

The turbojet engine is essentially a high-velocity gas generator, a machine in which the turbine

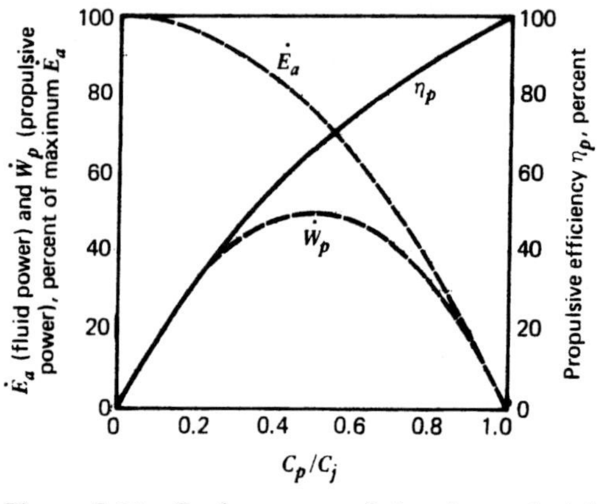

Figure 7.11 Performance of the theoretical jet engine.

drives only the compressor. Propulsion of the airplane is effected by the momentum change caused in the working fluid. In the turboprop engine, the turbine drives the compressor and the propeller, hence the energy in the gas discharged from the turbine is at a comparatively low level and accounts for only a minor contribution to the propulsion of the airplane.

The turbojet engine is ordinarily equipped with a convergent nozzle in which the gas is expanded from the turbine outlet section into the atmosphere. The turbojet engine normally operates with a sonic gas velocity at the nozzle outlet section and a pressure at this section that exceeds the atmospheric pressure. Supersonic gas velocities are generated by free expansion of the gas from the nozzle outlet section into the atmosphere. Free expansion of the gas is somewhat inefficient, but the simple construction of the convergent nozzle is a factor of considerable importance in the design and operation of the engine.

Figure 7.12 is a schematic representation of the exhaust nozzle for a turbojet engine. The minimum section in the gas stream is located very close to the outlet section of the nozzle. The pressure P_n at the nozzle outlet section exceeds the atmospheric pressure P_a. At section 6, the pressure in the fluid stream has dropped to the level of the atmospheric pressure.

The fluid velocity C at any point on the expansion path, either the reversible or actual path, is related to the temperature drop by

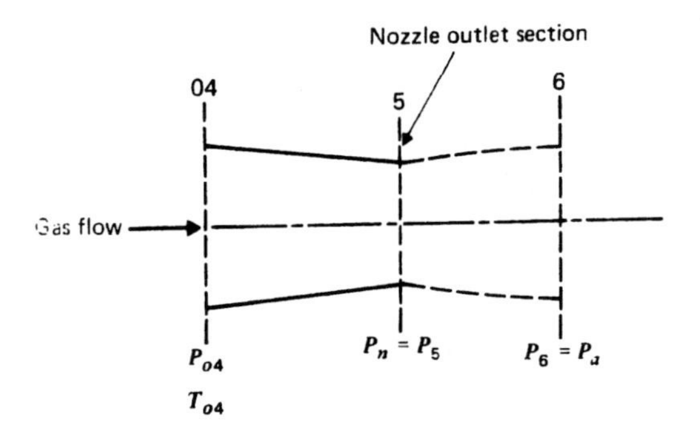

Figure 7.12 Turbojet engine exhaust nozzle.

$$\frac{C^2}{2} = c_p(T_{04} - T)$$

and the sonic gas velocity at any point is given by

$$C_a^2 = RTk$$

As the gas continues to expand, the fluid velocity C will attain, at some point along the path, a value equal to the sonic velocity C_a for the temperature T. This point occurs at the minimum flow area of the fluid stream; and, as noted above, the minimum flow area coincides virtually with the nozzle outlet section.

In Fig. 7.13, state 5 represents the fluid conditions at the nozzle outlet section. When C_5 equals C_a, a combination of the above two equations yields

$$\frac{T_5}{T_{04}} = \frac{2}{k + 1} \tag{7.14}$$

Temperature T_5 (state 5), at which the sonic velocity is generated, results from the actual expansion of the gas. Theoretically, the sonic velocity is attained at a like temperature $T_{5'}$ (state 5′) by reversible adiabatic expansion. However,

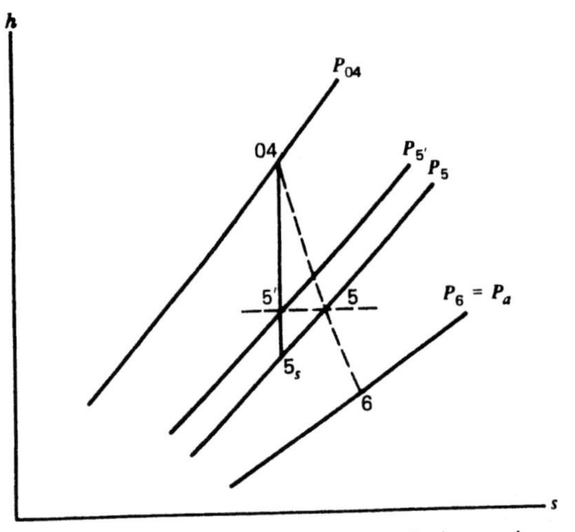

Figure 7.13 Gas expansion in a turbojet engine nozzle.

in comparison with the reversible process, a greater pressure drop is required to attain the sonic velocity by irreversible expansion. Thus, $P_5 < P_{5'}$.

The pressure P_5 is determined from the isentropic enthalpy change $(h_{04} - h_{5,s})$, which in turn is related to the actual enthalpy drop $(h_{04} - h_5)$ through the definition of the nozzle efficiency.*

$$\eta_n = \frac{h_{04} - h_5}{h_{04} - h_{5,s}} \qquad (7.15)$$

Supersonic velocities are generated outside the nozzle by expansion of the gas from pressure P_5 to the atmospheric pressure P_a. The irreversible expansion path 5–6 is shown in Fig. 7.13.

The mean axial velocity of the fluid stream beyond the nozzle outlet section is not readily evaluated because of the difficulty experienced in precisely defining the jet. Further consideration is given to this matter in the following section.

The convergent nozzle represents a practical device for expansion of the turbine exhaust gas. In order to improve the operating flexibility, convergent nozzles can be designed to incorporate variable geometry. The flow area at the nozzle outlet section can then be adjusted to conform to the flight requirements.

7.10 PERFORMANCE OF TURBOJET ENGINES

The turbojet engine is particularly effective for propulsion of aircraft designed for flight in the highest speed range, say, above a speed of 800 km/h. For a designated jet velocity, a comparatively high flight speed is conducive to achieving a high value of the propulsive efficiency. See Fig. 7.11.

* The nozzle efficiency is defined as the ratio of the kinetic energy generated in the actual nozzle to the kinetic energy generated by reversible adiabatic expansion between the same initial state and a commmon outlet pressure.

In a real engine, the mass flow rate of the working fluid is increased in the combustion chamber by the addition of fuel to the intake air. Because of this change, the engine thrust is usually taken as the difference between the gross thrust, developed at the exhaust nozzle, and the ram drag, imposed on the engine casing at the inlet section. The gross thrust F_g is developed by the momentum effect and the pressure difference at the nozzle outlet section. Then,

$$F_g = \dot{m}_g C_j + A_n(P_n - P_a) \qquad (7.16)$$

where

F_g = gross thrust applied to the engine casing by the gas

$\dot{m}_g$ = mass flow rate of the gas
C_j = velocity of the gas at the nozzle outlet section
A_n = nozzle area at the outlet section
P_n = gas pressure at the nozzle outlet section
P_a = atmospheric pressure

The ram drag is given by

$$F_r = \dot{m}_a C_a \qquad (7.17)$$

where

$\dot{m}_a$ = the airflow rate
C_a = the approach velocity of the air

The net force acting on the engine casing is equal to the difference between the gross thrust and the ram drag. Thus,

$$F_n = F_g - F_r \qquad (7.18)$$

The propulsive power is expressed by

$$\dot{W}_p = F_n C_p \qquad (7.19)$$

The equivalent jet velocity $C_{j.e}$ is determined from the combination of the actual jet velocity and the pressure difference effect created at the nozzle outlet section. Thus,

$$F_g = \dot{m}_g C_{j.e}$$
$$= \dot{m}_g C_j + A_n(P_n - P_a)$$
$$C_{j.e} = C_j + \frac{A_n}{\dot{m}_g}(P_n - P_a) \qquad (7.20)$$

The propulsive efficiency η_p is defined as the ratio of the propulsive power to the fluid power. Hence,

$$\eta_p = \frac{F_n C_p}{\frac{1}{2}(\dot{m}_g C_{j.e}^2 - \dot{m}_a C_a^2)} \qquad (7.21)$$

where

F_n = net thrust developed by the engine
C_p = airplane speed
$C_{j.e}$ = equivalent jet velocity
C_a = approach velocity of the air

The gas turbine thermal efficiency η_t for the engine is equal to the ratio of the fluid power to the rate at which energy is supplied by the fuel. Thus,

$$\eta_t = \frac{\frac{1}{2}(\dot{m}_g C_{j.e}^2 - \dot{m}_a C_a^2)}{\dot{m}_f(\text{LHV})} \qquad (7.22)$$

where

$\dot{m}_f$ = fuel rate
LHV = lower heating value of the fuel

The overall efficiency η_o of the engine in flight is equal to the ratio of the propulsive power to the rate at which energy is supplied by the fuel. Then,

$$\eta_o = \frac{F_n C_p}{\dot{m}_f(\text{LHV})} \qquad (7.23)$$

Also, $\eta_o = \eta_p \eta_t$.

Example 7.6

The following data are applicable to a turbojet engine in flight.

Speed of the airplane: 900 km/h (250 m/s)
Stagnation pressure and temperature of the gas at the turbine outlet section: 172.0 kPa, abs, and 970.0 K
Atmospheric pressure: 69.0 kPa, abs
Nozzle efficiency: 0.95
Airflow rate: 100 kg/s
Fuel/air ratio: 0.018
Lower heating value of the fuel: 43,930 kJ/kg

Calculate (a) the gross thrust and the net thrust, (b) the jet equivalent velocity, and (c) the propulsive, thermal, and overall efficiencies.

The pressure ratio, $P_a/P_{04} = 0.40$, is below the nozzle critical pressure ratio of about 0.54, hence the nozzle throat velocity is sonic.

1 Nozzle flow.
For $T = 970$ K, $k = 1.326$ (Table A.1 for 400 percent theoretical air)

$$T_5 = \frac{2}{k+1}T_{04} = \frac{2}{2.326}970.0$$
$$= 834.0 \text{ K} \quad \text{(first approximation)}$$

For $T_{av} = 902$ K, $k = 1.332$

$$T_5 = \frac{2}{2.332}970.0 = 831.9 \text{ K}$$

At $T_5 = 831.9$ K, $k = 1.339$

$$C_5 = C_{a.5} = \sqrt{RT_5k}$$
$$= \sqrt{287.2 \times 831.9 \times 1.339}$$
$$= 565.6 \text{ m/s}$$

$h_{04} = 29{,}859.7$ kJ/kgmol
$h_5 = 25{,}258.9$ kJ/kgmol (Table A.2)

$$\eta_n = \frac{h_{04} - h_5}{h_{04} - h_{5.s}}$$

$$0.95 = \frac{29859.7 - 25258.9}{29859.7 - h_{5.s}}$$

$$h_{5.s} = 25{,}016.8 \text{ kJ/kgmol}$$

$$T_{5.s} = 824.6 \text{ K}^*$$

For $T_{av} = 897.3$ K, $k = 1.332$

$$P_5 = P_{04}\left(\frac{T_{5.s}}{T_{04}}\right)^{k/k-1}$$

$$= 172.0\left(\frac{824.6}{970.0}\right)^{4.012}$$

$$= 89.65 \text{ kPa, abs}$$

$$\rho_5 = \frac{P_5}{RT_5} = \frac{89.65}{0.2872 \times 831.9}$$

$$= 0.3752 \text{ kg/m}^3$$

$$A_5 = \frac{\dot{m}_g}{\rho_5 C_5} = \frac{1.018 \times 100}{0.3752 \times 565.6}$$

$$= 0.4797 \text{ m}^2$$

2 Thrust.

$$F_g = \dot{m}_g C_j + A_n(P_n - P_a)$$

$$= (101.8 \times 565.6) + 0.4797 \cdot (89.65 - 69.0)10^3$$

$$= 67{,}484 \text{ N}$$

$$F_r = \dot{m}_a C_a = 100 \times 250$$

$$= 25{,}000 \text{ N}$$

$$F_n = F_g - F_r = 67484 - 25000$$

$$= 42{,}484 \text{ N}$$

3 Jet equivalent velocity.

$$C_{j.e} = \frac{F_g}{\dot{m}_g} = \frac{67484}{101.8} = 662.9 \text{ m/s}$$

4 Efficiency values.

$$\eta_p = \frac{F_n C_p}{\frac{1}{2}(\dot{m}_g C_{j.e}^2 - \dot{m}_a C_a^2)}$$

$$= \frac{42484 \times 250}{\frac{1}{2}[101.8(662.9)^2 - 100(250)^2]}$$

$$= 0.552$$

$$\eta_t = \frac{\frac{1}{2}\left(\dot{m}_g C_{j.e}^2 - \dot{m}_a C_a^2\right)}{\dot{m}_f(\text{LHV})}$$

$$= \frac{\frac{1}{2}\left[101.8(662.9)^2 - 100(250)^2\right]}{1.80(43930 \times 10^3)}$$

$$= 0.243$$

$$\eta_o = \frac{F_n C_p}{\dot{m}_f(\text{LHV})}$$

$$= \frac{42484 \times 250}{1.80(43930 \times 10^3)}$$

$$= 0.134$$

7.11 PERFORMANCE OF TURBOPROP ENGINES

In the medium-speed range for aircraft, 500 to 800 km/h, the turboprop engine operates with a comparatively high overall efficiency because of the high propulsive efficiency of the propeller. The thrust for propelling airplanes equipped with turboprop engines is developed principally by the propeller, with the exhaust gases contributing some 10 to 20 percent of the total thrust, depending upon the speed of the airplane.

The total power W_e developed by the engine is expressed as the net power delivered to the

* Because of the comparatively high value for the nozzle efficiency, the following calculation for temperature $T_{5.s}$ yields a sufficiently precise result.

$$\eta_n = \frac{T_{04} - T_5}{T_{04} - T_{5.s}}$$

$$0.95 = \frac{970.0 - 831.9}{970.0 - T_{5.s}}$$

$$T_{5.s} = 824.6 \text{ K}$$

shaft plus the fluid power, evaluated from the gain in the kinetic energy of the fluid in passing from the inlet to the exhaust section. The net shaft work W_n is calculated through use of Eq. 7.11.

$$\dot{W}_e = \dot{m}_a W_n + \frac{1}{2}\left(\dot{m}_g C_f^2 - \dot{m}_a C_a^2\right) \quad (7.24)$$

The jet velocity is considered to be subsonic in the above equation.

The propulsive efficiency η_p is equal to the ratio of the total propulsive power to the total power delivered by the engine.

$$\eta_p = \frac{\eta_s \dot{m}_a W_n + F_n C_p}{\dot{W}_e} \quad (7.25)$$

where η_s is the propeller efficiency. The first term in the numerator of Eq. 7.25 represents the propeller power, and the second term, the jet power.

An equivalent thrust is determined by

$$F_e = \frac{\eta_s \dot{m}_a W_n}{C_p} + F_n \quad (7.26)$$

The gas turbine thermal efficiency is defined as the ratio of the total power delivered by the engine to the rate at which energy is supplied by the fuel.

$$\eta_t = \frac{\dot{W}_e}{\dot{m}_f(\text{LHV})} \quad (7.27)$$

The overall efficiency η_o of the engine in flight is taken as the ratio of the propulsive power to the rate at which energy is supplied by the fuel.

$$\eta_o = \frac{F_e C_p}{\dot{m}_f(\text{LHV})} \quad (7.28)$$

7.12 THE ACTUAL CYCLE FOR THE GAS TURBINE ENGINE

The thermal efficiency of the gas turbine engine is influenced by a comparatively large number of factors. A detailed examination of the effect of all of these factors on the performance of the engine is however outside the scope of this text, particularly with respect to off-design operation.

Attention is now directed to the actual cycle that describes the operation of the simple, open-cycle, stationary gas turbine power plant. The theoretical, or Brayton-cycle, thermal efficiency is a function solely of the pressure ratio. Although the thermal efficiency of the theoretical cycle continues to increase with an increase in the pressure ratio, this trend is not generally observed for the actual cycle.

Five principal factors describe the cycle for the simple gas turbine engine, namely, the pressure ratio, the compressor inlet temperature, the turbine inlet temperature, the compressor efficiency, and the turbine efficiency. Some factors of secondary importance are the pressure, heat, combustion, and mechanical losses that contribute to a reduction in the thermal efficiency of the engine. Figure 7.14 shows the actual cycle for the engine.

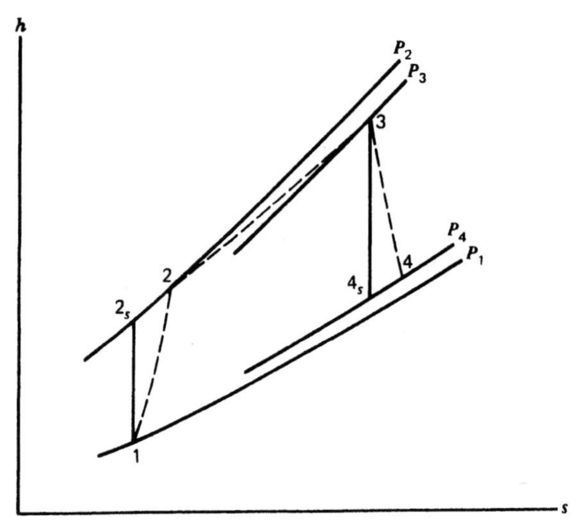

Figure 7.14 Actual gas turbine engine cycle.

Although the compressor and turbine efficiencies influence the performance of the engine, these factors are not operating variables that can be arbitrarily controlled. The machines are designed to achieve prescribed operating efficiencies that vary with changes in the load characteristics of the engine. In general, an increase in the efficiency of either machine, the compressor or turbine, will improve the thermal performance of the engine.

The compressor work is expressed by

$$W_C = \frac{c_p}{\eta_C} T_1 \left[1 - \left(\frac{P_2}{P_1} \right)^{k-1/k} \right]$$

For a prescribed pressure ratio, the compressor work is a function of the inlet air temperature T_1. Ordinarily, T_1 is essentially equal to the ambient air temperature. A reduction in the magnitude of T_1 causes a decrease in the compressor work, and consequently an increase is achieved in the net work of the engine. The compressor outlet temperature T_2 however decreases, hence for a prescribed combustion chamber outlet temperature T_3, the fuel/air ratio will increase.

The turbine work is expressed by

$$W_T = \eta_T c_p T_3 \left[1 - \left(\frac{P_4}{P_3} \right)^{k-1/k} \right]$$

For a prescribed pressure ratio, an increase in the turbine inlet temperature T_3 will cause an increase in both the turbine work and the net work of the engine. The turbine inlet temperature is limited by the heat-resistant properties of the materials used in the construction of the machine.

The pressure ratio prescribed for the engine cycle will be dependent, at least in part, upon the other four principal parameters, namely, the machine efficiencies and the compressor and turbine inlet temperatures. Figure 7.15 is a typical performance plot that shows the effect of a variation in the pressure ratio on the thermal efficiency of the engine cycle for a prescribed turbine inlet temperature.

The calculated values of the thermal efficiency of the engine cycle are based on representative values of the inlet air temperature and the compressor and turbine efficiencies. Appropriate allowances are made for the usual losses, namely, the combustion, heat, presssure, mechanical, and turbine leaving losses.

For prescribed turbine inlet temperatures of 1000 and 1300 K, the maximum value of the thermal efficiency of the cycle is attained for pressure ratios of 10 to 1 and about 20 to 1, respectively. For a higher turbine inlet temperature of 1600 K, the thermal efficiency continues to increase to a maximum value that is achieved for a pressure ratio in excess of 20 to 1.

The reason for the development of a peak in the thermal efficiency curve is not immediately obvious. An explanation however can be provided by examining the curves that show the variation with the pressure ratio for the fuel/air ratio and the work quantities, namely, the compressor, turbine, and net work.

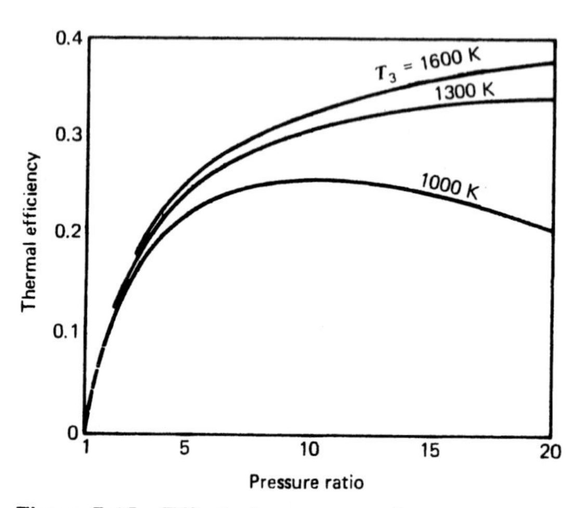

Figure 7.15 Effect of pressure ratio on the thermal efficiency of the gas turbine engine cycle.

7.13 COMBUSTION CHAMBERS

The objectives of the combustion chamber design are to achieve virtually complete combustion with a minimum pressure drop in the fluid stream. Many investigations have been conducted on the combustion reaction, and large quantities of data are available to the designer. Some of the significant operating characteristics of the gas turbine engine combustor are subsequently examined.

Natural gas burns readily in the combustion chamber of the gas turbine engine. Most engines however burn a liquid hydrocarbon fuel that is introduced into the combustion chamber in the form of a finely divided spray. Viscosity of the fuel is the most significant factor that influences the atomization of the fuel in the injection nozzle. Pressure atomization is achieved by forcing the fuel through the small orifices in the spray nozzle tip. Heavy fuel oils are preheated prior to atomization in order to reduce the viscosity and thereby ensure formation of a finely divided spray.

Before combustion is initiated, the fuel must be vaporized, mixed with air, and raised to the ignition temperature, which varies with the mixture ratio and the pressure. The combustion chamber design must provide for a transfer of energy from the high-temperature surfaces and combustion gases to the liquid fuel droplets. In addition to the time required for vaporization, there is a lag or delay period associated with the start of the combustion reaction in the gaseous mixture. Following ignition, the fuel must remain in the combustion chamber for a period of time sufficient for completion of the reaction.

The air entering the combustion chamber is generally divided by a metering device into the primary and secondary air streams. Ignition can only be established in a mixture of fuel and air that falls within certain limiting mixture ratios. The metering device directs the required amount of air to the primary combustion zone within the flame tube in order to provide, over wide ranges of operation, mixture ratios that are within the limits of inflammability. The stoichiometric fuel/air ratio for hydrocarbon fuels is approximately 1 to 15. However, to ensure adequate mixing of the fuel and air, principally to avoid cracking the fuel, an excess quantity of air, say, 15 to 20 percent, is ordinarily supplied to the primary combustion zone. The flame tube cap is designed to promote the necessary turbulence for adequate mixing of the air and vaporized fuel prior to ignition.

The velocity of flame propagation through a combustible mixture varies with the fuel/air ratio. Maximum velocity is achieved with a mixture that is slightly richer than the theoretical fuel–air mixture, that is, one corresponding to the stoichiometric ratio. The velocity of flame propagation is relatively low with respect to the inlet air velocity; hence after ignition occurs, the tendency exists for the flame to be blown from the combustion chamber. Stabilization of the flame is achieved by the admission of the air to the flame tube through swirl ports or passages, thus creating a vortex in the fluid. See Fig. 7.16. Higher rotational velocities in the region of the chamber axis than at the tube wall cause a reduction in the static pressure, with the pressure gradient decreasing toward the fuel nozzle outlet. Because of the pressure difference, a flow reversal into the stagnant region carries the high-temperature gases to the incoming fuel, thereby maintaining stable combustion.

Admission of a portion of the total air to the flame tube produces in the primary combustion zone the high temperatures required for high rates of combustion. Secondary air is gradually admitted, through the downstream swirl ports, to the primary gas stream in order to ensure completion of the combustion reaction. See Fig. 7.16. Farther downstream, additional secondary air is admitted to the dilution zone as a means to lowering the gas temperature to the limiting value set for admission to the turbine. Cooling of the combustion gases must not occur too rapidly, otherwise the reaction will be prematurely arrested as a result of low gas temperatures. Un-

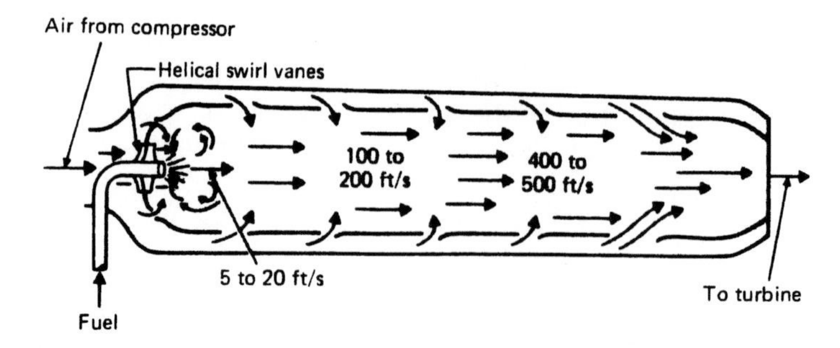

Figure 7.16 Schematic diagram for a gas turbine engine combustion chamber. (W. J. Hesse and N. V. S. Mumford, Jr., *Jet Propulsion for Aerospace Applications*, Pitman Publishing Corporation, New York, 1964.)

burned fuel components appearing in the engine exhaust gases result from inadequate mixing of the fuel and air, from the failure of some of the fuel particles to attain the ignition temperature, or from quenching prior to completion of the combustion reaction.

In the preceding discussion, attention was directed to the necessity for providing sufficient residence time for the fuel in the combustion chamber and for maintaining the sufficiently high temperatures that are required for establishing ignition and sustaining the combustion reaction. Turbulence is a third important requirement that must be observed in the design of a combustion chamber. Thorough mixing of the fuel and air, accomplished through turbulence, promotes complete combustion of the fuel. Briefly, adequate turbulence is achieved by proper location and design of the air ports in the liner and by establishing sufficiently high air velocities at the several points of admission to the gas stream.

Improper design or operation of the combustion chamber may cause a substantial temperature variation in the cross section of the gas stream at the outlet of the combustion chamber. As a result, "hot spots" can form in the turbine, and early failure of the first-stage nozzle vanes and rotor blades may occur. Adequate turbulence and proper flow control are essential to maintaining an acceptable temperature distribution in the gas stream discharged from the combustion chamber.

Because of the adverse effect on the turbine pressure ratio, the pressure drop in the combustion chamber should be held to a minimum consistent with the objectives of the design. The pressure loss is caused by wall friction and turbulent flow through ports and swirl vane passages and by the momentum change effected in the fluid. Because of the energy release, the velocity of the fluid increases during passage through the combustion chamber. An approximate evaluation of the pressure drop, owing to the momentum change, can be achieved by employing a theoretical combustion chamber or model in which the flow is described by a Rayleigh line process.

Present-day combustion chambers for gas turbine engines are designed primarily for burning a gaseous or liquid fuel. Natural gas is the predominant gaseous fuel, but a variety of waste and process gases can be fired in the engine. Aircraft turbine engines commonly burn JP-4, a fuel manufactured in petroleum refining. JP-4 and a similar commercial-grade fuel are widely used in industrial gas turbine engines, although heavy industrial oils are used to some extent. The aviation-grade turbine oils vaporize readily, and subsequent combustion is rapid and smokeless. Methanol can be burned effectively in a gas turbine engine, but the use of methanol would be contingent upon the availability of the fuel in sufficient quantity at a competitive price.

Figure 7.17 shows the constrution of a can-

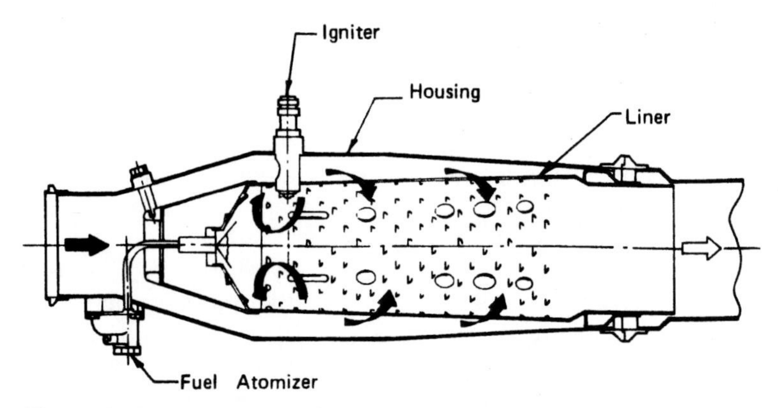

Figure 7.17 Can-type combustion chamber for a gas turbine engine. (Courtesy Garrett Turbine Engine Company.)

type combustion chamber. The engine is equipped with several combustion chambers of this type, each of which is provided with one or more pressure-atomizing nozzles. Installation of a can-type combustion chamber in the engine is shown in Fig. 7.18. A high energy release rate, kJ/h·m^3, can be achieved in the can-type combustion chamber, particularly for aircraft engine installations.

The annular-type combustion chamber is arranged for combustion of the fuel in a single or double annulus. The secondary air is admitted through openings in both walls of the annulus. Several fuel nozzles are installed in a concentric pattern at the inlet end of the combustion chamber.

The liner of the can-type and the walls of the annular-type combustion chambers are fabricated from heat-resistant steel. Metal temperatures are not exceptionally high, because one wall surface is exposed to the relatively cool secondary air. Because of a low stress level, comparatively thin sections are adequate for the construction of these components.

An industrial-type gas turbine engine can be equipped with a single combustion chamber as shown in Fig. 7.9. A machine arrangement of this kind is feasible where space limitations are

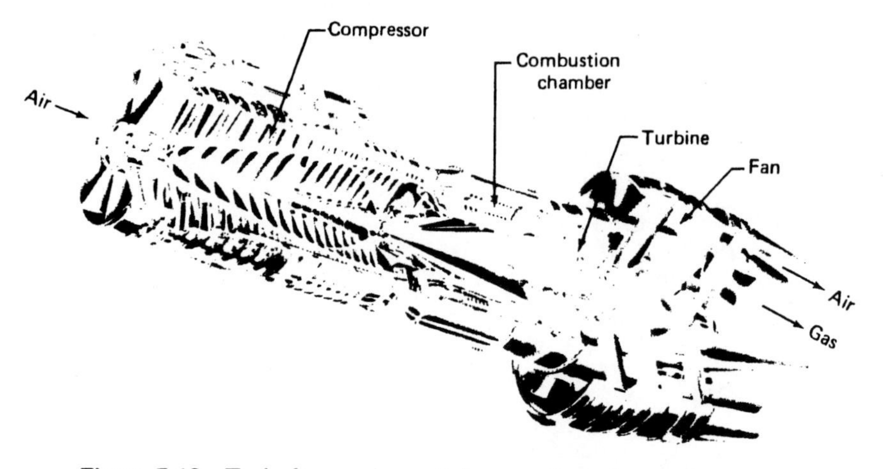

Figure 7.18 Turbofan engine, aft-fan configuration and can-type combustion chamber. (Courtesy The General Electric Company.)

not particularly stringent. In comparison with the can-type combustion chamber, the design of the single, industrial-type combustion chamber is directed to achieving a simple, rugged construction. Somewhat lower energy release rates are characteristic of these comparatively high-volume combustion chambers.

The development of an open-cycle, coal-fired gas turbine power plant was started in the United States circa 1945. Several difficult problems are encountered in constructing a coal-fired gas turbine engine. Turbine blade erosion by the coal ash is certainly a principal obstacle in this endeavor. Prior to entering the turbine, a substantial reduction must be achieved in the concentration of erosive dust present in the combustion gases. The dust collector, located between the combustion chamber and the turbine, should be capable of removing 99 percent of all ash particles larger than 10 microns if blade erosion is to be significantly reduced. The coal preparation and feeding equipment is complex, and at some point the coal must pass through an air lock. Other requirements for acceptable operation of a coal-fired gas turbine engine are a high combustion efficiency and a low pressure drop across the ash removal system.

The problems encountered in the direct firing of coal in a gas turbine engine can be largely eliminated by burning a low-energy gas derived from coal. Coal gasification was discussed in Chapter 2. The production of clean, low-energy gas in a chemically active, fluidized-bed gasifier would represent a good approach to reducing the dependence upon oil for the operation of industrial gas turbine power plants.

7.14 NO_x EMISSION

Because the gas turbine engine operates with a high percentage of excess air, combustion is essentially complete. The exhaust gas shows a low concentration of HC and CO. The emission of sulfur oxides is related directly to the sulfur content of the fuel. The NO_x emisson from the gas turbine engine is comparatively high, but still lower than the NO_x emission from an uncontrolled internal combustion engine. This high concentration of NO_x however creates a problem in the operation of the gas turbine engine and may retard future use of the engine in the transportation field and for power generation.

The 1970 Federal nitrogen oxides emission standard is 0.4 gram per mile for automotive vehicles. For the internal combustion engine, this limit is very difficult to achieve. In this respect, a number of proponents believe that the gas turbine engine has a greater potential for meeting the 0.4 g/mile limit if this NO_x emission standard is eventually imposed. See Section 4.7 for 1981 NO_x limit.

Many investigations are directed to achieving an effective reduction in the NO_x emission from the gas turbine engine. The only feasible approach to accomplishing this objective appears to be through a decrease in the formation of NO in the combustion chamber. Removal of NO_x from the engine exhaust gas by wet scrubbing is ineffective.

Oxides of nitrogen are formed at a high temperature in lean flames by the direct combination of atmospheric nitrogen and oxygen. The complete mechanism for the formation of these oxides is complex and not fully understood. NO is the predominant oxide of nitrogen formed in a typical gas turbine engine combustion chamber. Several reactions have been proposed to describe the formation of NO in a combustion reaction.

General acceptance has been given to the following reactions:

$$1 \quad N_2 + O \rightleftharpoons NO + N$$

$$2 \quad N + O_2 \rightleftharpoons NO + O$$

The principal source of oxygen atoms for reaction 1 is the dissociation of O_2. The oxygen atoms are essentially in thermal equilibrium with the oxygen molecules. Reaction 1 is the primary source of nitrogen atoms for reaction 2. The two reactions contribute approximately equal quantities of NO.

Several observations are pertinent to the formation of NO, irrespective of the mechanisms that may be adopted.

Because of insufficient time, equilibrium conditions are not established in the combustion reaction. The concentration of NO in the combustion gases is thus below the equilibrium concentration.

At the high temperatures, existing in flames, the NO level is sharply increased by an increase in temperature.

An increase in residence time will increase substantially the NO level.

The NO level will increase with an increase in the oxygen concentration.

The oxidation of NO to NO_2 is an exothermic reaction, and the rate at which it occurs is inversely proportional to the temperature. At room temperature, the oxidation of NO proceeds slowly. Most of the NO produced in the combustion reaction remains unaltered and is subsequently discharged from the engine. The total concentration of NO and NO_2 is ordinarily reported as NO_x.

An examination of the factors that influence the production of NO will disclose procedures that may possibly be employed to control the NO_x level in the combustion gases. Experimentation will be required to demonstrate the effectiveness of such measures.

The rate of NO formation is highly temperature dependent, thus a small change in the flame temperature will cause a significant change in the reaction rate. The maximum temperature in the primary zone of the combustion chamber can be readily decreased by introducing more air into this high-temperature zone. A temperature reduction can also be achieved by introducing an inert substance into the primary zone, such as recirculated, lower-temperature exhaust gases. Injection of a water spray into the primary zone, or water vapor into the combustion air, will effectively decrease the maximum temperature attained in the combustion chamber.

Temperature control achieved by increasing the quantity of air supplied to the high-temperature zone of the combustion chamber requires further examination. The additional air increases the oxygen concentration, which in turn tends to raise the NO level. This effect on the NO level is however counteracted by a considerably stronger effect, that is, the reduction in the NO level caused by the decrease in the maximum temperature of the reaction.

The NO level increases with increase in the residence time during which the gases are moving through the high-temperature zone of the combustion chamber. The residence time can be decreased by increasing the air velocity through the high-temperature zone or by decreasing the length of this zone. A decrease in the length of the zone can be accomplished by moving the air dilution holes upstream. The cooling effect achieved by air dilution causes a rapid drop in the gas temperature and the termination of NO production. Relocation of the air dilution holes however must not induce quenching of the flame prior to completion of the combustion reaction.

Further research and development will be required to reach an acceptable concentration of NO_x in the gases discharged from the gas turbine engine. While achieving a decrease in the NO level, the operating characteristics of the combustion chamber must not be adversely affected. A high combustion efficiency and low pressure losses should be maintained over a wide range of operation in the absence of any increase in the emission of HC and CO.

7.15 DISSOCIATION OF THE COMBUSTION GASES

The gas turbine engine operates generally within a temperature range in which the effects of the dissociation of the products of combustion are negligible, that is, turbine inlet temperatures are not much above 1100 C. Some high-performance aircraft engines however are currently operating with turbine inlet temperatures in excess of 1260 C. Still higher gas temperatures for future gas turbine engines may be anticipated as improved materials become available and more effective turbine cooling systems are developed.

Example 7.7

An examination will be made of the effect that dissociation of the combustion gases has on high-temperature operation of the gas turbine engine. The specifications for the cycle are

Compressor inlet temperature: 288.9 K

Fuel: liquid $C_{12}H_{26}$, T_f = 25 C

Compressor pressure ratio: 15 to 1

Compressor internal efficiency: 0.86

Turbine internal efficiency: 0.89

Air supplied: 200 percent of theoretical air

Neglect mechanical, heat, combustion, and pressure losses. Data selected from Table A.2 will be used for the cycle analysis. The properties for the combustion gas mixtures, listed in Table A.2, are based on prescribed fractions of CO_2, H_2O, O_2, and N_2; hence dissociation of the combustion products is not observed.

The following results are reported:

Compressor outlet temperature: 670.6 K

Compressor work: 392.63 kJ/kg air

Combustion chamber outlet temperature:

1794 K

Fuel/air ratio: 0.03338

Enthalpy change achieved by isentropic expansion: 1046.91 kJ/kg gas

Turbine work: 931.75 kJ/kg gas,
962.85 kJ/kg air

Cycle net work: 570.22 kJ/kg air

Cycle thermal efficiency: 0.387 (based on LHV of the fuel)

The effect of dissociation of the product gases on the performance of the cycle is examined through use of the NASA Lewis Research Center computer program devised by G. Sanford and B. J. McBride.[2]

For the combustion reaction, the following entries are made:

Fuel: liquid $C_{12}H_{26}$, T_f = 298.15 K

Inlet air temperature: 670.6 K

Equivalence ratio: 0.500

Pressure: 15 atm

Constant-pressure, adiabatic combustion assumed

The theoretical equilibrium composition of the product gases at the end of the combustion reaction is given by the following mole fractions:

CO	0.00001	NO	0.00282	O	0.00001
CO_2	0.06570	NO_2	0.00002	OH	0.00022
H_2O	0.07107	N_2	0.76034	O_2	0.09982

The corresponding temperature of the combustion gases is 1781 K. The total mole fraction for the primary nondissociated products (CO_2, H_2O, N_2, O_2) is 0.99693. The remaining mole fractions represent oxides of nitrogen (NO, NO_2) and the dissociated components (CO, OH, O).

Although a temperature of 1781 K represents, for current practice, a relatively high turbine inlet temperature, dissociation apparently is not present to a significant degree. A corresponding temperature of 1794 K was determined for a cycle analysis based on the Gas Tables.

The theoretical equilibrium composition will not in every case represent, to a fair degree, the actual composition of the gaseous mixture discharged from the combustion chamber. Chemical kinetics and residence time are factors that have a significant influence on the actual composition of the product gas mixture. For the preceding example, the theoretical equilibrium composition is probably a reasonably good representation of the actual composition.

The computer program was extended to include the isentropic expansion of the product gases to the atmospheric pressure. This process represents theoretical expansion in the turbine from 15 atmospheres to 1 atmosphere. Theoretical expansion may be considered for the products frozen in the initial composition or for the products maintained in continuous equilibrium.

For frozen expansion, the composition of the gaseous mixture at the end of expansion would not differ from the composition shown above for the initial state. When equilibrium is maintained during expansion, the degree of dissociation diminishes, and the mole fractions for the mixture at the terminal state are

CO	0.0	NO	0.00002	O	0.0
CO_2	0.06571	NO_2	0.0	OH	0.0
H_2O	0.07119	N_2	0.76179	O_2	0.10129

For the isentropic expansion process, the enthalpy change is

For frozen expansion: −1047.12 kJ/kg gas

For equilibrium expansion: −1050.89 kJ/kg gas

For this example, little difference is observed in the results reported for the two expansion processes.

The theoretical turbine work of 1047.12 or 1050.89 kJ/kg gas agrees closely with the value of 1046.91 kJ/kg determined from calculations based on the Gas Tables. Apparently, dissociation of the product gases will have a minor effect on the thermal efficiency of the gas turbine engine cycle for the currently observed operating temperatures.

The above example shows a decrease in the mole fraction of NO from 0.00282 to 0.00002 when chemical equilibrium is maintained during expansion of the gases. Such decrease will not be observed in the actual engine. Once the NO is formed in the engine, the concentration of this compound does not change as the temperature of the gaseous mixture decreases. At any designated temperature, the actual composition of the mixture will not agree with the equilibrium composition, particularly with respect to the NO. Again, the importance of chemical kinetics and residence time is apparent. In the actual engine, a small fraction of the NO is oxidized to NO_2 before the gases are discharged into the atmosphere. This slight decrease in the concentration of NO is however not a result of a shift in the equilibrium composition.

7.16 GAS TURBINE ENGINE APPLICATIONS

Aircraft gas turbine engines, including the turboprop machines, are extensively employed for propelling military, commercial, and private airplanes. Industrial gas turbine power plants are available for many different applications over a wide range in power output. Numerous marine installations can be cited. Application to rail transportation is in high-speed passenger service.

Despite a considerable amount of development, the automotive gas turbine engine has not moved into the production stage. The heavy-duty gas turbine engine appears to be a competitor of the Diesel engine for powering trucks, buses, and other mobile equipment; but applications in this area have been limited mainly to experimental vehicles. Production of gas turbine engines for installation in passenger automobiles evidently remains uncertain, despite occasional optimistic reports to the contrary.

Penetration of the automotive field could ultimately account for a substantial increase in the production of gas turbine engines. Several inherent characteristics of the gas turbine engine are considered to be advantageous with respect to automotive applications. The engine has good torque characteristics and operates quietly and smoothly. Unlike the internal combustion engine, the gas turbine engine can burn a wider variety of fuels with little difficulty, and cold-weather starting is not a problem. The gas turbine engine is a light-weight machine, requires no cooling system, and has a low emission of hydrocarbons and carbon monoxide.

The gas turbine engine developed for automotive service however has some disadvantages. The fuel economy is somewhat unsatisfactory, particularly when the engine is idling. Apparently, production costs for the automotive gas turbine engine will be comparatively high. The problem of high NO_x emission must be solved before large-scale production of the automotive engine can be undertaken.

The gas turbine engine is particularly well suited to electric power generation, in part, because of the constant-speed characteristic of the generator. Industrial generator sets are constructed for both mobile and stationary service. Gas turbine engine–electric generator sets are frequently installed in total-energy systems, particularly when fueled with natural gas. The energy in the turbine exhaust gas is utilized in processing, heating, and air conditioning applications, and some of these systems operate with an excellent overall economy.

A large number of gas turbine engine–electric generator sets are installed in electric power systems throughout the United States. These machines are integrated with steam and hydro units and normally operate for short periods of time while supplying peak power, or for longer periods of time while supplementing base load generation.

The gas turbine engine is inherently suited to peaking service. Start-up time, from a cold start to full load, is low, only a few minutes for some machines. Capital costs are comparatively low, and early delivery can usually be arranged. The gas turbine engine–electric generator set is often located at a strategic point in the system and operated by remote control. In comparison with the steam turbine-generator, the gas turbine power plant has a poorer heat rate, and maintenance costs are high, thus extended operation, that is, beyond peak periods, is ordinarily avoided. Projected installation of gas turbine engine–electric generator sets is clouded to some extent by the uncertain supply and high cost of natural gas and turbine oils.

Utility gas turbine engines are of two types, namely, heavy industrial and aircraft derivative. The machines in the latter category consist of one or more turbojet engines that function as gas generators and supply the motivating gas to the power turbine. The electric generator is driven by the power turbine. Because the turbojet engines and the power turbine are not mechanically coupled, the turbojet engines are operated at speeds above 3600 rpm, the limiting speed for the power turbine.

The heavy, industrial-type gas turbine engines are simple, open-cycle machines. In the event that gas turbine power is extended to electric generation in machines in the medium-capability range, regeneration will most likely be required. In Section 8.24, the combination of steam and gas power is examined under the heading of the "Combined Cycle."

Several mechanical arrangements are possible for the simple, open-cycle gas turbine engines. The directly coupled compressor–turbine arrangement is commonly employed for the constant-speed, heavy industrial-type machines. Where a variable-speed power takeoff is required, the turbine is usually constructed in two sections. The high-pressure turbine drives the compressor, often at constant speed, while the low-pressure turbine functions as the power turbine. The combustion gases are expanded in two steps successively through the high- and low-pressure turbines, which are not mechanically connected.

7.17 SOME ENGINE DESIGN CONCEPTS

A number of schemes are employed for improving in some manner the performance of the gas turbine engine. Higher thermal efficiencies are, in general, achieved by increasing the compressor pressure ratio and increasing the turbine inlet temperature. A substantial increase in the turbine inlet temperature has been accomplished over the past three decades, about 10 C per year, by improved methods for cooling the hot parts of the turbine and by developing materials capable of operating at higher temperatures.

The vanes in the first-stage nozzle ring are exposed to the highest-temperature gases in the turbine. The material used to fabricate these vanes must have excellent resistance to oxidation and corrosion and high resistance to thermal fatigue. The material must also possess casting and welding capability. A cobalt-base alloy, used to fabricate the vanes, meets these requirements.

The first-stage rotor blades are subjected to the most severe operating conditions in the turbine. Not only must the blade material withstand the effect of high temperature, but also the stresses caused by centrifugal force, the bending stresses, the superimposed vibratory stress, and on occasion thermal shock. A nickel-base superalloy is used for fabricating these blades.

Other parts of the turbine are subjected to less severe operating conditions and are fabricated from materials of appropriate qualities. Although several factors are involved in the selection of an appropriate material, the anticipated operating life and the maintenance costs are particularly significant.

If no provision is made for cooling the turbine components, the turbine inlet temperature is currently limited to about 1000 C. Inlet temperatures in the range of 1260 to 1320 C are achieved by fabricating the turbine components from superalloys and providing adequate cooling of these components. Not only are the superalloys expensive, but very large quantities of cooling air must be supplied by the compressor. An engine construction that incorporates air-cooled superalloy turbine components, while presently acceptable for high-performance aircraft applications, is not economically feasible for many other installations, particularly for an automotive gas turbine engine.

Substitution of ceramic materials for the superalloys is the objective of numerous metallurgical research and development programs. Ceramic components would not require air cooling, and turbine inlet temperatures could be increased to about 1375 C for gas turbine engines in all categories. An improvement in the thermal efficiency of the engine would be achieved because of the increase in the turbine inlet temperature and the elimination of the cooling air. For an engine of a designated size, the power output would be increased substantially by eliminating the diversion to the turbine cooling system of a portion of the air passing through the compressor.

An acceptable ceramic material must possess adequate stress characteristics and resistance to thermal shock for operation at a temperature level of 1375 C. Further, fabrication techniques must not be unduly complex. In comparison with the superalloys, ceramics are less dense, cheaper, and more resistant to corrosion from such contaminants as sodium and vanadium, present in low-cost oil fuels. Ceramics however are brittle, tend to break, and have poor impact resistance.

Many ceramic materials were screened, and silicon nitride (Si_3N_4) and silicon carbide (SiC) have been selected for further investigation. Fabrication of a variety of turbine components from these ceramics can be achieved by several processes, including molding, machining, and hot pressing. As an example, in one process, silicon turbine blades are formed by injection molding prior to the final nitriding.

Turbine blade cooling is a complex procedure because of temperature stress effects, required structural modifications, and thermal and aerodynamic losses. Air cooling is accomplished in a number of ways. Internal cooling provides for air flow through hollow blades or through radial holes or channels formed in solid blades. External cooling, or film cooling, is accomplished by the installation of hollow blades that are provided with leading- or trailing-edge slots. Effusion cooling, using a porous blade, and root cooling are other methods employed for turbine blade cooling. Various difficulties are encountered in devising an effective scheme for conveying the cooling air from the compressor to the point where the air escapes from the blade.

Several methods are available for liquid cooling of turbine blades. While liquid cooling is highly effective, a complex mechanical system must be contrived for handling the coolant.

Although the nozzle vanes are exposed to high-temperature gases, these components are, unlike the rotor blades, not subjected to severe stresses, hence vane cooling is not a critical design requirement. Turbine disks can be cooled by air blasts from strategically placed nozzles.

Load changes imposed on the gas turbine engine often create impaired flow conditions in the compressor, and as a result unstable operation and a poor fuel rate may be experienced. The performance of an axial-flow compressor can be improved to some degree by equipping the machine with adjustable stator vanes. These vanes are mounted on pivots and are rotated in order to correct the adverse flow effects caused by stalling. Compressor construction however is more complex because of the mechanism required for turning the vanes and may be economically justified only in a limited number of installations. Flow control can also be accomplished by equipping the compressor with adjustable inlet guide vanes, another form of variable geometry.

Improved performance of the turbojet engine can be accomplished by incorporating the twin-spool compressor design. Both the compressor and the turbine are constructed in two sections. The high-pressure turbine drives the high-pressure compressor, and the low-pressure turbine drives the low-pressure compressor through coaxial shafts. The two sections of the engine can operate at different speeds, an arrangement that is particularly advantageous with respect to the operation of the compressor.

A gas turbine engine–electric generator set, operating on a simple cycle, would ordinarily be a two-shaft machine, consisting of the compressor–turbine gas generator section and the power turbine. Various configurations have been proposed for the automotive gas turbine engine, including single-, two-, and three-shaft machine arrangements.

The overload capability of the gas turbine engine may be appreciably increased by water or steam injection. Water may be introduced at several points in the cycle, that is, at the compressor or at the combustion chamber. When injected at the compressor, the water is sprayed into the air stream at some point between stages. Water can be sprayed into the air stream prior to combustion, or into the gas stream subsequent to combustion. The thermal efficiency of the engine is essentially unaffected by water injection at the compressor, but is reduced when water is injected at the combustion chamber, prior to or subsequent to combustion. Pure water should be used in order to avoid formation of deposits in the engine. The quantity of water injected is typically 0.1 kg/kg air.

Steam injection into the downstream section of the combustion chamber will increase the engine output, but the effect on the thermal efficiency is dependent upon the source of the steam. A heat rate below the value achieved for non-overload operation can be anticipated when the steam is generated in a heat recovery boiler supplied by the turbine exhaust. The heat rate is however substantially higher when the steam is generated in a fuel-fired boiler. Either steam or water injection will provide, with good economy, the incremental power required for peak load operation.

The output of the gas turbine engine is affected to an appreciable degree by a change in the compressor inlet temperature because the compressor absorbs a substantial portion of the turbine work and the compressor work varies directly with the inlet air temperature for a prescribed pressure ratio. For example, an increase in the compressor inlet temperature from 27 C to 38 C is likely to effect a decrease of approximately 10 percent in the output of the engine. For those periods of high ambient temperature, the output of the engine could be restored to the 27 C capability by steam injection into the combustion chamber.

Supercharging is another method that can be employed for increasing the power output of the gas turbine engine. A forced-draft fan is used to increase the compressor inlet pressure. Intercooling is essential in order to eliminate the adverse effect of the increase in the compressor inlet temperature that results from compression of the intake air in the fan. Evaporative cooling, an effective means for air cooling, is achieved by spraying water into the evaporative cooler located between the fan and the compressor. The fan can be driven by an electric motor or, in a more complex system, by a turbine supplied with steam generated in a heat recovery boiler. The exhaust gas from the gas turbine engine is the energy source for the steam generation. Any method used for producing incremental power is normally adopted only if the approach to the problem can be economically justified.

7.18 THRUST AUGMENTATION

An important characteristic of a turbojet engine is the ability to achieve, for certain flight conditions, a higher-than-normal thrust. The augmented thrust is required for takeoff, high rate of climb, and enhancing the combat maneuverability of military aircraft. The increased thrust is maintained only for short periods of time, hence some increases in the specific fuel consumption, in kg/h per N, is acceptable while developing the incremental thrust.

A high ambient temperature causes a reduction in the thrust developed by the engine. Water injection into the compressor inlet is a method employed primarily for thrust restoration during takeoff. The water, or an alcohol–water mixture, evaporates during passage through the compressor and alters the operating characteristics of the engine. The net effect is an increase in the engine pressure and in the mass flow of the working fluid; and as a consequence, an increased thrust is achieved.

Thrust augmentation can also be accomplished by injecting water into the combustion chamber. In general, the overall effect is similar to that observed for water injection into the compressor inlet. Both methods may be employed simultaneously.

Afterburning is the most widely used method for achieving thrust augmentation in the turbojet engine. The afterburner is installed between the turbine and the nozzle. The requirements for the afterburner, with respect to the combustion characteristics and pressure losses, are generally similar to those for the primary combustion chamber.

Subsequent to discharge from the turbine, the combustion gases enter the afterburner where additional fuel is injected into the fluid stream. Because of the excess oxygen present in these gases, a substantial quantity of fuel can be burned to increase the temperature of the gas from T_4 to T_5, as shown in Fig. 7.19. Expansion of the gas in the turbine is terminated at state 4. For normal operation, with the afterburner turned off, expansion of the gas is continued, partly in the nozzle, to state 7 at atmospheric pressure. With the afterburner in operation, the gases are expanded from state 5 to state 6, again partly in the nozzle. The afterburner and nozzle are constructed to withstand high-temperature operation, with the maximum temperature T_5 attaining a level of approximately 1950 C.

For a prescribed pressure ratio, the isentropic enthalpy drop increases with increase in the initial temperture. Consequently, the increase in the nozzle inlet temperature from T_4 to T_5 will cause an increase in the nozzle outlet velocity, and as a result the gross thrust developed by the engine is increased.

A higher nozzle inlet temperature will, in addition, effect an increase in the specific volume of the gas, notably at the nozzle outlet section. A variable-area nozzle is required when the engine is equipped with an afterburner in order to accommodate the substantial change that occurs in the volume flow. Failure to provide for an increase in the area of the nozzle outlet section when the afterburner is turned on will cause a

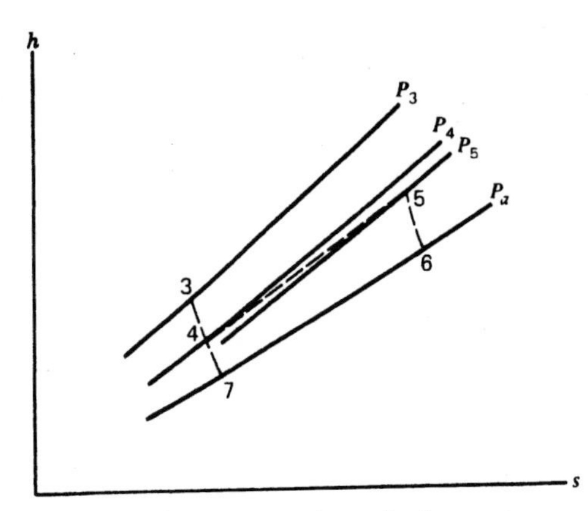

Figure 7.19 Turbojet engine afterburner operation h–s diagram.

change in the operating characteristics of the engine. A well-developed design allows the afterburner to be turned on and off with essentially no change observed in the operating characteristics of the basic engine. Significantly, the afterburner is operated only when the basic engine is developing maximum capability.

In contrast to the primary combustion chamber, dissociation of the product gases in the afterburner is responsible to a significant degree for the reduction of the actual gas temperature below the calculated theoretical value. Other factors contribute to this decrement, namely, low residence time, recirculation, and heat losses.

Example 7.8

An afterburner is added to the turbojet engine examined in Example 7.6. For the analysis of the engine, the airplane speed remains at 250 m/s. The afterburner operating data are

Fuel/air ratio: 0.041 (for the afterburner)

Nozzle efficiency: 0.92

Combustion efficiency: 0.87

Gas outlet temperature: 2000 K (nozzle inlet section)

Total pressure loss factor: 0.075

Property values selected from Tables A.1 and A.2 were used to evaluate the performance of the engine with the afterburner activated. The significant performance factors are compared with the corresponding factors reported in Example 7.6.

		Afterburner	
Performance Factor	*Units*	*Turned On*	*Turned Off*
Nozzle outlet velocity C_n	m/s	801.0	565.6
Gross thrust F_g	N	96,462	67,484
Net thrust F_n	N	71,462	42,484
Specific fuel consumption sfc	kg/h per N	0.297	0.153
Overall efficiency η_o		0.0689	0.134

The thrust augmentation ratio is 1.683.

7.19 THE TURBOFAN ENGINE

The development of the turbofan engine represents a particularly important advancement in aircraft propulsion. The design concept of the turbofan engine is shown schematically in Fig. 7.20. In principle, the fan is superimposed on the basic turbojet engine. A portion of the inlet air is compressed by the fan and subsequently expands in the annular fan nozzle. The remainder of the inlet air leaves the fan and enters the compressor. For the engine shown in Fig. 7.21, the high-pressure section of the turbine drives the high-pressure section of the compressor, and the low-pressure section of the turbine drives the fan and the low-pressure section of the compressor. The fan air completely bypasses the primary fluid stream.

A three-spool turbofan engine configuration employs three concentric shafts for driving the fan and the compressor elements. For example, the first turbine stage drives the high-pressure section of the compressor, while the second turbine stage drives the low-pressure section of the compressor. The fan is driven by the remaining three turbine stages.

The fan air may be conveyed in a duct to the aft end of the engine and expanded with the turbine exhaust gas in a common nozzle. The front-fan arrangement is employed for the engine shown in Fig. 7.21. In the aft-fan configuration, the fan is located aft of the primary turbine and performs a similar function in compressing the bypass air for subsequent expansion in the nozzle. See Fig. 7.18.

Fan action causes a momentum increase in the bypass air, and as a consequence a propulsive force is applied to the airplane. The velocity of the air discharged from the fan nozzle is somewhat lower, and closer to the airplane speed, than the velocity of the gas discharged from the primary nozzle. The fan thus achieves a higher propulsive efficiency than the primary section of the engine. See Section 7.8.

The most appropriate bypass ratio for an engine depends upon a number of factors. An important characteristic is the decrease in the thrust

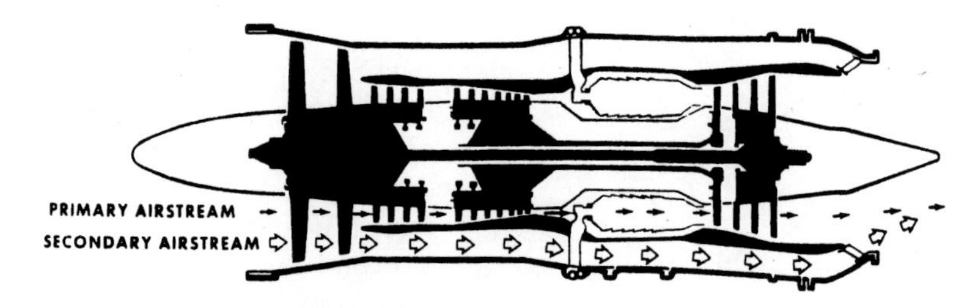

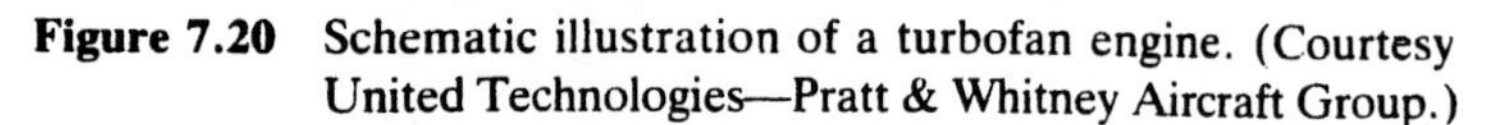

Figure 7.20 Schematic illustration of a turbofan engine. (Courtesy United Technologies—Pratt & Whitney Aircraft Group.)

specific fuel consumption, in kg/h per N, with an increase in the bypass ratio. Engines however are usually designed for a comparatively low bypass ratio in order to avoid an excessively large diameter for the fan rotor.

An increase in the bypass ratio will effect a decrease in the specific thrust N per kg/s of air. This trend is the result of shifting a larger portion of the total thrust developed by the engine to the fan air. On a unit mass basis, the fan air develops a lower thrust than the fluid in the primary section of the engine. Incidentally, the turbofan engine may be equipped with an afterburner.

Example 7.9

A comparison is made of the performance of a turbofan engine and a turbojet engine that operate under generally similar conditions.

Airplane flight speed: 210 m/s
Turbine inlet temperature: 1400 K

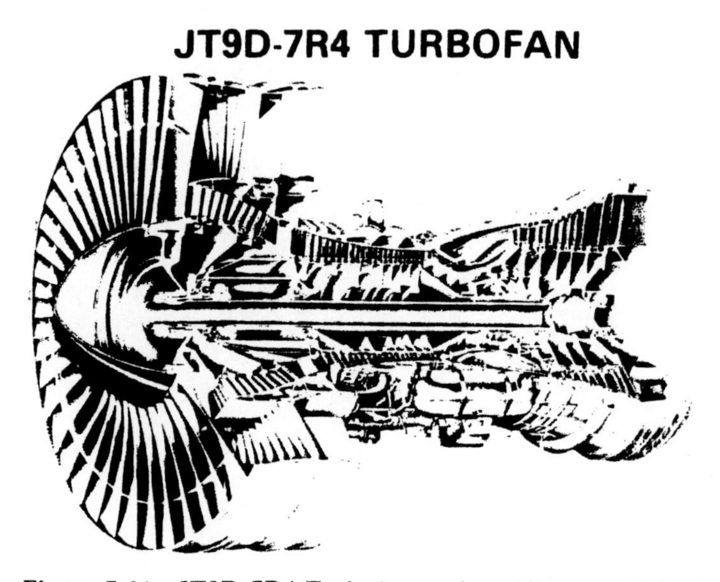

Figure 7.21 JT9D-7R4 Turbofan engine. (Courtesy United Technologies—Pratt & Whitney Aircraft Group.)

Turbofan engine

Compressor airflow: 50 kg/s

Fan airflow: 60 kg/s (bypass ratio 1.2 to 1)

Compressor pressure ratio: 15 to 1

Fan pressure ratio: 2.20 to 1

Turbojet engine

Compressor airflow: 60.377 kg/s

Compressor pressure ratio: 15 to 1

Appropriate efficiency values are prescribed. Calculations are based on the Gas Tables,[1] and the following results are reported.

Performance Factor	*Units*	*Turbofan Engine*	*Turbojet Engine*
Gross thrust F_g			
Fan	N	24.330	
Main jet	N	44.083	57.987
Total	N	68,413	57,987
Ram drag F_r	N	23,100	12,679
Net thrust F_n	N	45,313	45.308
Thrust specific fuel consumption (tsfc)	kg/h per N	0.09240	0.11159
Specific thrust	N per kg/s air	412	750
Overall efficiency η_o		0.1851	0.1532

In this example, the two engines develop the same net thrust, but in comparison with the turbojet engine, the turbofan engine operates with a lower thrust specific fuel consumption, and hence with a higher overall efficiency. The turbojet engine however develops a substantially higher specific thrust in comparison with the turbofan engine.

7.20 VARIATIONS OF THE BASIC GAS TURBINE ENGINE CYCLE

The basic gas turbine engine cycle can be modified to include multistage compression, two-stage expansion, and regeneration. Multistage compression provides for air compression in several stages, two to four, with cooling of the air between each compressor element. Compressor intercoolers are usually water cooled. Because of the inherently low fuel/air ratio, the combustion gases leaving the high-pressure turbine contain sufficient oxygen to support combustion in a second combustion chamber that precedes the low-pressure or reheat turbine.

The machine arrangement of the gas turbine engine provides an excellent opportunity to utilize the energy in the turbine exhaust gas for preheating the combustion air. Figure 7.22 shows the flow diagram for a gas turbine power plant equipped with a heat exchanger, commonly designated as a regenerator.

7.21 PRINCIPLES OF REGENERATION

Transfer of energy between the two fluid streams, gas and air, can for certain designs provide the means to achieving an appreciable increase in the thermal efficiency of the gas turbine engine. The increase in the thermal efficiency is a direct result of the reduction in the fuel/air ratio, which in turn is accomplished by heating the air subsequent to compression and prior to combustion. The energy extracted from the exhaust gas would otherwise be dissipated in the surroundings, or possibly utilized in a heat recovery boiler.

The principles of regeneration are disclosed by examination of the theoretical cycle that incorporates an internal exchange of heat. See Fig. 7.23. If temperature T_4 exceeds temperature T_2, heat can be transferred from the "exhaust gas" to the air. The gas temperature may be reduced to T_y as a limit, where $T_y = T_2$. A corresponding increase in the air temperature is observed, hence $T_x = T_4$, for a constant value of the specific heat c_p.

Temperature T_4 does not necessarily exceed temperature T_2. When temperature T_4 falls below temperature T_2, a reversed flow of heat occurs, that is, from the air to the gas, and as a consequence the thermal efficiency of the regenerative cycle is less than the thermal effi-

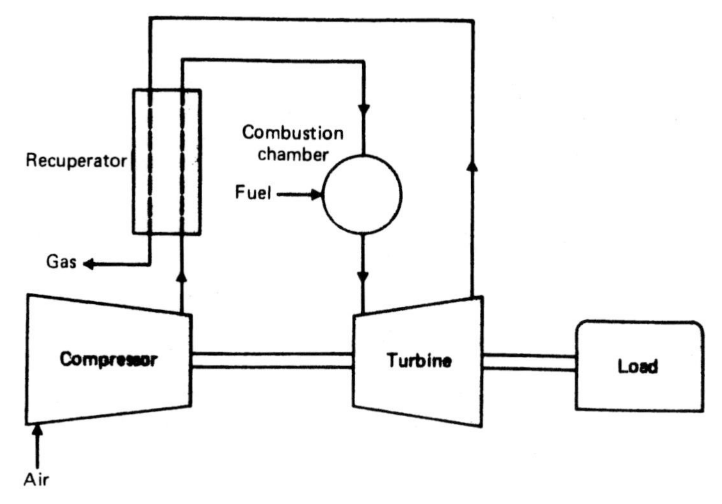

Figure 7.22 Flow diagram for a regenerative-cycle gas turbine power plant.

ciency of the corresponding Brayton cycle. Whether temperature T_4 is above or below temperature T_2 will be dependent upon the cycle pressure ratio and the temperature ratio T_3/T_1.

For the theoretical gas turbine engine regenerative cycle, $Q_s = c_p(T_3 - T_x)$ and $W_T = h_3 - h_4 = c_p(T_3 - T_4)$. Because $T_x = T_4$, $Q_s = W_T$. The thermal efficiency of the cycle is given by

$$(\eta_t)_{re} = \frac{W_T + W_C}{Q_s} = \frac{W_T + W_C}{W_T}$$

$$= 1 + \frac{W_C}{W_T}$$

$$= 1 + \frac{c_p(T_1 - T_2)}{c_p(T_3 - T_4)}$$

$$= 1 - \frac{T_2 - T_1}{T_3 - T_4}$$

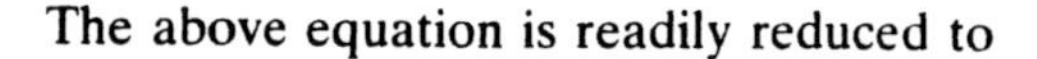

The above equation is readily reduced to

$$(\eta_t)_{re} = 1 - \frac{T_1}{T_3}\left(\frac{P_2}{P_1}\right)^{k-1/k} \tag{7.29}$$

Equation 7.29 indicates generally that the thermal efficiency of the theoretical regenerative cycle will, for a prescribed value of the initial temperature T_1, increase with increase in the maximum temperature T_3 of the cycle and decrease with increase in the cycle pressure ratio. Figure 7.24 illustrates the effect of these two parameters, that is, T_3 and P_2/P_1, on the thermal efficiency of the theoretical regenerative cycle.

The point at which reversed heat flow occurs is observed in Fig. 7.24. The crossover point for

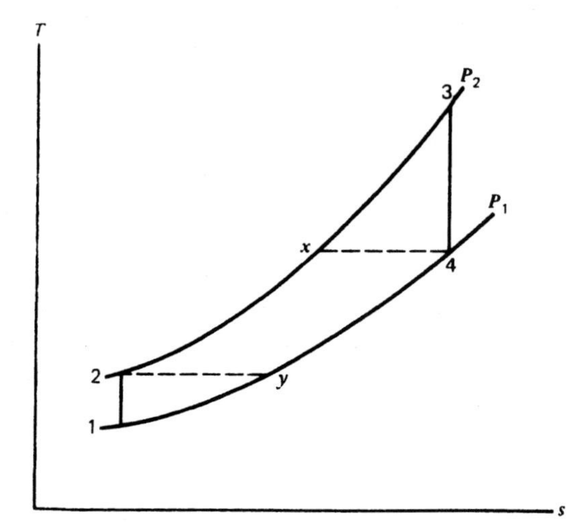

Figure 7.23 Gas turbine engine theoretical regenerative cycle.

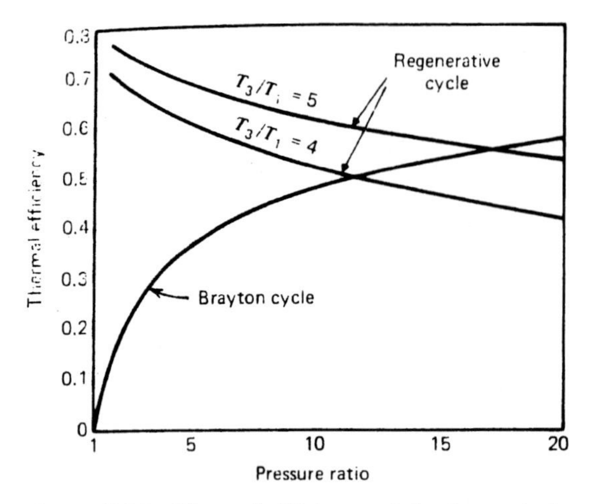

Figure 7.24 Thermal efficiency of the theoretical regenerative cycle for the gas turbine engine.

$T_3/T_1 = 5$ is established at a pressure ratio of approximately 17 to 1. At this point, the thermal efficiency values are alike for the Brayton and the theoretical regenerative cycles.

Equations 7.1 and 7.29 are combined to yield for the crossover point

$$\left(\frac{P_2}{P_1}\right)_{co} = \left(\frac{T_3}{T_1}\right)^{k/2(k-1)} \tag{7.30}$$

For $T_3/T_1 = 5$ and $k = 1.40$, $(P_2/P_1)_{co} = 16.72$. See Fig. 7.24.

7.22 REGENERATOR EFFECTIVENESS

The theoretical regenerative cycle is established for maximum heat recovery, that is, the temperature of the air is raised to the "exhaust gas" temperature, $T_x = T_4$, Fig. 7.23. Full heat recovery however is not practicable for an actual gas turbine power plant. The actual extent of the increase in the air temperature is designated by the regenerator effectiveness.

$$\eta_{re} = \frac{T_{a,o} - T_{a,i}}{T_{g,i} - T_{a,i}} \tag{7.31}$$

where

η_{re} = the regenerator effectiveness
$T_{a,o}$ = temperature of the air leaving the regenerator
$T_{a,i}$ = temperature of the air entering the regenerator
$T_{g,i}$ = temperature of the gas entering the regenerator

As $T_{a,o}$ approaches $T_{g,i}$, the value of η_{re} approaches 1 as a limit.

The design value of the regenerator effectiveness is determined in part from an economic analysis of the proposed system. A decrease in the power plant heat rate and hence lower fuel costs are achieved by the installation of a regenerator. The capital and maintenance costs charged to the regenerator however detract from the saving that results from the decrease in fuel consumption.

Design values for the regenerator effectiveness are typically in the range of 0.70 to 0.75. Higher values of η_{re} will improve the plant heat rate, but a substantial increase is required in the heat transfer surface of the regenerator.

Figure 7.25 shows the variation in the thermal efficiency, with respect to the pressure ratio, for

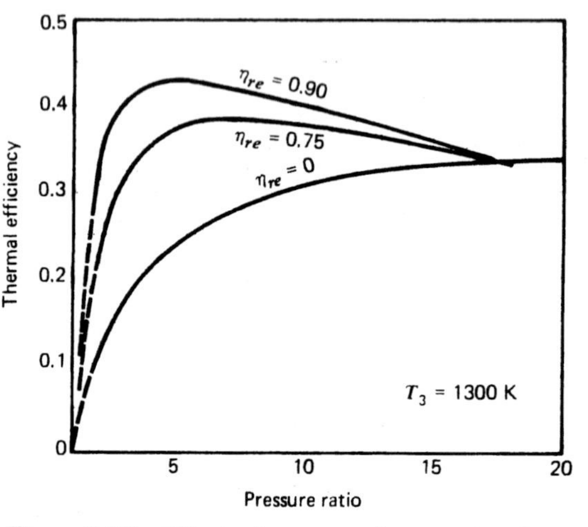

Figure 7.25 Effect of regeneration on the thermal efficiency of the gas turbine engine cycle.

a nonregenerative cycle and for two regenerative cycles for which $\eta_{re} = 0.75$ and 0.90, respectively. The optimum pressure ratios for the regenerative cycles are, respectively, 7 to 1 and 5 to 1, an indication that the greatest gain in the thermal efficiency, because of regeneration, is achieved in the region of low pressure ratios. Appropriate values for machine efficiencies, pressure losses, combustion losses, and mechanical losses were prescribed in the thermodynamic analysis of the three cycles.

The optimum pressure ratio for a power plant operating on a regenerative cycle is determined by an economic and thermodynamic analysis of the proposed design. In this respect, the turbine inlet temperature and the regenerator effectiveness are two important parameters. In general, a decrease in the value of η_{re} tends to cause an increase in the magnitude of the optimum pressure ratio of the cycle. An increase in the turbine inlet temperature has a similar effect.

7.23 HEAT EXCHANGER ENERGY TRANSFER

Two different types of heat exchangers are used for achieving heat transfer between the turbine exhaust gas and the combustion air. The regenerator consists of a ceramic or metallic element that rotates slowly through the gas and air streams. Energy is transferred from the gas to the rotor; and, following a brief period of storage, the energy is subsequently transferred to the air. A regenerator of this type is used for a high-performance automotive gas turbine engine and would most likely be used if regeneration is extended to aircraft turbine engines.

The recuperator, the second type of heat exchanger, is a shell-and-tube structure. Gas is usually directed through the tubes in order to facilitate cleaning, and the air flows across the tube bank in one or more passes. The term regenerator is commonly applied in a general sense to both types of heat exchangers.

Because of the adverse effect on the turbine pressure ratio, pressure losses in both the gas and air sides of the regenerator must be held at low levels. It is important to note that a pressure loss of 10 kPa, for example, on the gas side of the regenerator causes a much larger reduction in the turbine pressure ratio and in the plant thermal efficiency than a pressure loss of equal magnitude on the air side.

The temperature gradients for pure counterflow heat exchange are shown in Fig. 7.26. Normally, for the gas turbine engine recuperator, a small difference is observed between the magnitude of the terminal temperature differences ΔT_1 and ΔT_2. The mean temperature difference between the gas and air is then evaluated by

$$\Delta T_m = \frac{\Delta T_1 + \Delta T_2}{2} \qquad \text{(for counterflow)}$$

The heat transferred from the gas to the air is determined by

$$\dot{q} = U_o A_o F_c \, \Delta T_m \tag{7.32}$$

where

$\dot{q}$ = heat transfer rate
U_o = overall coefficient of heat transfer, based on the outside surface of the tubes
A_o = area of the external surface of the tubes
F_c = correction factor
ΔT_m = mean temperature difference between the gas and air streams, for counterflow of the fluids

Because the construction of the recuperator does not usually effect pure counterflow of the fluids, a correction factor F_c is inserted in the heat transfer equation in order to evaluate the real mean temperature difference $F_c \, \Delta T_m$.

The heat transferred between the two fluid streams is also equal to the heat transferred from the gas and to the air. Thus,

$$\dot{q} = \dot{m}_a c_{p,a}(T_{a,o} - T_{a,i})$$
$$= \dot{m}_g c_{p,g}(T_{g,i} - T_{g,o})$$

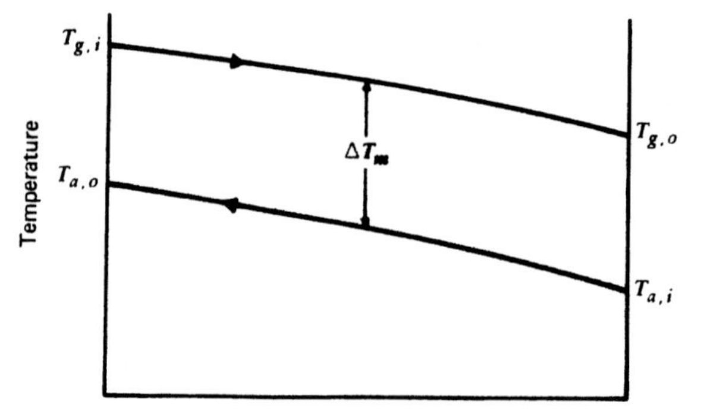

Figure 7.26 Temperature gradients for counterflow heat exchange.

A simplified equation used to determine an approximate area of the recuperator can be developed if ΔT_1 is assumed equal to ΔT_2. Thus,

$$A_o = \frac{\dot{m}_a c_{p.a}}{U_o F_c} \frac{\eta_{re}}{1 - \eta_{re}} \tag{7.33}$$

where

$\dot{m}_a$ = air flow rate
$c_{p.a}$ = specific heat of the air

An examination of Eq. 7.33 discloses the strong dependence of the heat transfer area A_o on the effectiveness of the recuperator. Above a value of $\eta_{re} = 0.75$, the quantity $\eta_{re}/(1 - \eta_{re})$ increases rapidly, thus causing a corresponding sharp increase in A_o.

Figure 7.25 indicates that the thermal efficiency of the engine cycle can be increased from approximately 0.340 to 0.385 by employing regeneration with $\eta_{re} = 0.75$ and to 0.430 with $\eta_{re} = 0.90$. These optimum efficiency values however are achieved at different pressure ratios.

Table 7.1 reports the significant characteristics of the three engine cycles. Values of A_o are determined from Eq. 7.33.

Although the increase in the recuperator effectiveness from 0.75 to 0.90 does achieve a substantial increase in the thermal efficiency of the engine cycle, a much larger recuperator is required. Ordinarily, the larger recuperator cannot be economically justified, and for some installations sufficient space may not be available.

Example 7.10

A gas turbine engine is equipped with a recuperator that has an operating effectiveness equal to 0.76. The air flow is 75 kg/s, and the fuel/air ratio is 0.01430. Air leaves the compressor at 240 C, and the gas leaves the turbine at 535 C. Gas flows through the tubes, and the cross flow of the air is through the tube bank. The tube O.D. is 15.0 mm, and the wall thickness is 1.0 mm. The heat transfer coefficients are: gas film 100

TABLE 7.1 The Influence of the Regenerator Effectiveness on the Characteristics of the Engine Cycle

η_{re}	*PR*	W_n *Ratio*	$\dot{m}_a$ *Ratio*	η_t	A_o *Ratio*
0	20	1	1	0.340	—
0.75	7	1.019	0.981	0.385	1
0.90	5	0.925	1.081	0.430	3.31

W/m²·K, air film 175 W/m²·K. Calculate the area of the recuperator heat transfer surface.

Temperature of the Gas Leaving the Recuperator

$$\eta_{re} = \frac{T_{a,o} - T_{a,i}}{T_{g,i} - T_{a,i}}$$

$$0.76 = \frac{T_{a,o} - 240}{535 - 240}$$

$$T_{a,o} = 464.2 \text{ C}$$

Enthalpy values for the combustion gas are selected from Table A.2 for combustion of the fuel with 400 percent theoretical air, although the actual percentage is somewhat higher.

$$T_{a,i} = 513 \text{ K}, \quad h_{a,i} = 516.81 \text{ kJ/kg}$$

$$T_{a,o} = 737 \text{ K}, \quad h_{a,o} = 753.70 \text{ kJ/kg}$$

$$T_{g,i} = 808 \text{ K}, \quad h_{g,i} = 845.25 \text{ kJ/kg}$$

$$\dot{m}_g(h_{g,i} - h_{g,o}) + \dot{m}_a(h_{a,i} - h_{a,o}) + \dot{Q} = 0$$

Assume $\dot{Q} = 0$

$$h_{g,o} = \frac{1}{1 + f/a}(h_{a,i} - h_{a,o}) + h_{g,i}$$

$$= \frac{1}{1.01430}(516.81 - 753.70) + 845.25$$

$$= 611.70 \text{ kJ/kg gas}$$

$$T_{g,o} = 596.0 \text{ K} = 323 \text{ C}$$

Mean Temperature Difference

$$\Delta T_1 = T_{g,o} - T_{a,i}$$

$$= 323 - 240 = 83 \text{ C}$$

$$\Delta T_2 = T_{g,i} - T_{a,o}$$

$$= 535 - 464 = 71 \text{ C}$$

$$\Delta T_m = 77 \text{ C}$$

$$F_c = 0.62^*$$

Overall Coefficient of Heat Transfer

$$R_o = \frac{D_o}{D_i}\frac{1}{h_i} + \frac{1}{h_0} = \frac{15.0}{13.0}\frac{1}{100} + \frac{1}{175}$$

$$= 0.01725 \text{ m}^2\cdot\text{K/W}$$

$$U_o = \frac{1}{R_o} = \frac{1}{0.01725} = 58.0 \text{ W/m}^2\cdot\text{K}$$

Area of Heat Transfer Surface

$$A_o = \frac{\dot{q}}{F_c\,\Delta T_m\,U_o} = \frac{\dot{m}_a(h_{a,o} - h_{a,i})}{F_c\,\Delta T_m\,U_o}$$

$$= \frac{75(753.70 - 516.81)10^3}{0.62 \times 77 \times 58.0}$$

$$= 6416 \text{ m}^2$$

7.24 THE ERICSSON CYCLE

A more complex gas turbine engine can be devised by basing the design of the machine on the Ericsson cycle rather than on the Brayton cycle. The Ericsson cycle, like the Carnot cycle, is a reversible cycle; hence for prescribed temperature limits, the thermal efficiency of the cycle is equal to the maximum theoretical value of $1 - (T_L/T_H)$. A Brayton cycle that operates within the same temperature limits will have a lower thermal efficiency.

The Ericsson cycle is depicted on the T–s plane in Fig. 7.27. In accordance with the concept of the reversible engine, isothermal heat transfer must be achieved for the exchange of energy between the heat source and the working fluid and between the working fluid and the heat sink. The heat transfer between the constant-pressure

* From R. A. Bowman, A. C. Mueller, and W. M. Nagle, "Mean Temperature Difference in Design," *Trans. ASME*, 1940.

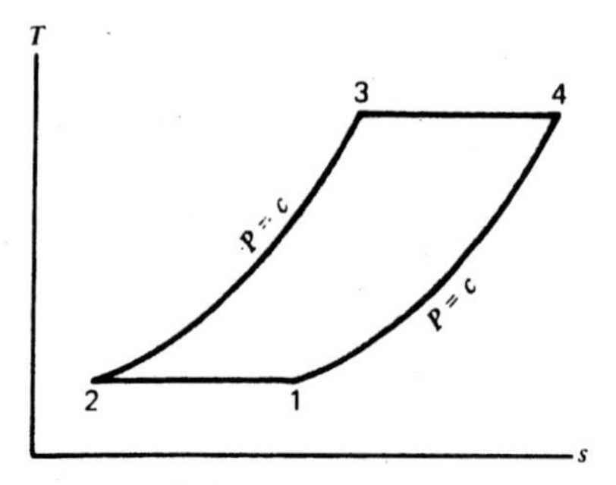

Figure 7.27 The Ericsson cycle.

processes 2–3 and 4–1 is effected internally by regeneration. Isothermal compression 1–2 and isothermal expansion 3–4 of the working fluid however are not practicable for a real engine.

An approach to isothermal compression and expansion can be achieved by utilizing a large number of compressor and turbine elements, combined with compressor intercooling and turbine reheating. In practice, the number of compressor elements ranges between two and four, while two turbine elements are ordinarily employed. In comparison with the simple gas turbine engine, a considerably more complex machine construction is required for multistage operation. In addition to the larger number of machine elements, the power plant must be equipped with a regenerator, one or more compressor intercoolers, and two combustion chambers.

Figure 7.28 shows a theoretical cycle for multistage operation including complete regeneration. Compression is achieved in three stages of like pressure ratio, hence the work quantities are of equal magnitude. Constant-pressure intercooling, *a–b* and *c–d,* is effected between the first and second and between the second and third compressor elements, respectively. For theoretical multistage compression, $T_d = T_b = T_1$. The primary and the secondary constant-pressure heating processes are indicated, respectively, by *x*–3 and *e–f*. Expansion of the working fluid is achieved in two stages, 3–*e* and *f*–4, of like pressure ratio.

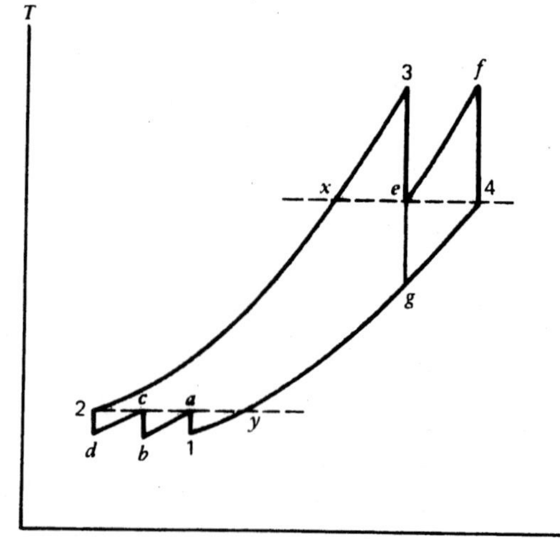

Figure 7.28 Theoretical gas turbine engine cycle for multistage compression and expansion and complete regeneration.

The principles of multistage compression were examined in Section 3.11. Briefly, the total work of compression is a minimum when the work quantities for the several stages are equal. Consequently, the stage pressure ratios are of like value, and the following relationships are established for the intermediate pressures.

For two-stage compression:

$$P_{ic} = \sqrt{P_2 P_1} \tag{7.34}$$

For three-stage compression:

$$P_{ic,1} = \sqrt[3]{P_2 P_1^2} \tag{7.35a}$$

$$P_{ic,2} = \sqrt[3]{P_2^2 P_1} \tag{7.35b}$$

For four-stage compression, the intermediate, or intercooler, pressures are determined by the procedure given in Section 3.11.

A similar analysis applied to theoretical two-stage expansion reveals that maximum total work is achieved when the work quantities for the two stages are equal and $T_f = T_3$. Because the pressure ratios are alike, the intermediate pressure is given by

$$P_{it} = \sqrt{P_3 P_4} \quad (7.36)$$

For a designated pressure ratio P_3/P_4, the total work for two-stage expansion exceeds the work for single-stage expansion because of interstage reheat. In the absence of reheat, expansion in the second stage would extend from state e to state g. The work of expansion W_{e-g} would be less than the work W_{f-4} achieved by expansion from state f to state 4.

The actual operation observed in a gas turbine power plant differs in many ways from the theoretical cycle. Pressure losses and irreversible compression and expansion are significant departures encountered in practice. Complete regeneration is not prescribed, and the turbine power outputs are dependent upon the arrangement of the compressor elements and hence are not likely to be of equal magnitude.

In the absence of regeneration, multistage compression, combined with interstage cooling, has little effect on the thermal efficiency of the power plant. Compressor work is reduced appreciably, but because of the decrease in the outlet temperature of the air, the fuel/air ratio must be increased in order to maintain the prescribed turbine inlet temperature. The reduction in the compressor work however leads to a substantial increase in the output of the power plant. The decrease in the compressor outlet temperature that results from multistage compression points to a definite need to incorporate regeneration into the design of the power plant in order to achieve the high thermal performance inherent in the Ericsson cycle.

Regeneration is essential for an additional reason. The turbine outlet temperature, for a prescribed inlet temperature and a prescribed pressure ratio, may be higher for two-stage expansion with reheat than for single-stage expansion. The heat recovery potentiality of the system increases with increase in the turbine outlet temperature.

Figure 7.29 shows the flow diagram for a gas turbine power plant. Provision is made for recuperation and multistage compression and expansion. A two-shaft machine arrangement is employed. The nominal rating of the power plant is 35 MW. The power plant cycle is depicted on the T–s plane in Fig. 7.30.

7.25 CLOSED-CYCLE GAS TURBINE POWER PLANTS

The theoretical and actual cycles for the open-cycle and the closed-cycle gas turbine power plants are quite similar. The main difference between these two types of power plants is observed in the mechanical construction and function of the heat transfer and combustion equipment. Heat rejection in an open-cycle power plant is achieved by discharging the exhaust gas into the atmosphere. For the closed-cycle power plant, heat rejection to the surroundings is accomplished in the precooler, a heat exchanger that is placed in the flow circuit between the low-pressure side of the recuperator and the compressor. The temperature of the working fluid would be reduced in a precooler from T_y to T_1, Fig. 7.30.

The precooler, similarly to an intercooler, is usually water cooled, although air cooling of these components may be effectively employed. The hot water discharged from the precooler and the intercoolers will provide, in a total energy system, a source of heat for material processing and for building heating and air conditioning.

Closed-cycle power plants are equipped with externally fired gas heaters that consist principally of a combustion chamber, and radiant and convective heat exchanger sections in which heat is transferred from the combustion gases to the working fluid. Because of external firing, a wide variety of fuels, particularly coal, can be burned in the closed-cycle gas turbine power plant. Significant economic gains may result from the absence of fuel restrictions. Overall gas-to-gas heat transfer coefficients are inherently of low magnitude, hence a relatively large heater is required.

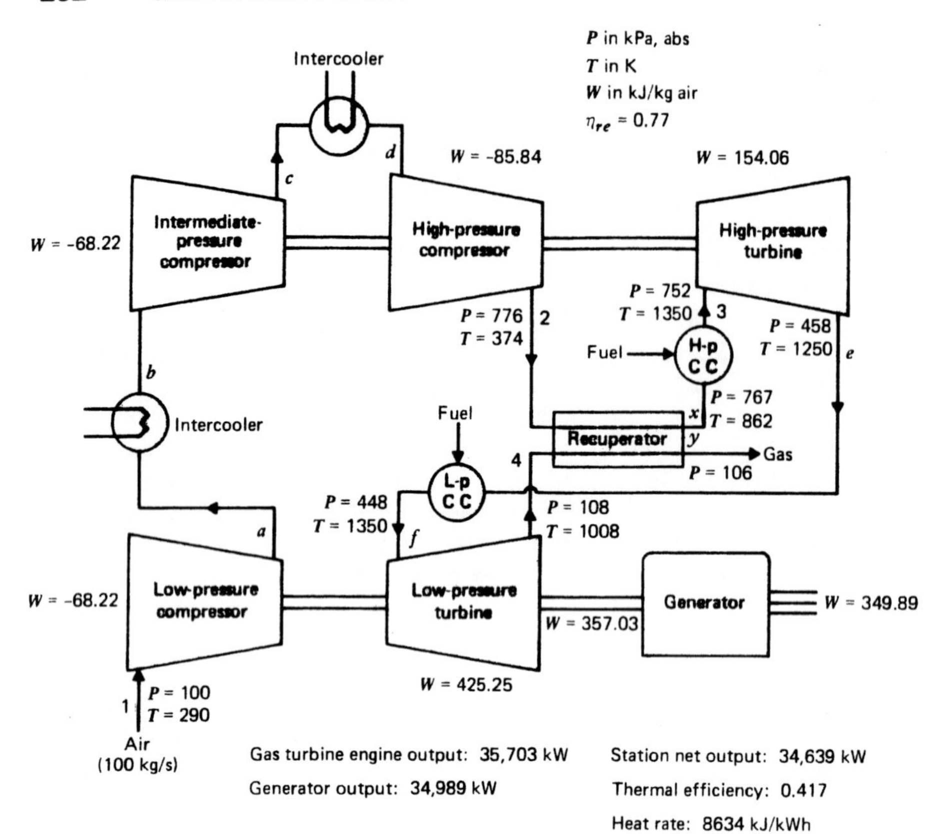

Figure 7.29 Flow diagram for a gas turbine power plant.

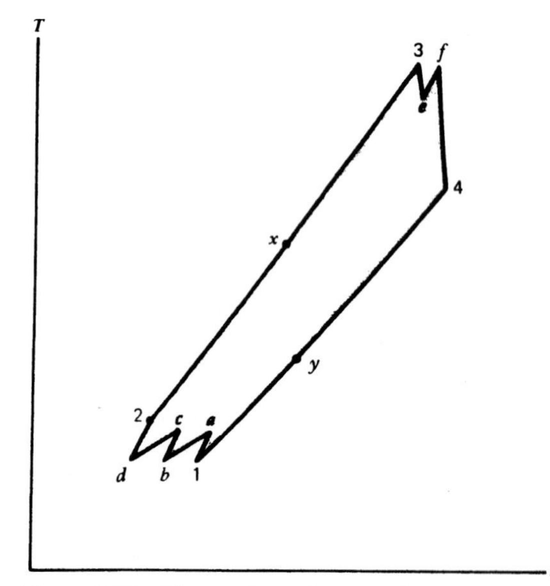

Figure 7.30 Thermodynamic cycle for the gas turbine power plant described by Fig. 7.29.

Because the working fluid is not required to enter the compressor at the pressure of the ambient air, as is the case for the open-cycle gas turbine engine, the minimum operating pressure for the closed-cycle power plant can be readily set substantially above the atmospheric pressure. High operating pressures cause low specific volumes for the working fluid, a factor of decided significance. Low gas volumes lead to a reduction in the flow areas; consequently, smaller components, that is, machines, heat exchangers, and piping, are typical of the closed-cycle power plant. Smaller heat exchangers are also a result of improved heat transfer coefficients that are attained at the higher gas pressures.

Continuous recirculation of the working fluid avoids fouling of compressor blades and heat transfer surfaces by suspended material in the atmosphere. Turbine blades and heat transfer surfaces are, in addition, free of fouling by residual material resulting from the combustion process. Reduced compressor, turbine, and heat exchanger performance, and restriction of flow passages because of such deposits, are consequently avoided in the closed-cycle power plant.

Regulation of the output from a closed-cycle gas turbine power plant operating at constant speed is accomplished by changing the pressure level in the system through a variation in the quantity of the working fluid in the cycle. Control valves are provided for shifting the working fluid between the cold-gas storage tanks and the active gas streams. This scheme for regulation of the power plant output results in improved part-load performance in comparison with the open-cycle power plant coupled to a constant-speed load, for which the load regulation is achieved through variation of the turbine inlet temperature or through flow control effected in the compound arrangements of compressors and turbines.

The open-cycle power plant must necessarily operate on air, while the closed-cycle system may be charged with any gas that demonstrates satisfactory operating characteristics. Increased plant thermal efficiencies and power outputs, for components of fixed dimensions, are possibilities that may result from the use of various gases other than air, although to date only air has been employed as the working fluid. Helium, carbon dioxide, neon, and nitrogen are some of the gases that have been investigated for possible use in the closed-cycle power plant. The gas-coooled nuclear reactor coupled with the closed-cycle gas turbine engine constitutes a mechanical system that exhibits excellent potentialities for high-capability electric power generation. In a plant of this design, helium is expected to perform satisfactorily as the reactor coolant and as the gas turbine engine working fluid.

Although the concept of the closed-cycle gas turbine power plant was advanced in 1939, only a small number of these plants have been constructed. As of 1974, seven closed-cycle power plants were in operation at various locations in Europe. Developmental work on the closed-cycle gas turbine power plant has been accomplished principally by Escher Wyss AG, a manufacturer of power equipment located in Switzerland.

Some typical design data for a closed-cycle gas turbine power plant are

Compressor inlet pressure: 800 to 1000 kPa

Pressure ratio: 4 to 1 (approximately)

Turbine inlet temperature: 700 C

The corresponding anticipated thermal efficiency, at the generator terminals, is 0.29 to 0.31.

Propulsion Supplement

The gas turbine engine is today the predominant means to propelling aircraft. Extensive investigations have been conducted on a variety of propulsion systems for application to space vehicles and to military projectiles and missiles. While these propulsion systems, like turbojet engines, are not power plants, they nevertheless are dependent upon some kind of energy conversion process in order to function. In the limited discussion presented in the following sections, only certain operating principles of the more common propulsion systems will be examined.

7.26 CHEMICAL ROCKET ENGINES

A turbojet engine is supplied with fuel from tanks installed in the aircraft, while the ambient air, taken on board in flight, provides the oxidizer. A chemical rocket engine, on the other hand, carries a complete supply of propellants, the fuel and the oxidizer. Both the turbojet engine and the chemical rocket engine derive the propulsive force by discharging a high-velocity jet of gas. Their performance capabilities however are very different, and hence applications of these engines are made to different kinds of vehicles.

Some of the important characteristics of rocket engines will now be examined. A rocket engine has no altitude limitation, hence it is used effectively to propel space vehicles. A slight improvement in the thrust is developed with an increase in altitude. The engine has no ram drag, and the flight speed is not limited. A rocket engine operates for a short period of time. Except for operation at extremely high flight speeds, the rocket engine performs with a low overall thermal efficiency.

Chemical rocket engines may be grouped in four main categories:

1 Liquid monopropellant rocket engines.
2 Liquid bipropellant rocket engines.
3 Solid propellant rocket engines.
4 Hybrid rocket engines.

A monopropellant may contain the oxygen required for combustion in its chemical or physical structure. Another type of monopropellant is a substance that is chemically unstable under certain conditions. Hydrogen peroxide is a monopropellant used to produce small jets of gas to correct the orientation or course of a space vehicle. The decomposition of hydrogen peroxide is an exothermic reaction described by

$$H_2O_2 \rightarrow (H_2O)_g + \frac{1}{2}O_2 + Q$$

The gases are generated at a temperature of about 425 C and are subsequently expanded through a nozzle.

Hydrazine (N_2H_4) is another monopropellant employed in small rocket engines that are used to maneuver space ships. With respect to decomposition into the component elements, hydrazine is thermodynamically unstable. Decomposition, an exothermic reaction, results in the release of a substantial quantity of energy.

A solid propellant rocket engine consists primarily of a combustion chamber and an attached nozzle. The combustion chamber is constructed in the form of a cylindrical case to which a thick layer of propellant is bonded. The hollow axial core is shaped to effect a substantially constant rate at which the fuel burns radially outward. Combustion starts and continues only on the exposed surface of the fuel. With a star-shaped core, for example, successive surface configurations, caused by radial burning, will have the same area as the initial surface, thus assuring a constant rate of gas generation. The propellant is a homogeneous mixture of the fuel, oxidizer, and a binding material that also burns. The high-temperature, high-pressure gases generated by the chemical reaction flow from the combustion chamber to the nozzle inlet and subsequently expand in the nozzle to produce a gas discharge velocity on the order of 1500 to 2500 m/s.[3] Solid propellant rocket engines are used primarily for propelling missiles.

Liquid bipropellant rocket engines have been extensively employed in space applications, both as the booster or first-stage engine and as the second-stage engine. Various kinds of propellants are used in these engines. Booster engines commonly burn a kerosene-type fuel with a liquid oxygen (LOX) oxidizer. A cryogenic high-energy propellant combination is achieved with liquid hydrogen (LH_2) and liquid oxygen. Cryogenic propellants cannot be stored indefinitely on board but must be loaded into the missile or vehicle shortly before launching.

The power output for a rocket engine is very high; hence for a liquid bipropellant engine, the fuel and oxidizer must be supplied to the combustion chamber at very high rates. One scheme provides for pressurizing the fuel and oxidizer tanks by a high-pressure inert gas, usually nitrogen. The pressure in the nitrogen supply tank is about 14 MPa, while the fuel and oxidizer tanks operate at about 3.5 MPa. The tank walls are thick and thus add to the mass of the vehicle. Gas pressurization can be effectively employed for short-duration operation, because small-capacity fuel and oxidizer tanks are adequate and do not significantly increase the mass of the system.

For long-range flight, pump pressurization of the rocket engine propellants is a superior method to gas pressurization. The liquid fuel and oxidizer are stored in light-weight tanks at low pressures and delivered to the combustion chamber at a pressure of 2 to 7 MPa by means of gas turbine-driven fuel and oxidizer pumps. The gas turbine is supplied with hot gases from a gas generator that also uses the main propellants. The exhaust from the gas turbine augments the thrust developed by the main combustion chamber and nozzle assembly.

The combustion chambers of the liquid bipropellant rocket engines operate at a pressure in the range of 2 to 7 MPa and achieve temperatures as high as 3000 C. Discharge gas velocities from the nozzle are in the range of 2500 to 4000 m/s, depending upon the propellant combination used to generate the gas.[3]

A liquid bipropellant rocket engine can be stopped and restarted in flight. This capability can also be acquired by a hybrid rocket engine that operates with a solid propellant, usually the fuel, and a liquid propellant, the oxidizer. In addition to the capability for intermittent operation, the hybrid engine can to some degree be throttled by varying the oxidizer flow rate.

7.27 ROCKET ENGINE PERFORMANCE

Thermodynamic calculations for the combustion process and the expansion of the gas through the nozzle are inherently complex if exactly carried out. The calculations can however be simplified by the introduction of certain empirical factors. Equilibrium conditions in the combustion chamber can be determined by the customary prescribed methods. Deviation from the equilibrium state is observed by applying an empirical combustion efficiency factor, defined as the ratio of the actual enthalpy change per unit mass of propellant to the enthalpy change for the reaction in attaining equilibrium. The properties of the gas entering the nozzle are thus defined. Values of the combustion efficiency are typically in the range of 0.95 to 0.98.

Nozzle flow is treated as the flow of an ideal gas. The value of the gas constant R thus does not change with an increase in the temperature of the gas, but the temperature dependence of the specific heats must be observed. An average value of the specific heat ratio k can be used, and various empirical factors are introduced.

The important determinations for the nozzle are the gas flow rate, the velocity of the gas at the outlet section, the thrust, and the specific impulse. There are various ways in which the calculations can be made. However, in the following analysis, the development of the pertinent equations will be based on three important parameters, namely, the nozzle inlet pressure and temperature and the nozzle outlet pressure. The principal sections for the nozzle are indicated in Fig. 7.31.

In a rocket engine, the gas is expanded in a De Laval nozzle, which consists of a converging and a diverging flow channel. The conditions at the throat of the flow channel consequently control the nozzle flow. Then, for reversible adiabatic flow,

$$T_t = \left(\frac{2}{k+1}\right)T_c \tag{7.37}$$

$$P_t = P_c\left(\frac{2}{k+1}\right)^{k/k-1} \tag{7.38}$$

where T_t and P_t are the temperature and the pressure at the throat section, and T_c and P_c are the stagnation temperature and pressure at the nozzle inlet section.

The nozzle flow rate is expressed by the continuity equation written for the throat section.

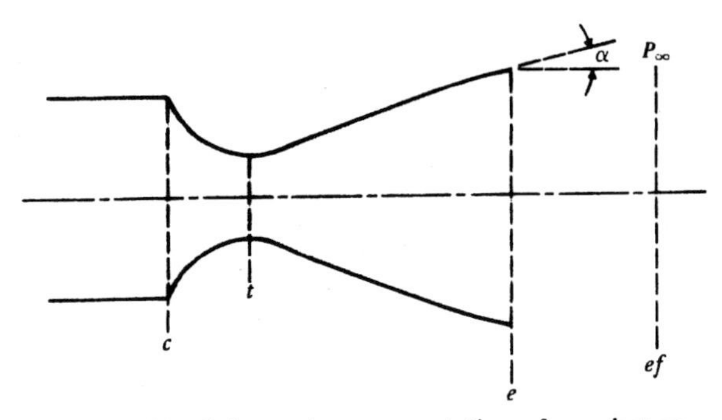

Figure 7.31 Schematic representation of a rocket engine.

$$\dot{m} = \rho_t A_t C_t \quad (7.39)$$

where the fluid velocity is given by

$$C_t = \sqrt{2c_p(T_c - T_t)} \quad (7.40)$$

and

$$c_p = \frac{kR}{k-1} \quad (7.41)$$

$$\rho_t = \frac{P_t}{RT_t} \quad (7.42)$$

The nozzle flow rate for reversible adiabatic expansion is now given by the following equation derived by combining Eqs. 7.37 to 7.42.

$$\dot{m}_i = A_t P_c \sqrt{\frac{k}{RT_c}\left(\frac{2}{k+1}\right)^{k+1/k-1}} \quad (7.43)$$

The flow rate for the actual nozzle differs from $\dot{m}_i$, determined for ideal gas flow in a reversible adiabatic expansion, because of a number of factors. Frictional effects cause a decrease in the gas velocity, while an increase in the fluid density causes the value of $\dot{m}$ to increase. See Eq. 7.39. An increase in the gas density can be attributed to a heat loss to the engine walls and to incomplete combustion. The flow rate $\dot{m}_a$ for the actual nozzle is predicted from

$$\dot{m}_a = z_d \dot{m}_i \quad (7.44)$$

where z_d, the coefficient of discharge, has a value of about 0.97 to >1.0. Values of z_d exceeding 1.0 result from actual gas densities that are higher than the theoretical gas density.

The theoretical gas velocity $(C_e)_{th}$ at the nozzle outlet section is expressed by

$$(C_e)_{th} = \sqrt{2c_p(T_c - T_e)} \quad (7.45)$$

and

$$T_e = T_c\left(\frac{P_e}{P_c}\right)^{k-1/k} \quad (7.46)$$

Substitution of Eqs. 7.41 and 7.46 into Eq. 7.45 yields

$$(C_e)_{th} = \sqrt{2\frac{kR}{k-1}T_c\left[1 - \left(\frac{P_e}{P_c}\right)^{k-1/k}\right]} \quad (7.47)$$

The actual gas velocity at the nozzle outlet section is less than $(C_e)_{th}$ because of fluid friction. The actual discharge velocity of the gas is predicted by application of a velocity coefficient z_c that, in turn, is related to the overall nozzle efficiency η_n, or $z_c = \sqrt{\eta_n}$. Values of η_n are ordinarily in the range of 0.90 to 0.96.

Rocket engine nozzles are constructed with relatively large angles of divergence. The half-angle of divergence α, Fig. 7.31, usually ranges from 10 to 18°. Because the thrust developed by the nozzle is a function of the axial velocity, a correction is introduced to take account of the nonaxial flow of gas that results from the nozzle divergence. The correction factor is given by

$$\lambda = \frac{1}{2}(1 + \cos\alpha) \quad (7.48)$$

The gas velocity at the nozzle outlet section, corrected for the effects of fluid friction and divergence of the jet, is then expressed by

$$C_e = z_c \lambda (C_e)_{th} \quad (7.49)$$

When the gas is not completely expanded within the nozzle, expansion continues outside the nozzle from the outlet section to the plane *ef*, Fig. 7.31, where the fluid pressure first drops to the ambient pressure P_∞. The nozzle thrust is determined from

$$F = \dot{m}_a C_e + A_e(P_e - P_\infty) \quad (7.50)$$

An effective nozzle discharge velocity is given by

$$C_{\text{eff}} = C_e + \frac{A_e}{\dot{m}_a}(P_e - P_\infty) \quad (7.51)$$

The specific impulse, an important parameter used to describe the performance of rocket engines, is defined as the impulse per unit mass of propellant. For steady-state operation, the engine develops a constant thrust, hence

$$I_{sp} = \frac{F\,\Delta\tau}{m_a} = \frac{F}{\dot{m}_a} \quad (7.52)$$

Values of I_{sp} are usually expressed as N per kg/s.

Typical values of I_{sp} for LH_2–LOX and kerosene–LOX propellants are, respectively, 3500 and 2500 N per kg/s. Corresponding values of k, the specific heat ratio, can be taken as 1.26 and 1.24, respectively. Values of I_{sp} for solid-propellant rocket engines are generally in the range of 2500 to 3000 N per kg/s.

7.28 VEHICLE ACCELERATION

Unlike a typical power plant, a rocket engine operates only for a relatively short period of time. The operation is terminated by shutting down the engine or when the fuel is completely consumed. The velocity of the vehicle is zero at takeoff. In flight however, the vehicle accelerates from an initial velocity U_i at $\tau = 0$, the time at which the engine fires. It is convenient to assume that the propellant burns at a constant rate.

The performance of the vehicle is first examined for constant-level flight in the absence of drag forces. The force or thrust acting on the vehicle because of the discharge of the propellant is

$$F_j = \dot{m}_p C_{eff} \quad \text{(Ref.: Eq. 7.16)}$$

where

$\dot{m}_p$ = propellant burning rate (constant)
C_{eff} = effective velocity of the jet (constant)

The constant force F_j is applied to the vehicle, which has an initial mass M_i, including the propellant, at time $\tau = 0$. At any subsequent time, the total mass of the vehicle is M, and the velocity is U. Then,

$$F_j = M\frac{dU}{d\tau}$$

where

$M = M_i - \dot{m}_p\tau$
τ = the thrust period

Then,

$$F_j \times \frac{d\tau}{M_i - \dot{m}_p\tau} = dU$$

Integration yields

$$-\frac{F_j}{\dot{m}_p}\ln(M_i - \dot{m}_p\tau)\,\Big|_i^f = U\,\Big|_i^f$$

and

$$-\frac{F_j}{\dot{m}_p}\ln\frac{M_i - \dot{m}_p\tau_f}{M_i - \dot{m}_p\tau_i} = U_f - U_i$$

but

$$\tau_i = 0 \qquad M_i - \dot{m}_p\tau_f = M_f$$

$$\frac{F_j}{\dot{m}_p} = C_{eff}$$

Thus,

$$U_f - U_i = C_{eff}\ln\frac{M_i}{M_f} \quad (7.53)$$

where

$U_f - U_i$ = change in the velocity of the vehicle that occurs during the thrust period
C_{eff} = effective velocity of the jet
M_i = initial mass of the vehicle, including the propellant
M_f = mass of the vehicle at the end of the thrust period

Equation 7.53 neglects the effect of gravitational and drag forces. The gravitational force is important at takeoff, but the magnitude of the force diminishes as the altitude of the vehicle increases. The value of the acceleration owing to gravity at an altitude h can be determined from

$$g = g_0\left(\frac{r_e}{r_e + h}\right)^2 \tag{7.54}$$

where g_0 is the acceleration owing to gravity at sea level on the surface of the earth and r_e is the radius of the earth. The standard value of 9.807 m/s^2 may be used ordinarily for g_0.

A long-range missile or space vehicle is usually launched from a vertical position with a subsequent change in the inclination angle θ occurring during the thrust period. See Fig. 7.32. The effect of gravity is to cause a reduction in the velocity change; and for a short period of time $\Delta\tau_b$, a constant value of g may be assumed. The reduction in the velocity change for the vehicle because of gravitional acceleration is given by $g\,\overline{\cos\theta}\,\Delta\tau_b$, where $\overline{\cos\theta}$ is the integrated mean value of $\cos\theta$ for the burning period $\Delta\tau_b$.

In passing through the earth's atmosphere, the vehicle encounters an aerodynamic drag. The drag, a retarding force, is expressed as a fundamental relationship involving the dynamic pressure $\frac{1}{2}\rho U^2$. Hence,

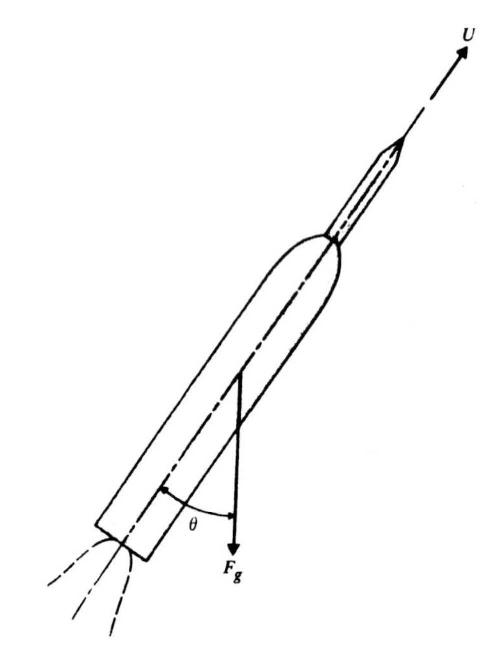

Figure 7.32 Rocket-powered vehicle traveling in a gravitational field.

$$F_D = C_D\left(\frac{1}{2}\rho U^2\right)A_f$$

where

- F_D = drag force acting on the vehicle
- C_D = drag coefficient
- ρ = local air density
- U = vehicle velocity
- A_f = vehicle frontal or projected cross-sectional area

The density of the air varies substantially with a change in altitude.

The drag coefficient is dependent upon the flight speed and the angle of attack of the vehicle, and upon the vehicle configuration and certain surface characteristics. A typical value of C_D is likely to be in the range of 0.1 to 0.2.[4] The aerodynamic drag is important to the flight analysis as the vehicle moves through the earth's atmosphere prior to reaching an upper region of rarefied air.

Drag causes a reduction in the velocity change achieved by the vehicle during the thrust period. For a small time interval $\Delta\tau_b$, a constant drag force F_D and a constant vehicle mass M can be assumed. Because of aerodynamic drag, the velocity change for the vehicle is then reduced by an amount equivalent to $(F_D/M)\,\Delta\tau_b$.

Equation 7.53 is expanded to include the effects of gravitational acceleration and aerodynamic drag. Thus, for a small time interval $\Delta\tau_b$,

$$U_f - U_i = C_{\text{eff}}\ln\frac{M_i}{M_f} - g\,\overline{\cos\theta}\,\Delta\tau_b - \frac{F_D}{M}\Delta\tau_b \tag{7.55}$$

There are many other aspects of rocket engine propulsion, not covered in the preceding discussion, that must be observed in the design and performance analysis of the engine and the vehicle in which it is installed.

7.29 RAMJET ENGINES

In the air-breathing engine category, the ramjet engine is the simplest propulsion device. The engine consists of three components, namely, the diffuser, the combustion chamber, and the nozzle. Compression of the air prior to combustion is accomplished by diffusion of the high-velocity stream of air approaching the inlet. Ramjet engines can operate at subsonic flight speeds, but superior performance is achieved at supersonic flight speeds because of the high pressure rise that is developed in the diffuser.

The free-stream air is first compressed adiabatically in passing through oblique and normal shocks. Compression through the normal shock reduces the velocity of the air to a value within the subsonic range, and subonic diffusion is subsequently effected in the divergent flow channel that extends to the combustion chamber.

The combustion chamber inlet Mach number is limited to about 0.25 to 0.3 in order to stabilize the flame. For a prescribed flight Mach number of 2.5, a high theoretical pressure increase is observed, that is, the ratio of the isentropic stagnation pressure to the static pressure is about 17 to 1. There are however substantial stagnation pressure losses incurred in the shocks, and further pressure losses occur in the subsonic diffuser section and with the flow of air through the flame holder.

Unlike the turbojet engine, combustion temperatures in the ramjet engine are not limited by the maximum temperature at which highly stressed rotating parts can safely operate. Thus, in the ramjet engine, the products of combustion enter the nozzle at a temperature in the range of 1650 to 1950 C, depending upon the provision made for cooling the engine walls.

Ramjet engines are used primarily for propulsion of military-type missiles and must operate at high flight speeds. These devices are comparatively simple, with the performance strongly influenced by the design of the supersonic–subsonic diffuser system. A significant disadvantage is encountered in the operation of the ramjet engine, namely, the inability to develop a static thrust. Because the engine cannot accelerate from a rest position, some means must be provided to launch a ramjet-equipped vehicle. Launching may be accomplished by a chemical rocket that can be detached from the vehicle when the ramjet engine begins to accelerate.

For a flight Mach number in the vicinity of 2.5, the specific impulse developed by a typical ramjet engine is about 20,000 N per kg/s.

PROBLEMS

7.1 A simplified analysis is applied to a gas turbine engine cycle that has a compression ratio of 8 to 1. The compressor and turbine inlet temperatures are, respectively, 300 and 1400 K. Values of gas (air) properties are obtained from the Air-Gas Tables. The compressor and turbine machine efficiencies are, respectively, 0.85 and 0.90. The gas-to-air mass ratio is 1.02 to 1. Calculate the thermal efficiency for the cycle and the thermal efficiency for the corresponding air standard Brayton cycle.

7.2 An airplane travels at a speed of 210 m/s. The ambient conditions are 40 kPa, abs, and 250 K. At the compressor entrance, the air velocity is 150 m/s. The inlet effectiveness is 0.93. Determine the static pressure and temperature at the compressor inlet section.

7.3 A gas turbine power plant employs three-stage compression with interstage cooling. The compressor machine efficiency is 0.86. The overall pressure ratio is 12 to 1. The compressor inlet conditions are 100 kPa, abs, and 295 K. Determine the compressor work and compare the result with the work required for single-stage compression between the prescribed overall pressure limits. Calculate the heat transferred in the intercoolers.

7.4 An airplane equipped with a turbojet engine travels at a flight speed of 290 m/s. The atmospheric pressure is 40 kPa, abs. The stagnation pressure and temperature at the nozzle inlet section are 100 kPa, abs, and 1000 K. The fuel/air ratio is 0.0185. The nozzle efficiency is 0.955. The net thrust developed by the engine is 70,000 N. Determine the air flow rate for the engine.

7.5 The flight speed of an airplane is in the supersonic range. The ambient pressure is 65 kPa, abs. The stagnation and static temperatures immediately downstream from the normal shock are, respectively, 377 and 340 K. Calculate the flight speed of the airplane and the stagnation pressure loss across the normal shock that occurs in the engine inlet. Normal-shock functions may be used in the calculations.

7.6 The thermal efficiency of a theoretical gas turbine engine regenerative cycle exceeds the thermal efficiency of a Brayton cycle by a quantity equal to 0.10. The cycles have equal pressure ratios. The maximum and minimum temperatures for the two cycles are, respectively, 1500 and 300 K. Determine the cycle pressure ratio. The air standard analysis may be applied.

7.7 The following data are applicable to a gas turbine power plant.

Shaft output: 5000 kW

Inlet air: 97 kPa, abs, and 30 C (ambient)

Inlet pressure loss: 2 kPa

Compressor pressure ratio: 5.5 to 1

Compressor machine efficiency: 0.84

Compressor mechanical efficiency: 0.98

Combustion chamber outlet temperature: 1000 C (stagnation)

Fuel: $C_{12}H_{26}$

Combustion efficiency: 0.97

Combustion chamber pressure loss: 3 percent of the inlet stagnation pressure

Turbine machine efficiency: 0.88

Turbine mechanical efficiency: 0.98

Turbine exhaust pressure: 100 kPa, abs (stagnation)

Determine the air flow rate and the power plant thermal efficiency.

7.8 The stagnation temperature at the inlet section of a turbojet engine nozzle is 920 K. The gas temperature at the nozzle outlet section is 820 K. Determine the Mach number at the nozzle outlet section.

7.9 The shaft work for a turbojet engine compressor is 510 kJ/kg air. The fuel/air ratio is 0.0185. The turbine exhaust stagnation conditions are 150 kPa, abs, and 970 K. The turbine internal efficiency is 0.89, and the mechanical efficiency is 0.98. Determine the stagnation pressure and temperature of the gas at the turbine inlet section.

7.10 At the compressor inlet of a turbojet engine, the stagnation pressure and temperature are 90.70 kPa, abs, and 301 K. The ambient air conditions are 70 kPa, abs, and 278 K. Calculate the aircraft flight speed and the effectiveness of the inlet.

7.11 A turbojet engine nozzle develops a gross thrust of 23,420 N. At the nozzle outlet section, $P = 70$ kPa, abs, and $T = 830$ K. The atmospheric pressure is 52 kPa, abs. Calculate the area of the nozzle outlet section and the gas flow rate.

7.12 The flight speed for an airplane powered by a turbojet engine is 245 m/s. The stagnation temperature at the turbine outlet section is 1100 K. The static pressure at the nozzle outlet is 89.6 kPa, abs. The atmospheric pressure is 70.0 kPa, abs. The fuel/air ratio is 0.0190, and the area of the nozzle outlet is 0.250 m^2. Calculate the net thrust developed by the turbojet engine.

7.13 An airplane equipped with turbojet engines operates at an altitude of 4570 m, where the air conditions are 430 mm mercury and −13 C.

Aircraft speed: 900 km/h

Airflow: 200 kg/s

Ram efficiency: 0.92

Compressor pressure ratio: 8 to 1

Compressor machine efficiency: 0.87

Compressor mechanical efficiency: 0.985

Combustion chamber outlet temperature: 1400 K

Fuel: $C_{12}H_{26}$

Combustion efficiency: 0.975

Combustion chamber pressure loss: 2.5 percent of the inlet stagnation pressure

Turbine machine efficiency: 0.91

Turbine mechanical efficiency: 0.985

Nozzle efficiency: 0.97

Determine the fuel/air ratio, the jet thrust, and the propulsive, thermal and overall efficiencies.

7.14 The engine described in Problem 7.13 is modified to operate as a turboprop engine. The aircraft flight speed is 630 km/h. The air flow rate is 100 kg/s. The stagnation pressure at the turbine outlet section is 76.5 kPa, abs. The efficiency of the exhaust nozzle is 0.98. The propeller efficiency is 0.83. Determine the fuel/air ratio, the total thrust, and the propulsive, thermal, and overall efficiencies.

7.15 Determine the air flow, in kg/s, to a turbojet engine that develops a net thrust of 100,000 N. The following data are applicable to the operation of the engine: aircraft flight speed 972 km/h, atmospheric pressure 55 kPa, abs; stagnation properties at the turbine outlet 150 kPa, abs, and 1000 K; fuel/air ratio 0.020; and nozzle efficiency 0.96.

7.16 The thermal efficiency for a gas turbine power plant is 0.210. A proposal is made to increase the thermal efficiency to 0.260 by installing a recuperator that has an effectiveness of 0.75. U_o = 48 W/m²·K, and F_c = 1.0. The plant rated output is 20,000 kW, and the plant annual factor is 0.55. The shaft work is 130 kJ/kg air when recuperation is employed. Fuel cost is $1.80 per million kilojoules. The capital cost of the recuperator is $220/m² heat transfer surface. The fixed charges are 20 percent of the capital cost. Estimate the net annual saving that may be anticipated from the installation of the recuperator.

7.17 Gas enters a recuperator at 750 K and leaves at 550 K. The air flow is 60 kg/s, and the fuel/air ratio is 0.020. The heat transfer area A_o = 920 m², and U_o = 115 W/m²·K; F_c = 1.0. Calculate the recuperator effectiveness.

7.18 A small gas turbine engine that drives an air compressor is supplied from a large pipeline with combustion gas at 420 kPa, abs, and 1000 K. The turbine exhaust stagnation pressure is 105 kPa, abs. The gas flow is 300 kg/min. Turbine efficiencies are: mechanical, 0.97 and internal, 0.85. At the compressor inlet, the pressure and temperature of the air are 100 kPa, abs, and 300 K, respectively. The air flow is 600 kg/min. Compressor efficiencies are: mechanical, 0.97 and internal, 0.80. Calculate the compressor discharge pressure.

7.19 A gas turbine power plant is designed for installation of a recuperator. The rated output of the power plant is 25,000 kW, and the plant annual factor is 0.60. The fuel cost is $1.00 per million kilojoules. The capital cost of the recuperator is $250/m² heat transfer surface, and the fixed charges are 20 percent of the capital cost; U_o = 55 W/m²·K; F_c = 0.95. For a recuperator effectiveness of 0.80, the thermal efficiency of the power plant is 0.30 and the air flow is 200 kg/s. For a recuperator effectiveness of 0.50, the thermal efficiency is 0.245 and the air flow is 165 kg/s. For no recuperation, the power plant thermal efficiency is 0.210. Determine whether installation of a recuperator is economically feasible.

7.20 Air enters a combustion chamber at P = 800 kPa, abs, and 585 K. The inlet air velocity is 65 m/s. The gas temperature at the outlet section is 1400 K; R = 0.287 kJ/kg·K. The pressure loss owing to fluid friction is equivalent to 120 percent of the dynamic pressure at the inlet section. Calculate the pressure drop attributed to fluid friction and acceleration of the fluid stream. The latter quantity may be determined from the

Rayleigh line one-dimensional compressible-flow functions.

7.21 The turbojet engine examined in Problem 7.13 is equipped with an afterburner that increases the gas temperature to 2100 K. The stagnation pressure at the nozzle inlet is 263.80 kPa, abs. The overall fuel/air ratio is 0.05873. The nozzle efficiency is 0.94. Calculate the increase in the net thrust produced by the afterburner.

7.22 A gas turbine power plant is comprised of three compressor units, a primary and a reheat turbine, and a recuperator. The high-pressure turbine drives the intermediate-pressure and the high-pressure compressors. The reheat turbine drives the low-pressure compressor and the electric generator. The shaft work for each compressor is −88.76 kJ/kg air. The inlet gas temperature for each turbine is 1350 K. The air leaves the recuperator at 793 K. The fuel is liquid dodecane, and the combustion chamber pressure loss is 3 percent of the inlet stagnation pressure. The combustion efficiency is 0.975. The gas stagnation pressure at the inlet of the high-pressure turbine is 950 kPa, abs, and 110 kPa, abs, at the outlet of the low-pressure turbine. The turbine internal and mechanical efficiencies are, respectively, 0.90 and 0.98. The generator efficiency is 0.96. Calculate the power plant net electrical output and the power plant thermal efficiency.

7.23 The pressure and temperature developed in the combustion chamber of a chemical rocket engine are 2.4 MPa, abs, and 3170 K. The atmospheric pressure is 55 kPa, abs. The nozzle outlet pressure is 85 kPa, abs. The throat area of the nozzle is 600 cm^2. The nozzle efficiency is 0.91; z_d = 0.98. The half-angle of divergence is equal to 12°; k = 1.25; R = 0.693 kJ/kg·K. Calculate the thrust and the specific impulse.

7.24 A space vehicle powered by a chemical rocket engine is traveling at an altitude of 800 km. The engine is shut down, and the total mass of the vehicle is 50,000 kg. The mean value of cos θ is 0.910. When the engine fires the propellant consumption rate is 200 kg/s, and the effective velocity of the jet is 3400 m/s. Determine the acceleration of the vehicle for a firing period of 30 s. (The mean diameter of the earth is 12,742 km.)

REFERENCES

1 Keenan, J. H., Chao, J., and Kaye, J. *Gas Tables*, John Wiley & Sons, Inc., New York, 1980.

2 Sanford, G. and McBride, B. J. *Computer Program for Calculation of Complex Chemical Equilibrium Compositions, Rocket Performance, Incident and Reflected Shocks, and Chapman-Jouguet Detonations*, NASA Lewis Research Center, Cleveland, Ohio, 1971.

3 Hesse, W. J. and Mumford, N. V. S., Jr. *Jet Propulsion for Aerospace Applications*, Pitman Publishing Corporation, New York, 1964.

4 Hill, P. G. and Peterson, C. R. *Mechanics and Thermodynamics of Propulsion*, Addison-Wesley Publishing Co., Reading, Mass., 1965.

BIBLIOGRAPHY

Dusinberre, G. M. and Lester, J. C. *Gas Turbine Power*, International Textbook Co., Scranton, Pa., 1958.

Hosny, A. N. *Propulsion Systems*, published by author, Columbia, S. Carolina, 1966.

Sawyer's Gas Turbine Engineering Handbook, Vol. I, *Theory and Design;* Vol. II, *Applications*, J. W. Sawyer, ed., Turbomachinery Publications, Inc., Norwalk, Conn., 1980.

Shepherd, D. G. *An Introduction to the Gas Turbine*, D. Van Nostrand Co., New York, 1949.

Sorensen, H. A. *Gas Turbines*, The Ronald Press Co., New York, 1951.

Vincent, E. T. *The Theory and Design of Gas Turbines and Jet Engines*, McGraw-Hill Book Co., New York, 1950.

Zucrow, M. J. *Aircraft and Missile Propulsion*, Vol. I and Vol. II, John Wiley & Sons, Inc., New York, 1958.

CHAPTER EIGHT

FOSSIL FUEL-FIRED STEAM POWER PLANTS

The steam-electric generating station is a large and highly complex energy conversion system. The power plant consists ordinarily of two main sections: (1) the steam generator; and (2) the turbine-generator, condenser, and feedwater heaters. In addition, a good deal of auxiliary equipment, such as pumps, fans, and conveyors, is required for the operation of the principal components of the power plant.

The plant is usually located relatively close to the load center in order to minimize electrical transmission losses. On the other hand, it may be economically feasible to locate the plant near a coal field at some distance from the load center. Higher transmission costs are incurred, but fuel transportation costs are substantially reduced. A power plant site adjacent to a natural source of cooling water will ordinarily be acquired, provided that there can be an unrestrictive use of an ample supply of water.

Figure 8.1 shows the principal components and the operating cycle for an elementary steam power plant. Deviations from the theoretical cycle are examined in later sections.

8.1 STEAM GENERATORS

The function of the steam generator, or boiler, is the conversion of liquid water into superheated steam through efficient combustion of the fuel. The design of the modern steam generator is the result of a continuous program of development that has been maintained in the steam power industry throughout the past century. Steam pressures and temperatures have gradually increased in response to the demand for improved thermal performance. Subcritical-pressure boilers are designed to operate at pressures that extend to about 20 MPa. For supercritical-pressure boilers design pressures range from 24 to 35 MPa.

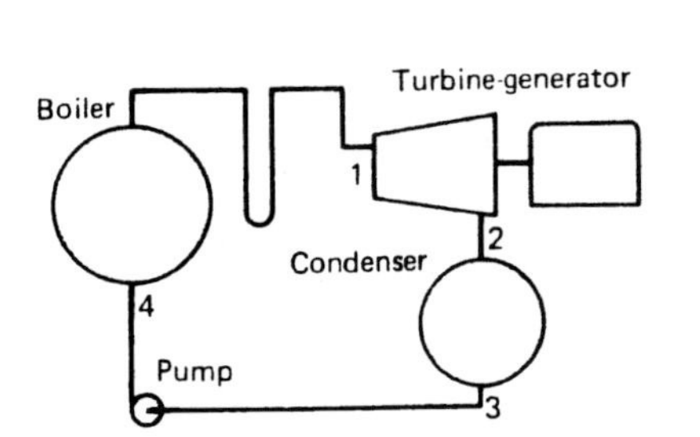

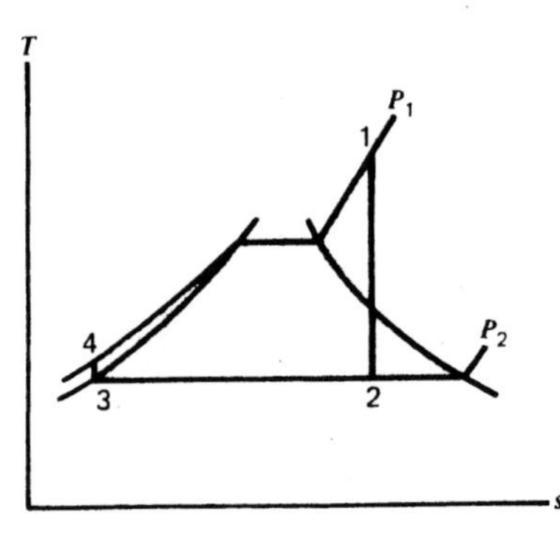

Figure 8.1 Basic steam power plant flow diagram and theoretical thermodynamic cycle.

Maximum steam temperatures are limited to approximately 540 C in nearly all of the large boilers installed in recent years in the United States.

Large, modern boilers are capable of generating from 0.5 to 4.5 million kilograms of steam per hour. Numerous problems in design and operation of steam generators have resulted from the necessity of burning fuels of inferior quality and observing environmental regulations that place limitations on the discharge of combustion gases and cooling water from the power plant.

Early boilers were of low capability, hence several boilers were required to supply steam to each turbine. Present-day power plant designs arrange for the combination of a single, large boiler with a single, large turbine-generator. This unit-type arrangement is particularly conducive to achieving high-economy operation.

A large steam generator consists generally of the following components: furnace, water walls, superheater, reheater, economizer, and air heater. The water walls are actually the vertical enclosing walls of the furnace. Certain auxiliary pieces of equipment are essential to the operation of the steam generator, namely, fans, pumps, soot blowers, and fuel handling and firing equipment.

Early boilers were essentially convective evaporating units that provided for little superheating of the steam. As the trend to higher steam pressures and temperatures continued, the superheater was required to absorb a greater share of the total energy released by combustion of the fuel. This requirement, coupled with the increased load on the water walls, placed the convective evaporating element in a position of minor importance. This element has been eliminated in the modern high-capability boilers. Thus, evaporation of the boiler water is accomplished in the furnace water walls and in the water-cooled wall enclosures, division walls, and boiler screen. The energy transfer to the water walls is effected principally by radiation from the furnace gases. Reduced furnace wall maintenance can be attributed to water cooling of the wall material.

In low-capability boilers, evaporation is achieved in a convective tube bank, and furnace water wall tubes may or may not be installed, depending upon the energy release rate. When used in these smaller units, the water wall tubes serve, in part, to protect the refractory wall material from the deleterious effect of high temperature.

Gas cooling, by transfer of energy to the furnace water walls, achieves a reduction in the temperature of the gas entering the superheater to a level where slag deposits and superheater tube corrosion can be controlled by soot blowers. In some installations, a slag or boiler screen is placed in front of the superheater in order to avoid excessive slagging of the superheater tube surface. Design procedures must necessarily observe proper matching of the evaporating sections and the superheater.

The reheat steam cycle was introduced into actual practice during the 1930 decade. Although the reheat cycle showed the expected superior operating characteristics in comparison with the nonreheat cycle, it did not initially possess sufficient advantages, principally economic, for gaining widespread adoption. Subsequently, significant changes in the economic factors that influence the trends in power plant design led to the inclusion of reheat in the basic steam power cycle.

Steam superheating in large power boilers is accomplished in both the superheater and the reheater. Although the superheater is usually a convective element, radiant superheaters, placed on the furnace wall, are also used in large boiler installations.

A wide variety of designs can be found upon examination of the boilers currently in use. However, in the limited scope of this text, consideration can be given only to a few of the significant types of steam generators.

8.2 BOILER DEVELOPMENT

The fire tube boiler is a low-capability steam generator used principally for producing steam for heating and processing applications. See Fig. A.1. In this type of boiler, the combustion gases pass through the tubes that are contained inside the drum. The fire tube boiler lacks vigorous circulation of the water within the drum; and, in addition, it is inherently unsuitable for high-pressure, high-capability applications. Despite these limitations, the fire tube boiler was used extensively at one time in heating plants and in stationary, marine, and locomotive power plants. Both the horizontal-drum and the vertical-drum arrangement can be employed for the construction of fire tube boilers.

Water tube boilers are not subject to the inherent limitations of the fire tube boiler. Positive circulation is promoted through the tubes, which are placed externally to the drum in an inclined or vertical position. See Fig. A.2. The external tube surface, over which the combustion gases pass, can be extended sufficiently to provide the area required for the transfer of energy.

Natural-circulation water tube boilers may be classified as

1 Longitudinal drum.
2 Cross drum.
   (a) Straight tube.
   (b) Bent tube.

Construction of the longitudinal-drum boiler places the tubes below the drum and parallel to the axis of the drum. The tubes are inclined at an angle of about 15° with the horizontal plane. The longitudinal-drum water tube boiler is limited because of structural characteristics to low-pressure, low-capability applications. Space available in this boiler is somewhat inadequate for installation of the superheater, hence only a modest amount of steam superheating can be accomplished.

The cross-drum boiler was developed as a means to surmounting the limitations on capability and pressure that are encountered in the longitudinal-drum design. In the cross-drum design, only the length of the drum is dependent upon the size of the boiler, that is, the width of the tube bank. See Fig. 8.2. The diameter of the drum need only be large enough to provide an adequate steam disengaging surface. An increase in the boiler capability is thus achieved by an increase in the width of the tube bank and a corresponding increase in the length of the drum.

The radiant-type boiler shown in Fig. 8.3 is typical of the high-capability, high-pressure steam generators currently operating in electric power stations. Steam generating capabilities for these boilers range from about 135,000 kg/h to an undetermined maximum exceeding 3.2×10^6 kg/h. The operating steam pressure is subcritical, usually 12.5 to 16.5 MPa. Steam superheat and reheat temperatures are usually 538 C. The radiant boiler can be fired with coal and lignite, oil, or natural gas. Boiler designs may provide for a pressurized furnace or a balanced-draft furnace. The induced-draft fan can be omitted for the pressurized-furnace design. Divided furnaces are often used in order to increase the furnace heat absorbing surface without increasing the furnace volume.

In the boiler shown in Fig. 8.3, evaporation of the boiler water occurs, for the most part, in the furnace water walls by absorption mainly of radiant energy from the combustion gases. The combustion gases pass successively from the furnace through the secondary superheater, reheater, primary superheater, economizer, and air heater.

The radiant boiler shown in Fig. 8.3 is fired with pulverized coal. The burners are installed on the front and rear walls of the furnace for opposed firing. Ash, in a dry form, is collected in the hopper installed at the bottom of the furnace. The radiant boiler shown in Fig. 8.4 is designed for coal firing in cyclone furnaces. Molten slag, discharged from the cyclone furnaces into the boiler furnace, is tapped into the slag tank. On contact with the water in the tank, the molten slag disintegrates and is subsequently removed for disposal.

The universal pressure boiler, a once-through design, is economically applicable within a pressure range of 14 to 28 MPa.[1] The water is pumped, in a single pass, successively through the several energy absorbing elements. There is no recirculation of water within the unit, hence a drum

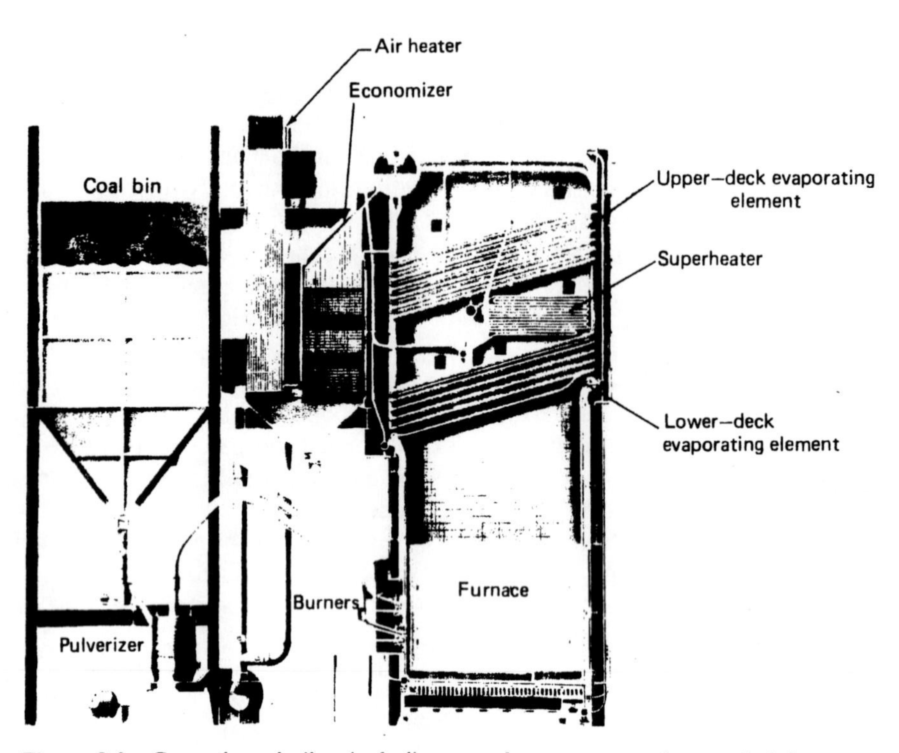

Figure 8.2 Cross-drum boiler, including superheater, economizer, and air heater; slag-tap furnace fired by a direct-firing pulverized coal system. (Courtesy Babcock & Wilcox.)

is not required for separation of the vapor from the liquid. The universal pressure boiler may be designed for operation at either subcritical or supercritcal steam pressures. The corresponding turbine throttle pressures are usually 16.5 and 24.1 MPa, gage, respectively. Boilers of this type are fired by oil or natural gas, or by coal in pulverized form or in a cyclone furnace. The steam generating capability of the universal pressure boiler ranges from about 135,000 kg/h to an undetermined maximum exceeding 4.5×10^6 kg/h. A steam flow of 4.2×10^6 kg/h corresponds to an electric generating capability of 1300 MW.

8.3 BOILER CIRCULATION

Natural circulation boilers depend upon fluid density differences for effecting the circulation of water through the evaporating elements. Analysis of the circulation process within a boiler is most complex, hence only a simplified treatment will be presented at this point in the discussion.

Figure 8.5 illustrates a single-tube circulation system. Energy absorbed at the surface of the riser is transmitted through the tube wall to the water, and subsequent evaporation produces a mixture of vapor and liquid water within the tube. The downcomer is filled with water that may be considered saturated and free of vapor bubbles.

When the circulation system is in equilibrium, the mass flow rate of the water in the downcomer is equal to the mass flow rate of the mixture of vapor and liquid water in the riser. Further, at section x–x, Fig. 8.5, the net pressure P_d of the fluid in the downcomer is balanced by the net pressure P_r of the fluid in the riser, $P_d = P_r$.

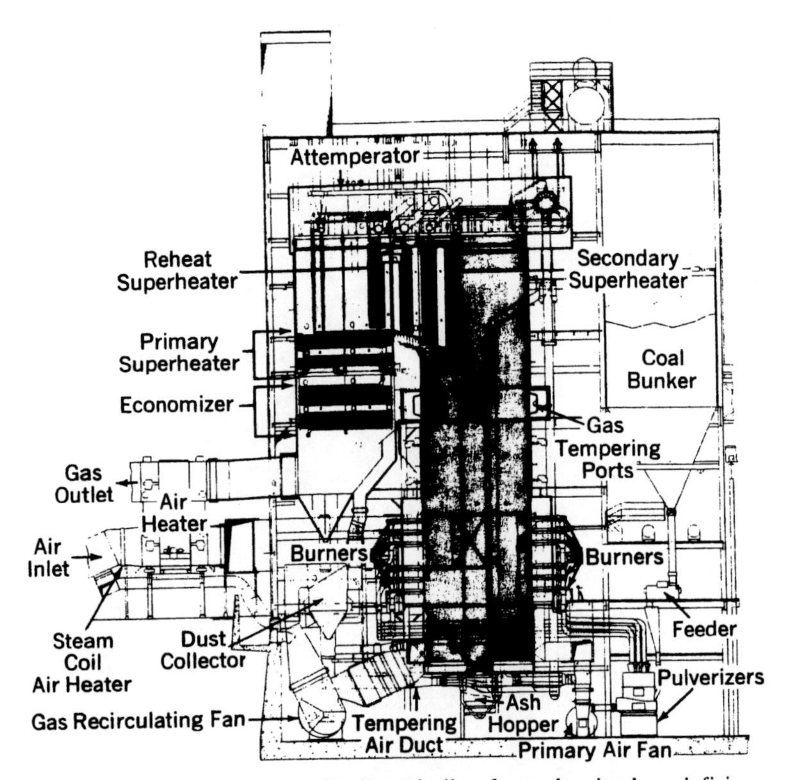

Figure 8.3 Carolina-type Radiant boiler for pulverized coal firing. Design pressure 19.82 MPa; primary and reheat steam temperatures 538 C, maximum continuous steam output 793,790 kg/h.[1] (Courtesy Babcock & Wilcox.)

$$P_d = P + g\rho_d z - \Sigma\, \Delta P_d$$

$$P_r = P + g\rho_r z + \Sigma\, \Delta P_r$$

Then,

$$gz(\rho_d - \rho_r) = \Sigma\, \Delta P_d + \Sigma\, \Delta P_r \qquad (8.1)$$

where

P = boiler drum pressure
z = height of the fluid columns
g = acceleration owing to gravity
ρ_d = density of the downcomer fluid
ρ_r = mean density of the riser fluid
$\Sigma\, \Delta P_d$ = pressure loss in the downcomer because of fluid friction and the bend, entrance, exit, and change-in-section losses
$\Sigma\, \Delta P_r$ = pressure loss in the riser because of fluid friction and the bend, entrance, exit, change-in-section, steam separator, and acceleration losses

Examination of Eq. 8.1 reveals that the sum of the circulation losses in the system is equal to the difference between the gravity pressures of the downcomer and riser fluids. This pressure difference actually causes the circulation of the fluid through the downcomer and the riser.

For a prescribed saturation pressure, an increase in the rate of energy absorption at the tube surface will be followed by an increase in the percentage of vapor in the two-phase mixture discharged from the riser. Because of the increase in the rate of steam formation, an increase is observed in the gravity pressure difference between the two fluid columns; and as a consequence, the circulation rate tends to increase. The relationship between the temperature difference, tube wall to liquid, and the circulation rate approaches a limiting condition beyond which

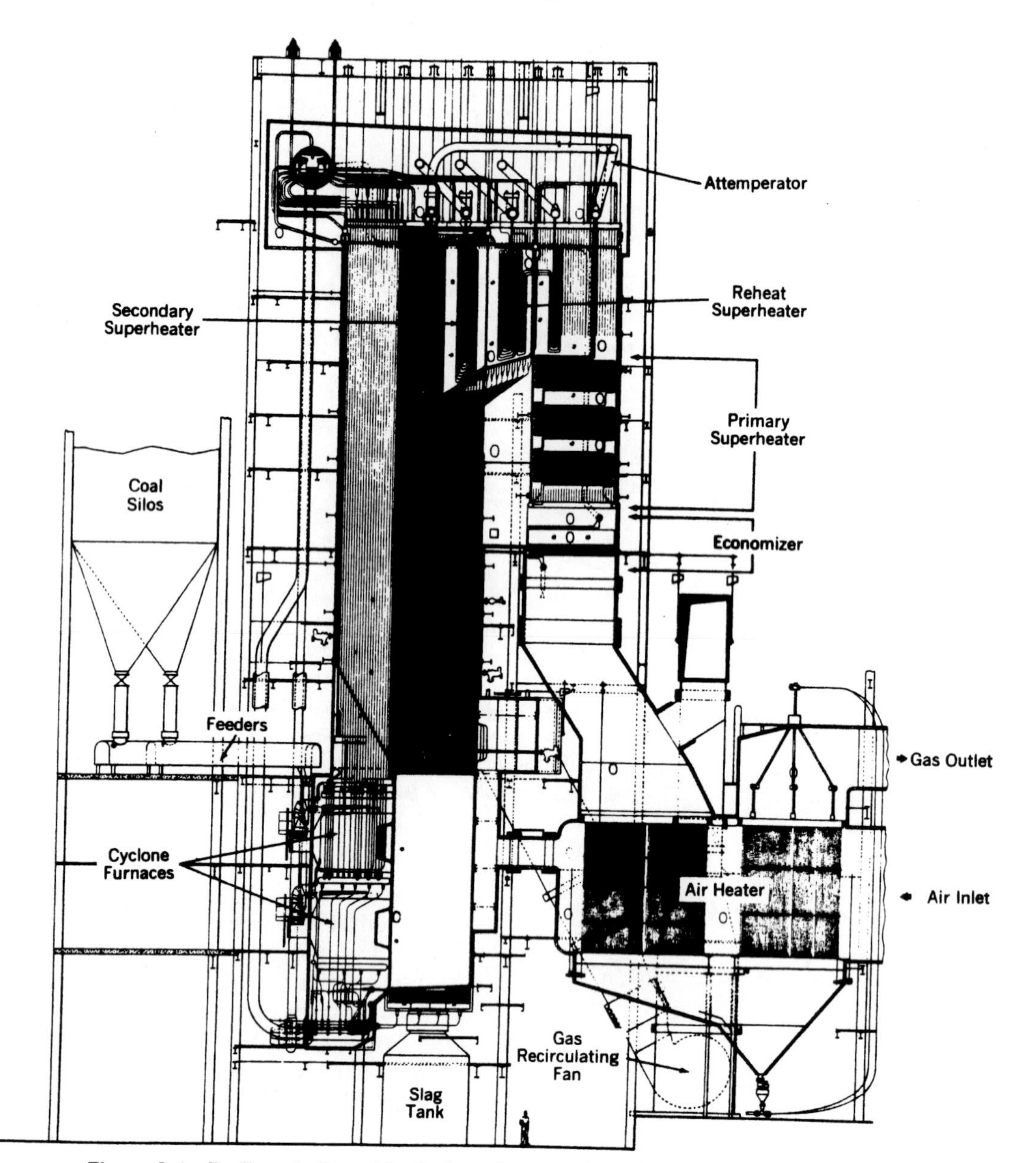

Figure 8.4 Radiant boiler with Cyclone furnaces (one wall) and bin system for coal preparation and feeding. Steam production: 566,990 kg/h at 13.10 MPa and 540 C, with reheat to 540 C.[1] (Courtesy Babcock & Wilcox.)

there is a decrease in the rate of steam production. The reduction in the rate of steam production with further increase in the temperature of the tube wall is attributed to two factors. First, fluid friction increases more rapidly than the increase in the gravity pressure difference between the two fluid columns. The second factor pertains to the nature of the boiling mechanism. In

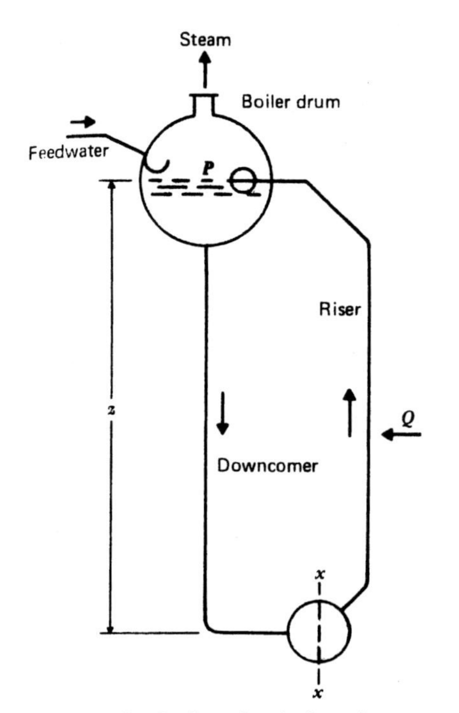

Figure 8.5 Boiler circulation element.

effect, a transformation occurs from nucleate boiling to partial nucleate and partial film boiling.

The characteristics of nucleate boiling are the formation and liberation of vapor bubbles on and from the heat transfer surface that remains in a wet state. The film conductance for nucleate boiling is high, that is, upward of 11,500 $W/m^2 \cdot K$. In film boiling, a layer of vapor forms on the heat transfer surface. The film conductance of the vapor is low, on the order of 60 $W/m^2 \cdot K$.

The transformation from nucleate to film boiling will occur at some point in the fluid above the lower end of the tube. An increase in the heat flux, for example, from moderate to high, will cause a downward shift in the point at which the transformation occurs. For a simplified analysis of the boiling phenomenon, the radiant energy transfer rate, in W/m^2, to the tube wall is assumed to be constant along the tube, with nucleate boiling established at the lower end and film boiling near the upper end. In order to balance the heat flow rate to the fluid, the tube wall temperature near the top of the tube will exceed the corresponding temperature at the bottom of the tube. In effect, at the top section of the tube, a greater temperature difference between wall and fluid is required to counteract the low film conductance. Actually, this inherent adjustment may not be possible without causing overheating and "blistering" of the tube metal. Tube failures of this kind can also result from faulty distribution of the water to the tubes in a bank. Thus, some of the parallel circuits are "starved" and hence cannot function properly.

The design of the riser and downcomer tubes in a boiler is a complex procedure that is dependent upon several factors, some of which are not readily evaluated. For a simplified riser design, two important criteria may be examined. One is the maximum allowable percent by volume of the vapor in the two-phase mixture leaving a heat absorbing riser circuit. Limiting values for this criterion are presented in Fig. 8.6 as a function of the operating pressure. Above a boiler pressure of 10.5 MPa, abs, a detailed analysis is required of the probable distribution of the heat fluxes throughout the furnace. The second criterion is the minimum allowable velocity of the water entering the tube. For risers that are essentially vertical and absorb energy on one side or completely around the tube, this minimum allowable velocity may be relatively low, 0.3 to 1.5 m/s. If the tube is slightly inclined with re-

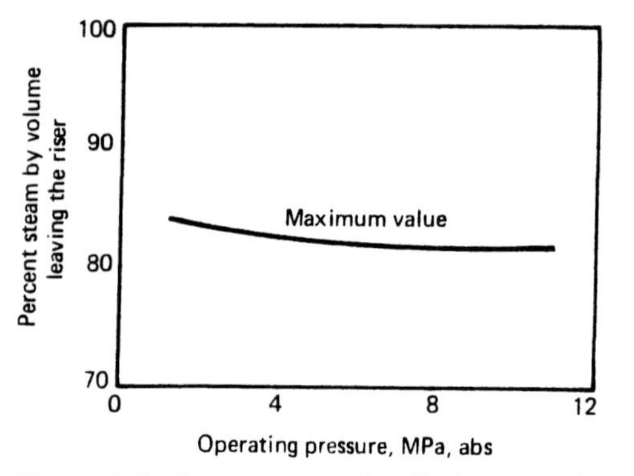

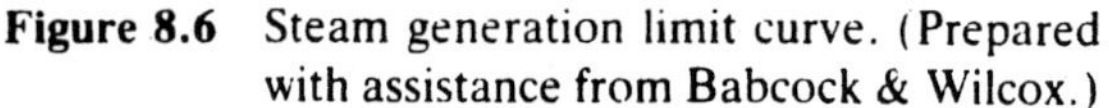
Figure 8.6 Steam generation limit curve. (Prepared with assistance from Babcock & Wilcox.)

spect to the horizontal and absorbs energy on the top side, for example, a tube on the furnace floor, a higher velocity, 1.5 to 3 m/s, is required.

Example 8.1

A furnace wall is constructed with closely spaced tubes, 75 mm O.D. and a wall thickness of 5.6 mm. The effective height of the tube is 15.25 m. The average operating steam pressure is 7000 kPa, abs. The inlet water velocity to the tube is 1.5 m/s. Determine the maximum allowable rate of energy absorption by the fluid in the tube. A tangential tube arrangement is assumed.

Tube transverse flow area: 0.0031969 m^2

Saturated water at 7000 kPa, abs:

$$v_f = 0.001351 \text{ m}^3/\text{kg}$$
$$v_g = 0.02737 \text{ m}^3/\text{kg}$$
$$h_{fg} = 1505.1 \text{ kJ/kg}$$

Flow rate:

$$\dot{m} = \frac{CA}{v_f}$$
$$= \frac{(3600 \times 1.5)0.0031969}{0.001351}$$
$$= 12{,}778 \text{ kg/h}$$

Maximum vapor fraction at discharge from the riser:

83 percent by volume (from Fig. 8.6)

Mass fraction of the vapor discharged from the tube:

$$x = \frac{1}{1 + \dfrac{V_l}{V_v}\dfrac{v_g}{v_f}}$$
$$= \frac{1}{1 + \dfrac{0.17}{0.83}\dfrac{0.02737}{0.001351}}$$
$$= 0.1942$$

Energy absorbed in the tube:

$$\dot{q} = x\dot{m}h_{fg}$$
$$= 0.1942 \times 12778 \times 1505.1$$
$$= 3{,}734{,}800 \text{ kJ/h}$$

Energy absorption rate:

Tube projected area

$$A_p = (\text{O.D.})L = 0.075 \times 15.25 = 1.14375 \text{ m}^2$$

$$\frac{\dot{q}}{A_p} = 3735 \div 1.14375$$
$$= 3265 \text{ MJ/h per m}^2 \text{ furnace wall area}$$

In general, natural boiler circulation is satisfactory for units that operate with steam pressures ranging up to 20 MPa. Within the higher range of pressures however, it is necessary to have considerable elevation in the boiler structure in order to achieve a differential head sufficient to promote adequate circulation. A few boiler designs, for various reasons, have incorporated controlled circulation of the boiler water.

In a controlled circulation boiler, the circulating pump moves the water from the boiler drum through the radiant evaporating elements. The mixture of steam and water discharged from the evaporating elements enters the drum where separation of the steam is achieved. The pressure drop provided for in the boiler design effects the flow of steam to and through the superheater. Feedwater is pumped through the economizer into the drum where it mixes with the recirculated boiler water.

The controlled circulation boiler does not differ significantly from the natural circulation boiler. Water flows through the downcomers to the circulating pumps, which in turn discharge into headers located at the bottom of the water walls. Flow from a header into the furnace wall tubes is regulated by inserting orifices of appropriate diameter into the tubes. These orifices effect a proper distribution of the water to the risers and thus ensure an adequate flow of water to all parallel circuits.

Several advantages are derived from the application of controlled circulation to steam boilers. Forced circulation is more positive in comparison with the flow established in a natural circulation boiler, and lower tube metal temperatures are experienced. Smaller furnace wall tubes can be used, for example, 40 mm O.D. against 90 mm O.D., and thus reduce fabrication difficulties.

At low boiler pressures, because of the high value of the gravity separating force, the steam readily separates from the water. With a substantial increase in pressure, there is a marked tendency to increase the carry-over of liquid with the steam leaving the boiler drum. The concentration of solids in the boiler water exceeds that of the relatively pure incoming feedwater. Thus, with considerable carry-over, the purity of the steam is reduced; and consequently, undesirable deposits may form on the internal surface of the superheater tubes and on the turbine blades.

Severe cases of poor steam separation result in priming, a condition that develops when slugs of water are carried over with the steam into the superheater and possibly into the turbine. Priming is particularly undesirable because of the thermal shock that is caused by the relatively cool slugs of water striking the high-temperature metallic surfaces.

Effective control of carry-over and priming has been achieved by installation of a steam separator in the boiler drum. Several different types of steam separators have been developed, utilizing the various principles of separating the liquid from the vapor.

8.4 FURNACE CONSTRUCTION AND OPERATION

Furnace walls are in contact with the high-temperature combustion gases on one side and the cool atmospheric air on the other side. The temperature difference created in the wall promotes heat losses, and the nonuniform temperature distribution tends to cause unequal wall expansion. In the construction of a furnace wall, it is essential to minimize heat losses, avoid disintegration of the refractory material, and permit expansion and contraction without buckling and cracking of the wall structure. Expansion joints are employed in some designs in order to allow for relative movement of the walls without opening large cracks through which air and gas can leak.

Initial water wall installations were employed as a means to reducing wall maintenance through partial cooling of the furnace. The advantages of completely cooled furnaces were soon recognized, and the fully protected refractory made possible the construction of high-capability boilers that operate with high energy release rates. As noted in an earlier section, the furnace water walls in high-capability boilers function as the main evaporating element.

Water wall panels, fabricated in the shop, consist of a single row of steel tubes connected by membrane bars that are securely welded to the tube surfaces on the center line of the tube. During erection of the boiler, the panels are assembled to form a gas-tight furnace enclosure. A high rate of heat transfer is achieved from the combustion gases to the continuous enclosing surface. For certain installations, a refractory lining is placed on a portion of the membrane wall in order to control the gas temperature within a particular zone.

Other types of furnace wall construction have been extensively employed. Bare tubes placed on the refractory wall in a tangent arrangement form a continuous energy absorbing surface. Tubes may be partially embedded in the refractory material or completely covered by the refractory material or by cast iron blocks. The tubes are covered in this manner in order to reduce the heat transfer or to provide protection against erosion or corrosion by the fuel ash.

Primary considerations in the design of boiler furnaces and the associated firing equipment are steam generating capability, stability, and operating flexibility. The furnace must be sufficiently large to allow essentially complete combustion of the fuel over a wide range of operation. Energy release rates, in kJ/h·m^3, and furnace temperatures must not approach such magnitudes as to cause rapid deterioration of the furnace structure and, as a consequence, materially increase boiler maintenance. Stable firing conditions are required, in conjunction with an absence of major operating difficulties associated with ash removal, deslagging of heat transfer surfaces, and air and gas leakage.

The requirements for a good furnace design must be observed in firing any specified fuel. Furnaces and equipment for firing gas, oil, or pulverized coal are generally similar in design, with special provisions included for handling slagging coals. These furnaces are often designed for firing the three different kinds of fuel in combination or separately, possibly with minor changes in the fuel firing equipment. Grate firing imposes narrow limitations on the design of the furnace and the fuel handling equipment, factors that reduce the flexibility in selecting acceptable coals.

8.5 MECHANICAL GRATE FIRING

Hand firing of coal is definitely excluded for all but the smallest furnaces. Boilers of moderate and high capability cannot be fired physically or economically by hand and require the uniformity of operation and control achieved by mechanical firing. The different types of equipment used for firing coal mechanically on grates are classified as

1 Chain grate and traveling grate stokers.
2 Underfeed stokers.
3 Spreader stokers.
4 Water-cooled, vibrating grate stokers.

The characteristics of coal that influence stoker selection are volatile matter content, caking qualities, ash content, and the softening and fusion temperatures of the ash.

Prior to the introduction of pulverized coal firing, circa 1920, coal was burned principally on the underfeed stoker and to a considerably lesser extent on the chain grate or traveling grate stoker. Because of the limitations on the available grate area, stoker appliations are restricted to boiler installations of comparatively low steam generating capability. Further, pulverized coal firing allows a somewhat greater flexibility in the selection of the fuel. Thus, the trend to pulverized coal firing was established, particularly with increase in the size of the newly installed boilers. The large coal-fired boilers currently erected in electric power generating stations are equipped solely for pulverized coal firing.

Among the several types of stokers listed above, the spreader stoker is most generally used in boilers that have a steam generating capability up to 175,000 kg/h. The spreader stoker is capable of firing a wide range of coals and a variety of by-product waste fuels, along with the advantages of simplicity, low capital cost, low maintenance, and ready response to load variations. A conservative rating for continuous firing is approximately 200 kg coal per hour for a square meter of grate area. Higher peak ratings may be carried for short periods of operation.

The operation of the spreader stoker is simple. See Fig. 8.7. Coal is fed from the hopper to the distributor by a feeder that regulates the quantity of fuel fired. The distributor, a rotor equipped with curved blades, revolves in the direction that achieves overthrow of the coal into the furnace. The distributors are designed to effect a uniform distribution of the fuel over the furnace area.

Combustion is most effective when the coal fired by the spreader stoker varies widely in size, that is, from very small particles to a maximum size of 2 cm. The fine particles burn in suspension, and the coarse pieces fall to the grate where combustion continues. A thin fuel bed, less than 10 cm in depth, is maintained on a stationary dumping grate or on a traveling grate. Rapid volatilization of the fuel in the furnace distills the tarry hydrocarbons from the coal before fusing or caking can occur. The absence of large masses of burning coal in the furnace permits

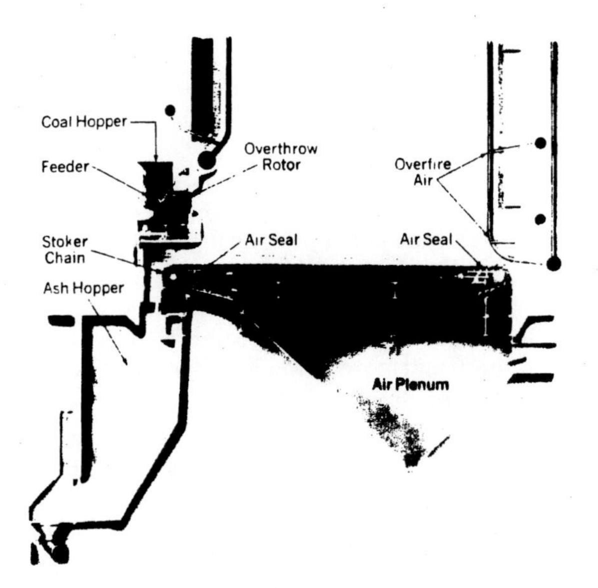

Figure 8.7 Traveling-grate spreader stoker with front ash discharge.[1] (Courtesy Babcock and Wilcox.)

rapid response to load changes. Primary combustion air is admitted to the furnace under the grate, and overfire air jets are used to promote turbulence and good mixing of the gases in the furnace and thereby ensure complete combustion.

When coal is fired by a spreader stoker, fine particles tend to leave the furnace as fly ash, especially at high firing rates. Because the combustible content of these particles is likely to be high, an excessive carbon loss can be incurred. Traps are customarily installed in the gas path in order to collect the particulate material and return it to the furnace for further burning.

8.6 PULVERIZED COAL FIRING

In the preceding section, note was taken of the two principal reasons why pulverized coal firing, rather than stoker firing, is the generally accepted method employed for burning coal in power boilers. One reason is the greater flexibility that can be exercised in the selection of the fuel. The other reason is related to boiler steam generating capability. Because of the physical limitations of the equipment, stoker installations are ordinarily restricted to steam generators that have a rated output of less than 175,000 kg steam per hour.

In the design of the pulverized coal-fired furnace, provision can be made for burning oil or gas. However, in any event, the conversion from coal firing can be readily accomplished. Various design-related factors, such as operating efficiency, firing problems, maintenance, and production costs, are of primary importance, although they are to a significant degree dependent upon the characteristics of the fuel selected.

In a system devised for pulverized coal firing, the coal is ground to a fine powder in a pulverizing mill and transported by a portion, 15 to 20 percent, of the preheated combustion air to burners installed in the furnace wall. Within the furnace, the radiant energy absorbed by the coal particles effects distillation of the volatile matter, which subsequently burns as a gas. The remaining coke particles burn with diffusion of oxygen through the layer of CO_2, CO, and N_2 that surrounds each particle. In order to achieve good combustion, the burners should provide for proper initial mixing of the fuel and air, followed by adequate turbulence to assist in the penetration of oxygen to the solid coke.

The formation of soot in the boiler furnace is a matter of some importance. If a deficiency of oxygen exists locally, the absorption of heat by the fuel molecules causes thermal decomposition of the hydrocarbon compounds. As a result of thermal cracking, hydrogen and amorphous carbon appear in the gaseous mixture. The particles of amorphous carbon are luminous in the high-temperature gaseous atmosphere and create a yellow flame, an indication of smoky combustion. In the absence of thermal cracking, the reaction is characterized by blue flame combustion. In practice, blue flame combustion is achieved only in the firing of gaseous fuels. These reactions occur very rapidly, and a relatively slight change in the firing conditions tends to effect a shift from one type of reaction to the other.

Thermal cracking is undesirable because the amorphous carbon burns with difficulty, a result of a slow reaction rate. Unless this carbon is maintained in a high-temperature atmosphere for a sufficient length of time, the particles will leave the combustion chamber in a partially burned state in the form of soot. It is important to note that soot particles are different from unburned carbon particles that are not a result of thermal cracking.

Two basic systems have been devised for preparing and handling pulverized coal. In the unit or direct system, the pulverized coal is transported directly from the mill to the burner. The indirect system provides for storage of the pulverized fuel at some point between the mill and the burner. Early systems were of the storage type; but gradually, despite some advantages, the storage system has been supplanted by the direct system in which the grinding and firing rates are inherently equal. The mill grinding capability for this system must be equal to the fuel demand at the maximum rate of firing, with the mills following the boiler "swings."

Coal pulverizing mills of various types are available; but all these mills grind the coal to a prescribed degree of fineness and ensure the rejection, and return for further grinding, of the oversize coal particles. Depending upon the mill design, the coal is processed by a combination of the basic grinding mechanisms, namely, crushing, impact, and attrition. The mill is supplied with crushed coal, and the product, finely ground coal, is transported to the burner in the primary air stream that flows through the mill.

The optimum requirements for pulverizing mills are low power consumption per ton of coal and a uniform grinding fineness over a wide range in the output of the mill. These mill characteristics are dependent upon certain fuel properties, namely, the moisture and ash contents and the grindability. Power is required to grind the inert ash, and an increase in the moisture content of the coal tends to cause a reduction in the grinding capability of the mill. The moisture content however is not especially signifiant, because the coal is dried during the grinding operation by the preheated primary air supplied to the mill for fluidizing the coal.

Coal differs considerably in the ease of grinding, as indicated by the grindability index, a property of the coal that is evaluated in the laboratory. The lower the grindability index, the harder the coal is to grind; that is, in comparison with the "softer" coals, a greater length of time is required and more power is consumed in grinding a ton of "hard" coal. Power requirements cover a wide range. Anthracite coal of low grindability requires about 40 kWh energy per metric ton for pulverization, while some bituminous coals require as little as 5 kWh per metric ton. In general, an increase in the mill capability factor will be experienced with an increase in the grindability index of the coal.

The degree of fineness to which the coal is ground influences the operating efficiency and the economy of the steam generating unit. High combustion efficiency is achieved when very finely ground coal is fired. This is particularly true for low-volatile coals, where extreme fineness speeds the combustion reaction. An increase in the amount of unburned carbon, hence reduced efficiency, results from coarse grinding. An increase in the degree of fineness of the coal however reduces mill capability and raises the power requirements. Thus, in achieving maximum operating economy, the prescribed degree of fineness will be dependent upon striking a balance between the two opposing effects, that is, the reduction in the combustion efficiency and the excess power required for grinding the coal.

The required degree of fineness varies with different coals. The more volatile coals achieve acceptable combustion with coarser grinding than the low-volatile coals. Except for fuels of very low volatile matter content, average practice prescribes grinding to a degree where 70 to 75 percent of the coal will pass through a No. 200 sieve. For anthracite coal, this figure is on the order of approximately 85 percent. Power expended in grinding the mineral matter in the coal is nonproductive.

8.7 FUEL BURNERS

A fuel burner may be designed for firing coal, oil, or gas, or for firing these three kinds of fuel in any combination. A gaseous fuel ordinarily requires little preparation prior to firing, except for possible removal of dust. A fuel oil is atomized in the burner before ignition is established. Atomization of the oil is usually effected mechanically or by the use of steam or air atomizers. The heavier oils are preheated prior to firing in order to reduce the viscosity and achieve proper dispersion of the oil into the air stream. Pulverized coal is conveyed from the mill to the burner in the preheated primary combustion air.

The typical amount of excess air for pulverized coal firing may be taken as 15 to 20 percent, while somewhat smaller values are observed for oil and gas firing, that is, 5 to 10 percent. Depending upon the type of stoker, the excess air for stoker firing of coal is roughly in the range of 15 to 50 percent.

For pulverized coal, oil, and gas firing, the combustion air is usually preheated to about 370 C. For stoker firing, the combustion air temperature is limited to about 175 C in order to avoid excessive heating of the stoker parts.

The furnace energy release rate is dependent upon a number of factors, including the type of furnace construction, the furnace operating characteristics, the fuel characteristics, and the quantity of excess air. For continuous operation in a water-cooled furnace, an average energy release rate of 1120 MJ/h per cubic meter of furnace volume can be maintained for stoker, oil, and gas firing, and 930 $MJ/h \cdot m^3$ for pulverized coal firing. Considerably higher energy release rates can be maintained during short periods of time for peak load operation.

A fuel burner is required to function in a manner that is conducive to achieving stable combustion, with low unburned fuel loss, over a reasonably wide range of operation. A self-contained burner conforms to these requirements independently of all other burners by means of proper adjustment and exact admission of air. On the other hand, the several burners may be operated collectively, with the furnace functioning as a mixing chamber and receiving at least a portion of the secondary combustion air.

The dry, pulverized coal arrives at the burner in a stream of preheated primary air. Good mixing of the coal and air has been achieved at the mill outlet. Ignition starts in the primary air stream where the fuel–air mixture is relatively rich. The preheated secondary air, at a temperature of 370 C, is directed into the primary stream of coal and air in order to cause turbulence and the resulting mixing of the fuel and air that is essential to achieving good combustion.

The circular-type burners are most frequently used for firing oil, gas, and pulverized coal. Maximum nozzle input per individual burner is 175 million kJ/h. Figure 8.8 depicts a dual-register, circular-type burner designed for firing pulverized coal. Turbulent combustion is promoted by the spin acquired by the secondary air in passing through the tangentially arranged vanes in the air register. Circular-type burners operate with an especially short, intense flame. The burners are installed in the front wall of the furnace, or in both the front and rear walls for opposed firing.

The vertical-type burner, used for firing pulverized coal, is installed in the roof of the furnance, or in a furnace arch, for firing the fuel downward into the furnace. The pulverized coal is carried into the furnace in the preheated primary air and directed into the secondary air that is admitted through ports placed in the furnace wall. The vertical flame travel is controlled by regulation of the primary and secondary air streams.

Medium- and high-volatile coals ignite readily, hence the secondary air can be supplied at the burner. Low-volatile coals, such as anthracite, do not ignite readily and require a rich mixture at the burner in addition to delayed contact with the secondary air. Vertical-type burners, described in the preceding paragraph, are usually employed for firing anthracite coal. Pulverized anthracite coal-fired installations are relatively

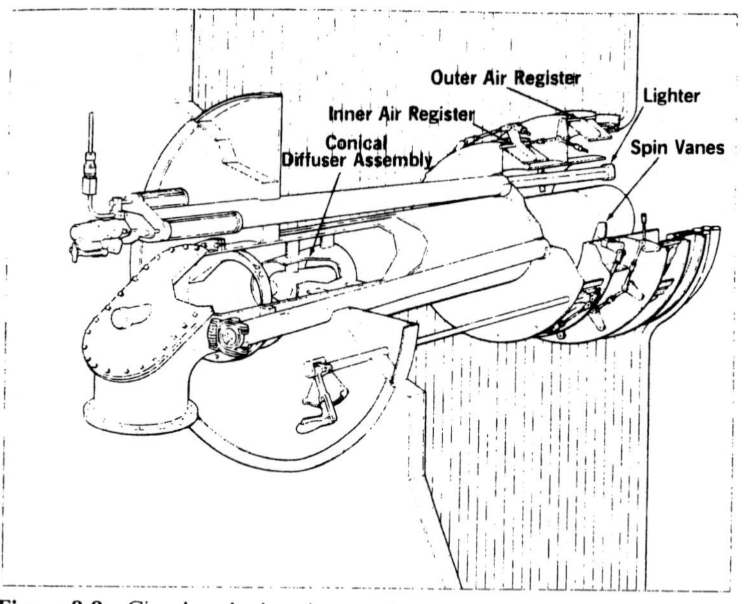

Figure 8.8 Circular dual-register pulverized coal burner. (Courtesy Babcock & Wilcox.)

uncommon, and furnace energy release rates are comparatively low, on the order of 560 to 670 MJ/h·m^3.

The tangential method of firing pulverized coal was developed as a means to creating intense turbulence through impingement of one flame on another. The burners are located in the four corners of the furnace and fire in a direction along a line that is tangent to a small circle lying in a horizontal plane at the center of the enclosure. Good mixing and high rates of combustion are achieved with this method of firing. The combustion gases sweep the furnace wall with a rotary motion and thus effect a relatively high rate of heat transfer by convection. Because of the location of the four sets of burners at the corners of the furnace, this type of firing requires a relatively complex system of piping for conveying the fuel and air to the furnace. Further, as in opposed firing, the burners must be closely regulated in order to achieve a balanced operation of the combustion system.

8.8 ASH REMOVAL

Successful firing of coal in pulverized form has been accomplished through the development of suitable furnaces and through the solution of certain difficult problems encountered in the operation of these furnaces. The greatest difficulties have occurred in firing coals that have low ash fusion temperatures and where, particularly under heavy sustained loads, the ash is removed in a molten state. The method of ash removal is thus introduced as a principal factor in the design of the furnace.

A large portion of the coal ash, probably about 50 percent, will be trapped in the slag pool. The remainder of the ash will leave the furnace and adhere to the convective heat transfer surfaces, be trapped in dry form at various points in the gas flow circuit, and be carried along in the combustion gases as fly ash.

When the average temperature of the gas leaving the furnace exceeds the ash initial deformation temperature, slag deposits will begin to build up on the boiler convective heat transfer surfaces. These deposits reduce the effectiveness of heat transfer and necessitate additional labor for surface cleaning.

In order to avoid excessive slagging of the boiler convective elements, the temperature of the gas leaving the furnace should be below the ash fusion temperature of the particular coal fired. Regulation of this gas temperature is accomplished by controlling the amount of cooling effected in the furnace. The quantity of energy absorbed in the furnace is influenced by the extent of the wall surface and the covering placed on the water wall tubes.

The energy absorbed in the furnace is normally a substantial portion of the energy released by combustion of the fuel. High gas temperatures are associated with high energy release rates, in kJ/h per m^3 furnace volume. In order to avoid excessive slagging of the convective heat transfer surfaces, energy release rates may necessarily be reduced through the use of a larger furnace for burning the fuel at a rate prescribed by the capability of the steam generating unit.

Slag builds up on the furnace walls to an equilibrium thickness, depending upon the load, and runs down the walls to collect on the furnace floor. By directing the flame from the burners in a sweep across the floor, the slag is readily maintained in a molten form for continuous tapping into a hydraulic sluicing disposal system. The furnace floor must be impervious to the liquid ash, and must be water cooled in order to avoid disintegration because of excessively high temperatures and because of erosion by the ash.

Certain coals, because of high ash fusion temperatures, can be burned in dry-bottom furnaces. Owing to the high ash fusion temperature, the particles of ash fall in a dry form to the bottom of the furnace. In order to facilitate removal of the ash, the furnace bottom is constructed in the form of a refractory-lined hopper. See Fig. 8.3. Cooling of the hopper surfaces is accomplished by placing water tubes on the refractory material. If any tendency exists for the ash particles to fall to the floor of the furnace in a plastic form, a slag screen consisting of horizontal water tubes may be placed at the top of the hopper. Ash particles are solidified in passing through the cool zone created by the screen. Only about 20 percent of the total ash is trapped in the dry bottom of the furnace; thus, a comparatively heavy load is placed on the dust collecting equipment that is required for all boiler installations.

When coal is fired on a chain grate, traveling grate, or underfeed stoker, most of the ash is discharged into a hopper located at the rear of the stoker. A portion of the coal fired by a spreader stoker is burned in suspension, hence the carry-over of particulate material, including ash, is relatively high. Ash and carbon particles are carried in the combustion gases that leave an oil-fired furnace, while a natural gas-fired installation is virtually free from carry-over of dust.

8.9 THE CYCLONE FURNACE

The development of the cyclone furnace is the most significant advancement achieved in burning coal since the introduction of pulverized coal firing. This type of furnace is capable of burning efficiently a variety of coals that are burned with difficulty on stokers and create, when fired in pulverized form, aggravated operating problems. In the latter application, the difficulty arises principally because of the characteristics of the ash.

For satisfactory firing in the cyclone furnace, the coal should have, at a temperature of 1425 C, a slag viscosity not in excess of 25 $N \cdot s/m^2$. Coals with a volatile matter content exceeding 15 percent on a dry basis should burn satisfactorily with air supplied to the furnace at a normal temperature of 370 C. A minimum ash content of 6 percent in the coal is required in order to establish a proper slag coating on the walls of the cyclone. Coals with an ash content as high as 25 percent, on a dry basis, can be burned satisfactorily in the cyclone furnace.

A cyclone furnace is constructed in the form of a horizontal, water-cooled cylinder that has a nominal internal diameter of 1.8 to 3 m. See Fig. 8.9. About 20 percent of the combustion air is supplied with the coal, as primary air, through the tangential inlet to the burner, located in the front wall of the furnace. The high-velocity secondary air is admitted tangentially, at approximately 90 m/s, to the furnace in order to increase the whirling action on the fuel. A small amount of air, up to about 5 percent, is admitted as tertiary air at the center of the burner. Excess air requirements for the cyclone furnace are low, normally 10 to 15 percent.

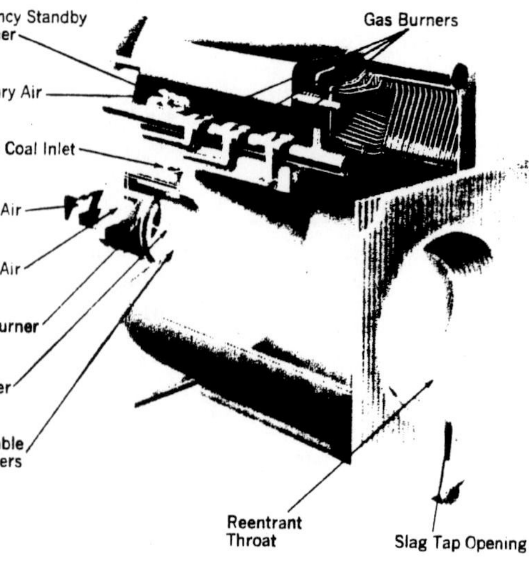

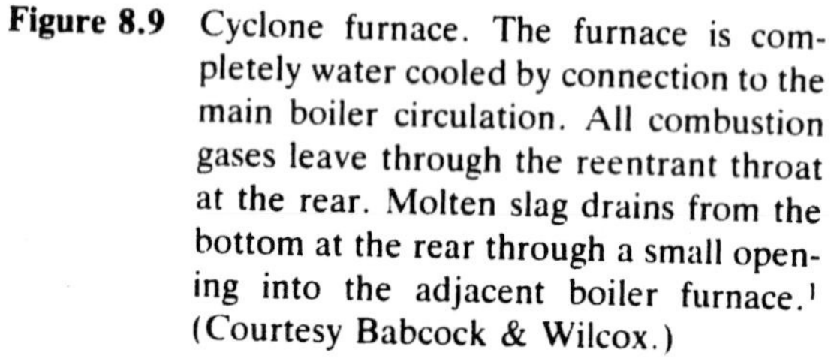

Figure 8.9 Cyclone furnace. The furnace is completely water cooled by connection to the main boiler circulation. All combustion gases leave through the reentrant throat at the rear. Molten slag drains from the bottom at the rear through a small opening into the adjacent boiler furnace.[1] (Courtesy Babcock & Wilcox.)

The energy release rates in the cyclone furnace are high, (17 to 30) $\times$ 10^6 kJ/h·m^3. Gas temperatures at the furnace outlet exceed 1650 C. The small quantity of fines in the coal is burned in suspension, while the heavier coal particles are thrown to the wall by centrifugal action and trapped in the thin layer of slag that covers the furnace wall. The high-velocity stream of air sweeping the furnace wall effectively achieves complete combustion of the embedded coal particles. Most of the coal ash is retained in the furnace as liquid slag that continuously drains from the wall into the slag tap. The carbon loss is low in this method of firing. Erosion of the boiler heat transfer surfaces by fly ash is not observed to any degree, because the particle size and concentration in the gas stream are both small.

In a typical arrangement, the cyclone furnace, in multiple units, is installed on the front wall of the cooling furnace. See Fig. 8.4. In order to achieve a reduction in the furnace width for large boilers, the cyclone furnaces are arranged for opposed firing. For such designs, the cyclone furnaces are installed on both the front and rear walls of the cooling furnace.

Because of a substantially greater air pressure drop encountered in firing coal in a cyclone furnace, the forced-draft fan power requirement is higher for this method of firing than for comparable pulverized coal firing. The power requirement for crushing coal in preparation for firing in a cyclone furnace is somewhat lower than the power required for pulverizing coal when burning bituminous coals and lignite of low grindability and low heating value. The reverse is true when firing bituminous coals of high grindability and high heating value.

The directly fired boiler furnace has two main functions, namely, combustion of the fuel and the transfer of energy from the combustion gases prior to the entrance of these gases into the boiler convective elements. The cyclone furnace concept assigns combustion of the fuel to the cyclone furnace and heat transfer to the cooling or boiler furnace. This method of firing coal exercises improved control over the combustion of certain coals that are burned with some difficulty in boiler furnaces. Incidentally, oil and gas can be fired in a cyclone furnace, although the cyclone furnace is intended primarily for the combustion of coal.

8.10 OIL AND GAS FIRING

Currently, oil and natural gas firing accounts for about 35 percent of the fuel burning capability of the electric power stations in the United States. Some of the furnaces in these power plants are designed for possible conversion to coal firing. Environmental constraints generally favor oil and gas firing, but the continued use of these fuels for steam generation could be adversely affected by uncertain supplies and escalating prices. Of greater significance however is the pressing need to reduce crude oil consumption in the United States. In the light of this need, conversion from oil firing to coal firing for electric power generation is likely to become mandatory in those plants where the conversion is feasible.

Certain advantages achieved by oil and gas firing can be cited. Neither fuel requires extensive handling or preparation for firing. Oil and gas firing avoids numerous operating problems, some severe, that are attributed to coal ash.

Oil storage at the power plant is usually provided. Depending upon the terms of the fuel contract, natural gas may be supplied on an interruptable basis, that is, gas delivery is cut off in the event of a short supply.

Gas burners are designed to mix the fuel with the combustion air and create the necessary turbulence for establishing good combustion. The gaseous fuel is readily introduced into the combustion zone through a perforated gas ring. Power plants normally burn residual oils that have high viscosity values, hence these oils are preheated prior to firing in order to expedite pumping and improve the atomization of the liquid fuel.

Compressed-air and pressure atomization of liquid fuels has been largely discontinued. The steam atomizing nozzle, a comparatively simple device, shows the best performance for firing the currently available oil fuels in water-cooled furnaces. Steam jets are directed across the fuel oil stream and effectively reduce the oil to a fine mist composed of small droplets. For operation at maximum capability of the boiler, the steam consumption for atomizing the fuel is less than 0.1 kg per kilogram of oil.

Furnaces designed for burning gas or oil are usually constructed with bare tube walls. In some designs, it may be necessary to cover the tubes in the area immediately surrounding the burners in order to stabilize combustion of the fuel. Block-covered water tubes form a flat furnace floor that is satisfactory for collecting ash and soot particles in dry form.

Certain of the fuel oils available today are somewhat lacking in high-quality combustion characteristics. These poorer-quality oils cause operating difficulties that develop from deposits of ash and slag on the heat absorbing surfaces. Deposits of oil ash usually contain complex sodium–vanadium compounds that are corrosive when in the molten phase and are the cause of high-temperature corrosion.

Steam generator efficiencies are inherently lower for gas and oil firing in comparison with coal firing under similar operating conditions. Fuel oils and natural gas contain substantially more hydrogen than coal. A high percentage of hydrogen in the fuel will produce a correspondingly high percentage of water vapor in the products of combustion, a factor that contributes to reduced boiler efficiency.

Tests have been conducted on firing a large power boiler with a fuel consisting of a mixture of coal and oil. The coal–oil mixtures were successfully blended up to a 50 to 50 mass ratio. Evaluation of the test results will indicate the technical and economic feasibility of firing the coal–oil fuel in boilers that are normally oil fired. Conversion from oil firing to coal–oil firing will contribute to the continued effort to reduce the consumption of petroleum products.

8.11 SUPERCRITICAL-PRESSURE BOILERS

Natural circulation-type boilers have been designed and constructed for operation at pressures ranging up to 20 MPa. Above this limiting pressure, natural circulation is ineffective. The Benson boiler, operating at 22.105 MPa, abs, the critical pressure of water, has been used extensively in Europe. In the United States, the trend to higher boiler pressures has been extended into the supercritical-pressure region.

The design of the supercritical-pressure steam generator provides for no internal circulation and no boiling of the working fluid. Water at a pressure somewhat higher than the outlet pressure of the unit is pumped in series flow through the several energy-absorbing elements of the steam generator. At some point in the flow circuit, the water becomes steam with no immediate increase in the specific volume. In passing through the energy-absorbing elements, the specific volume and the temperature of the working fluid increase from the inlet to the outlet section of the unit.

An increase in the turbine inlet pressure from 17 to 25 MPa will effect a reduction in the plant net heat rate of approximately 1.4 percent, for a superheat temperature of 538 C and a single steam reheat to 538 C. A further reduction of 1.2 percent in the plant net heat rate can be achieved by the inclusion of a second steam reheat to 538 C.

Unlike the subcritical-pressure boiler, the generation of steam in the supercritical-pressure boiler is not dependent upon internal circulation of the water, hence a drum is not included in the boiler structure for separation of the vapor from the liquid. The operation of the supercritical-pressure boiler is particularly sensitive to the impurities in the feedwater. Consequently, demineralization requirements for the make-up water are more exacting in comparison with the water treatment that is ordinarily adequate for subcritical-pressure boilers.

8.12 FURNACE HEAT TRANSFER

The furnace heat-absorbing surface is taken as the total enclosing surface, including the convective tube banks, minus the area of the uncooled refractory. Outlet gas temperatures are controlled by the quantities of energy released and absorbed in the furnace. Energy is transferred from the combustion gases to the heat-absorbing surfaces principally by radiation, while convective heat transfer is of secondary importance.

The analysis of furnace radiation is exceedingly complex. The character of the heat-absorbing surface is fixed by the type of furnace construction employed, while the condition of this surface is dependent upon the kind and thickness of the slag and ash deposits. Hydrocarbons, water vapor, carbon monoxide, and carbon dioxide emit and absorb radiant energy, while oxygen, nitrogen, and hydrogen are practically transparent to thermal radiation. Radiation from solid particles is an important contributing factor. The suspended solids consist of dust particles, ash, soot particles that result from thermal decomposition of hydrocarbons, and carbon particles produced in pulverized coal firing.

An analytical evaluation of the heat transfer in a boiler furnace is not easily achieved. The correlation of furnace heat transfer data is dependent to a considerable degree upon the introduction of experimentally determined factors. Not only is furnace geometry complex, but substantial variation is observed throughout the furnace in the combustion characteristics and in the energy exchange with the heat-absorbing surfaces. The variation in the fuel fired and in the quantity of excess air supplied are other factors that, in part, control furnace performance.

Boiler design requires a reasonably exact prediction of the gas temperature at the furnace outlet, because this parameter will influence the design of the boiler convective elements, particularly the superheater. For firing a particular fuel, the furnace exit gas temperature is a function of the furnace energy release rate, based on the water-cooled furnace enclosure surface. See Ref. 1, pages 4–16 to 4–19.

Evaluation of the furnace outlet gas temperature is illustrated by use of a simplified procedure, namely, the application of the modified Hudson–Orrok equation.

$$f_{ab} = \frac{1}{1 + \left[C \frac{\dot{m}_g}{\dot{E}_r} \left(\frac{\dot{E}_r}{A_p} \right)^{1/2} \right]} \tag{8.2}$$

where

f_{ab} = absorption fraction, defined as the ratio of the heat transferred from the gas to the energy released in the furnace
C = experimentally determined factor that is dependent upon the furnace design and geometry, the characteristics of the fuel fired, and the condition of the energy-absorbing walls
$\dot{m}_g$ = gas flow from the furnace, kg/h
$\dot{E}_r$ = energy release rate, MJ/h
A_p = projected area of the furnace energy-absorbing surface, m^2

The Hudson–Orrok equation is used principally for correlation of furnace test data. Furnace design would require a more complex procedure than the simplified approach represented by this equation.

Example 8.2

The temperature of the gas leaving a pulverized coal-fired furnace will be determined from the following operating data.

Coal higher heating value = 31,670 kJ/kg

Coal hydrogen fraction = 0.0501

Coal ash fraction = 0.082

Air supplied = 11.85 kg/kg coal

Furnace width = 9.2 m, depth = 7.6 m, height = 29.0 m

The fuel firing rate corresponds to an energy release rate of 950 MJ/h·m^3.

The temperature of the air entering the furnace is 367 C.

The unburned carbon loss and the radiation and unaccountable (R & U) loss are, respectively, equivalent to 1 and 2 percent of the higher heating value of the fuel.

Note: One-half of the R & U loss is assumed to result from additional unburned fuel, partially in the form of gaseous components.

1 Energy release.

The energy released in the furnace above room temperature, 25 C, and available for absorption in the furnace is equal to the lower heating value of the fuel minus the energy equivalent of the unburned fuel plus the preheat energy of the combustion air.

$$\begin{aligned}\text{LHV} &= \text{HHV} - 9m_H(h_{fg})_{\text{water, 25C}}\\ &= 31670 - (9 \times 0.0501 \times 2442.3)\\ &= 30{,}569 \text{ kJ/kg coal}\end{aligned}$$

$$\begin{aligned}\text{Energy released } E_r &= \text{LHV} - (f \times \text{HHV})\\ &\quad + m_a c_{p,a}(T - T_0)_a\\ &= 30569 - (0.02 \times 31670)\\ &\quad + [11.85 \times 1.023(367 - 25)]\\ &= 30569 - 633 + 4146\\ &= 34{,}082 \text{ kJ/kg coal}\end{aligned}$$

2 Fuel firing rate

$$\text{Furnace volume } V_f = 2027.7 \text{ m}^3$$

$$\begin{aligned}\text{Energy release rate } \dot{E}_r &= 2027.7 \times 950\\ &= 1{,}926{,}300 \text{ MJ/h}\end{aligned}$$

$$\begin{aligned}\dot{m}_f &= \frac{\dot{E}_r}{E_r}\\ &= \frac{1926300 \times 10^3}{34082}\\ &= 56{,}520 \text{ kg/h}\end{aligned}$$

3 Gas flow.

$$\begin{aligned}m_g &= m_f + m_a - m_{\text{ash}} - m_C\\ &= 1.0 + 11.85 - 0.08 - 0.01\\ &= 12.76 \text{ kg/kg coal}\end{aligned}$$

$$\begin{aligned}\dot{m}_g &= \dot{m}_f \times m_g = 56520 \times 12.76\\ &= 721{,}195 \text{ kg/h}\end{aligned}$$

Water content of the gas

$$\begin{aligned}m_w &= 9 \times m_H = 9 \times 0.0501\\ &= 0.451 \text{ kg/kg coal}\end{aligned}$$

$$\frac{0.451}{12.76} = 0.0353$$

$= 3.53$ percent water in the flue gas

4 Absorption fraction.

Reference 2 provides data for estimating the value of C contained in Eq. 8.2. For this example, $C = 0.0849$.

The projected area of the energy-absorbing surface is taken equal to the wall and roof areas of the furnace.

$$A_p = 1044.3 \text{ m}^2$$

$$f_{ab} = \frac{1}{1 + \left[C\frac{\dot{m}_g}{\dot{E}_r}\left(\frac{\dot{E}_r}{A_p}\right)^{1/2}\right]}$$

$$= \frac{1}{1 + \left[0.0849\frac{721,195}{1,926,300}\left(\frac{1,926,300}{1044.3}\right)^{0.5}\right]}$$

$$= 0.423$$

5 Gas temperature at the furnace outlet.

Energy in the gas

$$\dot{E}_g = (1 - f_{ab})\dot{E}_r$$
$$= 0.577 \times 1,926,300$$
$$= 1,111,475 \text{ MJ/h}$$

$$E_g = \frac{\dot{E}_g}{\dot{m}_g} = \frac{1,111,475}{721,195}$$
$$= 1.541 \text{ MJ/kg} = 1541 \text{ kJ/kg gas}$$

$$E_r = \frac{\dot{E}_r}{\dot{m}_g} = \frac{1,926,300}{721,195}$$
$$= 2.671 \text{ MJ/kg gas}$$

$$E_{ab} = f_{ab}E_r = 0.423 \times 2671$$
$$= 1130 \text{ kJ/kg gas}$$

The energy in the gas is equivalent to the enthalpy of the gas measured above 25 C. The temperature of the gas at the furnace outlet corresponds to $h = 1541$ kJ/kg, for a water vapor content of 3.5 percent. Thus, from Table 8.1,

$$T_{g,o} = 1593 \text{ K} = 1320 \text{ C}$$

The energy release rate can also be expressed as 1845 MJ/h per square meter of furnace energy-absorbing area, or the equivalent of 163 kilo-Btu/h·ft^2.

Figure 24, page 4–18, in Ref. 1 shows an approximate relationship between the furnace exit gas temperature and the energy release, expressed in kilo-Btu/h·ft^2 furnace area.

$$\text{For } \dot{E}_r = 163 \text{ kilo-Btu/h·ft}^2$$
$$T_{g,o} = 2520 \text{ F} = 1380 \text{ C}$$

The two values of 1320 C and 1380 C represent reasonably good agreement for the approximate methods employed for the determination of the furnace exit-gas temperature.

The effect on furnace performance caused by changes in the firing rate can be observed in Fig. 8.10. In order to simplify the calculations, the excess air, combustion efficiency, air preheat, and furnace conditions are assumed not to vary with changes in the firing rate. The increase in the furnace outlet-gas temperature with an increase in the boiler load is an important factor in superheater design and operation. A decrease in the furnace temperature, which accompanies a decrease in the boiler load, is significant with respect to maintaining stable combustion at light loads.

Preheated combustion air promotes higher furnace temperatures and improved combustion. Air preheat temperatures normally prescribed for firing fossil fuels are given in Section 8.7

8.13 FURNACE PERFORMANCE

The performance of a boiler furnace is measured by determining the combustion efficiency and the effectiveness of energy absorption. The combustion efficiency is defined as the ratio of the energy released in the furnace, excluding air preheat, to the higher heating value of the fuel.

Because the furnace energy-absorbing temperature, that of the water in the furnace walls, is substantially above the reference or room tem-

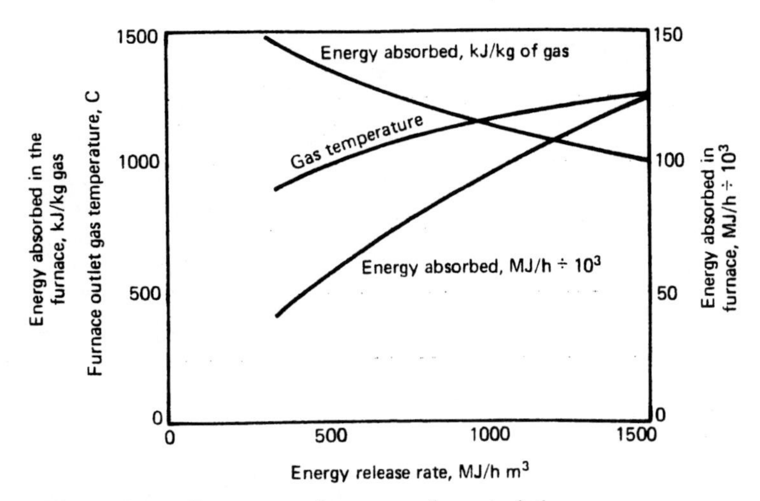

Figure 8.10 Furnace performance characteristics.

perature, a portion of the energy in the combustion gases is unavailable for absorption in the furnace. The availability ratio e_{av} is defined as the sensible energy in the gas above the energy-absorbing temperature divided by the energy released, including the air preheat. The numerator of this ratio, the available energy, is equal to the energy released minus the energy in the gas at the energy-absorbing temperature T_{ab}. The energy values are measured above room temperature.

$$e_{av} = \frac{E_r - h_{ab}}{E_r} = 1 - \frac{h_{ab}}{E_r} \quad (8.3)$$

The effectiveness of absorption, η_{ab}, is taken as the ratio of the energy absorbed in the furnace to the energy available.

$$\eta_{ab} = \frac{E_{ab}}{E_r - h_{ab}} = \frac{E_{ab}}{e_{av} E_r} \quad (8.4)$$

Example 8.3

Data taken from Example 8.2 will be used to evaluate the furnace performance factors. The energy-absorbing temperature T_{ab} is 280 C.

(a) Combustion efficiency.

$$\eta_B = \frac{\text{energy released, excluding air preheat}}{\text{HHV}}$$

$$= \frac{29936}{31670} = 0.945$$

(b) Availability ratio.
At 280 C, 3.5 percent water vapor

$$h_{ab} = 270 \text{ kJ/kg gas}$$

(from Table 8.1)

$$E_r = 2671 \text{ kJ/kg gas}$$

$$e_{av} = 1 - \frac{h_{ab}}{E_r} = 1 - \frac{270}{2671}$$

$$= 0.899$$

(c) Effectiveness of absorption.

$$E_{ab} = 1130 \text{ kJ/kg gas}$$

$$\eta_{ab} = \frac{E_{ab}}{e_{av} E_r} = \frac{1130}{0.899 \times 2671}$$

$$= 0.471$$

A low value of the effectiveness of absorption is not necessarily an indication of inferior performance. The absorption effectiveness of 0.471

TABLE 8.1 Enthalpy of Flue Gas, Above 25 C, in kJ/kg[a]

Temp. (K)	Percent by Mass of Water Vapor 0	2	4	6	8	10	12
300	2	2	2	2	2	2	2
400	103	105	106	108	110	112	113
500	206	210	213	217	220	224	228
600	312	317	323	328	334	339	344
700	421	428	435	443	450	457	465
800	532	542	551	560	569	579	588
900	646	658	669	681	692	703	715
1000	763	776	790	804	817	831	844
1100	882	897	913	929	945	961	977
1200	1002	1021	1039	1057	1075	1093	1112
1300	1125	1145	1166	1187	1207	1228	1249
1400	1248	1272	1295	1318	1342	1365	1388
1500	1374	1399	1425	1451	1477	1503	1529
1600	1500	1528	1557	1586	1614	1643	1672
1700	1627	1658	1690	1721	1753	1784	1816
1800	1755	1789	1824	1858	1892	1927	1961
1900	1884	1921	1958	1996	2033	2070	2108
2000	2013	2053	2094	2134	2175	2215	2256

[a] *Energy of evaporation of the water vapor is not included in the enthalpy values of the gas. Dry gas:* $\overline{m} = 30.215$.

for the furnace described in Example 8.3 defines the fraction of the available energy that is absorbed in the furnace and the fraction (1 − 0.471) of this energy that is not absorbed in order to provide a heat source for the downstream convective elements of the boiler. A proper division of the energy absorbed in the furnace and in the convective elements is highly important to boiler design.

8.14 SUPERHEATERS

Steam superheating is an accepted power plant practice, readily justified by improved economy and improved plant thermal efficiency. Superheating reduces the number of turbine stages that operate in the saturated steam region where droplet losses contribute to decreased machine efficiency. A reduction in blade erosion and in turbine maintenance will be achieved by a decrease in the moisture content of the steam.

Experience in the operation of superheaters has demonstrated the necessity of delivering to the superheater steam that contains minimum amounts of solids and liquid. As noted in Section 8.3, this objective has been accomplished by installing within the boiler drum a steam separator. Under normal conditions of separation, steam enters the superheater with a liquid content that does not exceed 0.25 percent.

Maximum steam temperatures at the superheater outlet are currently about 540 C. Because the resistance to heat flow through the superheated steam film is of significant magnitude, the temperature of the tube wall is somewhat higher than the bulk temperature of the steam. For metal temperatures that do not exceed 455 C, low-carbon steel can generally be used for fabrication of superheater tubes. When metal temperatures in excess of 455 C are encountered, an appropriate alloy steel is normally prescribed.

Superheaters are classified as convective and radiant types. Boilers are ordinarily equipped with convective superheaters, while the use of radiant superheaters is somewhat limited. The convective superheater is located adjacent to the furnace in order to absorb energy from the high-temperature gases discharged from the combustion zone. As previously noted, the temperature of the gas is controlled by the furnace energy absorption. In certain designs, the convective superheater is screened from the furnace radiation by a bank of evaporating tubes. The screen also protects the superheater against slagging. Slag particles are intercepted by the cooler tubes of the screen, where bridging is less serious than in the superheater that is constructed of closely spaced tubes, 50.8 or 63.5 mm O.D.

Convective superheaters in large power boilers are ordinarily split into two sections. See Figs. 8.3 and 8.4. Final energy absorption occurs in the secondary superheater placed adjacent to the furnace. Initial superheating is achieved in the primary section that is located in a zone of lower gas temperature. The principal reason for splitting the superheater is to provide for steam reheating. The superheating and reheating elements are arranged in a pattern that will achieve effective energy absorption and maintain steam temperatures at the prescribed levels over a relatively wide variation in the boiler load.

The tubes of the radiant superheater are placed on the furnace wall, where the energy transfer from the gases to the tube surface is effected principally by radiation. The superheater tube bank may replace a section of the water wall tubes. In another arrangement, the superheater tubes are placed in a plane behind the larger water wall tubes that serve as a screen and prevent overheating of the superheater tubes.

Radiant superheaters are occasionally used in combination with convective superheaters and arranged for series flow of the steam through the two elements. In certain designs, the boiler furnace is divided into two differentially fired sections. The walls of one furnace, used solely for steam superheating, are lined with radiant superheater tubes.

Modern power boilers provide for steam reheating. The reheater is usually a convective element located in the gas flow circuit. See Figs. 8.3 and 8.4. For a particular power plant design, the reheat steam temperature is ordinarily equal to the superheat steam temperature. Reheaters and superheaters are similar in construction and, in general, function in a like manner.

The general requirements of high efficiency and low maintenance in the operation of a power plant can, in part, be met through close control of the superheat and reheat steam temperatures. A reduction in the turbine efficiency will be experienced when the machine is operated with the steam temperatures below the design temperatures. On the other hand, steam temperatures that exceed the design values cannot be tolerated because of the adverse effect on the materials used in the construction of the boiler and the turbine.

Superheaters and reheaters are usually convective-type elements located in the gas stream. Figure 8.10 indicates that the gas temperature at the furnace outlet increases with an increase in the furnace firing rate. Further, the gas mass flow rate varies directly with the boiler load. The combined effect of these two factors is to cause the heat transfer rate to the convective elements to increase more rapidly than the increase in the steam flow rate. Thus, the net effect is an increase in the steam temperature with an increase in the boiler load. See Fig. 8.11. In the absence of temperature control, the design temperature of 540 C, for this particular illustration, is attained only for operation at 65 percent of the rated load. At higher boiler loads, the heat transfer surface is too large, while at lower loads the surface is inadequate with respect to maintaining the design steam temperature.

Several methods are available for the control of boiler superheat steam temperatures. The most effective methods are

Gas bypass.

Attemperation.

Excess air.

Gas recirculation.

Combination of radiant and convective superheaters.

Differentially fired divided furnace.

Burner selection.

Tilting burners.

Above the critical load, 65 percent of the rated boiler load in Fig. 8.11, steam temperatures may be controlled through bypassing a portion of the gas around the superheater tube bank. The prin-

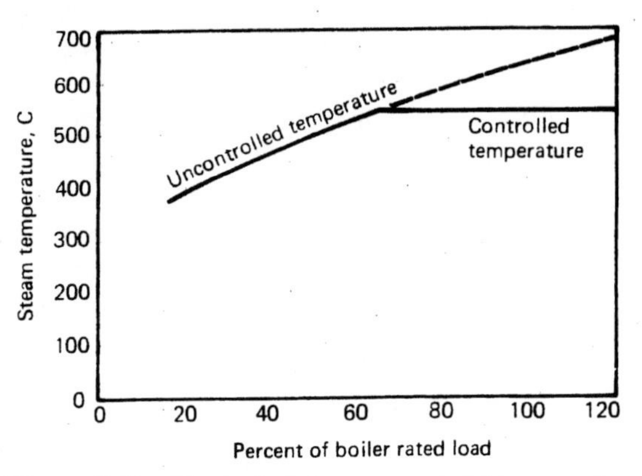

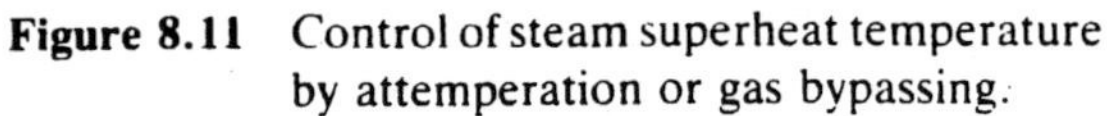

Figure 8.11 Control of steam superheat temperature by attemperation or gas bypassing.

cipal disadvantages of this method are the operating difficulties experienced with the movable dampers located in the high-temperature gas stream and the lack of quick response to load changes.

Effective temperature control can be achieved by attemperation of the steam at the point between the primary and the secondary superheaters. The controlled quantity of energy removed in the attemperator regulates the energy content, hence the temperature, of the steam at the outlet of the secondary superheater. One type of attemperator functions by injecting feedwater of high purity into the flowing steam. Subsequent evaporation of the water reduces the specific energy content of the steam passing to the secondary superheater. This type of attemperator may also be installed at the outlet of a single-element superheater.

Attemperation is also accomplished by transfer of energy from the steam to the boiler water or to the feedwater. A controlled quantity of steam flowing from the primary superheater is diverted through coils submerged in the boiler drum or through the tubes of a small feedwater heater. Subsequent to attemperation, the divided streams from the primary superheater recombine and pass to the secondary superheater for the final stage of energy absorption.

At fractional boiler loads, the outlet steam temperature from a convective-type superheater, installed in a drum-type boiler, can be increased by increasing the quantity of combustion air. Energy absorption in the furnace is reduced, and the gas flow through the superheater is increased. The resulting reduction in the boiler efficiency is offset by the higher turbine efficiency that is achieved by the increase in the steam temperature. This scheme for steam temperature control is not likely to be used for large power units.

Steam temperatures at the lighter boiler loads, for example, in Fig. 8.11 at less than 65 percent of the rated load, may be increased by recirculation of the flue gas from the economizer outlet to the bottom of the furnace. In general, furnace temperatures are lowered, hence the energy absorbed in the furnace is reduced. Recirculation of the flue gas also causes an increase in the mass flow of gas to the convective elements. The temperature of the gas entering the superheater is dependent upon a number of factors, thus the effect of flue gas recirculation on this characteristic is not readily apparent. Principally because of the increase in the mass flow of the gas, at a particular boiler load, the energy absorbed in the superheater will increase, and as a consequence an increase in the steam outlet temperature is achieved.

When the boiler load is increased, the rate of energy absorption by the furnace water walls does not increase as rapidly as the steam flow rate. Thus, a radiant superheater characteristically shows a decrease in the steam temperature with an increase in the boiler load. See Fig. 8.12. As previously noted, the convective superheater shows an opposite variation in the steam temperature with change in the boiler load. These opposing characteristics can be utilized in controlling the final steam temperature through a combination of properly proportioned superheaters of both types. As a result of series or parallel flow of the steam through the convective and radiant superheaters, the outlet steam temperature remains relatively constant within narrow limits over a wide range in the boiler output.

A divided-furnace boiler design is arranged for generation only of saturated steam in one section, while the superheater receives energy only

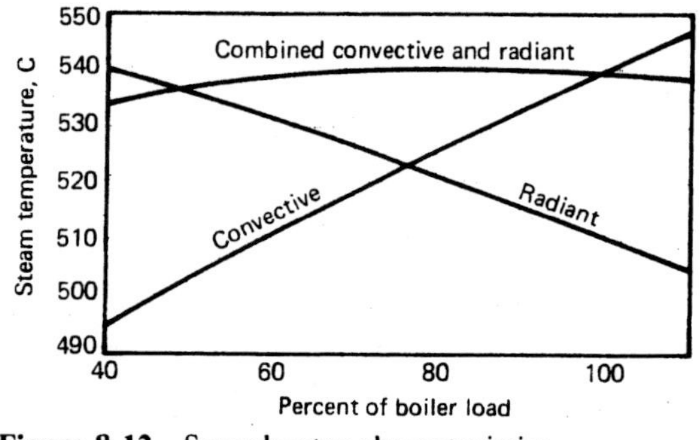

Figure 8.12 Superheater characteristics.

from the other section. The temperature of the superheated steam is regulated by altering the proportions of fuel fired in the two furnaces.

Burner positioning is a method used to decrease the furnace energy absorption and thus increase the energy in the gas that flows to the convective superheater and reheater. In principle, the burners are arranged in groups at different furnace levels. All of the burners are operated to achieve a high boiler output. As the boiler load is reduced, burners are removed progressively from service until at the lightest load only the burners in the top row remain in operation. The elevation of the firing zone at partial loads reduces furnace energy absorption through a decrease in the absorption effectiveness of the lower furnace zones. Tilting burners can be used to achieve the same results. The elevation of the firing zone is altered by a change in the direction of the flame travel within the furnace.

8.15 ECONOMIZERS AND AIR HEATERS

Steam generators are usually equipped with an economizer and an air heater. See Figs. 8.2 to 8.4. The economizer absorbs energy from the combustion gases for the purpose of preheating the feedwater prior to delivery of the water to the evaporating element. Gases discharged from the superheater and the reheater are cooled in the economizer and air heater to the lowest level of temperature consistent with acceptable power plant practice. The extent of the economizer surface is, in part, dependent upon the requirements for air preheating; that is, the flue gas leaving the economizer must contain sufficient energy for preheating the combustion air to the prescribed temperature level.

The location of the economizer in the boiler structure and the extent of the economizer surface will vary with the different boiler designs, particularly as these designs are influenced by the steam generating capability of the unit. Steaming economizers can be advantageously included in some boiler installations.

The energy-absorbing temperature is lower in a nonsteaming economizer than in the evaporating element, an important characteristic with respect to the effective transfer of energy from the gas to the water. Installation of the economizer is however justified only when it can absorb a prescribed quantity of energy at less cost than other types of heat transfer surface.

The air heater has the particularly important function of supplying preheated combustion air to the furnace. It may also be considered as a heat recovery element because of the low energy-absorbing temperature of the air, the receiver fluid.

The efficiency of a steam generating unit increases with a reduction in the outlet gas temperature or with the corresponding increase in the combustion air temperature. In general, for each 25 C increase in the air temperature, an increase of about 1 percent will be achieved in the boiler efficiency. Preheated air is also essential in pulverized coal-fired units for drying the coal during the grinding operation.

Because of the low overall coefficient of heat transfer, air heaters are relatively bulky structures. However, the differential pressure between the air and gas streams is small, hence a light-weight construction can be employed. Power plant air heaters are designed in accordance with either the recuperative or the regenerative principle of heat transfer.

The boiler shown in Fig. 8.4 is equipped with a tubular-type air heater, a recuperative-type heat exchanger. In order to facilitate removal of deposits from the tube surfaces, the gas is passed through the tubes, while the air flow is directed across the tube bank.

Figure 8.13 shows several arrangements for power plant tubular air heaters. Some of these arrangements are particularly applicable when the height of the air heater is limited. The upper center and upper right-hand figures show the air flowing across the tube bank in three passes, a flow pattern that improves the heat transfer characteristics of the recuperator.

For the two air heater arrangements shown in

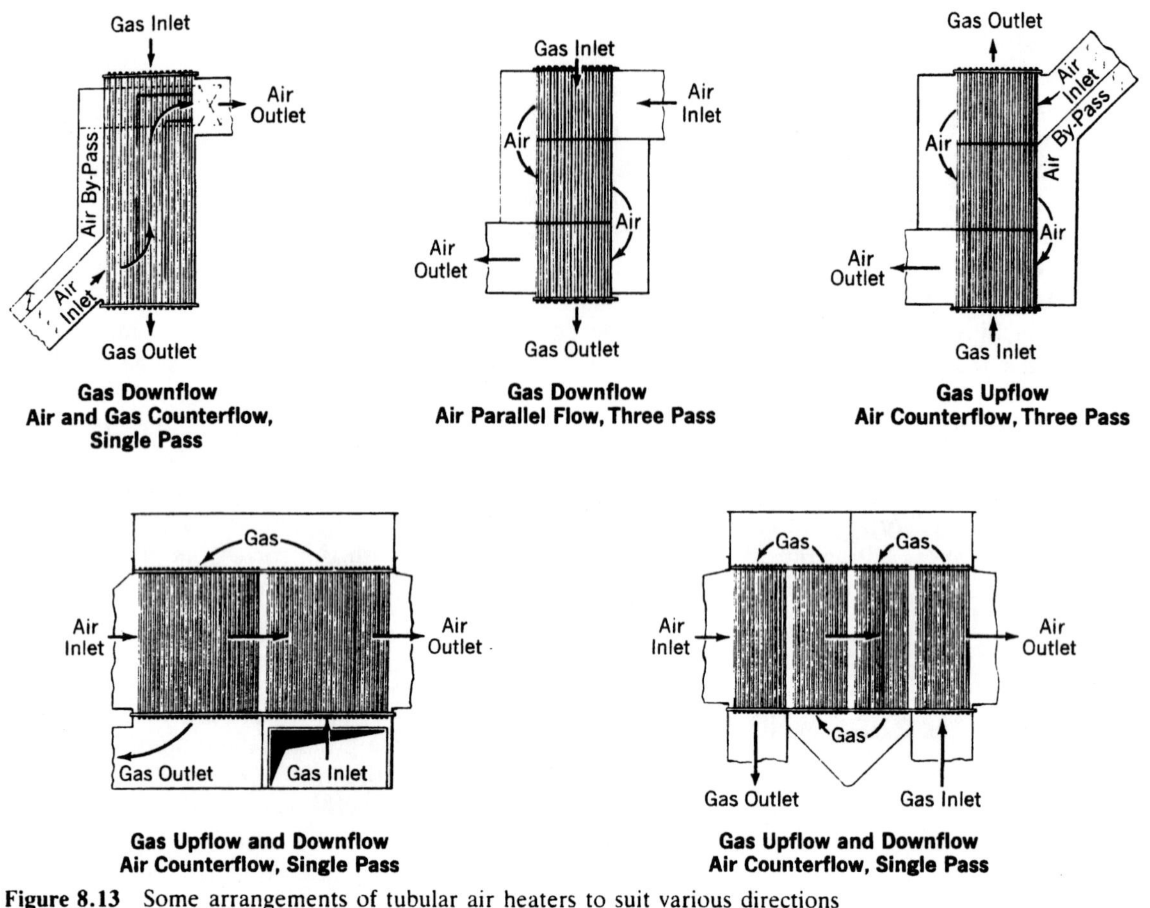

Figure 8.13 Some arrangements of tubular air heaters to suit various directions of gas and air flow.[1] (Courtesy Babcock & Wilcox.)

the upper center and upper right-hand sections of Fig. 8.13, counterflow of the gas and air, in comparison with parallel flow of the two fluids, achieves somewhat more effective heat transfer. Parallel flow of the fluids however yields, for prescribed gas and air temperatures, a higher minimum metal temperature, an important factor in avoiding condensation of water vapor contained in the flue gas. The minimum metal temperature can also be increased, when the air heater operates at light loads, by directing a portion of the inlet air through the cold-air bypass. In this case, there is an increase in the temperature of the gas leaving the heat transfer surface.

Regenerative-type air heaters are used extensively in power plant construction. In comparison with the tubular-type air heater, the regenerative-type air heater is, for similar operating conditions, considerably lighter and more compact. In one design, the rotor turns slowly at 3 rpm and moves the heat transfer elements, composed of crimped steel plates, through the separated streams of gas and air. See Fig. 8.14. Provision is usually made for counterflow of these two fluids. Energy is transferred from the gas to the plates and then, upon rotation of the plates into the air stream, from the plates to the air. The principal disadvantage encountered in this type of air heater is the leakage of air into the gas. Leakage occurs at the radial seals and through

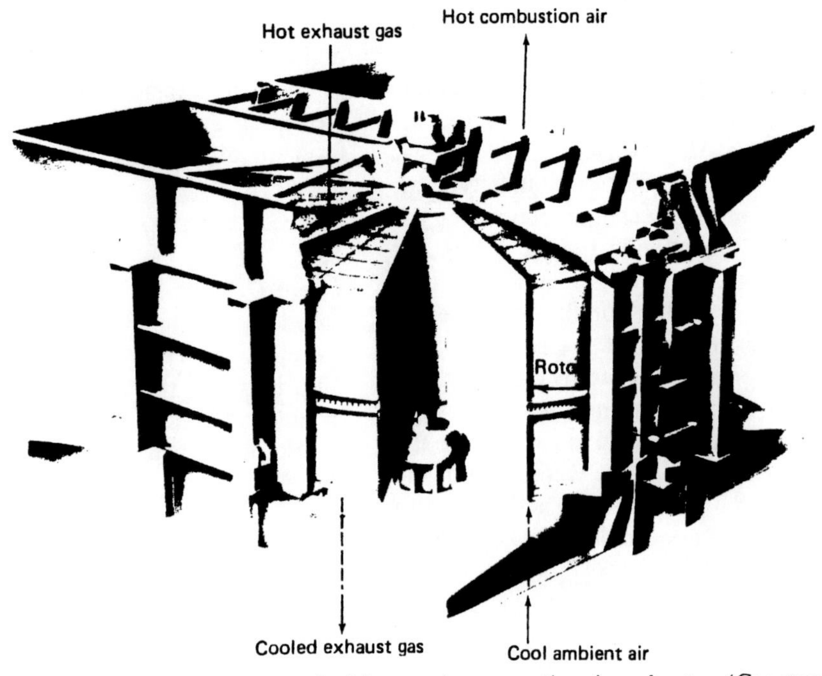

Figure 8.14 Ljungstrom® vertical-inverted regenerative air preheater. (Courtesy C-E Air Preheater.)

the annular space between the rotor shell and the housing. Leakage is also caused by entrainment of air in the rotor passages.

Other designs are employed for the construction of air heaters. A plate-type recuperative air heater provides for the flow of gas and air through alternate spaces between the plates.

In an alternative arrangement, the regenerative-type air heater is equipped with a stationary, cylindrical element comprised of the energy storage plates. The element is placed in the gas stream, and an air hood rotating across the element at about 1 rpm effects alternate flow of the air and gas through the passages formed by the plates.

Care must be exercised in operating the boiler, particularly at light loads, to avoid reducing the temperature of the combustion gases, in contact with the economizer and air heater surfaces, below the dew point temperature of the gas. Should this temperature reduction occur, the oxides of sulfur in the gas will combine with the condensed water vapor to form a weak but corrosive acid that is likely to attack the steel heat transfer surfaces.

Sulfur trioxide present in the flue gas causes an increase in the dew point temperature of the gas. The minimum metal temperature at which economizers and air heaters operate will be influenced by the sulfur content of the fuel and the method of firing the fuel. For an increase in the sulfur content of the fuel from 1 to 5 percent by mass, the minimum metal temperature increases approximately as follows[1]: residual oil firing, 115 to 125 C; natural gas firing, 105 to 125 C; coal, stoker fired, 105 to 125 C; coal, pulverized or burned in a cyclone furnace, 70 to 82 C.

8.16 BOILER HEAT TRANSFER

Energy is transferred from the combustion gases to the surface of the boiler elements by radiation and convection in a varying proportion. Except for the furnace elements, the principal means to transferring energy is by convection, although the screen tubes and the superheater will absorb a certain amount of energy radiated directly from the furnace gases. The several convective elements, including the superheater, reheater, evaporating section, and economizer, are arranged for normal, transverse flow of the gas through the tube bank. For a tubular-type air heater, as noted in the preceding section, the gas flows inside the tubes while the air is directed usually in a transverse flow over the tubes.

Heat transfer from the combustion gases to the tube surface for each of the several boiler convective elements is effected by the combined mechanisms of convection and intertube radiation. In comparison with the resistance of the gas film, the resistance of the water film in the screen tubes, in the evaporating section, and in the economizer is very small, and hence it is customarily neglected in the evaluation of U_o, the overall coefficient of heat transfer. The resistance of the steam film in the superheater and in the reheater and the resistance of the air film in the air heater are, on the other hand, relatively high and cannot be neglected.

The procedures used for the determination of the heat transfer film coefficients are somewhat involved and outside the scope of this text. Some representative values for the overall coefficient of heat transfer U_o for the boiler convective elements are

Screen	55 $W/m^2 \cdot K$
Superheater and reheater	50
Evaporating section	55
Economizer	75
Air heater	25

Variations in the values of U_o are observed for different boiler designs. The above values however are generally indicative of steam boiler performance.

Overall coefficients of heat transfer for boiler convective elements are of comparatively low magnitude, hence extensive heat transfer surfaces are required. Consequently, steam boilers are relatively large and bulky structures, but are operated effectively with a high degree of availability and reliability.

The heat transfer rate for a boiler convective element is determined by applying the basic heat flow equation

$$\dot{q} = U_o A_o \, \Delta T_m \tag{8.5}$$

where U_o is the overall heat transfer coefficient based on A_o, the area of the external surface of the element. The mean temperature difference ΔT_m between the two fluids is evaluated from the inlet and outlet temperatures of the two fluids by use of Eq. 8.6.

$$\Delta T_m = F_c \frac{\Delta T_1 - \Delta T_2}{\ln(\Delta T_1/\Delta T_2)} \tag{8.6}$$

Equation 8.6, with the omission of F_c, is derived for counterflow of the two fluids. The equation is also applicable to parallel flow. There are however many types of heat exchangers in which cross flow of the fluids occurs. For such cases, the counterflow log-mean temperature difference, $(\Delta T_1 - \Delta T_2)/\ln(\Delta T_1/\Delta T_2)$, is modified by the introduction of an appropriate cross flow correction factor F_c, where $F_c < 1$.

The terminal temperature differences ΔT_1 and ΔT_2 are identified in Fig. 8.15 for four types of heat exchangers. Figure 8.15a is applicable to the boiler screen and to the convective evaporating element. In both of these two sections of the boiler, the receiver fluid, water, is maintained essentially at a constant temperature. For this case, the cross flow correction factor F_c = 1.0.

The superheater, reheater, and economizer are designed for general counterflow or parallel flow of the two fluids, and Figs. 8.15b and 8.15c are, respectively, applicable to these two cases. The gas flows across the tube bank, but the receiver fluid, steam or water, makes several passes in flowing through the tubes. F_c = 1.0 for either counterflow or parallel flow through these elements.

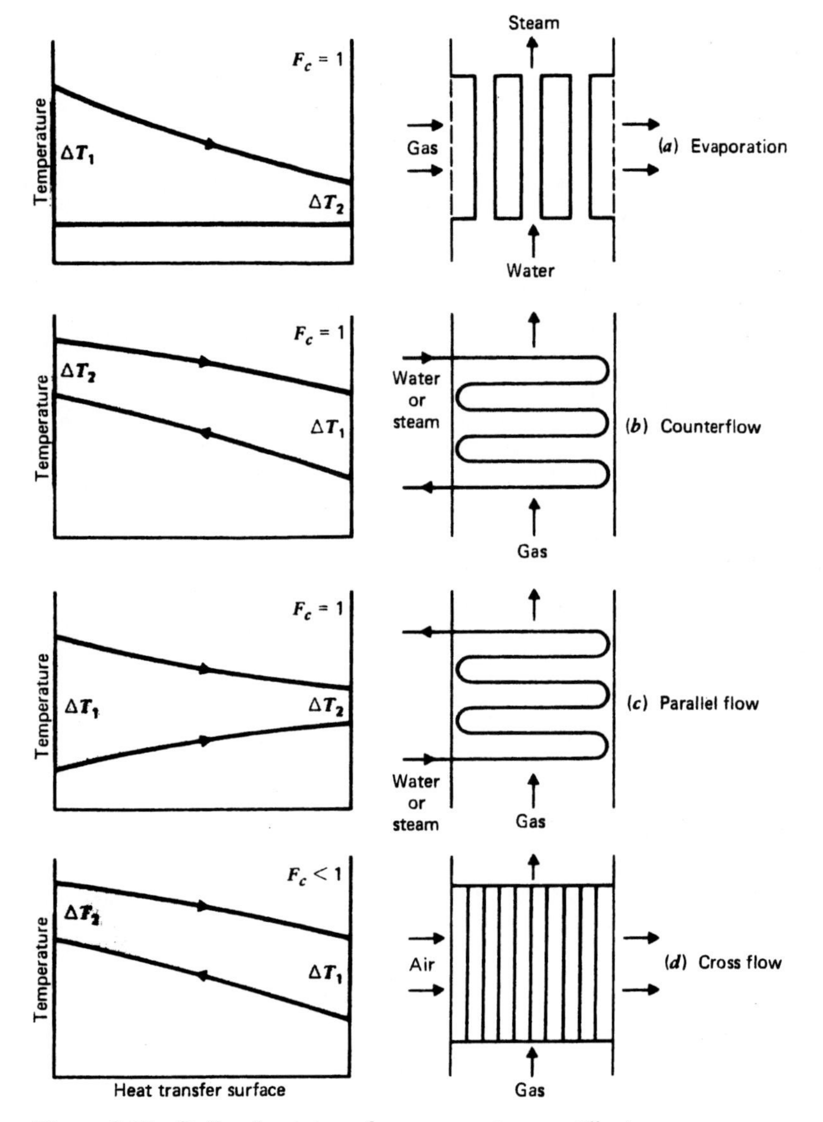

Figure 8.15 Boiler heat transfer temperature gradients.

Counterflow of the fluids is normally employed for achieving the most effective use of the heat transfer surface. In certain designs however a parallel-flow arrangement is used to obtain less variation in the metal temperature along the tube. For prescribed terminal temperatures of the source and receiver fluids and a prescribed heat flow rate, parallel flow, in comparison with counterflow, will show a lower value of ΔT_m, and hence a somewhat greater heat transfer surface will be required.

Figure 8.13 shows a number of arrangements for recuperative air heaters. Two of these arrangements will be examined. The upper center illustration depicts the gas flow through the tubes while the air flows across the tubes in three passes. For this type of air heater, three or four air passes will ordinarily be used. Either a parallel-flow or a counterflow arrangement will be used. See Fig. 8.13. Because the gas flows through the tubes in a single pass, $F_c = 1.0$. This would not be true if the gas were to flow through the tubes in two, or more, passes.

Insertion of the baffles in the air heater reduces the shell-side flow area, and consequently the velocity of the air is increased. The higher air velocity causes an increase in the air film heat transfer coefficient, and consequently an increase in U_o is achieved. Installation of the baffles thus improves the heat transfer rate in the air heater.

The schematic illustration in Fig. 8.15d shows the gas flowing through the tubes, say, for an air heater, while the air flows in a single pass across the tubes. Because of the cross flow of the air through the tube bank, $F_c < 1$. Values of F_c may be obtained from several sources, including Ref. 1, page 4–13.

Considerable variation can be observed in the value of F_c, depending upon the air heater effectiveness, which is defined as the ratio of the increase in the air temperature to the difference between the inlet gas temperature and the inlet air temperature. In general, for a prescribed amount of air preheat, any change in the gas temperature level that causes an increase in the heater effectiveness will, as a result, cause a decrease in the value of F_c.

Example 8.4

An economizer is arranged in 48 parallel loops, that is, there are 48 tubes in each horizontal row. The tube O.D. is 50 mm, and the tube length is 3.50 m. The unit is designed for counterflow of the gas and water.

Water temperatures: entering 112 C, leaving 147 C

Gas temperatures: entering 439 C, leaving 329 C

Gas flow: 134,000 kg/h

$U_o = 65 \text{ W/m}^2\cdot\text{K}$

Water vapor content of the gas: 6 percent by mass

Determine the number of tubes in each vertical row or loop.

Enthalpy values of the gas are selected from Table 8.1.

At 712 K, $h = 457$ kJ/kg gas (6 percent water vapor)

At 602 K, $h = 330$ kJ/kg gas

$$\dot{q} = \dot{m}_g(h_{g.i} - h_{g.o}) = 134{,}000(457 - 330) = 17.018 \times 10^6 \text{ kJ/h}$$
$$4.727 \times 10^6 \text{ W}$$

$$\Delta T_1 = T_{g.i} - T_{w.o} = 439 - 147 = 292 \text{ C} \quad \text{(see Fig. 8.15b)}$$

$$\Delta T_2 = T_{g.o} - T_{w.i} = 329 - 112 = 217 \text{ C}$$

$$\Delta T_m = \frac{292 - 217}{\ln(292/217)} = 252.6 \text{ C}$$

$$\dot{q} = U_o A_o \, \Delta T_m$$

$$4.727 \times 10^6 = 65 A_o \, 252.6$$
$$A_o = 287.90 \text{ m}^2$$

Tube area for a single horizontal row of tubes:

$$(A_o)_h = N_h \pi (\text{O.D.}) L = 48\pi 0.050 \times 3.50 = 26.39 \text{ m}^2$$

Number of tubes in a vertical row:

$$N_v = \frac{A_o}{(A_o)_h} = \frac{287.90}{26.39} = 10.91 \quad \text{say } 11$$

8.17 BOILER DRAFT LOSS

For a small boiler installatio· the natural draft created by a stack or chir y is sufficient to motivate the flow of air ı gas through the boiler. Substantially higher fluid pressure losses are encountered in power plant boilers, hence the movement of air and combustion gases is effected principally by fans with some assistance provided by the stack.

The power required for operation of the draft fan is directly related to the pressure drop, or draft loss, that occurs in the gas or air flowing through a boiler tube bank. The velocity of the fluid in the tube bank is particularly significant, because high velocities generally promote good heat transfer characteristics, but also cause increased pressure losses. In the design of a boiler, it is essential to establish an appropriate balance between these two effects through selection of the optimum fluid velocity.

The pressure loss in a gas stream flowing transversely through a tube bank is determined from

$$\Delta P_f = \frac{fNG^2}{(6.34 \times 10^8)\rho} \text{ cm water at 25 C*} \quad (8.7)$$

where

N = number of tube rows in the direction of flow
G = mass velocity of the fluid, kg/h·m², based on the minimum flow area
ρ = fluid density, kg/m³
f = friction factor

The friction factor f is dependent upon the Reynolds number and on the tube diameter, alignment, and the spacing in both directions. Correlation of the friction factor with test data is provided in chart form. See Fig. 8.16.

For an in-line tube arrangement, the gas density ρ, in Eq. 8.7, is evaluated at temperature T_d.

$$T_d = T_t \pm 0.9\, \Delta T_m \quad (8.8)$$

where

T_t = mean temperature of the tube wall
ΔT_m = mean temperature difference between the gas and the tube wall

and the plus sign stands for gas cooling by heat transfer to the tube surface and the minus sign, for the opposite direction of heat flow.

The additional pressure loss caused by a change in the direction of the gas flow is equivalent to $1.0h_v$ for a 90° bend and $1.35h_v$ for a 180° bend, where h_v is the velocity head expressed in cm water.

For the vertical flow of hot gases through tube banks, a correction is made for the "stack effect," or difference in the static head resulting from the buoyant effect of the column of hot gas.

$$\Delta P_s = \frac{Lg}{97.76}(\rho_a - \rho_g) \text{ cm water, at 25 C} \quad (8.9)$$

The net draft loss for gas flow through a convective tube bank is then given by

$$\Delta P = \Delta P_f \pm \frac{Lg(\rho_a - \rho_g)}{97.76} \text{ cm water, at 25 C} \quad (8.10)$$

* Equation 8.7 is a modification of the Fanning equation

$$\Delta P = 4f\rho\left(\frac{C^2}{2}\right)\frac{L}{D}$$

where ΔP is the pressure loss in a circular conduit. The friction factor f is a function of Reynolds number. The gas flow across a tube bank of N rows is considered to be equivalent to the gas flow through a series of N orifices. The length/diameter ratio L/D of the conduit is replaced by N, and $G/\rho = C$. Then,

$$\Delta P = \frac{2fNG^2}{\rho}$$

G is usually given in kg/h·m², and ρ = kg/m³. Then,

$$\Delta P = \frac{2fNG^2}{(3600)^2\rho} = \frac{fNG^2}{6.48 \times 10^6\,\rho} \text{ Pa}$$

At 25 C: 1 cm water = 97.78 Pa
Thus,

$$\Delta P = \frac{fNG^2}{6.34 \times 10^8\,\rho} \text{ cm water at 25 C}$$

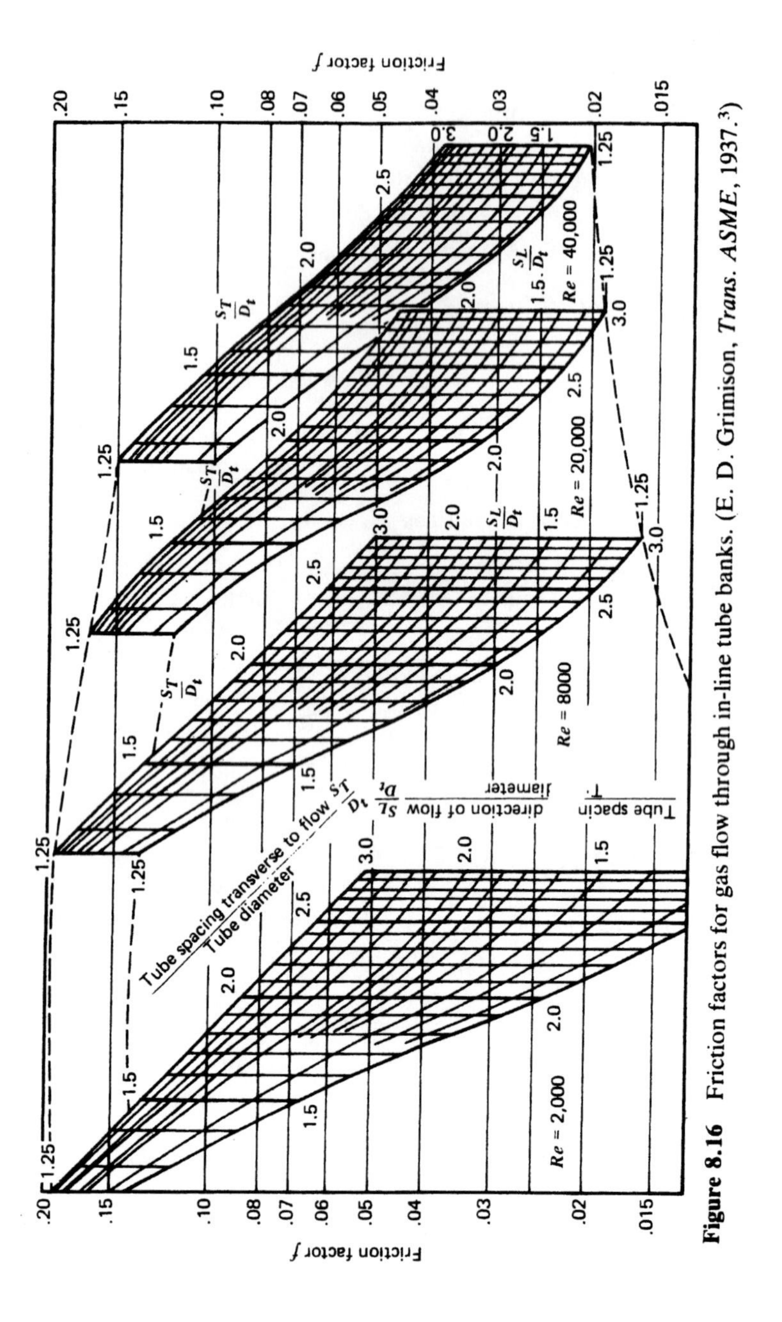

Figure 8.16 Friction factors for gas flow through in-line tube banks. (E. D. Grimison, *Trans. ASME*, 1937.[3])

where

ΔP_f = draft loss because of fluid friction, cm water
L = vertical distance between the points of measurement, m
g = acceleration owing to gravity, m/s^2
ρ_a = density of the ambient air, kg/m^3
ρ_g = average density of the gas, kg/m^3

For downward flow of the gas through the tube bank, the sign in Eq. 8.10 is positive. A negative sign is used for upward flow of the gas.

Example 8.5

Construction of the economizer examined in Example 8.4 provides for an in-line arrangement of the tubes in the direction of the gas flow. The tube center-to-center spacing in both directions is 75 mm, R_g = 0.2813 kJ/kg·K. Determine the draft loss for downward flow of the gas through the economizer.

The tube metal temperature is assumed to be equal to the water temperature. Then, $(T_t)_{mean}$ = 130 C. The mean gas temperature $(T_g)_{mean}$ = 384 C. Then,

$$T_d = T_t + 0.9\,\Delta T_m$$
$$= 130 + 0.9(384 - 130)$$
$$= 359 \text{ C } (632 \text{ K})$$
$$P_g = 96.5 \text{ kPa, abs (assumed)}$$
$$\rho_g = \frac{P_g}{R_g T_d} = \frac{96.5}{0.2813 \times 632}$$
$$= 0.5428 \text{ kg/m}^3$$

The minimum distance $(S_T - D_o)$ between adjacent tubes is 25 mm.

The minimum flow area is determined from

$$A_{min} = N_T L\,(S_T - D_o)$$
$$= 48 \times 3.50 \times 0.025$$
$$= 4.200 \text{ m}^2$$

$$G_{max} = \frac{\dot{m}_g}{A_{min}} = \frac{134000}{4.200}$$
$$= 31{,}905 \text{ kg/h·m}^2$$
$$= 8.862 \text{ kg/s·m}^2$$

The viscosity of the gas at temperature T_d is 2.960×10^{-5} N·s/m^2.

The viscosity of the flue gas was calculated by combining the viscosity values of the components of a typical gas mixture.

$$\text{Re} = \frac{G_{max} D_o}{\mu} = \frac{8.862 \times 0.050}{2.960 \times 10^{-5}} = 14{,}970$$

From Fig. 8.16,

For Re = 14,970,

$$\frac{S_T}{D_o} = 1.5, \quad \frac{S_L}{D_o} = 1.5$$
$$f = 0.079$$

$$\Delta P_f = \frac{fNG^2}{6.34 \times 10^8\,\rho}$$
$$= \frac{0.079 \times 11(31905)^2}{6.34 \times 10^8 \times 0.5428}$$
$$= 2.57 \text{ cm water}$$

Stack effect:

P_a = 97.85 kPa, abs
T_a = 22 C (assumed ambient conditions)

$$\rho_a = \frac{P_a}{R_a T_a} = \frac{97.85}{0.287 \times 295}$$
$$= 1.1557 \text{ kg/m}^3$$

$$L = N_L S_L = 11 \times 0.075 = 0.825 \text{ m}$$

$$\Delta P_s = \frac{Lg}{97.76}(\rho_a - \rho_g)$$
$$= \frac{0.825 \times 9.807}{97.76}(1.1557 - 0.5428)$$
$$= 0.051 \text{ cm water}$$

Total draft loss (for downward flow of the gas):

$$\Delta P = \Delta P_f + \Delta P_s = 2.57 + 0.05$$
$$= 2.62 \text{ cm water}$$

8.18 BOILER PERFORMANCE

Large steam generators are automatically controlled for achieving optimum conditions of operation. Although it is beyond the scope of this text to present a detailed discussion on boiler operation, the boiler control system can be briefly described. Steam flow from the boiler is controlled by the turbine governor that responds to load changes on the generator. A variation in the steam flow from the boiler causes changes in the fuel firing rate, air flow rate, and the feedwater flow rate. In this manner, the prescribed boiler operating pressure is maintained, in conjunction with the regulation of the superheat and reheat steam temperatures.

The absorption effectiveness η_{ab} for a single convective element in the boiler is determined by

$$\eta_{ab} = \frac{h_{g,i} - h_{g,o}}{h_{g,i} - h_{g,r}} \tag{8.11}$$

where

$h_{g,i}$ = enthalpy of the gas, above 25 C, entering the element

$h_{g,o}$ = enthalpy of the gas, above 25 C, leaving the element

$h_{g,r}$ = enthalpy of the gas, above 25 C, at the temperature of the receiving fluid entering the element

The numerator in Eq. 8.11 represents the energy absorbed in the element, while the denominator represents the energy available above the lowest absorbing temperature in the element.

Example 8.6

Determine the absorption effectiveness for the economizer examined in Example 8.4.

Gas entering the economizer:

$$T = 712 \text{ K}, \; h_g = 457 \text{ kJ/kg}$$

Gas leaving the economizer:

$$T = 602 \text{ K}, \; h_g = 330 \text{ kJ/kg}$$

Water entering the economizer:

$$T = 385 \text{ K}, \; h_g = 92 \text{ kJ/kg}$$

$$\eta_{ab} = \frac{h_{g,i} - h_{g,o}}{h_{g,i} - h_{g,r}}$$

$$= \frac{457 - 330}{457 - 92} = 0.348$$

The absorption effectiveness for the economizer cannot be arbitrarily prescribed, because the inlet and outlet gas temperatures are influenced, respectively, by the energy transferred in the upstream elements of the boiler and by the energy requirements for the downstream air heater.

The absorption effectiveness is a useful factor in design for comparing the performance of a proposed boiler element with that of a similar element in an existing boiler. In practice, a change in the absorption effectiveness of an element in an operating boiler could signify a decline in the boiler performance, possibly because of an accumulation of deposits, such as soot, dust, or scale, on the heat transfer surface.

The greater portion of the energy in the fuel fired to the steam generator is absorbed by the water, while the remainder of the energy, a loss, is not recovered. In order to analyze the performance of the steam generator, it is essential to subdivide this loss into a number of basic items, some of which are capable of reduction and some over which the station operators have little or no control. The steam generator energy balance, based on the higher heating value of the fuel, consists of eight items that represent the energy absorbed and the seven loss factors.

The steam generator losses result principally from incomplete combustion of the fuel, and from an outlet gas temperature that exceeds the inlet air temperature. If it were possible to lower the stack gas temperature to the level of the inlet air temperature, burn the fuel completely, and eliminate the external heat loss, the steam generator efficiency would closely approach 100 percent. In analyzing the loss items, it is helpful to consider each one as a quantity of energy that would be recovered if the above ideal conditions were achieved.

The steam generator energy balance consists of the following items.

1 Energy aborbed by the water.

$$e_1 = \frac{\dot{m}_s(h_s - h_w) + (\dot{m}_{rh}\,\Delta h_{rh})}{\dot{m}_f} \quad (8.12a)$$

where

$\dot{m}_s$ = steam flow at the superheater outlet, kg/h
$\dot{m}_{rh}$ = steam flow at the reheater outlet, kg/h
$\dot{m}_f$ = fuel firing rate, kg/h
h_s = enthalpy of the steam at the superheater outlet, kJ/kg
h_w = enthalpy of the entering feedwater, kJ/kg
Δh_{rh} = enthalphy change of the steam flowing through the reheater, kJ/kg

2 Energy in the dry stack gas.

$$e_2 = m_{dg}c_{p.g}(T_g - T_a) \text{ kJ/kg fuel} \quad (8.12b)$$

where

m_{dg} = mass of dry gas, kg/kg fuel
$c_{p.g}$ = specific heat of the dry gas, usually taken as 1.004 kJ/kg·K
T_g = outlet, or stack, gas temperature, C
T_a = inlet air temperature, C

3 Energy in the stack gas attributed to the moisture in the fuel.

$$e_3 = m_m(h_v - h_f) \text{ kJ/kg fuel} \quad (8.12c)$$

where

m_m = mass of water in the fuel, kg/kg fuel
h_v = enthalpy of superheated water vapor in the stack gas at T_g and the partial pressure of the water vapor, kJ/kg
h_f = enthalpy of liquid water at T_a, kJ/kg

Note: Because of the low partial pressure of the water vapor in the combustion gas mixture, h_v is virtually independent of pressure and can be evaluated at a representative pressure of 10 kPa, abs.

4 Energy in the stack gas attributed to the water vapor formed by combustion of the hydrogen in the fuel.

$$e_4 = 9m_H(h_v - h_f) \text{ kJ/kg fuel} \quad (8.12d)$$

where

m_H = net hydrogen in the fuel available for combustion, kg/kg fuel
h_v = enthalpy of superheated water vapor in the stack gas at T_g and the partial pressure of the water vapor, kJ/kg
h_f = enthalpy of liquid water at T_a, kJ/kg

5 Energy in the stack gas attributed to superheating the water vapor in the combustion air.

$$e_5 = m_a\omega c_{p.v}(T_g - T_a) \text{ kJ/kg fuel} \quad (8.12e)$$

where

m_a = mass of dry combustion air, kg/kg fuel
ω = specific humidity of the air, kg vapor per kg dry air
$c_{p.v}$ = specific heat of water vapor, usually taken as 1.88 kJ/kg·K
T_g = outlet, or stack, gas temperature, C
T_a = inlet air temperature, C

Items 2 to 5 represent the energy in the dry gas and water vapor that would be recovered by a reduction in the gas temperature from T_g to T_a. The magnitude of item 5 is comparatively small.

6 Loss attributed to incomplete combustion and indicated by the carbon monoxide in the stack gas.

$$e_6 = \frac{CO}{CO_2 + CO} m_{C.b}(23{,}560) \text{ kJ/kg fuel*} \quad (8.12f)$$

* Equation 8.12f is derived as follows:

$$12(CO_2 + CO) = \text{kg carbon burned per 100 mol dry gas}$$

$$\frac{12(CO_2 + CO)}{m_{C.b}} = \text{kg fuel fired per 100 mol dry gas}$$

$$28\ CO = \text{kg CO per 100 mol dry gas}$$

$$10{,}100 = \text{heating value of CO, kJ/kg}$$

$$\frac{28\,CO}{[12(CO_2 + CO)]/m_{C.b}} 10{,}100 = \frac{CO}{CO_2 + CO} m_{C.b} 23{,}560 \text{ kJ/kg fuel}$$

where

CO = carbon monoxide in the stack gas, mol/100 mol dry gas
CO_2 = carbon dioxide in the stack gas, mol/100 mol dry gas
$m_{C,b}$ = mass of carbon burned, kg/kg fuel

7. Loss attributed to unburned carbon.

$$e_7 = m_{C,ub}(32{,}750) \text{ kJ/kg fuel} \tag{8.12g}$$

where

$m_{C,ub}$ = mass of unburned carbon, kg/kg fuel
32,750 = heating value of carbon, kJ/kg

Items 6 and 7 represent the energy that would be recovered by complete combustion of the fuel.

8 R & U loss.

$$e_8 = (\text{HHV of the fuel}) - \sum_1^7 e \text{ kJ/kg fuel} \tag{8.12h}$$

The R & U item includes the loss attributed to heat transfer by radiation and convection from the steam generator and the loss resulting from unburned fuel components, namely, hydrogen, hydrocarbons, and carbon particles, that are not detected by the test procedure employed. In addition, the errors incurred in the evaluation of the other seven items appear in the R & U loss.

The heat loss from the steam generator is not readily measured, and a "standard" value can be used. Normally, the heat loss is equal to approximately 1 percent of the energy supplied by the fuel. In conducting a boiler test, precise measurements of some quantities are difficult to achieve. For acceptable test results, the unaccountable portion of the R & U loss should be within 1 to 2 percent.

The efficiency η_B of the steam generator is equal to the ratio of the energy absorbed by the water to the higher heating value of the fuel. Thus,

$$\eta_B = \frac{\dot{m}_s(h_s - h_w) + (\dot{m}_{rh}\,\Delta h_{rh})}{\dot{m}_f(\text{HHV})} \tag{8.13}$$

where

$\dot{m}_s$ = steam flow at the superheater outlet, kg/h
$\dot{m}_{rh}$ = steam flow through the reheater, kg/h
$\dot{m}_f$ = fuel firing rate, kg/h
h_s = enthalpy of the steam leaving the superheater, kJ/kg
h_w = enthalpy of the feedwater entering the boiler, kJ/kg
Δh_{rh} = enthalpy change of the steam flowing through the reheater, kJ/kg
HHV = higher heating value of the fuel, kJ/kg*

The principal objectives of the steam generator energy balance are the determination of the overall efficiency of the unit and the distribution of the various losses. An evaluation of each loss

* The thermal efficiency of a power plant or engine is defined generally as the ratio of the useful output to the corresponding energy supplied by the fuel. The energy supplied by the fuel is not an inherent quantity, and either the higher or lower heating value of the fuel is customarily used for the determination of the thermal efficiency. The difference between the higher and lower heating values of a fuel is equal to the energy of condensation of the water vapor that results from the combustion reaction. The combustion gases for thermal power plants and engines are discharged at temperatures above the condensation temperature of the water vapor. Because the exhaust gases contain water in the vapor phase, there is reason to charge the power-producing system only with the energy equivalent to the lower heating value of the fuel. Such practice has been adopted for internal combustion engines and gas turbine engines.

A similar concept may be extended to fossil fuel-fired steam power plants, but the established practice is to charge the power plant with the higher heating value of the fuel. The reason for using the higher heating value is primarily historic. The heating value of coal, a long-time power plant fuel, is determined in a bomb calorimeter and customarily reported as the higher heating value. In order to be consistent, the higher heating value of all fuels, solid, liquid, and gas, should be used in the determination of the thermal efficiency of steam power plants.

There is not a significant difference between a heating value determined at constant pressure and the corresponding heating value determined at constant volume.

In comparing the thermal efficiency values of different types of engines and power plants, it is important to observe the heating values on which the efficiency determinations are based.

will indicate the probability of improvement in performance that can be achieved through minor changes in design, or through changes in the operation of the unit. This is especially true for the initial test conducted subsequent to the erection of the boiler.

Boiler performance has a tendency to fall off somewhat with increased fouling of the heat transfer surfaces. Reduced performance is indicated by a gradual increase in the stack gas temperature.

In general, optimum performance of the boiler is achieved with a low outlet gas temperature, a minimum quantity of excess air, and a minimum quantity of combustible material discharged from the unit. Except for the effect caused by the fouling of the heat transfer surfaces, the outlet gas temperature is dependent upon the design of the unit and varies with the load on the boiler. Excess air should be held to a minimum quantity consistent with the practice of eliminating combustible components in the stack gas. Moisture losses, represented by items 3, 4, and 5, are inherent in the fuel and combustion air and hence can only be decreased indirectly by a reduction in the stack gas temperature.

An excessive amount of carbon in the refuse discharged from a stoker is principally the result of improper distribution and movement of the coal on the stoker and of faulty distribution of the primary air under the fuel bed. Improved operation of the stoker can reduce the carbon loss. In pulverized coal firing, the carbon loss is closely associated with the degree of fineness achieved in the grinding operation. This loss can also be controlled by modification of the operating procedure, including burner adjustment and air regulation.

Example 8.7

Determine the boiler energy balance and the boiler efficiency from the test data shown below. Steam reheating is not employed.

Steam flow: 200,000 kg/h at 4.0 MPa, abs, and 405 C

Feedwater conditions: 4.60 MPa, abs, and 162 C

Coal fired: 17,650 kg/h

Higher heating value of the coal: 33,330 kJ/kg

Refuse collected: 1382 kg/h

Combustible in the refuse: 19.2 percent

Combustion air: 14.0 kg dry air/kg fuel

0.013 kg water vapor/kg dry air

Combustion air temperature: 27 C

Stack gas temperature: 167 C

Barometric pressure: 96.5 kPa, abs

COAL ANALYSIS, MASS		ORSAT ANALYSIS VOLUMETRIC (PERCENT)	
C	0.8165	CO_2	14.0
H	0.0446	CO	0.2
O	0.0525	O_2	4.8
N	0.0127		
S	0.0104		
A	0.0633		
	1.000		

The total hydrogen and oxygen are reported in the coal analysis, wet basis.

Free moisture in the coal: 3.3 percent.

Dry gas

$$m_r = \frac{1{,}382}{17{,}650} = 0.0783 \text{ kg refuse/kg coal}$$

Wet gas
$$m_{wg} = m_f + (1 + \omega)m_a - m_r = 1.0 + (1.013 \times 14.0) - 0.078 = 15.104 \text{ kg/kg coal}$$

Water vapor contained in the gas
$$m_w = 9m_H + (m_w)_{air} = (9 \times 0.0446) + (14.0 \times 0.013) = 0.583 \text{ kg/kg coal}$$

Dry gas
$$m_{dg} = m_{wg} - m_w = 15.104 - 0.583 = 14.521 \text{ kg/kg coal}$$

Energy Balance

1 Energy absorbed in the boiler.

h_s at 4.0 MPa, abs, and 405 C = 3225.3 kJ/kg

h_w at 4.60 MPa, abs, and 162 C = 686.8 kJ/kg

$$e_1 = \frac{\dot{m}_s(h_s - h_w)}{\dot{m}_f}$$
$$= \frac{200{,}000(3225.3 - 686.8)}{17650}$$
$$= 28{,}764.9 \text{ kJ/kg coal}$$

2 Energy in the dry stack gas.

$$e_2 = m_{dg}c_{p,g}(T_g - T_a)$$
$$= 14.52 \times 1.004(167 - 27)$$
$$= 2040.9 \text{ kJ/kg coal}$$

3 Energy in the stack gas attributed to the moisture in the fuel.

h_v at 10 kPa, abs, and 167 C = 2815.8 kJ/kg

h_f at 27 C = 113.2 kJ/kg

$$e_3 = m_m(h_v - h_f)$$
$$= 0.033(2815.8 - 113.2)$$
$$= 89.2 \text{ kJ/kg coal}$$

4 Energy in the stack gas attributed to the water vapor formed by combustion of the hydrogen in the fuel.

$$\text{Net hydrogen} = 0.0446 - \frac{0.033}{9}$$
$$= 0.0409 \text{ kg/kg coal}$$

$$e_4 = 9m_H(h_v - h_f)$$
$$= 9 \times 0.0409(2815.8 - 113.2)$$
$$= 994.8 \text{ kJ/kg coal}$$

5 Energy in the stack gas attributed to superheating the water vapor in the combustion air.

$$e_5 = m_a\omega c_{p,v}(T_g - T_a)$$
$$= 14.0 \times 0.013 \times 1.88(167 - 27)$$
$$= 47.9 \text{ kJ/kg coal}$$

6 Loss attributed to incomplete combustion and indicated by the carbon monoxide in the stack gas.

Unburned carbon:

$$m_{C,ub} = m_r - m_{ash}$$
$$= 0.0783 - 0.0633$$
$$= 0.0150 \text{ kg/kg coal}$$

Carbon burned:

$$m_{C,b} = m_C - m_{C,ub}$$
$$= 0.8165 - 0.0150$$
$$= 0.8015 \text{ kg/kg coal}$$

$$e_6 = \frac{CO}{CO_2 + CO} m_{C,b}(23{,}560)$$
$$= \frac{0.2}{14.0 + 0.2} 0.8015 \times 23{,}560$$
$$= 266.0 \text{ kJ/kg coal}$$

7 Loss attributed to unburned carbon.

$$e_7 = m_{C,ub}(32{,}750)$$
$$= 0.0150 \times 32{,}750$$
$$= 491.2 \text{ kJ/kg coal}$$

8 R & U loss.

$$e_8 = \text{HHV} - \sum_1^7 e$$
$$= 33{,}330 - 32{,}695$$
$$= 635 \text{ kJ/kg coal}$$

Boiler Efficiency

$$\eta_B = \frac{m_s(h_s - h_w)}{\text{HHV}}$$
$$= \frac{e_1}{\text{HHV}} = \frac{28{,}765}{33{,}330} = 0.863$$

8.19 BOILER OPERATION

An extensive amount of auxiliary equipment is essential to the operation of the steam generator, namely, fuel measuring and handling equipment and mechanical and hydraulic systems for collecting and removing ash, dust, and soot. Facilities must be provided for proper chemical treatment of the boiler make-up water.

The flow of air and combustion gases through the boiler is achieved by the natural draft created by the stack and by mechanical draft developed by the forced- and induced-draft fans. The stack provides only a small portion of the draft required for the operation of a modern power boiler.

Furnace pressures, except for the pressurized designs, are maintained slightly below the atmospheric pressure in order to prevent leakage of gas into the boiler room. Forced-draft fans move the combustion air through the air heater to the stoker plenum chamber, or to the burner windbox, where a positive pressure is maintained. The induced-draft fan provides the pressure drop that causes the flow of gases from the furnace, through the several convective elements including the air heater, to the base of the stack. Prior to entering the stack, the combustion gases ordinarily pass through an element in which almost all of the particulate material is removed.

Some large boilers are equipped only with forced-draft fans. The entire unit operates with positive air and gas pressures, hence a gas-tight setting must be erected. Elimination of the induced-draft fan ensures lower capital and operating costs. Machine maintenance is reduced because fan erosion by dust-laden gas is avoided. Fan power requirements are reduced through substitution of cold-air compression for two-step compression of cold air and hot gas.

The formation of deposits on the heat transfer surfaces is important to the design and the operation of the boiler. Soot and ash deposited on the gas side and scale on the water side of a surface increase the resistance to heat flow and, under certain conditions, can cause an undesirable increase in the metal temperature. In addition, draft losses increase with the formation of deposits on the gas side of the boiler surfaces. Boiler operation is directed, in part, toward a reduction in the rate at which these deposits form. Gas-side surfaces are cleaned periodically during operation of the boiler, while scale is removed from the water-side surfaces at the time the boiler is shut down for maintenance. It is also essential to avoid a boiler water condition that will promote internal corrosion.

Soot and porous slag are readily removed from the boiler surfaces by steam- or air-operated blowers. These blowers, located at several points in the gas flow circuit, are commonly arranged for automatic sequential blowing. Retractable- or telescopic-type soot blowers are installed in the furnace and in other high-temperature zones of the boiler. The deposits of dense slag are usually removed by chilling the slag with water, either as a solid jet or mixed with steam or air.

Considerable attention has been directed to the cause and prevention of corrosion of the gas-side heat transfer surfaces in the boiler. Corrosion may be caused by the oxides of sulfur, an element present in coal and in most fuel oils. Certain metallic elements present in coal and oil ash are additional sources of corrosion.

Make-up water usually contains suspended material and dissolved salts and gases. The suspended material is customarily removed by filtration. Certain salts, principally those of calcium and magnesium, have a tendency to form a hard scale on the evaporating surfaces. In order to minimize the formation of such deposits, the make-up water is chemically treated in hot-process or zeolite softeners. The reaction may include precipitation, but in any event, the residual salts do not have a tendency to form a hard scale. Silica is especially troublesome in high-pressure operation, and particularly so in the supercritical-pressure boilers, because of the deposits that form on the surfaces of the superheater tubes and on the turbine blades. Silica removal is achieved in the water softening process.

A certain amount of water is discharged from the system, and steam is used for operation of auxiliary equipment, such as fuel-atomizing nozzles, air ejectors, and soot blowers. The make-up water, added continuously to the feedwater stream, compensates for the loss of steam and water from the system. Evaporation of the make-up water increases the concentration of solids in the boiler water, and blowdown is required in order to maintain this concentration below a prescribed limit. In principle, blowdown water, discharged as waste, is replaced with make-up water of lower solids concentration. As previously noted, carry-over of moisture from the steam drum to the superheater must be limited to about 0.25

percent. Solids contained in the water droplets, carried into the superheater, are a source of scale formation. Deposits on the tube wall occur when the droplets contact the surface and the liquid subsequently evaporates.

During shutdown for maintenance, boiler scale is removed either by chemical cleaning or by mechanical cutting or brushing. Certain soft deposits can be removed from the turbine blades by washing with water or saturated steam. The harder deposits are removed mechanically.

Certain dissolved gases, such as oxygen and carbon dioxide, cause internal corrosion. These gases are expelled from the system by deaeration of the feedwater in a deaerating heater or condenser. In a deaerating feedwater heater, the water, on contact with the steam, is raised to the boiling point where the solubility of the gases in water is essentially zero. Residual oxygen in the feedwater stream is removed by direct injection of sodium sulfite or hydrazine, chemicals that are effective oxygen scavengers. To guard against corrosion, the boiler water should be maintained at a minimum pH of 12 by addition of neutralizing chemicals.

8.20 CONDENSING SYSTEMS

Steam discharged from the turbine is directed into a condenser for two reasons. The condenser is operated at a high vacuum in order to create a low turbine exhaust pressure, ranging down to 12 mm mercury, abs. Turbines are ordinarily equipped with surface condensers that are indirect- or nonmixing-type heat exchangers. In the absence of mixing, the second function of the condenser can be realized, that is, the return of the condensate to the boiler. Because of the high steam flow, the condensate must be conserved, otherwise the operation of a large power boiler would be impracticable.

Barometric or jet-type condensers are often installed in low-capability power plants. These condensers are cheaper than the surface-type condenser, but because of the mixing of the steam and the cooling water in the condenser, the condensate cannot be directly returned to the boiler.

The shell-and-tube arrangement is used in the construction of the surface condenser. See Fig. 8.17. A large number of tubes, about 19 to 32 mm O.D., are installed between the tube sheets, each of which forms one side of a water box. In the single-pass design, the water enters at one box and flows through the tubes into the outlet box. For two-pass operation, the inlet box is divided in order to direct the water through one-half of the tubes. At the opposite end of the condenser, the flow is reversed for passage through the other half of the tubes.

Single-pass condensers require more cooling water but cause less resistance to water flow in comparison with the two-pass design. An electric power generating station, whenever feasible, is located adjacent to a natural source of cooling water sufficient for operation of a single-pass condenser.

The cooling water flow for a surface condenser is high, but the circulating pump is required only to develop sufficient head to accelerate the water and offset fluid friction in the conduits, piping, and condenser. The water side of the tubes is cleaned periodically in order to remove deposits of silt and other substances.

Surface condenser design provides for proper direction of the steam to all parts of the tube bank. The radial flow-type condenser, illustrated in Fig. 8.18, is designed for distribution of the steam in an arrangement that assures complete absence of inactive tube surfaces and a minimum pressure drop from the turbine exhaust to the air offtake.

Through cross flow and counterflow of the condensate and the steam, the condensate is continuously reheated with the resultant liberation of oxygen. Subcooling of the condensate is prevented by causing the liquid to drop through an active steam belt. Deaerated condensate can thus be delivered at the saturation temperature corresponding to the condenser pressure.

Various noncondensable gases are contained in the power plant steam/water fluid that is circulated through the system. The existence of these gases can be traced to several sources. The make-up water supplied to the system may contain a

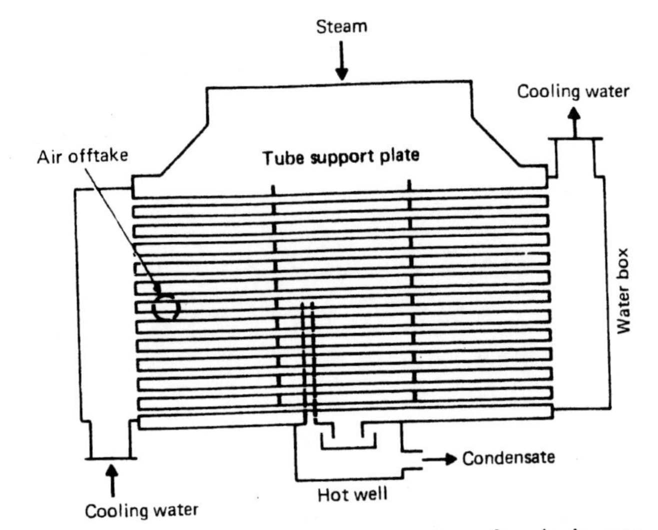

Figure 8.17 Schematic representation of a single-pass surface condenser.

relatively high fraction of dissolved noncondensable gases. In addition, noncondensable gases are released in internal chemical reactions, and noncondensable gases enter the system through air leaks where the internal pressure is below the atmospheric pressure. The noncondensable gases enter the condenser along with the turbine exhaust steam and will, unless continuously removed, quickly accumulate in the condenser and increase the turbine back pressure.

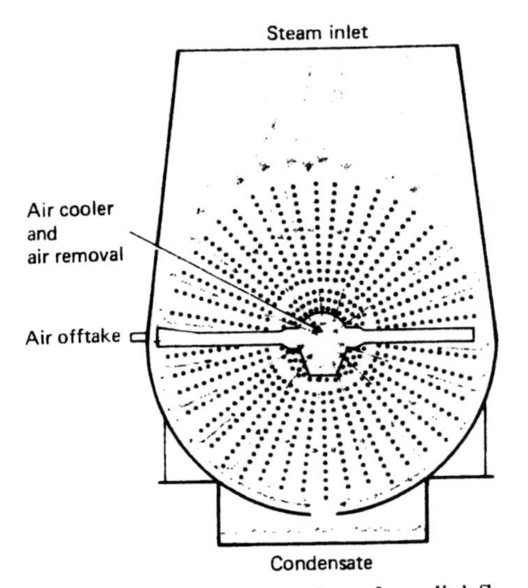

Figure 8.18 Cross-sectional view of a radial-flow surface condenser. (Courtesy Westinghouse Electric Corporation.)

The existence of noncondensable gases in the steam has a negative effect on the film heat transfer coefficient for condensing steam. Only a small quantity of these gases in the steam will cause a substantial reduction in the magnitude of this coefficient. Incidentially, the condensing film heat transfer coefficient for uncontaminated steam, determined experimentally, is within the general range of 8500 to 20,000 W/m^2·K. For industrial applications, a value of 11,500 W/m^2·K is commonly used.

The noncondensable gases are removed from the condenser by steam jet ejectors or by rotary vacuum pumps. In order to reduce the work expended in pumping these gases to the atmospheric pressure, the gas removal point is placed at the coldest region of the condenser.

The overall coefficient of heat transfer for the surface condenser is evaluated from the summation of the steam film, tube wall, and water film resistances. The conductance of the steam film can be taken at 11,500 W/m^2·K. The conductance of the water film can be determined through use of the Colburn equation.

$$h = 0.023\frac{k}{D}(\mathrm{Re})^{0.8}(\mathrm{Pr})^{0.33} \qquad (8.14)$$

where

- h = the conductance of the water film
- k = thermal conductivity of the water
- D = internal diameter of the tubes in the condenser
- Re = Reynolds number
- Pr = Prandtl number of the water

Fluid properties in Eq. 8.14 are evaluated at the average bulk temperature of the fluid.

Example 8.8

The cooling water flow to a steam surface condenser is 170 m³/min. The water velocity in the tubes is 2.0 m/s. The inlet water temperature is 27 C. The terminal temperature difference ttd is 5 C. The steam flow to the condenser is 134,000 kg/h. The energy transferred from the condensing steam is 2210 kJ/kg. Tubes: Admiralty metal, 22.22 mm O.D., 1.47 mm wall thickness. Cleanliness factor $F_c = 0.85$. Determine the required area of the heat transfer surface.

Cooling Water Temperature Increase

$$\dot{q}_s = \dot{m}_s \Delta h_s = 134{,}000 \times 2210$$
$$= 296.14 \times 10^6 \text{ kJ/h}$$
$$82.261 \times 10^6 \text{ W}$$

Density of the water = 996 kg/m³

c = 4.17 kJ/kg·K

$$\dot{q}_w = \dot{m}_w c(T_o - T_i)_w = \dot{q}_s$$

$$\dot{q}_w = (60 \times 170 \times 996)4.17(T_o - T_i)_w$$
$$= 296.14 \times 10^6$$

$$(T_o - T_i)_w = 7.0 \text{ C}$$

Mean Temperature Difference

$$T_{w.o} = T_{w.i} + (T_o - T_i)_w$$
$$= 27 + 7 = 34 \text{ C}$$

$$T_s = T_{w.o} + ttd$$
$$= 34 + 5 = 39 \text{ C}$$

$$\Delta T_1 = 12 \text{ C} \qquad \Delta T_2 = 5 \text{ C}$$

$$\Delta T_m = \frac{\Delta T_1 - \Delta T_2}{\ln(\Delta T_1/\Delta T_2)}$$

$$= \frac{12 - 5}{\ln\left(\frac{12}{5}\right)} = 8.0 \text{ C}$$

Steam Film Resistance

$$R_s = \frac{1}{h_s} = \frac{1}{11{,}500}$$
$$= 0.08696 \times 10^{-3} \text{ m}^2\text{·K/W} \qquad \text{(based on O.D.)}$$

Tube Wall Resistance

For Admiralty metal, k = 112.4 W/m·K

$$R_t = \frac{D_o}{D_m}\frac{L}{k}$$

$$= \frac{22.22}{20.75}\frac{1.47 \times 10^{-3}}{112.4}$$
$$= 0.01400 \times 10^{-3} \text{ m}^2\text{·K/W} \qquad \text{(based on O.D.)}$$

Water Film Resistance

At $T_{b.av}$ (30.5C): ρ = 996 kg/m³

c_p = 4.17 kJ/kg·K

(note table A.7)

μ = 0.795×10^{-3} N·s/m²

k = 0.617 W/m·K

$$D_i = 0.01928 \text{ m}$$

$$\text{Re} = \frac{\rho DC}{\mu}$$
$$= \frac{996 \times 0.01928 \times 2}{0.795 \times 10^{-3}} = 48{,}309$$

$$\text{Pr} = \frac{c_p \mu}{k}$$

$$= \frac{(4.17 \times 10^3)(0.795 \times 10^{-3})}{0.617}$$

$$= 5.373$$

Applying Eq. 8.14,

$$h = 0.023 \frac{k}{D}(\text{Re})^{0.8}(\text{Pr})^{0.33}$$

$$= 0.023 \frac{0.617}{0.01928}(48309)^{0.8}(5.373)^{0.33}$$

$$= 7163 \text{ W/m}^2\cdot\text{K}$$

$$R_w = \frac{D_o}{D_i}\frac{1}{h_w}$$

$$= \frac{22.22}{19.28}\frac{1}{7163}$$

$$= 0.16089 \times 10^{-3} \text{ m}^2\cdot\text{K/W}$$

(based on O.D.)

Total Resistance

$$R_o = R_s + R_t + R_w = 0.26185 \times 10^{-3} \text{ m}^2\cdot\text{K/W}$$

Overall Conductance

$$U_o = \frac{1}{R_o} = \frac{1}{0.26185 \times 10^{-3}}$$

$$= 3819 \text{ W/m}^2\cdot\text{K}$$

Applying a Cleanliness Factor

$$U_{o,c} = F_c \times U_o = 0.85 \times 3819 = 3246 \text{ W/m}^2\cdot\text{K}$$

Heat Transfer Surface

$$A_o = \frac{\dot{q}}{U_{o,c}\,\Delta T_m}$$

$$= \frac{82.261 \times 10^6}{3246 \times 8.0} = 3168 \text{ m}^2$$

The cleanliness factor allows for a reduction in the overall coefficient of heat transfer because of possible fouling of the water-side surface of the condenser tubes.

An increase in the water temperature causes a decrease in the viscosity of the water and consequently a decrease in the resistance of the water film. In turn, the decrease in the resistance of the water film effects an increase in the value of U_o for the condenser.

The overall coefficient of heat transfer used in the design of condensers for steam turbines can be obtained from *Standards for Steam Surface Condensers* published by the Heat Exchange Institute.[4] The design manual also provides data on the pressure loss in the water side of the condenser. Such data are useful in the determination of the pumping power.

Three principal objectives are observed in the design of a surface condenser. These interrelated objectives are (1) to maintain a low condensing temperature for the steam, (2) to avoid an excessively large heat transfer surface, and (3) to avoid high pumping costs. The condenser, in conjunction with the turbine, is designed in accordance with the requirement to achieve the most economical operation under normal conditions of loading.

The overall coefficient of heat transfer U_o is strongly influenced by the velocity of the water flowing through the tubes. An increase in U_o, caused by an increase in the water velocity, effects a decrease in the heat transfer area; and, as a consequence, the capital cost of the condenser and the annual fixed charges are reduced.

Fluid losses and pumping costs however increase sharply with an increase in the water velocity. The optimum water velocity is determined through an appropriate balance between the fixed charges and the pumping cost for a prescribed loading schedule for the condenser. See Fig. 8.19.

The temperature at which the steam condenses is influenced by the cooling water temperature. Consequently, both the inlet water temperature and the temperature increase of the water are important factors. Seasonal variations normally occur in the inlet water temperature, although some control of this temperature can be exer-

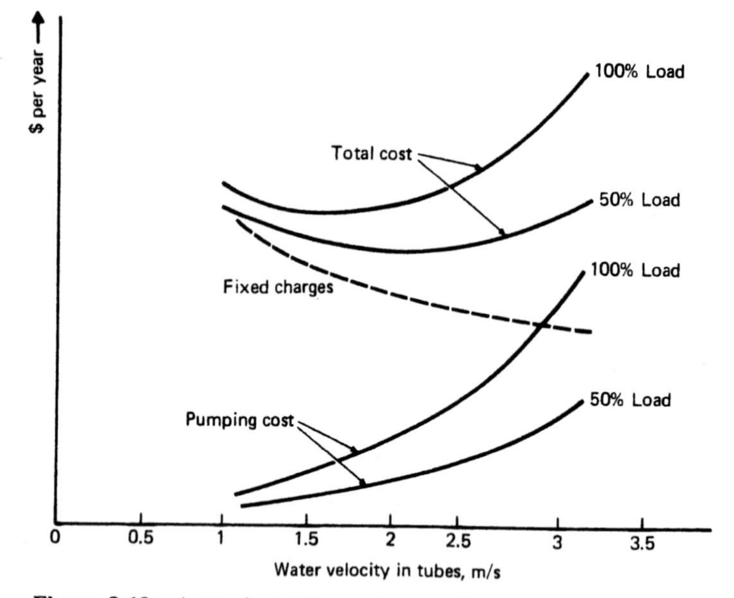

Figure 8.19 Annual cost for a steam surface condenser.

cised when the ultimate heat rejection from the power plant to the surroundings is achieved in a cooling tower. Appreciable temperature variations may occur in a natural body of water that is the source of the condenser cooling water. In general, a low inlet water temperature is highly desirable.

The temperature increase in the water passing through the condenser is regulated by the flow rate. This temperature increase can be reduced by increasing the cooling water-to-stream mass flow ratio, but higher pumping costs will be incurred. For the particular condenser, the increase in pumping power results from an increase in the mass flow rate of the water and from the increase in the fluid losses caused by the higher water velocity.

8.21 FEEDWATER HEATERS

Boiler feedwater is preheated by passage through a series of heaters supplied with steam extracted from the turbine. Except for a single deaerating heater, these feedwater heaters are of the surface or nonmixing type. In the closed heater, the water passes through a number of parallel circuits consisting of straight tubes or U-tubes. Provision for expansion is made by fastening both ends of the U-tube to the same tube sheet; or, when straight tubes are used, one of the tube sheets is installed in a floating head. Within the heater, the steam is directed by baffles in several cross flow passes over the tube bank, while the water, because of the tube arrangement, flows in two or more passes through the heater, which is constructed for horizontal or vertical mounting.

The basic design of the heater may be modified to include a drain-cooling section and a desuperheating section. The drain-cooling section, located at the bottom of the heater, is separated by a seal from the upper condensing section. Condensate from the upper section is directed by baffles through the cooling section to the outlet from the heater. In comparison with the discharge of condensate near the saturated-steam temperature, this provision for drain cooling achieves certain improvements in the performance of the heater. Because of subcooling below the saturation temperature, less flashing occurs when the condensate cascades to the next heater that is supplied with steam at a lower pressure. As a result, the loss in availability is reduced and the thermal efficiency of the steam cycle is improved. Also, corrosion–erosion effects are reduced because of decreased flashing of the condensate.

A desuperheating zone, arranged around the outlet end of the tubes, can be included in those heaters supplied with steam extracted from the superheat section of the turbine. Because of the higher temperature of the source fluid in this zone, the temperature of the feedwater leaving the heater exceeds the temperature that can be achieved by heat transfer from saturated steam.

The heat transfer characteristics of the closed feedwater heater are somewhat more complex than those of the surface condenser, because energy is not transferred in all heaters solely from saturated steam to the feedwater. Inclusion of drain-cooling and desuperheating sections will establish three zones in which the source fluid exists in different phases, namely, a superheated vapor, a two-phase saturated mixture, and a liquid. In the condensing section however, the performance of the feedwater heater does not differ significantly from that of the main condenser with respect to the water velocity and the overall coefficient of heat transfer.

For design conditions, a final terminal temperature difference of 3 C is commonly prescribed. The 3° difference is established between the saturated-steam temperature and the temperature of the water leaving the heater. A negative final terminal temperature difference can occur because of steam desuperheating.

The deaerating heater is an open- or mixing-type heat exchanger in which the extracted steam and the feedwater are in direct contact. The heater operates somewhat above atmospheric pressure and serves to expel the noncondensable gases from the feed stream through a vent condenser to the atmosphere. At the saturation state attained by the feedwater in the heater, the solubility of the gases in water is very low.

8.22 STEAM POWER CYCLES

The principal pieces of equipment in the steam power plant, namely, the boiler, turbine-generator, condenser, and feedwater heaters, are interconnected by the piping system to effect cyclic flow of the working fluid. For the high-capability electric power plants, the design of the system is based on the regenerative-reheat cycle. See Fig. 8.20. A low-capability power plant would omit the reheat and, in comparison, employ fewer feedwater heaters. Double steam reheat is used advantageously in the supercritical-pressure power plants.[5]

The superheat and reheat steam temperatures, respectively T_1 and T_3 in Fig. 8.20, are usually of like magnitude. Steam reheating 2–3 is achieved with some pressure drop, about 350 to 400 kPa, in the reheater and piping. Both superheating and reheating the steam improves the thermal efficiency of the power plant. Heat addition at the high temperature increases the available energy fraction of the heat supplied to the working fluid. Further, the thermal performance of the turbine is improved because of reduced moisture content of the steam expanding in the lower-pressure stages of the machine.

The improved thermal performance of the power plant achieved by high boiler pressures is realized in operating the subcritical-pressure systems at pressures ranging upward to 20 MPa and the supercritical-pressure systems in the range of 25 to 35 MPa. At the bottom of the cycle, condenser pressures are maintained at the lowest

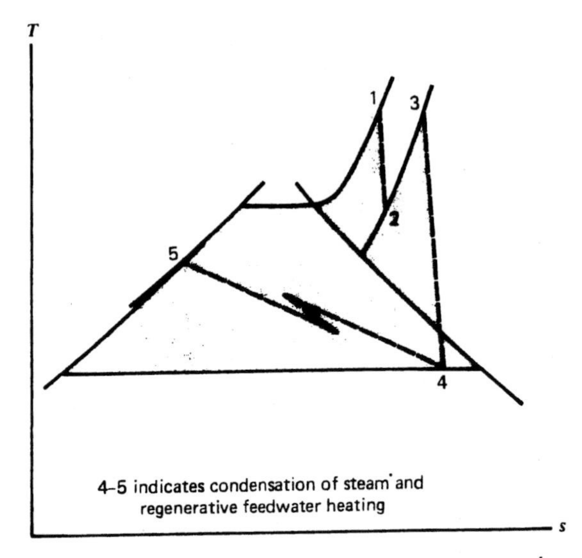

Figure 8.20 Regenerative-reheat steam power cycle.

level that can be economically attained with the available cooling medium.

Water, the commonly employed working fluid for the vapor cycle power plant, is significantly deficient in one respect, namely, the inherently high specific heat of liquid water. An appreciable amount of the heat supplied to the working fluid, above the condensing temperature, is transferred to the water in the liquid phase. An increase in the thermal efficiency of the cycle would be observed if the heat transfer to the liquid water were achieved by internal regeneration.

A hypothetical regenerative cycle can be devised by assuming an arrangement in which the water discharged from the condenser is directed, in counterflow with the steam, through a jacket placed around the turbine casing. The turbine is supplied with saturated steam. While the steam expands in the turbine, reversible heat transfer is achieved between the steam and the feedwater flowing through the jacket. Because of heat transfer, the feedwater enters the boiler at the temperature of the saturated steam. Thermodynamics textbooks demonstrate that, for operation between prescribed temperature limits, the hypothetical regenerative steam power cycle achieves a thermal efficiency equal to that of the Carnot cycle.

Heat transfer through the turbine casing, from the expanding steam to the feedwater, is impracticable; hence in practice the extraction cycle is used effectively for improving the thermal performance of the steam power plant. In an actual power plant, the feedwater, prior to entering the boiler, is heated in a series of heaters that are supplied with steam extracted at successive points from the turbine. Figure 8.21 shows the flow diagram for a four-stage extraction cycle.

In order to calculate the thermal efficiency of the regenerative-reheat cycle, the extracted steam flows from the turbine to the heaters must be determined. Because the flow rate of the working fluid is not the same through all components of the system, it is not particularly helpful to attempt to depict an extraction cycle on a *T–s* diagram. A flow diagram such as the one shown

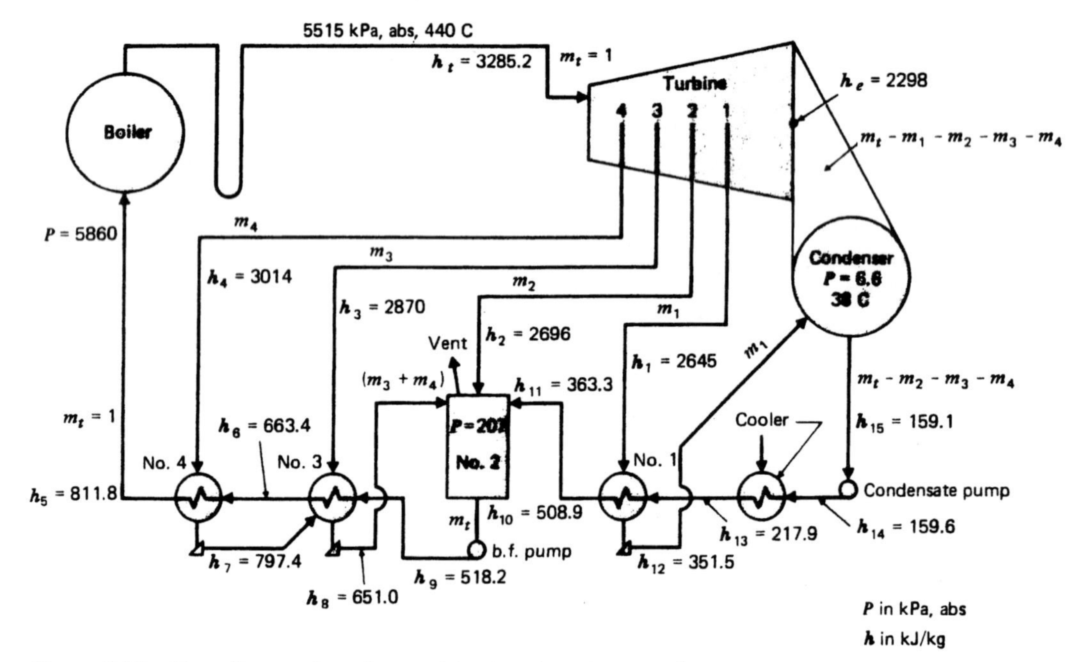

Figure 8.21 Flow diagram for a four-point extraction steam cycle.

in Fig. 8.21 has greater utility in analyzing the regenerative-reheat cycle.

Example 8.9

Determine the thermal efficiency for the extraction cycle shown in Fig. 8.21. A unit mass (m_t = 1) of steam enters the turbine and expands to the first extraction point (4), where a mass of steam (m_4) is extracted and flows to the No. 4 heater. The remainder of the steam ($1 - m_4$) expands to the second extraction point (3), where a mass of steam (m_3) is extracted and flows to the No. 3 heater. Subsequently, the steam continues to expand with extraction at points 2 and 1.

Heater No. 2 is a deaerating heater. Heater No. 3 and heater No. 4 are high-pressure closed heaters, and heater No. 1 is a low-pressure closed heater.

The generator and lubricating oil coolers and the steam jet air ejector condenser are represented by a single cooler. Steam used for air pumping is neglected in the simplified analysis.

1 Energy balance on the No. 4 heater.

$$m_4h_4 + m_th_6 - m_th_5 - m_4h_7 = 0$$

$$m_4(h_4 - h_7) + m_t(h_6 - h_5) = 0$$

$$m_4(3014 - 797.4) + 1(663.4 - 811.8) = 0$$

$(m_t = 1 \text{ kg})$

$$m_4 = 0.06695 \text{ kg}$$

2 Energy balance on the No. 3 heater.

$$m_3h_3 + m_4h_7 + m_th_9 - m_th_6 - (m_3 + m_4)h_8 = 0$$

$$m_3(h_3 - h_8) + m_4(h_7 - h_8) + m_t(h_9 - h_6) = 0$$

$$m_3(2870 - 651.0) + 0.06695(797.4 - 651.0) + 1(518.2 - 663.4) = 0$$

$$m_3 = 0.06102 \text{ kg}$$

3 Energy balance on the No. 2 heater.

$$m_2h_2 + (m_3 + m_4)h_8 + (m_t - m_2 - m_3 - m_4)h_{11} - m_th_{10} = 0$$

$$m_2(h_2 - h_{11}) + (m_3 + m_4)(h_8 - h_{11}) + m_t(h_{11} - h_{10}) = 0$$

$$m_2(2696 - 363.3) + (0.06102 + 0.06695) \cdot (651.0 - 363.3) + 1(363.3 - 508.9) = 0$$

$$m_2 = 0.04663 \text{ kg}$$

4 Energy balance on the No. 1 heater.

$$m_1h_1 + (m_t - m_2 - m_3 - m_4)h_{13} - m_1h_{12} - (m_t - m_2 - m_3 - m_4)h_{11} = 0$$

$$m_1(h_1 - h_{12}) + (m_t - m_2 - m_3 - m_4) \cdot (h_{13} - h_{11}) = 0$$

$$m_1(2645 - 351.5) + (1 - 0.04663 - 0.06102 - 0.06695)(217.9 - 363.3) = 0$$

$$m_1 = 0.05233 \text{ kg}$$

5 Turbine work.

$$\begin{aligned}
&m_t(h_t - h_4) \\
&\quad = 1(3285 - 3014) && = 271 \\
&(m_t - m_4)(h_4 - h_3) \\
&\quad = 0.9330(3014 - 2870) && = 134 \\
&(m_t - m_4 - m_3)(h_3 - h_2) \\
&\quad = 0.8720(2870 - 2696) && = 152 \\
&(m_t - m_4 - m_3 - m_2)(h_2 - h_1) \\
&\quad = 0.8254(2696 - 2645) && = 42 \\
&(m_t - m_4 - m_3 - m_2 - m_1)(h_1 - h_e) \\
&\quad = 0.7731(2645 - 2298) && = 268 \\
& && \overline{867}
\end{aligned}$$

W_T = 867 kJ/k steam supplied at the throttle

6 Pump work.

Condensate pump

$$W_p = (m_t - m_2 - m_3 - m_4)(h_{15} - h_{14}) = 0.8254\,(159.1 - 159.6) = -0.4 \text{ kJ}$$

Boiler feed pump

$$W_P = h_{10} - h_9 = 508.9 - 518.2$$
$$= -9.3 \text{ kJ/kg water}$$

Total pump work

$$= -9.7 \text{ kJ/kg water entering the boiler}$$

7 Heat supplied in the boiler.

$$Q_s = h_t - h_5$$
$$= 3285.2 - 811.8$$
$$= 2473.4 \text{ kJ/kg water}$$

8 Cycle thermal efficiency.

$$\eta_t = \frac{W_T + W_P}{Q_s} = \frac{867 - 10}{2473} = 0.347$$

The number of feedwater heaters incorporated into an actual steam power cycle depends, in part, upon the turbine throttle pressure and usually ranges between five and nine. The heaters are divided into two groups: the low-pressure and the high-pressure heaters. One of the low-pressure heaters is often a deaerating, or open-type, heater that discharges into the boiler feed pump.

A drain pump can be provided to pump the condensate discharged from a heater into the main feedwater stream. Installation of a drain pump for each heater however would represent generally an impracticable design.

Cascading the drain from a heater to the next heater in line eliminates the drain pump. Some flashing of the liquid will occur when the fluid enters the region of lower pressure in the adjacent downstream heater. Flashing of liquid into vapor is accompanied by a degradation of temperature and loss in availability. These adverse effects are largely diminished by the drain cooling section of the heater. Heater drain pumps are often completely eliminated by cascading the drains from the low-pressure heaters to the main condenser and the drains from the high-pressure heaters to the deaerating heater.

The modern, high-capability steam power plant operates on the regenerative-reheat cycle. Depending upon the characteristics of the particular cycle, the steam flow through the reheater will be less than the main flow from the superheater because of the diversion of some of the steam to one or two of the high-pressure feedwater heaters. Typical thermal efficiency values achieved by these power plants are 0.35 and 0.43, respectively, for subcritical-pressure and supercritical-pressure plants.

8.23 BINARY-VAPOR POWER CYCLES

Water is virtually the universal working fluid for the vapor power cycle. Ordinarily, water of sufficiently good quality is readily available, and conditioning for boiler make-up can be achieved economically. Certain adverse characteristics of water were noted in earlier sections and will be reviewed at this point.

The high specific heat of liquid water requires the application of regenerative feedwater heating in order to diminish the effect of transferring heat to the working fluid at temperatures substantially below the saturation temperature corresponding to the top pressure in the cycle. As the boiler pressure is increased, a greater proportion of the supplied heat is added to the water in the liquid phase, and a smaller proportion of the heat is added in achieving evaporation. Saturation pressures for water are comparatively high for the evaporation temperatures normally encountered in practice.

Some 50 years ago, when maximum boiler pressures were about 4.5 MPa, development of the mercury/steam binary-vapor power plant was undertaken in anticipation of improving the performance of electric generating stations. In principle, heat is transferred from the combustion gases to the mercury evaporating element and then to the steam superheater. Subsequent to expansion in the mercury turbine, the mercury is condensed in the mercury/steam heat exchanger. Steam produced in the heat exchanger is superheated, by the combustion gases, prior to expansion in the steam turbine. See Fig. 8.22.

For use in power plant cycles, mercury exhibits certain significant advantages, namely, the low specific heat in the liquid phase and the relatively low saturation pressures. At 540 C, the saturation pressure of mercury is only 1.277 MPa, abs. Containment of the mercury within a closed system can be accomplished without undue difficulty. Heat transfer from the solid boundary to pure mercury is adversely affected by the inherent spheroidal characteristics of the liquid. Improved contact with the heat transfer surface is achieved however by the use of titanium–magnesium-treated mercury. The maximum temperature of the mercury in the cycle is limited to approximately 510 C, because of the severe corrosion of all types of steel by mercury at temperatures in excess of this limiting value.

The first mercury power plant began commercial operation in 1928 in an electric generating station. A new mercury power plant was later installed in this station and commenced operation in 1949. Both the old and new mercury power plants were operated as topping units in conjunction with existing steam equipment. Three other mercury topping power plants were installed in an electric generating station in 1933

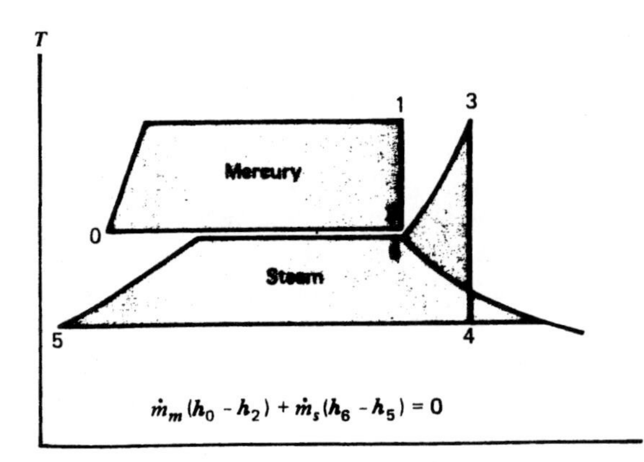

Figure 8.22 Theoretical mercury/steam binary-vapor power cycle.

and in two industrial power plants in 1933 and 1949. All of these power plants are located in the United States.

The mercury units installed in the Schiller station (Portsmouth, N.H.) and the associated steam equipment were designed and constructed as a complete power generating system.[6] Operation of the power plant began early in 1949. The system consists primarily of two 7500-kW mercury turbine-generators, two mercury boilers, a 25,000-kW steam turbine-generator, and four condenser-boilers. The throttle conditions for the mercury turbine are 779 kPa, gage, and 507 C; and for the steam turbine, the throttle conditions are 4137 kPa, gage, and 441 C. The anticipated net heat rate for the combined mercury/steam power plant was 9700 kJ/kWh.

The mercury/steam binary-vapor power plant is considered to be a successful development in the generation of electrical energy. No severe operating problems were experienced in these power plants. The incentive to construct additional mercury/steam power plants was curtailed by the improved thermal, and economic, performance achieved in the single-fluid steam power plants, through the trend to higher operating pressures and temperatures and, in part, through the construction of larger machines.

In addition to water and mercury, other fluids have been examined for use in the binary-vapor power plant. For the top fluid used in conjunction with water, a eutectic mixture of diphenyl and diphenyl oxide has been proposed. The mercury/benzene binary-vapor power cycle, suggested for possible adoption, would eliminate water as a working fluid. The low specific heat of liquid benzene, about 1.7 kJ/kg·K, should be viewed in the light of a comparatively low value of the energy of evaporation.

Liquid metals, particularly potassium and sodium, are used singly as working fluids in nuclear electric power plants devised for space applications. The feasibility of applying the potassium/steam binary-vapor cycle to electric power generation has been explored in some detail.[7,8] The potassium turbine would be supplied with potassium vapor at about 200 kPa, abs, and 830 C. The principal reason for considering this particular cycle is to operate the power plant at a high top temperature and thus achieve a high thermal efficiency, on the order of 0.54.[8]

The economic feasibility of a potassium/steam power plant must be demonstrated, particularly in the light of the competition from coal-fired and nuclear steam plants and from gas turbine/steam combined-cycle power plants. A considerable amount of developmental work can be anticipated before the potassium/steam binary-vapor cycle is applied to the construction of a high-capability power plant that will operate essentially trouble-free for extended periods of time.

8.24 GAS TURBINE/STEAM COMBINED-CYCLE POWER PLANTS

Over a period of several decades, the thermal performance of the steam power plant has been improved, mainly by constructing larger boilers and turbine-generators and by increasing the thermal efficiency of the steam power cycle. Increased thermal efficiency was accomplished by employing higher throttle pressures and temperatures, regenerative feedwater heating, and steam reheating. There are however economic limitations to continuing this approach to improving the power plant thermal peformance. For example, material costs increase when higher steam pressures and temperatures are prescribed, and inclusion of a second steam reheat will increase the power plant capital cost. The next step in achieving a reduction in the power plant heat rate would logically be directed to the gas side of the system.

The steam power plant is actually a two-fluid system, that is, energy is exchanged between the combustion gases and the water. The combustion reaction occurs essentially at atmospheric pressure, and very little change in pressure is observed as the gases flow successively through the several heat exchange components of the boiler prior to discharge to the atmosphere. Combustion and heat transfer processes have been improved, but in the conventional steam power plant the combustion gases do not participate directly in the process in which thermal energy is converted into mechanical energy.

The feasibility of combining gas and steam expansion in a power cycle has been extensively explored. Because steam generation involves the flow of large volumes of combustion gases, gas expansion is most appropriately accomplished in a gas turbine. Two general schemes are available for combining the steam and gas power cycles. In one arrangement, the hot exhaust gas from the gas turbine serves as the combustion medium for firing the steam boiler. In this system, combustion of the fuel is effected at two points in the cycle, namely, the gas turbine engine combustion chamber and the boiler furnace. See Fig. 8.23. In a modification of this system, an unfired steam boiler is employed.

The second approach combines the combustion process for the gas turbine engine with that of the boiler. See Fig. 8.24. In this system, all the fuel is burned in the pressurized boiler furnace.

The gain in the power plant thermal efficiency achieved by the combined cycle is inherently higher for Scheme B than for Scheme A. Scheme A has the advantage of independent operation of the two parts of the system, if required. If an unfired boiler is used in Scheme A, a light-weight, low first-cost type of construction is adequate for the boiler.

For a prescribed air flow rate, the mass flow rate of gas for the gas turbine is higher for Scheme B than for Scheme A, because all of the fuel for the combined system is fired upstream from the turbine inlet. Further, the energy equivalent of this incremental mass flow is more effectively utilized by expansion in the gas turbine in comparison with transfer to the steam cycle and subsequent expansion in the steam turbine. Principally because of the more effective manner in firing the fuel, for the two basic arrangements examined in this section, Scheme B will generally

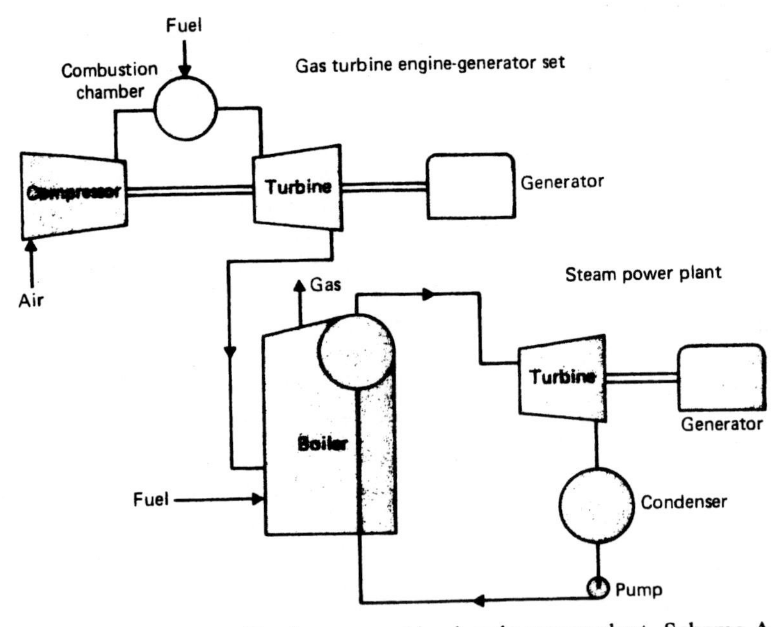

Figure 8.23 Gas turbine/steam combined-cycle power plant, Scheme A.

achieve the higher power plant thermal efficiency.

Development of the gas turbine/steam combined-cycle power plant for commercial applications has produced three variations of the basic arrangement designated as Scheme A. The three variations are described as (a) the unfired heat recovery cycle, (b) the supplemental-fired heat recovery cycle, and (c) the exhaust-fired cycle. These systems all employ gas turbine engines and

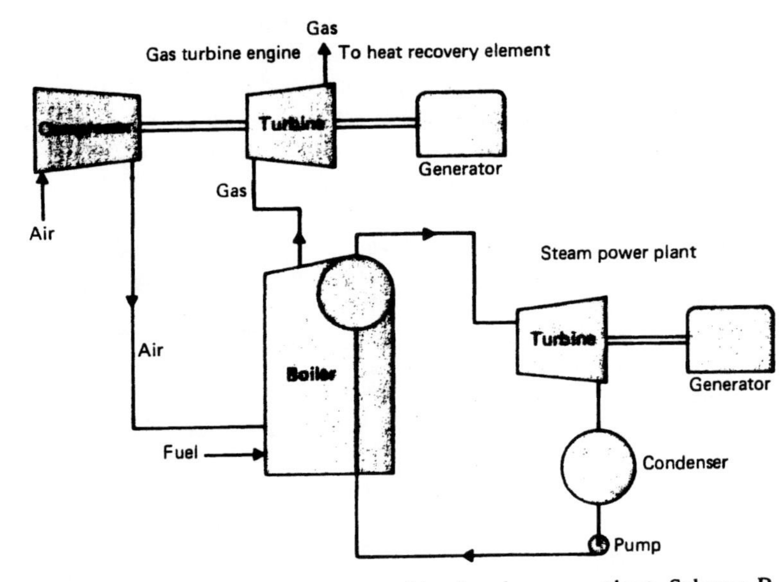

Figure 8.24 Gas turbine/steam combined-cycle power plant, Scheme B.

condensing steam turbine-generators but differ in the boilers and the proportion of the power plant output that is produced by the steam system. The portion of the power plant capability produced by the steam system is about 30 to 35 percent for the unfired heat recovery cycle, 35 to 60 percent for the supplemental-fired cycle, and 60 to 80 percent in the exhaust-fired cycle.[9]

The unfired combined cycle is the simplest of the several combined cycles currently used in commercial applications. The system is easily operated because all the fuel is fired in the gas turbine engine. The boiler usually consists of an evaporating section, superheater, and economizer. Steam extraction for feedwater heating is ordinarily not utilized. The exhaust gas however can be used for this purpose and thereby achieve a minimum stack gas temperature. Modest steam conditions are characteristic of this cycle and usually do not exceed 5860 kPa, gage, and 485 C.

The supplemental-fired heat recovery combined-cycle power plant is currently the system most likely to be selected for electric power generation. The boiler is essentially identical to the unfired heat recovery boiler, except for the supplemental-firing burners installed in the furnace. Because the furnace is lightly fired, refractory lining is adequate and water cooling of the furnace walls is not essential. A nonreheat steam turbine is used in this power plant with steam extraction for supplying the deaerating feedwater heater.

In comparison with the unfired system, the control of the supplemental-fired system is somewhat more complex. Of the two systems, the supplemental-fired heat recovery combined-cycle power plant has a slightly higher capital cost, expressed in dollars per net kilowatt of generating capability. Capital costs for the two systems are however substantially below the unit capital cost of a high-pressure, high-temperature reheat steam power plant.

The exhaust-fired combined-cycle power plant consists of a gas turbine engine combined with a conventional reheat steam power plant. In this system the air preheater and the forced-draft fan for the boiler are replaced by the gas turbine engine. Several power plants of this type have been installed in electric utility systems in the United States, notably the Oklahoma Gas & Electric Company Horseshoe Lake Unit No. 7, rated at 235 MW.

All of the combustion air for the exhaust-fired combined-cycle power plant is supplied through the gas turbine engine compressor. Because a low inlet air temperature is conducive to high thermal performance of the gas turbine engine, the customary air heater is omitted from the boiler structure. Stack gas cooling must consequently be accomplished by feedwater heating. Because both extracted steam and exhaust gas are source fluids for feedwater heating, the arrangement of the heaters is somewhat complex.

The net heat rate for the gas turbine/steam combined-cycle power plant is influenced, in part, by the proportion of the total power plant output that is produced by the steam turbine. See Table 8.2.

A conventional steam power plant operating on a reheat steam cycle with a turbine throttle pressure of 16.5 MPa has a heat rate of about 9500 kJ/kWh. In comparison with the conventional steam power plant, an improvement of 6 to 10 percent in the heat rate is achieved by the exhaust-fired combined-cycle power plant.

The boiler furnace in Scheme B, Fig. 8.24, is supercharged to several atmospheres. Boiler

TABLE 8.2 Heat Rates for Gas Turbine/Steam Combined-Cycle Power Plants, kJ/kWh[a]

Power Plant	Steam Turbine Capability (Percent)			
	0	35	60	80
Unfired	12,800	8600		
Supplemental fired		8600	9400	
Exhaust fired			8550	8950

The heat rate is based on the higher heating value of an oil fuel.

construction would necessarily depart significantly from the current practice followed in fabricating steam boilers in which the furnace and the gas side of the heat exchange elements operate at essentially atmospheric pressure. The Velox steam generator, developed sometime prior to 1930, is in certain respects a precursor to the present-day combined-cycle power plant. The combustion air in the Velox power plant is discharged from the compressor at a pressure of 150 kPa, gage, hence only moderate supercharging is effected. The gas turbine drives the compressor and the boiler circulation pumps, with auxiliary power provided, when needed, by the driving/starting motor. A number of Velox boilers have been installed in power plants located in Europe, but none in the United States.

Example 8.10

A theoretical combined cycle is devised by topping a superheat steam cycle with a gas turbine engine cycle. The exhaust from the gas turbine engine is used for steam generation in an unfired boiler that includes an economizer.

Gas turbine engine cycle.

Pressure ratio: 8 to 1

Compressor inlet temperature, T_1: 22 C (295 K)

Turbine inlet temperature, T_3: 1100 C (1373 K)

Fuel: $C_{12}H_{26}$, LHV = 44,085 kJ/kg

Steam cycle.

Turbine inlet conditions: 6 MPa, abs, and 480 C

Turbine exhaust pressure: 7 kPa, abs

Temperature of the gas entering the stack: 127 C

Air and gas properties are selected from Table A.2. Assume that combustion occurs with 285 percent theoretical air.

1 Gas turbine engine.
Turbine inlet: T_3 = 1373 K, h_3 = 1534.8 kJ/kg gas
Turbine outlet: T_4 = 830 K, h_4 = 878.6 kJ/kg gas
Compressor inlet: T_1 = 295 K, h_1 = 295.41 kJ/kg air
Compressor outlet: T_2 = 531 K, h_2 = 535.53 kJ/kg air
From combustion calculations:
$f/a = 0.02328$ 287 percent theoretical air

$$W_T = \left(1 + \frac{f}{a}\right)(h_3 - h_4) = 1.02328(1534.8 - 878.6) = 671.5 \text{ kJ/kg air}$$

$$W_C = h_1 - h_2 = 295.4 - 535.5 = -240.1 \text{ kJ/kg air}$$

$$W_n = W_T + W_C = 671.5 - 240.1 = 431.4 \text{ kJ/kg air}$$

$$\eta_t = \frac{W_n}{(f/a)(\text{LHV})} = \frac{431.4}{0.02328 \times 44085} = 0.420$$

2 Steam cycle.
At 6 MPa, abs, and 480 C: h_1 = 3374.0 kJ/kg steam
At 7 kPa, abs:

$$h_{2,s} = 2115.5 \text{ kJ/kg}$$
$$h_{f,2} = 162.6 \text{ kJ/kg}$$
$$v_{f,2} = 0.001007 \text{ m}^3/\text{kg}$$

Pump work:

$$W_P = -v_{f,2}(P_1 - P_2) = -0.001007(6000 - 7) = -6.0 \text{ kJ/kg water}$$

Cycle net work:

$$W_n = (h_1 - h_{2,s}) + W_P$$
$$= (3374.0 - 2115.5) - 6.0$$
$$= 1252.5 \text{ kJ/kg steam}$$

Thermal efficiency of the steam cycle:

$$\eta_t = \frac{W_n}{(h_1 - h_{f,2}) + W_P}$$
$$= \frac{1252.5}{(3374.0 - 162.6) - 6.0}$$
$$= 0.391$$

3 Combined cycle.
Gas entering the boiler:

$$T = 830 \text{ K}, h = 878.6 \text{ kJ/kg gas}$$

Gas leaving the boiler: $T = 400$ K, $h = 407.8$ kJ/kg gas

Heat transferred from the gas:

$$\Delta h_g = 878.6 - 407.8 = 470.8 \text{ kJ/kg gas}$$

Steam leaving the boiler: $h = 3374.0$ kJ/kg

Water leaving the pump:

$$h = h_{f,2} - W_P = 162.6 - (-6.0) = 168.6 \text{ kJ/kg}$$

Heat transferred to the water:

$$\Delta h_w = 3374.0 - 168.6 = 3205.4 \text{ kJ/kg}$$

Ratio of steam flow to gas flow:

$$\frac{\dot{m}_s}{\dot{m}_g} = \frac{\Delta h_g}{\Delta h_w} = \frac{470.8}{3205.4}$$
$$= 0.1469 \text{ kg steam/kg gas}$$

Steam turbine work:

$$W_{st} = \left(1 + \frac{f}{a}\right)\frac{\dot{m}_s}{\dot{m}_g}(W_n)_{st}$$
$$= (1 + 0.02328)0.1469 \times 1252.5$$
$$= 188.3 \text{ kJ/kg air}$$

Gas turbine engine work: $W_{gt} = 431.4$ kJ/kg air
Total work:

$$W_{\text{total}} = W_{gt} + W_{st}$$
$$= 431.4 + 188.3$$
$$= 619.7 \text{ kJ/kg air}$$

Heat supplied to the combined cycle:

$$Q_s = \frac{f}{a}(\text{LHV}) = 0.02328 \times 44085$$
$$= 1026.3 \text{ kJ/kg air}$$

Combined cycle thermal efficiency:

$$\eta_t = \frac{W_{\text{total}}}{Q_s} = \frac{619.7}{1026.3} = 0.604$$

Steam turbine capability:

$$\frac{W_{st}}{W_{\text{total}}} = \frac{188.3}{619.7} = 0.304$$

Example 8.10 indicates the improvement in the power plant thermal efficiency that is achieved by application of the gas turbine engine/steam combined cycle. In this theoretical analysis, the thermal efficiency of the combined cycle exceeds by an appreciable increment the thermal efficiency of either the independent gas turbine engine cycle or the independent steam power cycle. For an actual unfired combined-cycle power plant, with 30 percent steam turbine capability, a thermal efficiency of about 0.40 can be anticipated.

The gas turbine engine/steam combined-cycle power plant demonstrates, because of the inherently high thermal efficiency, excellent potentialities for expanded use in electric power generation, provided that the dependence upon petroleum fuels and natural gas can be avoided. Currently, the most feasible approach to achieving this independence appears to be through the large-scale production of low-energy gas manufactured in coal conversion plants. Because of the comparatively high cost of transporting low-energy fuel, the coal conversion plants must be located in close proximity to the power plant, or cluster of power plants, in which the fuel is burned. The distance over which it is feasible to transport the fuel will be controlled by the overall operating economy of the entire system, including the conversion of coal and the generation of electric power.

The General Electric Company is developing a water-cooled gas turbine that would operate initially in a combined-cycle power plant.[10,11] The ultrahigh-temperature gas turbine engine (UHT-GT) is expected to deliver 180 MW and operate at a temperature of 1540 C. Cooling will be accomplished with water that circulates through the engine via small channels close to the surface of the stationary nozzles and the rotating blades, thus limiting blade temperatures to 650 C. Because blade temperatures are not excessively high, the water-cooled gas turbine can be constructed primarily of nickel-based superalloys rather than the more advanced high-temperature materials currently under development. The cooling water used to cool the rotating blades escapes from the blade tips and is collected in a stationary shroud. Following passage through a heat exchanger, the water is recycled. The water consumption for a UHT-GT combined-cycle power plant is estimated to be about one-half of the amount of water consumed in a steam turbine power plant of the same generating capability.

A thermal efficiency of about 0.50 is anticipated for the UHT-GT combined-cycle power plant for base-load and midrange power production. A comparable value of 0.43 is achieved in a high-capability supercritical-pressure steam power plant. This gain in thermal efficiency, although significant, is anticipated for a combined-cycle power plant that would burn oil or natural gas. The UHT-GT engine however is capable of burning low-grade fuels, particularly low-energy gas derived from coal. A UHT-GT combined-cycle power plant could eventually generate electrical energy from coal with an anticipated thermal efficiency of 0.45, a figure that includes the negative effect of the energy losses associated with coal gasification and sulfur removal.

While a thermal efficiency of 0.45 does not exceed by a wide margin the thermal efficiency of 0.43 achieved in high-performance steam power plants fired directly with coal, the advantages of burning sulfur-free and ash-free fuel are particularly significant. There would be no need to install equipment for SO_2 removal from the stack gases, and operating problems related to coal ash would be eliminated. Removal of dust from the gas stream can be accomplished with a minor capital outlay.

The application of pressurized fluidized-bed combustion of coal in a gas turbine engine/steam combined-cycle power plant would achieve a number of advantages. The system would burn coal directly while achieving a high thermal efficiency and effectively controlling the emission of gaseous pollutants. A wide variety of coals, including those with high sulfur content, would be burned at high pressure in a bed of limestone. Because the sulfur in the coal is largely absorbed in the bed, the waste gases would be almost free of sulfur dioxide. Control of particulate material however would be required. Steam is generated in tubes submerged in the fluidized bed, thus the combustor functions as a gas turbine engine combustion chamber and as a supercharged steam boiler.

8.25 INDUSTRIAL POWER PLANTS

In addition to the utility electric generating stations, there are many industrial steam power plants that operate primarily to supply steam for processing a wide variety of products. These power plants may also generate electric power for onsite use. Excess power generated by an industrial power plant can usually be sold to the local utility under an energy interchange agreement.

When process steam is not required, electrical energy can usually be purchased from a utility at a cost below that for electrical energy generated in an industrial power plant. Incidentally, determination of the individual costs for producing electric power and steam in an industrial power plant may require a somewhat complex procedure.

Steam power plants are widely used for propelling large ships. Marine boilers are oil fired, and the steam is used chiefly for operating the main propulsion engines. Some steam is used by the auxiliary turbine-generators for production of electric power.

Because of their widely diverse characteristics, no simple classification can be made of industrial steam power plants. The principal differences are observed in the steam generating capability, in steam pressures and temperatures, in the arrangement of the heat transfer surfaces, and in the kind of fuel fired in the furnace.

Industrial power plants burn a wide variety of fuels other than coal, oil, and natural gas. Many of these fuels are by-products or waste materials, such as wood refuse, coke, oil refinery refuse, bagasse, blast furnace gas, and spent liquor from wood-pulping digesters. Coal is burned on stokers or in pulverized fuel systems. A practical lower limit on pulverized coal firing corresponds to a steam output of about 100,000 kg/h.[1] Stoker firing can be extended to a steam generating capability of about 200,000 kg/h.

Turbines installed in industrial steam power plants commonly operate noncondensing at a prescribed exhaust pressure. The exhaust steam and the steam obtained by controlled extraction from the turbine provide the heat source for the processing operations. Turbine throttle pressures and temperatures for large machines extend to 12.5 MPa and 540 C. Low boiler pressures are not uncommon when steam is generated solely for heating.

8.26 SUMMARY

Fossil fuel-fired steam power plants will for many years generate the major portion of the electrical energy produced in the United States. Because of uncertain supplies and high cost of natural gas and fuel oil, greater dependence must be placed upon coal as the predominant fossil fuel for steam-electric power generation. Difficulties experienced in plant siting will continue, but this problem may be alleviated to some degree in the light of a need to expand the electric generating capability throughout the country. Control of air and water quality can be accomplished generally with reasonably good results, but the implementation of the controls must necessarily increase the production cost of electrical energy and aggravate to some extent the power plant operating problems.

During the past several decades, the trend has been to increase the size of the largest turbine-generators that operate as the base load machines. In later years, these machines attained a generating capability in the range of 1000 to 1300 MW. Some reversal in the trend to larger machines is now observed with the installation of a number of turbine-generators rated at 500 to 600 MW. Selection of the generating capability is dependent primarily on economic factors.

Rapidly increasing power plant construction costs are a matter of considerable concern. During the years 1964, 1965, and 1966, the construction cost for large fossil fuel-fired power plants continued to decline appreciably to a low value of about $100 per kilowatt of generating capability. Since 1966, construction costs have increased sharply, principally because of inflation, additional equipment required for environmental protection, and extended construction time. The interest paid on the invested capital during the construction period is by no means a small item.

The capital cost of a 1000-MW steam-electric coal-fired power plant, for a 1982 commercial operation date, is likely to be about $800 per kilowatt of generating capability. In 1990, the capital cost for a 1300-MW coal-fired power plant will probably exceed $1000 per kilowatt.

PROBLEMS

8.1 The closely spaced tubes in a furnace wall are 13.5 m long, and the outside diameter is 75 mm. The tube wall thickness is 5.6 mm. The water inlet velocity to the tubes is 1.5 m/s. The operating pressure is 10 MPa, abs. The energy input to the furnace wall is 500 kW/m^2. Determine whether the steam generation is within acceptable limits.

8.2 A pulverized coal-fired furnace has an energy absorbing surface equivalent to 1150 m^2. Coal, fired at the rate of 65,360 kg/h, has a higher heating value of 31,380 kJ/kg. The combustion efficiency is 0.95. The air and gas flows are, respectively, 11.73 and 12.60 kg/kg coal. The flue gas contains 4 percent water vapor. The air preheat temperature is 347 C. The value of C for Eq. 8.2 is 0.085. The boiler energy absorbing temperature is 330 C. Calculate the absorption fraction, availability ratio, and effectiveness of absorption for the furnace.

8.3 A boiler generates 200,000 kg steam per hour. At the superheater outlet, the steam pressure and temperature are 7.0 MPa, abs, and 500 C. The steam enters the superheater at 7.10 MPa, abs, with 0.25 percent moisture. The gas temperatures entering and leaving the superheater are, respectively, 1045 and 860 C. The overall coefficient of heat transfer U_o = 50 W/m^2·K. Calculate the superheater heat transfer area for (a) counterflow and (b) parallel flow of the steam and gas.

8.4 Determine the absorption effectiveness for the superheater described in Problem 8.3. The water vapor content of the flue gas is 6 percent.

8.5 A small boiler consists of a noncooled furnace, a convective evaporating element, a superheater, and an economizer. Steam leaves the boiler at 1.80 MPa, abs, and 350 C, and the water enters the economizer at 160 C. The boiler efficiency is 0.82. The higher heating value of the coal is 31,635 kJ/kg, and the total hydrogen content of the coal is 4.9 percent. The unburned fuel loss is equivalent to 4 percent of the HHV. The fuel is fired with 14.0 kg air per kg coal. The refuse is 0.12 kg/kg coal. Calculate the quantity of steam generated per kilogram of coal and the temperature of the gas leaving the economizer.

8.6 A surface condenser is constructed with 4000 tubes arranged for two-pass operation. The tube length is 6 m, and the O.D. and wall thickness are 25 mm and 1.25 mm, respectively. The water velocity in the tubes is 2 m/s. The cooling water enters the condenser at 15 C and leaves at 25 C. The value of U_o is 2800 W/m^2·K. The steam enters the condenser with a quality of 0.90, and the condensate is saturated. Calculate the condensing temperature of steam and the steam flow rate, in kg/h.

8.7 Steam is expanded in a turbine from the initial state (P = 7.0 MPa, abs, and T = 500 C) to an exhaust pressure of 7.5 kPa, abs. The quality of the exhaust steam is 0.87. At the single extraction point, the steam conditions are 175 kPa, abs, and x = 0.99. The feedwater leaves the mixing-type heater at P = 175 kPa, abs, saturated. Calculate the thermal efficiency of the cycle.

8.8 A combined gas turbine engine/steam power plant is arranged in accordance with Scheme B (Fig. 8.24). Feedwater heating by extracted steam is not employed. The gas turbine engine pressure ratio is 6 to 1. Air enters the compressor at 19 C, and the gas enters the turbine at 857 C. The steam turbine inlet conditions are 10 MPa, abs, and 500 C, and the exhaust pressure is 7.5 kPa, abs. The gas leaves the economizer at 177 C. The machine efficiencies are: gas turbine 0.90, steam turbine 0.90, and gas turbine engine compressor 0.85. Mechanical and generator efficiencies are, respectively, 0.98 and 0.96. The fuel, liquid dodecane, is fired with 15 percent excess air. The combustion efficiency is 0.975. Calculate the power plant thermal efficiency and the total electrical output of the plant.

8.9 Determine the improvement in the thermal efficiency that can be achieved by topping a steam cycle with a mercury vapor cycle. Steam enters the turbine at 3.5 MPa, abs, and 500 C, and exhausts at 7.5 kPa, abs. The throttle conditions for the mercury turbine are 500 C, x = 1.00, and the exhaust temperature is 260 C. The mercury/steam heat exchanger produces dry, saturated steam. The machine efficiency for both turbines is 0.90. The pump work may be neglected.

8.10 A superheater design is based on the following data. Inlet steam 3.10 MPa, abs, and x = 1.0. Outlet steam 3.0 MPa, abs, and 350 C. Inlet gas temperture 1040 C. The water vapor content of the gas is 6 percent. Steam flow 115,000 kg/h; U_o = 48 W/m^2·K. The prescribed absorption effectiveness is 0.230. Determine the superheater heat transfer area.

8.11 A furnace wall is constructed with closely spaced tubes 14.0 m long. The tube O.D. is 75 mm, and the tube wall thickness is 6.0 mm. The boiler operating pressure is 8 MPa, abs. The energy input to the furnace wall is 475 kW/m^2. Calculate the minimum velocity for the water entering the tubes.

8.12 A furnace is designed to burn pulverized coal. The energy absorbing area is equivalent to 1250 m^2. The energy release rate is 2.525 × 10^6 MJ/h. The furnace absorption fraction is 0.42, and C = 0.085. The water content of the flue gas is 3 percent. Calculate the gas flow and the temperature of the gas leaving the furnace.

8.13 The firing rate for the furnace described in Problem 8.12 remains unchanged. The outlet gas temperature is to be reduced to 1500 K by increasing the area of the energy absorbing surface. Determine the required area of this surface.

8.14 Determine the availability ratio and the absorption effectiveness for the furnace described in Problem 8.12. The boiler pressure is 10 MPa, abs.

8.15 A boiler generates steam at a rate of 250,000 kg/h. The gas flow rate is 305,000 kg/h. Steam enters the superheater dry and saturated at 3.60 MPa, abs, and leaves at 3.50 MPa, abs, and 400 C. The gas enters the superheater at 1050 C. The gas contains 6 percent water vapor. U_o = 50 W/m^2·K. Calculate the area of the heat transfer surface for (a) counterflow and (b) parallel flow of the two fluids. Determine for (a) and (b) the maximum tube wall temperature. The resistance of the steam film is equal to 1/20 of the gas film resistance.

8.16 The superheater described in Problem 8.15, arranged for counterflow, is fabricated from 65-mm-O.D. tubes 5.50 m long. The horizontal center-to-center spacing is 155 mm, and the vertical spacing is 85 mm. There are 12 tubes in a vertical row. The gas pressure is 99 kPa, abs. The gas viscosity is 4.37 × 10^{-5} N·s/m^2, and R = 0.287 kJ/kg·K. The ambient conditions are 101 kPa, abs, and 20 C. Calculate the draft loss through the superheater for both upward and downward flow of the gas.

8.17 Calculate the absorption effectiveness for the superheater described in Problem 8.15. The gas leaves the superheater and enters the convective evaporating element. The gas leaves this element at 400 C, and the water enters the element at 215 C. U_o = 55 W/m^2·K. Calculate for the evaporating element the absorption effectiveness, the evaporation rate, and the area of the heat transfer surface.

8.18 A steam generator produces 300,000 kg steam per hour. The gas flow is 365,000 kg/h. The water enters and leaves the economizer at 105 and 140 C, respectively. The inlet and outlet gas temperatures are 436 and 330 C. The tube O.D. is 50 mm, and the tube center-to-center spacing is 75 mm in both directions. The tube length is 3.20 m. G_{max}, the gas flow through the minimum section, is 9.80 kg/s·m^2. U_o = 75 W/m^2·K. The mean gas pressure is 100 kPa, abs. R = 0.286 kJ/kg·K. The viscosity of the gas is 3.03 × 10^{-5} N·s/m^2. The ambient conditions are 1 atm and 27 C. The economizer is arranged for counterflow of the two fluids and downward flow of the gas. Calculate the number of tubes in a horizontal row and in a vertical row, and determine the draft loss for the economizer.

8.19 The gas flows from the economizer described in Problem 8.18 to an air heater arranged for cross flow of the air through the tube bank.

The gas outlet temperature is 210 C. The water vapor content of the gas is 6 percent. The air enters the heater at 25 C. The air flow is 345,000 kg/h. The tube O.D. is 50 mm, and the tube length is 4.60 m. U_o = 25 W/m^2·K; F_c = 0.93. Each row of tubes in line with the gas flow contains 76 tubes. Calculate the number of tubes in each transverse row and the absorption effectiveness of the air heater.

8.20 A steam generator is fired with coal that has the following ultimate analysis: C 0.807, H 0.045, O 0.024, N 0.011, S 0.018, ash 0.062, moisture 0.033. The flue gas analysis is CO_2 13.1, CO 0.2, O_2 6.0, N_2 80.7 percent. The actual air supplied is 14.5 kg/kg coal. The air inlet conditions are 100 kPa, abs, 25 C, and 55 percent relative humidity. The flue gas is discharged at 195 C. The higher heating value of the coal is 33,265 kJ/kg. The steam generation rate is 250,000 kg/h, and the fuel firing rate is 23,210 kg/h. The refuse, collected at the rate of 1903 kg/h, contains 25 percent carbon. The steam leaves the unit at 10.0 MPa, abs, and 500 C, and the feedwater enters the unit at 180 C. Determine for the steam generator the energy balance and the efficiency.

8.21 A closed feedwater heater is supplied with steam extracted from the turbine. The temperature of the saturated steam in the heater is 90 C, and the enthalpy of the entering steam is 2645.0 kJ/kg. The feedwater enters the heater at 52 C, and the terminal temperature difference is 3 C. The condensate is saturated. The feedwater flow is 90,000 kg/h. U_o = 3260 W/m^2·K. Calculate the steam flow to the heater and the area of the heat transfer surface. The heater is constructed for four water passes. The tube O.D. is 18 mm, and the tube wall thickness is 1.25 mm. The water velocity is 1.5 m/s. Calculate the length of the tubes.

8.22 The following data are applicable to a sodium/steam binary-vapor power cycle. The throttle conditions for the sodium turbine are 9.080 kPa, abs, and 838 C (h = 5651.4 kJ/kg, s = 8.6805 kJ/kg·K). The turbine exhausts at 282 C (h_g = 5189.8; h_f = 757.5 kJ/kg; s_g = 11.3671; s_f = 3.3468 kJ/kg·K). The sodium/steam heat exchanger produces dry, saturated steam at 5.5 MPa, abs. Steam enters the steam turbine at 5.5 MPa, abs, and 450 C, and exhausts at 7.5 kPa, abs. The machine efficiency for both turbines is 0.90. Calculate the heat supplied, the heat rejected, and the thermal efficiency for the binary-vapor power cycle.

8.23 A Diesel engine has a shaft output of 2000 kW. The brake thermal efficiency is 0.38 based on the lower heating value of the fuel (44,086 kJ/kg). The air/fuel ratio is 20 to 1. The exhaust gas temperature is 430 C, and c_p = 1.100 kJ/kg·K. The exhaust gas is used to generate steam in a heat recovery boiler. The gas leaves the boiler at 255 C, and dry, saturated steam is produced at a pressure of 1.50 MP, abs. The steam is used in a Rankine cycle engine that exhausts at 15 kPa, abs. The turbine machine efficiency is 0.90. Calculate the power output and the thermal efficiency for the combined power plant.

8.24 A gas turbine engine operates with a compressor pressure ratio of 10 to 1. The inlet air conditions are 100 kPa, abs, and 300 K. The compressor and turbine machine efficiencies are, respectively, 0.87 and 0.91. The turbine inlet temperature is 1380 K, and the turbine exhaust stagnation pressure is 108 kPa, abs. The combustion efficiency is 0.98, and the combustion chamber pressure loss factor is 0.03. The engine burns liquid dodecane. Mechanical and generator efficiency values are, respectively, 0.98 and 0.95. The turbine exhaust gas flows to an unfired boiler where steam is generated at 7.0 MPa, abs, and 500 C. The gas leaves the boiler at 375 C. The steam expands in a Rankine cycle engine to 5 kPa, abs. The steam turbine machine efficiency is 0.90. Calculate the thermal efficiency for the combined-cycle power plant, and determine the electrical output for an air flow of 100 kg/s.

8.25 The working substances for a binary-vapor power cycle are mercury and Freon-12. The fluids are dry and saturated at the start of expansion, and the turbine machine efficiency is

0.90. The mercury cycle operates between 540 and 120 C, and the Freon-12 cycle, between 110 and 25 C. Calculate the thermal efficiency of the binary-vapor power cycle and discuss the characteristics of this particular cycle. Determine the thermal efficiency for a corresponding steam power cycle. The turbine inlet conditions are 10 MPa, abs, and 540 C, and the exhaust steam temperature is 25 C. Pump work may be neglected in this problem.

8.26 A reheat steam power cycle includes three low-pressure extraction stages. The drain from heater No. 2 is cascaded to heater No. 1, and the combined drain from the two closed heaters is cascaded to the main condenser. Heater No. 3 is a deaerating heater that operates at a pressure of 175 kPa, abs, and discharges to the boiler feed pump. The steam conditions are

Turbine throttle: 16.650 MPa, abs, and 538 C (h = 3399 kJ/kg)

Entering reheater: 4.165 MPa, abs, and 342 C (h = 3068 kJ/kg)

Leaving reheater: 3.758 MPa, abs, and 538 C (h = 3532 kJ/kg)

First extraction point: h = 2805 kJ/kg

Second extraction point: h = 2728

Third extraction point: h = 2596

Entering condenser: h = 2396

Condenser pressure: 7 kPa, abs

The feedwater conditions are

Entering No.1 heater: h = 163 kJ/kg

Entering No. 2 heater: h = 265

Entering No. 3 heater: h = 416

No. 1 heater drain: h = 186

No. 2 heater drain: h = 288

Leaving boiler feed pump: h = 507

Draw the flow diagram and calculate the thermal efficiency for the regenerative-reheat cycle. Calculate the thermal efficiency for the corresponding reheat cycle that operates without feedwater heating.

REFERENCES

1 *Steam—Its Generation and Use,* The Babcock and Wilcox Company, New York, 1978.

2 Reid, W. T., Cohen, P., and Corey, R. C. "An Investigation of the Variation in Heat Absorption in a Pulverized-Coal-Fired Water-Cooled Steam-Boiler Furnace," *Trans. ASME,* **70,** July 1948.

3 Grimison, E. D. "Correlation and Utilization of New Data on Flow Resistance and Heat Transfer for Cross Flow over Tube Banks," *Trans. ASME,* **59,** Oct. 1937.

4 *Standards for Steam Surface Condensers,* 6th ed., Heat Exchange Institute, New York, 1970.

5 Chait, I. L. and Fiehn, A. J. "Design of the Supercritical 325-MW Unit Addition to Genoa Station No. 3," *Combustion,* Sept. 1972.

6 "The Mercury Power Plant from South Meadow to Schiller," General Electric Co. Ger-246. Reprinted from *Power Generation,* March 1950.

7 *Mechanical Engineering,* Nov. 1975, page 20.

8 Fraas, A. P. "A Potassium-Steam Binary Vapor Cycle for Better Fuel Economy and Reduced Thermal Pollution, ASME Paper No. 71-WA/Ener-9, Nov.–Dec. 1971.

9 Tomlinson, L. O. "Comparing Combined Cycle Plants," *Gas Turbine International,* Nov.–Dec. 1972.

10 *Gas Turbine International,* Nov.–Dec. 1975, page 8.

11 Alff, R. K., Manning, G. B., and Sheldon, R. C. "The High Temperature Water Cooled Gas Turbine in Combined Cycle with Integrated Low Btu Gasification," *Combustion,* Mar. 1978.

BIBLIOGRAPHY

Barnard, W. N., Ellenwood, F. O., and Hirshfeld, C. F. *Heat-Power Engineering,* Parts II and III, John Wiley & Sons, Inc., New York, 1935.

Combustion Engineering, Combustion Engineering, Inc., New York, 1966.

Gaffert, G. A. *Steam Power Stations,* McGraw-Hill Book Co., New York, 1946.

Potter, P. J. *Power Plant Theory and Design,* The Ronald Press Co., New York, 1959.

Skrotski, B. G. A. and Vopat, W. A. *Power Station Engineering and Economy,* McGraw-Hill Book Co., New York, 1960.

Zerban, A. H. and Nye, E. P. *Power Plants,* International Textbook Co., Scranton, Pa., 1956.

CHAPTER NINE

NUCLEAR POWER PLANTS

Nuclear energy is the most recent form of energy to become directly available to mankind and represents a timely supplement to the energy obtained from fossil fuels. In the light of dwindling supplies of fossil fuels and increasing costs of these fuels, construction of nuclear power plants should seemingly be vigorously supported throughout the world. In a general sense, this support is observed, although various individuals and groups are attempting to halt plant construction, principally because of a belief that nuclear power generation represents a hazardous practice.

At the start of 1979, 72 nuclear power plants were operating in the United States and generating about 13 to 14 percent of the electrical energy produced in the country. By the year 1985, an additional 75 nuclear power plants, currently under construction, could be in operation and generating as much as 20 percent of the total electrical energy. As of January 1, 1979, 94 nuclear power plants were under construction in the United States, with an additional 30 plants in the planning stage.

A relatively large number of concepts for construction of nuclear power reactors have been proposed. In the United States, three types of power reactors have attained commercial status, namely, pressurized water, boiling water, and high-temperature gas reactors. In this group, the pressurized water reactor is the leading type of power reactor, while the high-temperature gas reactor has had only limited application.

9.1 NUCLEAR ENERGY

The fusion process, in which energy is released by the fusion of light nuclei, remains in the laboratory stage of development. Energy conversion achieved by controlled fission of a heavy nucleus is commercially applicable to electric power generation. A present-day nuclear power plant incorporates the nuclear reactor in a conventional steam power cycle. Essentially, the reactor replaces the fossil fuel-fired furnace as a heat source for the working fluid.

Reactor physics is highly complex, and only the most significant principles pertinent to power generation systems are examined in this text. Calculations for the energy released in a nuclear reaction are in general indicative of the conversion and do not represent precise determinations.

Release of energy by means of a variety of chemical reactions has been observed and employed by man for many years. These reactions result from changes that occur in the outer region of the atom, leaving the nucleus unchanged. The discovery of radioactivity was a precursor to a series of investigations of the emission phenomena involving the atomic nucleus. These experiments ultimately disclosed the possibility of releasing tremendous quantities of energy in reactions characterized by alterations of the nucleus of the atom.

The phenomenon of radioactivity was first observed by Henri Becquerel in 1896. He discovered accidentally that the fogging of a photographic plate was caused by an emission of some kind from a compound containing uranium. Subsequent investigations in the field of radioactivity by Pierre and Marie Curie led to the discovery of a new element, radium, a substance of far greater radioactivity than uranium. The discovery of radioactivity was highly significant because of the nature of the phenomenon. A radioactive substance was observed to be capable of spontaneous emission of energy in the absence of any transfer of energy to the substance.

Naturally occurring isotopes are stable in most chemical elements, but elements of atomic number 84 and above have no stable isotopes, and generally emit alpha particles on radioactive decay. The alpha particle is identical to the nucleus of the helium atom $^{4}_{2}He$. These particles are highly ionizing and have little penetrating ability. Alpha particles carry a positive charge.

Among the radioactive elements, the character of emission differs with respect to energy level and the number of emitting rays. Through spontaneous atomic disintegration, many nuclides decay into other unstable nuclides and thus create a decay chain that continues until a stable isotope is formed. In addition to alpha particle emission, two other types of emission are observed in radioactive decay, namely, beta ray and gamma ray emission.

Beta particles are actually high-energy electrons, and the beta ray carries a negative charge. The gamma ray carries no charge and is electromagnetic radiation of very short wavelength. Gamma rays have great penetrating ability and can traverse material for considerable distances. Emission of alpha and beta particles effects a decrease in mass and transmutation of the element to another element of different chemical characteristics. Radioactive decay is important in reactor operation.

The decrease in mass is particularly significant, because in accordance with the law of conservation of mass and energy, a corresponding release of energy must occur. The phenomenon of mass decrease is observed in the formation of the atomic nucleus from the component particles.

The nucleus of the atom consists of a number of protons that are positively charged particles and a varying number of neutrons that are electrically neutral. Incidentally, for an electrically neutral atom, the number of electrons is equal to the number of protons. The mass of the proton is 1.007277 amu (atomic mass unit), and the mass of the neutron is 1.008665 amu, based on an atomic mass of carbon equal to 12.000000 amu. The electrons add little to the mass of the atom, because the mass of the electron is 1/1836 of the mass of the proton. See Table 9.1. The number of protons contained in a nucleus is indicated by the atomic number Z, and the total number of protons and neutrons is indicated by the mass number A. A nucleon is either a proton or a neutron. Hydrogen $^{1}_{1}H$, the basic element, has a nucleus that consists of a single proton.

TABLE 9.1 Significant Mass Values (Atomic Mass Units)

Particle	*Symbol*	*Mass (amu)*
Proton	$^{1}_{1}p$	1.007277
Neutron	$^{1}_{0}n$	1.008665
Electron	e^{-}	0.000549
Alpha particle	$^{4}_{2}\alpha$	4.001505
Hydrogen atom	$^{1}_{1}H$	1.007825
Helium atom	$^{4}_{2}He$	4.002603
Lithium atom	$^{7}_{3}Li$	7.016004
Carbon atom	$^{12}_{6}C$	12.000000
Deuteron	$^{2}_{1}H$	2.013553
Deuterium atom	$^{2}_{1}H$	2.014102
Triton	$^{3}_{1}H$	3.015501
Tritium atom	$^{3}_{1}H$	3.016050

Source: *Handbook of Physics*, by E. U. Condon and H. Odishaw.[1] Copyright 1958, 1967, by McGraw-Hill, Inc. Used with permission of McGraw-Hill Book Company.

For all the elements, except hydrogen ^1_1H, the mass of the nucleus is less than the total mass of the component nucleons. The decrease in mass that occurs with the formation of a nucleus effects a release of energy in accordance with the law of conservation of mass and energy, stated quantitatively by $\Delta E = \Delta mc^2$. The term c denotes the velocity of light in vacuum.

The nucleons are bound together by a force system of the nucleus that favors attraction rather than repulsion.* The nucleus has a greater stability than the component nucleons prior to combination. Transfer to a state of greater stability is in effect a transfer to a state of lower energy level, hence the release of energy.

The iron nucleus $^{56}_{26}Fe$ has the greatest binding energy per nucleon (8.7857 MeV); consequently, the iron nucleus has the greatest stability. See Fig. 9.1. For mass numbers exceeding 56, the binding energy per nucleon decreases rather uniformly. For mass numbers less than 56, the binding energy per nucleon also decreases, but with some irregularity. The discrepancy of the helium nucleus is especially noteworthy in this respect.

Figure 9.1 is also a representation of the relative stability of the nuclei. Any nuclear transformation in the direction of iron from either side is a move to greater stability accompanied by a release of energy. The energy release may be achieved in a reaction that transforms a heavier nucleus into a lighter nucleus for $A > 56$, or generally a lighter nucleus into a heavier nucleus for $A < 56$.

The energy released in the formation of a nucleus is termed the binding energy, because this quantity of energy is required to decompose the nucleus into the constituent nucleons. The protons and neutrons are bound together in the nucleus by the forces of attraction existing among these particles. Separation of the nucleons against the opposing forces requires work and hence a transfer of energy to the nucleus.

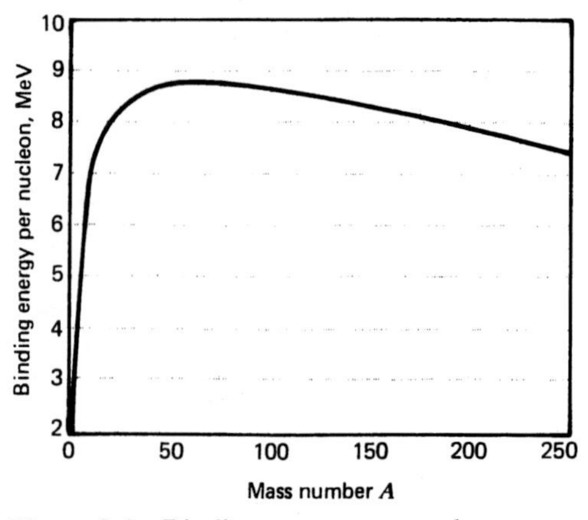

Figure 9.1 Binding energy per nucleon.

The binding energy of the helium nucleus will be calculated from the mass decrease that occurs in the formation of the nucleus from the component nucleons. The mass of the helium atom ^4_2He is given as 4.002603 amu. The mass of the nucleus is equal to the mass of the atom minus the mass of the two electrons.

$$4.002603 - (2 \times 0.000549)$$
$$= 4.001505 \text{ amu} \qquad \text{(mass of the helium nucleus)}$$

The mass of the constituent protons (2) and neutrons (2) prior to formation of the nucleus is equal to

$$(2 \times 1.007277) + (2 \times 1.008665) = 4.031884 \text{ amu}$$
$$\text{Mass decrease} = 4.031884 - 4.001505$$
$$= 0.030379 \text{ amu}$$
$$= 0.007595 \text{ amu per nucleon}$$
$$\text{Binding energy} = 0.007595 \times 931$$
$$= 7.071 \text{ MeV per nucleon}^\dagger$$

Alpha particles are emitted from the nucleus of a radioactive isotope at a relatively high velocity, approximately 1⁄10 the speed of light. These particles are heavy and possess little penetrating ability—a few centimeters in air—and rapidly

* The strong nuclear interaction, one of the four natural forces, is the force that holds the nucleus together against the electric repulsion of the protons.

† ΔE (MeV) = 931 Δm (amu)
MeV = million electron volts
1 eV = 1.60210×10^{-19} J

experience a decrease in velocity and energy. Disintegration effected by emission of alpha particles results in transmutation of the element. For example,

$$^{238}_{92}U \longrightarrow {}^{4}_{2}\alpha + {}^{234}_{90}Th$$

Through emission of an alpha particle, uranium 238 is converted to thorium 234. The nucleus of the parent isotope loses two protons and two neutrons. In addition, two low-energy electrons are emitted from the atom. Incidentally, the alpha particle will, on attaining a low level of energy, attract two free electrons to form an electrically neutral helium atom.

Emission of beta particles is also observed in ratioactive decay. The beta particle has the same mass and charge as the electron. The beta particles, emitted at a velocity approaching the speed of light, possess considerable penetrating ability. The nucleus contains no free electrons, hence beta particles are believed to originate in the spontaneous conversion of a neutron into a proton and an electron, plus a very small particle called a neutrino. The electron is emitted while the proton remains in the nucleus. Beta emission from thorium 234 is representative of this reaction. Thus,

$$^{234}_{90}Th \xrightarrow{\beta} {}^{234}_{91}Pa$$

While the atomic number increases from 90 to 91, the mass number remains at 234. The protactinium nucleus is of a chemical species different from that of the thorium nucleus.

Following emission of an alpha or beta particle, the remaining nucleus is still in an excited state and, in order to reach the ground state, emits the excess energy in the form of gamma radiation about 10^{-15} s later. Light rays or x-rays are emitted from the electrons, while electromagnetic radiation in the form of gamma rays is emitted from the nucleus. The highly penetrating gamma rays have physical characteristics similar to x-rays but have greater energy and shorter wavelength.

Many different nuclear reactions may be achieved by bombardment with accelerated particles. Of particular interest are the experiments conducted late in the 1930 decade on the neutron absorption of uranium. These experiments ultimately demonstrated that the absorption by the nucleus of the uranium isotope ^{235}U of a slowly moving neutron, traveling at a velocity within the thermal range, produces an unstable compound nucleus. The unstable structure may fission, and stability is achieved in the formation of two nuclei of medium weight. The product nuclei, except for a very small percentage of symmetrical divisions, are of unequal weight and range from zinc to gadolinium, a span of 35 elements and a large number of isotopes. The lighter and heavier fragments fall, respectively, into groups with mass numbers ranging approximately from 80 to 110 and from 125 to 155.

Fission of the uranium 235 nucleus produces, in addition to the two medium-weight nuclei, an assortment of particles including several neutrons. A large quantity of energy, about 200 MeV, is released with each fission. Most of this energy exists in the form of kinetic energy of the fission fragments and the neutrons. A small fraction of the energy is emitted as beta and gamma radiation. The total mass of the fission products is less than the combined mass of the original nucleus and the incident neutron. As in the several reactions cited above, the energy released is quantitatively equivalent to the mass decrease.

The release of 200 MeV per fission represents a tremendous quantity of energy, certainly in comparison with the combustion reaction on the basis of equal masses of fuel. Incidentally, the energy release for the combustion of one atom of carbon with oxygen to produce CO_2 is only 4.0 eV.

The uranium isotope ^{235}U is the predominant nuclear fuel used in present-day power reactors. A typical fission reaction is represented by

$$^{235}_{92}U + {}^{1}_{0}n \rightarrow {}^{137}_{56}Ba + {}^{97}_{36}Kr + 2{}^{1}_{0}n + Q \quad (9.1)$$

The mass of the ^{235}U nucleus and the incident neutron is greater than the total mass of the product nuclei, barium 137 and krypton 97, and the two product neutrons. The mass reduction for this reaction is 0.2032 amu, and the equivalent energy is 189 MeV. Q is the energy released in the reaction.

The two fragments that result from the fission of the uranium nucleus have like charges and, consequently, because of repulsion, move apart at extremely high velocities. The fission fragments penetrate the reactor materials to a depth of about 0.015 mm, and the kinetic energy of the fragments, a very large quantity, is converted into thermal energy.

The electrical energy associated with the separation of the nuclei can be calculated approximately from a Coulomb equation. The calculated value of 200 MeV is essentially in agreement with the energy equivalent of the mass difference, about 200 MeV. Thus, electrical repulsion between the fission fragments causes the conversion of mass energy into kinetic energy of the product particles.

The barium–krypton reaction shown above represents only one of several types of reactions that can occur in the fission of the ^{235}U nucleus. Barium 137 is a stable isotope, while krypton 97 is unstable. The decay of the unstable isotopes in a reactor will be examined at a later point.

Another possible fission reaction is given by

$$^{1}_{0}n + ^{235}_{92}U \rightarrow ^{236}_{92}U^* \rightarrow ^{140}_{54}Xe + ^{94}_{38}Sr + 2^{1}_{0}n + Q \quad (9.2)$$

When the neutron enters the ^{235}U nucleus, the structure is transformed into the unstable compound $^{236}U^*$ nucleus that immediately fissions into xenon 140 and strontium 94. Both of these product nuclei are unstable and subsequently undergo successive beta decay as shown below.

$$^{140}_{54}Xe \xrightarrow[16\ s]{\beta} ^{140}_{55}Cs \xrightarrow[1.1\ m]{\beta} ^{140}_{56}Ba \xrightarrow[12.8\ d]{\beta} ^{140}_{57}La \xrightarrow[40\ h]{\beta} ^{140}_{58}Ce \quad (9.3)$$

$$^{94}_{38}Sr \xrightarrow[1.3\ m]{\beta} ^{94}_{39}Y \xrightarrow[20\ m]{\beta} ^{94}_{40}Zr \quad (9.4)$$

The half-life time for each beta decay is shown in the preceding equations. Beta decay causes successive changes in the chemical species of the atom. For example, the ejection of the first beta particle changes xenon 140 to cesium 140 as the number of protons in the atom is increased from 54 to 55. The cerium 140 and the zirconium 94 nuclei are stable and the decay is thus terminated.

The mass values for the several particles are

$^{1}_{0}n$	1.0087 amu
$^{235}_{92}U$	235.0439
$^{140}_{58}Ce$	139.9054
$^{94}_{40}Zr$	93.9063
β	0.0005

Initial mass

$^{1}_{0}n$	1.0087 amu
$^{235}_{92}U$	235.0439
	236.0526 amu

Final mass

$2^{1}_{0}n$	2.0174 amu
$^{140}_{58}Ce$	139.9054
$^{94}_{40}Zr$	93.9063
$6\ \beta$	0.0030
	235.8321 amu

Mass decrease = 0.2205 amu

Energy equivalent = 205 MeV

Following penetration of the incident neutron, the heavy nucleus, for example, ^{235}U, divides into two very highly excited fragment nuclei that contain all of the neutrons. The fragments are rich in neutrons and must reduce the neutron number. Within an exceedingly short period of time, the fragments expel the excess, or prompt, neutrons and simultaneously emit gamma radiation as they drop to states of lower energy. In the lower-energy state, continued instability of the fragment nuclei leads to beta decay and neutron decay.

When uranium 235 nuclei fission in the reactor, about 99 percent of the neutrons are emitted immediately. The remainder of the neutrons are released in delayed reactions. In a typical fission reaction, uranium 235 fissions into bromine and lanthanum:

$$^{1}_{0}n + ^{235}_{92}U \rightarrow ^{87}_{35}Br + ^{147}_{57}La + 2^{1}_{0}n + Q \quad (9.5)$$

The bromine isotope $^{87}_{35}Br$ is unstable and decays into krypton 87 by emission of a beta particle. Thus,

$$^{87}_{35}Br \xrightarrow{\beta} ^{87}_{36}Kr \quad (9.6)$$

Krypton 87 is also unstable and decays by two means: (1) normal beta emission to form rubidium $^{87}_{37}Rb$, a stable isotope, and (2) the emission of a neutron to form the stable krypton isotope $^{86}_{36}Kr$.

$$1 \quad ^{87}_{36}Kr \xrightarrow{\beta} ^{87}_{37}Rb \quad (9.7)$$

$$2 \quad ^{87}_{36}Kr \xrightarrow{^{1}_{0}n} ^{86}_{36}Kr \quad (9.8)$$

The neutron emitted in reaction 2 is designated as a delayed neutron.

The highly excited nucleus $^{87}_{35}Br$ produced by fission contains more neutrons than is acceptable for stability. The beta decay of $^{87}_{35}Br$ is a precursor to the expulsion of the neutron in a secondary reaction. Thus, the delay time for the neutron emission is controlled by the beta decay half-life time of 55.6 s for $^{87}_{35}Br$. This subsequent neutron decay represents the source of the delayed neutrons observed in the reactor.

Reaction 1 can be written as

$$^{87}_{36}Kr \longrightarrow ^{87}_{37}Rb + \beta + \nu + Q \quad (9.9)$$

The neutron decays into a proton, a beta particle, and a neutrino ν. The neutrino is a particle of zero rest mass and does not carry a charge. Despite these characteristics, the neutrino carries away a fraction of the energy of the reaction. The energy Q released by the reaction is shared by the beta particle and the neutrino. The beta particles are emitted at extremely high speeds in the range of 120,000 to 240,000 km/s.

A reaction often leaves the nucleus in an excited state. The subsequent drop of the nucleus from a higher energy state to a lower, more stable state is accompanied by gamma radiation, in the absence of any chemical change in the substance. Unlike alpha and beta particle emission, gamma rays, as noted earlier, are electromagnetic radiation of exceedingly high frequency and therefore possess great penetrating capability.

The following energy distribution is representative of an average fission of about 200 MeV.

Kinetic energy of the fission fragments	170 MeV
Kinetic energy of the neutrons	5
Beta rays and gamma rays	15
Energy of the neutrinos associated with beta decay	10
	200 MeV

9.2 NEUTRON EMISSION

When the ^{235}U nucleus is fissioned by the incident neutron, the number of product neutrons will probably be one, two, or three, depending upon the manner in which the nucleus fissions. Average values of the number of product neutrons (see Table 9.2) are highly significant with respect to a sustained reaction.

A steady-state chain reaction can only be sustained when, for each fission, a product neutron effects in turn a successive fission of another nucleus. Because of certain losses, not all of the product neutrons will cause a successive fission

TABLE 9.2 Neutrons Produced in Thermal Neutron Fission

Nucleus	*Average Number of Neutrons Produced per Fission*	
Uranium 235	2.47	(2.43)
Plutonium 239	2.91	(2.90)
Uranium 233	2.55	(2.51)

Source: Reference 2. Figures in parentheses are from reference 4.

reaction. For this reason, the excess number of neutrons, 1.47 for uranium 235, is a highly important characteristic. In the absence of an adequate number of excess neutrons, a nuclear power reactor would not function.

Neutrons ejected by fission are designated as "fast," because speeds of about 16,000 km/s are initially attained. These neutrons have an initial kinetic energy on the order of 2 MeV. A fast neutron in collision with a uranium atom is more likely to lose kinetic energy than to be captured or to produce fission. Provision must then be made in thermal reactor design for reducing the speed of a sufficiently large number of fast neutrons. A material designated as a moderator is introduced into the reactor for this purpose. Successive collisions of the neutrons with the nuclei of the moderator material cause the neutrons to lose kinetic energy. Eventually, the neutrons possess only thermal energy, about $\frac{1}{40}$ eV. Thereafter, the average kinetic energy of the neutrons remains unchanged with further collisions because the nuclei of the moderator material have only thermal motion.

Fast neutrons have little ability to fission either ^{235}U or ^{238}U and pass from the fuel rod, some to be reduced subsequently to slow neutrons in the moderator. Slow neutrons, on the other hand, have a high probability of producing fission in ^{235}U but cannot cause fission in ^{238}U.

A possible neutron balance in a natural uranium reactor that is just critical is indicative of the manner in which a steady-state chain reaction is sustained. A representative group of 1000 fast neutrons produced by fission of uranium is selected for examination. See Table 9.3. An average number of 2.5 neutrons is assumed to be released for each fission.

The product fast neutrons, in Table 9.3, total 1000, hence the chain reaction is sustained in a steady state. Of the total 1000 neutrons emitted by fission, 400 neutrons effect subsequent fission, a number adequate for maintaining steady-state operation of the reactor. The production of a relatively large number of excess neutrons above those neutrons required for sustaining the chain reaction is a factor of great importance in reactor design. The neutron balance provided in Table 9.3 shows the large number of neutrons not available for subsequent fission because of resonance capture and leakage from the reactor.

TABLE 9.3 Neutron Balance in a Natural Uranium Reactor

		Product Neutrons (Fast)
Disposition of 1000 fast neutrons		
Resonance capture by ^{235}U and ^{238}U	100	
Leakage from the reactor	192	
Fission of ^{235}U	4	10
Fission of ^{238}U	4	10
Transformation by the moderator to slow neutrons	700	
	1000	
Disposition of 700 slow neutrons		
Fission of ^{235}U	382	955
Fission of plutonium	10	25
Captured by fission products	23	
Captured by reactor materials	2	
Resonance capture by ^{235}U and ^{238}U	208	
Leakage from reactor	75	
	700	1000

Source: Reference 3.

In an inelastic neutron process, the kinetic energy of the system, including the target nucleus and the incident neutron, is not conserved. The neutron actually enters the nucleus of the atom, and as a consequence the stability of the nucleus is disrupted. In an effort to regain stability, an internal readjustment occurs in the structure of the new nucleus that has been formed by penetration of the incident neutron. There are three general ways in which equilibrium can be established, including fission. Inelastic scattering is the term applied to the process that ejects a neutron for each neutron absorbed. The energy of the new neutron is lower than that of the incident neutron.

Neutron absorption is the third inelastic process in which equilibrium is established, in this case by a release of energy and particles. In the reactor, neutron absorption, or resonance capture, is particularly significant because the process results in a loss of neutrons. Resonance capture of an incident slow neutron by ^{235}U is represented by

$$ {}_0^1n + {}_{92}^{235}U \rightarrow {}_{92}^{236}U + \gamma + Q \qquad (9.10) $$

This reaction has a cross section of 107 barns for thermal neutrons. The cross section is a measure of the probability of the occurrence of a reaction. Further discussion of this term is presented at a later point. A cross section of 107 barns is low in comparison with a value of 545 barns for the fission of ^{235}U by thermal neutrons. Incidentally, the product ^{236}U is of no value, but the gamma radiation and the associated energy release cannot be ignored.

The fuel for a uranium reactor consists principally of ^{238}U, whether the reactor is charged with natural or enriched uranium. Resonance capture of neutrons by the ^{238}U nuclei is a highly significant reaction because of the subsequent reactions that occur. The initial absorption of a neutron is represented by

$$ {}_0^1n + {}_{92}^{238}U \rightarrow {}_{92}^{239}U + \gamma + Q \qquad (9.11) $$

This reaction is similar to the reaction described by Eq. 9.10.

The product ^{239}U is unstable, and the proportion of protons and neutrons is readjusted by beta decay in two steps. The first step produces neptunium 239.

$$ {}_{92}^{239}U \rightarrow {}_{93}^{239}Np + \beta + \nu + Q \qquad (\tau = 23 \text{ min}) \qquad (9.12) $$

The second step transforms the neptunium 239 into plutonium 239.

$$ {}_{93}^{239}Np \rightarrow {}_{94}^{239}Pu + \beta + \nu + Q \qquad (\tau = 2.3 \text{ days}) \qquad (9.13) $$

The conversion of ^{238}U to ^{239}Pu is particularly important because ^{239}Pu can undergo fission, with a cross section of 730 barns, by thermal neutrons.

Plutonium can be used for fueling reactors, but the substance is radioactive and highly toxic and must be carefully handled. The unstable plutonium isotope ^{239}Pu has a very long half-life, about 24,000 years. Through emission of an alpha particle, ^{239}Pu is transformed to ^{235}U.

$$ {}_{94}^{239}Pu \rightarrow {}_{92}^{235}U + {}_2^4\alpha \qquad (9.14) $$

Both ^{238}U and ^{235}U are naturally radioactive and degenerate to the stable lead isotopes ^{206}Pb and ^{207}Pb in 14 and 11 steps, respectively. Alpha and beta particles are emitted, and additional energy is released in the form of gamma rays and neutrinos.

9.3 CONVERSION AND BREEDING

Some of the excess neutrons produced in the reactor can be used to manufacture, from fertile materials, certain other nuclei that fission with slow neutrons. When the reactor produces more than one new fissionable nucleus for each nucleus that is fissioned, the process is termed breeding. The quantity of fissionable fuel in a breeder reactor thus increases with the operation of the reactor. The artificially produced fuel can be removed for fueling other reactors.

Conversion is the term applied to the production of a new fuel in a quantity that is less than the amount consumed. The new fissionable fuel produced in the converter reactor is consumed in the reactor to augment the fuel charge.

The transmutation processes employed for breeding and converting fertile substances are similar. Two fertile isotopes considered for these processes are thorium 232 and uranium 238. The latter process was examined in the preceding section.

Thorium 232 is transformed to uranium 233, an isotope that is fissionable by slow neutrons. The transformation is achieved by neutron resonance capture and successive beta decay in the following steps.

$$ {}_{0}^{1}\mathrm{n} + {}_{90}^{232}\mathrm{Th} \rightarrow {}_{90}^{233}\mathrm{Th} + \gamma + Q \qquad (9.15)$$

$$ {}_{90}^{233}\mathrm{Th} \rightarrow {}_{91}^{233}\mathrm{Pa} + \beta + \nu + Q \qquad (9.16)$$

$$(\tau = 23 \text{ min})$$

$$ {}_{91}^{233}\mathrm{Pa} \rightarrow {}_{92}^{233}\mathrm{U} + \beta + \nu + Q \qquad (9.17)$$

$$(\tau = 27.4 \text{ days})$$

Conservation of neutrons becomes a matter of considerable importance in the operation of a breeder or converter reactor. The average number of neutrons emitted per fission of uranium 235 may be taken as 2.5. If a reactor is required to manufacture as much fissionable materials as it consumes, one neutron is needed for sustaining the chain reaction, and one neutron is used in generating new fuel. Only 0.5 neutron remains for leakage or capture by nonfission processes. In general, the number of neutrons available for generation of new fuel is equal to 1.5 minus the number lost because of leakage and capture.

Neutron leakage, necessarily held to a minimum, is controlled by instituting appropriate design procedures. The loss of neutrons to the reactor structural materials can be reduced by selection of materials of low neutron cross section. The cross section of a substance is, in part, a function of the energy of the incident neutron. Thus, the probability of neutron capture by nonfission processes can be greatly reduced by elevating the energy of the incident neutron to higher values.

9.4 URANIUM ENRICHMENT

Naturally occurring uranium contains 0.72 percent of the ^{235}U isotope. Reactors are ordinarily charged with slightly enriched uranium in which the ^{235}U fraction ranges between 2 and 5 percent. Uranium enrichment requires some kind of process for separation of the ^{235}U isotope. Because the ^{235}U and the ^{238}U isotopes have similar chemical properties, a nonchemical separation process must be employed.

Several different methods have been examined for ^{235}U separation. Gas separation in large high-speed centrifuges has been successfully demonstrated in the developmental stage, and a production-scale plant is under construction. So far, the gaseous diffusion method has been used for effecting uranium enrichment.

Uranium exists naturally in the form of various compounds. One common mineral is uraninite (uranium oxide, U_3O_8). Most of the commercial-grade ore in the United States is mined in the western part of the country and contains on the average 2.5 kg U_3O_8, or equivalent, per metric ton of ore. The ore is processed for extraction of the uranium in the uranium mill, producing an end product known in the industry as "yellowcake." The yellowcake, a crude uranium concentrate that contains from 70 to 90 percent U_3O_8, is further refined and converted into uranium hexafluoride (UF_6).

The principles of gas kinetics are the basis of the method used for separation of the uranium isotopes by gaseous diffusion. Gaseous uranium hexafluoride is a mixture of two gases, namely, $^{238}UF_6$ and $^{235}UF_6$. The molecules of the lighter $^{235}UF_6$ gas have higher velocities and strike the walls of the containing vessel more frequently than the molecules of the heavier $^{238}UF_6$ gas. If the walls of the containing vessel are actually screens or barriers that have microscopic passages, the molecules of the lighter gas will diffuse through the barrier in a proportion greater than the concentration of the light molecules in the gas mixture. Because the holes in the barrier are very small and only sufficiently large to allow passage of individual molecules, bulk flow of the gas through the barrier does not occur.

The gas outside the barrier now has a higher concentration of $^{235}UF_6$ molecules than the initial gas mixture, and the gas inside the containing vessel has a lower concentration of $^{235}UF_6$ molecules. The degree of separation attained in this manner in a single step is very small, and the process must be repeated a great many times in order to achieve the enrichment required in a fuel intended for a power reactor.

A gaseous diffusion plant consists of hundreds of stages arranged for cascade flow. Each diffusion stage is constructed in the form of a chamber divided into two zones by a finely porous, thin barrier. System operation establishes a pressure difference between the two zones of the chamber. The gaseous mixture consisting of $^{235}UF_6$ and $^{238}UF_6$ is admitted to the higher-pressure zone. A portion of the mixture diffuses into the lower-pressure zone, and the remainder, slightly depleted in the ^{235}U isotope, flows from the higher-pressure zone downward to the next stage. The portion of the mixture that diffuses through the barrier is slightly enriched in the ^{235}U isotope and is pumped upward to the next stage in the cascade. A suitably enriched mixture is ultimately discharged from one end of the cascade, and a mixture depleted in the ^{235}U isotope is discharged from the other end.

Because many stages are required, the gaseous diffusion plants are large and complex. Each diffusion stage is equipped with a pump and a gas cooler. Chemical processing and pump power requirements contribute substantially to the operating cost for uranium enrichment by gaseous diffusion. The material used for fabrication of the barriers must resist corrosion and clogging of the microscopic passages by the process gas. The technology for gaseous diffusion is well established, and the system has demonstrated high reliability.

Chemical processing is employed to convert the enriched UF_6 to uranium dioxide (UO_2). The raw UO_2 is refined in a series of metallurgical processes and then compacted under high pressure to form small cylindrical pellets. Fabrication is completed by sintering the pellets at a high temperature in a reducing atmosphere and subsequently grinding them to a prescribed diameter. The dimensions of the pellets vary somewhat with the reactor design. A pellet diameter of about 8 mm is typical, and the length of the pellet is ordinarily somewhat greater than the diameter.

Uranium enrichment by gas centrifuge will apparently achieve certain advantages. In comparison with the gaseous diffusion process, the gas centrifuge will require one-tenth as much power and less heat will be rejected.

9.5 REACTOR DESIGN

The neutrons produced by nuclear fission have an extremely high initial velocity corresponding to about 2 MeV kinetic energy. Three general types of reactors are classified in accordance with the neutron velocity that is prescribed for effecting fission. If most of the fissions in the reactor are initiated by very-high-speed neutrons with energies above 1000 eV, the term "fast reactor" is applied. In an "intermediate reactor," the neutron energies are in the range of 0.05 to about 1000 eV. Most of the fission in a "thermal reactor" is caused by neutrons that have energy values below 0.05 eV, that is, average values in the vicinity of 0.025 eV.

In order to effect proper fission of the fuel, neutron velocities must be reduced to an appropriate level for the particular type of reactor. Neutrons that escape absorption by nuclei lose kinetic energy by elastic collisions with nuclei and particles until equilibrium is attained with the surroundings. A material, termed a moderator, is placed in the thermal reactor for this express purpose. Fast-moving neutrons diffuse out of the fuel into the moderator where an effective reduction in the velocity of the neutrons is achieved. Subsequently, the now slowly moving neutrons diffuse back into the fuel where they have a greater probability of causing fission in the ^{235}U and are less likely to be caught by the ^{238}U.

Reactors can be designed to achieve a chain reaction by fast neutrons. However, in a reactor fueled with natural uranium, fission induced by fast neutrons will not sustain a chain reaction. Early reactors were fueled with natural uranium, hence reactor development has been directed principally to designs utilizing fission by thermal neutrons.

The moderator must have little capability for absorbing neutrons and should have a low atomic weight. The reduction in the velocity of the fast-moving neutrons is achieved in elastic collision with the nuclei of the moderator. The neutrons experience a greater reduction in kinetic energy on each collision with a light-weight nucleus than would be the case in collision with relatively heavy nuclei. The nuclei of several elements can function effectively as moderators.

Hydrogen is a highly effective moderator because the nucleus consists of a proton, which is about the same size as the neutron. The introduction of hydrogen into the reactor would be as a component of water, which is cheap and readily available. The hydrogen nucleus however has a tendency to absorb a neutron, a process that results in the formation of a deuteron and gamma ray emission. Heavy water, which contains deuterium, is a satisfactory moderator with respect to neutron absorption, but the substance is costly to manufacture.

Lithium and boron are strong absorbers of neutrons and consequently would be unsuitable to serve as moderators. Graphite, a good moderator, is structurally weak; and beryllium, also satisfactory, is costly and hence its use is limited. Although helium has a light-weight nucleus, the concentration of the nuclei is low, since the helium would be present in the reactor in the gaseous phase. The moderating characteristics of helium would consequently be poor because of the relatively long distance the neutrons must travel between collisions with the helium nuclei.

Microscopic cross section is the term applied to the probability that a particular interaction between a nucleus and a moving particle will occur. For a specified reaction, the microscopic cross section is the effective area presented by a stationary nucleus to the approaching particle. The unit for the microscopic cross section is the barn, a quantity established on a nuclear scale of magnitude and equal to 10^{-24} cm^2. The microscopic cross section for a reaction is dependent upon the particular type of the reaction and the energy of the incident particle. The cross section is designated by σ, with an appropriate subscript used to denote the general type of the reaction; for example, σ_f is the cross section for fission. Incidentally, the microscopic cross section is not necessarily related to the geometric cross-sectional area of the nucleus.

The total cross section σ_t for a neutron approaching a fissionable atom is expressed by

$$\sigma_t = \sigma_c + \sigma_s + \sigma_f \tag{9.18}$$

where

σ_c = the capture cross section, a measure of the area presented for absorption without fission

σ_s = the equivalent area of the nucleus that will scatter or deflect the neutron

σ_f = the area that is presented for a neutron to strike and cause fission to occur (This characteristic is peculiar to only a few of the many nuclei.)

Also,

$$\sigma_a = \sigma_c + \sigma_f$$

where

σ_a = the absorption cross section, the area presented for absorption with or without fission

The term critical signifies that the fission chain reaction is being maintained at a constant rate. If the fission rate is increasing, the reactor is described as supercritical. If the fission rate is decreasing, the chain reaction is not maintained, and the reactor becomes subcritical.

The size of the reactor core is particularly significant. The production of neutrons by fission will be dependent upon the volume of the core. Neutron leakage occurs at the boundary of the core, hence the extent of the core surface in relation to the volume of the core becomes important to the design of the reactor. The ratio of volume to surface, V/S, establishes a characteristic linear dimension that is indicative of the core size. There will be a unique size of the core at which the reactor is exactly critical. Below this size, a chain reaction cannot be sustained because of excessive leakage of neutrons from the core. Above the critical size, the neutron density, and thus the reactor power, could increase indefinitely.

A reflector is placed around the reactor core to return many of the neutrons that tend to escape from the core. A material of low atomic weight is used for the reflector in a thermal reactor for the purpose of assisting in the reduction of the neutron velocity. For the fast reactor, the reflector should be made of a material that has a high atomic weight. On collision with a body of greater mass, the neutron will be reflected with little loss in kinetic energy. Graphite can be used as a reflector in a thermal reactor; and for a water reactor, the water around the core serves as both the reflector and the moderator.

The reactor must be adequately shielded in order to trap the unintended and biologically harmful emission consisting of alpha and beta particles, gamma rays, and neutrons. Neutrinos are not known to interact with any material through which they pass, hence there is no requirement to shield against neutrino emission from the reactor. Alpha and beta particles travel relatively short distances and are therefore easily caught in the reactor structure. Gamma rays and thermal neutrons travel appreciable distances and have intermediate ranges of penetration. Reactor shielding is designed principally to intercept the neutron and gamma ray emissions, usually by large masses of concrete placed around the reactor vessel. Steel and lead are also effective shielding materials, although the latter is comparatively expensive. Fast neutrons can be moderated prior to possible escape from the reactor.

9.6 REACTOR CONTROL

The control system for a power reactor has two main functions, namely, regulating the reactor output and providing adequate measures to protect the reactor from failure. The thermal output of the reactor must be maintained constant at prescribed levels for finite periods of time. Actually, it is not unusual for such time periods to be of extensive duration. The reactor output however must quickly be adjusted in response to changes in the load imposed on the turbine-generator.

Any mechanical device or machine can fail in some manner. Fortunately, most such failures are of minor importance. Presumably, for some reason the reactor output could increase to a level that would place the reactor in a condition of incipient failure. The reactor control system will however sense the power buildup and quickly shut down the reactor before damage to the structure can occur.

Reactor control is decidedly a complex subject, and design procedures for control systems are not simply treated. There are however certain aspects of reactor control that can be examined at this point, and some significant observations will be made.

The reactor power is fixed by the rate at which the fissions occur. The fissioning rate, in a particular reactor core, is proportional to the quantity of fissionable material present in the core and to the average neutron density. Because the quantity of fissionable material contained in most cores will decrease at a slow rate, the power level of the reactor is controlled by regulating the neutron density.

When the reactor is critical, a chain reaction is achieved and steady-state operation is maintained. If 2.5 neutrons are emitted per fission on the average, the neutron production is taken as 1.5, since one of the emitted neutrons must be used to achieve a subsequent fission. Because of leakage and nonfissionable absorption, 1.5 neutron do not participate in the fission process. Thus, for a steady-state chain reaction,

$$\text{Neutron production} = \begin{matrix}\text{neutron}\\\text{leakage}\end{matrix} + \begin{matrix}\text{neutron}\\\text{absorption}\end{matrix}$$

An increase in the neutron production, without

a corresponding increase in the neutron leakage and absorption, would increase the neutron density. As a result of this change, the fission rate would increase and a rise in the reactor temperature would be observed. The neutron density increases in accordance with the compound interest principle, hence, unless multiplication is halted at some prescribed level, the reactor core would ultimately be destroyed by melting.

For a fixed thermal output, the reactor control system must then maintain the neutron production at a constant level. In order to change the reactor output, the neutron production is altered briefly. For example, to increase the reactor power output, the neutron fission rate must be increased to a higher level. As soon as the prescribed output is attained, the neutron multiplication rate is again reduced to zero. Thus, neutron production must remain constant, except for a short period of time when the reactor power output is changing in either direction.

The effective neutron multiplication factor k_{eff} is defined as the ratio of the neutron population in each generation to the population in the immediately preceding generation. In addition, k_{eff} is defined by

$$k_{eff} = \frac{\text{neutron production}}{\text{neutron leakage + neutron absorption}} \tag{9.19}$$

For a thermal reactor, k_{eff} is evaluated as the product of a number of factors pertaining to neutron production, absorption, and capture and to neutron retention within the core.

The reactivity ρ is defined as the ratio of the excess of the effective multiplication factor to the effective multiplication factor. Thus,

$$\rho = \frac{k_{eff} - 1}{k_{eff}} \tag{9.20}$$

The value of $k_{eff} = 1.0$ corresponds to a constant power level in the reactor. If the value of k_{eff} is less than 1, the chain reaction is not sustained and will ultimately die down.

The neutron lifetime, designated by τ^*, is defined as the average time between successive neutron generations. τ^* is the mean time that elapses between the production of neutrons by fission and the subsequent fission effected by these neutrons or loss of the neutrons from the reaction. Only prompt neutrons are covered by this definition.

The number of neutrons existing in a unit volume is designated by n. The time rate of change of the neutron density is given by $\partial n/\partial \tau$. Then,

$$\frac{dn}{d\tau} = \frac{n\rho}{\tau^*} \quad \text{or} \quad \frac{dn}{n} = \frac{\rho}{\tau^*}\, d\tau$$

Integration of the above equation yields

$$\ln \frac{n}{n_0} = \frac{\rho}{\tau^*}\tau \tag{9.21}$$

where n and n_0 represent, respectively, the neutron density at time τ and $\tau = 0$.

An equivalent form of Eq. 21 is given by

$$n = n_0 e^z \tag{9.22}$$

where $z = \frac{\rho}{\tau^*}\tau$

Example 9.1

A reactor is operating at a low power of 1 W. The reactor then becomes supercritical with $k_{eff} = 1.0015$. The average neutron lifetime is 0.0001 s, a value that is applicable to prompt neutrons. Determine the reactor power level at the end of 1 s.

$$\rho = \frac{k_{eff} - 1}{k_{eff}}$$

$$= \frac{1.0015 - 1}{1.0015} = 0.0014978$$

$$z = \frac{\rho}{\tau^*}\tau$$

$$= \frac{0.0014978}{0.0001} 1 = 14.978$$

$$\frac{n}{n_0} = e^z = e^{14.978}$$

$$= 3.198 \times 10^6$$

The reactor power is proportional to the neutron density, hence within a period of 1 s, the reactor power is increased from 1 W to 3.198 MW.

If in the above example k_{eff} is taken as 1.002, the power level at the end of one second is 466 MW.

The extremely rapid increase in the reactor power in relation to the increase in the effective multiplication factor would seemingly create a severe problem in the development of a control system capable of quick response to changes in the reactor operation. The difficulty arises actually from the extremely short average neutron lifetime of the prompt neutrons. Fortunately, a natural phenomenon, that of the delayed neutrons, alters to a highly significant degree the magnitude of τ^*.

The atomic fragments produced simultaneously with the emission of the prompt neutrons by fission of the uranium 235 nucleus are in a highly unstable condition. Some of these fragments emit additional neutrons after a short time delay. See Section 9.1. About 0.75 percent of all of the neutrons produced from fission of ^{235}U are delayed for an average time of about 10 s. Although the delayed neutrons constitute only a small fraction of the total number of neutrons produced by fission, the presence of these neutrons causes an increase in the average neutron lifetime to 0.1 s. This time period of 0.1 s exceeds by a factor of 1000 the average neutron lifetime of 0.0001 s that would be applicable if all of the neutrons were emitted promptly at fission.

The presence of the delayed neutrons greatly decreases the multiplication rate of the neutrons. As a consequence of this lower multiplication rate, sufficient time is available for the control system to regulate the reactor power output and maintain stable operation. Adjustment in the neutron production can be made with sufficient rapidity to contain any sudden or abnormal increase in the neutron multiplication rate.

Example 9.2

Calculate the time required for the reactor power to double for (a) assumed fission with prompt neutrons and (b) actual fission with prompt and delayed neutrons. $k_{eff} = 1.002$.

$$\rho = \frac{k_{eff} - 1}{k_{eff}}$$

$$= \frac{1.002 - 1}{1.002} = 0.0019960$$

$$\frac{n}{n_0} = 2 = e^z \qquad z = 0.69315$$

(a) $\tau^* = 0.0001$ s

$$\tau = \frac{z\tau^*}{\rho}$$

$$= \frac{0.69315 \times 0.0001}{0.0019960} = 0.0347 \text{ s}$$

(b) $\tau^* = 0.1$ s

$$\tau = \frac{z\tau^*}{\rho}$$

$$= \frac{0.69315 \times 0.1}{0.0019960} = 34.73 \text{ s}$$

The effect of the presence of the delayed neutrons on the neutron multiplication rate can be observed in this example.

Critical operation of the reactor is achieved when the production of neutrons equals the leakage plus the absorption. Production can be altered by changing either the neutron absorption or the neutron leakage. Thermal reactor control is usualy accomplished by changing the neutron absorption rate and thus effecting a change in the neutron multiplication rate.

Regulation of neutron absorption is achieved by altering the quantity of neutron absorbing material present in the reactor. Control rods composed of a substance that has a high-thermal-neutron absorption cross section are moved in the reactor to appropriate positions. The control rods are moved singly or in groups; and when the rods are fully inserted into the reactor core, the chain reaction is terminated. Control rods are fabricated from various metals but contain appropriate quantities of substances such as cadmium and boron that have high absorption cross sections for thermal neutrons.

Control rods perform different functions. Certain groups of rods function as safety rods and effect reactor shutdown. Coarse regulation of the reactor power level is accomplished by the shim rods. Fine changes in the reactor power level are effected by a regulator rod. Positive operation of the safety rods can be ensured in an arrangement that provides for fall of the rods into the reactor core. The rods are held above the core by magnetic latches and when released drop under the influence of gravity. In an alternative arrangement, the control rods enter at the bottom and move upward through the reactor core.

In practice, the reactor is designed for excess reactivity, that is, it is larger than the indicated critical size. Certain of the fission products formed in the reactor core as a result of the continued reaction strongly absorb neutrons and hence interfere with the fission process. These absorbing substances are called poisons. Thus, to prevent the reactor from ultimately becoming subcritical because of the formation of these poisons, a supercritical design is developed. The excess reactivity also serves to compensate for fuel consumption and for possible errors in design, in addition to providing for operating flexibility.

A reactor has a negative temperature coefficient when the multiplication factor decreases slightly with an increase in the reactor temperature. If for some reason the temperature of such a reactor increases, the multiplication factor will decrease and, in turn, cause the temperature to drop. The original value of the multiplication factor will be attained when the reactor temperature drops to the initial level. A reversed sequence will take place when the temperature initially decreases. The negative temperature coefficient has a stabilizing influence on the reactor operation.

When the reactor is developing a constant power output, relatively slight variations, including temperature changes, will occur. Such variations are compensated by the movement of the control rods in the reactor core.

A uranium 235 reactor becomes prompt critical when the effective multiplication factor k_{eff} attains a value of 1.0075. At this k_{eff} value, a chain reaction is sustained on prompt neutrons alone, that is, the delayed neutrons are not needed to sustain the chain reaction. If k_{eff} exceeds a value of 1.0075, an extremely rapid exponential multiplication of the reactor power level will be experienced because of the very short average neutron lifetime. A control system must therefore ensure against k_{eff} exceeding a value of 1.0075.

Example 9.3

Calculate the time required for the reactor power to double if only prompt neutrons participate in the reaction. $k_{eff} = 1.0076$; $\tau^* = 0.0001$ s.

$$\rho = \frac{k_{eff} - 1}{k_{eff}}$$

$$= \frac{1.0076 - 1}{1.0076} = 0.0075427$$

$$\frac{n}{n_0} = 2 = e^z \qquad z = 0.69315$$

$$\tau = \frac{z\tau^*}{\rho}$$

$$= \frac{0.69315 \times 0.0001}{0.0075427} = 0.00919 \text{ s}$$

Example 9.3 illustrates the dangerous condition that will develop, if the effective multiplication factor is allowed to reach or exceed a value of 1.0075. Because the reactor must be designed with excess reactivity, as noted above, there exists an inherent capability of attaining a "runaway" condition. The control system however will prevent such an occurrence.

9.7 TYPES OF REACTORS

Although many different kinds of nuclear reactors demonstrate potentialities for power production, two general classifications are recognized, namely, the heterogeneous and the homogeneous types. Present-day nuclear power reactors are in the heterogeneous class.

The core of the heterogeneous-type reactor consists in part of a large number of fuel rods. The coolant is circulated around the fuel rods to remove the thermal energy produced by nuclear fission. An adequate number of control rods are inserted in the core among the fuel rods for regulating the reactor power level. After leaving the core, the coolant flows either directly to the steam turbine or through a heat exchanger. In the latter system, energy is transferred in the heat exchanger from the coolant to a secondary fluid that ordinarily is the working fluid for the vapor power cycle.

The homogeneous-type reactor, in one design, is charged with a fissionable salt of uranium, for example, uranyl sulfate (or uranyl nitrate), dissolved in the moderator that is either H_2O or D_2O, light or heavy water. The solution is critical in the core, and the thermal energy generated at this point is removed by circulating the solution through a heat exchanger and transferring the energy to a secondary fluid. The secondary fluid would normally be the working fluid in a power cycle. With respect to the reactor fuel, another concept provides for the circulation of a thorium-and-uranium slurry in water.

Although development of the homogeneous reactor has, to date, been restricted to experimental systems, several advantages can be cited for the homogeneous design. Unlike the heterogeneous reactor, there is an absence of problems encountered in the transfer of heat at very high fluxes from the fuel through the cladding to the coolant. The absence of structural materials in the core reduces the capture of neutrons in nonfissionable reactions. The homogeneous reactor is inherently more stable with respect to power surges. Further, fission products can be removed and new fuel added continuously, with no need to shut down the reactor. Component maintenance in the homogeneous reactor power plant is however a distinct problem. The circulating fuel is intensely radioactive, and difficulties because of erosion and corrosion are experienced.

A sufficiently high pressure must be maintained in an aqueous fuel reactor to prevent boiling. High-temperature operation in the absence of boiling can be achieved by dissolving the fuel in a liquid metal, such as bismuth, or in a molten salt, for example, lithium fluoride–beryllium fluoride. Again, these concepts have been applied only to experimental reactors.

9.8 REACTOR POWER

The core of the heterogeneous reactor consists of a large number of slender fuel rods. Each fuel rod, depending upon the length, will contain 100 to 200 small-diameter, about 10-mm, cylindrical fuel pellets. Because of the small cross-sectional area of these pellets in comparison with that of the core, the radial variation in the neutron flux in the rod may be considered to be negligible. There is however in the rod a longitudinal variation in the neutron flux.

The highest fission rate in the reactor core occurs essentially at the geometric center of the composite structure. Thus, the neutron flux and hence the heat generation or power density attain generally maximum values at the center of the cylindrical core. The radial power distribution decreases from the peak value on the center line to lower values at the outer limits of the core. In a similar manner, the power density decreases longitudinally from a peak value at the midpoint of a fuel rod to lower values at the core end sections. In conclusion, the neutron flux and the heat generation therefore peak in both axial and radial directions within the reactor core.

Because of the activity at the boundary of the core, the neutron density is not zero at the outer physical limits of the core. In order to evaluate the power density of the reactor, it is convenient to extend hypothetically the boundary of the core, both radially and longitudinally. Then, in addition to the actual length L and radius R of the core, the corresponding extrapolated dimensions L_e and R_e are established. At the hypothetical boundary, produced by extrapolation, the neutron flux is assumed to be negligible.

The specific energy release rate for a nuclear reaction is given by

$$\dot{q}_v = E_f \Sigma_f \phi \quad (9.23)$$

where

$\dot{q}_v$ = specific energy release rate, MeV/s·cm³
E_f = energy per reaction, MeV/fission
Σ_f = effective macroscopic fission cross section, cm^{-1}
ϕ = neutron flux, neutrons/s·cm²

Since the effective macroscopic fission cross section is equal to $\sigma_f N_{ff}$,

$$\dot{q}_v = E_f \sigma_f N_{ff} \phi \quad (9.24)$$

where

σ_f = effective fission microscopic cross section, cm²
N_{ff} = density of the fissionable fuel, nuclei/cm³

In Eq. 9.24, σ_f ordinarily designates the effective fission microscopic cross section for ^{235}U for thermal neutrons. Some fission by fast neutrons will occur in a power reactor. Fission by fast neutrons is observed by inserting the term C_{fn} in Eq. 9.24. Thus,

$$\dot{q}_v = E_f \sigma_f N_{ff} \phi C_{fn} \quad (9.24a)$$

C_{fn} is equal approximately to 1.08 for commercial pressurized water reactors.[4]

The total energy generated in the reactor core is about 200 MeV per fission. Not all of this energy is however transferred to the coolant, thus the effective heat recovery is a somewhat smaller quantity, that is, about 180 MeV per fission.

The fuel density N_{ff} for a uranium oxide fuel is given by

$$N_{ff} = 2.372 f \times 10^{22} \quad {}^{235}\text{U nuclei/cm}^3 \quad (9.25)$$

In this equation, f represents the mass fraction of ^{235}U in the fuel. Thus, for 3-percent-enriched UO_2, $f = 0.03$.

The evaluation of the effective fission microscopic cross section σ_f is a somewhat involved procedure. Pertinent data for the calculation of σ_f are available in Ref. 5.

Example 9.4

Determine for a light-water-moderated uranium reactor the specific energy release rate for the following conditions.

$$\phi = 10^{13} \text{ neutrons/s·cm}^2$$
$$f = 0.035$$
$$\sigma_f = 577 \text{ barns} = 577 \times 10^{-24} \text{ cm}^2$$

From Eq. 9.25,

$$N_{ff} = 2.372 f \times 10^{22} = 0.035 \times 2.372 \times 10^{22} = 8.302 \times 10^{20} \text{ nuclei/cm}^3$$

From Eq. 9.24a,

$$\dot{q}_v = E_f \sigma_f N_{ff} \phi C_{fn} = 180(577 \times 10^{-24})(8.302 \times 10^{20})10^{13} \times 1.08 = 9.312 \times 10^{14} \text{ MeV/s·cm}^3$$

Since 1 MeV = 1.602×10^{-13} J,

$$\dot{q}_v = (9.312 \times 10^{14})(1.602 \times 10^{-13}) = 149.2 \text{ W/cm}^3$$

9.9 HEAT GENERATION IN A SINGLE FUEL ROD

While the radial variation in the neutron flux in a single fuel rod is negligible, the longitudinal variation in the neutron flux cannot be ignored. The longitudinal variation in the neutron flux is readily observed when $\dot{q}_v$ can be approximated by a simple function of z, the coordinate that corresponds with the geometric axis of the fuel rod. For a more complex variation in $\dot{q}_v$, the fuel rod can be subdivided into several segments, and each segment is examined in accordance with the local conditions.

Figure 9.2 depicts a sinusoidal variation in the

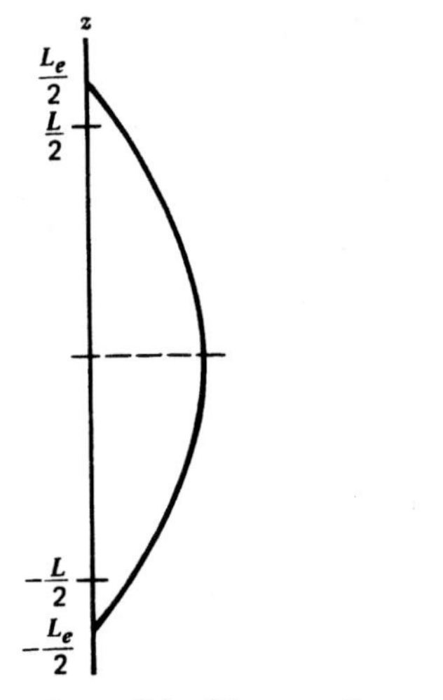

Figure 9.2 Neutron flux variation in a fuel rod.

neutron flux that has a maximum value at the center of the rod. The value of the neutron flux drops to zero at the extrapolated ends of the rod, a distance $L_e/2$ in each direction from the center point. The variation in $\dot{q}_v$ with z is then given by the cosine function

$$\dot{q}_{v,z} = \dot{q}_{v,c} \cos \frac{\pi z}{L_e} \tag{9.26}$$

where

$\dot{q}_{v,c}$ = the specific energy release rate at the center point of the rod
$\dot{q}_{v,z}$ = the specific energy release rate at a distance z from the center point

The total heat resulting from fission and transferred from the rod is determined from

$$\dot{q} = \int_{-L/2}^{L/2} A_t \dot{q}_{v,c} \cos \frac{\pi z}{L_e} \, dz$$

where

A_t = the transverse area of the fuel pellet
L = the actual length of the fuel rod

Integration of the above equation yields

$$\dot{q} = \frac{2}{\pi} A_t L_e \dot{q}_{v,c} \sin \frac{\pi L}{2L_e} \tag{9.27}$$

When extrapolation of the length of the fuel rod can be neglected, $L_e = L$, and Eq. 9.27 is reduced to

$$\dot{q} = \frac{2}{\pi} \dot{q}_{v,c} A_t L \tag{9.28}$$

In this equation, $\dot{q}_{v,c}$ is the amplitude and $(2/\pi)$ $\dot{q}_{v,c}$ is the average height of the cosine function.

The quantity $\dot{q}_{v,c}$ will vary with the position of the fuel rod in the reactor core. Thus, the fuel rods near the center of the core have the highest value of $\dot{q}_{v,c}$. Because of discontinuities in the core, the magnitude of $\dot{q}_{v,c}$ will diminish somewhat irregularly from the center to the outer limits of the core.

Example 9.5

The diameter of the pellets in a fuel rod is 10 mm. The length of the rod is 380 cm. At the center of the rod, $\dot{q}_{v,c}$ = 149 W/cm³. (See Example 9.4.) Determine the thermal power for the fuel rod.

Applying Eq. 9.28,

$$\dot{q} = \frac{2}{\pi} \dot{q}_{v,c} A_t L$$
$$= \frac{2}{\pi} \times 149 \times \frac{\pi}{4} \times 380$$
$$= 28{,}310 \text{ W}$$

9.10 HEAT GENERATION IN THE REACTOR CORE

A heterogeneous reactor core consists of a large number of fuel rods each of which is surrounded by the coolant. Light water serves as the coolant in most power reactors and, in addition, as the moderator.

The highest power density exists at the center of the cylindrical core, the point at which the neutron flux has a maximum value. All of the fuel rods have a maximum neutron flux in the middle plane of the core where $z = 0$. These observations are applicable to a normal neutron flux distribution, that is, one not distorted by the presence of control rods in the core.

The heat generation rate for a heterogeneous reactor core is equal to the product of the total volume of the fuel rods (nA_tL) and the average specific energy release rate ($K\dot{q}_{v,co}$). Thus,

$$\dot{q}_t = KnA_tL\dot{q}_{v,co} \qquad (9.29)$$

where

$\dot{q}_t$ = heat generation rate in the reactor core
n = number of fuel rods in the reactor core
A_t = transverse area of a single fuel rod
L = height of the reactor core (active length of the fuel rods)
$\dot{q}_{v,co}$ = specific energy release rate at the geometric center of the core
K = a characteristic of the reactor

The transverse area A_t of either the fuel pellet or the fuel rod may be used in Eq. 9.29, provided that the appropriate value of $\dot{q}_{v,co}$ is employed.

The heat generated in the reactor core includes the heat generated in the solid fuel and in other reactor materials. The latter quantity may be taken approximately as 5 percent of the heat generated in the fuel.

A theoretical analysis of the reactor heat generation shows a value of 0.289 for K in Eq. 9.29.[5] Examination of the operating data for two reactors discloses K values of 0.368 and 0.416.

Example 9.6

Fuel rod data for a power reactor are: diameter = 10.92 mm, length = 365.76 cm, number = 42,640. The maximum specific energy release rate is 0.61667 kW/cm^3. K = 0.3685. Power generated in the fuel and cladding is 97.3 percent of the total reactor power. Determine the reactor thermal power.

$$A_t = 0.9369 \text{ cm}^2$$

$$\begin{aligned}\dot{q}_t &= KnA_tL\dot{q}_{v,co}\\ &= 0.3685 \times 42640 \times 0.9369\\ &\quad \times 365.76(0.61667 \times 10^{-3})\\ &= 3320 \text{ MW(t)}\end{aligned}$$

Note: MW(t) = thermal power, MW.

$$\begin{aligned}\text{Total reactor power} &= \frac{3320}{0.973}\\ &= 3413 \text{ MW(t)}\end{aligned}$$

The reactor thermal power can be determined by employing the average heat generation rate per unit length of fuel rod. Thus,

$$\dot{q}_t = nL\dot{q}_{l,\text{av}} \qquad (9.29a)$$

9.11 REACTOR CORE CONSTRUCTION

In this section, only the heterogeneous-type core is examined. Although the fuel elements may be fabricated in a variety of shapes, for example, a plate, the slender circular rod is the form ordinarily used.

The small cylindrical UO_2 fuel pellets are encased in the thin-walled tube, about 10 mm O.D., usually fabricated from a zirconium alloy (Zircaloy). The term cladding is applied to this containment tube and does not refer to a metallic deposit on the pellets. Initially, the Zircaloy tube is capped and welded at one end. Following insertion of the fuel pellets, the tube is charged with helium gas and then capped and welded at the other end. A radial clearance is provided between the fuel pellets and the cladding. The tube is designed to accommodate differential thermal expansion effects and to avoid failure because of radial volumetric growth of the fuel.

The cladding has several functions, including the protection of the UO_2 fuel from corrosion and erosion by the coolant and providing mechanical strength for the structurally weak oxide fuel. An important function of the cladding is the confinement of the fission products, particularly those in the gaseous phase. A plenum is provided at the top of the tube for collection of the fission gases. Containment of the fission products appreciably reduces the radioactivity of the coolant and prevents escape of the radioactive products prior to reprocessing the fuel. Operating experience has demonstrated that a limited amount of cladding failure can be tolerated.

A fuel assembly consists of a square array of 200 to 300 fuel rods held in place by a support fixture at each end of the assembly. See Fig. 9.3. The coolant enters the assembly through holes in the end plate and flows between and parallel to the fuel rods. Fuel assemblies are about 20 cm square and include several guide tubes for the control rods. Depending upon the rating of the reactor, a relatively large number of fuel assemblies, 200 to 250, constitute the reactor core. Fuel assemblies can be removed for possible maintenance or for replenishing the reactor fuel inventory.

Metallurgical research has been an important phase of reactor development, largely because of the effect of radiation on material properties. The materials that compose the reactor structure are continuously exposed to gamma radiation and bombarded by high-intensity streams of high-energy particles. In a high-power reactor, the cumulative effect of this high-level bombardment over an extended period of time can cause many changes in the physical and chemical properties of those materials. Some of the effects are embrittlement, distortion, reduced ductility, and a change in the thermal conductivity. The gaseous fission products, a significant source of radiation damage, cause swelling in metals and alloys. A properly prepared oxide fuel however can contain a large amount of fission gas in the absence of appreciable swelling. In addition to the fission gases, the UO_2 matrix is capable of containing other fission products.

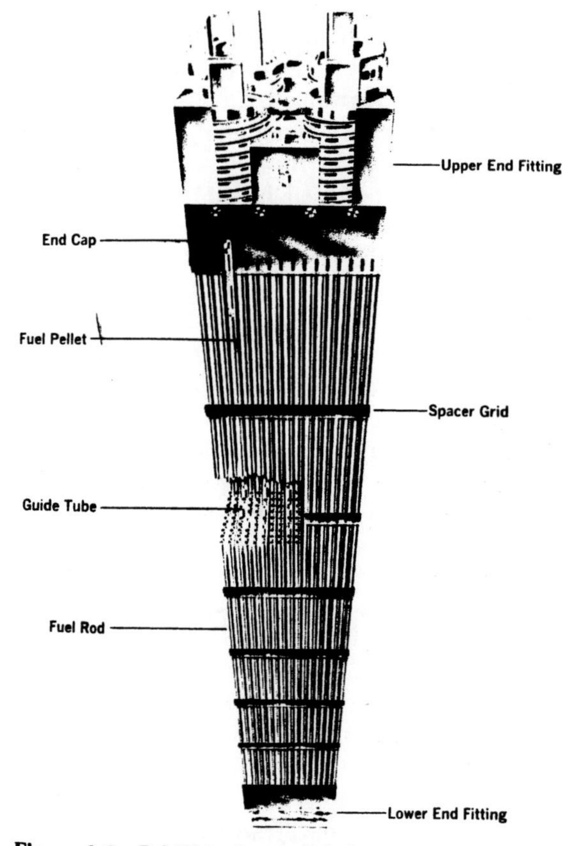

Figure 9.3 B&W fuel assembly for pressurized water reactor for electric power generation. (Courtesy Babcock & Wilcox.)

Cruciform-shaped control rods are fabricated from a neutron absorbing material such as hafnium or a cadmium alloy. Another type of construction consists of Inconel tubes filled with boron carbide pellets that provide the neutron absorbing capability. The guide tubes serve to position properly the control rods when the rods are inserted into the fuel assembly. Fuel rods are omitted, as necessary, in the lattice for the purpose of providing slots for inserting the control rods.

Control of the axial power distribution and shaping the radial power distribution in the core are achieved by arranging the control rods in an appropriate pattern and by providing for the independent movement of the individual control element assemblies. When the control rods are inserted into the reactor core, a distortion of the axial power distribution will be effected. As a consequence, the peak power density normally observed at the center of the core is displaced from the core center to a region ahead of the end of the control rod. Any section of the core into which a control rod penetrates will experience a reduction in the power density because of neutron absorption by the control rod material.

In general, zirconium and stainless steel are acceptable structural materials from the standpoint of resisting the effects of corrosion, high temperature, and radiation. Zircaloy, as noted earlier, is used in a thermal reactor for fuel cladding and for fabrication of other parts of the reactor. In addition to a low neutron absorption characteristic, Zircaloy maintains under reactor conditions adequate strength and resistance to interaction with water at temperatures that range up to 345 C. The reactor vessel lining, the thermal shield, and other parts of the reactor are fabricated from stainless steel. Because of the comparatively high operating temperatures, stainless steel is the proposed cladding material for the fast breeder reactor.

Coolants used in commercial heterogeneous power reactors are helium, water, and sodium. An acceptable primary coolant must satisfy four principal requirements with respect to purity, corrosion, leakage, and heat transfer. It is essential to maintain the coolant at a high level of purity because certain impurities aggravate corrosion. Pure water, unlike sodium, is not highly radioactive. Impurities in the water however can become radioactive, hence removal of foreign matter by demineralization is required. Pure water is highly corrosive, and sodium has a tendency to become corrosive at temperatures in excess of 540 C. Proper water treatment will assist in limiting corrosion within the reactor and the connected equipment. The inert characteristic of helium precludes corrosion, but, as is common with other coolants, impurities must be continuously removed from the gas. Leakage of the coolant from the system is prevented by employing appropriate constructional techniques.

With respect to the transfer of energy from the fuel rods, the coolant should preferably have a high film coefficient of heat transfer and a high thermal capacity. Water has a high thermal capacity, but the film resistance to heat flow is comparatively high. In addition, the water must be maintained under a high pressure in order to prevent boiling. The thermodynamic properties of sodium are generally acceptable for reactor applications. The film resistance to heat flow is low, while moderate system pressures are sufficient to preclude formation of vapor. Helium has a low film coefficient of heat transfer, a characteristic of fluids in the gaseous phase.

A small percentage, less than 1 percent, of the fission neutrons is produced in secondary reactions a few seconds after the initial fission process has occurred. A somewhat larger fraction of the energy released in the reactor results from the added decay of the fission products. Thus, immediately after shutdown of the reactor, continued circulation of the coolant for a period of time is necessary in order to remove the energy released in delayed reactions.

Several requirements must be observed in fabricating the fuel assemblies. The form of the assembly should promote effective removal of the thermal energy by the coolant. Local hot spots in the fuel rods must be avoided to prevent melting of the cladding. Optimum designs achieve a high surface-to-volume ratio for the fuel assembly. Pressure drops in the coolant circuit, for flow through the core, should not be excessively high. The form of the fuel assembly should contribute to the development of a rigid structure, particularly with respect to maintaining flow channels of uniform width. The flow channels are usually very narrow, hence a slight amount of warping can produce considerable relative variation in the channel width.

The film heat transfer coefficient for sodium is high, hence little difficulty is experienced in achieving effective removal of energy from the fuel rods. However, because of the poorer heat transfer characteristics of water and gases, the channels that carry these fluids should be designed for highly turbulent flow as a means to improving the transfer of heat from the surface of the fuel rods to the coolant.

9.12 REACTOR HEAT TRANSFER

The limitations on the power output of the reactor are not nuclear but are imposed by the rate at which energy can be removed from the core. An adequate flow of the coolant around each fuel rod must be assured. The heat generated within the fuel is transferred by conduction to the surface of the pellet, then through the gap between the pellet and the cladding, and successively through the cladding and the coolant film. The temperature profile for a fuel rod is shown in Fig. 9.4.

Because of the substantial axial variation in $\dot{q}_v$ for a single fuel rod, the heat transfer from the rod is most conveniently determined for a short length l. Since both the diameter and the length of the fuel element are small, the neutron flux is assumed not to change in either the axial or the radial direction. Consequently, the heat flow from the element of limited length is considered to be radially uniform and constant in the axial direction.

A cross section of the fuel pellet is shown in Fig. 9.5. The basic conduction equation describes the heat flow through a thin cylindrical shell of thickness dr. Thus,

$$\dot{q}_r = -k_f A \frac{dT}{dr}$$

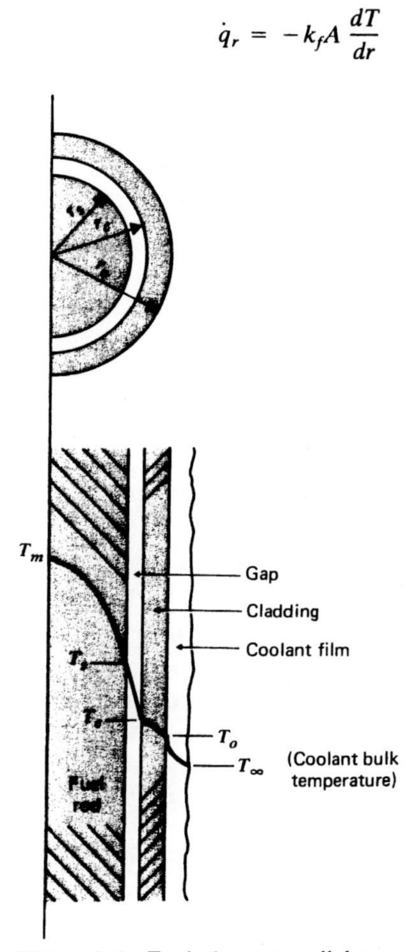

Figure 9.4 Fuel element radial temperature profile.

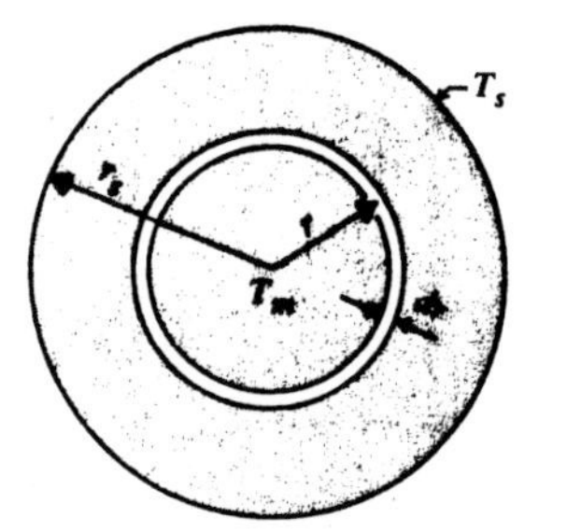

Figure 9.5 Fuel pellet cross section.

where

k_f = the thermal conductivity of the fuel, assumed to be temperature independent
A = the heat transfer area, equal to $2\pi rl$
dT/dr = the temperature gradient in the fuel

All the heat generated within the cylindrical body of radius r and volume V_r is transferred through the area A. Then,

$$\dot{q}_r = V_r \dot{q}_v = \pi r^2 l \dot{q}_v$$

The above two equations are now combined:

$$-k_f 2\pi rl \frac{dT}{dr} = \pi r^2 l \dot{q}_v$$

or

$$-\frac{k_f}{\dot{q}_v} dT = \frac{r dr}{2}$$

The boundary conditions for the element are $T = T_m$ at $r = 0$, and $T = T_s$ at $r = r_s$. Integration of the preceding equation yields

$$\frac{k_f}{\dot{q}_v}(T_m - T_s) = \frac{r_s^2}{4} \tag{9.30}$$

where T_m and T_s are, respectively, the temperatures at the center and at the surface of the cylindrical fuel element.

All the heat generated within the solid cylinder is transferred from the surface A_s. Thus, the heat flow $\dot{q}_s$ from the surface of the cylinder is given by

$$\dot{q}_s = \pi r_s^2 l \dot{q}_v \tag{9.31}$$

Equations 9.30 and 9.31 are combined to eliminate $\dot{q}_v$ and yield

$$\dot{q}_s = 4\pi k_f l(T_m - T_s) \tag{9.32}$$

Equation 9.32 shows that the heat generated per unit length of the fuel rod is solely a function of the temperature drop $(T_m - T_s)$ from the center to the surface of the pellet. The temperature drop however is a function of $\dot{q}_v$ and the radius of the fuel pellet as shown by Eq. 9.30. Evidently an increase in the diameter of the fuel pellet, for prescribed values of $\dot{q}_v$ and T_s, would cause an increase in T_m. The diameter of the fuel pellet is in part limited by the maximum allowable temperature of the fuel.

The surface A_s of the pellet through which the heat is transferred is equal to $2\pi r_s l$. Thus, substitution of $l = A_s/2\pi r_s$ into Eq. 9.32 yields

$$\dot{q}_s = \frac{2A_s k_f}{r_s}(T_m - T_s) \tag{9.33}$$

The thermal conductivity k_f of the fuel is usually evaluated at the mean temperature of the fuel. The accumulation of fission products inside the fuel will in time reduce the magnitude of k_f.

Significant variations are observed in the radial dimensions of the gap that exists between the fuel pellets and the cladding. The gap represents the clearance required to allow ready insertion of the fuel pellets into the containment tube. Variations in the pellet diameter and in the diameter of the cladding are expected when practical manufacturing processes are employed. Heat transfer across the gap is improved by the helium gas with which the tube is charged during fabrication. Incidentally, the thermal conductivity of helium is substantially higher than that of air. The thermal conductivity of the gap fluid however decreases as fission product gases are released into the gap during burnup of the fuel. Convective heat transfer through the thin layer of gas in the gap is comparatively small and hence may be neglected.

The basic equation is now written for the heat conduction through the gas in the gap.

$$\dot{q}_g = -k_g A_g \frac{dT}{dr}$$

where $A_g = 2\pi rl$. Substitution for A_g and subsequent integration yield

$$\dot{q}_g = \frac{2\pi l k_g}{\ln(r_c/r_s)}(T_s - T_c) \tag{9.34}$$

The numerator and the denominator are each multiplied by $r_c - r_s$, where r_c and r_s are, respectively, the outer and the inner radial dimensions of the cylindrical gap. See Fig. 9.4. Then, $r_c - r_s$ is equal to the gap clearance Δr_g.

$$\dot{q}_g = \frac{2\pi(r_c - r_s)lk_g}{(r_c - r_s)\ln(r_c/r_s)}(T_s - T_c)$$

But $$\frac{2\pi(r_c - r_s)l}{\ln(r_c/r_s)} = \overline{A}_g$$

where $\overline{A}_g$ is the log mean area of the gap through which the heat is transferred. Then,

$$\dot{q}_g = \frac{k_g \overline{A}_g}{\Delta r_g}(T_s - T_c) \tag{9.35}$$

where T_s and T_c are, respectively, the temperature at the surface of the fuel pellet and the temperature at the inner surface of the cladding.

Because the fuel rod gap is in the form of a thin-walled cylinder, the heat transfer area $\overline{A}_g$ can be evaluated as the arithmetical mean area of the gap. Consequently, $\overline{A}_g = \pi l(r_c + r_s)$.

The above procedure is followed in developing an equation for the conduction of heat through the cladding. Thus,

$$\dot{q}_c = \frac{k_c \overline{A}_c}{\Delta r_c}(T_c - T_o) \tag{9.36}$$

where

$\overline{A}_c = \pi l(r_o + r_c)$, the arithmetical mean heat transfer area of the cladding
k_c = the thermal conductivity of the cladding
Δr_c = the thickness of the cladding
T_c = the inner surface temperature of the cladding
T_o = the outer surface temperature of the cladding

Figure 9.6 shows one configuration of the unit flow channels that comprise in part the reactor core. At the core wall, the form of the unit flow channel deviates from the configuration shown in Fig. 9.6, and deviations are caused by the replacement of a fuel rod by a control rod guide tube. Because of a number of factors, the coolant does not flow uniformly through the many unit flow channels. Variations in the dimensions of the channels can be expected as a result of tolerances that are allowed in the manufacture of the fuel rods and in the fabrication of the fuel assemblies. Distribution of the coolant to the fuel assemblies will not be exactly uniform.

In order to evaluate the heat transfer characteristics in any unit flow channel, the geometry of the channel must be described and the velocity, pressure, and temperature of the coolant known. Several heat transfer equations are available for predicting the coolant film conductance h_o. The heat transfer through the coolant film is given by

$$\dot{q} = h_o A_o (T_o - T_\infty) \tag{9.37}$$

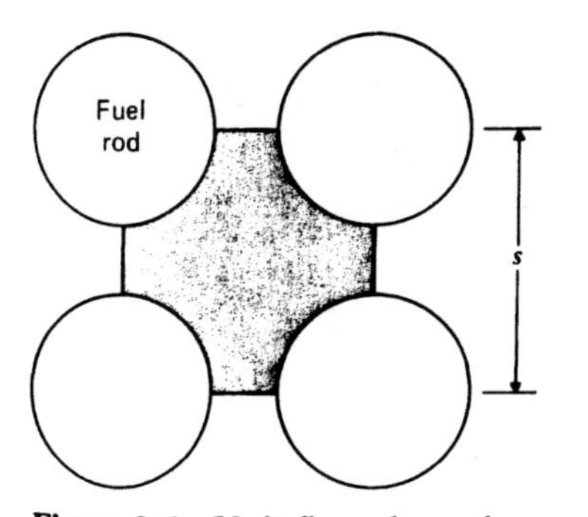

Figure 9.6 Unit flow channel.

where

h_o = the coolant film conductance based on A_o
$A_o = 2\pi r_o l$, the area of the outside surface of the cladding
T_o = the outer surface temperature of the cladding
T_∞ = the bulk temperature of the coolant

Equations 9.33, 9.35, 9.36, and 9.37 are arranged in the following forms:

$$T_m - T_s = \dot{q}_s \frac{r_s}{2A_s k_f}$$

$$T_s - T_c = \dot{q}_g \frac{\Delta r_g}{k_g \overline{A}_g}$$

$$T_c - T_o = \dot{q}_c \frac{\Delta r_c}{k_c \overline{A}_c}$$

$$T_o - T_\infty = \dot{q}_o \frac{1}{h_o A_o}$$

Since $\dot{q}_s = \dot{q}_g = \dot{q}_c = \dot{q}_o$ under steady-state conditions, a summation can be made of the above four equations:

$$T_m - T_\infty = \dot{q}_s \left[\frac{r_s}{2A_s k_f} + \frac{\Delta r_g}{k_g \overline{A}_g} + \frac{\Delta r_c}{k_c \overline{A}_c} + \frac{1}{h_o A_o} \right]$$

and

$$\dot{q}_r = \frac{T_m - T_\infty}{\dfrac{r_s}{2A_s k_f} + \dfrac{\Delta r_g}{k_g \overline{A}_g} + \dfrac{\Delta r_c}{k_c \overline{A}_c} + \dfrac{1}{h_o A_o}} \tag{9.38}$$

The above equation is multiplied by A_o/A_o. Thus,

$$\dot{q}_r = \frac{A_o(T_m - T_\infty)}{\dfrac{A_o}{A_s}\dfrac{r_s}{2k_f} + \dfrac{A_o}{\overline{A}_g}\dfrac{\Delta r_g}{k_g} + \dfrac{A_o}{\overline{A}_c}\dfrac{\Delta r_c}{k_c} + \dfrac{1}{h_o}} \tag{9.39}$$

where

$\dot{q}_r$ = the heat transferred, per unit time, from a prescribed length of fuel rod
T_m = the temperature at the center of the fuel pellet
T_∞ = the bulk temperature of the coolant
k_f = the thermal conductivity of the fuel
k_g = the thermal conductivity of the gas in the gap
k_c = the thermal conductivity of the cladding material
r_s = fuel pellet radius
Δr_g = gap width
Δr_c = cladding thickness
h_o = coolant film conductance
A_o = outer area of the cladding
A_s = outer area of the fuel pellet
$\bar{A}_g$ = mean area of the gap
$\bar{A}_c$ = mean area of the cladding

Equation 9.39 is particularly applicable to the determination of the conductive heat transfer through a composite cylinder. An area ratio can be conveniently replaced by a ratio of the corresponding radii. For example, $r_o/\bar{r}_g = A_o/\bar{A}_g$, where $\bar{r}_g$ is the mean radius of the gap.

The terms in the denominator of Eq. 9.39 represent the several resistances to heat flow. The temperature drop through each section of the composite structure is proportional to the respective resistance.

Example 9.7

Determine the thermal power for a section of a reactor fuel rod that operates under the following conditions.

Maximum temperature of the fuel: 2210 C
Coolant temperature: 320 C
Fuel rod O.D.: 10.925 mm
Fuel pellet diameter: 9.400 mm
Cladding thickness: 0.675 mm
Thermal conductivity
Fuel: 2.127 W/m·K
Helium: 0.303 W/m·K
Cladding: 16.430 W/m·K
Coolant film conductance: 43,700 W/m²·K

The calculations are based on Eq. 9.39.

Cladding: I.D. = 9.575 mm $\quad \bar{r}_c = 5.125$ mm
$\Delta r_c = 0.675$ mm
Gap: $\bar{r}_g = 4.744$ mm $\quad \Delta r_g = 0.0875$ mm
$A_o = 343.219 \times 10^{-6}$ m²/cm length

$$\dot{q}_r = \frac{A_o(T_m - T_\infty)}{\frac{r_o}{r_s}\frac{r_s}{2k_f} + \frac{r_o}{\bar{r}_g}\frac{\Delta r_g}{k_g} + \frac{r_o}{\bar{r}_c}\frac{\Delta r_c}{k_c} + \frac{1}{h_o}}$$

$$\frac{r_o}{r_s}\frac{r_s}{2k_f} = \frac{5.4625}{4.7000}\frac{4.7000 \times 10^{-3}}{2 \times 2.127}$$
$$= 1.2841 \times 10^{-3} \text{ m}^2\cdot\text{K/W}$$

$$\frac{r_o}{\bar{r}_g}\frac{\Delta r_g}{k_g} = \frac{5.4625}{4.7440}\frac{0.0875 \times 10^{-3}}{0.303}$$
$$= 0.3325 \times 10^{-3} \text{ m}^2\cdot\text{K/W}$$

$$\frac{r_o}{\bar{r}_c}\frac{\Delta r_c}{k_c} = \frac{5.4625}{5.1250}\frac{0.675 \times 10^{-3}}{16.430}$$
$$= 0.0438 \times 10^{-3} \text{ m}^2\cdot\text{K/W}$$

$$\frac{1}{h_o} = \frac{1}{43700}$$
$$= 0.0229 \times 10^{-3} \text{ m}^2\cdot\text{K/W}$$

$$\Sigma R = 1.6833 \times 10^{-3} \text{ m}^2\cdot\text{K/W}$$

$$\dot{q}_r = \frac{A_o(T_m - T_\infty)}{\Sigma R}$$
$$= \frac{343.219 \times 10^{-6}(2210 - 320)}{1.6833 \times 10^{-3}}$$
$$= 385.36 \text{ W/cm length}$$

$$\frac{T_m - T_s}{T_m - T_\infty} = \frac{R_f}{\Sigma R}$$

$$\frac{2210 - T_s}{2210 - 320} = \frac{1.2841 \times 10^{-3}}{1.6833 \times 10^{-3}}$$

$$T_s = 768 \text{ C}$$

A similar procedure is used to calculate the remaining interface temperatures. Then, the temperatures are

Center of the fuel pellet:	2210 C
Surface of the fuel pellet:	768 C
Cladding inner surface:	395 C
Cladding outer surface:	346 C
Coolant:	320 C

The temperature drops are

Fuel:	1442 C
Gap:	373 C
Cladding:	49 C
Coolant film:	26 C
	1890 C

The Colburn equation is used generally to correlate the heat transfer data for the turbulent flow of water or gas parallel to the fuel elements.

$$\mathrm{Nu} = 0.023\ \mathrm{Re}^{0.8}\ \mathrm{Pr}^{1/3}$$

For all-liquid-water conditions in the reactor, a modified Colburn equation is recommended.[4] Thus,

$$h_o = 0.030 \frac{k}{D_e}\left(\frac{\rho D_e C}{\mu}\right)^{0.8}\left(\frac{c_p \mu}{k}\right)^{1/3} \qquad (9.40)$$

where

h_o = convective film heat transfer coefficient
k = thermal conductivity of the fluid
ρ = density of the fluid
μ = dynamic viscosity of the fluid
c_p = specific heat of the fluid
C = fluid velocity
D_e = equivalent diameter of the flow channel

In this form of the Colburn equation, Eq. 9.40, all of the fluid properties are evaluated at the bulk temperature of the water.

The equivalent diameter D_e for a unit flow channel is determined by

$$D_e = 4\,\frac{\text{area of the flow channel}}{\text{portion of the perimeter that is a solid boundary}}$$

$$D_e = 4\,\frac{S^2 - (\pi/4)D_o^2}{\pi D_o} \qquad (9.41)$$

where

S = the fuel rod spacing in a square array (Fig. 9.6)
D_o = the fuel rod O.D.

For turbulent flow of liquid metals, the Colburn equation is not applicable because of the exceptionally high thermal conductivity of metallic fluids. The film heat transfer coefficient, for the flow of liquid metals outside and parallel to the fuel elements, may be evaluated generally for a constant heat flux by

$$\frac{hD_e}{k} = 5.8 + 0.02\mathrm{Pe}^{0.8} \qquad (9.42)$$

The Peclet number Pe is equivalent to the product of the Reynolds and the Prandtl moduli. Thus,

$$\mathrm{Pe} = D_e G c_p / k$$

where G = the mass velocity, the product of ρC. Equation 9.42 is based on investigations conducted by R. A. Seban.[6]

The physical properties of liquid metals do not vary significantly across the flow channel because a small difference exists between the temperature of the heat transfer surface and the bulk temperature of the fluid. This small temperature difference is related to the high film conductance that is characteristic of liquid metals. Viscosity is not a significant property with respect to liquid–metal convective heat transfer.

Example 9.8

Liquid sodium at a bulk temperature of 700 C flows through a 12.5-mm-diameter hole in a graphite block. Surface temperature of the graphite is 730 C. Sodium velocity is 4.6 m/s. Sodium properties at 700 C are

$$c_p = 1.255 \text{ kJ/kg·K}$$
$$k = 59.67 \text{ W/m·K}$$
$$\rho = 778.5 \text{ kg/m}^3$$

Determine $\dot{q}/A$.

$$G = \rho C = 778.5 \times 4.6$$
$$= 3581 \text{ kg/s·m}^2$$
$$D_e = D_i = 0.0125 \text{ m}$$

$$\text{Pe} = \frac{D_e G c_p}{k}$$
$$= \frac{0.0125 \times 3581 \times (1000 \times 1.255)}{59.67}$$
$$= 941$$

$$h = \frac{k}{D_e}[5.8 + 0.02\text{Pe}^{0.8}]$$
$$= \frac{59.67}{0.0125}[5.8 + 0.02(941)^{0.8}]$$
$$= 50{,}530 \text{ W/m}^2\text{·K}$$

$$\dot{q}/A = h(T_s - T_b)$$
$$= 50530 \times 30$$
$$= 1516 \times 10^3 \text{ W/m}^2$$
$$= 1516 \text{ kW/m}^2$$

Circulation of the coolant through the reactor core, heat exchangers, and interconnecting piping is achieved by pumping. In order to avoid an adverse effect on the thermal performance of the power plant, pumping power is held within acceptable limits by controlling the magnitude of the fluid losses. Fluid power $\dot{W}_f$ is determined from $\dot{W}_f = \dot{V}\,\Delta P$, where $\dot{V}$ is the volume flow rate and ΔP is the pressure increase developed by the pump. The power input to the pump $\dot{W}_p$ exceeds $\dot{W}_f$ because of the machine losses.

From the standpoint of heat removal and pumping power, water, both light and heavy, is superior to other coolants. The $\dot{W}_p/\dot{q}$ ratios for liquid metals and gases are, respectively, about 5 and 100 times the value of $\dot{W}_{P}/\dot{q}$ for water.[5]

9.13 BOILING HEAT TRANSFER

Figure 9.7 represents generally the transfer of heat to water flowing in a heated channel, such as a unit flow channel in a reactor core. The several regimes shown on this diagram have been described in detail in the heat transfer literature. At this point only the most significant characteristics of the generalized curve will be examined.

The portion of the curve extending from *a* to *b* represents convective heat transfer, in the absence of boiling, to liquid water. From *b* to *c*, the heat flux is still small, and a few vapor bubbles form but quickly collapse. This regime is defined as subcooled-nucleate or local boiling. The bulk of the water does not reach the saturation temperature until the temperature difference ΔT_c is established.

As the temperature difference continues to increase toward ΔT_d, nucleate boiling continues with increased bubble formation. The regime *c–d* represents nucleate boiling. High heat flux rates, with a moderate temperature difference, are promoted by the agitation and turbulence that develop as a result of bubble formation.

The heat flux at point *d*, designated as the critical heat flux, is the maximum heat flux that is attained with nucleate boiling. An increase in ΔT beyond the value of ΔT_d causes a departure

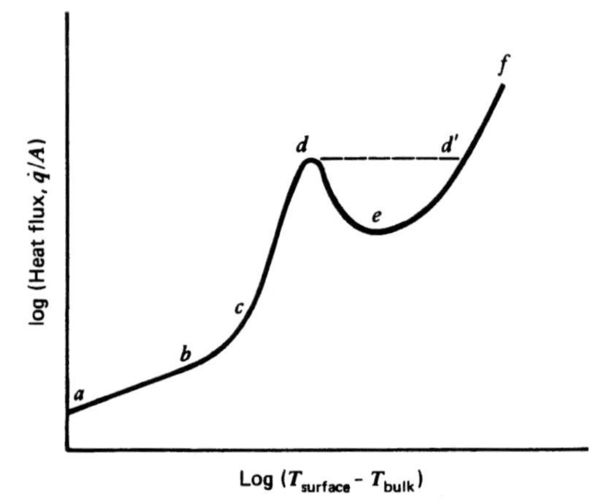

Figure 9.7 Typical boiling curve for water flowing in a heated channel.

from nucleate boiling. In the regime extending from d to e, the bubbles become more numerous and interfere with the flow of water to the surface. The bubbles coalesce to form a film of superheated vapor on portions of the surface. Within this regime, d to e, the magnitude of $\dot{q}/A$ tends to decrease with an increase in ΔT because of the comparatively low film conductance of superheated vapor. Boiling in this range is unstable as a result of the periodic formation and collapse of the film.

The regime extending from e to f represents stable film boiling, and an increase in ΔT produces a characteristic increase in $\dot{q}/A$. However, in this regime much higher temperature differences are observed, and thermal radiation from the surface becomes a factor of increasing significance as the surface temperature moves into the high range.

When the local heat flux in a boiler tube or reactor channel exceeds the critical heat flux, at point d in Fig. 9.7, the surface temperature can increase very quickly, along a horizontal line, to $T_{d'}$ at a point in the film boiling regime. The metal temperature at d′ may be sufficiently high to cause the heating surface to burn out or melt. Burnout does not necessarily occur at d', because burnout can occur at other points on the boiling curve depending upon the heating surface material and the operating conditions. The burnout heat flux is taken as the heat flux that corresponds to the surface temperature at burnout. The temperature at d' often is sufficiently high to cause burnout, hence in this case the burnout heat flux and the critical heat flux are of like magnitude.

Certain aspects of water boiling in reactor cores are of interest. Subcooled boiling is a surface phenomenon, because the bubbles that form in the liquid tend to collapse near the surface. This type of boiling is observed when the bulk of the liquid is at a temperature below the saturation temperature for the system pressure. Saturated boiling occurs when the bulk of the liquid is at the saturation temperature. For either case, the surface temperature must, for bubble formation, be higher than the saturation temperature. Thus, in the vicinity of the surface, the liquid is superheated, while the bulk of the liquid will be at or below the saturation temperature, depending upon the type of boiling.

Under certain conditions of high loading, subcooled boiling is established in liquid water-cooled reactors. Subcooled boiling is promoted as a means to achieving a high film heat transfer coefficient. The temperature of the cladding at the outer surface exceeds the saturation temperature for the system pressure, while the coolant bulk temperature remains below the saturation temperature.

In order to protect the reactor from burnout during subcooled boiling, the design procedure must include a prediction of the value of $(\dot{q}/A)_{bo}$, the burnout heat flux. Equation 9.43, developed by Jens and Lottes, correlates the results derived from experimental data obtained on subcooled water flowing upward through a vertical tube.[7]

$$\left(\frac{\dot{q}}{A}\right)_{bo} = C\left(\frac{G}{10^6}\right)^m (T_{sat} - T_b)^{0.22}\ \text{kJ/h·m}^2 \qquad (9.43)$$

where T_{sat} and T_b are, respectively, the saturation temperature and the bulk temperature of the water at the burnout point, C. The values of C and m, selected from Table 9.4, are dependent upon the system pressure. The correlation provided by Eq. 9.43 extends over a range in G from 4.69×10^6 to 38.1×10^6 kg/h·m^2 and a subcooling range of 3.1 to 90.6 C. The equation is not applicable to saturated boiling because for this condition, $T_{sat} - T_b = 0$.

Example 9.9

A pressurized water reactor operates at a pressure of 16 MPa, abs. Water leaves the core at

TABLE 9.4 Constants for Equation 9.43

Water Pressure (MPa, abs)	$C \times 10^{-6}$	m
3.5	8.193	0.160
7.0	5.231	0.275
14.0	2.603	0.500
21.0	1.022	0.725

344.80 C. $G = 12.20 \times 10^6$ kg/h·m². Determine $\dot{q}/A$ at burnout.

$$T_{sat} = 347.44 \text{ C}$$

Applying Eq. 9.43,

$$C = 2.151 \times 10^6 \qquad m = 0.564$$

$$\left(\frac{\dot{q}}{A}\right)_{bo} = C\left(\frac{G}{10^6}\right)^m (T_{sat} - T_b)^{0.22}$$
$$= (2.151 \times 10^6)(12.20)^{0.564} \cdot (347.44 - 344.80)^{0.22}$$
$$= 10.92 \times 10^6 \text{ kJ/h·m}^2$$

Usually, only a carefully controlled minimum amount of local or subcooled boiling would be acceptable in the design of a pressurized water reactor. However, near the outlet end of the core, nucleate boiling may develop in the hotter channels, and under extreme conditions, the DNB (departure from nucleate boiling) may be reached. These abnormal operating conditions can result from reduced coolant flow or occur during power transients.

The burnout heat flux for subcooled boiling, predicted from experimental data, is compared with the design heat flux in order to evaluate the margin of safety that may be anticipated in the operation of the reactor.

Saturated flow boiling occurs in the upper portion of the boiling water reactor. Unlike subcooled flow, the bubbles that form on the heating surface do not collapse, or they are detached from the surface and enter the mainstream of the flowing fluid. The bulk temperature of the water is equal to the saturation temperature, hence heat transfer from the bubbles to the liquid, and subsequent collapse of the bubbles, do not occur. The quality of the liquid water–steam mixture leaving the reactor core is normally 15 percent.

For design conditions, a relationship is established for boiling water reactors between the burnout heat flux and two parameters, namely, the mass velocity and the quality of the fluid. See Fig. 9.8. For a system pressure other than 6.895 MPa, abs, the values of $(\dot{q}/A)_{bo}$ selected from Fig. 9.8 are reduced by 0.724×10^6 kJ/h·m² per 1 MPa increase in pressure, and vice versa for a pressure decrease.

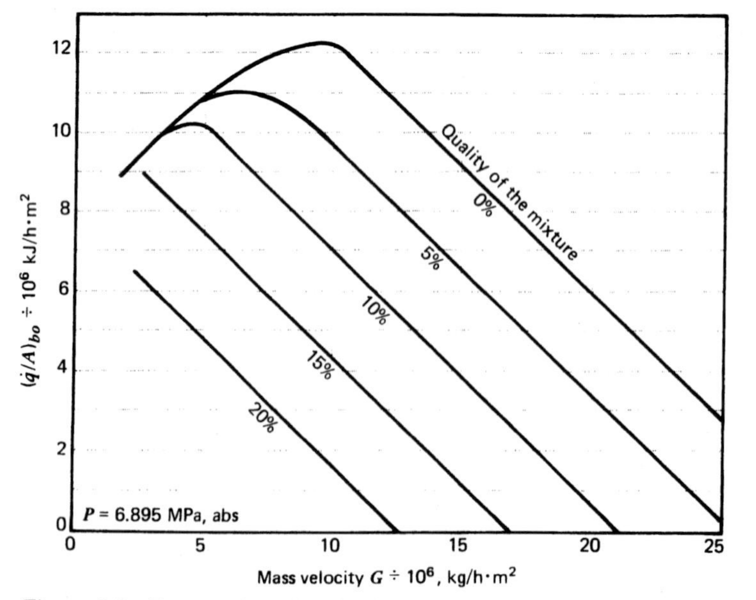

Figure 9.8 Burnout heat flux for boiling water reactors. (E. Janssen and S. Levy, General Electric Company.[8])

Example 9.10

The coolant leaves the core of a boiling water reactor at a pressure of 7.25 MPa, abs, and with a quality of 15 percent. $G = 6.35 \times 10^6$ kg/h·m² around the fuel rods. Determine the burnout heat flux for the specified operating conditions.

From Fig. 9.8, for $G = 6.35 \times 10^6$ kg/h·m² and $x = 15$ percent,

$$(\dot{q}/A)_{bo} = 6.52 \times 10^6 \text{ kJ/h·m}^2$$

Correction for pressure increase is

$$\Delta(\dot{q}/A)_{bo} = (-0.724 \times 10^6)(7.25 - 6.895)$$
$$= -0.257 \times 10^6 \text{ kJ/h·m}^2$$

At $P = 7.25$ MPa, abs,

$$(\dot{q}/A)_{bo} = (6.52 - 0.26) \times 10^6$$
$$= 6.26 \times 10^6 \text{ kJ/h·m}^2$$

Example 9.10 illustrates the comparatively low burnout heat flux encountered in saturation boiling. The possibility of fuel rod failure is avoided by designing the core for a relatively low maximum heat flux, say, 4×10^6 kJ/h·m².

9.14 HOT CHANNEL CONCEPT

A reactor can be designed with a high degree of precision, but the manufacturing process will not produce an exact duplicate of the design model. Although exact dimensions can be specified, variations in the actual dimensions of the fuel rods, cladding, and other parts of the core are experienced because of allowable manufacturing tolerances. Flow variations develop because of failure to achieve the prescribed distribution of the coolant to the several fuel assemblies as required in the design. In addition, the long slender fuel rods have a tendency to bow and alter the flow channel cross section. For a number of reasons, the neutron flux distribution in the core will not conform exactly to the pattern developed by the application of reactor physics. There are discontinuities in the core, such as water holes, that further distort the neutron flux distribution.

Because of the differences between the design and the actual operating conditions throughout the core, temperatures in the fuel and in the cladding will deviate from the temperatures predicted by application of design techniques. The direction and magnitude of a local temperature deviation will be influenced by the collective effect of the variations from the design conditions prescribed for a particular point in the reactor core. A temperature in excess of the predicted temperature can be a matter of concern because of the possible development of a hot spot. Reactor design consequently must include an analytical investigation of the probability of hot spots developing in the core. This procedure is followed to ensure safe operation of the reactor and to limit core maintenance which, at best, is accomplished under difficult circumstances.

The nuclear reactor, unlike the boiler furnace, is charged with sufficient fuel for many hours of operation. The nuclear energy could be readily released at an extremely high rate, but damage to the reactor structure would very likely be experienced. The power output of the reactor is limited by the temperature developed in the fuel and in the cladding. The temperature within the fuel must remain by a safe margin below the melting point, 2840 C for UO_2, and burnout of the cladding must be avoided.

The "hot channel" concept has been established as a means to investigating the probability of excessively high temperatures occurring in the reactor core. The hot channel is devised as a hypothetical region in the core where the worst conditions prevail. Here, all the factors that influence heat generation are at their maximum values, and all the factors that promote heat removal are at their minimum values.

Hot spot factors are generally classified in accordance with the physical phenomenon that causes the deviation, namely, nuclear or mechanical. The designer customarily evaluates the individual factors, under each classification, from statistical data.

An evaluation of the effectiveness of the thermal design of a reactor core can be achieved by the determination of the enthalpy rise hot channel factor $F_{\Delta h}$, which is defined as the ratio of the enthalpy rise in the hot channel to the enthalpy rise of the coolant in the average channel. $F_{\Delta h}$, in turn, is equal to the product of several subfactors, and the lower each subfactor becomes, the closer the core outlet temperature can be to the maximum permissible reactor coolant temperature. In a typical 1960 nuclear power plant, the hot channel factor was about 3.4, actually a very conservative value. Because of improvements in core design and more realistic evaluations of the subfactors, the hot channel factor has been reduced to about 1.1.

9.15 PRESSURIZED WATER REACTORS

Almost all the power reactors constructed in the United States are classified as water reactors and are either pressurized or boiling reactors. Light water serves as both the coolant and as the moderator. Good operating records have been achieved by both types of water reactors.

Figure 9.9 shows a flow diagram for a pressurized water reactor power system. The water, under sufficient pressure to prevent bulk boiling, is circulated through the reactor and the steam generator. Steam produced in the steam generator, by transfer of heat from the pressurized water, is piped to the turbine. In a conventional steam power system, steam is extracted from the turbine for supplying the high-pressure and the low-pressure feedwater heaters.

A pressurizer is provided for establishing and maintaining the pressure of the reactor coolant within prescribed limits. The typical pressurizer is a vertical vessel equipped with a spray nozzle located at the top and electric heaters installed at the bottom of the vessel. The water level in the vessel is normally maintained at about the midpoint. Density changes in the reactor coolant

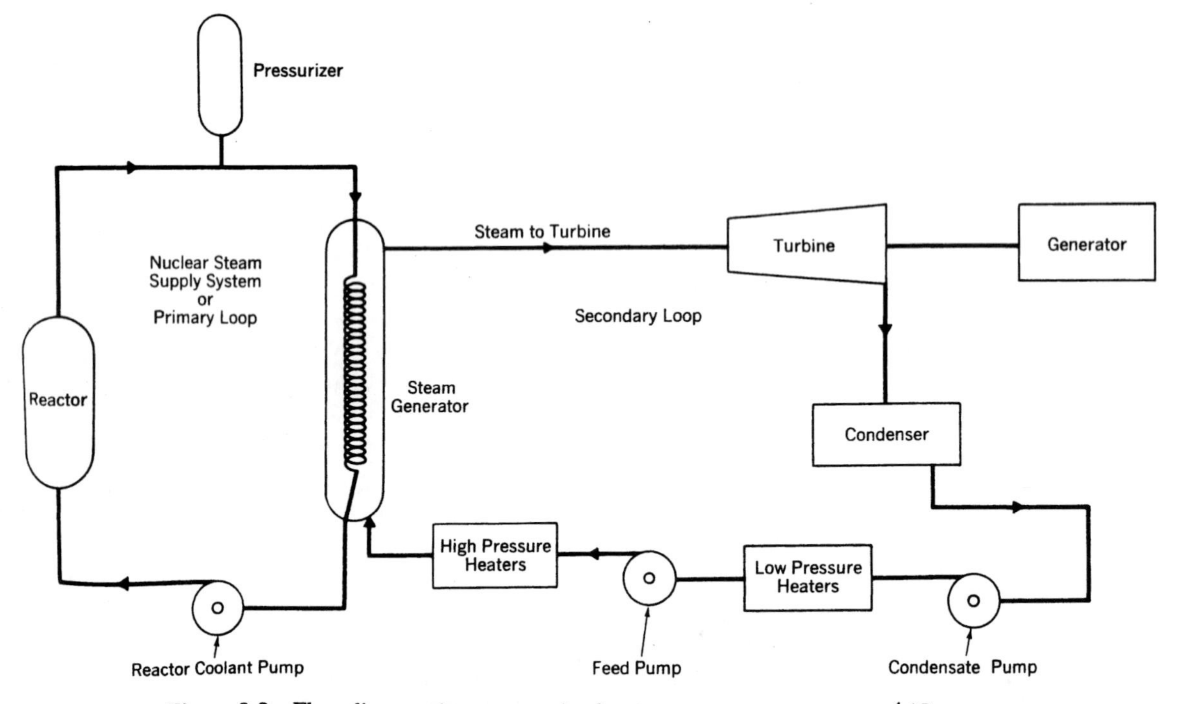

Figure 9.9 Flow diagram for a pressurized water reactor power system.[4] (Courtesy Babcock & Wilcox.)

that occur during operation of the system are compensated by either evaporating water or condensing vapor in the pressurizer.

A portion of the coolant is continuously removed from the main stream and passed through a purification demineralizer. Impurities are removed by ion exchange, and the coolant is subsequently returned to the main stream. Purification is essential in order to limit the radioactivity buildup of the long-lived impurities. Chemicals are added to the return flow from the purifier for effecting reactivity, pH, and oxygen control.

Figure 9.10 illustrates the construction of the reactor vessel and the internal components. A thermal shield is installed to absorb radiation emitted from the core and thus reduce the absorption of gamma and neutron energy in the wall of the pressure vessel. A number of control rods and the associated drive system are installed in the reactor vessel. A control rod drive line consists of a drive, guide tubes, and one control rod or control rod assembly. Only one control rod drive line is shown in Fig. 9.10.

Design and operating data for the pressurized water reactor shown in Fig. 9.10 are presented in Ref. 4.

The nuclear power system usually includes at least two steam generators of the shell-and-tube design. The steam generator is ordinarily placed in a vertical position, and either U-tubes or straight tubes are installed in the shell. The reactor coolant flows through the tubes, and steam is generated on the outside surface of the tubes. Circulation of the pressurized water through the reactor and the steam generators is effected by the reactor coolant pumps. A moderate increase,

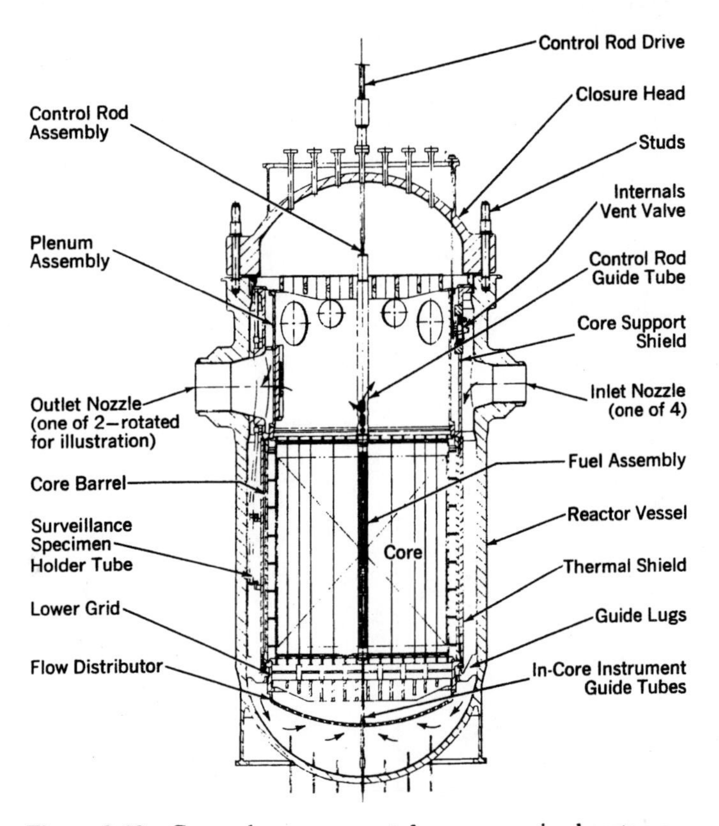

Figure 9.10 General arrangement for a pressurized water nuclear power reactor.[4] (Courtesy Babcock & Wilcox.)

30 to 55 C, is achieved in the temperature of the coolant in passing through the reactor, hence a high volume flow of the coolant is required for removal of the heat generated in the core.

The temperature of the reactor coolant entering the steam generator is somewhat higher than the saturation temperature of the generated steam, hence a modest amount of superheating, say 20 C, can be achieved.

It is essential in designing the reactor core to ensure against burnout. The value of $\dot{q}/A$ along the average channel is multiplied by an appropriate value of the hot channel factor $F_{\Delta h}$ in order to obtain the hot channel $\dot{q}/A$ variation. See Fig. 9.11. The burnout heat flux $(\dot{q}/A)_{bo}$ can be determined by use of Eq. 9.43, or from experimental data if such data are available. A minimum burnout ratio, say, 1.5, is customarily prescribed as a safety measure. If the minimum burnout ratio is too low, the design would be altered in some manner to improve the margin of safety. The value of $(\dot{q}/A)_{\max}$ is the maximum heat flux that can be tolerated at any place in the core.

The reactor core design should also be examined for possible local boiling and for the possibility of developing an excessively high temperature at the center of the fuel pellets. A limited amount of melting within the fuel pellets is not regarded as a particularly alarming occurrence.

The point in the channel where local or subcooled boiling begins is determined, in part, by use of the Jens and Lottes equation.[7]

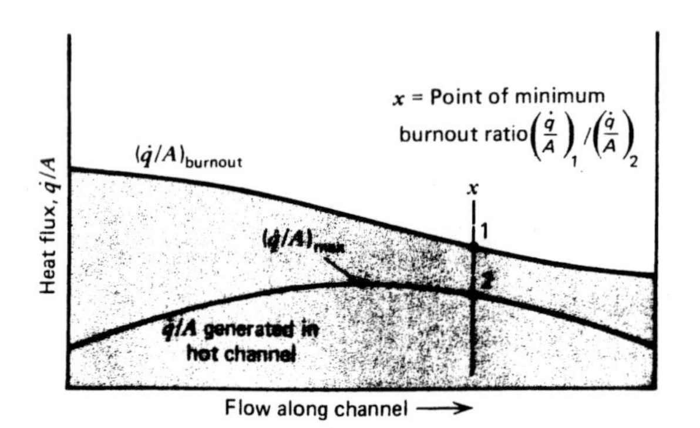

Figure 9.11 Pressurized water reactor minimum burnout ratio.

$$T'_w - T_{sat} = 18.16\left(\frac{\dot{q}/A}{10^6}\right)^{1/4} \Big/ e^{P/6.2} \quad (9.44)$$

where

T'_w = wall (cladding surface) temperature, C, at which local boiling begins
T_{sat} = saturation temperature of the water, C
$\dot{q}/A$ = heat flux, kJ/h·m²
P = water pressure, MPa, abs

The difference between the surface temperature T_w and the bulk temperature T_b of the water is determined from

$$T_w - T_b = \frac{\dot{q}/A}{h_o} \quad (9.45)$$

where h_o is the water film heat transfer coefficient evaluated by Eq. 9.40, the modified Colburn equation.

Local boiling occurs when the wall temperature T_w calculated from Eq. 9.45 is higher than the wall temperature T'_w for local boiling determined by Eq. 9.44.

Example 9.11

Determine whether local boiling occurs in a pressurized water reactor when operating under the following conditions.

P = 15.0 MPa, abs (T_{sat} = 342.2 C)
T_b = 318 C k = 0.521 W/m·K
μ = 0.867 × 10^{-4} N·s/m²
c_p = 6.085 kJ/kg·K
(From ASME Steam Tables)

$\dot{q}/A$ = 5.10 × 10^6 kJ/h·m²
G = 11.2 × 10^6 kg/h·m²
D_e = 13.35 mm

Substitution of numerical values into Eq. 9.44 yields

$$T'_w - T_{sat} = \frac{18.16(5.10)^{0.25}}{e^{2.419}}$$

$$= 2.4 \text{ C}$$

$$T'_w - 342.2 = 2.4 \qquad T'_w = 344.6 \text{ C}$$

$$\frac{GD_e}{\mu} = \frac{11.2 \times 10^6 \times 0.01335}{0.867 \times 10^{-4} \times 3600}$$

$$= 479{,}046$$

$$\frac{c_p\mu}{k} = \frac{6.085 \times 0.867 \times 10^{-4} \times 1000}{0.521}$$

$$= 1.013$$

Using Eq. 9.40,

$$h_o = 0.030 \frac{k}{D_e}\left(\frac{GD_e}{\mu}\right)^{0.8}\left(\frac{c_p\mu}{k}\right)^{1/3}$$

$$= 0.030 \frac{0.521}{0.01335}(479046)^{0.8}(1.013)^{1/3}$$

$$= 41{,}176 \text{ W/m}^2\cdot\text{K} \qquad 148{,}234 \text{ kJ/h}\cdot\text{m}^2\cdot\text{K}$$

$$T_w - T_b = \frac{\dot{q}/A}{h_o}$$

$$= \frac{5.10 \times 10^6}{148234} = 34.4 \text{ C}$$

$$T_w = 318 + 34 = 352 \text{ C}$$

Since $T_w > T'_w$, local or subcooled boiling will occur.

9.16 BOILING WATER REACTORS

The cores of the boiling water and the pressurized water reactors are basically similar. For these two types of reactors, ordinary or light water functions as the coolant and the moderator. The fuel for the BWR is also UO_2, and the fuel pellets, rods, and fuel assemblies for both types of reactors are similarly fabricated. The heat flux $\dot{q}/A$ for the BWR and the heat flux for the PWR are essentially of the same magnitude.

The operating pressure of about 15 MPa in the PWR will suppress bulk boiling for the normal coolant temperature. In the BWR, on the other hand, the reactor operating pressure is maintained at a somewhat lower value, about 7 MPa, in order to establish bulk boiling within the core. During the early stages of the development of the water-cooled and water-moderated reactor, the boiling process was believed to affect adversely the operation of the reactor. Tests conducted on experimental reactors however demonstrated the feasibility of the boiling water design and further showed that this type of reactor has inherently safe and stable operating characteristics.

The moderating property of water diminishes with an increase in temperature, as a result of the reduction in the density of the fluid. When boiling occurs, the moderating capability of water is further reduced because of the comparatively low density of the liquid–vapor mixture. The resulting decrease in reactivity has a tendency to shut down the reactor, a phenomenon that contributes to safe operation of the reactor. The void fraction in the core of the boiling water reactor however is not sufficiently great to cause serious deterioration of the moderating capability of the water in a liquid–vapor binary phase.

Figure 9.12 illustrates the flow diagram for the direct-cycle reactor system. The steam generated in the reactor flows directly to the turbine. Elimination of the heat exchanger, which is used in the PWR system, simplifies the mechanical arrangement and avoids a loss in availability, inherent in the exchange of energy between the reactor coolant and the turbine working fluid. High- and low-pressure feedwater heaters are incorporated into the steam cycle.

In the BWR power system, saturated steam is delivered to the turbine, and as a consequence the liquid content of the expanding steam is relatively high. Figure 9.12 shows the installation of a mechanical moisture separator in the crossover piping between the high- and low-pressure sections of the turbine. Removal of some of the liquid from the two-phase fluid achieves an increase in the specific enthalpy of the steam.

The steam leaving the reactor is radioactive.

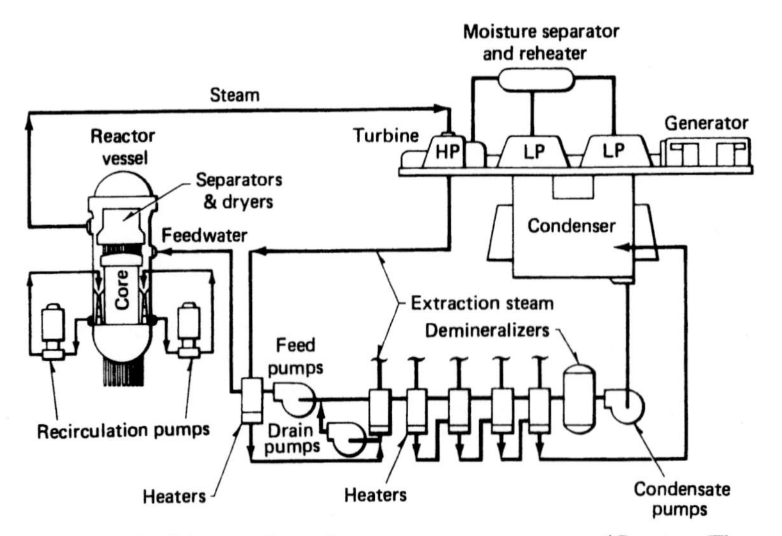

Figure 9.12 Direct-cycle nuclear reactor power system. (Courtesy The General Electric Company.)

This characteristic however is not detrimental to the operation of the system. Impurities are continuously removed from the feedwater by the demineralizers. Pure water however becomes slightly radioactive in the reactor because of neutron absorption by the several isotopes of oxygen. The predominant reaction is the conversion of ^{16}O to ^{16}N. The ^{16}N isotope has a short half-life of about 7 s in decaying to ^{16}O with beta and gamma emissions. When the system is shut down, the radioactivity in the piping, turbine, and feedwater system quickly disappears and maintenance work is safely accomplished.

The main feed flow from the turbine enters the reactor vessel above the core and mixes with the recirculated flow. See Fig. 9.13. The combined water flow enters the core at the bottom and flows upward through the fuel channels. The liquid entering the core is subcooled because of the mixing of the recirculated flow with the relatively cool return flow from the turbine.

Recirculation within the reactor vessel is accomplished by a series of jet pumps located around the periphery of the core. The jet pumps, which have no moving parts, create a continuous internal circulation path for a major portion of the core coolant flow. The driving flow for the jet pumps is created by the external recirculation pumps that handle about one-third of the core flow. The recirculation pump power for the BWR is only a small fraction of the pumping power required for the PWR.

The BWR can function with natural circula-

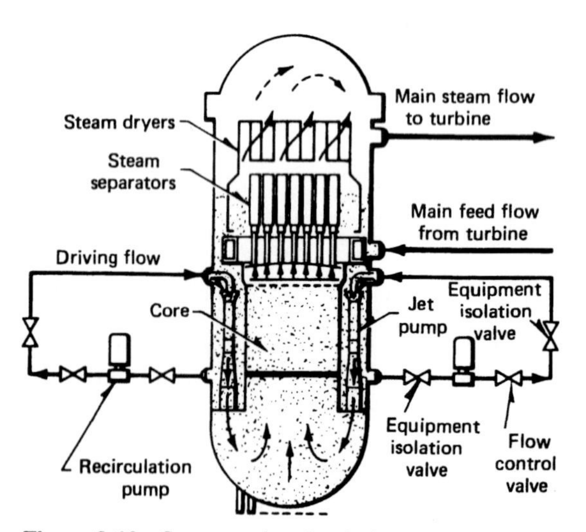

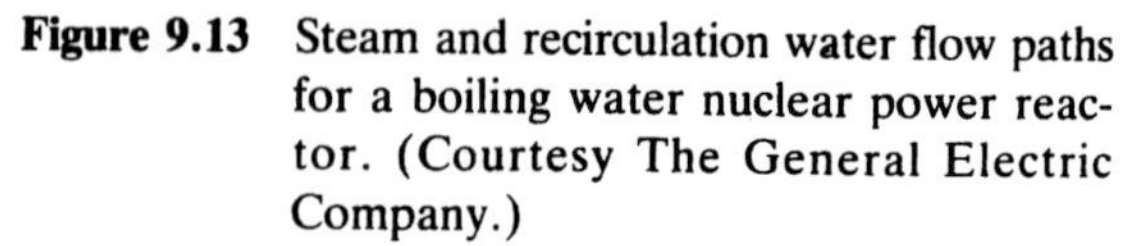

Figure 9.13 Steam and recirculation water flow paths for a boiling water nuclear power reactor. (Courtesy The General Electric Company.)

tion and deliver about one-third power. This capability is clearly advantageous in the event that a "black restart" is required. Further, the reactor can operate at reduced load should power to the external recirculating pumps fail or some other mishap occur in the recirculation system.

A low-quality mixture of liquid and vapor is discharged from the top of the core. A typical value for the average quality of this mixture is about 15 percent. The vapor flowing from the region immediately above the core is directed successively through the steam separator assembly and the steam dryer assembly. Most of the liquid entrained in the vapor is removed mechanically, first by centrifugal action in the separators and finally by impingement on the dryer vanes.

Changes up to 25 percent in the power output of the reactor can be achieved by altering the recirculation flow rate without changing the position of the control rods. A change in the recirculation flow rate causes a change in the core void fraction, and, as a consequence, the reactivity is altered.

The recirculation ratio is defined as the ratio of the recirculation flow rate to the rate at which saturated vapor is produced in the core. An average quality of 15 percent for the vapor–liquid mixture at the core outlet corresponds to a recirculation ratio of 5.67.

The axial variation in the heat flux generated in the fuel rods is generally similar for the boiling water and the pressurized water reactors. The burnout heat flux in the BWR, on the other hand, remains at a high value until vapor formation starts. Because of the mixing of the relatively cool feedwater with the recirculated saturated liquid, subcooled water enters the core. The temperature of the water increases during upward flow in the core until the saturation temperature is attained. Nucleate boiling begins at some point before the water reaches the saturation temperature. Bulk boiling with steam generation is subsequently established. As the fluid moves upward in the core, the quality of the mixture increases with the continued formation of vapor.

9.17 NUCLEAR POWER PLANT STEAM TURBINES

The steam supplied to the turbine in a nuclear power plant is under a moderate pressure of about 7 MPa. For the BWR power plant, the steam is saturated, while limited superheat, about 20 C, can be provided in a PWR power system. Such steam conditions are sharply different from the 20 MPa, 540 C total temperature conditions that are typical of the fossil fuel-fired, subcritical-pressure power plant.

Because of the lower throttle pressure and temperature inherent in the design of the power system, the thermal efficiency of the nuclear power plant is considerably below the thermal efficiency of a modern fossil fuel-fired steam power plant. The thermal efficiency values range, respectively, from 0.30 to 0.32 and from 0.34 to 0.40+. The characteristics of the thermal power cycle strongly influence the overall performance of the power plant.

The liquid content of the steam expanding in a turbine should ordinarily not exceed 12 percent if severe erosion of the blades is to be avoided. In fossil fuel-fired power plants, steam superheating and reheating substantially reduces blade erosion. Nuclear power plant turbines, on the other hand, are supplied with saturated or with slightly superheated steam for the BWR and PWR power plants, respectively. Certain measures however can be taken in the design of nuclear power plant turbines to minimize effectively the erosion of the blade surfaces caused by the impact of water droplets that are formed in the expanding steam.[9]

The extent of the erosion damage is dependent upon the quantity of liquid water in the steam and upon the fluid velocity. A reduction in the fluid velocity can be achieved by designing the turbine for operation at 1800 rpm, rather than at the more conventional speed of 3600 rpm. The 1800-rpm machines are inherently larger structures than the machines that operate with the same generating capability at the higher speed of 3600 rpm.

A number of materials capable of resisting erosion are available for fabrication of nozzle vanes, blades, and other parts of the turbine subjected to water droplet impingement and high-velocity flow over the metallic surfaces.

Liquid removal from the expanding fluid is an effective means to controlling the moisture content of the steam. A certain number of turbine stages are designed for moisture extraction. Radial grooves are milled into the surface of the moving blades on the convex side at the leading edge. Water tends to collect in these grooves and is thrown by centrifugal action into a circumferential channel in the turbine casing. The extracted water is drained to the feedwater heating system, and the collecting channel may, in addition, serve as a steam extraction belt.

Separation of liquid water from the steam can be effectively accomplished in the crossover between the high- and low-pressure sections of the turbine. In addition, at this point the steam may be reheated by a portion of the steam generated in the BWR core or in the PWR steam generator. The reheat steam is diverted from the main supply line at a point upstream from the turbine throttle. Water removed in the moisture separator and the condensate from the reheater are piped to the feedwater heating system.

Improved thermal performance of the PWR and the BWR power plants would be realized if the steam temperature at the turbine throttle were increased to 540 C, a common temperature for fossil fuel-fired electric generating stations. A change in the steam conditions at the turbine throttle from 7 MPa, gage, saturated to 7 MPa, gage, 540 C would effect a reduction of 20 percent in the plant heat rate for a nuclear power system. Steam superheating in a nuclear power plant could be accomplished either in a nuclear reactor or in a separate, fossil fuel-fired superheater.

Steam superheating in a fossil fuel-fired unit is not a particularly desirable application. A more complex installation will result, and two different kinds of fuel must be supplied. The advantages of the nuclear design would be offset to some degree by burning the additional fossil fuel. The designer must observe two sets of environmental regulations, and the operating problems are aggravated when two different fuel-burning systems are combined.

Nuclear superheating has been explored in a number of demonstration-type nuclear superheating reactors. The 62.5-MW Pathfinder Atomic Power Plant, located at Sioux Falls, South Dakota, is the first commercial power plant in the world to generate superheated steam with nuclear energy. Criticality was first achieved in March 1964. The reactor is a controlled-recirculation boiling water reactor provided with an integral nuclear superheater. Currently, commercial light-water nuclear power plants are not constructed for nuclear superheating of the steam.

Substantial improvement in the thermal efficiency of future commercial nuclear power plants could be achieved by the application of nuclear superheat and nuclear reheat. A thermal efficiency of 0.427 is anticipated for a nuclear power plant that operates with steam conditions at 14 MPa, gage, and 540 C/3.85 MPa, gage, and 540 C.[10]

9.18 GAS-COOLED REACTORS

Commercial nuclear power reactors installed in the United States, with a few exceptions, employ water as the cooling medium. The use of a gaseous cooling medium is however entirely practicable. The efficacy of gas cooling has been demonstrated in a number of gas-cooled reactors placed in operation since 1960. Helium is the most commonly used gaseous coolant, although limited use has been made of nitrogen and carbon dioxide.

The typical high-temperature gas-cooled reactor (HTGR) employs a graphite moderator and helium for the coolant. Helium is chemically inert and transparent with respect to neutron interaction. In addition to functioning as the moderator, graphite is used for the core structural material and for the material employed in fabricating the fuel elements. Neutrons are not lost through capture by the metallic fuel cladding and by other metallic elements that are present in the core of the water-moderated reactors. Unlike most materials, the strength of graphite increases with an increase in temperature. The maximum strength of graphite is attained at about 2480 C, a temperature that is well above the reactor operating temperature range.

The HTGR power plant is a two-fluid system. Helium is the primary coolant and water is the working fluid in the steam power cycle. When water functions as the reactor coolant, the critical temperature of water imposes a limit on the coolant outlet temperature. Unlike water, helium exists in the reactor in the gaseous phase far above the critical state. The temperature of the helium leaving the reactor is consequently not limited because of any inherent characteristic of the gas.

The gas-cooled reactor is designed for high-temperature operation so that the heat source fluid can enter the steam generator at a substantially elevated temperature. The helium temperature is sufficiently high to produce high-pressure, high-temperature steam, including reheating, for operation of the turbine-generator. As a consequence of improving the steam power cycle, a thermal efficiency of 0.39 can be achieved in the HTGR power plant. A thermal efficiency of 0.39, in contrast to a value of about 0.31 for the PWR and BWR power plants, compares favorably with the thermal efficiency values reported for the high-capability, fossil fuel-fired steam power plants.

The first commercial HTGR power plant constructed in the United States is located at Peach Bottom, Pennsylvania. The power plant has a generating capability of 40 MW and began commercial operation in June 1967. The 330-MW Fort St. Vrain Nuclear Generating Station is the second HTGR power plant constructed in the United States. The station, located in the vicinity of Platteville, Colorado, began commercial operation in 1974.

Certain design data for the Fort St. Vrain Station steam generating system are of interest.[11] The primary coolant system is comprised of a reactor and two coolant loops, each of which contains one steam generator and two helium circulators. The helium enters the steam generator at 775 C and leaves at 4.73 MPa, abs, and 394 C. Steam is produced at 17.32 MPa, abs, and 540 C. The reheat steam conditions are 4.14 MPa, abs, and 539 C.

The employment of a gaseous medium for reactor cooling suggests coupling a gas turbine engine with the HTGR.[12] In this arrangement, the helium gas, circulating at a high pressure in a closed system, functions as the reactor coolant and as the engine working fluid. The gas turbine engine, while producing power, also acts as the helium circulator. The nuclear reactor essentially replaces the combustion chamber in the conventional gas turbine engine. The HTGR/GT power plant is inherently less complex than the two-fluid HTGR/steam cycle power plant, and the gas turbine engine is a much smaller machine than the steam turbine of the same output.

9.19 BREEDER REACTORS

The light-water reactor is an ineffective user of uranium fuel. Only 2 percent of the energy available from uranium is used in the light-water reactors. The bulk of the available energy in the processed uranium ore remains unused in the depleted uranium, an end product of the enrichment process. By the year 1985, the quantity of depleted uranium accumulated in supplying enriched fuel for light-water reactors is expected to be about 375,000 metric tons.

During the operation of the light-water reactor, some of the ^{238}U is converted to fissionable ^{239}Pu. The plutonium can be recovered in processing the spent fuel taken from the reactor. More than 275 metric tons of high-commercial-value plutonium are expected to be available by the year 1990.

The breeder reactor is expected to overcome certain deficiencies of the light-water reactor. Power plant thermal efficiencies in the range of 0.38 to 0.40 are anticipated, and consequently, a more effective utilization of the fuel is achieved while a substantially smaller quantity of heat is rejected to the surroundings.

The commercial value of the recycled plutonium is higher when used in a breeder reactor than in a light-water reactor. A most important function of the breeder reactor is the conversion of fertile uranium 238 into fissionable plutonium. Consequently, the breeder reactor is capable of utilizing as much as 75 percent of the energy available from uranium.[14]

The concept of the breeder reactor is by no means new. The first electric power produced from a nuclear reactor was generated in the first fast breeder reactor. This event occurred in 1951, and the EBR-I had a generating capability of 0.2 MWe. Subsequently, a number of experimental and demonstration fast breeder reactor power plants have been constructed, or are scheduled for construction, in the United States, Japan, and in several European countries. Present efforts in the United States for achieving electric power generation in a breeder reactor power plant are directed principally to the 350-MWe Clinch River Fast Demonstration Plant.

The intent of the breeder reactor is to utilize plutonium for the fuel and to convert simultaneously ^{238}U into fissionable plutonium. These reactions are more effectively achieved by fast, or high-energy, neutrons than by thermal neutrons. The average fission energy in a fast breeder reactor is typically 0.15 MeV, in comparison with 0.03 eV for the neutrons in a light-water reactor. Plutonium is more effectively fissioned in a fast reactor than in a thermal reactor because not as much parasitic absorption of neutrons in plutonium occurs for fast neutrons as for thermal neutrons. Further, approximately twice as much ^{238}U is fissioned directly by fast neutrons in the fast reactor as in the thermal neutron reactors.

The fuel for the light-water reactor must be periodically replenished if continued operation is expected. With continued operation of the fast reactor, the quantity of fissionable fuel in the reactor continues to increase. Eventually, the amount of fissionable fuel in the reactor will be equal to twice the initial charge of fissionable material. Thus, the fast breeder reactor maintains its own supply of fissionable fuel and, in addition, produces sufficient fissionable material to fuel another reactor.

The time required for a breeder reactor to produce sufficient fuel to replenish its own plutonium needs and those of a similar reactor is designated as the doubling time. The doubling process is similar to earning interest on invested money, and the doubling time for a particular breeder will depend upon the manner in which the bred plutonium is handled. In general, the doubling time ranges from 12 to 20 years.[15]

Production of fast neutrons in the breeder reactor necessitates the removal of moderating materials, such as light water in the LWR, that can slow the neutrons by scattering collisions and consequently decrease the energy spectrum toward the thermal range. Coolants used for a fast reactor must be a very-low-density gas or a substance of high atomic weight in comparison with the mass of the neutron, and thus have little or no moderating effect on the neutrons. A number of substances, including steam, helium, carbon dioxide, and a variety of liquid metals, have been examined for possible use as coolants in fast reactors. In this respect, sodium and helium demonstrate the best potentialities.

The liquid metal fast breeder reactor (LMFBR) is cooled by circulating liquid sodium through the core. Sodium reacts vigorously with water to form sodium hydroxide, sodium oxide, and hydrogen gas, hence sodium containment must be ensured through proper construction of the circulation system. When sodium flows through a neutron flux, a neutron reaction causes the formation of two radioactive isotopes, namely, sodium 22 and sodium 24. Sodium 24, the principal activity, emits gamma rays and has a half-life of 15 h. The main problem experienced with sodium 22 is a long half-life of 2.6 years.

Because of the chemical and neutron reactivity of sodium, the LMFBR power plant is constructed with two sodium loops. Energy is transferred from the reactor coolant, that is, the so dium in the primary loop, to the sodium in the secondary loop and finally, in a second heat exchanger, to water in the power loop. The sodium in the secondary loop is not radioactive.

The primary loops must be heavily shielded because of the high radioactivity of the sodium 24. Contact maintenance of the equipment in the primary loops can be performed by delaying the work for about 10 days in order to allow the sodium 24 to decay to a sufficiently low level. There is however an added problem caused by the long half-life of the sodium 22, although in comparison with sodium 24 the activity of this isotope is less intense.

The core for the LMFBR differs in several respects from the typical core for the light-water reactors.[16] Fewer control rods are required for the LMFBR, and the volume of the core is smaller, approximately 80 percent of the LWR core. Other differences are observed in the fuel composition, coolant, and the reactor structural materials. The fuel pins for the LMFBR are smaller in diameter, about 6.5 mm. Stainless steel is used for the fuel cladding and for the fabrication of most of the structural components of the core.

Plutonium used to fuel the fast breeder reactor may be in the form of an oxide, metal, carbide, or nitride. The early FBR plants will employ a mixed oxide fuel consisting of fissionable PuO_2 mixed with fertile UO_2. This mixture is selected because of good burnup experience and an extensive in-pile experience. Physical properties of the mixed oxide fuel and the UO_2 fuel are quite similar, and similar processes are employed in the fabrication of the fuel pins. The melting point of Pu–UO_2 is 2700 C, in comparison with 2880 C for UO_2.

Protective measures are required in the manufacture of the plutonium fuel, in part because of the toxic property of this element. Further, extensively irradiated plutonium fuel can be radioactive after reprocessing. Remote or semi-remote handling of the plutonium increases the manufacturing cost of the fuel.

The fuel material in the FBR contains 15 percent fissionable isotopes, in contrast to a typical 3 percent for the LWR. The substantial fuel enrichment of 15 percent is essential in the fast breeder reactor, because parasitic capture of neutrons by the core materials is high in this type of reactor.

Plutonium used to fuel the FBR is obtained by reprocessing the spent fuel removed from light-water reactors. The plutonium acquired in this manner from a PWR has the following isotopic composition: ^{239}Pu 0.61, ^{240}Pu 0.22, ^{241}Pu 0.13, and ^{242}Pu 0.04. The fissionable isotopes in the FBR would ordinarily be ^{239}Pu and ^{241}Pu.[17]

The core of the FBR is surrounded by radial and axial blanket regions that are composed of fuel pins manufactured from depleted or natural uranium. In the radial blanket, the depleted uranium extends throughout the entire length of the fuel pins. The axial blanket is formed by placing depleted uranium fuel in the upper and lower ends of the fuel pins and the Pu–U mixed oxide fuel in the center portion of these multipurpose fuel pins. The reactor core is formed by the center portion of the composite fuel pins.

Because of the high temperature that can be achieved in a sodium-cooled reactor, steam can be generated at conditions comparable to those of a modern fossil fuel-fired power plant, that is, up to 24 MPa, gage/540 C/540 C. Consequently, the thermal efficiency (0.40) of the FBR steam cycle is substantially higher than the thermal efficiency of the PWR steam cycle.

The sodium-cooled fuel elements can operate, without difficulties developing from boiling burnout conditions, with a heat flux in excess of 11×10^6 kJ/h·m^2. On the other hand, the heat flux in the PWR is limited by avoiding unstable boiling.

Several types of gas-cooled fast breeder reactors (GCFR) have been proposed during the past two decades, but of this number, only two are receiving serious consideration. One of the two leading types is based on a conservative design and employs stainless-steel-clad, mixed plutonium and uranium oxide fuel rods. Helium is used for reactor cooling and for transferring energy to the indirect steam cycle.

Construction costs for the GCFR/steam cycle power plant are likely to be somewhat below the comparable costs for the LMFBR plant, principally because of the omission of the intermediate sodium loop. The thermal efficiency of the GCFR plant is expected to be slightly lower than that of the LMFBR plant, 0.38 compared with 0.40 for the latter.[18]

The second type of GCFR power plant that is receiving serious consideration would incorporate certain advanced reactor designs and a direct helium gas turbine power cycle.[20] The GCFR/GT power plant tentatively represents the simplest and most economical nuclear power plant that can be designed for operation on a thermal power cycle and employing nuclear fission. The predominant characteristics of this power plant are fuel breeding and the single fluid that functions both as the reactor coolant and as the working fluid in the power cycle. Typically, the power plant would be comprised of the helium-cooled breeder reactor coupled to three gas turbine engine-generators.

9.20 FUSION REACTORS

Development of the fusion reactor is an advanced step in the program directed to the application of nuclear energy to electric power generation. The design principles for the fusion reactor are completely different from those of the various types of fission reactors. The reactor operating temperature is perhaps the most significant difference observed between fission and fusion reactors. In a fission reactor, the temperature of the uranium oxide fuel is normally maintained somewhat below the melting temperature of 2880 C. Fusion of the light nuclei however can only be achieved at extremely high temperatures.

Light-weight nuclei may be combined in a variety of reactions to form a heavier nucleus with a release of energy. For example, combination may be effected between two deuterium nuclei, or between tritium and hydrogen. The latter reaction is expressed by

$$^3_1\text{H} + ^1_1\text{H} \rightarrow ^4_2\text{He} + \text{energy}$$

The energy release for this reaction is approximately 19.8 MeV.

With respect to the energy released in a single nuclear reaction, fission is quantitatively superior to fusion on the order of about 10 to 1. On an equal mass basis however, the energy release ratio is approximately 7 to 1 in favor of the fusion reaction.

The fuel for the fusion reactor is deuterium, a stable, heavy isotope of hydrogen present in natural water. Deuterium exists in combination with oxygen as D_2O, a compound designated as heavy water. Natural hydrogen is composed of two isotopes, namely, deuterium ^2_1H (0.015 atomic percent) and the light isotope ^1_1H (99.985 percent). If both of the hydrogen isotopes are assumed to be combined with ^{16}O, the mass concentration of heavy water occurring in natural water is about one part in 6000. The deuterium can be separated from ordinary hydrogen at an extremely low cost in comparison with the value of the energy that, in principle, can be released by the fusion process.

TABLE 9.5 World Fuel Resources

Coal, oil, natural gas	
Proved[21]	34×10^{18} kJ
Estimated total[21]	227×10^{18} kJ
Uranium: estimated total (< \$65/kg)[21]	
LWR	2.5×10^{18} kJ
Breeder	173×10^{18} kJ
Deuterium	
D–T reaction[a]	216×10^{18} kJ
D–D reaction	1.2×10^{28} kJ

[a] *If the deuterium in the ocean water is used in the deuterium–tritium reaction, the limiting factor would be the world reserve of lithium, the material used to manufacture tritium. Both lithium 6 and lithium 7 are used for this purpose. The lithium reserve however can be stated in terms of lithium 6, because it is the least abundant lithium isotope. The measured, indicated, and inferred lithium resources in the United States, Canada, and Africa total 9.1 × 10^6 metric tons of lithium metal, including 675,000 metric tons of lithium 6 (7.42 atom percent of the total).[22] From this amount of lithium 6, the fusion energy obtainable in the D–T reaction at 3.19 × 10^{-12} J/atom would be 216 × 10^{18} kJ (205 Q). The calculated value of 216 × 10^{18} kJ is based on an assumption that the burnout rates for ^{6}Li and ^{7}Li are equal.*

An examination of the data presented in Table 9.5 will disclose the significance of the vast quantity of deuterium that exists in the ocean water of the world. The successful development of a commercial-type fusion reactor fueled with deuterium will provide the means to utilizing a virtually unlimited source of energy for the production of electric power. The incentives for supporting the development of the fusion reactor are unquestionably high.

There are several possible reactions between the nuclei of light elements that can be the basis for controlled fusion. Four such reactions involving deuterium are[23]

$$^{2}_{1}H + ^{2}_{1}H \rightarrow ^{3}_{2}He + ^{1}_{0}n \qquad 3.2\ \text{MeV} \qquad (9.46a)$$

$$^{2}_{1}H + ^{2}_{1}H \rightarrow ^{3}_{1}H + ^{1}_{1}p \qquad 4.0\ \text{MeV} \qquad (9.46b)$$

$$^{2}_{1}H + ^{3}_{1}H \rightarrow ^{4}_{2}He + ^{1}_{0}n \qquad 17.6\ \text{MeV} \qquad (9.46c)$$

$$^{2}_{1}H + ^{3}_{2}He \rightarrow ^{4}_{2}He + ^{1}_{1}p \qquad 18.3\ \text{MeV} \qquad (9.46d)$$

The selection of the most promising reaction from the above group is made on the basis of the requirements for establishing a sustained fusion reaction. Three criteria are observed: (1) the deuterium fuel must be maintained at a very high temperature, between 100 million and 5 billion degrees Kelvin (10 to 500 keV), depending upon the fuel cycle; (2) the resulting ions, the plasma, must be held in a configuration where their density is on the order of 10^{15} ions/cm^3; and (3) for fusion to occur at these temperatures and densities, the confinement time must be on the order of tenths of a second. These requirements are truly formidable, but the ultimate development of the fusion reactor is believed to be possible.

A high plasma temperature is indicative of the high velocity and the corresponding high kinetic energy acquired by the nuclei. The kinetic energy must be sufficiently high to overcome the repulsion of the nuclei, and, as a consequence, the collisions become so severe as to cause the nuclei to fuse. A high particle density increases the probability that the encounter between the nuclei will occur. Finally, sufficient time must be provided to enable the fusion reaction to occur and become self-sustaining.

A significant relationship exists between the plasma density and the confinement time. An increase in the plasma density promotes a shorter confinement time; and, alternatively, if the ion density is reduced, the confinement time must be increased. In this respect, the deuterium–tritium reaction is of particular interest. For the D–T reaction at 10 keV, a change in the plasma density from 10^{20} to 10^{12} particles/cm^3 increases the mean reaction time from about 10^{-4} s to 10^{4} s. For a plasma density of 10^{15} particles/cm^3, the mean reaction time is somewhat less than 1 s, that is, approximately on the order of 0.7 s.

Fusion occurs most easily between deuterium and tritium, the heavy and the heavy-heavy isotopes of hydrogen, respectively. Deuterium–tritium reactions are self-sustaining above a temperature of 5 keV (50×10^6 K). The corresponding temperature for the deuterium–deuterium reaction is 50 keV (500×10^6 K).

The two D–D reactions described by Eqs. 9.46a and 9.46b occur with equal probability. Not only is the ignition temperature high for these two reactions, but the energy yields are low. The ignition temperature of the plasma is defined as the temperature above which the fusion energy release exceeds the natural radiation losses that represent an inescapable plasma energy loss.

The energy yield, 18.3 MeV, for the deuterium–helium reaction described by Eq. 9.46d is comparatively high. A high self-sustaining temperature of 100 keV however is required for the D–He reaction. The corresponding temperature for the D–T reaction is considerably lower, 5 keV, with almost the same energy yield, 17.6 MeV. The quantity 17.6 MeV represents the heat release per fusion in the reactor.

Unlike deuterium, tritium does not occur abundantly in nature because it decays radioactively, by beta decay, to helium 3 with a half-life of 12.3 years. Tritium however may be produced in a lithium "breeding blanket" that surrounds the plasma core of the fusion reactor. The breeder concept for the production of tritium is essential to establishing an acceptable fuel economy for the fusion reactor.

The deuterium–tritium reaction produces a neutron and helium 4. The energy distribution for the products is 14.1 MeV for the neutron and

3.5 MeV for the helium 4. Neutrons react with the two isotopes of lithium, ^{6}Li and ^{7}Li, as shown by

$$^1_0n + ^6_3Li \rightarrow ^3_1H + ^4_2He + 4.78\ MeV \qquad (9.47a)$$

$$^1_0n + ^7_3Li \rightarrow ^3_1H + ^4_2He + ^1_0n - 2.47\ MeV \qquad (9.47b)$$

The first reaction, Eq. 9.47a, is exothermic, and the reaction energy augments considerably the total energy of the cycle. The second reaction, Eq. 9.47b, however is endothermic. The energy level of the product neutron (2.1 MeV) is lower than that of the incident neutron (14.1 MeV).

Natural lithium is composed of 7.42 percent ^{6}Li and 92.58 percent ^{7}Li. ^{6}Li enrichment is proposed as a means to increasing the lithium yield, because the ^{6}Li isotope is burned out at a more rapid rate than the ^{7}Li isotope.

From a material standpoint, the fusion reactor is in a favorable position. Deuterium and lithium, the input materials, are relatively inexpensive, and helium, the by-product from the reaction, is a valuable commodity. In addition to the primary nuclear reactions, there are secondary reactions that occur in the materials of the surrounding structures. Some of these reactions breed neutrons; but, for the most part, they are neutron absorbing, hence deleterious. Despite the neutron losses, an ample supply of tritium is assured. Apparently, it will be possible to develop a high overall tritium breeding ratio and thus achieve a low fuel doubling time.

The nuclear reactions pertinent to the fusion process apparently can be achieved in the absence of insurmountable difficulties. The principal problem encountered in developing the fusion reactor concerns the design and construction of the equipment that can effectively contain the plasma. Extensive laboratory-type experiments have been conducted for about two decades in several countries in an effort to develop a plasma containment device that subsequently can be extrapolated to a larger scale for incorporation into a fusion reactor of commercial capability. These containment devices are highly complex, but the operating principles can be examined.

At the typically high reaction temperature, the gas is completely ionized, and the plasma is an excellent electrical conductor. Confinement of the conducting fluid free from any material walls can be achieved, at least in principle, by strong magnetic fields. The charged deuterium and tritium ions, and the electrons that neutralize the medium, are constrained from escaping, because each particle can move only along magnetic field lines. Actually, the particle moving in a magnetic field must follow a helical path around the magnetic field lines, and, as a consequence, the particle is in effect partially immobilized by the field.

Four approaches to fusion power are listed in Table 9.6. The low-density closed system is considered to represent the most promising approach to fusion power. The Russian device, called the T-3, has demonstrated that hot (10^7 K), dense ($2 \times 10^{13}/cm^3$) plasma can be contained in a tokamak for 10 ms or even longer. This performance falls short of the reactor parameter (10^8 K, $10^{14}/cm^3$, and 1 s). Experiments conducted in the United States demonstrated that high temperatures and high densities can be achieved by compression of a tokamak plasma. In a recent experiment, a strong magnetic field held a plasma for 20 ms at a temperature of 60 $\times 10^6$ K. Four high-power, neutral-beam injectors pumped additional energy into the hot plasma.

The collective experiments indicate that the techniques employed would be capable of elevating tokamak plasmas to ignition temperatures in a larger facility. Much more investigative work however remains before a fusion power demonstration plant can be constructed.

The confinement chamber of the tokamak is constructed in the form of a torus that has a metal conducting wall. The external current-carrying windings, placed on the wall, produce a toroidal magnetic field. A magnetic transformer driven by an external power supply induces an axial current in the plasma. The induced current, in

TABLE 9.6 Four Approaches to Fusion Power[24]

	Popular Name	*Typical Density (cm^{-3})*	*Typical Burning Time (s)*
Open systems	Magnetic mirror	10^{13}	10
Low-density closed systems	Tokamak	10^{14}	1
High-density closed systems	Theta pinch	10^{16}	0.01
Laser pellet systems	Laser fusion	10^{24}	10^{-9}

turn, creates a poloidal field in the torus. The combination of the two fields produces a typical helical-shaped tokamak confinement field.

9.21 FUSION POWER REACTORS

The form that the fusion power reactor will eventually take is a matter of interest. Figure 9.14 shows a schematic representation of a possible configuration for the deuterium–tritium fusion reactor. For a tokamak design, this illustration represents a cross section of the torus. The plasma is contained inside an evacuated tube that typically may have a diameter of 4 m. The surrounding vacuum wall, through which the 14-MeV neutrons from the plasma pass, is maintained at a temperature of 500 to 1000 C, depending upon the wall coolant and upon the kind of wall material that can be developed. The vacuum wall also intercepts radiation from the plasma, supports vacuum forces, and must withstand bombardment by the neutrons as well as some charged particles.

Outside the vacuum wall are two concentric regions, namely, the lithium breeding moderator and the magnet shield. Tritium is manufactured in the lithium blanket; and in the blanket, the kinetic energy of the 14-MeV neutrons produced by fusion is converted into thermal energy. The thickness of the lithium moderating region, which must be about 1 m, is controlled by the nuclear cross sections. Large superconducting magnets are placed outside the magnet shield, which is about 1 m thick. The diameter of the magnets may be 7 to 8 m and must be designed to resist high magnetic forces. These magnets will operate in a cryogenic environment and consume very little power.

Because of the large number of unsolved problems, many years will pass before a demonstration fusion reactor power system can be built. Significant advantages for the fusion power plant may be cited. The supply of deuterium, the basic fuel, is virtually inexhaustible and can be obtained at a negligible cost. Large quantities of radioactive wastes will not be produced in these power plants, and helium, the reaction product, is nontoxic, nonnoxious, and nonradioactive, and has a high market value. Fusion power plants will be inherently very safe to operate, and the radioactivity associated with the operation of the plant will be relatively low.

The inherent safety of the fusion reactor can be attributed to two characteristics of the device. The basic fusion reaction yields no radioactive products, and the inherent nature of the fusion reactor absolutely precludes any kind of nuclear excursion. Although there are no radioactive waste products, the inner portions of the reactor become highly activated and represent a potential hazard should maintenance or overhauling be required. It is also essential to guard against leakage of tritium from the plant, because of the radioactive characteristics of this isotope.

High energy conversion efficiencies, on the order of 0.60 or higher, are predicted for the fusion power plant. Of particular significance, because of environmental impact, will be the cor-

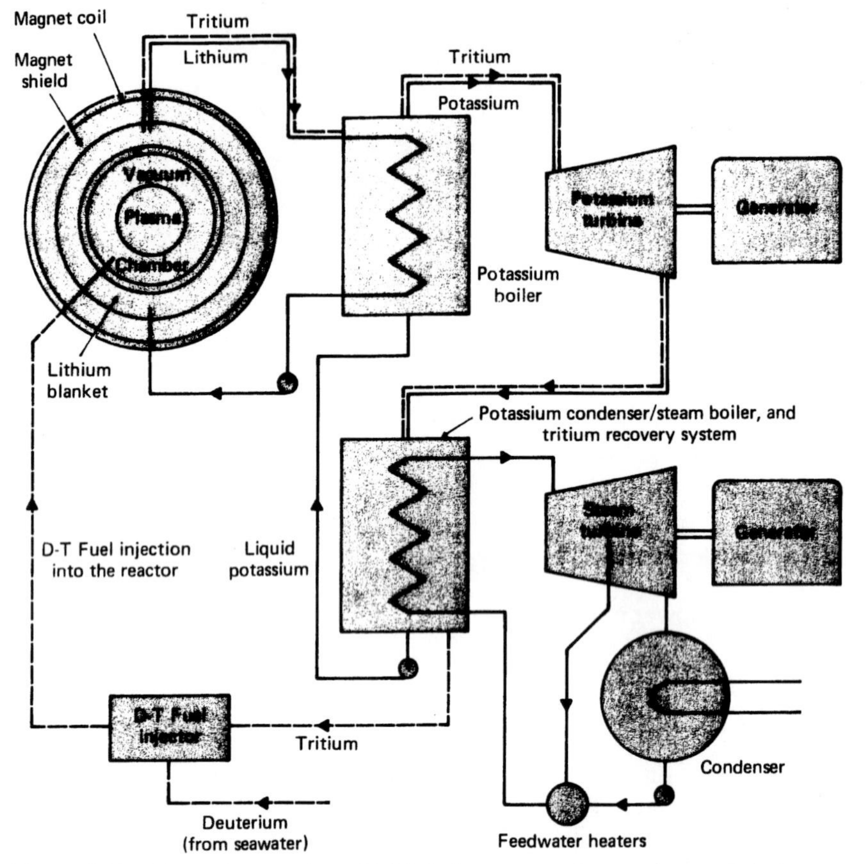

Figure 9.14 Deuterium–tritium (D–T) fusion power plant conceptual design.

responding low heat rejection per kilowatt of electrical energy generated.

Figure 9.14 shows the flow diagram for a proposed fusion power plant. The binary-vapor power cycle consists of a potassium topping cycle and a conventional steam cycle. Included in the cycle is a tritium recovery system. The molten lithium, or lithium salt, is pumped from the blanket through the potassium boiler and returned to the blanket. Tritium, produced in the blanket, combines with the potassium, and the two fluids are expanded in the potassium turbine. Following expansion, the potassium vapor is condensed, by transfer of energy to the water, in the potassium–steam heat exchanger. In this component, the tritium is recovered and subsequently mixed with deuterium for injection into the reactor.

The material consumption in a fusion power plant is very small. Table 9.7 lists the material flow for a 2030-MW fusion power plant.

Nonfissile neutron multiplication is achieved by placing a layer of beryllium adjacent to the vacuum chamber that encloses the plasma. Natural beryllium consists entirely of ^{9}Be. The reaction is represented by

$$ {}^{1}_{0}n + {}^{9}_{4}Be \rightarrow 2\,{}^{4}_{2}He + 2\,{}^{1}_{0}n \qquad (9.48) $$

TABLE 9.7 Fuel Reactions[26]

Tritium burnup	0.527 kg/day
Deuterium burnup	0.352 kg/day
Lithium burnup	1.104 kg/day
Beryllium burnup	0.362 kg/day
Helium production	10.2 m^3(STP)/day

Fuel costs are essentially negligible for the fusion power plant. Operation and maintenance costs are quite high because of an assumed frequent replacement of the first wall, that is, every five years. From an overall standpoint, the system is capital intensive. In comparison with the fusion power plant, the fuel costs for the fission-type power plant are high, but capital costs are expected to be somewhat lower.

9.22 SUMMARY

While nuclear power plants cause no air pollution, the relatively high heat rejection from the light-water reactor power plants is a negative factor. Development of the HTGR and the FBR power plants will to some degree alleviate this problem through an improvement in the power plant thermal efficiency.

Stringent safety measures are observed in the design, construction, and operation of a nuclear power plant. Despite such precautions, considerable opposition exists in the United States to the construction of additional nuclear power plants and even to the continued operation of those plants currently producing electrical energy. The opposition cites a number of equipment failures that have occurred in nuclear power plants and fear the occurrence at some time of a failure of catastrophic proportions. The proponents of nuclear power, on the other hand, claim that the likelihood of reactor accidents occurring is much lower than that of many nonnuclear accidents of about the same degree of severity. Further, the proponents point to the need for continued construction of nuclear power plants in order to ensure an adequate electric power generating capability in the years ahead.

The rapid escalation of construction costs for nuclear power plants is a matter of considerable concern. From 1965 to 1974, the capital cost for a nuclear power plant increased from about $200/kW to about $560/kW, although not at a uniform rate. During the same period of time, the construction schedule was extended from five to eight years to eight to 12 years, or from an average 6.3 years to an average 9.4 years.[27] An increase in the length of the construction schedule is particularly disturbing, because of the resulting increase in the amount of interest paid during construction. The 1980 construction cost for a nuclear power plant is in the vicinity of $1000/kW.

Two of the major problems the electric utility industry faces today are a severe lack of suitable power plant sites and the absence of standardized construction that would be both cost and time saving. The offshore floating nuclear power plant concept offers solutions to both of these problems.[29]

The nuclear power plant and the supporting platform would be constructed in accordance with a standardized design in a shipyard-type facility employing assembly line techniques. The completed power plant is subsequently towed to the prepared offshore site and connected to the mainland power grid by underwater cables.

A cost comparison of a floating nuclear power plant and an onshore power plant is complex, but offshore power plants may prove to be less costly to construct than the land-based plants. Offshore power plants may also have a lower environmental impact. The environmental constraints encountered in the dissipation of the rejected heat into the cold ocean water should not be as severe as the restrictions that must be observed when inland water or the atmosphere serves as a heat sink.

PROBLEMS

9.1 The power output of a reactor is increased from 1900 MW(t) to 2000 MW(t) within a period of 20 s. Determine the value of the effective neutron multiplication factor.

9.2 The power output of a reactor is to be increased from 1800 MW(t) to 3600 MW(t). A value of 1.005 is prescribed for the effective neutron multiplication factor. Is the increase in the power output achieved within a reasonable period of time?

9.3 Estimate the energy release rate, per cm length, for a uranium oxide fuel pellet that has an O.D. of 9.5 mm. The fuel is 2.5 percent enriched. The effective microscopic cross section is 577 barns, and the neutron flux is 10^{13} neutrons/s·cm^2. The effective heat recovery is 182 MeV per fission. $C_{fn} = 1.07$.

9.4 The diameter and actual length of a fuel rod are, respectively, 10 mm and 350 cm. At the center of the rod, the energy generation rate is 100 W/cm^3. The extrapolated length of the fuel rod is assumed to be 2 percent greater than the actual length. Estimate the thermal power of the fuel rod.

9.5 A reactor core is designed to generate 3000 MW(t). The height of the core is 360 cm, and the diameter of the fuel rods is 10.5 mm. The power generated in the fuel and cladding is 97 percent of the total reactor power. The maximum specific heat generation rate is 580 W/cm^3. The value of the K factor for the reactor is 0.385. Calculate the number of fuel rods required for generation of the prescribed thermal output.

9.6 The dimensions of a fuel rod are: pellet diameter 9.500 mm, outer diameter of the cladding 11.050 mm. The bulk temperature of the coolant is 575 K. The temperature of the outer surface of the cladding is 600 K, and the surface temperature of the fuel pellet is 1077 K. The coolant film conductance is 45,000 W/m^2·K. The thermal conductivity of the fuel is 2.130 W/m·K. Calculate the maximum temperature of the fuel.

9.7 A nuclear power reactor operates at a nominal power output of 2000 MW(e). Actually, there is a 2-percent oscillation of the power output over a period of 2.5 min. Determine the effective neutron multiplication factor.

9.8 The dimensions of a uranium oxide fuel element are: diameter 9 mm, length 12 mm. The fuel enrichment is 3 percent. The effective microscopic cross section is 577 barns. The effective heat recovery is 180 MeV per fission. The power generated in the single fuel pellet is 9.50 W. Calculate the neutron flux.

9.9 A fuel rod assembly contains 200 rods, each 360 cm long. The diameter of the fuel pellets is 9.40 mm. At the center of the rods, the energy release rate is 300 W/cm^3. Calculate the thermal power generated in the fuel rod assembly.

9.10 A boiling water reactor operates at a pressure of 7.60 MPa, abs. The flow rate for the water entering the core is 55×10^6 kg/h, and the steam generation rate is 8.0×10^6 kg/h. The core flow area is 9.30 m^2. Determine the burnout heat flux in the vicinity of the outlet section of the core.

9.11 The flow through the reactor core described in Problem 9.10 is reduced to 36×10^6 kg/h. Assume no change in the rate of steam generation. Determine the burnout heat flux at the core outlet section.

9.12 A reactor core contains 40,000 fuel rods each 365 cm long. The fuel rod diameter is 10.90 mm. The maximum specific thermal output for the core is 615 W/cm^3. $K = 0.360$. The power generated in the fuel and cladding is 97.25 percent of the total reactor power. The coolant (water) flow rate is 63×10^6 kg/h. The inlet water temperature is 300 C, and the water pressure is 15.5 MPa, abs. Determine the outlet water temperature.

9.13 Water enters a reactor core at 300 C and leaves at 330 C. The water velocity is 4.60 m/s. The fuel rod O.D. is 10.95 mm, and the center-to-center spacing of the rods in both directions is 14.45 mm. At the average bulk temperature of the water, the fluid properties are: thermal conductivity 0.5222 W/m·K, specific heat 6.090 kJ/kg·K, viscosity 0.8670×10^{-4} N·s/m^2, density 690 kg/m^3. The fuel pellet diameter is 9.40 mm, and the cladding I.D. is 9.60 mm. Thermal conductivity values are: cladding 16.5, helium 0.299, and fuel 1.85 W/m·K. The maximum temperature of the fuel is 2500 C. Determine the energy release in the fuel rod per centimeter of length.

9.14 A pressurized water reactor operates at a pressure of 17.25 MPa, abs. At a particular point in a channel, the bulk temperature of the water is 350.0 C. The corresponding mass velocity is 11.5×10^6 kg/h·m^2, and the heat flux is 6.25×10^6 kJ/h·m^2. Determine the probability of burnout occurring.

9.15 A graphite block in a power reactor is cooled by sodium flowing through holes that have a diameter of 11.5 mm. The height of the block is 30 cm. The sodium enters the block at 745 C and leaves at 775 C. The surface temperature of the block is 785 C. The fluid properties are: density 764.6 kg/m^3, specific heat 1.247 kJ/kg·K, thermal conductivity 57.77 W/m·K. Calculate the velocity of the sodium.

9.16 A pressurized water reactor operates at a pressure of 16.5 MPa, abs. At a particular section in the core, the heat flux is 5.40×10^6 kJ/h·m^2. The mass velocity of the water is 12.2×10^6 kg/h·m^2. The equivalent diameter of the channel is 0.0134 m. The fluid properties are: specific heat 6.092 kJ/kg·K, viscosity 0.8572×10^{-4} N·s/m^2, thermal conductivity 0.524 W/m·K. Determine the bulk temperature of the water at which local boiling will occur.

REFERENCES

1. Condon, E. U. and Odishaw, H. *"Handbook of Nuclear Physics,"* McGraw-Hill Book Co. New York, 1967.
2. Kramer, A. W. *Nuclear Energy*, Power Engineering, Barrington, Ill.
3. Shoupp, W. E. "The Physics of Nuclear Power," *Westinghouse Engineer*, Sept. 1954.
4. *Steam—Its Generation and Use*, The Babcock & Wilcox Co., New York, 1978.
5. El-Wakil, M. M. *Nuclear Heat Transport*, International Textbook Co., Scranton, Pa., 1971.
6. *Liquid Metals Handbook*, Atomic Energy Commission, Department of the Navy, June 1952.
7. Jens, W. H. and Lottes, P. A. "Analysis of Heat Transfer, Burnout, Pressure Drop, and Density Data for High Pressure Water," Argonne Natl. Lab. Rept. ANL-4627, Argonne, Ill., May 1951.
8. Janssen, E. and Levy, S. "Burnout Limit Curves for Boiling Water Reactors," General Electric Co. Report APED-3892, San Jose, Calif., April 1962.
9. Spencer, R. C. and Miller, E. H. "Performance of Large Nuclear Turbines," *Combustion*, Aug. 1973.
10. Gruenwald, K. H. and Bezella, W. A. "Milestone: Nuclear Superheat at Pathfinder Plant," *Atomics*, Sept./Oct. 1963.
11. Schuetzduebel, W. G. and Hunt, P. S. "A Review of Design Criteria for the Steam Generator at the 300-MW(e) High Temperature Gas-Cooled Reactor Plant of the Public Service Company of Colorado," *Combustion*, June 1970.
12. Goodjohn, A. J. and Kenton, R. D. "The High-Temperature Gas-Cooled Reactor—An Advanced Nuclear Power System for the 1980s," ASME Paper No. 73-Pwr-8, New York, Sept. 1973.
13. Keller, C. and Schmidt, D. "Industrial Closed-Cycle Gas Turbines for Conventional and Nuclear Fuel," ASME Paper No. 67-GT-10, New York, March 1967.
14. Taylor, J. J. and Hawes, D. J. "Liquid-Metal Fast-Breeder Reactors," Presented before the Northwest Public Power Association, Vancouver, B.C., July 16, 1970.
15. "The Who, What, Why, and Where of Fast Breeder Reactors," *Combustion*, June 1972.
16. "Technical Review of and Cost Data for Reactor Concepts," *Combustion*, June 1971.
17. Jacobi, W. M. "Westinghouse Liquid Metal Fast Breeder Reactor Demonstration Plant Design," *Combustion*, June 1971.
18. Ash, E. B. "Unique Features of a Sodium-Cooled Fast-Breeder Reactor," *Combustion*, June 1970.
19. Dee, J. B. and Melese-d'Hospital, G. B. "Gas Cooled Fast Breeder Reactor Designs," Parts 1 and 2, *Mechanical Engineering*, Jan.–Feb. 1972.

20. Lys, L. A., Brogli, R. and Helbling, W. "Parametric Studies of Gas Cooled Fast Reactors with Closed Cycle Gas Turbines," *Atomkernenergie*, March–April 1969.
21. Boxer, L. W. "Uranium Resources, Production and Demand," *Combustion*, Dec. 1971.
22. Hubbert, M. K. "The Energy Resources of the Earth," *Scientific American*, Sept. 1971.
23. Seaborg, G. T. "Fission and Fusion—Developments and Prospects," *Combustion*, Dec. 1970.
24. Dean, S. O. "Fusion Power: Types, Status, Outlook," *Power Engineering*, March 1974.
25. Fraas, A. P. and Rose D. J. "Fusion Reactors as Means of Meeting Total Energy Requirements," ASME Paper No. 69-WA/Ener-1, New York, Nov. 1969.
26. Mills, R. G. "Problems and Promises of Controlled Fusion," ASME Paper No. 74-WA/NE-10, New York, Nov. 1974.
27. Olds, F. C. "Power Plant Capital Costs Going Out of Sight," *Power Engineering*, Aug. 1974.
28. Olds, F. C. "Capital Cost Calculations for Future Power Plants," *Power Engineering*, Jan. 1973.
29. Collier, A. R. and Nichols, R. C. "Floating Nuclear Power Plants for Offshore Sitting," *Combustion*, June 1973.

BIBLIOGRAPHY

Ash, M. *Nuclear Reactor Kinetics*, McGraw-Hill Book Co., New York, 1979.

Bell, G. I. and Glasstone, S. *Nuclear Reactor Theory*, Van Nostrand Reinhold Co., New York, 1970.

Burcham, W. E. *Nuclear Physics: An Introduction*, McGraw-Hill Book Co., New York, 1963.

El-Wakil, M. M. *Nuclear Energy Conversion*, The American Nuclear Society, La Grange Park, Ill., 1978.

El-Wakil, M. M. *Nuclear Heat Transport*, The American Nuclear Society, La Grange Park, Ill., 1979.

Etherington, H. *Nuclear Reactor Handbook*, McGraw-Hill Book Co., 1958.

Glasstone, S. and Sesonske, A. *Nuclear Reactor Engineering*, Van Nostrand Reinhold Co., Princeton, N.J., 1981.

Graves, H. W., Jr. *Nuclear Fuel Management*, John Wiley & Sons, Inc., New York, 1979.

Knief, R. A. *Nuclear Energy Technology*, Hemisphere Publishing Co., Washington, D.C., 1981.

Lamarsh, J. R. *Introduction to Nuclear Engineering*, Addison-Wesley, Reading, Mass., 1975.

CHAPTER TEN

POWER PLANTS AND THE ENVIRONMENT

Air pollution, caused by certain constituents of the exhaust gases, is the principal environmental disturbance attributed to internal combustion engines and to gas turbine engines. Although the total heat rejected by these engines is a significant quantity, the dispersion of these power plants, many of low capability, effectively eliminates an appreciable increase in the local temperature of the atmosphere to which the heat is rejected.

Heat rejection from a nuclear or fossil fuel-fired power plant, on the other hand, has become a major problem in power plant siting and operation. These power plants however contribute little to water pollution. Nuclear power plants must incorporate provisions for shielding the environment from radioactivity and for safe storage or disposal of spent fuel. Fossil fuel-fired power plants are required to operate with restricted emission of CO, CH, NO_x, SO_x, and particulate material.

The coal-fired steam-electric power plants consume a large fraction, about 75 percent, of the coal mined in the United States, and consequently are indirectly involved in the environmental impact caused by coal mining. Pit mining of coal produces huge piles of spoil material, mostly rock, that present an unsightly appearance. Leaching of acid chemicals from the spoil material is objectionable because the ultimate disposition of these chemicals would be in aquifers, streams, and rivers. Untreated acid mine water should not be discharged into local waterways. Huge gashes in the surface of the earth result from strip coal mining. In past years, strong resentment developed toward this method of mining because of the ugly-appearing terrain that remained from the operation. Today, governmental regulations require replacement of the overburden and, on occasion, restoration of the original earth contour.

The necessity for maintaining a minimum impact on the environment has created numerous problems for the designer and the operator of stationary power plants, most of which are located in electric generating stations. In this chapter, the solutions to these problems are explored, and the effects on power plant capital and operating costs are examined.

10.1 AIR QUALITY CONTROL

Prior to the latter part of the 1960 decade, the selection of a site for a power plant was not a particularly burdensome undertaking. The site selected was usually reasonably close to the load center and preferably on the outskirts of a town or city. The plant would ordinarily have access to an ample supply of cooling water and to suitable facilities for transportation of fuel. Provision would be made for collecting particulate material prior to discharge of the effluent gases into the atmosphere.

Today, site selection is a far more complex undertaking. A considerable amount of time and money is expended in preparing environmental impact statements. Opposition to locating a power plant in a particular area may come from individuals or organized groups who believe continued construction of power plants is no longer essential or propose that the plant be built elsewhere. Restrictions on the use of water for cooling and other applications will most likely be a major factor in site selection. The manner in which land may be used is becoming more restrictive throughout the country and will necessarily influence power plant planning. Federal, state, and local air pollution control regulations are pertinent factors that must be considered in power plant location and operation.

Of the total emission of atmospheric pollutants from all sources in the United States, substantial quantities of three classes of substances are emitted by power plants. The estimated percentage distribution of emissions assigned to thermal power plants is: SO_x 45, NO_x 30, particulates 20, and miscellaneous 5. CO and HC are each less than 1 percent.[1] Installation of pollution control equipment will increase power plant capital costs. Further, control of the emission of pollutants will generally add to the power plant operating problems.

The Environmental Protection Agency (EPA) has defined the danger levels for the five most common air contaminants, identified in the preceding paragraph. They are SO_x, NO_x, CO, HC, and particulates. At the danger level, the health of a person might be harmed significantly during periods of high air pollution. Air quality control regions throughout the United States have been designated by the EPA. The individual states are responsible for setting air quality standards and enforcing compliance with those standards in areas within these regions.

Two standards are applied in air pollution control. An emission standard regulates the amount, emission rate, or concentration of a pollutant emitted from a stack. An ambient concentration standard regulates the concentration of a pollutant at ground level. No simple relationship exists between the two standards because of meteorological and topographical influences.

The quantity of a pollutant and the point of discharge into the atmosphere are under the control of the individual power plant. The degree of pollution however is measured at ground level and hence may be influenced by the discharge of the pollutant from several sources. A tall stack does not necessarily ensure widespread dispersion, or low concentration, of the pollutants emitted from the power plant. Wind conditions and the topography of the surrounding area can cause a downdraft that produces a local high concentration of the polluting substance.

Stack emission standards have been prescribed by the EPA for fossil fuel-fired electric generating units constructed or modified after 1970. The emission limits, in $lb_m/10^6$ Btu, are

SO_2	coal 1.2 (0.52), oil 0.8 (0.34)
NO_x	coal 0.7 (0.30), oil 0.3 (0.13), gas 0.2 (0.086)
Particulates	coal, oil, gas 0.1 (0.043)

The quantities in parentheses are expressed in $kg/10^6$ kJ. The quantities shown above represent discharge limitations. More stringent limitations may be applied locally if ground conditions so require. Units constructed prior to 1970 are subject only to state emission regulations and therefore must comply with a wide range of control measures.

Local conditions may permit the application of emission standards to a group of power plants rather than to a single stack. In this manner, pollution is effectively controlled while some of the power plants in the group can be operated with less stringent emission regulation.

The Clean Air Act as amended in 1977 required the EPA to review the source performance standards for utility boilers. Numerous factors were considered in conducting the review, and new emission standards were adopted. The standards set in 1971, under the Clean Air Act of 1970, stipulated a reduction of up to 75 percent in sulfur oxides emission from power plants burning high-sulfur coal. No controls were required for plants burning low-sulfur coal. The revised standards, for power plants built subsequent to September 18, 1978, require a minimum reduction in the SO_x emission of 90 percent for plants burning high-sulfur coal and 70 percent for plants burning low-sulfur coal.

Example 10.1

The annual average thermal efficiency for a 1000-MW pulverized coal-fired steam power plant is 0.37, and the annual plant factor is 0.95. The coal analysis is C 0.7545, H 0.0495, O 0.0396, N 0.0131, S 0.0371, ash 0.1062. The HHV of the coal is 31,522 kJ/kg. Determine the annual production of sulfur dioxide in the stack gas and the quantity of SO_2 that must be removed in order to conform to the clean air requirements. Complete conversion of the sulfur to SO_2 is assumed.

Coal Fired

$$\dot{m}_f = \frac{\dot{W}}{\eta_t(\text{HHV})} = \frac{10^6 \times 3600}{0.37 \times 31522}$$

$$= 0.3087 \times 10^6 \text{ kg/h} \qquad \text{(at the rated load)}$$

$$\dot{m}_f = 0.95 \times 8760 \times 0.3087 \times 10^6$$

$$= 2569 \times 10^6 \text{ kg/year}$$

Sulfur Dioxide Production

$$1.998 \text{ kg } SO_2/\text{kg sulfur}$$

$$1.998 \times 0.0371 = 0.0741 \text{ kg of } SO_2/\text{kg coal}$$

$$\dot{m}_{SO_2} = (2569 \times 10^6)\, 0.0741$$

$$= 190.4 \times 10^6 \text{ kg/year}$$

$$= 190{,}400 \text{ metric tons/year}$$

Sulfur Dioxide Removed from the Flue Gas

(a) SO_2 limit: 0.52 kg/10^6 kJ

$$\frac{0.52}{10^6}(\text{HHV})\dot{m}_f = \frac{0.52}{10^6}\, 31522(2569 \times 10^6)$$

$$= 42.1 \times 10^6 \text{ kg/year emitted}$$

$$SO_2 \text{ removed} = SO_2 \text{ produced} - SO_2 \text{ emitted}$$

$$= 190.4 \times 10^6 - 42.1 \times 10^6$$

$$= 148.3 \times 10^6 \text{ kg/year}$$

(b) 90 percent minimum SO_2 reduction

$$SO_2 \text{ removed} = 0.90 \times 190.4 \times 10^6$$

$$= 171.4 \times 10^6 \text{ kg/year}$$

Carbon dioxide and carbon monoxide discharged from fossil fuel-burning equipment contribute to air pollution, but in different ways. Carbon monoxide is toxic, and the concentration of this gas is limited in congested areas. The carbon dioxide content of dry air is about 0.035 percent by volume, and the increase in CO_2 is approximately 4 percent per year. The latter figure is the annual percentage increase and not the increase in percentage points. Despite the vast quantities of fossil fuels burned each year, the oxygen content of the atmosphere remains essentially unchanged.

An increase in the carbon dioxide content of the atmosphere should, according to one point of view, produce an increase in the mean temperature of the earth's surface and the adjacent atmosphere. This temperature increase is attributed to the increased "greenhouse" effect of the carbon dioxide. Carbon dioxide is essentially transparent to the short-wavelength solar radiation but absorbs, in pronounced bands, the longer-wavelength terrestrial radiation. Carbon dioxide present in the atmosphere thus tends to diminish the cooling effect of terrestrial radiation, and a gradual increase in atmospheric carbon dioxide should cause a long-term warming trend.

The release of energy by combustion reactions should also contribute to the temperature increase. On the other hand, the increase in the quantity of particulate matter in the atmosphere should cause a decrease in the mean temperature of the earth by reducing the net gain in solar energy.

Actual temperature measurements do not indicate a sustained change in temperature in either direction. For the period 1900 to 1940, earth temperatures increased. Since 1950 however, the warming trend has been reversed, and the earth is now experiencing a period of decreasing temperatures. Climatic changes are probably controlled by a large number of factors. Combustion of fossil fuels may or may not be a contributing factor to a significant degree.

10.2 SO_2 CONTROL

The control of sulfur oxides discharged from power plants has proved to be a major operating problem. Oxides of sulfur are produced by combustion of the sulfur in the fuel, hence the implementation of air quality standards forced many power plants to burn low-sulfur fuels instead of the usual high-sulfur coal or oil. Fuel costs can be substantially higher for these substitute low-sulfur fuels. The price of low-sulfur oil is high, and the charge for transporting low-sulfur western coals to power plants located in other regions of the country adds to the cost of the fuel. Operating problems can develop when the western coals are burned in furnaces designed to burn coals of different origin, principally because of differences in the ash content and in the characteristics of the ash.

Many methods have been proposed and investigated for reducing the SO_2 effluent from coal- and oil-fired power plants. One method is directed to the source of the SO_2, that is, to the sulfur in the fuel. The sulfur content of coal and oil can be reduced by mineral beneficiation and chemical processing. These clean fuels of low ash and low sulfur content are more costly than the untreated original fuels, hence other methods of SO_2 control may prove to be economically superior.

Limestone injection is one of the methods investigated for possible application to SO_2 control. Limestone, or dolomite, is preferably added to the fuel, coal, or oil to react with the sulfur oxides formed in the combustion process. The limestone–sulfur dioxide reaction produces a solid compound that remains in the ash or is removed from the flue gas stream. Limestone may be injected at some point in the furnace, if not added to the fuel. Test results indicate that 20 to 25 percent removal of SO_2 can be achieved in a conventional furnace by the addition of limestone or dolomite.[2] The use of other additives for SO_2 removal has been explored.

The application of fluidized-bed combustion to power plant boilers is currently under investigation, in part, because a high combustion efficiency and a reduction in the size of the boiler structure can be achieved. As a consequence, a lower capital cost for the steam generating equipment may be anticipated. Further, fluidized-bed combustion offers an excellent possibility for achieving substantial SO_2 removal. Studies conducted primarily at atmospheric pressure demonstrated that the fluidized-bed combustion system is capable of removing 90 percent of the SO_2 from the flue gas and can also reduce NO_x emissions up to 50 percent. Combustion efficiencies as high as 99 percent were measured.[3]

Fluidized-bed combustion is accomplished by burning the coal in a bed of limestone that is maintained in a fluid state by the combustion air and the gases produced by the combustion of the fuel. At the temperature established in the bed, the limestone is calcined to form CaO, which in turn reacts with the SO_2 and oxygen present in the bed.

$$CaO + SO_2 + 1/2O_2 \rightarrow CaSO_4 \qquad (10.1)$$

When used on a once-through basis, the consumption of limestone is high, hence regeneration of the $CaSO_4$ in a separate fluidized-bed reactor is under investigation. The reaction with a reducing gas at a temperature of 1100 C is represented by

$$CaSO_4 + CO \rightarrow CaO + SO_2 + CO_2 \qquad (10.2)$$

The regenerated CaO is returned to the fluidized-bed combustion system as a means to reducing the consumption of limestone. The SO_2 is a marketable by-product, although conversion to elemental sulfur or to sulfuric acid may enhance the market value of the recovered sulfur dioxide.

The operating temperature of the fluidized bed may be maintained in the proximity of 815 to 925 C. Several advantages are derived from this comparatively low operating temperature, including reduced NO_x emission. Coals that have low ash-fusion temperatures are burned with less difficulty than in a conventional furnace because slagging problems are largely eliminated.

At the present time, the predominant approach to SO_2 control is directed to scrubbing the flue gas prior to discharge into the stack. Various processes are being investigated, but lime or limestone scrubbing appears to be the most feasible process. Scrubbers would be installed in new power plants and retrofitted to existing plants.

The experience with flue gas desulfurization systems has to date been generally disappointing. These systems have the potential for reliable operation, but the average availability has proved to be poor. The principal operating problems are caused by erosion, corrosion, plugging, and scaling. The effectiveness of the SO_2 removal is however not a problem. The systems currently in operation are capable of achieving the prescribed design effectiveness, that is usually in the range of 0.70 to 0.90.[5]

The basic sulfur dioxide absorber is a tower 11 m in diameter and 22 m high. One tower is required for each 150 MW of generating capability. Briefly, a water slurry of an alkali reagent such as limestone is passed down through the packed absorber tower to react with the SO_2 in the rising flue gas to form a slurry that contains calcium sulfite, calcium sulfate, and nonreacted limestone. Particulate matter in the flue gas is removed ahead of the packed tower by means of a wet scrubber or by an electrostatic precipitator. The limestone slurry from the absorber tower can be utilized in the scrubber as a reagent and removes about 20 percent of the SO_2, as well as the fly ash.

The end product of the flue gas treatment is a slurry that contains up to 15 percent solid material. Disposal of the dewatered slurry as land fill is a feasible arrangement, and water recycling reduces the requirement for make-up water. The volume of the disposed solid material is roughly equal to 25 percent of the original volume of the coal.

The flue gas desulfurization (FGD) systems installed in fossil fuel-fired power plants are usually of the "throwaway" type. The product sulfate salts are worthless and are consequently discarded. A more complex process however is available for treating the waste slurry as a means to producing marketable products and reducing substantially the quantity of material transported to the disposal site.

Example 10.2

The power plant described in Example 10.1 is equipped with a wet scrubber for removal of the sulfur dioxide from the flue gas. The SO_2 reacts with limestone supplied to the tower. Determine the quantity of limestone required for the operation of the scrubber and the amount of the product calcium sulfate.

A simplified reaction is assumed. Thus,

$$CaCO_3 + SO_2 + 1/2O_2 \rightarrow CaSO_4 + CO_2$$

$$\begin{aligned} m_{CaCO_3} &= \frac{\overline{m}_{CaCO_3}}{\overline{m}_{SO_2}} m_{SO_2} \\ &= \frac{100.09}{64.06} 171.4 \times 10^6 \\ &= 267.8 \times 10^6 \text{ kg/year} \\ &= 267{,}800 \text{ metric tons/year} \end{aligned}$$

$$\begin{aligned} m_{CaSO_4} &= \frac{\overline{m}_{CaSO_4}}{\overline{m}_{SO_2}} m_{SO_2} \\ &= \frac{120.14}{64.06} 171.4 \times 10^6 \\ &= 321.4 \times 10^6 \text{ kg/year} \\ &= 321{,}400 \text{ metric tons/year} \end{aligned}$$

The maximum amount of sulfur that can be recovered from the calcium sulfate is 85,800 metric tons per year, that is, 90 percent of the sulfur originally in the coal.

Capital costs for wet-process flue gas desulfurization equipment are high, about equal to the capital cost of the entire boiler installed in a new unit. These systems contribute to high draft losses and require disposal of large quantities of sludge. Dry scrubbing of the flue gas is viewed as an alternative to wet scrubbing as a means to effecting a cost reduction and improving the operation of the flue gas cleaning equipment.[6]

The utility of a two-stage combined dry scrubber/SO_2 absorber has been investigated in a number of test installations. In the first stage, a calcium or sodium compound is introduced, in a solution or slurry, into a spray dryer where the sulfur in the flue gas reacts with the sorbent material. The water in the droplets is evaporated by the thermal energy in the flue gas to produce a dry powder mixture of sulfate and sulfite, with some residual unreacted alkali. In the second stage, the solid material is removed from the flue gas with a fabric (bag) filter. More than 25 percent of the total SO_2 conversion efficiency is contributed by the fabric filter as the SO_2 combines with the unreacted sorbent material that enters the fabric material.

In addition to SO_2 removal, the bag filter in the two-stage system removes the particulates contained in the flue gas leaving the boiler. See Section 10.4.

Test results obtained for the two-stage flue gas cleaning system are regarded as impressive, and a limited number of commercial power plant applications could justifiably be undertaken.

The development of an effective flue gas desulfurization system is essential if power plants are to burn the high-quality high-sulfur coals that are in abundant supply in the eastern region of the United States. Substitute fuels, such as low-energy gas produced by coal conversion, are not likely to be economically competitive with coal burned directly in boiler furnaces. Fluidized-bed firing of coal in steam boilers is in the early stage of development and ultimately may not prove to be technically and economically superior to conventional coal firing coupled with flue gas desulfurization.

10.3 NO_x CONTROL

Reduction in the emission of nitrogen oxides from power plant boilers has in recent years received considerable attention. Certain difficulties are encountered in reducing the formation of NO in the furnace, hence development of an acceptable process for removal of nitrogen oxides from the flue gas would represent a major achievement in power plant practice. An example of this approach is the investigation of the effectiveness of an aqueous alkaline scrubbing process for NO_x absorption.

The formation of thermal NO inside a boiler furnace is influenced by several factors. The principal factor is the flame temperature, which in turn is influenced by the quantity of excess air, combustion air temperature, extent of furnace cooling, and the furnace wall deposits. Heat transfer to the furnace wall has a cooling effect on the flame. An increase in the quantity of excess air will lower the flame temperature but increase the oxygen concentration, and thus have a dual effect on the formation of NO. In this respect, a reduction in the flame temperature tends to decrease the formation of NO, while an increase in the oxygen concentration has an opposite effect.

Thermal NO is formed in the furnace by the fixation of atmospheric nitrogen supplied by the combustion air. Another source of NO is the chemically bound nitrogen in the fuel. The formation of NO from the fuel nitrogen involves complex reactions that are not exactly defined. Fuel nitrogen is a particularly significant source of NO in coal-fired installations. The average nitrogen content of bituminous coal and lignite is approximately 1.4 percent, although in some coals the nitrogen content is in the range of 2.0 to 2.5 percent. In comparison with furnaces designed for oil and gas firing, the coal-fired furnace is much larger, that is, the energy released per unit furnace volume is relatively low. Consequently, flame temperatures are lower in coal-fired furnaces, a factor that offsets the effect of the higher nitrogen content of the fuel.

The temperature dependence of the formation of thermal NO is shown in Table 10.1.

The formation rate of thermal NO increases very rapidly with increase in the flame temperature, which affects, to a substantially lesser de-

TABLE 10.1 Influence of Temperature on the Formation of Thermal NO

Temperature (C)	*Time to Form 500 ppm NO (s)*	*NO at Equilibrium (ppm)*
1095	1370	180
1315	16.2	550
1540	1.1	1380
1760	0.117	2600
1980		4150

Source: Reference 7.

gree, the increase in the equilibrium concentration of NO. Usually, in a combustion chamber operating at atmospheric pressure, the NO is formed in the flame near the adiabatic temperature, about 1925 C. This temperature will vary somewhat with the fuel composition and the firing conditions.

Because of insufficient residence time, the quantity of thermal NO formed is usually substantially below the equilibrium concentration corresponding to the maximum temperature achieved in the combustion chamber. As the combustion gases cool, the formation rate of NO decreases, and subsequently, at some point in the gas stream, the NO concentration will exceed the equilibrium value for the local temperature. Equilibrium conditions however are not established because there is virtually no dissociation of the NO. Once formed, the NO remains in the gas stream.

Correlation of experimentally acquired data on the formation of thermal NO_x is a somewhat complex procedure.[8] The most significant factor in the correlation is the relative furnace cooling in the primary burning zone. The primary burning zone is defined as that portion of the flame that generally is above 1540 C, or the quenching temperature of NO_x. The burning area energy release rate (BAER) is evaluated for this zone in terms of kJ/h per m^2 of heat absorbing surface.

The thermal NO_x emission from coal-fired units is appreciably higher than the emission of NO_x from gas- and oil-fired units. See Fig. 10.1. The incremental difference is especially large when the coal-fired units operate with high to medium slagging. NO_x emission is lower for opposed-fired and turbo-fired furnaces than for front-fired furnaces.

Exact data are not available for NO_x formation from fuel nitrogen. The portion of fuel nitrogen that is converted to NO_x is shown to be a function of the percentage of nitrogen in the fuel and increases as the amount of nitrogen in the fuel decreases. However, for a range of 0.2 to 1.0 percent nitrogen in the fuel, the portion converted to NO_x remains constant.[8] The converted portion of fuel nitrogen is reported by various investigators to range from 18 to 80 percent. Rawdon and Sadowski[8] used a conversion fraction of 0.21 for the fuels examined, which for the most part had a nitrogen content within the range of 0.2 to 1.0 percent.

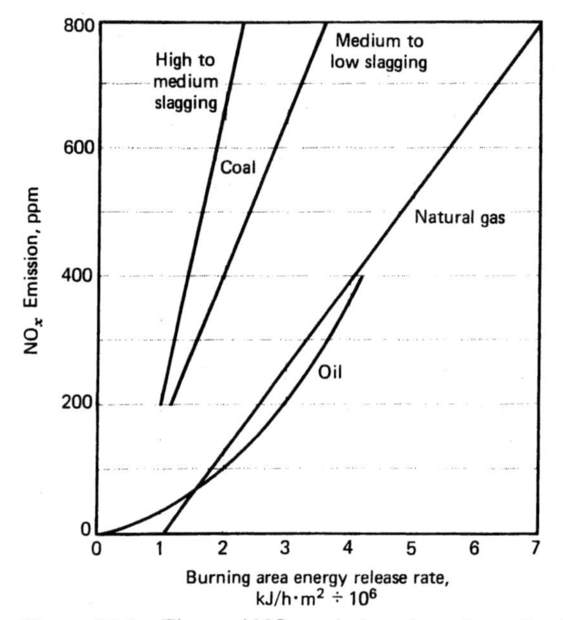

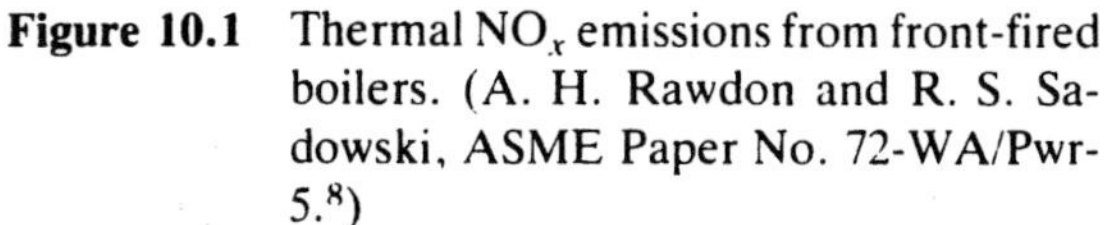
Figure 10.1 Thermal NO_x emissions from front-fired boilers. (A. H. Rawdon and R. S. Sadowski, ASME Paper No. 72-WA/Pwr-5.[8])

Example 10.3

The emission of NO_x from a typical coal-fired power plant is determined from the following data.

HHV of the fuel: 31,660 kJ/kg

Energy released: 34,105 kJ/kg coal

Coal fired: 56,461 kg/h

Dry flue gas: 12.30 kg/kg coal

Burning zone radiant surface area: 585.29 m^2

Nitrogen content of the fuel: 1.2 percent

Medium to low slagging conditions are assumed.

(a) Thermal NO_x

$$\text{BAER} = \frac{\dot{m}_f E_r}{A_{ab}} = \frac{56461 \times 34105}{585.29} = 3.29 \times 10^6 \text{ kJ/h·m}^2$$

NO_x emission = 735 ppm (from Fig. 10.1)

The following equation, derived from basic relationships, is used to express the NO_x emission in terms of kg/10^6 kJ.

$$NO_x = \left[\frac{46.0\, m_{dfg}}{\overline{m}_{dfg}(\text{HHV})}\right] NO_x\text{ (ppm) kg/10}^6\text{ kJ}$$

where

46.0 = molal mass of NO_2, kg/kgmol

$\overline{m}_{dfg}$ = molal mass of the dry flue gas, kg/kgmol

m_{dfg} = dry flue gas, kg/kg fuel

HHV = higher heating value of the fuel, kJ/kg

$$NO_x = \left[\frac{46.0 \times 12.30}{30 \times 31660}\right] 735 = 0.438 \text{ kg/10}^6 \text{ kJ}$$

(b) Fuel NO_x

Assume that 25 percent of the N_2 in the fuel is converted to NO_x.

$$NO_x = \frac{\overline{m}_{NO_2}}{\overline{m}_{N_2}} \times \text{conversion ratio} \times N_2 \text{ in fuel} = \frac{46.0}{28.0} 0.25 \times 0.012 = 0.00493 \text{ kg/kg fuel}$$

$$NO_x = \frac{10^6}{\text{HHV}} \times (NO_x \text{ kg/kg fuel}) \text{ kg/10}^6 \text{ kJ}$$

$$NO_x = \frac{10^6}{31660} 0.00493 = 0.156 \text{ kg/10}^6 \text{ kJ}$$

(c) Total NO_x emission = 0.438 + 0.156 = 0.594 kg/10^6 kJ

The emission of NO_x determined in this example is substantially above the limiting value of 0.30 kg/10^6 kJ established for coal-fired power plants.

Because NO formation is strongly temperature dependent, the control of NO_x emission is most effectively achieved through furnace temperature regulation. Two methods are mainly employed for achieving this regulation, namely, flue gas recirculation and two-stage combustion.[9] Other possible methods for NO_x control, observed in the following paragraph, are less effective.

Reduced furnace temperatures may be achieved by increased furnace cooling, but this method, available for new power plants, is very costly. NO_x emission may be diminished to some extent by decreasing the quantity of excess air, but as a result, unburned fuel losses are likely to increase. A reduction in the combustion air temperature will decrease furnace temperatures, however an adverse effect on the thermal performance of the boiler will be experienced. This method of NO_x emission control is available to a limited extent for new installations and would require a greater proportion of energy absorption in the economizer.

Flue gas recirculation is an effective method for accomplishing a substantial reduction, up to 90 percent, in the emission of thermal NO_x, but it is relatively ineffective in reducing the NO_x formed from the fuel nitrogen. About 15 to 20 percent of the flue gas is recirculated to the furnace and admitted with the combustion air. Flame temperatures are reduced as a consequence of the increased mass flow of gas in relation to the fuel flow. The application of flue gas recirculation to existing units is very difficult and is most appropriately employed for new construction. Because of the required additional ductwork and fans, capital costs are high. The system is easily controlled and is well suited to pulverized-coal firing.

Gas temperature control can be achieved by two-stage combustion or off-stoichiometric firing, two somewhat similar combustion modification methods that have proved to be highly effective in reducing the emission of NO_x. Two-stage combustion is accomplished by operating all of the burners with a fuel-rich mixture. The oxygen content of the fuel-rich flame is very low, hence the formation of NO is restricted by the paucity of oxygen. Additional air is admitted through over-fire air ports at the point where the cooling gases begin to leave the highly radiative section of the furnace. The additional air is of sufficient quantity for establishing the correct overall air/fuel ratio essential to completing the combustion process. Flame temperatures at this point are comparatively low, and little NO is formed despite the relatively high oxygen content of the combustion gases.

Off-stoichiometric firing provides for operating some of the burners fuel rich, while the remainder of the burners are operated air rich or admit only air. Similarly to two-stage combustion, this method of firing achieves a substantial reduction in NO formation. The fuel-rich burners provide a reducing atmosphere in which the flame is initially developed, and the air-rich burners provide the air dilution essential to completing the combustion of the fuel in a comparatively low-temperature flame.

Existing units are easily modified for two-stage combustion or off-stoichiometric firing with relatively low capital cost. System control is more difficult to achieve in comparison with flue gas recirculation, and because of incomplete combustion, firing problems are experienced in coal-burning units. Some tests have demonstrated that off-stoichiometric firing can reduce the NO_x emission by a factor of 5 or 6.

Acid rain and acid fog are terms that are appearing in various reports and news releases with increasing frequency. Most rain is naturally slightly acidic, but an increase in this acidity is observed in some regions, particularly in the eastern United States and in certain areas of Europe. The increased acidity is attributed to the presence in the atmosphere of oxides of sulfur and nitrogen, two pollutants discharged from fossil fuel combustion systems. Sulfur dioxide and nitrogen oxides react with oxygen and moisture to form sulfuric and nitric acids. Acid rain is defined as rain with an acidity lower than pH 5.6. On the pH scale, the neutral point is 7.0, and pH 0.0 indicates that a substance is highly acidic.

Various effects have been ascribed to the increased acidity of rain. The acidity of some surface and subsurface waters could be increased with a deleterious effect on aquatic life. Acid rain may possibly damage vegetation and cause corrosion of various materials. Studies conducted to date on the effects of acid rain have proved generally to be inconclusive. It is not presently clear how increasing or decreasing the emission of sulfur and nitrogen oxides will affect the acidity of rain. Despite this uncertainty, the "specter" of acid rain will quite likely be raised whenever a proposal is made to construct a coal-fired power plant.

10.4 PARTICULATE CONTROL

Discharge of particulate material into the atmosphere is undesirable for a number of reasons. The finely divided material is a source of dirt and can be biologically harmful, for example, in causing respiratory and eye irritation. Further, particulate material participates in smog formation, that is to say, NO_2 + particulates + photochemical oxidants + sunlight = smog.

Three types of devices, namely, precipitators, bag filters, and Venturi scrubbers, are employed for removal of the particulate material from the combustion gases prior to discharge of the gases into the atmosphere.

The electrostatic precipitator has a high coefficient of fly ash removal from the flue gas. The unit is a relatively large structure, arranged for horizontal gas flow, and consists of a number of channels formed by vertical metallic plates, usually spaced 200 mm apart.[10] These plates function as the collecting electrode surfaces. The discharge electrodes are usually circular steel wires, about 2.5 mm in diameter, suspended in the channels and spaced about 150 mm apart.

The collector plates are grounded, and a strong field is created by the high voltage applied to the discharge electrodes. The corona field creates ions among the gas molecules, and the ions in turn impart an electric charge to the solid particles suspended in the gas stream. The charge on the particles is similar to the charge on the discharge electrodes, thus the field causes a force to act on the particles and move them to the collecting plates. The dc power supply for fly ash collection delivers an operating voltage of 40 to 55 kV.

The ash particles collect on the plates and are dislodged by rapping and fall into the collecting hopper. Proper rapping is essential in order to avoid any appreciable reentry of the particles into the gas stream. The thickness of the dust layer is usually within a range of 3 to 12 mm. Subnormal operation of the precipitator is observed to occur when the dust resistivity exceeds a critical value of about 2×10^{10} ohm·cm.

The formation of SO_3 in the combustion gases from the sulfur in the coal is particularly significant with respect to the operation of the precipitator.[11] The electrical conductivity of the fly ash particles is closely associated with the adsorption of SO_3 on the particle surfaces. Before SO_2 emission controls were imposed, most of the coal burned in the United States had a relatively high sulfur content. As a consequence of the high sulfur content, sufficient SO_3 was formed to maintain the resistivity of the fly ash below the critical value of 2×10^{10} ohm·cm.

The resistivity of fly ash can be empirically correlated with temperature and the sulfur content of the fuel as shown in Fig. 10.2. This figure however represents a simplified correlation because of ignoring the effects on resistivity caused by the presence of moisture and ash in the coal.

Precipitators have been located mainly in the gas stream at a point downstream from the air heater and thus operate at a comparatively low inlet gas temperature in the range of 115 to 175 C. If a low-sulfur (0.5 to 1.0 percent) coal is fired and the precipitator operating temperature is in the vicinity of 150 C, the resistivity of the fly ash deposit will be well above the critical value of 2×10^{10} ohm·cm. Some difficulty can also be experienced when medium-sulfur coal is burned under similar operating conditions. The resistivity of the dust layer can be reduced by the addition of a small quantity of SO_3 to the flue gas ahead of the precipitator. Sparking is reduced and the precipitator effectiveness is improved.

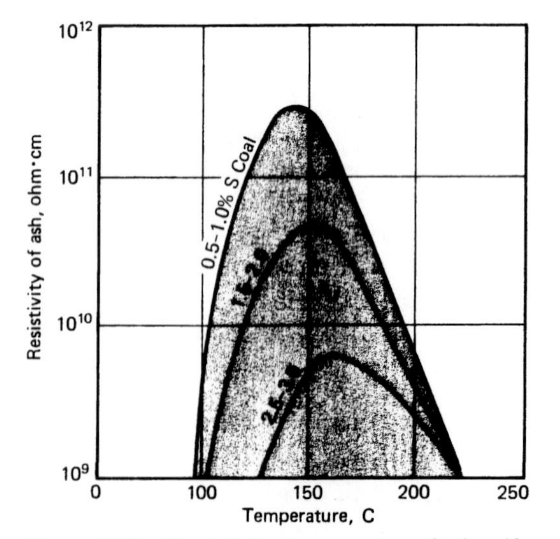

Figure 10.2 Resistivity–temperature–fuel sulfur relationships for eastern bituminous coals. (A. B. Walker, *Combustion*, Nov. 1974.[11])

Figure 10.2 discloses the sharp reduction in resistivity that will result from a comparatively small increase in the precipitator operating temperature. Higher operating temperatures have been achieved in a number of installations by locating the precipitator between the economizer and the air heater. Thus, in an operating temperature range of 320 to 480 C, problems resulting from high ash resistivity are eliminated.

Electrostatic precipitators are highly effective in the removal of particulate material from the flue gas. Test results show that a precipitator effectiveness of 0.9999 can be achieved, that is, 99.99 percent of the entering particulate material is removed from the flue gas. In order to increase the effectiveness of the precipitator from 0.99 to 0.9999, for example, an approximate increase of 65 percent in the collecting area must be provided. The federal air quality control standards limit the particulate emission to 0.043 kg/10^6 kJ. In order to comply with this limitation, the effectiveness of the collector must be in the 0.99+ range.

A precipitator must be designed exactly to conform to the operating characteristics of the boiler and to the characteristics of the fuel fired in the furnace. An inadequate or improper design will lead to reduced effectiveness and low reliability. Corrosion can be a problem when the precipitators are operated "cold." Interior surfaces should remain above the SO_3 dew point temperature. In this respect, properly applied insulation can reduce the tendency for condensation to occur.

The required collecting area for hot precipitators is in the range of 70 to 100 m^2 for a gas flow of 100 m^3/min (actual) in order to achieve 99 percent particulate removal. If 99.8 percent particulate removal is prescribed, the required collecting area is in the range of 80 to 120 m^2 for a gas flow of 100 m^3/min (actual).

The required collecting area for a cold precipitator is roughly in the same range as that for a hot precipitator when based on the actual gas flow rate, m^2 area per m^3/min actual gas flow. However, for a prescribed power plant rating, the gas flow rate for the cold precipitator is about 65 percent of the gas flow rate for the hot precipitator, based on gas temperatures of 150 and 370 C, respectively. Thus, for a particular power plant design, the collecting area for a cold precipitator would be about 65 percent of the collecting area for a hot precipitator. It is important to observe that the required collecting area is related to the volumetric gas flow rate.

Particulate removal can be accomplished by the use of bag filters. The baghouse is a metal shell in which a number of filter bags are suspended. The filter bags are typically about 9 to 10 m high and about 30 cm in diameter. Gas flow through the filter cloth is either to or from the interior of the bag. The dust that collects on the bag fabric is periodically dislodged by a short reversed pulse of compressed air. The effectiveness of particulate removal from the gas stream is not significantly influenced by the ash chemistry or by the quantity and the particle size distribution of the ash in the flue gas. A collection effectiveness of 99.8 percent is not unusual for a fabric filter.

In comparison with precipitators, capital costs for bag filters tend to be lower, but somewhat higher operating costs for labor and maintenance can be experienced. Maintenance costs are related chiefly to corrosion problems and material replacement for filter bags. Corrosion is aggravated by acid condensation that can even be a problem when a low-sulfur fuel is burned.

Mechanical dust collectors, for example, cyclone separators, are used on occasion in series with other dust removal systems. In an upstream position, the mechanical collector provides erosion protection and reduces the heavy particle loading on the downstream system. The addition of a mechanical collector to a precipitator increases the capital cost as well as the operating cost of the dust removal system. Increased operating costs are attributed primarily to the high draft losses that are experienced in effecting the mechanical separation of the particulate material.

The capital cost of the flue gas cleaning system

may be taken as about 20 percent of the cost of the entire power plant. Of this total amount, SO_2 control can account for 65 to 75 percent, while particulate control would account for the remaining 25 to 35 percent.[13] Some 2 to 5 percent of the net power generated by the station is consumed in operating the flue gas cleaning equipment. Installation of the flue gas cleaning system will consequently increase power generation costs because of the increase in the fixed charges and because of the additional operating costs for energy and maintenance.

The fraction of the ash, in the fuel, that leaves the boiler in the flue gas varies with the kind of fuel and the method of firing the fuel. The comparatively small quantity of ash contained in fuel oil is largely deposited inside the furnace and other sections of the boiler. When coal is burned on a traveling-grate, chain-grate, or underfeed stoker, most of the ash is discharged into the ash hopper, and the carry-over of the ash in the flue gas is a minor fraction of the total ash in the coal. Incidentally, some of the sulfur in the coal will be detected in the ash deposited in the hopper.

When coal is fired by a spreader stoker, a substantial portion of the coal is burned in suspension. The carry-over of particulate material in the flue gas is thus high in comparison with that observed in burning coal on other types of stokers. Collection equipment is normally installed for separating and returning the coarse carbon-bearing particles to the furnace for further burning, while discharging the fine material to the ash removal system.

In pulverized coal firing, combustion is achieved by burning the fuel solely in suspension, hence an inherent tendency exists for a high carry-over of ash in the flue gas leaving the boiler. The carry-over is about 80 percent of the ash in the coal for dry ash firing and about 50 percent for firing the fuel in a slag tap furnace. When coal is burned in a cyclone furnace, the fly ash loading in the flue gas is reduced to 20 to 30 percent of the ash in the coal (Ref. 1, Chapter 8).

Ash disposal in coal-fired power plants is usually not a problem. The cinders discharged from power plants equipped with stokers find ready use in the construction industry. Fly ash, bottom ash, and slag discharged from pulverized coal-fired power plants have customarily been hauled from the plant and disposed as land fill. The use of fly ash in the manufacture of commercial products has been investigated, and some success has been achieved in using fly ash as an ingredient of concrete.

Example 10.4

The power plant described in Example 10.1 is equipped with a hot electrostatic precipitator for removal of the particulates from the flue gas that enters the precipitator at 98.2 kPa, abs, and 330 C. The unburned carbon is 0.008 kg/kg coal, and 15 percent of the particulate material is trapped in the furnace bottom. The pulverized coal is fired with 17 percent excess air. The prescribed precipitator collecting area is 90 m^2 for a gas flow of 100 m^3/min, a value that is expected to achieve the removal of 99 percent of the particulates from the flue gas. The gas velocity in the precipitator is 2.25 m/s. Develop a preliminary design for the precipitator.

Combustion Calculations Report

Air supplied: 12.157 kg/kg coal

Flue gas: 13.043 kg/kg coal

$\overline{m}_g = 29.86$ kg/kgmol $\quad R_g = 0.278$ kJ/kg·K

Precipitator Flow Area

$$\dot{m}_g = \dot{m}_f \times \text{g/f} = 85.75 \times 13.043 = 1118.4 \text{ kg/s}$$

$$\dot{V}_g = \dot{m}_g \frac{R_g T_g}{P_g} = 1118.4 \frac{0.278 \times 603}{98.2} = 1909.1 \text{ m}^3/\text{s}$$

$$A_x = \frac{\dot{V}_g}{C_x} = \frac{1909.1}{2.25}$$
$$= 848.5 \text{ m}^2$$

The prescribed width-to-depth ratio w/d is 1.5. Then,

$$d = 23.784 \text{ m} \qquad w = 35.676 \text{ m}$$

Collecting Plate Data

Plate spacing: $s = 0.200$ m

Number of channels:

$$\frac{w}{s} = \frac{35.676}{0.200} = 178.4 \sim 178$$

Collecting area:

$$\dot{V}_g = 114{,}546 \text{ m}^3\text{/min}$$
$$A_c = 90 \frac{114546}{100} = 103{,}091 \text{ m}^2$$
$$\text{Collecting area per plate} = 2dL$$
$$A_c = 2ndL$$
$$103{,}091 = 2 \times 178 \times 23.784L$$
$$L = 12.175 \text{ m} \qquad \text{(plate length)}$$

Particulate Removal

$$m_p = 0.1142 \text{ kg/kg coal (total)}$$
$$m_p \text{ to flue gas} = 0.85 \times 0.1142$$
$$= 0.09707 \text{ kg/kg coal}$$
$$m_p \text{ to stack} = 0.01 \times 0.09707$$
$$= 0.0009707 \text{ kg/kg coal}$$
$$0.0009707 \frac{10^6}{31{,}522} = 0.0308 \text{ kg}/10^6 \text{ kJ}$$

The particulate discharge of 0.0308 kg/10^6 kJ from the power plant is below the limiting value of 0.043 kg/10^6 kJ. See Section 10.1.

10.5 NUCLEAR POWER PLANT SHIELDING

Nuclear reactions create several types of radiation, most of which are present in the power reactor and in the surrounding shield. These different forms of radiation result from both primary and secondary reactions and involve complex physics.

The particles pertinent to reactor shielding can be separated into two classes, namely, charged and neutral particles.

CHARGED	NEUTRAL
Fission fragments	Neutrinos
Alpha particles	Neutrons
Deuterons	Gamma rays (electromagnetic radiation)
Tritons	
Protons	
Beta rays (+ or −)	

A charged particle in moving through a substance will, because of its charge, interact electrically with the atoms of the substance. Because of this strong interaction, the charged particles lose energy very rapidly and can be stopped in relatively thin layers of material. Consequently, the heavy fission fragments, although possessing most of the energy released in the fission reaction, are normally arrested in the fuel elements. Other heavy charged particles, such as alpha particles and protons, have longer ranges. The range of one of these particles however is very much less than the thickness of the ordinary structural components of the reactor.

The short-range alpha particles cannot penetrate the skin, but they are harmful if ingested or otherwise deposited within the body. Thus, a substance that emits alpha particles would be classified as a radiopoison.

Because of a small mass, electrons have comparatively long ranges for a particular energy level. An electron of about 5 MeV energy has a range of 2.6 cm in water and 0.33 cm in lead. Hence, for the energy levels normally encountered in a reactor, electrons can be arrested rather easily at low cost. Under certain conditions, electrons will lose energy by radiating gamma rays that may be much more difficult to stop than the original electron.

Almost 10 percent of the reactor power is in the form of gamma and neutron radiation. Heat is generated in a material through which this radiation passes. A steel thermal shield is placed just inside the reactor vessel in order to reduce the energy flux of the emitted radiation to an acceptable level. The thermal shield, in particular, protects the reactor vessel from excessive neutron damage.

While the internal shielding is adequate for protection of the reactor vessel, the gamma and neutron radiation external to the reactor vessel is capable of inflicting serious biological damage. A heavy external shield is thus required for absorbing the highly penetrating gamma and neutron radiation. Protection of the operating personnel from this harmful radiation is usually achieved by erecting around the reactor vessel a biological shield constructed of concrete about three meters thick. Absorption of the radiative energy within the concrete causes considerable heat generation, thus requiring water cooling of the external shield.

The reactor building, a prestressed reinforced concrete structure, houses the reactor vessel, steam generators, coolant pumps, and other operating equipment. The walls of the building are sufficiently thick to provide external shielding against radioactivity within the structure that might be caused by a leak of some kind in the system.

Equipment is provided for remote handling of the fuel elements during shutdown of the reactor for refueling. Because of residual activity within the core following shutdown, provision is made for continued reactor cooling to guard against excessively high temperatures developing in the fuel elements. Many other safeguards are employed in a nuclear power plant for protection of the operating personnel and for preventing the occurrence of any harmful or alarming activity in the area external to the power plant.

Neutrinos carry away about 5 percent of the reactor power, but pose no shielding problem. A neutrino can travel through matter for an enormous distance before collision with another particle will occur, hence it cannot cause biological damage, nor can the neutrino flux be attenuated by shielding in order to generate heat for augmentation of the reactor power. A. W. Kramer appropriately states, "It is impossible to shield against *nothing* and the neutrino comes about as close to nothing as anything one can think of."[15]

10.6 NUCLEAR WASTE

Power reactors are periodically shut down, normally at yearly intervals, in order to refuel the reactor and accomplish the usual plant maintenance. About one-third of the fuel assemblies are replaced each year, and the other two-thirds are relocated in the core in order to utilize most effectively the fissionable material in each assembly group. The spent fuel assemblies are removed from the core and transported to a storage facility under carefully guarded conditions. Underwater storage is facilitated in racks designed to prevent the occurrence of a chain reaction. Provision must also be made for removing, during a limited period of time, the heat generated by radioactive decay.

When the decay heat reaches a sufficiently low level, the spent fuel assemblies can be shipped to a fuel reprocessing plant. Considerable care must be exercised in transporting the spent fuel. The shielded casks in which the spent fuel assemblies would be shipped are specially designed for handling radioactive material and for resisting damage during transit.

Continued nuclear power development is contingent upon establishing a safe, economical, and effective method for the disposal of the radioactive waste material. Retrievable storage of the reactor spent fuel assemblies would be highly uneconomical because significant quantities of valuable fuel components, namely, plutonium and unburned uranium, would not be recovered and recycled, although recovery at a later date is possible. Reactor operation produces different types of nuclear wastes, but the greatest care must be exercised in handling and storing the high-level waste that results from reprocessing the spent fuel for recovery of most, 99.5 percent, of the uranium and plutonium.

Currently, there are no facilities in the United States available for the commercial reprocessing of spent fuel removed from light-water reactors. Fuel reprocessing plants however are expected to be in operation early in the 1980 decade. Initially, the unburned uranium will be returned for reactor fueling. Commercial reprocessing and recycling of plutonium in the United States is presently in doubt, but this question may be resolved when a national energy plan is adopted. Plutonium extracted from the spent fuel taken from light-water reactors will be used subsequently to fuel fast breeder reactor (FBR) power plants.

At the reprocessing plant, the spent fuel rods are chopped and the material dissolved in processes carried out in a sealed room behind heavy walls. The valuable unburned fuel is separated from the waste material by chemical treatment and recovered for recycling. The remaining waste material is in a highly purified liquid form.

Initially, the most dangerous components of the high-level waste are the fission products, notably cesium 137 and strontium 90, each with half-lives of about 30 years. In ten half-lives (300 years), the radiation, initially intense, from the fission fragments will dwindle to the level of background radiation. Subsequently, the radiation from the several transuranic isotopes becomes important, with the main activities in the isotopes of plutonium and americium. The actinide isotopes emit weak radiation, but some of them have long half-lives, particularly plutonium 239 (24,000 years) and americium 241 (458 years). A very long period of time will consequently elapse before the radioactive actinides decay sufficiently for the radiation to drop to the level of background radiation.

Provision must be made for storing the high-level nuclear waste, undisturbed, for hundreds of thousands of years during the time that the radiation danger persists. Storage of the liquid radioactive waste is not desirable, because of possible leakage, excessive bulk, corrosion problems, and other problems related to containment. The current, most promising method for handling this waste provides for reducing the liquid waste to a solid that is subsequently sealed in a relatively insoluble borosilicate glass. The final form of the waste material is a black ceramic clinker that is chemically inert. It does not dissolve in water or react with air. In subsequent steps, the solid material is encased in a metal tube that is placed in a thick-walled shipping cask and transported to a final storage facility. The shipping cask is designed to survive all conceivable accidents that may occur during transit.

There is every reason to believe that the high-level nuclear waste contained in the canisters can be safely stored in a well-protected underground or aboveground facility. The volume of waste is relatively small, about 2 m^3 for a power reactor operating at full output for a year. About 10 individual canisters would contain this quantity of solid waste.

In the final storage facility, all individuals must be excluded from the vicinity of the containers of the radioactive waste material for a period of 350 years. Adequate protection can be achieved by storing the waste behind an appropriate shield. At the end of the 350-year period, the waste material will still contain residual quantities of radiopoisons that must remain isolated from food and water, but otherwise routine storage is adequate.

Of particular significance is the very small volume of nuclear waste material that is produced, hence no extensive storage capability is required. Waste material may be stored permanently as in a salt mine, rock cavern, or ice cap, or in a retrievable form. Because of heat generation in the radioactive waste, the need for cooling must be examined in the design of any storage facility.

It is essential to examine at this point the consequences of a breakdown that could conceivably occur in an underground waste storage facility. Under severe corrosive stresses, the stainless steel canisters have an anticipated life of 10 to 40 years. Failure of a canister would expose the glass in which the radioactive waste is sealed. Recent investigations disclose that glass, if subjected to certain conditions of high temperature and high pressure, can become unstable in the presence of water. Should the glass disintegrate because of the severe environment, migration of the radioactive waste is a likely event.

Improved stability in the storage of radioactive waste can apparently be accomplished through use of a high-density solid known as a cermet, which is a combination of a ceramic and a metal. The waste is fixed in microscopic particles that are uniformly dispersed through and permanently bound within a metal alloy matrix. The cermet has excellent mechanical strength and a very high thermal conductivity that ensures against an excessive temperature increase in the solidified mass. In addition, the waste volume can be reduced in the transformation from the initial form to the cermet. The volumetric reduction is particularly high for certain compositions of the waste.

Conceptual designs have been developed for ground-surface storage of commercial radioactive waste.[17] The solid waste, in one scheme, would be recieved at the storage area in sealed, stainless steel canisters 0.3 m in diameter and 3 m long. One to three "as-received" canisters are subsequently sealed in a mild steel cask. Shielding against gamma and neutron radiation is accomplished by surrounding the cask with a heavy concrete sleeve approximately 1 m thick. The shielded, sealed casks would be individually placed in a vertical position on concrete support pads that rest on the ground. The heat generated in the waste material will be removed in cooling the canister by convective circulation of air between the concrete sleeve and the sealed steel cask. Only a modest storage area, perhaps 500 hectares, would be required to accommodate the nuclear waste produced by power reactors up to the year 2000.

Low-level radioactive waste material is produced in various nuclear power plant operations. Most of this material, which is found in a wide variety of forms, remains radioactive for a considerable length of time, that is, up to several hundred years. Disposal of this waste is through shallow land burial. Low-concentration migration of radionuclides has been detected at some of the burial sites, but such migration does not pose any present significant threat to public health and safety.

Facilities for transportation and storage of nuclear power plant wastes have been devised to withstand the effects of all conceivable accidents and natural catastrophes. Do nuclear wastes, in storage or in transit, create a hazard because of possible sabotage or acts of terrorism? The consequences of such acts should be examined.

In the event of sabotage, the heavy, inert solid material spilled from a ruptured canister would not disperse through the atmosphere, and the spill could easily be contained in a cleanup operation. It is virtually impossible to manufacture an explosive device from the waste material, because this material does not have the proper composition for such use. The penetrating radiation is a real deterrent to theft of the waste material—personal risk would be great. Further, virtually all of the fissionable material is recovered from the spent nuclear fuel, including the plutonium produced by transmutation of uranium 238.

10.7 POWER PLANT HEAT DISSIPATION

The heat rejected from a steam-electric power plant is equivalent to approximately 60 percent of the energy in the fuel for a fossil fuel-fired plant and 68 percent for a light-water reactor power plant. For all of the electric generating stations in the United States, the rejected heat represents a vast quantity of energy. In addition, all the generated electrical energy is ultimately converted into heat. Prior to 1965, the local and global effects of heat generation received, for the most part, little attention. In the light of present-day concern over the deterioration of the environment, heat generation and the dissipation of heat into the surroundings are no longer regarded as matters of minor importance, in part because of the ever-increasing consumption of fossil and nuclear fuels.

The conversion of electrical energy into heat presents in itself no particular problem because of the widespread dissipation of the heat. When all of the heat sources are considered, namely, all thermal power plants, various kinds of heating processes and systems, and certain natural heat sources, a gradual increase in the mean temperature of the earth could be anticipated. Such increase in temperature however has not been observed. The very-long-range effect of this heat liberation on the earth's temperature remains uncertain and for the present time causes no alarm.

The heat dissipation from the steam-electric generating stations creates the main problem related to heat generation, because of the comparatively high concentration of this energy. Little heat is rejected directly to the atmosphere from a nuclear power plant, while in a typical fossil fuel-fired power plant, 10 to 15 percent of the input energy is discharged directly into the atmosphere. Consequently, a light-water reactor nuclear power plant discharges about 50 percent more heat to the steam condensing system than a fossil fuel-fired power plant of a comparable generating capability. It is this heat rejection to the condensing system that creates a major problem in acquiring a site for a proposed electric generating station.

In earlier years, steam power plants were usually designed for once-through flow of the cooling water, with some exceptions noted for low-capability plants. In a typical once-through cooling system, the water is taken from a natural source, pumped through the condenser, and returned to the source in a piping arrangement that avoids appreciable mixing of the warm outflow with the cooler inlet flow. In order to achieve a low steam condensing temperature, the increase in the temperature of the cooling water is usually held to within a range of 5 to 8 C. For a high-capability turbine-generator installation, the cooling water flow rate is a particularly large quantity. For example, a 1000-MW unit would require roughly a condenser cooling water flow of 130×10^6 kg/h (130×10^3 m^3/h) for a water temperature increase of 8 C. The cooling water flow rate is cited for a fossil fuel-fired power plant, thus the flow rate would be substantially higher for a light-water nuclear reactor power plant that has the same generating capability.

Once-through cooling represents the most economical method for dissipating into the surroundings the heat rejected by a steam power plant. The quantity of water available at the plant site must be sufficiently large to achieve condensation of the steam without incurring a substantial rise in the cooling water temperature. For a number of reasons, power plant sites that have once-through cooling capabilities are becoming increasingly more difficult to acquire.

In addition to the difficulty experienced in acquiring a suitable power plant site near an adequate supply of cooling water, the restrictions on the use of water will prove in many cases to be a greater obstacle to utilizing once-through cooling. Federal and state regulations limit heat rejection from a power plant in terms of the increase in the temperature of the water. The Water Pollution Control Act of 1965 specifies a maximum allowable increase of 5 F (2.8 C) in the water temperature for open rivers and streams. For a large power plant, the 5 F increment would require a condenser flow rate that exceeds the average flow in most rivers in the United States.

The limit on the rise in the water temperature is established in order to avoid profound effects on the aquatic ecosystems of the heat-receiving water bodies. As the water temperature is increased, the maximum oxygen-carrying capacity of the system is reduced while the biological and biochemical oxygen demands within the system are simultaneously increased. The life cycle of the various aquatic organisms, for example, fish, can be altered, perhaps with unfortunate consequences. The effect of heated water on aquatic organisms has however not proved to be deleterious in every case. The self-cooling ability of a river can be an important factor in water utilization. Given a sufficient run, energy will be effectively transferred from the warm river water to the surroundings by radiation, convection, and evaporation.

10.8 UTILIZATION OF REJECTED HEAT

Two questions related to steam-electric power plants are frequently raised. Why is so much of the energy in the fuel wasted? Why cannot the rejected heat be "put to good use" rather than be dissipated into the surroundings? Because of the thermodynamic limitations imposed by the second law, most of the rejected heat is unavailable energy that is fixed by the characteristics of the steam power cycle and cannot be properly classed as waste heat. The balance of the rejected heat results from the losses inherent in the energy conversion processes. The means that can be employed to reducing the magnitude of the rejected heat were examined in earlier sections of the text. Possible use of the rejected heat will now be investigated.

In order to effect in the turbine maximum conversion of thermal energy into mechanical energy, the steam should be expanded to the lowest temperature that can be economically achieved by the power plant cooling system. A condenser pressure of 25 mm Hg, abs, will establish a steam condensing temperature of 25.7 C. A corresponding temperature of about 21 C may be assumed for the cooling water leaving the condenser. The power plant cooling system thus typically discharges a very large quantity of low-temperature water. Possible use for this warm water has been explored by numerous investigators.

Applications to agriculture have been investigated in a variety of experimental programs, including crop spraying with warm irrigation water and heating the soil by warm water circulated through pipes buried in the ground. Other possible uses for warm water are heating aquaculture ponds, greenhouses, and poultry houses. Such applications however would require water at temperatures somewhat in excess of 20 C.

Exhaust steam or steam bled from the turbine can be used for district heating and cooling, producing hot water, and for many different industrial and commercial applications. Actually, these practices are well established, but the possibility exists for considerable expansion in district heating and cooling. Other suggestions include sidewalk and roadway heating for removal of snow and ice. Both steam and hot water can serve as heat sources, but for most schemes the fluid must arrive at the point of application at a temperature of about 95 C or higher. Evidently, to obtain an exhaust steam temperature of 100 to 105 C, the turbine must operate with essentially an atmospheric exhaust pressure.

The dissipation of the rejected energy over a wide area through a variety of heating and cooling processes appears to offer a solution to the problem of thermal pollution of water bodies. Actually, an obstacle of significant degree arises in the incompatability of the electrical load with the heating and cooling loads. The annual variation in the electrical load is small in relation to the variation in the heating load, which is heavy during the winter months, light during the spring and fall, and virtually zero in the summer. Significant differences are observed in the heating loads throughout the country. Further, the daily variations in the electrical and heating loads are not coincidental. The absence of a summer heating load however is compensated by a cooling load that, throughout the country, will vary from light to very heavy.

Any method employed for widespread heat dissipation must necessarily be integrated into the design of the individual power plant. Evidently, the economic feasibility of the overall operation must be established. The extent of the piping system is an important factor from the standpoint of both the capital cost and the heat loss. Not all power plants are located in areas that have a sufficiently high population density for supporting heating and cooling of various types of buildings.

Little use is likely to be found for the warm cooling water normally discharged from the condenser, because of the high volume and low temperature of this water. Consequently, turbine exhaust steam, or bled steam, must be used for regional heating and cooling applications. A substantial loss in available energy occurs when the turbine outlet pressure is increased from 25 mm to 760 mm Hg, abs. For example, steam at typical inlet conditions is assumed to expand reversibly and adiabatically through the intermediate- and low-pressure turbines. For a condenser pressure of 25 mm Hg, abs, the theoretical work is 1370 kJ/kg steam. When the exhaust pressure is increased to 760 mm Hg, abs, the theoretical work is reduced to 900 kJ/kg. The reduction in the theoretical work indicates that an appreciable decrease in the thermal efficiency of the power plant will be experienced because of the increase in the turbine exhaust pressure.

While the production of electrical energy per unit quantity of fuel is reduced because of the higher turbine exhaust pressure, there is a financial return on the steam used for regional heating and cooling. Again, the economic feasibility of dual-purpose operation must be determined for each proposed installation. Incidentally, in cooling applications, steam would be supplied for the operation of absorption refrigeration systems that require only a comparatively small amount of electrical energy for fluid pumping.

A steam turbine can be designed to operate at any prescribed exhaust pressure, but the performance of the machine will deteriorate if the exhaust pressure at the last stage varies appreciably from the design value. This characteristic of the machine must be observed in any proposed installation that would utilize the turbine exhaust steam for heating or cooling applications.

10.9 POWER PLANT COOLING SYSTEMS

Increased electrical generating capability in the United States must for the next two to three decades be accomplished mainly through the construction of coal-fired and nuclear power plants. A suitable site for each power plant must be found, and, in addition to other requirements, the site must have the potentialities for establishing an effective means for heat rejection. The magnitude of the siting problem is comprehended when the number of new power plants is identified. Construction of some 350 power plants is anticipated during the 15-year period 1980 to 1995. Dependence upon once-through cooling will diminish because of the restrictions on the use of water and the general inadequate supply of water for this purpose. Consequently, the power plant planner must turn with increasing frequency from once-through cooling to alternative methods for dissipating the heat rejected from steam-electric generating stations.

The various systems used for steam power plant cooling are

1 Once-through.
2 Cooling lakes.
3 Spray ponds.
4 Mechanical-draft wet towers.
5 Natural-draft wet towers.
6 Dry cooling towers.

Each of these systems has certain advantages and disadvantages that should be examined in the selection of the most appropriate cooling system for a particular power plant site. The environmental impact, operating problems, and plant economy are relevant factors in this selection.

Cooling lake systems resemble once-through cooling systems, except for recirculation of the cooling water in the lake. Construction costs are reasonable, provided the soil in the basin has a sufficiently low permeability. Make-up water is not likely to be excessive, because rainfall and local runoff will compensate for evaporation losses. The amount of land required for construction of a cooling lake is somewhat extensive, that is, 0.4 to 2 hectares per megawatt of installed generating capability. Localized fogging and icing can in some installations prove to be a problem.

Spray pond cooling involves the discharge of the warm condenser water into the ambient air above the surface of the water in the pond. A fine spray is produced by the atomizing nozzles, and the small droplets are cooled by direct contact with the ambient air. Most of the heat transfer from the droplets is effected by evaporation. In comparison with a cooling lake, a spray pond requires less area, but evaporation losses are higher, and drift increases the water losses. Spray pond performance is strongly dependent upon wind speed and direction. Occurrence of localized fogging and icing is a possibility, particularly so because of drift. In large-area ponds, the spray has a tendency to become ineffective because the air near the center of the pond is very humid. The power required for achieving proper atomization of the spray water is an operating cost that cannot be ignored.

A mechanical-draft wet tower can be arranged to move the ambient air in cross flow through the condenser water as it descends in the tower. See Fig. 10.3a. Air enters at each side of the tower and flows horizontally toward the center of the structure. After passing through the drift eliminators, the air flows vertically through the open center section of the tower and is subsequently discharged through the axial-flow, induced-draft fans located at the top of the tower. The water enters at the top of the tower and falls through the packing that breaks the water into small droplets in order to increase the area of contact with the air and thus improve the heat transfer between the two fluids. Except for the water lost by evaporation, the cooled water is collected in a sump at the bottom of the tower and subsequently returned to the condenser. The drift eliminators, installed adjacent to the packing, remove the entrained water droplets from the main air stream. In a counterflow arrangement, the air enters through the side of the tower and flows vertically through the fill and drift eliminators to the inlet of the draft fans. See Fig. 10.3b.

Mechanical-draft wet towers are readily controlled, and the ambient relative humidity has a minimal effect on the tower performance. Power is required for moving the air and water, although the pumping head is usually low. A wet tower requires only a small fraction of the water needed for once-through cooling. In once-through cooling however, all the water that flows through the condenser can be returned to the source. On the other hand, the water supplied from the source to the tower is lost through evaporation that is essential to the cooling process.

Natural-draft wet towers employ a chimney effect for moving the ambient air. These "hyperbolic" towers range in size from 75 to 100 m in diameter and a height between 100 and 150 m. Air enters through a peripheral section at the bottom of the tower and, in the counterflow arrangement, flows upward through the descending water. See Fig. 10.3d. The heat exchanger section occupies a small portion of the tower, with the remainder of the structure used to promote the chimney effect. Tower construction can also be arranged for cross flow of the air and the water as shown in Fig. 10.3c. Natural-draft wet towers are most effective in areas where the ambient relative humidity is high, a condition that produces a comparatively strong buoyancy effect.

In comparison with the mechanical-draft tower, the natural-draft tower has a higher capital cost, but maintenance costs are lower, and no fan power is required. There is some objection to the great height of the natural-draft tower, although the high-level discharge of the plume reduces possible localized fogging and icing. Exact control of the outlet water temperature is difficult to achieve in the natural-draft tower.

Dry cooling towers are receiving some attention despite certain disadvantages associated with the operation of this type of cooling unit. The condenser water flows inside finned tubes, and heat is transferred through the tube wall to the cooling air. Either an induced-draft fan or a natural-draft stack is used to effect air flow through the heat exchanger. The dry cooling tower requires no source of water, hence there is no evaporative loss and fogging, mist formation, and icing are avoided. High capital costs and high maintenance costs are characteristic of dry cooling towers, and large volumes of air must be circulated through the tower. The conventional

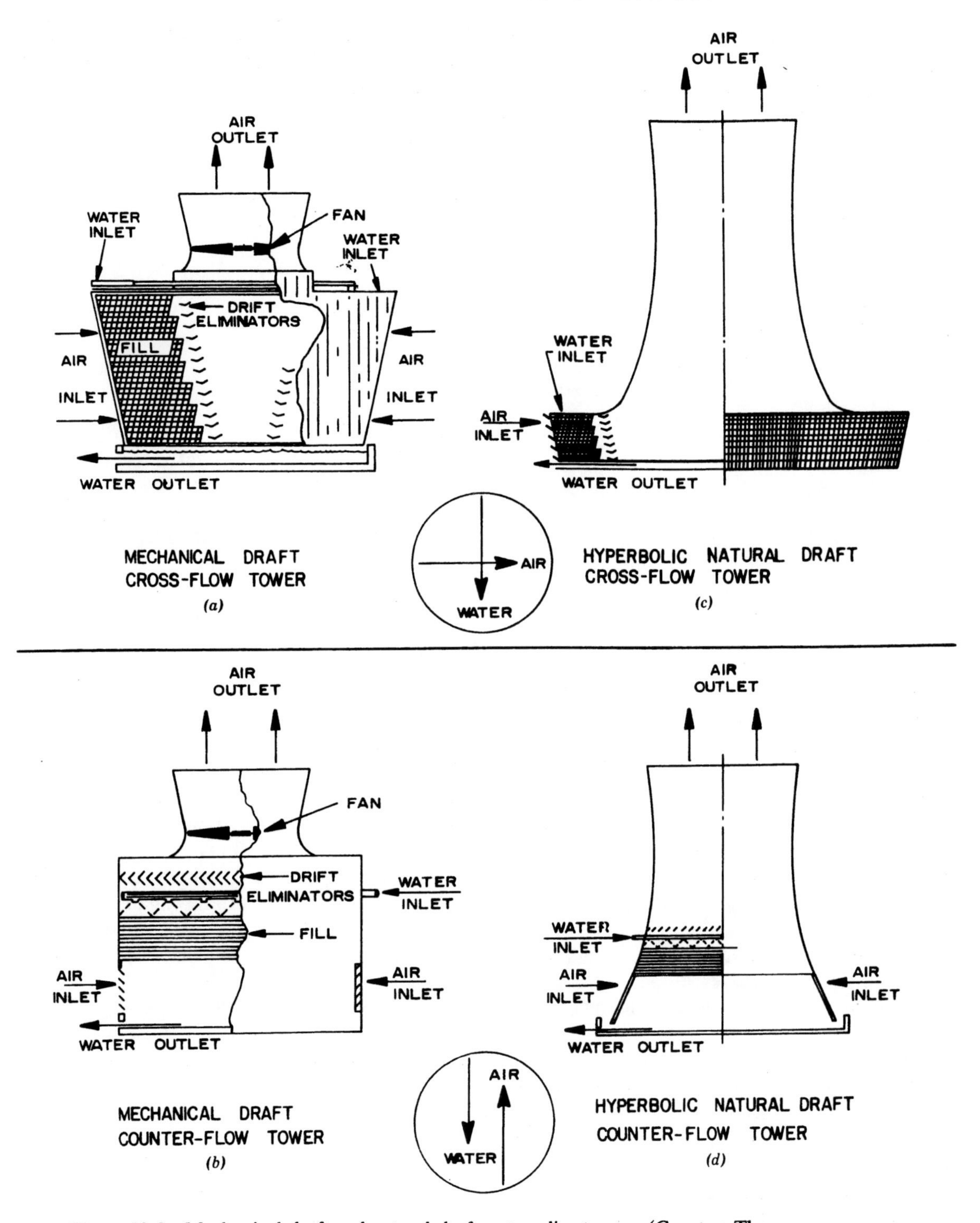

Figure 10.3 Mechanical-draft and natural-draft wet cooling towers. (Courtesy The Marley Company.)

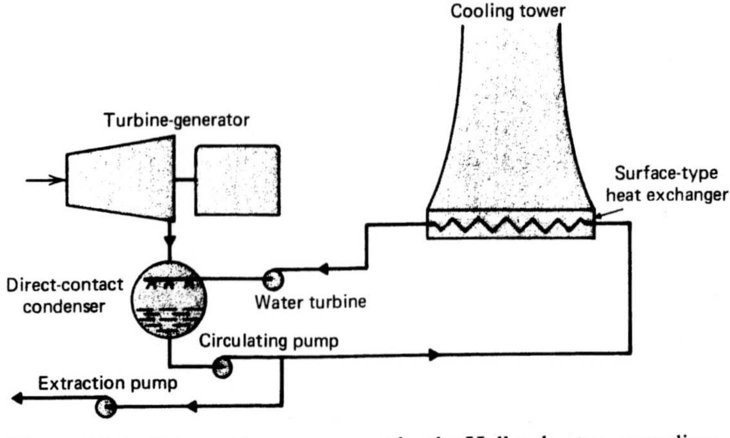

Figure 10.4 Schematic arrangement for the Heller dry-tower cooling system.

low turbine exhaust pressure cannot be economically achieved through the use of dry cooling towers.

The Heller dry cooling tower system, used principally in Europe, incorporates air-cooled heat exchangers and jet or spray condensers. The turbine exhaust steam is condensed by direct contact with the cooling water. The major portion of the condenser outflow is directed to the heat exchangers in the cooling tower, while the remainder of the water is pumped to the boiler, that is, as feedwater. See Fig. 10.4. In comparison with a surface condenser, a spray condenser is less complex, occupies less space, and can be installed and maintained at a lower cost. The closed water system used with a dry tower allows the use of a spray condenser without contamination of the boiler water. Of particular significance is the low terminal temperature difference, about 1 C, achieved in a spray condenser. On the other hand, the terminal temperature difference for a surface condenser should exceed 3 C.

The plume of warm, moist air discharged from a wet cooling tower can be reduced or eliminated by designing a tower for wet/dry cooling. In one arrangement of this type of tower, the air flows initially through the finned tube heat exhanger, where dry cooling is achieved, and then through the wet cooling section. The quantity of water vapor discharged from the tower is controlled by varying the proportion of heat rejected in the wet and dry sections. Figure 10.5 shows the construction of a parallel path wet/dry cooling tower. This type of tower is particularly effective in areas where the supply of water for evaporative cooling is limited or only periodically available.

Intermingling of the cooling tower plume with the power plant stack gases is a highly undesirable occurrence when an acid condensate is formed. Serious corrosion problems can result from this event. Possible mixing of the stack gases with the wet tower plume may be minimized by discharging the two fluid streams at different elevations.

The detrimental effect of operating a turbine at a higher-than-conventional back pressure is demonstrated by the increase in the turbine heat rate. An increase in the turbine exhaust pressure from 38 mm to 89 mm Hg, abs, causes an increase in the Rankine cycle heat rate of 7.1 percent for a typical (7 MPa, 284 C) LWR power plant and 3.6 percent for a typical (25 MPa, 538 C/538 C) fossil fuel-fired power plant. For the two different exhaust pressures, the corresponding heat rejection temperatures are 33.0 and 48.8 C.[22] The typical condenser pressure achieved with a wet evaporative cooling system will not exceed 75 mm Hg, abs.

Discontinuance of once-through cooling will require for most future steam-electric power plants the installation of cooling towers. Because of the

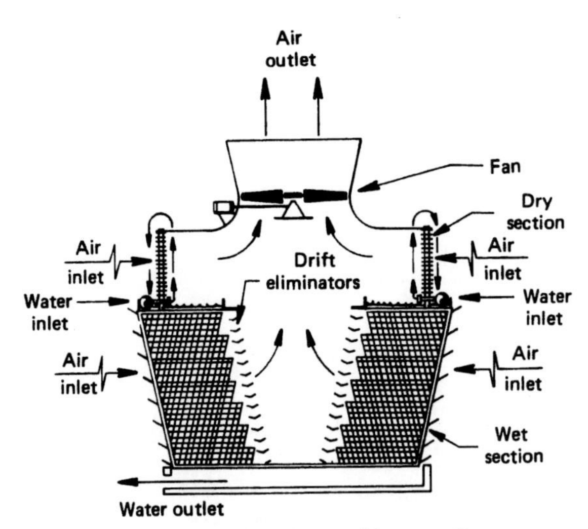

Figure 10.5 Parallel-path wet/dry cooling tower. (Courtesy The Marley Company.)

inherently higher heat rejection temperature, the use of cooling towers will result in increased heat rates, particularly during the summer months. The consequences of this change to cooling towers will be increased consumption of fuel and increased heat rejection to the surroundings. These two detrimental effects should be regarded with considerable concern when an increase of 6 percent in the heat rate of future power plants is a reasonable assumption in the event once-through cooling is superseded by wet tower cooling.

The discharge of water vapor from wet cooling towers may reach objectionable levels for a high concentration of generating capability at a single power plant site. In addition to fogging and icing problems, the dispersion of water vapor into the ambient air may create a highly humid atmosphere in the area surrounding the power plant. An inadequate supply of water and an excessively high discharge of water vapor are two factors that can contribute to a decision to employ dry rather than wet cooling towers.

Dry tower cooling however imposes a considerably more severe penalty on the power plant heat rate. Condenser pressures will be high, that is, in the range of 200 to 500 mm Hg, abs. The application of dry tower cooling, in comparison with once-through cooling, will increase the power plant heat rate by 12 percent or more.

Higher capital costs will be incurred when power plants are equipped with cooling towers. Estimated incremental capital costs, that is, the added investment over a conventional once-through cooling system, for fossil fuel plants are: mechanical-draft wet tower $7/kW, natural-draft wet tower $11/kW, and mechanical-draft dry tower $45/kW. These incremental capital costs are increased by about 50 percent for nuclear light-water power plants. Still higher incremental capital costs can be anticipated for retrofitting cooling towers to existing power plants that currently employ once-through cooling.

The energy consumed in operating cooling towers, in comparison with once-through cooling, effects an increase in the production cost of electrical energy, say, 0.4 to 0.6 mills/kWh for wet towers. For dry towers, the incremental cost is considerably higher, perhaps 2 to 2.5 mills/kWh.

10.10 COOLING TOWER DESIGN

A cooling tower is designed for operation at a prescribed set of conditions, hence adequate controls must be provided in order to facilitate off-design operation. The most severe ambient conditions for operating the tower will occur during the summer months, and reference to meteorological data will provide local wet-bulb and dry-bulb temperatures pertinent to the design of the tower. The wet-bulb temperature is particularly significant with respect to the design of a wet tower, because most of the cooling is accomplished by evaporating a portion of the circulated condenser water.

In a commercial wet cooling tower, the temperature T_{cw} of the cold water leaving the tower cannot drop to the wet-bulb temperature T_{wb} of the entering air. This temperature difference ($T_{cw} - T_{wb}$) is designated as the "approach." Further, the temperature $T_{a,2}$ of the outlet air cannot reach the temperature T_{hw} of the hot water entering the tower.

A cooling tower is a part of the overall cooling system for the power plant, hence it is designed in conjunction with the condenser. The design of the two main components, condenser and cooling tower, will be influenced by the parameters prescribed for the optimum design of the entire power plant.

The following parameters are specified for a basic wet tower design.

1 Cooling water flow rate.
2 Steam condensing temperature T_{sat}.
3 Temperature of the hot water T_{hw}.
4 Temperature of the cold water T_{cw} (approach).
5 Temperature of the outlet air $T_{a,2}$.

The design of the cooling tower is largely empirical, but application of the first law can be illustrated. A schematic view of a wet cooling tower is shown in Fig. 10.6. The energy equation written for the wet cooling tower is

$$\dot{m}_w h_{hw} + \dot{m}_a h_{a,1} + \omega_1 \dot{m}_a h_{v,1} - [\dot{m}_w - \dot{m}_a(\omega_2 - \omega_1)]h_{cw} - \dot{m}_a h_{a,2} - \omega_2 \dot{m}_a h_{v,2} = 0 \tag{10.3}$$

where

$\dot{m}_w$ = hot water flow rate
$\dot{m}_a$ = dry air flow rate
h_{hw} = enthalpy of the hot water
h_{cw} = enthalpy of the cold water
$h_{a,1}$ = enthalpy of the dry air entering the tower
$h_{a,2}$ = enthalpy of the dry air leaving the tower
$h_{v,1}$ = enthalpy of the water vapor in the inlet air
$h_{v,2}$ = enthalpy of the water vapor in the outlet air
ω_1 = specific humidity of the inlet air
ω_2 = specific humidity of the outlet air

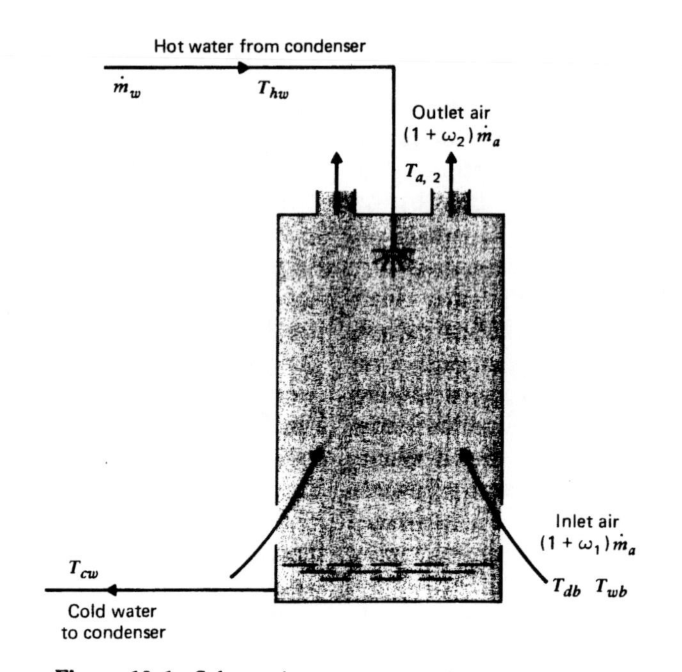

Figure 10.6 Schematic arrangement for a wet cooling tower.

Example 10.5

Calculate the air flow rate and the make-up water flow rate for a wet cooling tower installed in a 500-MW steam power plant. The following operating data are applicable to the analysis.

Temperature of the hot water: 43 C
Ambient air
  Dry-bulb temperature: 24 C
  Wet-bulb temperature: 18 C
  Relative humidity: 0.564
  Barometric pressure: 101.35 kPa, abs
Temperature of the air leaving the tower: 35 C
Approach: 10 C
Flow rate of the water entering the tower: 590,000 kg/min
The air is assumed to leave the tower saturated.

Calculated values of the specific humidity are

Air entering the tower: 0.01057 kg vapor/kg dry air
Air leaving the tower: 0.03657 kg vapor/kg dry air

Enthalpy values of the water vapor are

At the tower inlet: 2545.4 kJ/kg vapor
At the tower outlet: 2565.3 kJ/kg vapor

Enthalpy values of the water are

Hot water: $h = 180.10$ kJ/kg water
Cold water: $T_{cw} = T_{wb} + (\text{approach}) = 18 + 10 = 28$ C

$$h = 117.43 \text{ kJ/kg water}$$

From Eq. 10.3,

$$\dot{m}_a c_{p,a}(T_{a,1} - T_{a,2}) + \dot{m}_a(\omega_1 h_{v,1} - \omega_2 h_{v,2}) + \dot{m}_w h_{hw} - [\dot{m}_w - \dot{m}_a(\omega_2 - \omega_1)]h_{cw} = 0$$

$$\dot{m}_a \times 1.004(24 - 35) + \dot{m}_a[(0.01057 \times 2545.4) - (0.03657 \times 2565.3)] + (590000 \times 180.10) - [590000 - \dot{m}_a(0.03657 - 0.01057)]117.43 = 0$$

$$\dot{m}_a = 493{,}670 \text{ kg/min}$$

The corresponding volume flow rate, determined at a temperature of 24 C and a partial pressure of 99.66 kPa, abs, is 422,450 m³/min. Thus, the air flow to the tower is

$$\dot{V}_a = 422{,}450 \text{ m}^3\text{/min} \quad \text{based on ambient conditions}$$

The make-up water is equivalent to the water evaporated in the tower. Hence,

$$(\dot{m}_w)_{evap} = \dot{m}_a(\omega_2 - \omega_1) = 493670(0.03657 - 0.01057) = 12{,}835 \text{ kg/min}$$

The make-up water flow rate is equal to 2.18 percent of the inlet water flow rate.

The approach is a particularly significant cooling tower design parameter.[23] For a prescribed wet-bulb temperature, constant water flow rate, and constant range $T_{hw} - T_{cw}$, a reduction in the approach will effect a decrease in the temperature of both the hot and the cold water. Steam turbine performance is consequently improved, but the tower size and cost, and the fan power, increase substantially.

The approach may be held constant and the range decreased. As a result, the hot water temperature decreases and a lower condensing temperature is achieved. However, the water flow rate, tower cost, and fan power increase.

If the wet-bulb temperature and the hot water temperature remain constant and the range is increased, the approach is reduced. As a consequence, the water flow rate is reduced, but an increase is observed in the tower cost and in the fan power.

PROBLEMS

10.1 A power plant burns a Pennsylvania High-Volatile A bituminous coal [S (sulfur) = 1.3 percent, HHV = 31,635 kJ/kg]. Determine whether SO_2 control is required in order to conform to the EPA stack emission standards for plants constructed after 1970. Repeat the determination for a Wyoming High-Volatile B bituminous coal (S = 0.5 percent, HHV = 30,125 kJ/kg).

10.2 A 600-MW power plant burns a Pennsylvania Low-Volatile bituminous coal (S = 1.8 percent, HHV = 33,265 kJ/kg). The annual plant factor is 0.70, and the thermal efficiency is 0.39. Assume that the coal is burned in a fluidized bed. Sufficient limestone is added to the furnace to reduce the emission of SO_2 to the limiting value prescribed by the EPA stack emission standards. If the limestone is used on a once-through basis, what is the annual consumption of limestone and the annual production of calcium sulfate?

10.3 Explain the control of NO_x emission by (a) flue gas recirculation and (b) by two-stage combustion.

10.4 A steam turbine exhausts to a surface condenser designed for an inlet water temperature of 16 C, a cooling water temperature increase of 10 C, and a terminal temperature difference of 4 C. When the supply of river water is insufficient for once-through cooling, the condenser cooling water is circulated through a dry cooling tower. The operating temperatures for the tower are: inlet water 68 C, outlet water 58 C, inlet air 32 C, and outlet air 62 C. The steam turbine inlet conditions are 15 MPa, abs, and 500 C. Estimate the reduction in the specific turbine work that occurs when the operation is shifted from once-through cooling to dry tower cooling.

10.5 (a) When the system described in Problem 10.4 operates with dry tower cooling, the heat rejection rate is 910×10^6 kJ/h. The specific heat of the air is 1.017 kJ/kg·K. The overall coefficient of heat transfer for the water-to-air heat exchanger is 60 W/m^2·K. Calculate the water flow rate, kg/min, the air flow rate, m^3/min, and the area of the heat exchanger heat transfer surface.
(b) The pressure drop on the air side of the cooling tower heat exchanger is 5.5 cm water. The fan static efficiency is 0.60. Calculate the fan shaft power.

10.6 The cooling water flow from a power plant surface condenser is 300,000 kg/min. The prescribed temperature of the water leaving the condenser is 45 C. The design air conditions are: 27 C dry-bulb temperature, 21 C wet-bulb temperature, and 100 kPa, abs, barometric pressure. The prescribed approach for a wet tower design is 9 C, and the air leaves the tower at 37 C (saturated). Calculate the air flow rate, based on the tower inlet conditions, and determine the make-up water flow rate.

10.7 A steam power plant employs a Heller dry cooling tower system. The steam flow rate to the jet condenser is 300,000 kg/h, and the enthalpy of the exhaust steam is 2340 kJ/kg. The temperature of the water entering the tower is 60 C. The ratio of the return flow from the tower to the steam flow is 40 to 1. The ambient air temperature is 25 C, and the air leaves the tower at 45 C. The water-to-air heat exchanger is arranged for counterflow of the fluids, and U_o = 55 W/m^2·K. Calculate the air flow rate and the heat transfer area for the heat exchanger.

10.8 The pulverized coal firing rate for a boiler is 68,000 kg/h. The coal HHV is 31,635 kJ/kg. The heat release rate is 34,000 kJ/kg coal, and the dry flue gas is 14.3 kg/kg coal. The burning zone radiant surface area is 732 m^2. Medium slagging conditions are observed in the furnace. The nitrogen content of the coal is 1.1 percent. Estimate the NO_x emission, in kg/10^6 kJ.

10.9 A steam turbine-generator produces 5000 kW with a steam rate of 6.0 kg/kWh. The enthalpy of the exhaust steam is 2325 kJ/kg, and the enthalpy of the condensate is 226 kJ/kg. The cooling water enters and leaves the condenser at 30 and 50 C, respectively. A cooling pond serves as the heat sink. The pond area required for heat dissipation is determined from $A = \dot{m}_w c_w (T_1 - T_2)/\dot{m}_v h_{fg}$, where $\dot{m}_w$ is the cooling water flow rate, c_w is the specific heat of the water, T_1 is the temperature of the water entering the pond, T_2 is the final temperature of the water in the pond, $\dot{m}_v$ is the evaporation rate per unit pond area, and h_{fg} is the energy of evaporation of the water at temperature T_1. For the prescribed operating conditions, the evaporation rate for the pond is 0.350 kg/h·m^2. Calculate the pond area required for the heat dissipation.

10.10 A 350-MW electric power plant burns a coal that has a HHV of 30,705 kJ/kg. The theoretical air for the coal is 329 kg/10^6 kJ. The coal rate is 0.417 kg/kWh, and the fuel is burned with 125 percent theoretical air. The refuse collected is 0.13 kg/kg coal. The flue gas enters the precipitator at 330 C and 100 kPa, abs. The precipitator is required to remove 99.8 percent of the particulate material from the flue gas. The gas velocity through the precipitator is 2 m/s. Estimate the area of the precipitator collecting surface, and calculate the gas flow area.

REFERENCES

1 Patterson, W. D. "Progress in Satisfying Environmental Requirements," *Combustion*, Oct. 1974.

2 Land, G. W. "Trials of Additives for Sulfur Dioxide Removal in Industrial Plants," *Combustion*, Dec. 1969.

3 Hoke, R. C., Ruth, L. A., and Shaw, H. "Combustion and Desulfurization of Coal in a Fluidized Bed of Limestone," ASME Paper No. 74-PWR-6, New York, Sept. 1974.

4 Albrecht, P. E. and Lieberman, J. A. "Reliability of Flue Gas Desulfurization Systems," *Power Engineering*, June 1975.

5 Andrews, R. L. "Current Assessment of Flue Gas Desulfurization Technology," *Combustion*, Oct. 1977.

6 Estcourt, V. F., Grutle, R. O. M., Gehri, D. C., and Peters, H. J. "Tests of a Two-Stage Combined Dry Scrubber/SO_2 Absorber Using Sodium or Calcium," *Combustion*, Nov. 1978.

7 Aghassi, W. J. and Cheremisinoff, P. N. "NO_x Control in Central Station Boilers," *Power Engineering*, June 1975.

8 Rawdon, A. H. and Sadowski, R. S. "An Experimental Correlation of Oxides of Nitrogen Emissions from Power Boilers Based on Field Data," ASME Paper No. 72-WA/Pwr-5, New York, Nov. 1972.

9 Bell, A. W., deVolo, N. B., and Breen, B. P. "Nitric Oxide Reduction by Controlled Combustion Processes," Presented at the Western States Section/Combustion Institute Meeting, Berkeley, Calif., April 20–21, 1970.

10 White, H. J. "Electrostatic Precipitators," *Combustion*, March 1953.

11 Walker, A. B. "Experience with Hot Electrostatic Precipitators for Fly Ash Collection in Electric Utilities," *Combustion*, Nov. 1974. Also presented at the 1974 American Power Conference.

12 Reigel, S. A. "Reverse Pulse Baghouses for Industrial Coal-Fired Boilers," *Power Engineering*, Aug. 1974.

13 Troupe, J. S. "Handwriting on the Power Plant Wall—Flue Gas Treatment," *Combustion*, Oct. 1978.

14 Rittenhouse, R. C. "R & D in Generation," *Power Engineering*, Dec. 1975.

15 Kramer, A. W. Note on Reactor Shielding. *Power Engineering*, Sept. 1958.

16 Hammond, R. P. "Nuclear Power Risks," *American Scientist*, March–April 1974.

17 Pittman, F. K. "Management of Commercial Radioactive Waste," Presented at the 1974 Thermal Power Conference, Washington State University, Pullman, Wash., Oct. 1974.

18 Budenholzer, R. J., Hauser, L. G., and

Oleson, K. A. "Selecting Heat Rejection Systems for Future Steam-Electric Power Plants," *Combustion*, Oct. 1972.

19 Glicksman, L. R. "Thermal Discharge from Power Plants," ASME Paper No. 72-WA/Ener-2, New York, Nov. 1972.

20 Steele, B. L. "Selection of Plant Cooling Source(s)," ASME Paper No. 75-IPWR-6, New York, May 1975.

21 Rittenhouse, R. C. "The Revolution in Water Management," *Power Engineering*, Aug. 1975.

22 Jaske, R. T. "A Future for Once-Through Cooling," *Power Engineering*, Feb. 1972.

23 Bloebaum, O. E. "A Practical Approach to Cooling Tower Sizing Related to Thermal Power Generation," Presented at the 1971 Thermal Power Conference, Washington State University, Pullman, Wash., Oct. 1971.

BIBLIOGRAPHY

Faith, W. L. *Air Pollution Control*, John Wiley & Sons, New York, 1959.

Fowler, J. M. *Energy and the Environment*, McGraw-Hill Book Co., New York, 1975.

Glasstone, S. and Jordan, W. H. *Nuclear Power and Its Environmental Effects*, The American Nuclear Society, La Grange Park, Ill., 1980.

Obert, E. F. *Internal Combustion Engines and Air Pollution*, Harper & Row, Publishers, New York, 1973.

CHAPTER ELEVEN

REFRIGERATION AND AIR CONDITIONING

Mechanical refrigeration is highly important to many industrial and commercial operations, and essential to air conditioning and food preservation. A substantial amount of energy is consumed in the operation of refrigeration equipment, particularly so during the summer months when the air conditioning load is heavy. Because of the general need to conserve energy, cooling systems should be designed to achieve efficient operation within the economic limits prescribed for a particular application.

Two different systems are used principally for achieving mechanical refrigeration, namely, the gas or vapor compression system and the absorption system. A third system, installed primarily aboard ships, employs the principle of water vapor refrigeration.

Somewhat allied to air conditioning are the numerous processes used in industry for drying a great variety of products. Refrigeration, air conditioning, and drying are mechanical applications dependent upon electrical and thermal energy sources for their operations.

11.1 GAS COMPRESSION REFRIGERATION SYSTEMS

The operation of the theoretical gas compression refrigeration system is based on the reversed Brayton cycle, indicated by 12_s34_s1 in Fig. 11.1. The actual cycle is described by 12341. Subsequent to essentially adiabatic compression 1–2, the gas is cooled, with a small pressure drop, to temperature T_3 by transfer of heat to the surroundings. Adiabatic expansion is effected in the change from state 3 to state 4. Refrigeration is achieved by the process 4–1 that occurs with a transfer of heat to the gas. Because the work of expansion is less than the work of compression, a net transfer of work to the system is required.

In comparison with the theoretical cycle, reduced performance of the actual refrigeration system is caused by irreversible compression and expansion of the working fluid and by miscellaneous heat, pressure, and mechanical losses.

Example 11.1

Operating data for a gas compression refrigeration system are

Working fluid: air, $c_p = 1.004$ kJ/kg·K, $k = 1.40$

Air temperature at the start of compression: 220 K

Air temperature at the start of expansion: 300 K

Cycle pressure ratio: 6 to 1

Compressor efficiency: 0.85

Expander efficiency: 0.90

Air circulation rate: 25 kg/min

The miscellaneous losses are neglected.

Determine the refrigeration capability and the coefficient of performance for the system.

The gas temperatures achieved by reversible adiabatic compression and expansion are, respectively,

$$T_{2,s} = 367.1 \text{ K} \qquad T_{4,s} = 179.8 \text{ K}$$

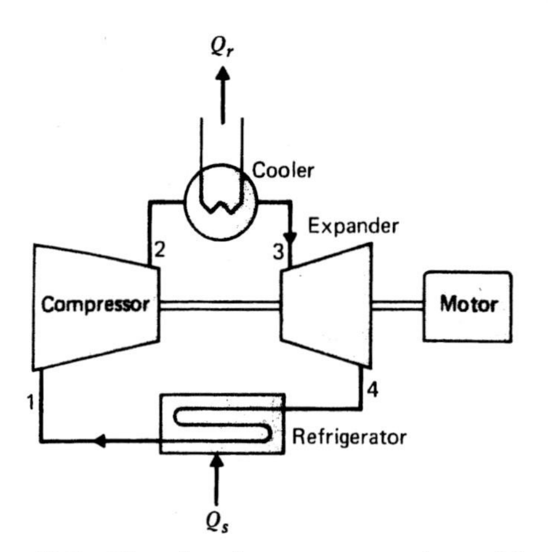

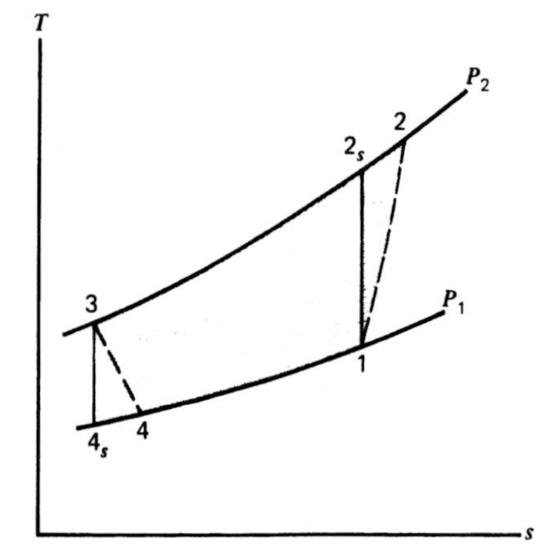

Figure 11.1 Closed-cycle gas compression refrigeration system.

Then,

$$W_C = \frac{h_1 - h_{2,s}}{\eta_C} = \frac{c_p(T_1 - T_{2,s})}{\eta_C}$$

$$= \frac{1.004(220 - 367.1)}{0.85}$$

$$= -173.8 \text{ kJ/kg air}$$

$$W_E = \eta_E(h_3 - h_{4,s})$$

$$= \eta_E c_p(T_3 - T_{4,s})$$

$$= 0.90 \times 1.004(300 - 179.8)$$

$$= 108.6 \text{ kJ/kg air}$$

$$W_n = W_C + W_E$$

$$= -173.8 + 108.6$$

$$= -65.2 \text{ kJ/kg air}$$

Gas temperature at the end of expansion:

$$W_E = h_3 - h_4 = c_p(T_3 - T_4)$$

$$108.6 = 1.004(300 - T_4) \qquad T_4 = 191.8 \text{ K}$$

$$Q_s = c_p(T_1 - T_4)$$

$$= 1.004(220 - 191.8) = 28.3 \text{ kJ/kg air}$$

Refrigeration capability:

$$Q_s = \dot{m}_a Q_s$$

$$= 25 \times 28.3 = 707.5 \text{ kJ/min}$$

Coefficient of performance:

$$\text{cop} = \frac{Q_s}{W_n} = \frac{28.3}{65.2} = 0.434$$

The gas compression refrigeration system operates with a characteristically low coefficient of performance. Consequently, except for certain gas-cooling applications, other more effective refrigeration systems are in general use today. Early closed-cycle, air compression refrigeration systems operated with reciprocating machines and were used in a variety of commercial installations where leakage of a toxic refrigerant could create a hazardous situation. Open-cycle systems have been used, but these systems are susceptible to icing problems.

High machine efficiencies are essential to the operation of the gas compression refrigeration system as a means to improving the coefficient of performance. Subsequent to compression, the temperature of the gas should be reduced to the lowest level consistent with good practice. For a prescribed pressure ratio, a reduction in the temperature of the gas at the inlet to the expander causes a reduction in the work of expansion, but an increase in the refrigeration effect is observed, and as a consequence an increase is achieved in the coefficient of performance.

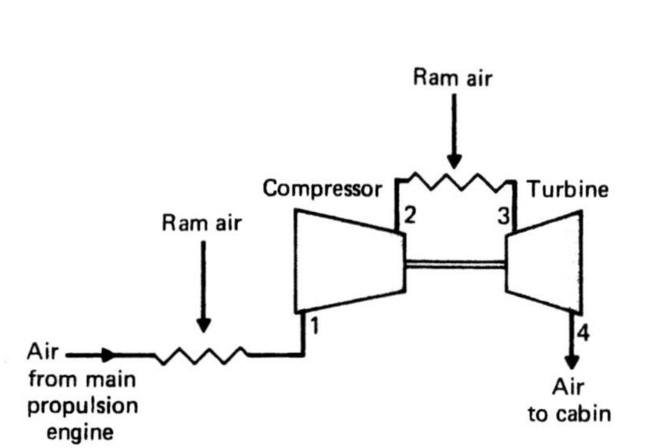

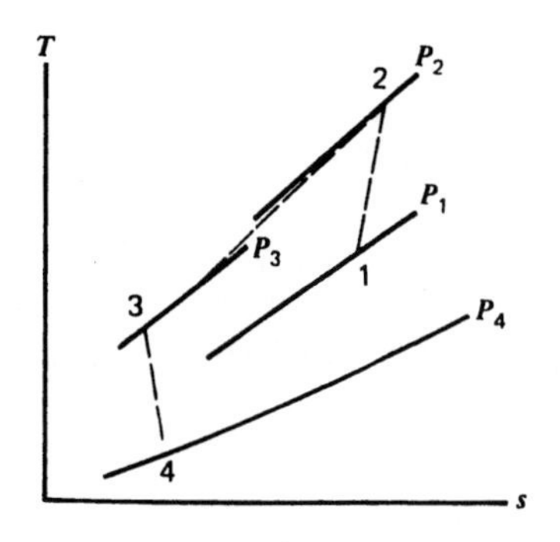

Figure 11.2 Aircraft cabin cooling system.

Figure 11.2 shows the flow diagram and the thermodynamic cycle for an air compression refrigeration machine devised for aircraft cabin cooling. The compressor and turbine are small, high-speed machines keyed to a common shaft. Air piped from the compressor of a main propulsion engine is cooled by ram air, prior to entering the refrigeration system compressor at pressure P_1. In a second heat exchanger, ram air is used to cool the system air before it expands in the turbine to the cabin pressure P_4. The pressure drop $P_2 - P_3$ is comparatively small. The low-temperature air discharged from the turbine provides the means to cooling and ventilating the interior of the airplane. Because $P_3/P_4 > P_2/P_1$, the cooling unit operates without a work input.

11.2 VAPOR COMPRESSION REFRIGERATION SYSTEMS

The flow diagram and the thermodynamic cycle for the vapor compression refrigeration system are shown in Fig. 11.3. Compression of the vapor from state 1 to state 2 will be essentially adiabatic when accomplished in high-speed rotating compressors. Large reciprocating compressors are ordinarily water cooled, and compression is nonadiabatic. The expansion process 5–6 is effected by some kind of throttling device, usually a valve or a capillary.

In Fig. 11.3, the numbers on the flow diagram correspond with the numbers on the T–s diagram. In addition to the transfer of heat to the

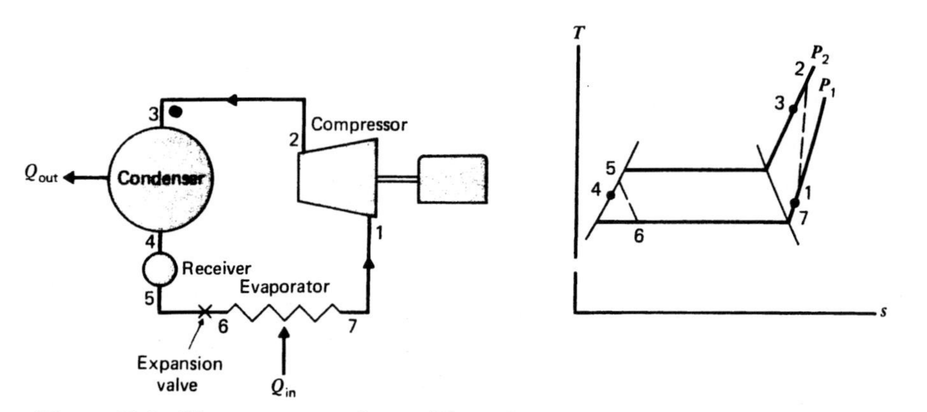

Figure 11.3 Vapor compression refrigeration system.

evaporator and from the condenser, heat is also transferred between the surroundings and other parts of the system, namely, the receiver and various sections of the piping.

The system pressures are important operating parameters. The high pressure in the system controls the saturation temperature of the refrigerant in the condenser. This condensing temperature must be several degrees above the temperature of the cooling medium in order to effect heat transfer from the system. The low pressure in the system controls the temperature of the refrigerant in the evaporator. In order to achieve heat transfer, the evaporating temperature of the refrigerant must be several degrees below the temperature of the heat source.

The difference between the two system pressures should be maintained as small as practicable in order to minimize the compressor work and ensure an acceptable coefficient of performance.

Example 11.2

The following test data are representative of a vapor compression refrigeration system charged with R-12 refrigerant.

Condenser pressure: 700 kPa, abs
Evaporator pressure: 140 kPa, abs
Fluid temperature at
- Compressor inlet: −3 C
- Compressor outlet: 58 C
- Condenser inlet: 40 C
- Condenser outlet: 22 C
- Expansion valve inlet: 27 C
- Evaporator outlet: −13 C

Compressor displacement P.D.: 11.35 m³/min
Compressor volumetric efficiency: 0.82
Heat removed in the compressor: 605 kJ/min

Determine the system refrigeration capability, compressor power, and the coefficient of performance. See Fig. 11.3.

1 System refrigeration capability.

$$v_1 = 0.13074 \text{ m}^3/\text{kg at 140 kPa, abs, and } -3 \text{ C}$$

$$\dot{m} = \frac{\eta_v(\text{P.D.})}{v_1} = \frac{0.82 \times 11.35}{0.13074} = 71.19 \text{ kg/min}$$

Refrigeration effect:

$$h_6 = h_5 = 61.607 \text{ kJ/kg at 700 kPa, abs, and 27 C}$$

$$h_7 = 183.036 \text{ kJ/kg at 140 kPa, abs, and } -13 \text{ C}$$

$$Q_s = h_7 - h_6 = 183.036 - 61.607 = 121.429 \text{ kJ/kg}$$

Refrigeration rate:

$$\dot{Q}_s = \dot{m}Q_s = 71.19 \times 121.429 = 8644.5 \text{ kJ/min*}$$

2 Compressor power.

$$h_1 = 189.049 \text{ kJ/kg at 140 kPa, abs, and } -3 \text{ C}$$

$$h_2 = 220.432 \text{ kJ/kg at 700 kPa, abs, and 58 C}$$

$$\dot{W} = \dot{m}(h_1 - h_2) + Q = 71.19(189.049 - 220.432) - 605 = -2839 \text{ kJ/min} \quad -47.32 \text{ kW}$$

3 Coefficient of performance.

$$\text{cop} = \frac{\dot{Q}_s}{\dot{W}} = \frac{8644}{2839} = 3.04$$

4 System energy balance.

$$h_3 = 207.580 \text{ kJ/kg at 700 kPa, abs, and 40 C}$$

$$h_4 = 56.758 \text{ kJ/kg at 700 kPa, abs, and 22 C}$$

* The capability of a refrigeration machine is usually expressed in tons of refrigeration. One ton = 200 Btu(IST)/min, a quantity derived from the energy required to melt one ton of ice at 32 F during a 24-h period. Then, one ton of refrigeration = 211.0 kJ/min. Thus, the capability of this machine is 8644.5/211.0 = 41.0 tons of refrigeration.

Energy Balance

		Gain (kJ/kg)	Loss (kJ/kg)
Compressor	(1–2)	39.881	8.498
Line	(2–3)		12.852
Condenser	(3–4)		150.822
Line and receiver	(4–5)	4.849	
Expansion valve	(5–6)	—	—
Evaporator	(6–7)	121.429	
Line	(7–1)	6.013	
		172.172	172.172

For the compressor, the energy gain and the energy loss are equivalent to the work input and the heat transferred to the cooling medium, respectively.

Numerous applications of low-temperature refrigeration can be cited. A commercial low-temperature installation will usually require some kind of multistage operation.

For vapor compression refrigeration systems, the lower limit on the evaporating temperature is controlled by the vapor pressure and the freezing temperature of the refrigerant. Very low vapor pressures introduce problems in compressing the vapor through a high pressure ratio. A reduction in the work required for compression of the vapor can be accomplished by multistage operation or by employing a cascade system. The cascade system is comprised of a series of single-stage refrigeration units that are charged with a series of refrigerants of progressively lower boiling points. The evaporator of the first unit is used to cool the condenser of the second unit, and so forth, moving down the cascade.

11.3 REFRIGERANTS

The fluorinated hydrocarbons known by the trade names Freon and Genetron are widely used as refrigerants in vapor compression refrigeration systems. The use of ammonia in these systems is limited, because of its toxicity, to commercial and industrial applications. Ammonia is used in the absorption refrigeration system. Refrigerants are selected on the basis of thermodynamic properties and other characteristics, namely, chemical stability, corrosive action, toxic effects, fire and explosion hazards, miscibility with lubricating oil, and cost.

With respect to the thermodynamic properties, attention is directed to the condensing and evaporating pressures for the normal range of the operating temperatures. Moderate condensing pressures are desirable in order to avoid heavy construction in certain parts of the system. Excessively low evaporating pressures are undesirable because of the difficulty encountered in sealing against air leakage into the system. A high critical temperature is essential to the condensing process. The circulation rate is dependent, in part, upon the latent energy of vaporization, hence this property should have a relatively high value. A low specific volume for the vapor is desirable when reciprocating compressors are used, but this characteristic is not especially significant for compression in centrifugal machines. Fluid viscosities and heat transfer coefficients influence the design of the piping and the heat exchangers.

R-12 (dichlorodifluoromethane, CCl_2F_2) is a widely used refrigerant. This chemically stable and essentially nontoxic substance performs satisfactorily in various types of refrigeration systems where the evaporator pressure is maintained above or slightly below the atmospheric pressure. The relatively low specific volume of the vapor at normal intake conditions is particularly significant with respect to the use of reciprocating compressors.

At a specified temperature, the vapor pressure for R-11 (trichloromonofluoromethane, CCl_3F) is substantially below the corresponding vapor pressure for R-12. Consequently, R-11 would be used in those systems that operate with comparatively high evaporator temperatures, particularly in air conditioning applications. This refrigerant has a relatively high specific volume, hence compression is ordinarily effected in centrifugal machines.

11.4 HEAT PUMPS

The vapor compression refrigeration system can be used for space heating or for other applications that require heat sources. When the system is used in this manner, a stream of air, or another fluid, flows over the condenser surface and transfers the rejected energy to a heat receiver. Energy taken from the heat source is transferred to the working fluid in the evaporator, "pumped" or elevated to a higher temperature, and in a final step rejected from the system in the condenser. Pumping is achieved by an expenditure of mechanical energy, that is, by the work input to the compressor.

In order to achieve an acceptable operating economy, the compressor work should be minimized by maintaining a low condensing pressure and a high evaporating pressure. External factors however control these pressures. The condensing pressure, hence the temperature, must be sufficiently high to establish the temperature difference required for heat transfer between the working fluid and the heat receiver, for example, the warm air circulated through the building. The usual sources of heat are the ambient air, earth, and natural bodies of water. Well water, because of a relatively constant temperature, is an excellent heat source for the heat pump.

The temperature of the heat source, and consequently the evaporator temperature, have a pronounced effect on the thermodynamic performance of the heat pump. For a prescribed value of the rejected heat Q_r, a reduction in the evaporator temperature will cause an increase in the compressor work W_C, and as a consequence, the coefficient of performance Q_r/W_C of the heat pump decreases. In this respect, the use of the ambient air as a heat source creates a problem in the operation of the heat pump. During very cold weather, low evaporator temperatures must be maintained, hence the coefficient of performance is reduced, and the operating economy of the heat pump is likely to be unsatisfactory.

Example 11.3

A comparison will be made of the cost of the energy required for heating a building by means of a conventional oil-fired heating unit and by use of a heat pump. The ambient air is the heat source for the heat pump, which is charged with R-12.

Fuel oil cost: $3.00/10^6 kJ

Efficiency of the oil-fired heating plant: 0.70

Cost of electrical energy: 3¢/kWh

Outdoor temperature: 0 C

Building temperature: 21 C

Building heat loss: 75,000 kJ/h

The heat pump will be operated with an evaporator temperature of −5 C, and a condensing temperature of 60 C (P_{sat} = 1.5259 MPa, abs). The isentropic efficiency of the compressor, including the effect of mechanical losses, is 0.78. The motor efficiency is 0.95. Line losses are neglected. The working fluid enters the compressor dry and saturated at the evaporating temperature. The temperature of the liquid R-12 entering the expansion valve is 30 C. The heat pump cycle is shown in Fig. 11.4.

1 Compressor.

$$h_1 = 185.243 \text{ kJ/kg}$$
$$h_{2,s} = 216.698 \text{ kJ/kg}$$

Electrical energy input to the motor:

$$W = \frac{h_1 - h_{2,s}}{\eta_C \eta_e}$$
$$= \frac{185.243 - 216.698}{0.78 \times 0.95} = -42.449 \text{ kJ/kg R-12}$$

2 Heating effect.

$$h_4 = h_3 = 64.539 \text{ kJ/kg}$$

Energy transferred to the evaporator from the ambient air:

$$Q_s = h_1 - h_4$$
$$= 185.243 - 64.539 = 120.704 \text{ kJ/kg}$$

The energy available for heating the building is obtained from two sources, namely, the electri-

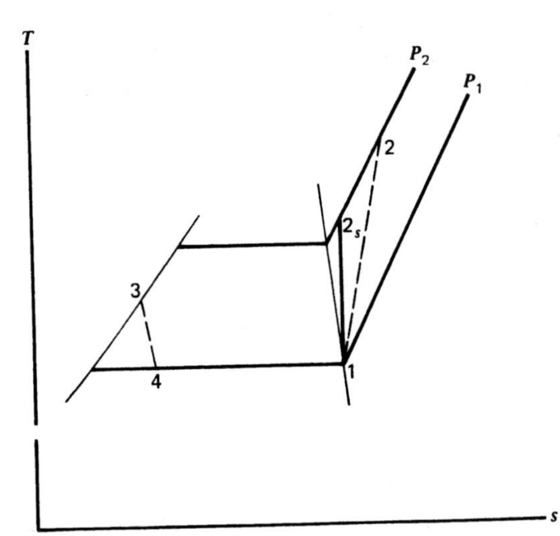

Figure 11.4 Heat pump thermodynamic cycle.

cal energy supplied to the motor and the energy transferred from the ambient air.

$$(Q_s + Q_r) - W = 0$$

$$Q_r = W - Q_s$$

$$= -42.449 - 120.704$$

$$= -163.153 \text{ kJ/kg R-12}$$

3 Operating cost for the heat pump.

$$\dot{m} = \frac{\dot{Q}}{Q_r} = \frac{75000}{163.153}$$

$$= 459.69 \text{ kg/h} \qquad \text{(R-12 circulation rate)}$$

Cost of electrical energy:

$$\dot{W} = \frac{\dot{m}W}{3600} = \frac{459.69 \times 42.449}{3600} = 5.42 \text{ kW}$$

$$5.42 \times 3 \text{ ¢} = 16.26 \text{ ¢/h}$$

4 Cost for oil firing.

$$\frac{75000}{0.70}\frac{300 \text{ ¢}}{10^6} = 32.14 \text{ ¢/h}$$

5 Cost for direct electric heating.

$$\frac{75000}{3600} \times 3 \text{ ¢} = 62.50 \text{ ¢/h}$$

The heat loss from a building may be assumed to vary directly with the difference between the interior temperature of the building and the ambient air temperature. If the outdoor air temperature of 0 C, prescribed in this example, is decreased to −30 C, the heat loss from the building is increased to 182,140 kJ/h. For the altered operating conditions, the following results are reported.

Evaporator temperature: −35 C

h_1 = 171.784 kJ/kg $h_{2,s}$ = 224.592 kJ/kg

Input to the compressor motor: −71.266 kJ/kg R-12

Energy transferred to the evaporator: Q_s = 107.245 kJ/kg

Heating effect: Q_r = −178.511 kJ/kg

Compressor power: 20.20 kW

Heating costs
- Heat pump: 60.60 ¢/h
- Oil firing: 78.06 ¢/h
- Direct electric: 151.78 ¢/h

The competitive advantage of the heat pump diminishes with a decrease in the evaporator operating temperature, which in turn is caused by a decrease in the temperature of the heat source. This reduced effectiveness of the heat pump is indicated by the decline in the coefficient of performance. Thus, for operation of the heat pump in this example, at an ambient air temperature of 0 C,

$$\text{cop} = \frac{Q_r}{W} = \frac{-163.153}{-42.449} = 3.84$$

For operation of the heat pump at the reduced ambient air temperature of −30 C,

$$\text{cop} = \frac{Q_r}{W} = \frac{-178.511}{-71.266} = 2.50$$

It is important to note that the above example represents two sets of "design" calculations and does not pertain to different operating conditions for a particular heat pump. The influence of the outdoor temperature on the performance of a typical heat pump will now be examined.

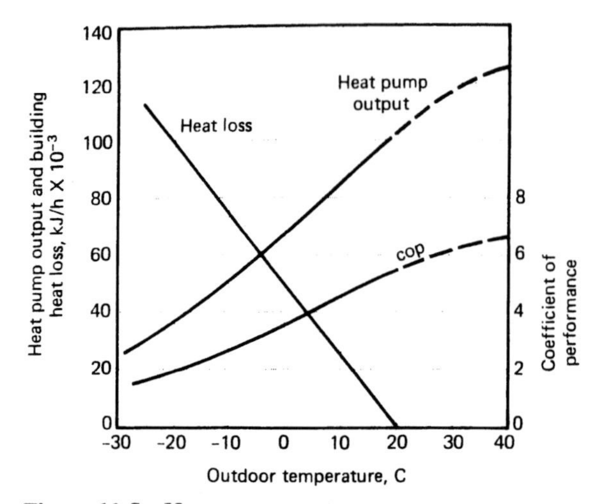

Figure 11.5 Heat pump performance chart.

Figure 11.5 illustrates the performance of a heat pump used for space heating. The building heating system is designed for a heat loss of 80,000 kJ/h, based on a prescribed indoor temperature of 20 C and an outdoor temperature of −12 C. The heat pump is rated at 5 kW, the output of the electric motor.

When the outdoor temperature exceeds −4 C, the heat pump operates intermittently and supplies all of the heat required for maintaining the building temperature at 20 C. For an outdoor temperature below −4 C, the heat pump operates continuously but cannot supply all of the required heat, and supplemental heat is consequently provided by energizing the resistance heating elements.

In the event of very low outdoor temperatures, the heat pump operation may not be economically practicable, because of a low coefficient of performance. The heat pump operation is then discontinued when the coefficient of performance drops below a prescribed limiting value, say, 1.25.

Capital costs for the heat pump are high in comparison with similar costs for the direct electric and fossil fuel heating systems. Supplemental direct electric heating is employed as a means to minimizing the system capital cost. The extent of this supplemental heating is dependent principally upon energy costs and the local climatic conditions. The same factors are used to determine whether a building should be equipped with a heat pump or an alternative heating system.

During the summer months, the heat pump may be operated as a refrigeration machine in conjunction with an air conditioning system. The heat pump/refrigeration machine should have the capability of handling the building heating and cooling loads. The investment cost of the unit is normally charged against both services, that is, heating and cooling of the structure.

A heat pump can be used to augment the heat obtained from a solar energy collector. Under certain conditions, the temperature of the energy storage fluid may not be sufficiently high for space heating. The comparatively low-temperature fluid would however be a very good heat source for a heat pump. The economy of a complex system of this kind must be closely examined, in particular with respect to the capital costs.

11.5 WATER VAPOR REFRIGERATION SYSTEM

For certain applications, water can function effectively as a refrigerant in a vapor compression refrigeration system. A simple system of this kind is shown schematically in Fig. 11.6. The booster pump maintains a very low pressure in the flash chamber, or evaporator, by continuous removal of water vapor and noncondensable gases. The returning water, which is at a temperature of 6 to 8 C above the evaporator temperature, partially flashes into vapor on entering the tank. Simultaneously, the temperature of the remaining liquid drops to the evaporating temperature.

For adiabatic operation, the energy equation for the evaporator can be expressed by

$$\dot{m}_{w,i}h_{w,i} + \dot{m}_{mk}h_{mk} - \dot{m}_{w,o}h_{w,o} - \dot{m}_v h_{v,1} = 0$$

In a closed, chilled-water loop, operating with steady-state conditions, $\dot{m}_{w,i} = \dot{m}_{w,o}$. Also, $\dot{m}_{mk} = \dot{m}_v$, that is, the make-up water flow rate is equal to the rate at which the vapor is removed

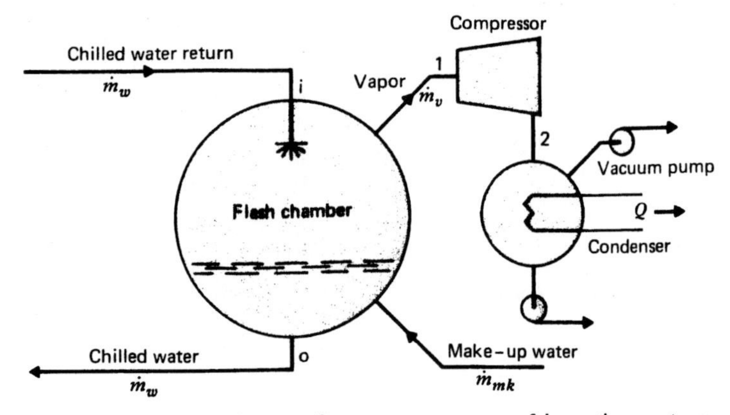

Figure 11.6 Flow diagram for a water vapor refrigeration system.

from the evaporator. Then,

$$\dot{m}_w(h_{w,i} - h_{w,o}) - \dot{m}_v(h_{v,1} - h_{mk}) = 0 \quad (11.1)$$

where

$\dot{m}_w$ = chilled water circulation rate
$\dot{m}_v$ = vapor flow rate
$h_{w,i}$ = enthalpy of the chilled water entering the evaporator
$h_{w,o}$ = enthalpy of the chilled water leaving the evaporator
$h_{v,1}$ = enthalpy of the vapor leaving the evaporator
h_{mk} = enthalpy of the make-up water

The vapor discharged from the flash chamber carries over to the booster pump entrained liquid droplets. The liquid content of the vapor is normally less than 5 percent by mass.

The refrigeration capability of the system is determined from

$$\dot{Q}_s = \dot{m}_w(h_{w,i} - h_{w,o}) \quad (11.2)$$

A very large volume of vapor is produced at the low pressure in the flash chamber, hence high-capability centrifugal or steam jet pumps are required for removal of the vapor. The pump discharge pressure must be of sufficient magnitude to establish a vapor saturation temperature in excess of the temperature of the condenser cooling water. Incidentally, if the booster pump were required to compress the vapor to the atmospheric pressure, the energy expended in pumping would be much greater than the energy consumed in pumping the vapor to the condenser pressure.

Example 11.4

A water vapor refrigeration system supplies chilled water at the rate of 680 kg/min. The water leaves the evaporator at 4 C and returns at 13 C. The make-up water is supplied at 17 C. The quality of the vapor leaving the evaporator is 0.97. The isentropic efficiency of the centrifugal compressor is 0.70. The compressor mechanical efficiency and the motor efficiency are, respectively, 0.98 and 0.95. The condenser pressure is 5.0 kPa, abs, achieved with a water outlet temperature of 30 C. Determine the refrigeration system performance.

Enthalpy values for the water are

$$h_{w,i} = 54.60 \text{ kJ/kg}$$
$$h_{w,o} = 16.78$$
$$h_{mk} = 71.38$$
$$h_{v,1} = 2433.3 \text{ kJ/kg at 4 C},$$
$$x = 0.97 \qquad (P_{sat} = 0.820 \text{ kPa, abs})$$

1 Refrigeration capability.

$$\dot{Q}_s = \dot{m}_w(h_{w,i} - h_{w,o})$$
$$= 680(54.60 - 16.78) = 25{,}718 \text{ kJ/min}$$

2 Vapor discharged from the evaporator.

$$\dot{m}_w(h_{w,i} - h_{w,o}) - \dot{m}_v(h_{v,1} - h_{mk}) = 0$$
$$25718 - \dot{m}_v(2433.3 - 71.4) = 0$$
$$\dot{m}_v = 10.89 \text{ kg/min}$$

3 Compressor power.

The vapor is compressed from 4 C, $x = 0.97$, at the evaporator outlet, to $P = 5.0$ kPa, abs, the pressure maintained in the condenser.

For reversible adiabatic compression: $h_{v,2} = 2694.1$ kJ/kg

$$\dot{W}_s = \dot{m}_v(h_{v,1} - h_{v,2})$$
$$= 10.89(2433.3 - 2694.1) = -2840 \text{ kJ/min}$$

Motor input:

$$\dot{W}_e = \frac{\dot{W}_s}{\eta_C \eta_m \eta_e}$$
$$= \frac{-2840}{0.70 \times 0.98 \times 0.95} = -4358 \text{ kJ/min}$$

4 Coefficient of performance.

The power required to operate the condensate pump and the mechanical vacuum pumps is taken as 6 percent of the compressor power. Then,

$$\dot{W}_P = 0.06 \times \dot{W}_e$$
$$= 0.06(-4358) = -261 \text{ kJ/min}$$

$$\dot{W}_{total} = \dot{W}_e + \dot{W}_P$$
$$= -4358 - 261 = -4619 \text{ kJ/min}$$
$$77.0 \text{ kW}$$

$$\text{cop} = \frac{\dot{Q}_s}{\dot{W}_{total}} = \frac{25718}{4619} = 5.57$$

The estimated coefficient of performance for a R-12 vapor compression refrigeration system that operates between the same evaporating and condensing temperatures as the water vapor system, 4 and 32.9 C, respectively, is approximately 5.7. Calculations were based on compression in a reciprocating machine.

Installation of a water vapor refrigeration system would be particularly advantageous where steam, available for pumping, can be provided at a cost below that of the equivalent electrical energy. Vapor and gas pumping would be accomplished by steam jet ejector pumps. Water vapor refrigeration cannot be employed for low-temperature applications, because the minimum evaporating temperature must be a few degrees above the freezing temperature of water, thus a lower limit of about 2 C is established.

11.6 THE ABSORPTION REFRIGERATION SYSTEM

The power required for operation of the vapor compression refrigeration system is a significantly large quantity, particularly for high-capability or low-temperature systems. If the working fluid could be compressed in the liquid phase, the work input would be a small fraction of the energy required for vapor compression between prescribed pressure limits. Liquid compression can be accomplished by "absorbing" the refrigerant in a secondary fluid, or "carrier" liquid, and then pumping the mixture from the low-pressure to the high-pressure section of the system. A satisfactory combination of this kind can be achieved with ammonia and water. At some point in the high-pressure section of the system, the ammonia is expelled from the solution.

Figure 11.7 shows the flow diagram for a commercial-type absorption refrigeration system. Ammonia vapor flows from the evaporator to the absorber where contact is made with the weak liquor returning from the generator. In the absorber, the ammonia vapor is readily absorbed by the low-temperature aqua. Because this reaction is accompanied by liberation of energy, the absorber is water cooled for most effective operation, which occurs at a comparatively low

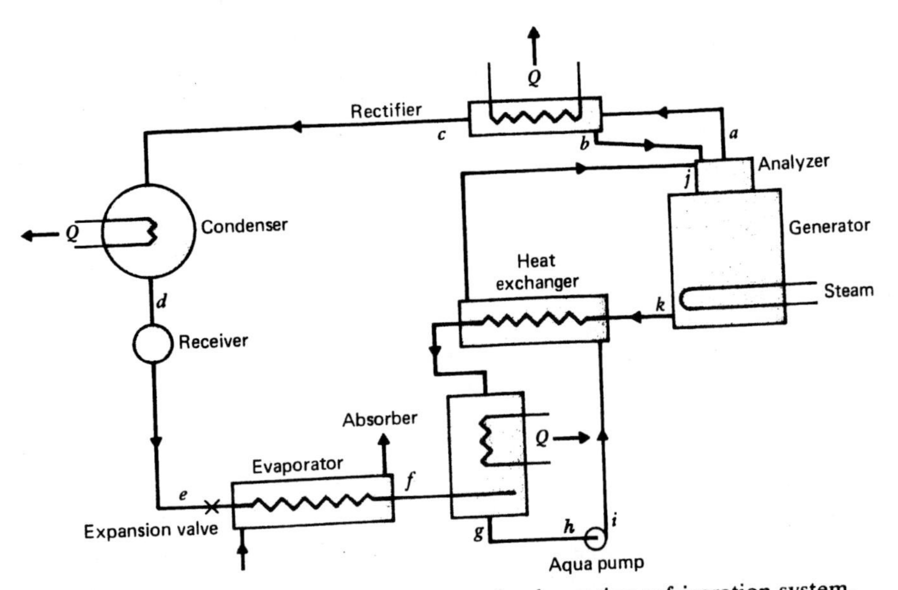

Figure 11.7 Flow diagram for an ammonia absorption refrigeration system.

temperature, about 25 C. The resulting strong liquor is pumped from the absorber to the generator where a temperature of 100 to 150 C is maintained by steam heating. At an elevated temperature, within this range, the solubility of ammonia in water is low. The heat, added in the generator, is required for separation of some of the ammonia from the solution. The vapor discharged from the generator is a mixture of ammonia vapor and water vapor. The remaining components of the system, namely, the condenser, expansion valve, and evaporator, are similar to the corresponding pieces of equipment installed in the vapor compression system.

The vapor that enters the condenser should be virtually anhydrous ammonia, hence almost all of the water vapor must be removed from the mixture discharged from the generator. The analyzer, usually incorporated into the generator, partially fulfills this function. The hot vapor mixture from the generator rises through the falling spray or sheets of the cooler strong liquor returning, through the heat exchanger, from the absorber. The interchange of energy in the analyzer, resulting from direct contact between the two fluid streams, causes a substantial portion of the water vapor to condense from the outflowing mixture. In addition, as the strong liquor is heated, there is a tendency to distill off some of the ammonia that subsequently flows from the element.

After leaving the analyzer, the vapor mixture enters the rectifier, a water-cooled heat exchanger, where nearly all of the water vapor is condensed. The drip from the rectifier is customarily piped back to the analyzer. The combined effect of the analyzer and the rectifier is to deliver essentially anhydrous ammonia to the condenser.

The ammonia/water mass ratio changes at several points in the system. The concentration of the aqua ammonia solution defines the quantity of ammonia, as a mass fraction, that is present in the solution.

A heat exchanger is placed between the absorber and the generator as shown in Fig. 11.7. This element provides for an interchange of energy between the hot weak liquor and the cool strong liquor. Improved operating economy is thus achieved by reducing both the steam supplied to the generator and the cooling water supplied to the absorber.

Because ammonia is strongly absorbed by water, low-pressure operation of the absorber can readily be maintained. It is thus possible to establish wide-range operation of the evaporator with respect to pressure and temperature. Control of the load on the ammonia absorption refrigeration system is effected by regulating the heat supplied to the generator.

The power requirements for the absorption refrigeration system are comparatively low, but a substantial quantity of thermal energy is supplied in the form of steam. Because of the influence of local factors, it is difficult to compare generally the overall operating costs for the vapor compression and the absorption refrigeration systems. Generally, the absorption refrigeration system shows superior economy when a supply of relatively cheap steam is available. For certain installations however, it may be economically feasible to use a gas-fired generator for operation of an absorption refrigeration system.

In an absorption refrigeration system, a decrease in the evaporator pressure effects only minor changes in the pumping power and the refrigerant flow rate. Thus, unlike in the vapor compression system, the coefficient of performance and the refrigeration capability of the absorption system remain relatively constant with changes in the evaporator temperature. These are important characteristics with respect to achieving low-temperature refrigeration.

An increase can be expected in the utilization of solar energy for cooling buildings. In this respect, the absorption refrigeration system has reasonably good potentialities, because the system can operate effectively with a low-level energy source.

Example 11.5

The prescribed operating conditions for an ammonia absorption refrigeration system are listed below. (The letters in parentheses refer to Fig. 11.7.)

	Pressure (kPa, abs)	*Temperature (C)*
Condenser	1170	
Evaporator	210	
Weak aqua leaving the generator (k)	1200	105
Strong aqua entering the analyzer (j)	1200	94
Vapor leaving the analyzer (a)	1200	99
Drip leaving the rectifier (b)	1190	49
Vapor leaving the rectifier (c)	1190	49
Aqua leaving the condenser (d)	1170	25
Ammonia vapor leaving the evaporator (f)	210	−12
Strong aqua leaving the absorber (g)	200	27
Strong aqua at the pump inlet (h)	175	27
Strong aqua at the pump outlet (i)	1225	

Steam is supplied to the generator at 150 kPa, abs, $x = 1.0$.

Aqua pump:

Internal efficiency = 0.65

Combined motor efficiency and pump mechanical efficiency = 0.93

Refrigeration rate: 22,500 kJ/min

The performance of the system will be evaluated by analyzing the several components that comprise the system. See Fig. 11.7.

An equilibrium concentration exists in the fluid at a number of points in the system. The concentration x and the enthalpy of the fluid, selected from Chart A.1 in the Appendix, are tabulated below for each of these points.

	x	*h (kJ/kg)*
Weak aqua leaving the generator (k)	0.323	277
Strong aqua entering the analyzer (j)	0.373	205
Vapor leaving the analyzer (a)	0.943	1539
Vapor leaving the rectifier (c)	0.9975	1346
Aqua (drip) leaving the rectifier (b)	0.665	19

The strong aqua leaves the absorber at 27 C, $x = 0.373$ (point g). At these conditions, $h = -128$ kJ/kg.

1 Evaporator.

For the evaporator, values of the enthalpy for anhydrous ammonia are selected from the ammonia tables.

Entering: liquid ammonia at 25 C, $h_e = 298.8$ kJ/kg
Leaving: ammonia vapor at 210 kPa, abs, and − 12 C

$$h_f = 1436.2 \text{ kJ/kg}$$

Ammonia circulation rate:

$$\dot{m} = \frac{\dot{Q}_s}{h_f - h_e}$$

$$= \frac{22500}{1436.2 - 298.8} = 19.782 \text{ kg/min}$$

2 Absorber.

The control volume includes the absorber, heat exchanger, and aqua pump. (Neglect the small amount of water in the ammonia leaving the evaporator.)

Entering the control volume:

$$\dot{m}_f = 19.782 \text{ kg/min (ammonia)}$$

$$\dot{m}_k \text{ (weak aqua)} \quad x_k = 0.323$$

Leaving the control volume:

$$\dot{m}_j \text{ (strong aqua)} \quad x_j = 0.373$$

Ammonia mass balance:

$$\dot{m}_f + x_k \dot{m}_k = x_j \dot{m}_j$$

$$19.782 + 0.323\dot{m}_k = 0.373\dot{m}_j$$

Total mass balance:

$$\dot{m}_f + \dot{m}_k = \dot{m}_j$$

$$19.782 + \dot{m}_k = \dot{m}_j$$

Then,

$$\dot{m}_k = 248.07 \text{ kg/min} \quad \text{(weak aqua)}$$

$$\dot{m}_j = 267.85 \text{ kg/min} \quad \text{(strong aqua)}$$

The specific volume of the aqua ammonia is not precisely equal to the sum of the volumes of the component fluids. At the conditions ordinarily observed in the absorption refrigeration system, about 16 percent of the ammonia liquid in the solution is absorbed within the volume occupied by the water. Thus,

$$v_{sol} = (1 - x)v_w + 0.84xv_a$$

where

v_{sol} = specific volume of the solution
v_w = specific volume of the water
v_a = specific volume of the liquid ammonia

3 Aqua pump power.

At 27 C $v_w = 0.001003 \text{ m}^3\text{/kg}$

$$v_a = 0.001667 \text{ m}^3\text{/kg}$$

$$v_{sol} = (1 - x_h)v_w + 0.84x_h v_a$$
$$= (1 - 0.373)0.001003 + (0.84 \times 0.373 \times 0.001667)$$
$$= 0.001151 \text{ m}^3\text{/kg}$$

$$\dot{W}_P = \frac{\dot{V}_{sol}(P_h - P_i)}{\eta_P} = \frac{\dot{m}_h v_h (P_h - P_i)}{\eta_P}$$

$$= \frac{267.85 \times 0.001151(175 - 1225)}{0.65}$$

$$= -498.0 \text{ kJ/min}$$

4 Absorber heat transfer.

The control volume is defined in **2**.

$$h_f = 1260 \text{ kJ/kg at 210 kPa, abs, and } -12 \text{ C}$$

$$\dot{m}_f h_f + \dot{m}_k h_k + \dot{Q} - \dot{m}_j h_j - \dot{W} = 0$$

$$(19.782 \times 1260) + (248.07 \times 277) + \dot{Q} - (267.85 \times 205) - (-498.0) = 0$$

$$\dot{Q} = -39{,}299 \text{ kJ/min}$$

5 Rectifier heat transfer.

Ammonia flow rate at point c $= 19.782$ kg/min

Total flow rate at point c $= \dfrac{19.782}{0.9975} = 19.832$ kg/min

Total mass balance:

$$\dot{m}_a - \dot{m}_b - \dot{m}_c = 0$$

$$\dot{m}_a - \dot{m}_b - 19.832 = 0$$

Ammonia mass balance:

$$x_a\dot{m}_a - x_b\dot{m}_b - x_c\dot{m}_c = 0$$

$$0.943\dot{m}_a - 0.665\dot{m}_b - 19.782 = 0$$

Then,

$$\dot{m}_a = 23.718 \text{ kg/min}$$

$$\dot{m}_b = 3.886 \text{ kg/min}$$

The energy balance is represented by

$$\dot{m}_a h_a + \dot{Q} - \dot{m}_b h_b - \dot{m}_c h_c = 0$$

$$(23.718 \times 1539) + \dot{Q} - (3.886 \times 19) - (19.832 \times 1346) = 0$$

$$\dot{Q} = -9734 \text{ kJ/min}$$

6 Condenser heat transfer.
From the ammonia table:

$$h_c = 1524.2 \text{ kJ/kg}$$

$$h_d = 298.8 \text{ kJ/kg}$$

Neglect the water carried by the ammonia to the condenser. Thus,

$$\dot{m} = 19.782 \text{kg/min}$$

$$\dot{Q} = \dot{m}(h_d - h_c) = 19.782(298.8 - 1524.2) = -24{,}241 \text{ kJ/min}$$

7 Generator and analyzer heat transfer.

$$\dot{m}_b h_b + \dot{m}_j h_j + \dot{Q} - \dot{m}_a h_a - \dot{m}_k h_k = 0$$

$$(3.886 \times 19) + (267.85 \times 205) + \dot{Q} - (23.718 \times 1539) - (248.07 \times 277) = 0$$

$$\dot{Q} = 50{,}234 \text{ kJ/min}$$

8 System energy balance.

	Gain (kJ/min)	*Loss (kJ/min)*
Generator	50,234	
Rectifier		9,734
Condenser		24,241
Evaporator	22,500	
Absorber		39,229
Aqua pump	498	
	73,232	73,204

9 System performance.
Aqua pump:

$$\text{Fluid power } \dot{W}_P = -498 \text{ kJ/min}$$

$$\text{Motor input } \dot{W}_e = \frac{\dot{W}_P}{\eta_m \eta_e} = \frac{-498}{0.93} = -535 \text{ kJ/min} \quad (8.92 \text{ kW})$$

Steam consumption:

$$Q_{\text{steam}} = h_{fg} \text{ at 150 kPa, abs} = 2226.5 \text{ kJ/kg}$$

$$\dot{m}_{\text{steam}} = \frac{\dot{Q}_{\text{gen}}}{Q_{\text{steam}}} = \frac{50234}{2226.5} = 22.56 \text{ kg/min} \quad 1354 \text{ kg/h}$$

Coefficient of performance:

$$\text{cop} = \frac{\dot{Q}_{\text{evap}}}{\dot{Q}_{\text{gen}} + \dot{W}_e} = \frac{22500}{50234 + 535} = 0.443$$

The coefficient of performance for the absorption refrigeration system is inherently low, typically about 0.5. The higher coefficient of performance ordinarily achieved in a vapor compression refrigeration system is offset usually by the higher cost of the electrical energy required for operation of the compressor.

Adequate cooling of the absorber is essential to achieving a low evaporator temperature in the absorption refrigeration system. With a cooling water temperature of about 13 C, an evaporator temperature of −60 C can be readily achieved in a single-stage ammonia absorption system. In a two-stage absorption system, one stage is used to cool the absorber of the low-temperature stage.

Low-capability absorption refrigeration systems, for example, domestic-type refrigerators, are operated solely through the supply of thermal energy. The pump is omitted, and a third fluid, hydrogen, is added to the system and confined to the low-pressure section, that is, the evaporator and absorber. Although the total pressure is uniform throughout the system, the partial pressures of the refrigerant are maintained at the customary "high" and "low" levels. The several components of the system are arranged for thermal circulation of the refrigerant and for gravity flow of the liquid from the condenser to the evaporator.

An absorption refrigeration system designed to produce chilled water is commonly employed in air conditioning applications where a minimum refrigerant temperature of about 2 C is acceptable. In this system, a lithium bromide solution is the absorbent and water is the refrigerant. The chilled water flows from the evaporator to the external load and then back to the evaporator. Thermal energy for operating the system is usually supplied by low-pressure steam or hot water.

Figure 11.8 is a cross-sectional view of a commercial absorption refrigeration system designed to produce chilled water. The refrigerant (water) is sprayed over the surface of the evaporator tubes, and subsequent evaporation of the refrigerant removes heat from the chilled water. The refrigerant, in the vapor phase, is absorbed by a spray of lithium bromide solution in the water-cooled absorber. The dilute solution collects in a sump and is pumped to the steam-heated concentrator where the refrigerant is separated from the solution. Subsequently, the concentrated solution is returned to the absorber and the refrigerant vapor flows to the condenser. The condensed refrigerant flashes through an orifice into the evaporator where the cycle is repeated. System pressures are low, namely, $\frac{1}{10}$ atmosphere in the concentrator and condenser, and $\frac{1}{100}$ atmosphere in the evaporator and absorber.

11.7 AIR CONDITIONING

The requirements for air conditioning are complete control throughout the year of the dry-bulb temperature, relative humidity, and distribution of the air. Other requirements are often stipulated, such as control of dust, pollen, or contaminating gases and vapors. Winter heating and humidification, or summer cooling, rather than all-year air conditioning may be entirely adequate for a particular installation.

Air may be conditioned for human comfort without regard to the economy, for example, a residential application. The economy however cannot be ignored in the consideration of commercial and industrial installations, since improved comfort contributes to increased business activity and individual productivity. Many industrial operations are performed more effectively when the humidity and temperature of the air are precisely controlled.

Figure 11.9 represents schematically an air conditioning unit. During the heating season, the cold outside air passes through a preheater, mixes with the recirculated air, and flows through the filter for removal of dust. Operation of the preheater is ordinarily discontinued when outdoor temperatures are in a moderate range. The temperature of the water supplied to the air washer is adjusted to humidify or dehumidify the mixture of outside and recirculated air, depending upon the vapor content of the recirculated air. Following passage through the washer, the air is heated in the main heat exchanger to the prescribed outlet temperature and then conveyed to the various parts of the building.

Heating and humidification of the air cause the quantity and the properties of the vapor to change as shown in Fig. 11.10a. Vapor preheating is represented by the path 1–2. The second process, 2–3, occurs in the spray chamber and the main heater. For a constant total pressure of the air–water vapor mixture, an increase in the vapor pressure causes a decrease in the partial pressure of the dry air.

The ratio of the outside air to the recirculated air is dependent upon a number of factors, including the intended use of the building. For

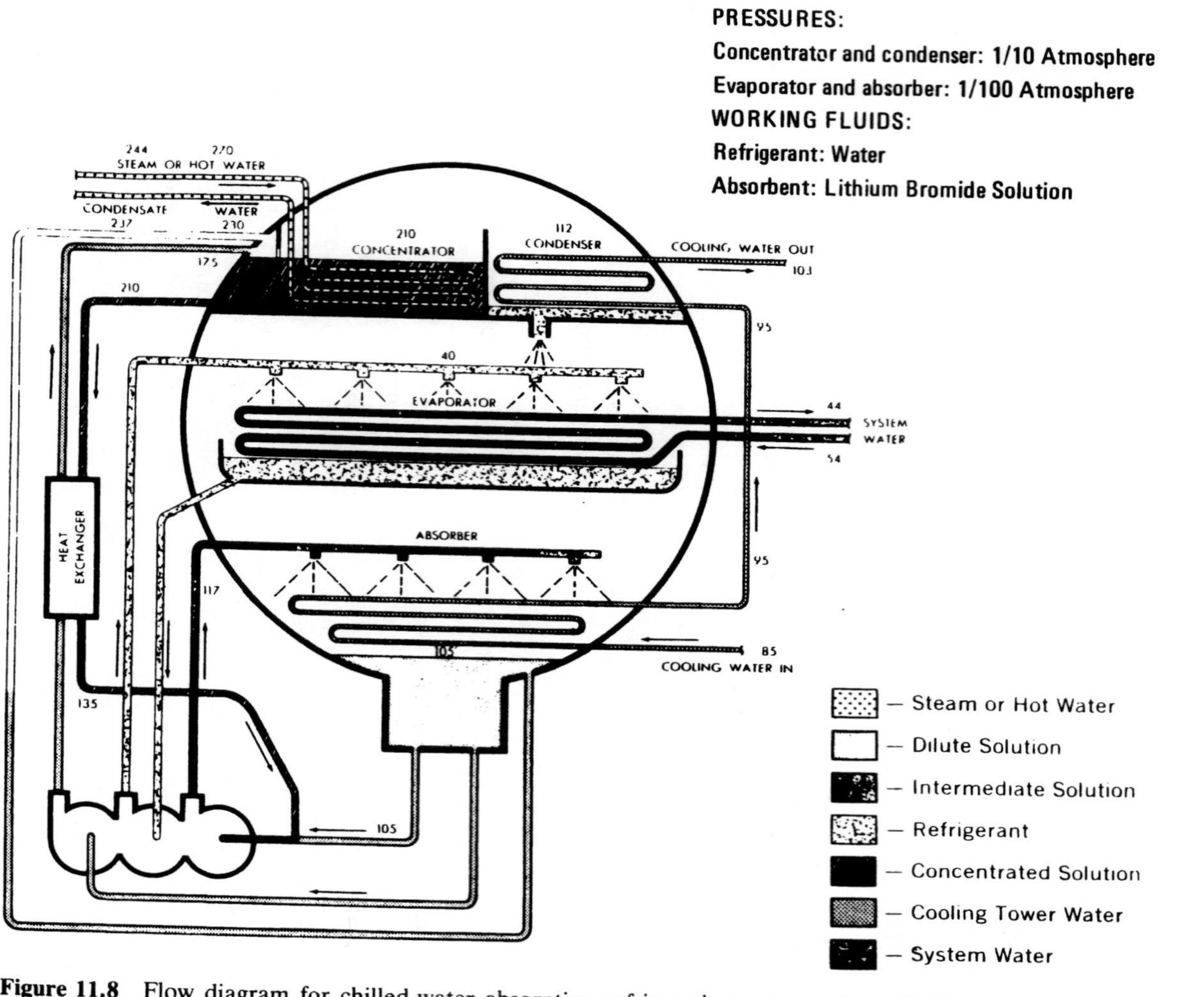

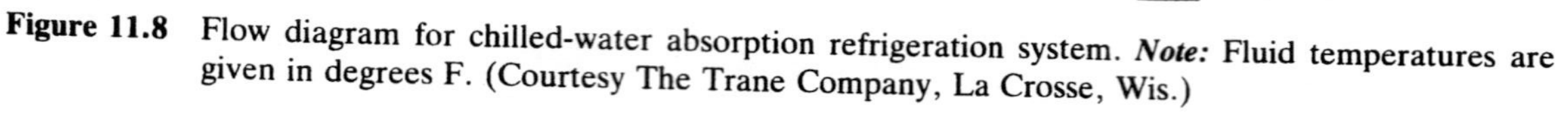

Figure 11.8 Flow diagram for chilled-water absorption refrigeration system. ***Note:*** Fluid temperatures are given in degrees F. (Courtesy The Trane Company, La Crosse, Wis.)

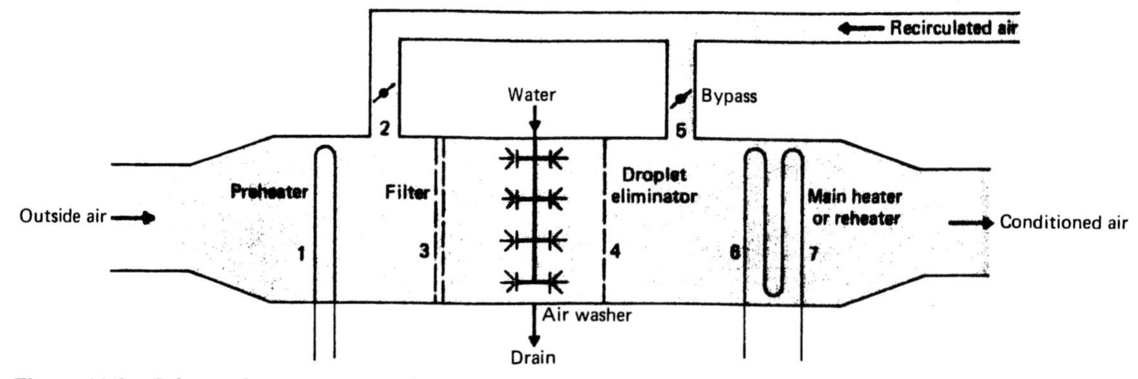

Figure 11.9 Schematic arrangement for an air conditioning system.

public and commercial buildings, it is customary to supply generally outside air at a minimum rate of about 0.3 to 0.4 m^3/min for each occupant. Higher rates are required for certain rooms because of above-normal air contamination. Air that contains obnoxious odors, toxic gases, or suspended material is normally not recirculated but exhausted directly from the building.

The quantities of heat and water supplied to a heating unit are determined by application of the first law to a designated control volume. See Fig. 11.11. Thus, for a single stream of fluid the energy equation is given by

$$\dot{Q} = \dot{m}_a[(h_{a,2} - h_{a,1}) + \omega_2 h_{v,2} - \omega_1 h_{v,1} - (\omega_2 - \omega_1)h_{f,3}] \quad (11.3)$$

where

h_a = enthalpy of the dry air
h_v = enthalpy of the water vapor
ω = specific humidity of the air–vapor mixture
$h_{f,3}$ = enthalpy of the liquid water supplied at temperature T_3
$\dot{m}_a$ = mass flow rate of the dry air

Example 11.6

Operating data for a humidifying heater are:

Outside air: 250 kg/min at −7 C, RH = 0.40

Recirculated air: 550 kg/min at 21 C, RH = 0.50

Outlet air: 49 C, RH = 0.10

Water inlet temperature: 27 C

Barometric pressure: 99.50 kPa, abs

Determine the rates at which energy and water are supplied to the heater.

1 Calculated values of the specific humidity.

	P_v (kPa, abs)	ω (kg vapor/kg dry air)*
Outside air (o)	0.1358	0.000850
Recirculated air (r)	1.2430	0.007869
Conditioned air (c)	1.1736	0.007424

Note: The specific humidity of the recirculated air is higher than the specific humidity of the conditioned air, because of the vapor gain contributed by the load.

* The specific humidity ω is determined from

$$\omega = 0.622 \frac{P_v}{P_a - P_v}$$

where $0.622 = R_a/R_w$, P_v is the vapor pressure, and P_a is the total pressure of the air–vapor mixture. The vapor pressure is determined from

$$P_v = \phi P_{v,\text{sat}}$$

where ϕ is the relative humidity and $P_{v,\text{sat}}$ is the saturated vapor pressure at the temperature of the air–vapor mixture.

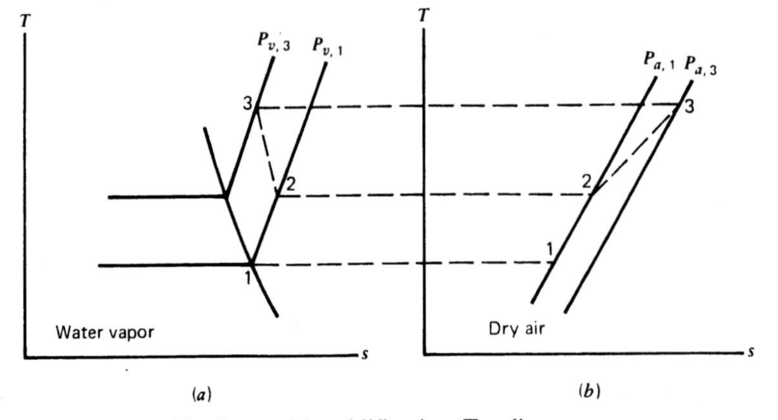

Figure 11.10 Heating and humidification T–s diagram.

2 Water consumption.

	$\dot{m}_a$ (kg/min)	$\dot{m}_v$ (kg/min)
Outside air (o)	249.788	0.2123
Recirculated air (r)	545.706	4.2942
Conditioned air (c)	795.494	5.9057

$$\Delta \dot{m}_v = (\dot{m}_v)_c - (\dot{m}_v)_o - (\dot{m}_v)_r$$
$$= 5.9057 - 0.2123 - 4.2942$$
$$= 1.3992 \text{ kg/min}$$

3 Energy input.
Enthalpy values are

h_v at -7 C = 2488.4 kJ/kg
h_v at 21 C = 2539.9
h_v at 49 C = 2590.3
h_f at 27 C = 113.2

The energy equation for the heater, based on Eq. 11.3, is

$$(\dot{m}_a)_o c_{p,a}(T_o - T_c) + (\dot{m}_a)_r c_{p,a}(T_r - T_c)$$
$$+ (\dot{m}_v h_v)_o + (\dot{m}_v h_v)_r + (\Delta \dot{m}_v h_f)_w + \dot{Q}$$
$$- (\dot{m}_v h_v)_c = 0$$

$$249.79 \times 1.004(-7 - 49) + 545.71$$
$$\times 1.004(21 - 49) + (0.2123 \times 2488.4)$$
$$+ (4.2942 \times 2539.9) + (1.3992 \times 113.2) + \dot{Q}$$
$$- (5.9057 \times 2590.3) = 0$$

$$\dot{Q} = 33{,}089 \text{ kJ/min}$$

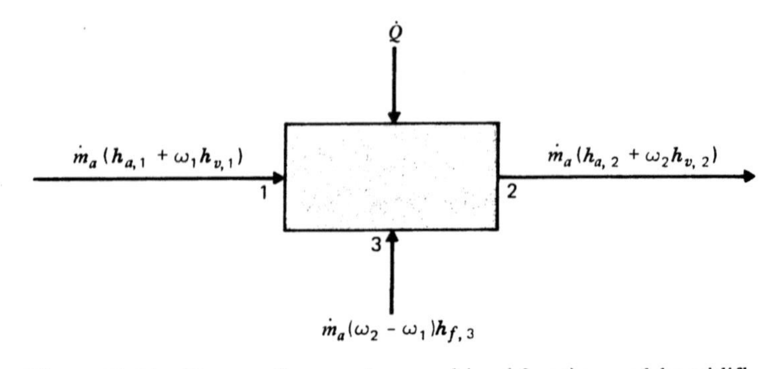

Figure 11.11 Energy diagram for combined heating and humidification.

During the warm periods of the year, cooling and dehumidification are the principal requirements for air conditioning. In order to maintain a comfortable indoor temperature, the energy input to the building must be continuously removed, together with the heat liberated by the occupants, lighting fixtures, and various pieces of equipment. It is also necessary to remove the water vapor discharged into the air by the occupants and by certain equipment, such as steam kettles. A high discharge of water vapor will impose a relatively heavy dehumidifiction load on the system.

In a typical cooling and dehumidification operation, the hot, humid outside air mixes with the recirculated air, and the combined fluid steam passes through the spray of refrigerated water. See Fig. 11.9. An adequate quantity of refrigerated water must be supplied for cooling the air–vapor mixture and condensing a portion of the vapor. The temperature of the spray water will control the dehumidifier outlet temperature, and the air leaving the dehumidifier is not entirely saturated. The required dehumidification often delivers air at a temperature too low for comfort when distributed to the room. Thus, after passing through the droplet eliminator, the dehumidified air is reheated in a steam-heated or gas-fired heat exchanger. See Fig. 11.12.

Bypassing a portion of the recirculated air around the dehumidifying washer will ordinarily improve the operating economy of the air conditioning system. The bypass arrangement is shown in Fig. 11.9.

Cooling and dehumidification of the outside air entering the system may impose a comparatively heavy load on the dehumidifier. Consequently, because of economic considerations, a minimum quantity of outside air, consistent with the fresh air requirement, is introduced into the air conditioning system.

Dehumidification and cooling may be achieved by passing the air through a finned-tube-type evaporator. Because only a portion of the air–vapor mixture is in direct contact with the cold surface of the evaporator, the entire air mass will not leave the dehumidifier in a saturated state.

The heat removed from the air–vapor mixture in the dehumidifier is determined by application of the first law. See Fig. 11.13. The energy equation for a single fluid stream is given by

$$\dot{Q} = \dot{m}_a[(h_{a,2} - h_{a,1}) + \omega_2 h_{v,2} - \omega_1 h_{v,1} + (\omega_1 - \omega_2)h_{f,2}] \quad (11.4)$$

Example 11.7

An auditorium seating 2500 persons is to be conditioned to 22.2 C, RH = 0.60, that is, the air enters the auditorium at this temperature and relative humidity. The air requirements are 0.85 m^3/min per occupant, based on the auditorium

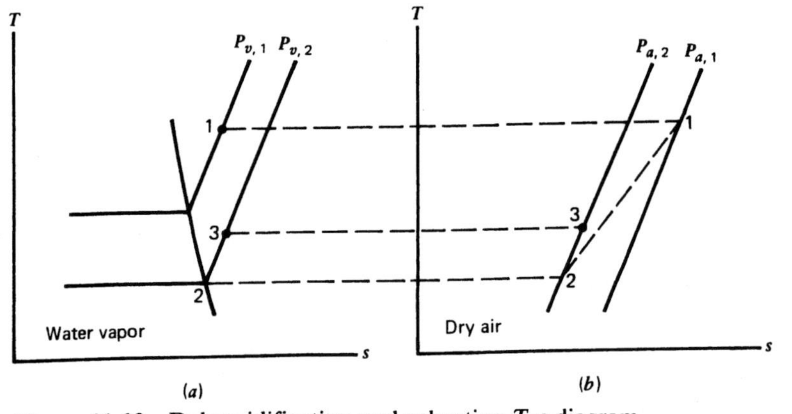

Figure 11.12 Dehumidification and reheating T–s diagram.

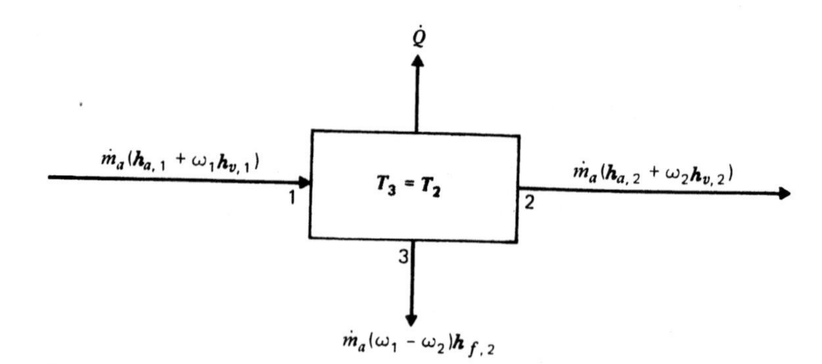

Figure 11.13 Energy diagram for cooling and dehumidification.

inlet conditions. One-third of this air is outside air supplied to the air conditioning system. The recirculated air leaves the auditorium at 27 C, RH = 0.49. Thirty-five percent of the recirculated air is directed through the bypass duct and mixes with the air leaving the washer. The ambient air conditions are 100.0 kPa, abs, 35 C, and RH = 0.55. Determine the refrigeration and heating requirements for the air conditioning unit. The system arrangement is illustrated by Fig. 11.9.

1 Specific humidity.

	P_v *(kPa, abs)*	ω *(kg vapor/kg dry air)*
Outside air	3.092	0.01985
Conditioned air	1.606	0.01015
Recirculated air	1.746	0.01105

2 Dry air flow.

$$\dot{V} = 2500 \times 0.85$$
$$= 2125 \text{ m}^3/\text{min} \quad \text{(air–vapor mixture)}$$
$$P_v = 1.606 \text{ kPa, abs}$$
$$P_a = P_b - P_v$$
$$= 100.0 - 1.606 = 98.394 \text{ kPa, abs}$$

$$\dot{m}_a = \frac{P_a\dot{V}}{R_aT}$$
$$= \frac{98.394 \times 2125}{0.287 \times 295.4} = 2466.2 \text{ kg/min}$$

The dry air flow is subdivided into three streams:

Outside air: 822.1 kg/min

Recirculated air:
- To bypass 575.4 kg/min
- To washer 1068.7 kg/min

3 Dehumidifier.

The specific humidity of the air leaving the washer is determined from a vapor balance. Thus,

$$\dot{m}_{a,4}\omega_4 + \dot{m}_{a,5}\omega_5 = \dot{m}_{a,6}\omega_6$$
$$1890.8\omega_4 + (575.4 \times 0.01105) = 2466.2 \times 0.01015$$
$$\omega_4 = 0.00988 \text{ kg vapor/kg dry air}$$

(assume a saturated mixture)

and

$P_v = 1.5630$ kPa, abs $\quad T_{dp} = 13.7$ C

$h_g = 2526.5$ kJ/kg $\quad h_f = 57.4$ kJ/kg

4 Heat removed in the dehumidifier from the outside air.

A modified form of Eq. 11.4 is used to determine $\dot{Q}$.

$$\dot{Q} = \dot{m}_a[c_{p,a}(T_4 - T_1) + \omega_4 h_{v,4} - \omega_1 h_{v,1} + (\omega_1 - \omega_4)h_{f,4}]$$

$T_1 = 35$ C $\quad T_4 = 13.7$ C

$\omega_1 = 0.01985 \quad \omega_4 = 0.00988$ kg vapor/kg dry air

$h_{v,1} = 2565.3 \quad h_{v,4} = 2526.5$

$\dot{m}_a = 822.1$ kg/min $\quad h_{f,4} = 57.4$ kJ/kg

Substitution of these values into the above equation yields

$$\dot{Q} = -38{,}452 \text{ kJ/min}$$

5 Heat removed in the dehumidifier from the recirculated air.

$$\dot{Q} = \dot{m}_a[c_{p,a}(T_4 - T_2) + \omega_4 h_{v,4} - \omega_2 h_{v,2} + (\omega_2 - \omega_4)h_{f,4}]$$

$T_2 = 27$ C $\quad \omega_2 = 0.01105$ kg vapor/kg dry air

$h_{v,2} = 2550.8$ kJ/kg $\quad \dot{m}_a = 1068.7$ kg/min

Substitution of the appropriate values into the above equation yields

$$\dot{Q} = -17{,}645 \text{ kJ/min}$$

6 Required refrigeration.

$$\dot{Q}_{\text{total}} = -(38452 + 17645) = -56{,}097 \text{ kJ/min}$$

7 Reheater energy transfer.

The air leaving the dehumidifier is heated to a temperature of 22.2 C. The energy equation is given by

$$\dot{Q} = \dot{m}_a[c_{p,a}(T_7 - T_4) + \omega_4(h_{v,7} - h_{v,4})]$$

$T_4 = 13.7$ C $\quad T_7 = 22.2$ C

$h_{v,4} = 2526.5 \quad h_{v,7} = 2542.1$ kJ/kg

$\omega_4 = 0.00988$ kg vapor/kg dry air

$\dot{m}_a = 1890.8$ kg/min

These values are substituted into the above equation. Thus,

$$\dot{Q} = 16{,}428 \text{ kJ/min}$$

On passing through the reheater, there is, in effect, a decrease in the energy of the bypass air.

$$\dot{Q} = \dot{m}_a[c_{p,a}(T_7 - T_5) + \omega_5(h_{v,7} - h_{v,5})]$$

$T_5 = 27$ C $\quad \omega_5 = 0.01105$ kg vapor/kg dry air

$h_{v,5} = 2550.8$ kJ/kg $\quad \dot{m}_a = 575.4$ kg/min

The appropriate values are substituted into the above equation. Thus,

$$\dot{Q} = -2828 \text{ kJ/min}$$

The net transfer of energy in the reheater is

$$\dot{Q}_{\text{net}} = 16428 - 2828 = 13{,}600 \text{ kJ/min}$$

If all of the recirculated air is directed to the dehumidifying washer, the energy requirements are

Dehumidifier: −62,925 kJ/min

Reheater: 20,427 kJ/min

A comparison of the above energy quantities with the corresponding energy quantities determined in Example 11.7 discloses the advantage of bypassing a portion of the recirculated air around the dehumidifier. The refrigeration and the reheating requirements are both reduced. The system operating conditions may allow sufficient bypassing of the recirculated air to eliminate the need for air reheating.

11.8 DEHUMIDIFICATION

The dehumidifying washer is ordinarily appropriate for a comparatively large air conditioning system. The air leaving the dehumidifier will usually not be completely saturated. If two rows of spray nozzles are installed in the washer, so as to create a curtain of water that extends completely across the flow channel, the quantity of air that bypasses the spray will be small, somewhat less than 5 percent. The water spray will remove some particulate matter from the air, but removal of dust by air washing is by and large not effective.

Dehumidification is also achieved by passing the air over refrigerated coils. Finned tubes are used, and for effective operation of the dehumidifier, several rows of tubes are installed. In passing through the dehumidifier, some of the air does not come in contact with the cold surface of the tubes, hence the air leaving the dehumidifier is not completely saturated. For a moderate velocity, 100 to 150 m/min, at the dehumidifier inlet section, and a deep tube bank, say, six to eight rows of tubes, the quantity of bypass air will be small, perhaps about 2 percent.

Figure 11.14 shows a dehumidification process in which the discharged air is nonsaturated, state 2. The air that enters the dehumidifier is at state 1, and the air that comes in contact with the cold surface, or the cold water spray, is at state c. State 2 is determined by adiabatic mixing of the bypass air at state 1, with the dehumidified air at state c. The line 1–2 however does not represent the actual path followed by the air. State 2′ represents the final condition of the air for the corresponding theoretical case, in which saturated air is discharged from the dehumidifier. The change in state from 2 to 3 is achieved by reheating the air.

Example 11.8

Air at 35 C and 60 percent relative humidity enters a dehumidifier in which the surface temperature of the coils is maintained at 10 C. The atmospheric pressure is 101.0 kPa, abs. Ten percent of the entering air bypasses the coils. Determine the state of the air discharged from the dehumidifier and the heat transfer in the dehumidifier. (Ref.: Fig. 11.14.)

At State 1

$$P_{v,\text{sat}} \text{ at } 35\text{ C} = 5.628 \text{ kPa, abs}$$

$$P_v = \phi P_{v,\text{sat}} = 0.60 \times 5.628 = 3.377 \text{ kPa, abs}$$

$$\omega_1 = 0.622 \frac{P_v}{P_a - P_v} = 0.622 \frac{3.377}{101.0 - 3.377} = 0.02152 \text{ kg/kg dry air}$$

$$h_{v,1} = 2565.3 \text{ kJ/kg}$$

At State c

$$P_v = P_{v,\text{sat}} = 1.2276 \text{ kPa, abs}$$

$$h_{v,c} = 2519.8 \text{ kJ/kg}$$

$$\omega_c = 0.622 \frac{1.2276}{101.0 - 1.2276} = 0.007653 \text{ kg/kg dry air}$$

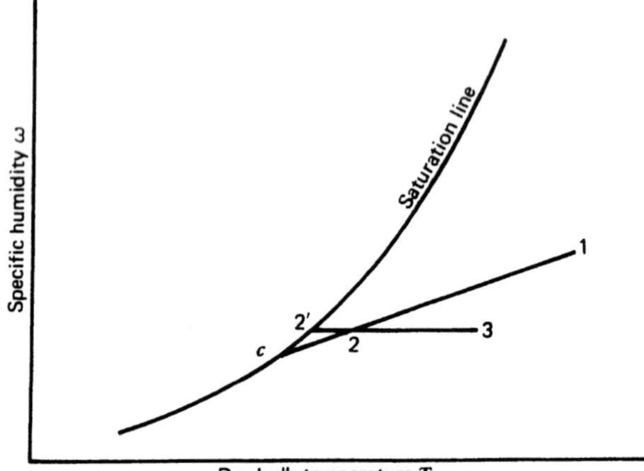

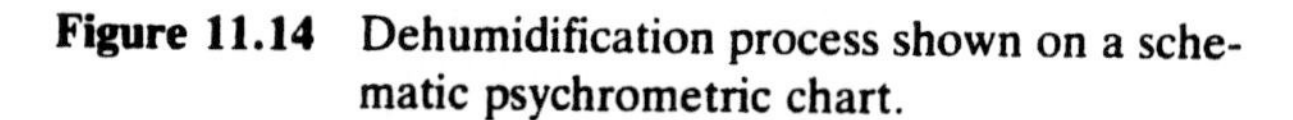

Figure 11.14 Dehumidification process shown on a schematic psychrometric chart.

At State 2

$$\omega_2 = 0.1\omega_1 + 0.9\omega_c$$
$$= (0.1 \times 0.02152) + (0.9 \times 0.007653)$$
$$= 0.009040 \text{ kg/kg dry air}$$

The energy equation for adiabatic mixing is

$$(h_{a,2} + \omega_2 h_{v,2}) = m_1(h_{a,1} + \omega_1 h_{v,1}) + m_c(h_{a,c} + \omega_c h_{v,c})$$

$$m_1 c_{p,a}(T_2 - T_1) + m_c c_{p,a}(T_2 - T_c) + \omega_2 h_{v,2} - m_1\omega_1 h_{v,1} - m_c\omega_c h_{v,c} = 0$$

$$0.1 \times 1.004(T_2 - 35) + 0.9 \times 1.004(T_2 - 10) + (0.009040 h_{v,2}) - (0.1 \times 0.02152 \times 2565.3) - (0.9 \times 0.007653 \times 2519.8) = 0$$

T_2 is determined by a trial-and-error solution.

Substitute $T_2 = 12.5$ C

and

$$h_{v,2} = 2524.4 \text{ kJ/kg}$$

Then,

$$R = -0.0556$$

Substitute $T_2 = 12.6$ C

and

$$h_{v,2} = 2524.5 \text{ kJ/kg}$$

Then,

$$R = 0.0457$$

Select $T_2 = 12.6$ C

Relative Humidity at State 2

$$\omega_2 = 0.622 \frac{P_v}{P_a - P_v}$$

$$0.009040 = 0.622 \frac{P_{v,2}}{101.0 - P_{v,2}}$$

$$P_{v,2} = 1.4469 \text{ kPa, abs}$$

$$P_{v,\text{sat}} \text{ at } 12.6 \text{ C} = 1.4591 \text{ kPa, abs}$$

$$\phi = \frac{P_{v,2}}{P_{v,\text{sat}}} = \frac{1.4469}{1.4591} = 0.992$$

Dehumidifier Heat Transfer

$$h_{a,1} + \omega_1 h_{v,1} + Q - h_{a,2} - \omega_2 h_{v,2} - (\omega_1 - \omega_2)h_{f,c} = 0$$

$$c_{p,a}(T_1 - T_2) + \omega_1 h_{v,1} - \omega_2 h_{v,2} - (\omega_1 - \omega_2)h_{f,c} + Q = 0$$

$$1.004(35 - 12.6) + (0.02152 \times 2565.3) - (0.009040 \times 2524.5) - (0.02152 - 0.00904)42.01 + Q = 0$$

$$Q = -54.35 \text{ kJ/kg dry air*}$$

11.9 ADIABATIC SATURATION

A nonsaturated air–vapor mixture in direct contact with liquid water will, on flowing through an ideally insulated chamber of adequate length, attain a saturated state at a temperature designated as the temperature of adiabatic saturation. The process assumes complete mixing and adequate contact between the air–vapor mixture and the liquid water. The phenomenon of adiabatic saturation is not restricted to air and water. Only these two fluids however are important to this discussion.

Vapor produced by evaporation of the water diffuses into the nonsaturated air–vapor stream. The energy required for evaporation is transferred from the air–vapor mixture to the liquid water. Equilibrium is established between the mass transfer rate and the heat transfer rate. During flow through the chamber, the temperature of the air–vapor mixture decreases toward

* For an alternative solution, the energy equation is applied to the air that comes in contact with the cold surface of the tubes, that is, $m_c = 0.9$.

$$m_c[c_{p,a}(T_1 - T_c) + \omega_1 h_{v,1} - \omega_c h_{v,c} - (\omega_1 - \omega_c)h_{f,c}] + Q = 0$$

$$0.9[1.004(35 - 10) + (0.02152 \times 2565.3) - (0.007653 \times 2519.8) - (0.02152 - 0.007653)42.01] + Q = 0$$

$Q = -54.39$ kJ/kg dry air that enters the dehumidifier

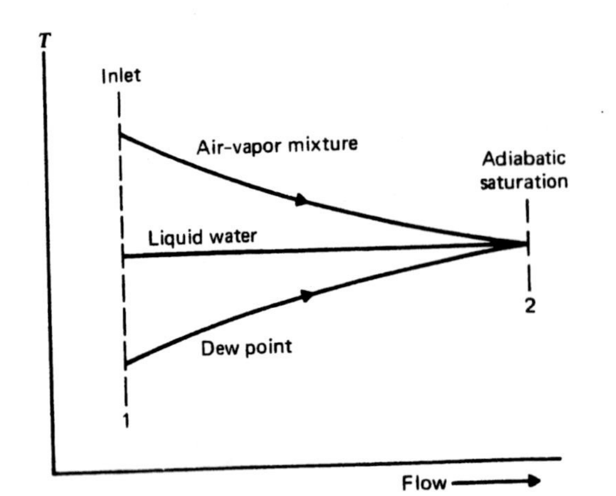

Figure 11.15 Temperature gradients for adiabatic saturation.

the adiabatic saturation temperature as a limiting value. See Fig. 11.15.

The decrease in the temperature of the air–vapor mixture, from T_1 to T_2, is also shown in Fig. 11.16. The transfer of vapor to the mixture causes an increase in the partial pressure of the vapor and an increase in the specific humidity of the air–vapor mixture. As a consequence of the increase in the vapor pressure, the dew point temperature of the mixture will increase to the adiabatic saturation temperature as a limit.

The energy equation for adiabatic saturation is given by

$$h_{a,1} + \omega_1 h_{v,1} + (\omega_2 - \omega_1)h_{f,2} - h_{a,2} - \omega_2 h_{v,2} = 0 \qquad (11.5)$$

where

- h_a = enthalpy of the dry air, kJ/kg (see Fig. 11.17)
- h_v = enthalpy of the water vapor, kJ/kg
- $h_{f,2}$ = enthalpy of the liquid water, kJ/kg
- ω = specific humidity of the air–vapor mixture, kg vapor/kg dry air
- $\omega_2 - \omega_1$ = liquid water evaporated and transferred to the air–vapor mixture, kg/kg dry air

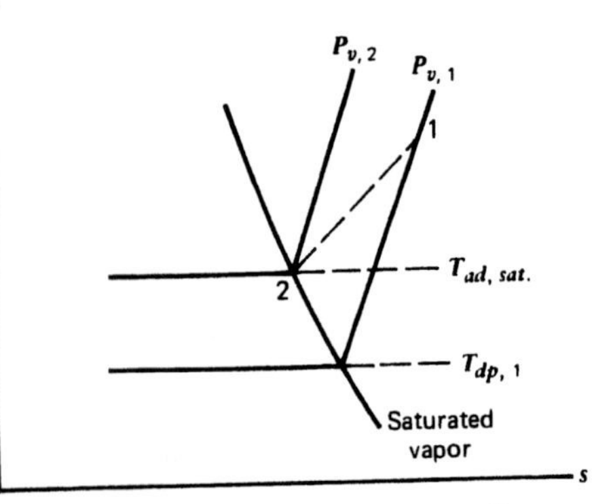

Figure 11.16 Adiabatic saturation T–s diagram.

The water supplied to the system is assumed to be at the adiabatic saturation temperature.

The pressure of the water vapor in the air–vapor mixture is commonly determined from the measured wet-bulb temperature of the mixture. A psychrometer is used to measure the wet-bulb temperature that is used subsequently in an appropriate equation to calculate the partial pressure of the water vapor. Carrier's equation, a semiempirical equation, may be used for this purpose.

$$P_v = P_{wb} - \frac{(P_a - P_{wb})(T_{db} - T_{wb})}{1546 - 1.44T_{wb}} \qquad (11.6)$$

where

- P_v = vapor pressure of the air–vapor mixture
- P_{wb} = saturated vapor pressure corresponding to the wet-bulb temperature
- P_a = total pressure of the air–vapor mixture
- T_{db} = dry-bulb temperature, C
- T_{wb} = wet-bulb temperature, C

Pressures are expressed in consistent units.

The wet-bulb temperature and the temperature of adiabatic saturation are in close agreement for the air–water vapor mixtures normally encountered in air conditioning applications. Such agreement may not be observed for mixtures other than air and water vapor.

It can be demonstrated from an energy balance that the adiabatic saturation temperature

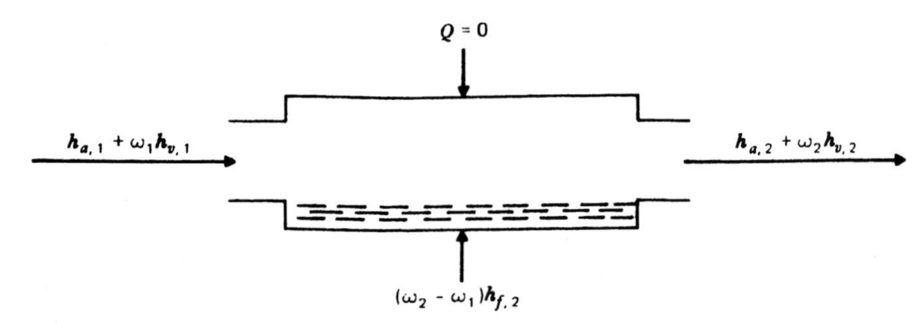

Figure 11.17 Energy diagram for adiabatic saturation.

of the air–vapor mixture remains constant as the vapor content of the mixture increases during passage through the adiabatic saturator. Thus, adiabatic saturation is achieved at a constant adiabatic saturation temperature, and hence for the air–water vapor mixture at a constant wet-bulb temperature. The observation of the constant wet-bulb temperature is particularly important to the use of the psychrometric chart.

Example 11.9

Air enters an adiabatic saturator at $P_a = 100.0$ kPa, abs, $T_{db} = 38$ C, and $\phi = 0.15$. Determine the adiabatic saturation temperature of the air by use of Eq. 11.5.

1 Specific humidity of the inlet air.

$$\text{At 38 C, } P_{v,\text{sat}} = 6.632 \text{ kPa, abs}$$

$$P_v = \phi P_{v,\text{sat}}$$

$$= 0.15 \times 6.632 = 0.995 \text{ kPa, abs}$$

$$\omega_1 = 0.622 \frac{P_v}{P_a - P_v}$$

$$= 0.622 \frac{0.995}{100.0 - 0.995}$$

$$= 0.006251 \text{ kg/kg dry air}$$

$$h_{v,1} = 2570.7 \text{ kJ/kg}$$

2 Adiabatic saturation temperature.

A trial-and-error solution will be used. The final trial only is shown.

$$\text{Assume } T_{ad,\text{sat}} = 19.1 \text{ C}$$

$$P_{v,2} = 2.212 \text{ kPa, abs (saturated)}$$

$$\omega_2 = 0.622 \frac{2.212}{100.0 - 2.212}$$

$$= 0.01407 \text{ kg/kg dry air}$$

$$h_{v,2} = 2536.4 \text{ kJ/kg}$$

$$h_{f,2} = 80.19 \text{ kJ/kg}$$

A modified form of Eq. 11.5 will be used for the solution.

$$c_{p,a}(T_1 - T_2) + \omega_1 h_{v,1} + (\omega_2 - \omega_1)h_{f,2} - \omega_2 h_{v,2} = 0$$

$$1.004(38 - 19.1) + (0.006251 \times 2570.7) + (0.01407 - 0.00625)80.19 - (0.01407 \times 2536.4) = R$$

$$R = -0.015$$

The assumed value of $T_{ad,\text{sat}} = 19.1$ C is acceptable.

3 Wet-bulb temperature.

Applying Eq. 11.6 to the condition of the aiı at the inlet section,

$$\text{Assume } T_{wb} = 19.1 \text{ C, then } P_{wb} = 2.212 \text{ kPa, abs}$$

Applying Eq. 11.6,

$$P_v = P_{wb} - \frac{(P_a - P_{wb})(T_{db} - T_{wb})}{1546 - 1.44T_{wb}}$$

$$= 2.212 - \frac{(100.0 - 2.212)(38 - 19.1)}{1546 - (1.44 \times 19.1)}$$

$$= 2.212 - 1.217 = 0.995 \text{ kPa, abs}$$

The calculated value, $P_v = 0.995$ kPa, abs, is equal to the value of $P_{v,1} = 0.995$ kPa, abs, determined in 1. Thus, $T_{wb} = 19.1$ C.

The state of an air–water vapor mixture is usually defined by prescribing, or measuring, two of three interrelated properties, namely, the dry-bulb temperature, the wet-bulb temperature, and the relative humidity. The psychrometric chart is particularly useful for solving, with sufficient precision, numerous problems in air conditioning applications. See Fig. 11.18. Incidentally, the adiabatic saturation temperature is taken equal to the wet-bulb temperature.

Several processes are illustrated in Fig. 11.18, namely:

0–1 Adiabatic saturation—evaporative cooling
0–2 Sensible heating
0–3 Sensible cooling
0–4 Cooling and dehumidification
0–5 Heating and humidification

An examination of Eq. 11.5 shows that if the energy of the liquid water is neglected, the enthalpy of the air–water vapor mixture at the adiabatic saturation state is equal to the enthalpy of the mixture entering the adiabatic saturator. Thus, a line of constant wet-bulb temperature on the psychrometric chart is also essentially a line of constant enthalpy.

Some psychrometric charts show both lines of constant wet-bulb temperature and lines of constant enthalpy. The slopes of these lines deviate slightly. See Fig. 11.18. It may be observed that the enthalpy of the mixture decreases slightly as the dry-bulb temperature increases at a constant wet-bulb temperature and constant total pressure. Two states, one a saturated state, are selected at a common wet-bulb temperature. The difference between the enthalpy of the mixture at the saturated state and the enthalpy of the mixture at the nonsaturated state is equal to the product of the enthalpy of liquid water, at the wet-bulb temperature, and the difference between the respective specific humidities.

The principle of adiabatic saturation is important to cooling and drying processes. When the relative humidity of the ambient air is low, say, about 10 percent, evaporative cooling can be particularly effective. Air is directed by a fan through a device supplied with water and arranged to add the water to the air by evaporation. The process tends to approach that of adiabatic saturation, but the outlet air is not necessarily saturated.

The evaporative cooler is only effective in a region where the summertime relative humidity

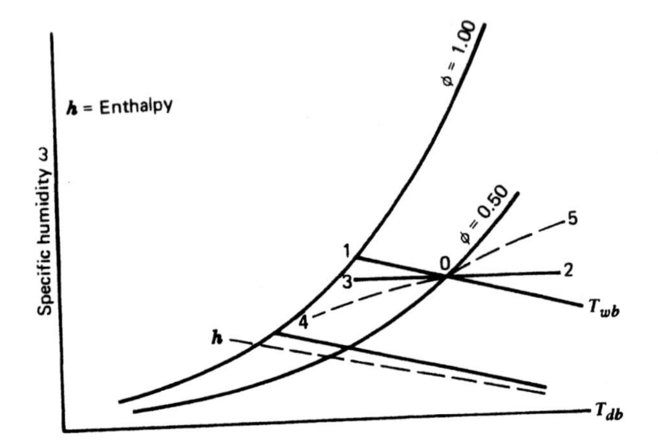

Figure 11.18 Schematic psychrometric chart.

is low. Compared with an air conditioning system that includes a vapor compression refrigeration machine, the evaporative cooler is considerably lower in first cost and consumes substantially less electrical energy. Care must be exercised in operating the evaporative cooler to avoid an excessively high humidity in the building.

A second important application is that of the adiabatic drier. Atmospheric air is usually heated to an appropriate temperature and subsequently blown through a drying tunnel or shed. A variety of wet products is dried in this fashion by evaporation of most of the liquid in the material into the flowing air. Depending upon the operating characteristics of the system, the air may not leave the drier in a saturated state. It is important to note that heat is not supplied to the products in the drying tunnel, but only to the incoming air.

PROBLEMS

11.1 A closed-cycle air compression refrigeration system is rated at 10,000 kJ/min. Following compression, the gas is cooled to a temperature 8 C above the ambient temperature of 30 C. The gas enters the "cold room" at −71 C. The gas circulation rate is 163 kg/min. The compressor and expander machine efficiencies are, respectively, 0.84 and 0.91. Mechanical losses are neglected. Determine the coefficient of performance and the power input to the machine.

11.2 The operating pressures for a vapor compression refrigeration system are 200 and 1200 kPa, abs. The refrigerant (ammonia) temperatures are: entering the compressor 0 C, leaving the compressor 135 C, entering the condenser 120 C, leaving the condenser and entering the expansion valve 25 C, and leaving the evaporator −10 C. The isentropic compressor efficiency, based on the shaft input, is 0.765. Calculate the compressor work and heat transfer, refrigeration effect, and coefficient of performance. Write the energy balance, and determine the power requirement for a refrigeration capability of 20,000 kJ/min.

11.3 A water vapor refrigeration system is operated by a centrifugal compressor. The compressor isentropic efficiency is 0.72, and the mechanical efficiency is 0.985. The evaporator is maintained at a pressure of 1.0 kPa, abs. The condenser pressure is 10 kPa, abs. The vapor leaves the evaporator with a liquid content of 2.5 percent. The maximum power input to the compressor is 50 kW. The chilled water returns to the evaporator at 14 C. The make-up water is supplied at 20 C. Determine the maximum circulation rate, in kg/min, for the chilled water.

11.4 A building is heated by a warm air heating system. The heat loss from the building is 500,000 kJ/h. The building interior is maintained at 20 C and 50 percent relative humidity. The warm air leaves the heater at 60 C. Thirty percent of the air supplied to the heater is outside air at −23 C and 10 percent relative humidity. The remainder of the air is recirculated from the building interior. The humidity gain in the building can be neglected. The barometric pressure is 100 kPa, abs. Calculate the rates at which the heat and water are supplied to the heater.

11.5 A vapor compression refrigeration system is charged with R-12 refrigerant. The condensing and evaporating temperatures are, respectively, 30 and −25 C. The refrigerant enters the expansion valve at 25 C and enters the compressor at −10 C. The compressor isentropic efficiency is 0.78, and adiabatic compression is assumed. The refrigerant flow rate is 50 kg/min. Calculate the refrigeration capability, in kJ/min, and the coefficient of performance.

11.6 An aircraft compressor-turbine cooling unit is supplied from the main engine with compressed air at 170 kPa, abs, and 318 K. Air enters the turbine, which drives the unit compressor, at 273 K and leaves the turbine at 96.5 kPa, abs. The turbine inlet pressure is equal to the compressor discharge pressure. The cabin temperature is 20 C. The compressor and turbine machine efficiencies are, respectively, 0.80 and 0.88. Calculate the turbine inlet pressure and the cooling effect, in kJ/kg air.

11.7 Air enters a convention hall at 22 C and 50 percent relative humidity and at the rate of 3400 m^3/min. Thirty-five percent of this air is supplied from the outside, while the remainder of the air is recirculated from the hall through the dehumidifier. The ambient conditions are 33 C, 55 percent relative humidity, and P_{bar} = 101 kPa, abs. The air leaves the hall at 27 C and 40 percent relative humidity. Determine the refrigeration requirement and the energy requirement for reheating the air, in kJ/min.

11.8 Repeat Problem 11.7 for an alternative arrangement in which 50 percent of the recirculated air is directed through the bypass around the dehumidifying washer.

11.9 The compressor installed in a vapor compression refrigeration system is comprised of two single-acting cylinders. The bore and stroke are, respectively, 100 mm and 140 mm, and the compressor speed is 900 rpm. The compressor volumetric efficiency is 0.86. The refrigerant, ammonia, enters the expansion valve at 24 C and leaves the evaporator at 150 kPa, abs, and 0 C. The ammonia enters the compressor at 125 kPa, abs, and 10 C. Calculate the refrigeration capability, in kJ/min.

11.10 The compressor for the system described in Problem 11.9 discharges the ammonia at 1000 kPa, abs, and 168 C. Heat is removed from the ammonia by the compressor jacket water at the rate of 272 kJ/min. Determine the compressor power and the system coefficient of performance.

11.11 A building is heated by a heat pump that is charged with Freon-12. The pressure in the condenser is maintained at 1.6 MPa, abs, and the refrigerant enters the expansion valve at 45 C. The compressor overall isentropic efficiency is 0.76, and the motor efficiency is 0.94. The system heat source is the outside air. The cost of the electrical energy is 4 cents per kilowatt hour. Calculate the heating cost for two different outdoor temperatures.

(a) For an outdoor temperature of 5 C, the heat loss from the building is 50,000 kJ/h. The refrigerant leaves the evaporator at 300 kPa, abs, and 5 C.

(b) For an outdoor temperature of −15 C, the heat loss is 112,500 kJ/h. The refrigerant leaves the evaporator at 150 kPa, abs, and −15 C.

11.12 Determine for the conditions prescribed in Problem 11.11 the cost for heating the building by an oil-fired unit that has an operating efficiency of 0.72. The fuel cost is $3.10 per 10^6 kJ.

11.13 The evaporating temperature for a water vapor refrigeration system is 5 C, and the vapor leaves the evaporator with a liquid content of 3.5 percent. The condenser pressure is 10 kPa, abs. The chilled water returns to the evaporator at 10 C. The make-up water is supplied at 22 C. The chilled-water circulation rate is 1000 kg/min. The overall isentropic efficiency of the motor-driven compressor is 0.69. The power required for operating the mechanical pumps is taken as 6 percent of the compressor power. Calculate the refrigeration capability, in kJ/min, the power input to the motors, and the system coefficient of performance.

11.14 A small foundry is heated by a warm-air heating system. The building interior is maintained at 18 C, and the outdoor conditions are −12 C and 45 percent relative humidity. The building heat loss is 100,000 kJ/h. The air leaves the heater at 62 C, and no air recirculation is provided. Calculate the rate at which energy is supplied in the heater.

11.15 A quantity (100 m^3) of air enters a dehumidifier at 100 kPa, abs, 35 C, and 55 percent relative humidity. The air leaves the spray-type dehumidifier at 13 C. Calculate the heat transfer in the dehumidifier.

11.16 A room in a textile mill is supplied with air at 27 C and 60 percent relative humidity. The room dimensions are 60 × 30 × 6 m. The room air is changed 10 times per hour. The ambient air conditions are T_{db} = 15 C, T_{wb} = 7 C, and P_b = 97.0 kPa, abs. Water is supplied to the heater at 20 C. Calculate the quantities of energy and water supplied, per minute, in the heater.

11.17 An adiabatic drier is supplied with air that is heated to 60 C. The ambient air conditions are 20 C dry-bulb temperature, 9 C wet-bulb temperature, and 100 kPa, abs, barometric pressure. At the drier outlet, the dry-bulb temperature is 4 C above the wet-bulb temperature. The wet-bulb temperature of the air is assumed to remain constant during flow through the drier. The drying rate is 50 kg water per hour. Calculate the air flow to the heater, in m^3/min, based on the ambient conditions, and the energy supplied to the heater, in kJ/h.

11.18 Air enters a dehumidifier at 32 C. The atmospheric pressure is 100.50 kPa, abs, and the measured wet-bulb temperature is 26 C. The surface temperature of the coils is 7 C. The air bypass fraction is 12 percent. Determine the dry-bulb temperature and the relative humidity of the air leaving the dehumidifier. Calculate the dehumidifier heat transfer.

11.19 An evaporative cooler is used to maintain the interior temperature of a building at 25 C. The building heat gain is 90,000 kJ/h. The ambient conditions are P_b = 100.2 kPa, abs, T_{db} = 37 C, and ϕ = 0.10. Determine the air flow rate to the cooler, in m^3/min, based on atmospheric conditions. The air is assumed to leave the cooler in a saturated state.

11.20 Repeat Problem 11.20 for operation of the cooler with the discharge of nonsaturated air at a dry-bulb temperature of 18 C.

BIBLIOGRAPHY

ASHRAE Handbook 1981 Fundamentals, American Society of Heating, Refrigerating and Air-Conditioning Engineers, Inc. Atlanta, Ga.

Fanger, P. O. *Thermal Comfort*, McGraw-Hill Book Co., New York, 1972.

Jennings, B. H. *The Thermal Environment*, Harper & Row, Publishers, New York, 1978.

Jordan, R. C. and Priester, G. B. *Refrigeration and Air Conditioning*, Prentice-Hall, Inc., Englewood Cliffs, N.J., 1956.

Pita, E. G. *Air Conditioning Principles and Systems: An Energy Approach*, John Wiley & Sons, Inc., New York, 1981.

Sparks, N. R. and Dillio, C. C. *Mechanical Refrigeration*, McGraw-Hill Book Co., New York, 1959.

CHAPTER TWELVE

DIRECT ENERGY CONVERSION

The possibility of achieving on a commercial scale the direct conversion of chemical or thermal energy into electrical energy has attracted the interest of many investigators. Direct conversion of energy would eliminate the dependence upon a thermal engine and notably for the photovoltaic cell and the fuel cell free the conversion process from the limitations imposed by the second law of thermodynamics. Some of the more significant direct energy conversion systems are examined in this chapter. The operating characteristics of the various systems will be investigated, along with the inherent limitations on the application to commercial-scale electric power generation.

Magnetohydrodynamic Power

12.1 THE MAGNETOHYDRODYNAMIC POWER PLANT

In a fossil fuel-fired steam electric power plant, the generation of electrical energy is accomplished in three main steps, namely, the conversion of the chemical energy in the fuel into thermal energy in a combustion reaction, the conversion of thermal energy into mechanical energy in a steam power cycle, and finally the conversion of mechanical energy into electrical energy in the generator. The magnetohydrodynamic (MHD) power generator omits the intermediate step by converting thermal energy directly into electrical energy. Although the boiler and turbine are eliminated, the MHD power plant is not necessarily less complex than the conventional fossil fuel-fired electric generating station. MHD power plants however operate at high temperatures and thus provide a means to improving the performance of the present-day electric power generating plant.

The conventional electric generator coupled, for example, to a steam turbine functions by creating a rotating magnetic field in the stationary conductors. The MHD electric generator utilizes an electrically conducting gas that flows through the magnetic field. In both of these generators, an induced emf causes a current to flow in the conductor, one solid and the other gaseous, and in an external circuit.

The principle of MHD power generation was discussed over 100 years ago by Michael Faraday during his attempt to establish the law of motional electromagnetic induction. Throughout the past century, only sporadic investigations were made of devices based on magnetohydrodynamic interaction. The approach to electric power generation was instead directed to the electromechanical energy converter, a rotating machine consisting basically of copper windings and an iron-cored magnetic circuit.

In recent years, interest in the MHD generator has intensified, with current research and development conducted on an international scale in some six or seven countries. The Russian investigators have constructed an experimental power plant in which the MHD generator has a design power output of 25 MW.

In the United States, a program has been established to design and construct a full-scale demonstration-type MHD power system that will be operational circa 1990. The intention is to demonstrate commercial feasibility of electric power generation in an open-cycle MHD/steam cycle power loop rated at approximately 1000 MW(t).

Because of characteristics unlike those of conventional power plants, the development of the MHD generator is dependent upon the solution to numerous problems that are related, in part, to very high operating temperatures. Effective operation of the MHD power generator requires the attainment of high electrical conductivity in the gas at comparatively low temperatures.

Because of the high temperature of the effluent gas, the MHD generator is employed as the topping unit in a combination with another power generating system. Current proposals anticipate topping the steam power cycle with the MHD generator. The MHD generator is readily fired with gas or oil, but most developments for commercial power generation are directed to coal firing. In addition to the coal-burning capability, the MHD/steam cycle power plant will have a high overall thermal efficiency and a reduced environmental impact. Thermal efficiency values for the MHD/steam cycle power plant are expected to be in the range of 0.50 to 0.60. The comparatively low heat rejection from the power plant reduces the demand for cooling water. The operating conditions of the MHD generator are conducive to the removal of sulfur dioxide and nitrogen oxides from the effluent gases.

There are two basic types of the MHD generator. One type employs an ionized gas or plasma for the working fluid, and the other type uses a liquid metal. The ionized gas is used in either an open or closed cycle. Schemes proposed for closed-cycle operation are principally dependent upon heating the gas in a nuclear reactor, contingent upon development of reactors capable of high-temperature operation, that is, in the vicinity of 2400 C. The closed-cycle liquid-metal system has been explored in some detail, but the thermal efficiency of this system is inherently lower than that of the open-cycle system. The major effort today is directed to the development of the coal-fired open-cycle MHD power generator because it offers the best possibility for the generation of electrical energy on a commercial scale in a system that utilizes the magnetohydrodynamic principle.

12.2 OPERATING PRINCIPLES OF THE MAGNETOHYDRODYNAMIC GENERATOR

Figure 12.1 is a schematic representation of the MHD power generator that will serve to illustrate the system operating principles. The hot, electrically conducting gas is expanded in the diverging channel. Electrodes are placed in the

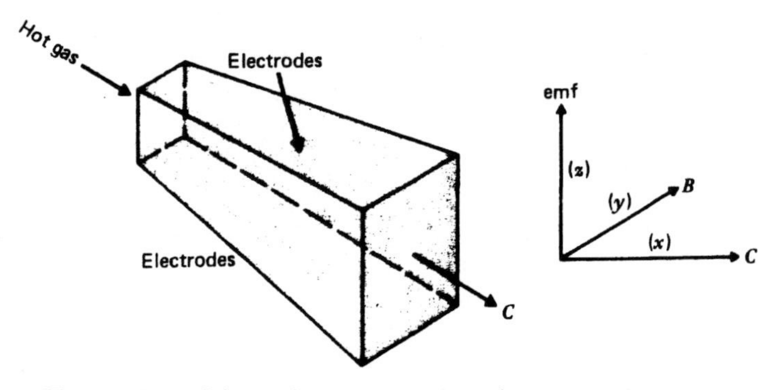

Figure 12.1 Schematic representation of a magnetohydrodynamic (MHD) generator duct.

upper and lower channel walls, and the field coils are installed outside the electrodes. The channel walls are cooled to protect the structural materials from the high-temperature effects of the flowing gas.

The direction of the magnetic field, along the y-axis, is perpendicular to the direction of flow, along the x-axis. Interaction of the conducting gas with the magnetic field creates an induced emf in the gas that drives an electric current through the gas, the electrodes, and the external load. The direction of the emf is along the z-axis, causing the current to flow from the lower electrode through the gas to the upper electrode.

The current flowing through the gas creates an induced field that interacts with the applied magnetic field. The net effect of this field interaction is to produce a force that opposes the motion of the gas. The retarding force, in turn, is counteracted by the force that is applied to the gas to effect expansion of the gas in the generator channel. In this manner, the MHD generator converts thermal energy into electrical energy.

In the MHD generator, electromotive forces are actually induced both transverse to and along the direction of the flowing fluid. The transverse emf is caused by the Faraday induction between the moving fluid and the magnetic field, as described above. The axial emf is caused by the Faraday interaction that results from the movement of the electric charge (current) in a direction perpendicular to the fluid flow and the magnetic field. The axial emf is designated as the Hall effect, and the axial field is known as the Hall field. Under the conditions encountered in a MHD generator that is designed for practical operation, the Hall effect is always significant and cannot be ignored.

Because of the Hall effect, MHD generators may be designed for extraction of electrical energy by employing either the transverse or Hall emf. In principle, the design is directed to the manner in which the electrodes are installed in the walls of the generator and connected to the external load.

The electrical conductivity is an important property of the working fluid in a MHD generator. Thermal ionization of a combustion gas mixture is readily achieved by maintaining the gas at a high temperature since the electrical conductivity of the gas is strongly temperature dependent. The electrical conductivity of combustion products increases from 1 to 100 mho/m with a temperature increase from 2200 to 3100 K.[2] For a number of reasons, the electrical conductivity of the gas can be much lower than that of copper. Among these reasons is the very large magnetic field, on the order of 30,000 to 100,000 gauss, envisioned for commercial MHD generators.

The electrical conductivity of the combustion gas is enhanced by the addition of a small quantity of an easily ionized impurity called seed. Potassium in a compound such as K_2CO_3 or KOH is a convenient and inexpensive seed material. The use of cesium as a seed material would improve the electrical conductivity of the gas to a substantial degree. However, in comparison with potassium, cesium is considerably more costly, and its use is contingent upon economical and highly effective recovery.

The fluid inlet temperature for a steam or gas turbine is limited because of the hot, highly stressed moving parts that operate with close tolerances. Such temperature limitations are not imposed on the MHD generator. The stationary walls of the MHD generator are water cooled and hence operate at temperatures well below the temperature of the working fluid. The MHD generator channel can be constructed from readily available materials capable of withstanding the moderate stresses that develop because of thermal effects and internal pressure.

The capability for high-temperature operation is an important advantage to MHD power generation, as noted earlier. Further, in the MHD generator, the energy conversion process is associated with a high-volume flow of gas. Fluid and thermal losses are proportional to the surface area of the flow channel. The MHD generator is thus a device that is inherently suited to the generation of power on a very large scale.

12.3 MAGNETOHYDRODYNAMIC GENERATOR ANALYSIS

The char-fired MHD generator is proposed as an alternative to the coal-fired MHD generator. Char is a fuel derived from coal by the removal of a large portion of the volatile constituents. Relatively large quantities of char could be made available in the future as a by-product of conversion of coal into liquid and gaseous fuels. Char is a particularly good fuel for a MHD open-cycle power plant because of a low hydrogen content, which in turn leads to superior electrical conductivity in the product gases when alkali seed is added. Incidentally, the thermodynamic characteristics of the coal-fired MHD system are somewhat inferior to those of the char-fired MHD system, and as a consequence higher operating temperatures are required when firing coal.

A combustion system comprises the first component of a MHD/steam cycle power plant. The fuel, coal or char, is burned with air to produce a high-temperature working fluid. Subsequent to discharge from the combustion system, the product gases flow through the MHD generator and the heat recovery components. See Fig. 12.2. Steam generation, air preheating, and seed recovery are the primary functions of the heat recovery system. In order to achieve the best heat rate for the power plant, the overall system should be designed for maximum air preheat.

A simplified analysis of the MHD generator can be made by employing data selected from Ref. 4 for a system in which char is burned with oxygen. The thermodynamic properties of the products of combustion, useful for the evaluation of the system performance, are determined by computer calculations for equilibrium composition. Figure 12.3 is a Mollier chart for the products resulting from the combustion of char and oxygen in a stoichiometric ratio, with seeding by dry cesium carbonate of an amount that yields 0.7 percent by mass of cesium in the gas mixture. For this particular chart, the pressure ranges from 0.9 to 8 atm and the temperature, from 2200 to 2700 K.

Prescribed operating conditions for the char–oxygen MHD power generating system are

Combustor outlet: $P_0 = 8$ atm

$T_0 = 2650$ K

Then, from Fig. 12.3, $h_0 = -5179$ kJ/kg

Prescribed design parameters for the generator are

Gas flow rate: 1000 kg/s

Generator inlet velocity: $C_1 = 700$ m/s

Ratio of the generator heat loss to generator power: $\lambda = 0.05$

Derived quantities for the generator are

INLET	OUTLET
$P_1 = 5.04$ atm	$P_2 = 0.9$ atm
$T_1 = 2554$ K	$T_2 = 2270$ K
$h_1 = -5424$ kJ/kg	$h_2 = -6079$ kJ/kg
$s_1 = 7.615$ kJ/kg·K	$s_2 = 7.697$ kJ/kg·K
$\rho_1 = 0.9655$ kg/m³	$\rho_2 = 0.1973$ kg/m³
$A_1 = 1.48$ m²	$A_2 = 8.93$ m²
$C_1 = 700$ m/s (prescribed)	$C_2 = 567.5$ m/s
$\sigma_1 = 5.30$ mho/m	$\sigma_2 = 3.99$ mho/m
$B_1 = 7$ tesla	$B_2 = 3.15$ tesla

Length of generator: 19.76 m

Diffuser outlet: $P_3 = 1.204$ atm

$h_3 = -5918$ kJ/kg

The expansion of the working fluid in the generator is irreversible because of internal losses. The efficiency of the MHD generator is determined by $(h_{01} - h_{02})/(h_{01} - h_{02,s})$.

The electrical energy generated in the MHD generator is determined by applying the energy equation. Then,

$$\dot{W} = \dot{m}_g[h_1 + (C_1^2/2) + Q - h_2 - (C_2^2/2)]$$

$$= \dot{m}_g(h_0 - h_3 + Q)$$

Since $Q = -\lambda W$,

$$\dot{W} = \dot{m}_g \frac{h_0 - h_3}{1 + \lambda}$$

$$= 1000 \frac{-5179 - (-5918)}{1 + 0.05}$$

$$= 703{,}810 \text{ kW}$$

$$= 703.8 \text{ MW}$$

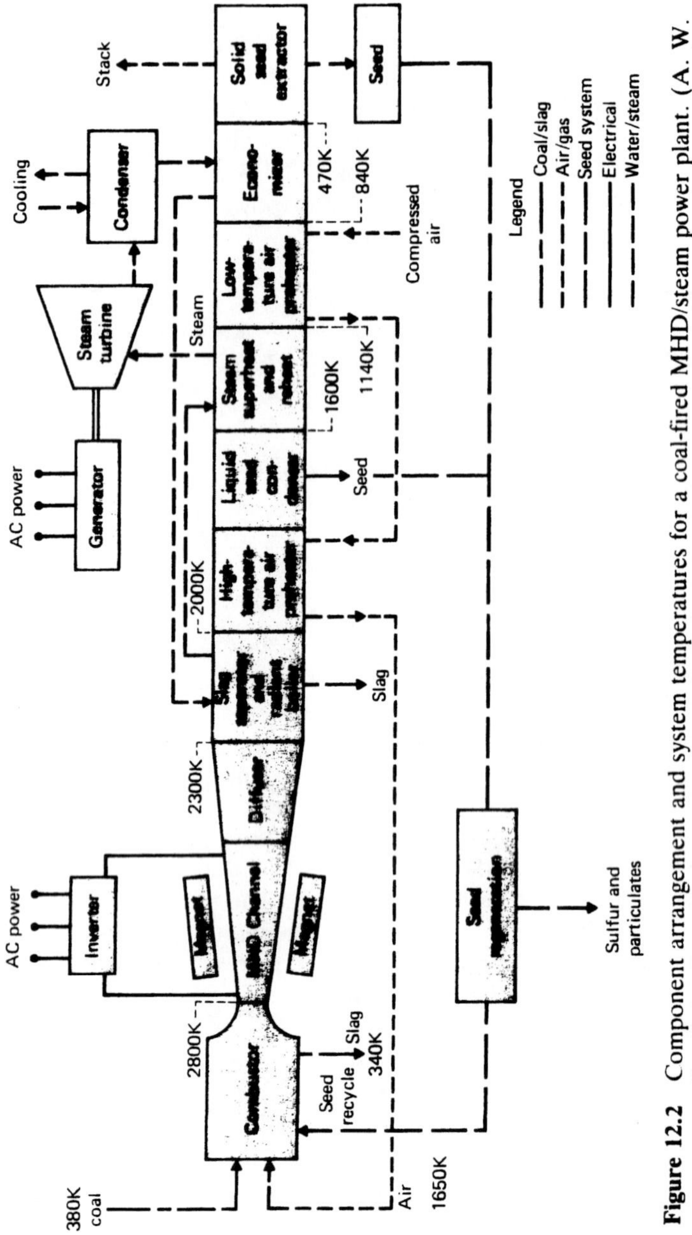

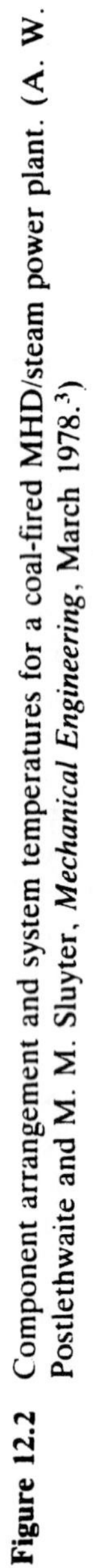
Figure 12.2 Component arrangement and system temperatures for a coal-fired MHD/steam power plant. (A. W. Postlethwaite and M. M. Sluyter, *Mechanical Engineering*, March 1978.[3])

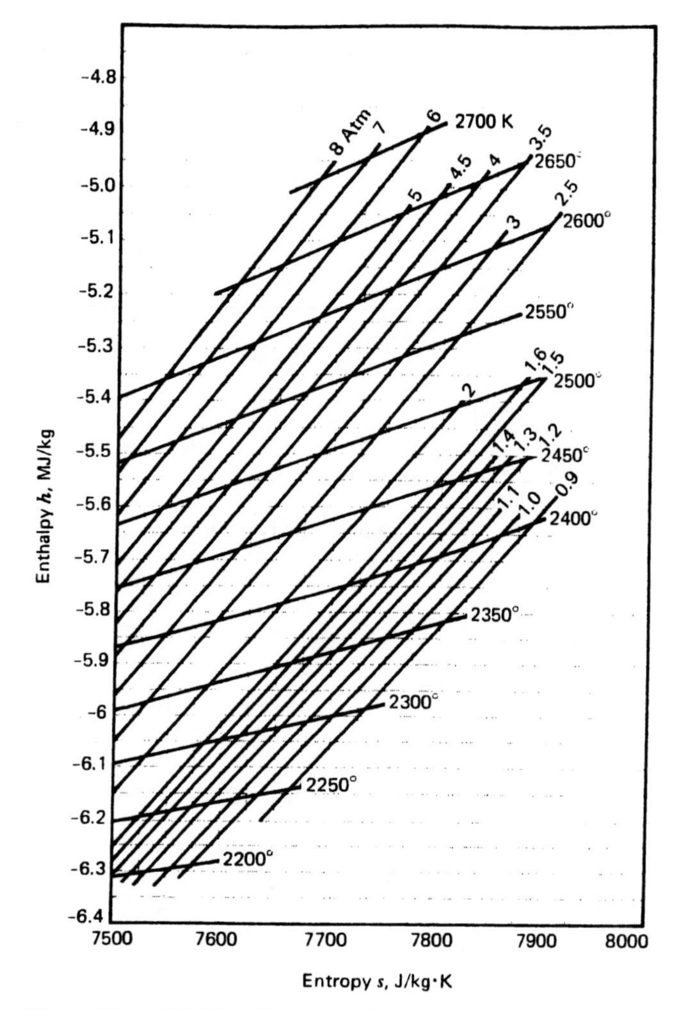

Figure 12.3 Mollier diagram: char and oxygen products; 0.7 percent cesium seeding by dry Cs_2CO_3; $\Phi = 1$. (S. Way, *Combustion*, May 1971.[4])

where h_0 and h_3 are essentially stagnation properties.

Two important physical characteristics of the MHD generator are the volume and the length of the diverging flow channel. Attention is again directed to Fig. 12.1, and a segmented arrangement of the electrodes is prescribed. The induced electrical potential is given by the cross product of the velocity of the fluid conductor and the magnetic flux density. Thus,

$$E_{\text{ind}} = C \times B \text{ m/s} \times \text{V·s/m}^2 = \text{V/m}$$

The total induced voltage drives the current against the resistance of the external load and against the internal resistance of the flowing fluid. The generator parameter n is defined as the ratio of the load resistance to the sum of the load resistance R_e and the generator internal resistance R_g. Then,

$$n = \frac{R_e}{R_e + R_g} = \frac{V}{V^*} = \frac{Ez}{E_{\text{ind}}z} = \frac{E}{E_{\text{ind}}}$$

where

E = the voltage gradient measured at the electrodes, V/m
z = the constant spacing between the electrodes, m
V = Ez, the operating voltage, V

The current density is

$$J = \sigma(E_{ind} - E)$$
$$= \sigma(E_{ind} - nE_{ind})$$
$$= \sigma CB(1 - n) \quad \text{A/m}^2$$

where σ is the fluid conductivity, mho/m.

The generator power is

$$\dot{W} = VJA$$

where A = the electrode area, m^2; $A = xy$, where x = the length and y is the width of the electrodes.

$$V = Ez = nE_{ind}z = nCBz$$

Thus,

$$\dot{W} = (nCBz)\sigma CB(1 - n)xy$$

and

$$\dot{W}_{MHD} = V\sigma C^2B^2n(1 - n) \qquad (12.1a)$$

where

$\dot{W}_{MHD}$ = generator power output, W
V = volume of the generator, m^3
σ = gaseous electrical conductivity, mho/m
C = gas velocity in the generator duct, m/s
B = magnetic induction, tesla or weber/m^2, V·s/m^2
n = ratio of the load resistance to the sum of the load resistance and the generator internal resistance

The value of n is not arbitrarily assumed but is selected in the design of the generator to yield an acceptable generator conversion efficiency. Values of n are generally in the range of 0.7 to 0.9.

The pressure drop, per unit length, in the flowing gas can be expressed by

$$\frac{dP}{dx} = -JB \quad \text{N/m}^2 \text{ per m}$$

Then,

$$dx = -\frac{dP}{JB} = -\frac{dP}{\sigma CB^2(1 - n)}$$

and by integration,

$$L = \frac{P_1 - P_2}{\sigma CB^2(1 - n)} \qquad (12.1b)$$

where L = the length of the generator duct, m

Example 12.1

The volume and the length of the above MHD generator duct will be determined by application of Eqs. 12.1a and 12.1b. Average values of σ, C, and B will be used in the calculations.

$$\sigma = 4.645 \text{ mho/m} \quad \text{(A/V per m)}$$
$$C = 634 \text{ m/s}$$
$$B = 5.075 \text{ Wb/m}^2 \quad (\text{V·s/m}^2)$$
$$n = 0.72 \quad \text{(assumed)}$$

$$\dot{W}_{MHD} = 703.8 \times 10^6 \text{ W}$$

1 Generator duct volume.

$$\dot{W}_{MHD} = V\sigma C^2B^2n(1 - n)$$
$$703.8 \times 10^6 = 4.645V(634)^2 \cdot (5.075)^2 0.72(1 - 0.72)$$
$$V = 72.598 \text{ m}^3$$

2 Generator duct length.

$$L = \frac{P_1 - P_2}{\sigma C B^2(1 - n)}$$

$$= \frac{(5.04 - 0.9)101325}{4.645 \times 634(5.075)^2(1 - 0.72)}$$

$$= 19.75 \text{ m}$$

12.4 MAGNETOHYDRODYNAMIC GENERATOR OPERATION

The cost of the oxygen supplied to the char–oxygen system is an item of substantial proportion. Because of the economic impact of this cost item, the combustion of the fuel with oxygen is not likely to represent a practicable approach to MHD power generation. Char and coal however can be burned with air, and sufficiently high combustion temperatures can be attained through a high level of air preheat. Figure 12.4 is applicable to the combustion of char with air.

Removal of nitrogen oxides from the effluent gases of the MHD generator is receiving appropriate attention because of the air quality control regulations that apply to thermal-electric generating stations. Both the atmospheric nitrogen and the fuel nitrogen will contribute to NO_x emission. The MHD combustion system may be operated to produce a minimum concentration of NO_x by establishing fuel-rich conditions in the combustor and slow cooling rates in the radiant boiler.

Removal of sulfur oxides from the effluent gases in the MHD/steam cycle power plant apparently can be achieved with minimum difficulty. The alkali seed material effectively captures almost all the sulfur contained in the coal. Separation of the sulfur and particulates is accomplished in the seed regeneration component as shown in Fig. 12.2.

The MHD generator produces direct current, hence for commercial power distribution an inverter must be employed to change the direct current to alternating current. A substantial portion of the gross power output of the MHD/steam cycle power plant is consumed in compressing the combustion air for the MHD generator. Power is also required for energizing the magnets installed on the walls of the MHD generator. The energy consumed for this purpose may reach a substantial amount unless superconducting magnets are used.

There is no assurance that the MHD generator will begin, say, by the year 2000, to contribute significantly to generation of power on a commercial scale. Solution to the various technical problems may be achieved, but there still remains the need to demonstrate whether the MHD/steam cycle power plant will be economically competitive with other types of power plants.

Thermoelectric Power

12.5 THE THERMOELECTRIC EFFECT

A thermoelectric effect was discovered in 1821 by Thomas J. Seebeck, the German physicist. When a body of thermoelectric material is maintained at a uniform temperature, the positive and negative electrical charges are uniformly distributed throughout the material. If one surface of the body, for example, the end of a rod, is heated, the charge distribution is altered. The positively charged ions in the crystals remain fixed, while the negatively charged electrons tend to move to the cool end of the rod.

The negative charges are carried by the free electrons, which are in effect certain of the orbital electrons that can pass readily between adjacent atoms. The heating effect causes an increase in the kinetic energy of these electrons; and as a consequence, they begin to move by diffusion toward the cooler end of the rod. The higher concentration of electrons at the cooler end results in a gradient of electrical charge, and a potential difference is established between the hot and the cold ends of the rod. An electric current may then be made to flow in an external circuit or load.

Accumulation of negatively charged electrons

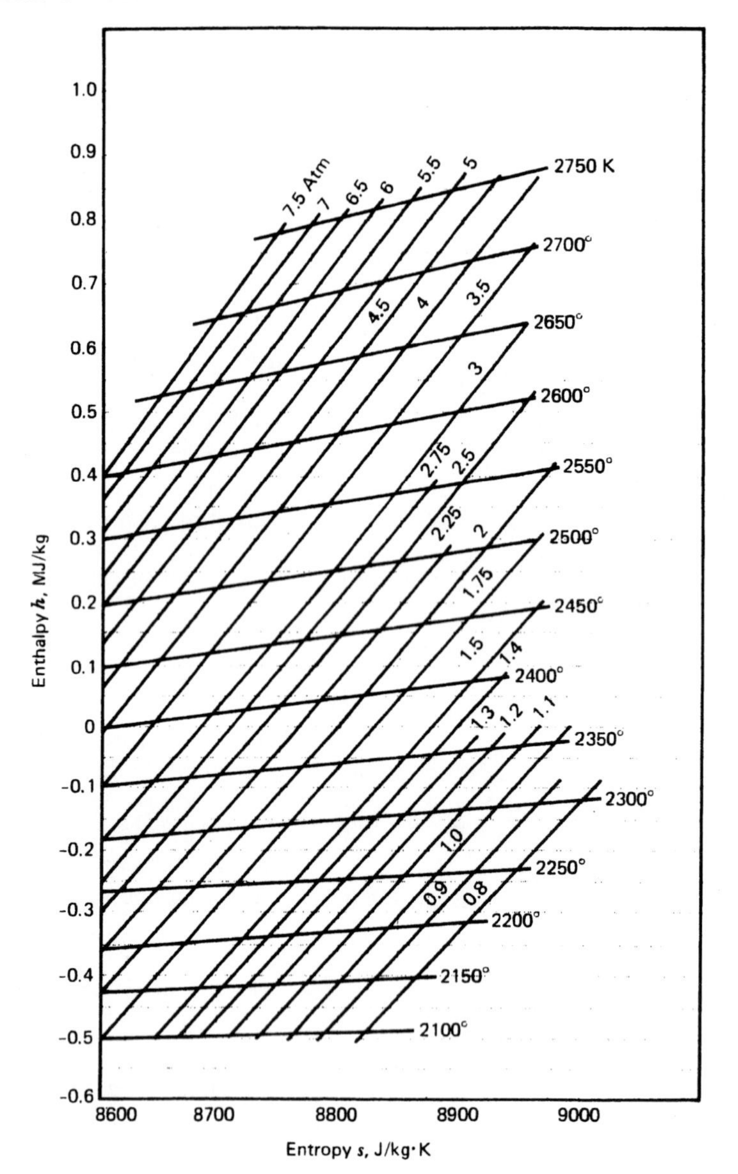

Figure 12.4 Mollier diagram: Φ = 0.95 char and air products; air with 1 percent moisture; 0.7 percent cesium seeding by dry Cs_2CO_3. (S. Way, *Combustion*, May 1971.[4])

at the cooler end of the rod causes that end to be negatively charged with respect to the hot end. As a consequence of this negative charge, electrons flowing from the hot end of the rod are repelled at an increasing rate until equilibrium conditions are established, at which point the electron flow from both ends of the rod will be equal. The cool end of the rod however is still negatively charged with respect to the hot end. Characteristically, the lower the number of free electrons in a substance, the greater will be the accumulation of electrons at the cool end of the rod and, in turn, the higher will be the voltage generated for a prescribed temperature difference between the two ends of the rod.

An important factor in the advancement of thermoelectric technology is the ability to adjust the number of free electrons in semiconductor materials. The importance of the control of the number of free electrons is attributed to two basic relationships, namely, the output voltage of any thermoelectric material is inversely proportional to the number of free electrons in that material, and the conductivity of the material is directly proportional to the number of free electrons. The significance of these two relationships can be illustrated through an examination of the thermoelectric characteristics of insulators and metals.

Insulators containing 10^{10} free electrons per cubic centimeter generate Seebeck (output) voltages of about 10,000 microvolts per degree Kelvin temperature difference between the hot and cold ends.[5] However, because of the extremely high internal resistance of an insulator, the current is low, and little power is generated. On the other hand, metals have a higher density of free electrons and produce Seebeck voltages of about 5 microvolts per degree Kelvin temperature difference, but they have an extremely low internal resistance. The combination of low voltage and high current however results in little power output.

Apparently, an acceptable thermoelectric material would have conductivity and Seebeck voltage characteristics between those of insulators and those of metals. Efficiency curves indicate that the optimum electron density is about 10^{19} free electrons per cubic centimeter. This value is well within the range of good-conducting semiconductors and will produce about 175 microvolts per degree Kelvin temperature difference. Acceptable values of the thermoelectric efficiency can be achieved with a number of materials, for example, lead telluride (PbTe), a commonly used semiconductor material.

In constructing thermoelectric generators, it is essential to observe that each thermoelectric material has an operating temperature range within which best performance is achieved. An upper limit on the operating temperature is reached when the material becomes "intrinsic." The intrinsic state is observed when the heat input causes both positive and negative charges to occur in equal numbers, which reduces the magnitude of the output voltage generated.

The approach to high-temperature operation is the creation of a material that has the proper conductivity and does not become intrinsic. While semiconductors become intrinsic at temperatures approaching 1250 K, many insulators do not become intrinsic at this temperature level and can be modified to acquire good thermoelectric characteristics. It is possible to decrease the electrical resistivity of a material that is normally classified as an insulator by the introduction, in small quantity, of a metallic additive.

Three different types of semiconductors can be manufactured by controlling the physical makeup of the crystal lattice. The designations "*n*-type," "*p*-type," and "intrinsic" are employed. In the *n*-type semiconductor, the lattice has a small excess number of negatively charged electrons, while in the *p*-type semiconductor the lattice has a deficiency of electrons. This deficiency creates "holes" that are free to move and provide the effect of positively charged particles. Thus, the electrons are available to function as negative charge carriers and the holes, as positive charge carriers. If the lattice contains no charge carriers or an equal number of positive and negative charge carriers, the semiconductor is classified as intrinsic.

12.6 BASIC THERMOELECTRIC GENERATORS

Figure 12.5 illustrates a basic thermoelectric generator constructed from two dissimilar metals. For many years, the only practical application for an arrangement of this kind was in instru-

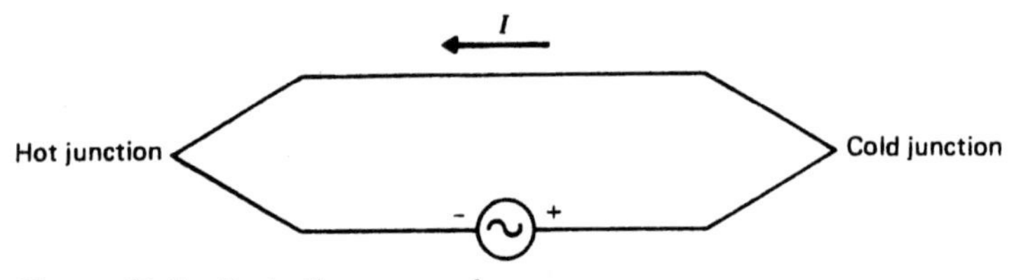

Figure 12.5 Basic thermocouple.

mentation with the device employed as a thermocouple for temperature measurements. The reason for this limited application can be attributed principally to the low energy conversion efficiency that is achieved when metals are used in the construction of a thermoelectric power generator.

Figure 12.6 shows a section of semiconductor material arranged in contact with a heat source and a heat sink. The migration of the negative electrical charges is toward the cold end of the material. In addition to having a relatively low number of free electrons and a low electrical resistivity, a good semiconductor material should have a low thermal conductivity. A temperature gradient is established in the semiconductor; and unless the material has a low thermal conductivity, an excessively large fraction of the heat supplied to the device would be wastefully conducted from the heat source directly to the heat sink. A low thermal conductivity for the semiconductor material thus contributes to an improvement in the energy conversion efficiency of the thermoelectric device.

When an *n*-type semiconductor and a *p*-type semiconductor are combined as shown in Fig. 12.7, both the positive and the negative terminals are cold-side conductors. This arrangement has an important advantage, namely, that there is no heat leakage from the hot-side conductor through the external circuit as is the case for the device shown in Fig. 12.6. The direction of the electron flow and the direction of the conventional current are shown on the diagram.

To generate a sufficiently high voltage in a thermoelectric power device, a number of semiconductors are arranged in an array that produces an additive voltage effect. See Fig. 12.8. The arrangement alternates the *n*-type and the *p*-type semiconductors. Multiple hot junctions and cold junctions are fabricated in joining the two types of semiconductors, and the terminal connections are made on the cold side of the array.

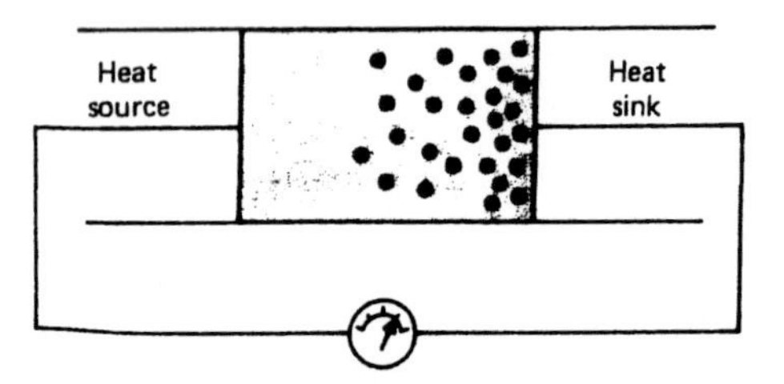

Figure 12.6 Electron concentration in a thermoelectric material.

12.7 THERMOELECTRIC GENERATOR PERFORMANCE

For prescribed hot and cold junction temperatures, the performance of a thermoelectric gen-

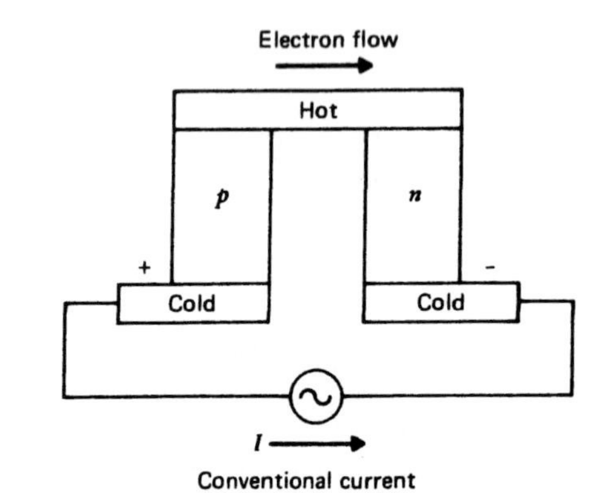

Figure 12.7 Thermoelectric couple.

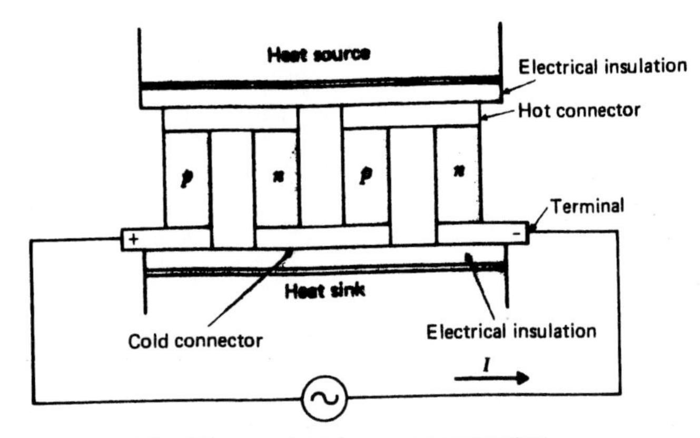

Figure 12.8 Thermoelectric power generator.

erator is measured by three parameters, namely, the Seebeck coefficient, the electrical conductivity, and the thermal conductivity. The three parameters are combined to provide a figure of merit Z for the material, as follows:

$$Z = \frac{S^2\sigma}{k} \tag{12.2}$$

where

S = the Seebeck coefficient, V/K
σ = the electrical conductivity, mho/m
k = the thermal conductivity, W/m·K

The dimension of Z is 1/K.

The system losses are influenced by both the electrical conductivity and the thermal conductivity of the material. A high electrical conductivity and a low thermal conductivity are essential to achieving a satisfactory value of Z.

The material figure of merit for a junction is defined by

$$Z = \frac{(S_1 + S_2)^2}{\left[\left(\frac{k_1}{\sigma_1}\right)^{1/2} + \left(\frac{k_2}{\sigma_2}\right)^{1/2}\right]^2} \tag{12.3}$$

The subscripts 1 and 2 designate the properties of the two different materials that form the junction. For semiconductor thermocouples, the same basic material is customarily used to fabricate both the p-type and the n-type elements, hence $k_1 \cong k_2$. The p and n properties are acquired by impurity doping that is arranged to give $S_1 \cong S_2$ and $\sigma_1 \cong \sigma_2$. As a consequence of these equalities, the material figure of merit for a junction can be given by $Z = S^2\sigma/k$.

A second figure of merit M takes account of the temperature range T_h to T_c utilized by the material.

$$M = Z\frac{T_h + T_c}{2} = Z\overline{T} \tag{12.4}$$

where $\overline{T}$ is the average operating temperature of the thermoelectric generator; M is a dimensionless quantity.

For a thermoelectric generator operating between two temperatures, T_h and T_c, the maximum conversion efficiency is given by

$$\eta_{mc} = \eta_C \frac{(1 + Z\overline{T})^{1/2} - 1}{(1 + Z\overline{T})^{1/2} + \frac{T_c}{T_h}} \tag{12.5a}$$

where η_C is the Carnot cycle thermal efficiency expressed by $1 - (T_c/T_h)$. Z is taken as the average value, $(Z_{hot} + Z_{cold})/2$.*

The variation of η_{mc}/η_C with M or $Z\overline{T}$ is shown in Fig. 12.9. Evidently, a good thermoelectric material will have a high M value.

* The derivations for Eqs. 12.5a and 12.5b are available from several sources, including Angrist, S. W. *Direct Energy Conversion*, Allyn and Bacon, Inc., Boston, 1976.

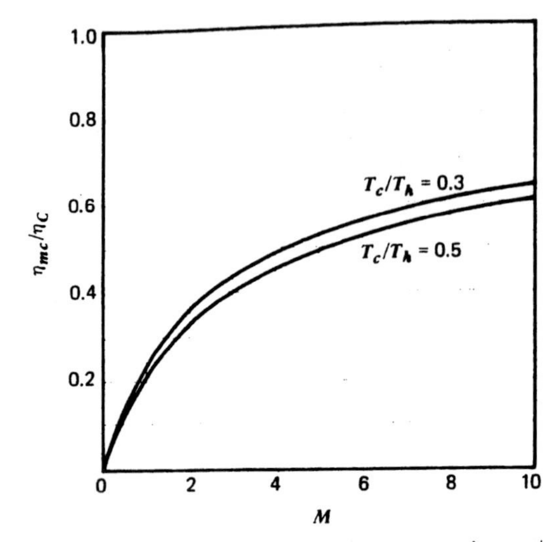

Figure 12.9 Variation of the efficiency ratio η_{mc}/η_C with M for a thermoelectric generator.

In order to achieve the maximum value of the conversion efficiency, Eq. 12.5a, the ratio of the load resistance R_L to the internal resistance R_G of the generator is required to equal $(1 + Z\overline{T})^{1/2}$. The maximum possible power output however is obtained with $R_L = R_G$. Thus, in achieving the maximum conversion efficiency of a thermoelectric generator, the power output will be slightly less than the maximum possible value. If the thermoelectric generator is operated for maximum power, the conversion efficiency will drop slightly below the optimum value given by Eq. 12.5a.

The conversion efficiency for maximum power is given by

$$\eta_{mp} = \frac{\Delta T}{T_h} \frac{1}{\dfrac{4}{ZT_h} + 2 - \dfrac{1}{2}\dfrac{\Delta T}{T_h}} \tag{12.5b}$$

Table 12.1 lists the values of the figure of merit Z for several semiconductor materials. In this group, the highest value of Z is for a modification of bismuth telluride (Bi_2Te_3). This material however has a comparatively low operating temperature and cannot be used at temperatures above 525 to 575 K. The commonly used lead telluride

TABLE 12.1 Figures of Merit for Some Typical Semiconductors

Material	*Type*	*Temperature (C)*	*Figure of Merit Z*
Bi_2Te_3	*p*	25	2.5×10^{-3}
Bi_2Te_3	*n*	25	2.5×10^{-3}
PbTe	*n*	450	1.3×10^{-3}
ZnSb	*p*	175	1.4×10^{-3}
GeTe	*p*	450	1.7×10^{-3}
MnTe	*p*	900	0.4×10^{-3}
$CeS_{1.4}$	*n*	1100	1.8×10^{-3}
$AgSbTe_2$	*p*	400	1.8×10^{-3}
InAs	*n*	700	0.7×10^{-3}

Source: Snyder, P. E. "Chemistry for Thermoelectric Materials," *Chemical and Engineering News*, March 13, 1961.[6] Reprinted with permission of The American Chemical Society.

(PbTe) has, in comparison with bismuth telluride, a somewhat higher temperature limit, that is, about 775 K, but a lower Z value.

The whole thermoelectric circuit delivers maximum power when connected to an external load that has a resistance equal to that of the circuit. Then, with proper matching of the thermoelement with the load, $R = R_o$, where R is the resistance of the thermoelement and R_o is the resistance of the external load.

The output power for a thermoelement is expressed by

$$\dot{W} = I^2 R_o$$

and

$$I = \frac{S\,\Delta T}{R + R_o} = \frac{S\,\Delta T}{2R}$$

Thus,

$$\dot{W} = \frac{S^2(\Delta T)^2}{(2R)^2} R = \frac{S^2(\Delta T)^2}{4R}$$

where

$$R = \frac{\rho L}{A}$$

or

$$\dot{W} = \frac{S^2(\Delta T)^2 A}{4\rho L}$$

The power output $\dot{W}$ delivered to the external load by each element of the thermoelectric generator is then determined with satisfactory precision from

$$\dot{W} = \frac{S^2 A(\Delta T)^2 \sigma}{4L} \qquad \text{(maximum power)} \qquad (12.6)$$

where

A = the cross-sectional area of the element
S = the Seebeck coefficient
σ = the electrical conductivity of the thermoelectric material
L = the length of the element
ΔT = the temperature gradient across the element

12.8 THERMOELECTRIC POWER SYSTEMS

Thermoelectric power generators have been developed principally for application to space vehicles and to military equipment, and for service at remote terrestrial sites and in undersea installations. These generators have been fueled with radioisotopes and fossil fuels and operate with an overall thermal efficiency of approximately 5 percent. The power output is low, that is, about 1 kW, and manufacturing costs for these power generators are relatively high. An important advantage achieved by the thermoelectric generator is the absence of moving parts.

Equation 12.5 represents the maximum energy conversion efficiency of the thermoelectric elements. In other words, the equation indicates the maximum fraction of the thermal energy entering the hot junction that is delivered as electrical energy to the external load.

Ordinarily, 90 percent or more of the heat that enters the thermoelectric elements is simply conducted through the elements to the cold conductor and consequently does not participate in the energy conversion process. The heat that must be removed from the cold junction consists not only of the energy conducted from the source but, in addition, the heat generated by the I^2R losses in the material and some heat that the electric current pumps from the hot junction to the cold junction by the Peltier effect.

On the cold side of the generator, it is essential to establish a low thermal resistance at the interface between the cold junctions of the thermoelectric elements and the cold conductor. In comparison with free-convection devices, improved air cooling can be accomplished by employing forced-convection heat exchangers, which characteristically produce higher heat transfer coefficients. The disadvantages to forced-convection air cooling are the introduction of a moving part, namely, the fan, and a reduction in the power output of the generator. The power required for driving the fan must be diverted from the gross power output of the thermoelectric elements.

For fossil fuel-fired thermoelectric generators, an important design objective is to achieve an effective transfer of heat from the combustion gases to the hot junctions of the elements. In this respect, the stack losses introduce a serious problem. The gases leave the combustion chamber at a temperature essentially equal to the temperature of the hot junctions and consequently carry from the system an appreciable amount of energy. This heat loss, in the absence of any recovery, will substantially degrade the overall thermal efficiency of the system, defined as the ratio of the net electrical energy output to the energy in the fuel.

The burner efficiency η_B is defined as the ratio of the heat transferred through the combustion chamber walls to the lower heating value of the fuel. The maximum overall thermal efficiency of the system is equal to the product of the burner efficiency and the maximum conversion efficiency of the thermoelectric elements. The burner efficiency decreases with an increase in the stack gas temperature, as indicated generally by $\eta_B = 1.0 - 0.00045T$, where T is the stack gas temperature, in C.

With possible minor exceptions, commercial application of the thermoelectric power generator appears to be unlikely in the foreseeable future because of low power capability, high capital cost, and low thermal efficiency. Discovery of cheaper and more abundant materials with superior thermoelectric properties could reduce the production cost and possibly improve the thermal performance of the generator to a significant degree.

Although the possibilities have been examined, there is little likelihood of the thermoelectric generator replacing the steam turbine-generator in large electric generating stations, or operating as a topping or a bottoming unit in conjunction with a steam turbine. The temperature of the exhaust gas from an internal combustion engine varies widely depending upon the engine expansion ratio, load, and speed. For an exhaust gas temperature in the range of 400 to 700 C, a thermoelectric generator could conceivably be used to recover some of the energy in the exhaust gas. The economics of this application of the thermoelectric generator would be a crucial factor, but the concept is attractive because in a number of ways it would be superior to an internal combustion engine/steam turbine combined-cycle power plant.

Example 12.2

A thermoelectric generator operates with a cold-junction temperature of 295 K and a hot-junction temperature of 725 K. The properties of the semiconductor materials are

At 295 K,

	n-TYPE	*p*-TYPE
S (V/K)	-160×10^{-6}	200×10^{-6}
σ (mho/cm)	960	200
k (W/cm·K)	0.0123	0.00717

At 725 K,

	n-TYPE	*p*-TYPE
S (V/K)	-132×10^{-6}	237×10^{-6}
σ (mho/cm)	640	250
k (W/cm·K)	0.0142	0.00749

Determine the maximum conversion efficiency of the generator and the corresponding thermal efficiency.

1 Material figure of merit for the junctions. For the cold junction:

$$Z = \frac{(S_1 + S_2)^2}{\left[\left(\frac{k_1}{\sigma_1}\right)^{1/2} + \left(\frac{k_2}{\sigma_2}\right)^{1/2}\right]^2}$$

$$= \frac{[(160 + 200) \times 10^{-6}]^2}{\left[\left(\frac{0.0123}{960}\right)^{0.5} + \left(\frac{0.00717}{200}\right)^{0.5}\right]^2}$$

$$= 1.416 \times 10^{-3} \quad 1/\text{K}$$

A similar calculation for the hot junction yields

$$Z = 1.313 \times 10^{-3} \quad 1/\text{K}$$

The average value of Z for the two junctions is 1.364×10^{-3} 1/K.

2 Maximum conversion efficiency.

$$\eta_C = \frac{T_h - T_c}{T_h} = \frac{725 - 295}{725} = 0.593$$

$$\overline{T} = 510 \text{ K}$$

$$(1 + Z\overline{T})^{1/2} = [1 + (0.001364 \times 510)]^{0.5} = 1.302$$

$$\eta_{mc} = \eta_C \frac{(1 + Z\overline{T})^{1/2} - 1}{(1 + Z\overline{T})^{1/2} + \frac{T_c}{T_h}}$$

$$= 0.593 \frac{1.302 - 1}{1.302 + \frac{295}{725}}$$

$$= 0.105$$

3 Thermal efficiency
Burner efficiency:

$$\eta_B = 1.0 - 0.00045T$$

$$= 1.0 - (0.00045 \times 452) = 0.80$$

$$\eta_t = \eta_B \times \eta_{mc}$$

$$= 0.80 \times 0.105 = 0.084$$

The thermal efficiency of the thermoelectric generator under actual operating conditions will be somewhat below 0.080, say, about 5 percent.

Example 12.3

A thermoelectric generator is fabricated from *p*-type and *n*-type semiconductor materials (PbTe) that have the same S, σ, and k properties in the operating range of 295 to 725 K. At the mean temperature of 510 K,

$$S = 187 \times 10^{-6} \text{ V/K}$$
$$\rho = 1.64 \times 10^{-3} \text{ ohm·cm}$$
$$k = 0.0146 \text{ W/cm·K}$$

The length of the thermoelements is 1.0 cm, and the cross-sectional area is 1.20 cm^2. Determine the maximum power and the corresponding conversion efficiency. Calculate the output voltage for the thermoelectric couple and the current established in the circuit.

1 Maximum power.
Applying Eq. 12.6,

$$\dot{W} = \frac{S^2A(\Delta T)^2\sigma}{4L}$$
$$= \frac{(187 \times 10^{-6})^2 1.20(430)^2 610}{4 \times 1.00}$$
$$= 1.183 \text{ W} \quad \text{(per thermoelement)}$$

$$\frac{\dot{W}}{A} = \frac{1.183}{1.20} = 0.986 \text{ W/cm}^2$$

2 Conversion efficiency for maximum power.

$$Z = \frac{S^2\sigma}{k} = \frac{S^2}{\rho k}$$
$$= \frac{(187 \times 10^{-6})^2}{(1.64 \times 10^{-3})0.0146}$$
$$= 1.460 \times 10^{-3} \text{ 1/K}$$

Applying Eq. 12.5b,

$$\eta_{mp} = \frac{\Delta T}{T_h} \frac{1}{\dfrac{4}{ZT_h} + 2 - \dfrac{1}{2}\dfrac{\Delta T}{T_h}}$$
$$= \frac{430}{725} \frac{1}{\dfrac{4}{(1.460 \times 10^{-3})725} + 2 - \dfrac{1}{2}\dfrac{430}{725}}$$
$$= 0.1082$$

3 Output voltage.
The open-circuit voltage is equal to the product of the Seebeck coefficient, for both legs of the couple, and the temperature difference across the legs.

$$V_{oc} = 2S\,\Delta T = 2(187 \times 10^{-6})430 = 0.1608 \text{ V}$$

Total resistance:

$$R_t = R_L + R_G$$

where $R_L = R_G$ for maximum power

Voltage across the load:

$$V_L = \frac{R_L}{R_t} V_{oc} = \frac{R_L}{2R_L} V_{oc} = \frac{1}{2} 0.1608 = 0.0804 \text{ V}$$

The output voltage of the thermoelectric generator is increased by the addition of thermoelectric couples.

4 Current.

$$\dot{W} = 2 \times 1.183 \text{ W} \quad \text{(for the thermoelectric couple)}$$

$$I = \frac{\dot{W}}{V_L} = \frac{2.366}{0.0804} = 29.43 \text{ A}$$

An alternative calculation shows for a single leg of the couple.

$$R = \frac{\rho L}{A}$$
$$= \frac{(1.64 \times 10^{-3})1.0}{1.20} = 0.001367 \text{ ohm}$$

$$I^2 = \frac{\dot{W}}{R} = \frac{1.183}{0.001367}$$

$$I = 29.42 \text{ A}$$

5 Performance of the thermoelectric couple for operation at maximum conversion efficiency.

From Eq. 12.5a,

$$\eta_{mc} = 0.1102$$

$$Z = \frac{S^2}{\rho k}$$

$$= \frac{(187 \times 10^{-6})^2}{(1.64 \times 10^{-3})0.0146} = 1.460 \times 10^{-3} \text{ 1/K}$$

$$\overline{T} = 510 \text{ K} \qquad (1 + Z\overline{T})^{1/2} = 1.321$$

$$R_G = 2R = 2 \times 0.001367 = 0.002734 \text{ ohm}$$

$$R_L = R_G(1 + Z\overline{T})^{1/2}$$

$$= 0.002734 \times 1.321 = 0.003612 \text{ ohm}$$

$$R_t = R_L + R_G$$

$$= 0.003612 + 0.002734 = 0.006346 \text{ ohm}$$

$$I = \frac{V_{oc}}{R_t} = \frac{0.1608}{0.006346} = 25.34 \text{ A}$$

$$\dot{W} = I^2 R_L = (25.34)^2 \times 0.003612$$

$$= 2.319 \text{ W} \qquad \text{(per couple)}$$

$$\dot{W} = 1.160 \text{ W} \qquad \text{(per thermoelement)}$$

12.9 THERMOELECTRIC REFRIGERATION

The generation of an electric current by establishing a temperature difference across a thermoelectric material suggests that a countereffect can be created by passing an electric current through the material. This important discovery was made in 1834 by Jean Peltier, a French artisan. Peltier observed that when an electric current was passed through a junction of two dissimilar materials, a heating or cooling effect would occur at the junction, depending upon the magnitude and direction of the current.

An emf exists at the common surface of the two dissimilar materials. When the electric current is made to flow against this emf, heating in excess of that caused by the ohmic resistance of the conductor will result. When the current flows in the direction of the emf, a cooling effect can occur. In the first case, a definite heating effect is observed at the junction because the Joule (ohmic) heating and the Peltier effects are additive. In the second case, a difference between the Joule heating and the Peltier effects is observed, and by properly adjusting the current, a cooling effect can be created.

Incidentally, a third thermoelectric effect was discovered by Lord Kelvin (William Thomson) in 1856. In analyzing the performance of a thermoelectric energy converter, the Thomson effect, because of its minor importance, is however not considered.

It is interesting to note that Peltier failed to understand the meaning of his discovery. Four years later, in 1838, Emil Lenz presented a correct explanation of Peltier's observations. Some seven decades however were to elapse before the Peltier effect was employed in a refrigeration system. The thermoelectric refrigeration units developed in recent years are, like the thermoelectric power generators, intended for special applications and currently have little commercial utility.

12.10 ANALYSIS OF THERMOELECTRIC COOLING DEVICES

The basic equation used in the design of Peltier devices is the expression for the determination of the net heat removed from the cold junction, namely,

$$\dot{Q}_c = SIT_c - \frac{1}{2} I^2 \rho m - \frac{k}{m} \Delta T \qquad (12.7)$$

where both legs of the couple are assumed to have the same length L and the same cross-sectional area A, and

where

S = the Seebeck coefficient
I = the current
T_c = the temperature of the cold junction
ρ = the electrical resistivity of the thermoelectric material
k = the thermal conductivity of the thermoelectric material
m = the length-to-area ratio L/A
ΔT = the temperature difference between the hot and the cold junctions, $T_h - T_c$

In Eq. 12.7, the Seebeck coefficient, the electrical resistivity, and the thermal conductivity are evaluated, respectively, from

$$S = \bar{S}_p - \bar{S}_n$$
$$\rho = \bar{\rho}_p + \bar{\rho}_n$$
$$k = \bar{k}_p + \bar{k}_n$$

The bar denotes an average value of the property over the range between the hot-junction and the cold-junction temperatures.

The three terms on the right-hand side of Eq. 12.7 represent, respectively, the Peltier cooling effect, one-half of the Joule heat that flows back to the cold junction, and the heat transferred by conduction to the cold junction because of the temperature difference between the two junctions.

The heat transfer $\dot{Q}_c$ given by Eq. 12.7 is reduced, to some degree, by the heat conducted through the insulation from the hot junction to the cold junction. In order to account for this heat leakage, a quantity $-\dot{Q}_i$ is inserted on the right-hand side of Eq. 12.7.

$$\dot{Q}_i = k_i A_i (T_h - T_c)/L$$

where k_i is the thermal conductivity of the insulation and A_i is the effective heat transfer area of the insulation. The magnitude of $\dot{Q}_i$ will be dependent, in part, on the kind and the quantity of insulation used in fabricating the cooling device.

The design method most often used for Peltier cooling devices provides for selecting a geometry for a given application and optimizing this design. The design is commonly optimized for the prescribed geometry by determining the current that will pump the maximum amount of heat. This current, designated as the optimum current I_o, is determined by differentiating the quantity $\dot{Q}_c$ in Eq. 12.7 with respect to I and setting the result equal to zero. Thus,

$$\frac{d\dot{Q}_c}{dI} = ST_c - I\rho m = 0$$

and

$$I_o = \frac{ST_c}{\rho m} \tag{12.8}$$

The maximum temperature difference that can be produced across the couple occurs when the current is equal to I_o and the heat removed from the cold reservoir is equal to zero. Thus, for Eq. 12.7, $\dot{Q}_c = 0$ and $I = I_o$. Then,

$$\Delta T_{\text{max}} = \frac{S^2 T_c^2}{2\rho k} \tag{12.9}$$

The heat removed from the cold reservoir by the optimum current is determined by substituting the equivalent of I_o, Eq. 12.8, into Eq. 12.7. Thus,

$$(\dot{Q}_c)_{I,o} = \frac{S^2 T_c^2}{2\rho m} - \frac{k}{m}\Delta T$$

This equation is readily reduced to

$$(\dot{Q}_c)_{I,o} = \frac{k}{m}(\Delta T_{\text{max}} - \Delta T) \tag{12.10}$$

The construction of a thermoelectric cooling device is represented schematically in Fig. 12.10.

Example 12.4

A thermoelectric cooler is to be designed for operation with a cold conductor temperature of

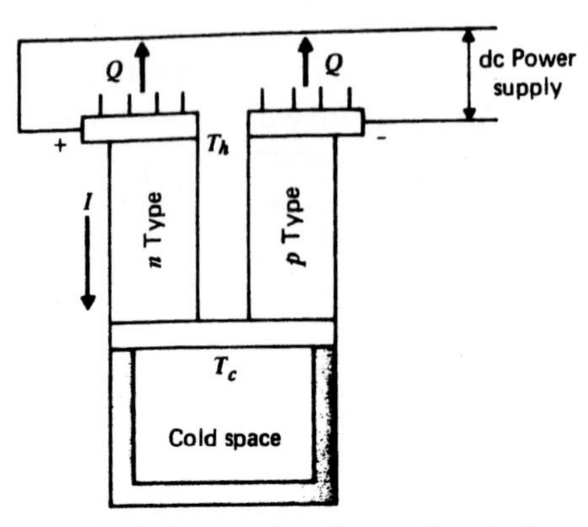

Figure 12.10 A thermoelectric cooling device.

270 K and a hot conductor temperature of 300 K. $A = 0.250\ \text{cm}^2$; $L = 1.05$ cm. For the thermoelectric material,

$$\bar{S}_n = -190 \times 10^{-6}\ \text{V/K}$$

$$\bar{S}_p = 190 \times 10^{-6}\ \text{V/K}$$

$$\bar{\rho}_n = \bar{\rho}_p = 1.05 \times 10^{-3}\ \text{ohm·cm}$$

$$\bar{k}_n = \bar{k}_p = 0.0125\ \text{W/cm·K}$$

Determine $\dot{Q}_c$ and the coefficient of performance for a design in which I is optimized.

$$S = \bar{S}_p - \bar{S}_n = [190 - (-190)] \times 10^{-6} = 380 \times 10^{-6}\ \text{V/K}$$

$$\rho = \bar{\rho}_p + \bar{\rho}_n = (1.05 + 1.05) \times 10^{-3} = 2.10 \times 10^{-3}\ \text{ohm·cm}$$

$$k = \bar{k}_p + \bar{k}_n = 0.0125 + 0.0125 = 0.0250\ \text{W/cm·K}$$

1 Current.

$$m = \frac{L}{A} = \frac{1.05}{0.250} = 4.20\ \text{cm}^{-1}$$

$$I_o = \frac{ST_c}{\rho m} = \frac{(380 \times 10^{-6})270}{(2.10 \times 10^{-3})4.20} = 11.63\ \text{A}$$

2 Cooling rate.

$$\Delta T_{\max} = \frac{S^2 T_c^2}{2\rho k}$$

$$= \frac{(380 \times 10^{-6})^2(270)^2}{2(2.10 \times 10^{-3})0.0250} = 100.25\ \text{K}$$

$$(\dot{Q}_c)_{I,o} = \frac{k}{m}(\Delta T_{\max} - \Delta T)$$

$$= \frac{0.0250}{4.20}(100.25 - 30) = 0.418\ \text{W}$$

3 Power input.

$$R = \frac{\rho L}{A} = \rho m$$

$$= (2.10 \times 10^{-3})4.20 = 0.00882\ \text{ohm}$$

$$\dot{W} = I^2 R$$

$$= (11.63)^2\ 0.00882 = 1.193\ \text{W}$$

4 Coefficient of performance.

$$\text{cop} = \frac{\dot{Q}_c}{\dot{W}} = \frac{0.418}{1.193} = 0.350$$

In reviewing the preceding design method, Crosby et al.[7] show that Eq. 12.10 does not represent the maximum amount of heat pumping for the current I_o. For an improved design concept, the heat pumping is optimized with respect to the length-to-area ratio m.

The quantity $\dot{Q}_c$ in Eq. 12.7 is differentiated with respect to m, and the result is set equal to zero. Then,

$$\frac{d\dot{Q}_c}{dm} = -\frac{1}{2}I^2\rho + \frac{k\,\Delta T}{m^2} = 0$$

and

$$m_o = \frac{1}{I}\left(\frac{2k\,\Delta T}{\rho}\right)^{1/2} \tag{12.11}$$

When I is set equal to I_o, the following relationship is established:

$$\frac{m_o}{m_{I,o}} = \left(\frac{\Delta T}{\Delta T_{\max}}\right)^{1/2} \tag{12.12}$$

Optimizing the design for m rather than for I is particularly significant. For prescribed values of the current and the cross-sectional area of the element, the length of the couple legs will be reduced and more heat will be pumped. Further, less thermoelectric material is required and a higher coefficient of performance is achieved.

Example 12.5

The thermoelectric cooler described in Example 12.4 is redesigned for the optimum value of m. $A = 0.250\ \text{cm}^2$; $I = 11.63$ A.

1 Optimum value of m.

$$m_o = m_{I,o}\left(\frac{\Delta T}{\Delta T_{\max}}\right)^{1/2}$$

$$= 4.20\left(\frac{30}{100.25}\right)^{0.5} = 2.298\ \text{cm}^{-1}$$

2 Cooling rate.

When the design of the thermoelectric device is optimized for m, Eq. 12.7 is used to determine the cooling rate. Thus,

$$\dot{Q}_c = SIT_c - \frac{1}{2}I^2\rho m - \frac{k}{m}\Delta T$$

$$= (380 \times 10^{-6})11.63 \times 270 - {}^1\!/_2(11.63)^2(2.10 \times 10^{-3})2.298 - \frac{0.0250}{2.298}30 = 0.541\ \text{W}$$

3 Power input.

$$R = \rho m = (2.10 \times 10^{-3})2.298 = 0.00483\ \text{ohm}$$

$$\dot{W} = I^2R = (11.63)^2 \times 0.00483 = 0.653\ \text{W}$$

4 Coefficient of performance.

$$\text{cop} = \frac{\dot{Q}_c}{\dot{W}} = \frac{0.541}{0.653} = 0.828$$

Examples 12.4 and 12.5 are indicative of the influence of certain parameters on the design of a thermoelectric cooler. Somewhat different results would be observed for an actual design of one of these devices.

When the design is optimized for m rather than for I, the length of the couple legs is reduced, and as a consequence an increase is experienced in the cooling rate. Actually, the heat conduction from the hot to the cold junction increases, but a reduction of greater magnitude occurs in the Joule heating. Further, the power input to the couple is reduced because of the decreased electrical resistance of the shorter legs, and the net overall effect is an improvement in the coefficient of performance of the thermoelectric cooler.

Characteristically, the design optimized for m shows that the second and third terms on the right-hand side of Eq. 12.7 are of like magnitude. Thus, for this case,

$${}^1\!/_2 I^2\rho m = k\,\Delta T/m$$

Thermionic Power

12.11 THERMIONIC EMISSION

The working concept of the thermionic energy converter is somewhat similar to that of the various conventional thermal engines, for example, the steam power plant. Heat is supplied to the thermionic element at a high temperature, and a portion of this heat is rejected from the element at a low temperature. The difference between Q_s and Q_r is the portion of the supplied heat that is converted, in principle, directly into electrical energy through the flow of electrons in an electric circuit. The conversion of thermal energy into electrical energy by a thermionic converter is accomplished without employing any moving parts.

Thermionic emission was investigated by Thomas Edison and contemporary workers during the 1880–1890 decade. Other investigations were made in subsequent years, but development of the thermionic energy converter only reached a significant stage within the past two decades. Thermionic devices of comparatively low power capability were developed principally for military and space applications. For thermionic energy converters used in space, the heat source is either solar energy or nuclear energy, while a land-based converter would operate on a fossil or nuclear fuel.

Figure 12.11 illustrates the operating principle of the thermionic energy converter. The device in its simplest form consists of two flat-plate electrodes enclosed in a plasma or vacuum. The electrodes, designated as the cathode and the anode, are fabricated from dissimilar materials. Heat is transferred from the heat source to the cathode in order to raise the temperature of this electrode sufficiently to cause electrons, because of thermal agitation, to "boil off" and flow to the anode. The electrons are collected on the anode, which is maintained at a temperature below that of the cathode. When the two electrodes are connected externally, the electrons flow in the external circuit from the anode to the cathode.

Some discussion of the electron flow is essential to describing the operating principle of the thermionic energy converter. The electrons that move in the outermost orbits of the atoms in the electrodes are not strongly bound to the nuclei. These electrons consequently can move about randomly from an orbit around one nucleus to an orbit around another nucleus and thus constitute the free electrons of the electrodes.

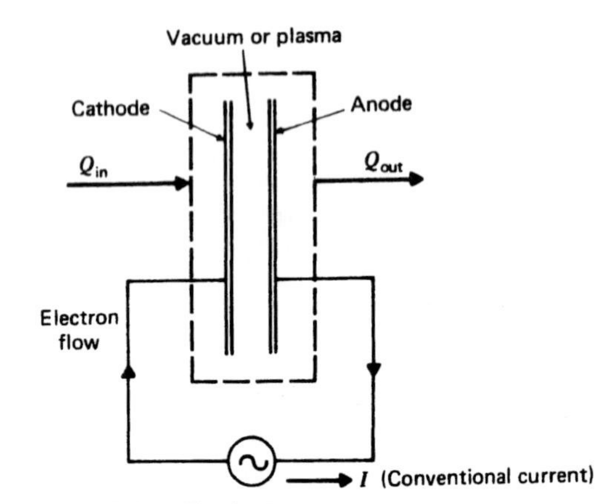

Figure 12.11 Basic thermionic energy converter.

The kinetic energy of the free electrons can be distributed to certain discrete energy levels that differ from one another by a very small but finite amount of energy. A limited number of free electrons can be contained in each energy level although, similarly to an electron orbit, an energy level can be empty, partly filled, or completely filled. The highest level of energy corresponds to the Fermi level, and the electrons in this level have the highest energy of the free electrons.

The electrons in the thermionic energy converter have different Fermi levels, that is, the cathode has a low Fermi level and the anode has a high Fermi level. In other words, the free electrons in the anode Fermi level possess more energy than the free electrons in the cathode Fermi level.

The energy that is required to cause a free electron in the cathode Fermi level to escape from the cathode body and come to rest, just outside the cathode surface, is called the cathode work function. There is a corresponding energy quantity, related to the escape of a free electron from the anode body, designated as the anode work function.

When a sufficient amount of heat from a high-temperature source is supplied to the cathode, high-energy free electrons in the Fermi level will acquire enough energy to escape from the cathode body and surface. The amount of energy gained by the electrons is equal to the work function of the cathode plus a quantity of kinetic energy. If the electrons pass the interelectrode gap between the cathode and the anode, without a loss in energy, the electrons will lose energy as they strike the surface and enter the anode body. The lost energy is equal to the kinetic energy of the electrons plus the energy corresponding to the anode work function. A quantity of energy equivalent to the work function is required to cause the electron to pass through the anode surface. The lost energy is rejected as heat from the low-temperature anode.

Figure 12.12 shows the relative magnitudes of

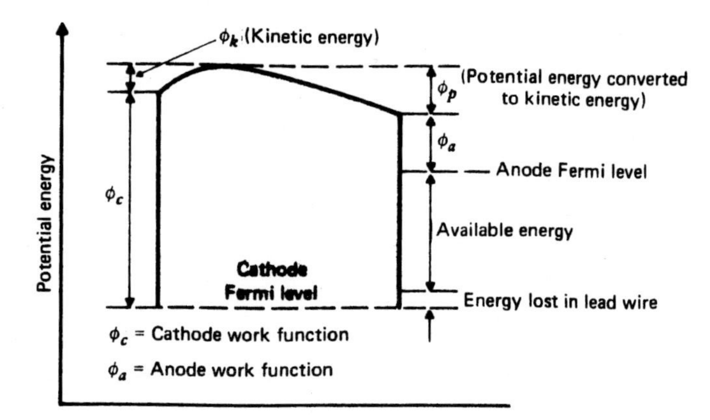

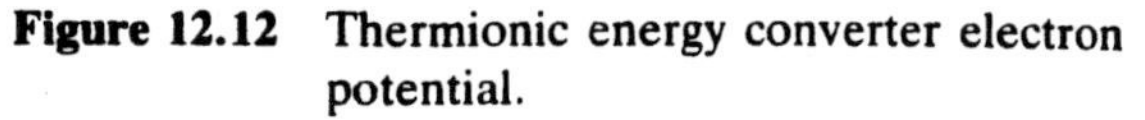
Figure 12.12 Thermionic energy converter electron potential.

the several energy quantities pertinent to the thermionic energy converter. Because the cathode work function is larger than the anode work function, a net amount of energy becomes available for performing work once the electrons have entered the anode and acquired the status of free electrons in the anode Fermi level. The available energy is a measure of the power that is produced by the thermionic energy converter when the electrons flow from the cathode, through the interelectrode gap, to the anode and back to the cathode through the external circuit, which consists of the load and the lead wire. The overall effect is the conversion of heat directly into electrical energy.

In common with any flow of electrons in a vacuum, a space charge is formed in the interelectrode gap. The space charge, a mutual electrical repulsion of the electrons, inhibits the electron flow from the emitter. The effect of the space charge can be minimized by restricting the interelectrode spacing to very small distances, that is, less than 0.01 mm. Converters constructed in this manner are known as vacuum or close-spaced diodes. A plasma diode avoids the close electrode spacing by introducing a gas that can be readily ionized to provide positive ions in the interelectrode space. When the proper number of ions are present, the space charge barrier is eliminated and the heat losses in the converter are reduced. The operating lifetime of both type of diodes is limited by problems related to fabrication techniques and material failures.

12.12 PERFORMANCE OF THERMIONIC ENERGY CONVERTERS

The thermal efficiency of the thermionic energy converter is defined as the power developed across the load divided by the total input heat rate. No losses are assumed in the interelectrode gap. The anode temperature is assumed sufficiently low to obviate any electron flow in the interelectrode gap from the anode to the cathode.

The converter output is equal to the cathode work function, minus the anode work function and the loss in the lead wire. $W_{out} = \phi_c - \phi_a - \phi_l$. The total heat input is equal to the sum of the cathode work function, the kinetic energy of the electron, and the losses that consist of the radiation loss between the cathode and the anode and of the heat conduction along the lead wire. $Q_{in} = \phi_c + 2kT_c + Q_{r+l}$. Then

$$\eta_t = \frac{\phi_c - \phi_a - \phi_l}{\phi_c + 2kT_c + Q_{r+l}} \tag{12.13}$$

where

ϕ_c = cathode work function, eV (electron volt)
ϕ_a = anode work function, eV
ϕ_l = energy loss in the lead wire, eV
k = Boltzmann constant, 1.38054×10^{-23} J/K
T_c = cathode temperature, K
Q_{r+l} = radiative and conductive heat loss

Note: 1 eV = 1.60210×10^{-19} J

Equation 12.13 can be simplified by neglecting the Joule losses, that is, the energy losses because of the electrical resistance in the electrodes and in the lead wires, and by neglecting the heat conduction in the external leads. Thus, $\phi_l = 0$, and $Q_{r+l} = Q_r$. Q_r is the energy radiated from the cathode to the anode.

The terms ϕ_c, ϕ_a, and $2kT_c$ are divided by e, the elementary charge, which has a value of 1.60210×10^{-19} C or A·s (coulomb). Thus, $\phi_c/e = V_c$ (in volts). The value of $k/e = 0.86171 \times 10^{-4}$ V/K.

The current J_{max} is introduced into Eq. 12.13 and yields

$$\eta_t = \frac{J_{max}(V_c - V_a)}{J_{max}\left(V_c + 2\dfrac{k}{e}T_c\right) + \dot{Q}_r} \tag{12.14}$$

The saturation current, expressed in amperes per cm², is calculated from the Richardson–Dushman equation:

$$J_{max} = AT_c^2 \exp\left(-\frac{\phi_c}{kT_c}\right) \tag{12.15}$$

where A, the Richardson–Dushman constant, is equal to 120 A/cm² T_c^2; and J_{max} is the maximum electron current that an emitting surface can produce per unit area at temperature T_c.

In order to express the work function in volts, e is inserted into Eq. 12.15 to yield

$$J_{max} = AT_c^2 \exp(-11605V_c/T_c) \text{ A/cm}^2 \tag{12.16}$$

In order to establish consistent units, the radiation loss $\dot{Q}_r$ in Eq. 12.14 is expressed in W/cm² emitter surface:

$$\dot{Q}_r = \sigma F_e(T_c^4 - T_a^4) \tag{12.17}$$

The value of σ, the Stefan–Boltzmann constant, is 5.6697×10^{-12} W/cm²·K⁴, and

$$F_e = \frac{1}{\dfrac{1}{e_c} + \dfrac{A_c}{A_a}\left(\dfrac{1}{e_a} - 1\right)} \tag{12.17a}$$

where e_c and e_a are, respectively, the emissivity values for the cathode and anode surfaces, and A_c and A_a are the corresponding surface areas.

Equation 12.14 is indicative of the performance of a thermionic energy converter. A more complex procedure however must be used in an actual design when evaluating the thermal performance of a thermionic device.

Example 12.6

A thermionic energy converter operates with a cathode temperature of 1150 K and an anode temperature of 520 K. V_c = 1.72 volts and V_a = 1.12 volts. Emissivity values are 0.11 for the cathode and 0.08 for the anode. A_c/A_a = 1.0. Calculate the maximum current and the corresponding power and thermal efficiency for the converter.

1 Maximum current.

$$J_{max} = AT_c^2 \exp\left(-\frac{11605V_c}{T_c}\right)$$
$$= 120(1150)^2 \exp\left(-\frac{11605 \times 1.72}{1150}\right)$$
$$= 4.60 \text{ A/cm}^2$$

2 Power.

$$\dot{W} = J_{max}(V_c - V_a)$$
$$= 4.60(1.72 - 1.12) = 2.76 \text{ W/cm}^2$$

3 Heat supplied.

$$F_e = \frac{1}{\dfrac{1}{e_c} + \dfrac{1}{e_a} - 1}$$

$$= \frac{1}{\dfrac{1}{0.11} + \dfrac{1}{0.08} - 1} = 0.0486$$

$$\dot{Q}_r = \sigma F_e(T_c^4 - T_a^4)$$
$$= (5.6697 \times 10^{-12})0.0486[(1150)^4 - (520)^4]$$
$$= 0.462 \text{ W/cm}^2$$

$$\dot{Q}_s = J_{\max}\left(V_c + 2\frac{k}{e}T_c\right) + \dot{Q}_r$$
$$= 4.60[1.72 + 2(0.86171 \times 10^{-4})1150] + 0.46$$
$$= 8.82 + 0.46 = 9.28 \text{ W/cm}^2$$

4 Thermal efficiency.

$$\eta_t = \frac{\dot{W}}{\dot{Q}_s} = \frac{2.76}{9.28} = 0.297$$

The theoretical thermal efficiency of the thermionic energy converter is equal to the Carnot cycle thermal efficiency for operation between the prescribed temperatures T_c and T_a, the cathode and the anode temperatures, respectively, of the thermionic device. For the prescribed temperatures in Example 12.6, $T_c = 1150$ K and $T_a = 520$ K, the Carnot cycle thermal efficiency is equal to 0.548.

Because of the neglect of some of the losses inherent in the device, the calculated thermal efficiency in Example 12.6 is high in comparison with the thermal efficiency achieved by a practical thermionic energy converter. The thermal efficiency of an actual thermionic device will be dependent upon the emitter and the collector temperatures and upon the system characteristics, and will range generally from 10 to 15 percent. With improved materials available for fabricating the emitter and the collector, an increase in the conversion efficiency to about 20 to 25 percent may be possible.

The thermionic energy converter is a simple device, and, unlike the conventional power plants, the thermal efficiency is independent of the size of the device. A low-voltage, high direct-current output is characteristic of the thermionic energy converter. A practical converter operates with a cathode temperature in the range of 1200 to 2000 K and an anode temperature less than 1000 K. The converter produces a current density of about 5 to 10 A/cm^2 at 0.5 to 2 V.

For a prescribed collector temperature of 1000 K, an increase in the emitter temperature from 1800 K to 2500 K will increase the conversion efficiency by a factor of about 1.5, while the power density (in W/cm^2) is increased by a factor of about 3. A significant reduction in the conversion efficiency and in the power density will be experienced with an increase in the collector temperature for a prescribed emitter temperature.

12.13 THERMIONIC ENERGY CONVERTER CONSTRUCTION

Various configurations are employed in fabricating thermionic devices. Two common arrangements are closely spaced flat plates and concentric cylinders. The latter arrangement is used, although not exclusively, for nuclear thermionic energy converters, with the inner cylinder, a fuel rod, functioning as the emitter.

The emitter–collector spacing is an important factor in the design of the thermionic energy converter. An increase in the distance between the emitter and the collector causes the conversion efficiency and the output power density to decrease. For example, an increase in the emitter–collector spacing from 0.125 mm to 0.500 mm would cause a substantial decline in the output power density and the conversion efficiency, perhaps on the order of 40 and 15 percent, respectively. While a very small spacing is desirable with respect to the thermal performance of the thermionic energy converter, a minimum spacing of 0.25 mm would likely be prescribed when consideration is given to the practical fabrication limitations and reliability requirements. In comparison with the emitter–collector spacing effect, the cathode-to-anode area ratio has a relatively small effect on the performance of the thermionic energy converter.

Suitable materials for fabrication of the cathode of a thermionic device are tungsten, molybdenum, and tantalum. Silver oxide, tungsten oxide, and copper are suitable anode materials. Cesium vapor is most frequently used in a plasma-type converter for reducing the space charge, although rubidium and potassium can also be employed for this purpose.

In order to function effectively, the thermionic energy converter must operate in a high-temperature range; as a consequence, heat is rejected at a relatively high temperature. As a means to conserving this rejected heat, the thermionic energy converter is inherently suited to function as a topping power unit, for example, operating in conjunction with a steam power plant.

12.14 THERMIONIC TOPPING POWER PLANT

The feasibility of topping a coal-fired steam power plant by a thermionic energy converter has been explored by the Office of Coal Research.[8] The block diagram for the proposed system is shown in Fig. 12.13. The steam power plant follows a conventional design that incorporates regenerative feedwater heating.

Pulverized coal is fired in the thermionic converter furnace with the combustion air preheated to 1260 C. The combustion gases pass over the thermionic converters, raising the emitter temperature to about 1510 C, and leave the furnace at a temperature of about 1700 C. Subsequent to discharge from the furnace, the combustion gases flow successively through the final air heater, the economizer, and a conventional rotary air heater. The gases enter the stack at a temperature of 150 C. The final air heater is designed and fabricated for high-temperature operation.

In the proposed system, the final air heater is an important element, because a high air preheat temperature is essential to establishing the high furnace temperature that is required for operating the thermionic emitter at a sufficiently elevated temperature. The gas and air temperatures for this heater are shown in Fig. 12.13.

The steam generated in the economizer leaves the element at a state above the critical state of water. Subsequently, the steam is superheated to 538 C in passing over the thermionic collector surface, which is maintained at about 650 C. Following expansion in the high-pressure turbine, the steam is reheated from 288 C to 538 C through contact with the thermionic collector surface.

For the system depicted in Fig. 12.13, 20 percent of the total power is generated in the thermionic energy converter and the balance in the steam plant. There are several ways in which the dc output from the thermionic converter may be conditioned. The most promising approach appears to involve the use of a dc acyclic motor coupled to one of the turbine-generator shafts. This comparatively simple arrangement transmits the thermionic converter power directly to the input shaft of the ac generator and hence into the station electrical distribution system. The acyclic motor is a low-voltage, high-current (typically 150,000 A) dc machine capable of operation at power levels in the range of several hundred megawatts, and at speeds that allow direct coupling to the shaft of an 1800-rpm generator. The anticipated efficiency of an acyclic motor in the 200-MW class is 98 percent or higher.

In order to demonstrate the improvement in thermal performance that may be derived from thermionic topping, a comparative study was conducted on the T.V.A. Bull Run power plant.[8] The study prescribes a modular arrangement for the thermionic elements. Each thermionic module is rated at 20 kW electrical output and consists of an internal series string of 25 separate diodes, each of which delivers about 0.7 V and 1140 A.

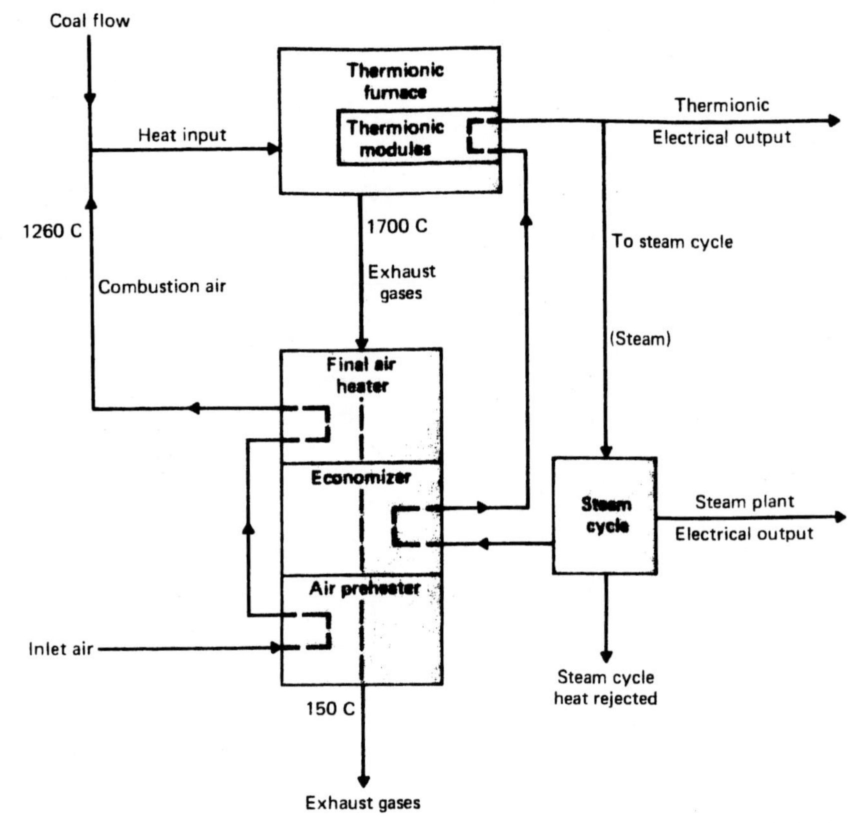

Figure 12.13 Block diagram for a thermionic topping power system. (Office of Coal Research, U.S. Department of Interior.[8])

The thermionic module is protected against the corrosive effects of the fossil fuel combustion gases by a silicon carbide envelope. Heat is transferred from the combustion gases through the envelope to a heat transfer medium that, in turn, delivers the thermal energy to the emitter structure. Collector cooling is accomplished by the flow of the single-phase superheat and reheat steam. A conversion efficiency of 15 percent is predicted; and on this basis, the heat rate for the diode module is 24,000 kJ/kWh.

The application of the thermionic topping unit to the Bull Run power plant is based on the installation of 10 thermionic furnace modules. In each furnace, there will be installed 1120 thermionic modules, 817 for superheating and 303 for reheating the steam. The overall total number of thermionic modules for the entire power plant is 11,200.

The comparative study conducted on the Bull Run power plant indicates that thermionic topping would increase the station output and thermal efficiency from 914 MW at 41.3 percent to 1139 MW at 50.6 percent. Thermionic topping is estimated to result in lower electrical generating capital costs, and an appreciable reduction in the annual fuel cost is anticipated.

Application of thermionic topping to nuclear power plants as well as to fossil fuel plants is a possibility. An analysis conducted by Engdahl et al.[9] shows that thermionic topping applied as an addition to a 450-MW boiling water reactor (BWR) nuclear power plant will increase the output to 832 MW and raise the thermal efficiency from 32 to 45 percent. A significant reduction in the capital cost of the power plant would be achieved, and appreciably lower net power generating costs are anticipated.

Before thermionic topping can be realized, the operating capability of the thermionic energy converter must be demonstrated. The converter must have an acceptable operating lifetime and function properly under the severe conditions encountered in the furnace, caused principally by the high operating temperatures and the corrosive ash. Development of improved materials for converter fabrication will contribute significantly to achieving commercial adaptation of the thermionic energy converter.

The second part of the overall program conducted by the Office of Coal Research involved the design, construction, and test of several thermionic converters. Of particular significance was the development of an emitter envelope capable of protecting the converter against the combustion gas and coal slag environment. Eighteen converters were tested, and power levels up to 133 W were obtained.

Photovoltaic Power

12.15 SOLAR CELLS

One important phase of solar energy utilization is the direct conversion of solar energy into electrical energy by means of the photovoltaic effect. The first photovoltaic cells capable of operating with practical conversion efficiencies, that is, above 1 percent, were manufactured at the Bell Laboratories early in the 1950 decade. By the year 1955, field tests had been conducted on small panels of cells mounted on the tops of telephone poles for the purpose of supplying power to telephone amplifiers. The most significant application of the silicon solar cell has occurred in space exploration, where the cells are used to produce electrical energy for the operation of various devices installed in space vehicles and artificial satellites. The solar cell can presently be regarded as an appropriate energy converter, for a limited number of applications, when a small amount of power will suffice and when high capital costs, in dollars per kilowatt, are acceptable.

A wide variety of substances exhibit a measurable photovoltaic effect. Only semiconductors however are known to have the capability of generating electric power at a level sufficiently high for practical applications. Among the semiconductor materials commonly used are silicon, germanium, cadmium sulfide, and gallium arsenide.

The silicon solar cell is most often employed for the solar-electric direct energy conversion process because of a superior conversion efficiency and a comparatively longer life. The batch-type process for manufacturing the single-crystal silicon solar cell is time consuming, requires precise temperature control, and depends upon the services of highly skilled technicians. The manufacture of the typical cell begins by pulling silicon crystals from a melt and subsequently sawing the material into wafers 1 mm thick. The next step, shaping and polishing the thin wafers, is an exacting operation.

The conventional silicon solar cell is a *n*/*p*-type semiconductor device, 2 × 2 cm in area and 0.25 to 0.35 mm thick. The cell has a boron *p*-type base with a phosphorus-diffused, *n*-type light-sensitive top layer. Following polishing and surface preparation, the thin *n*-type layer is formed by diffusing phosphorus, under the influence of high temperature, into the boron base material.

The thickness of the solar cell is a design factor of some significance. A 2-ohm·cm *n*/*p*-type silicon solar cell, which has a conversion efficiency of 0.114 at a thickness of 0.40 mm, would have a conversion efficiency of about 0.096 at a thickness reduced to 0.10 mm.[10] In general, the power-to-mass ratio increases with a decrease in the thickness of the solar cell.

Earlier solar cells were of the *p*/*n* type, that is, the positive surface faced the sun. These cells were susceptible to rapid deterioration from destructive solar radiation. Subsequent research demonstrated that greatly superior radiation resistance could be achieved by reversing the cell and placing the negative surface in the sun-facing position.

Research and development in the field of photovoltaic energy conversion are directed to producing an inexpensive, long-life, and reliable solar cell. Among the currently active projects are those that seek to develop materials of improved characteristics, for example, a cell made from oxidized gallium arsenide. Some of the investigations are concerned with crystal growth and other manufacturing techniques. Progress has been reported in developing continuous-type processes for producing silicon crystals. The newly developed processes are expected ultimately to reduce by a large factor the manufacturing cost of solar cells.

In full sunlight, the open-circuit voltage of the silicon solar cell measures about 0.6 V. Individual silicon solar cells manufactured for terrestrial use are available with conversion efficiencies that reach 15 percent. An array consisting of rectangular cells in this class would have a somewhat lower conversion efficiency, that is, above 12 percent, based on the total area of the array.[11] Higher efficiencies are forecast with improved manufacturing techniques.

Various investigations are currently directed to improving the conversion efficiency of thin-film solar cells. The conversion efficiencies of thin-film silicon solar cells and thin-film cadmium sulfide solar cells are currently between 9 and 10 percent.

Tests have demonstrated that a polycrystalline semiconductor has a lower photovoltaic conversion efficiency than a single-crystal semiconductor fabricated from the same material. The single-crystal form is however more costly to produce than the polycrystalline form of the material.

An improvement in the conversion efficiency of the solar cell is anticipated with the development of an advanced concept for cell construction. The new cell consists of a layer of $Al_{1-x}Ga_xAs$ *n*-type semiconductor on a GaAs *p*-type semiconductor base. In the *n*-type layer the Al concentration decreases from 80 percent at the cell surface to zero percent at the *n*/*p* junction, that is, x increases correspondingly from 0.2 to 1.0. The variation in the Al and Ga proportions in the *n*-type region is expected to create a superior photon absorption characteristic for the cell.

Research groups have achieved impressive results in the development of this advanced GaAs-type solar cell by obtaining a terrestrial conversion efficiency above 23 percent, the highest reported value. The theoretical conversion efficiency for this cell is 26 to 28 percent, depending upon the design characteristics of the cell, primarily the thickness and composition of the Al–Ga–As layer.[11] For a number of reasons however, somewhat lower conversion efficiencies will be obtained in practice. Incidentally, the maximum conversion efficiency for a single-crystal silicon solar cell achieved under ideal conditions is about 22 percent.

The cost of silicon solar cells has declined in recent years to a current (1980) price in the range of \$7 to \$12 per peak watt for flat-plate photovoltaic modules. The Department of Energy is aiming to reduce (in constant dollars) the cost of solar cell modules to \$3 per peak watt by 1982 and to 70¢ per peak watt by 1986. Development of thin-film solar cells is directed to reducing the manufacturing cost to less than 30¢ per peak watt by 1986. A peak watt represents the maximum power output of the solar cell under bright noontime sunlight. It is important to observe that the cost to produce the solar cells is only a part of the total cost to construct the solar-electric power plant.

12.16 THE PHOTOVOLTAIC EFFECT

The silicon photocell is fabricated by forming a junction between two different materials in a single crystal of silicon, an element that is normally a poor electrical conductor. Addition of a trace of an element, such as arsenic or phosphorus, that contains one more electron than silicon renders the silicon conductive. In this material, designated as *n*-type silicon, conduction is ef-

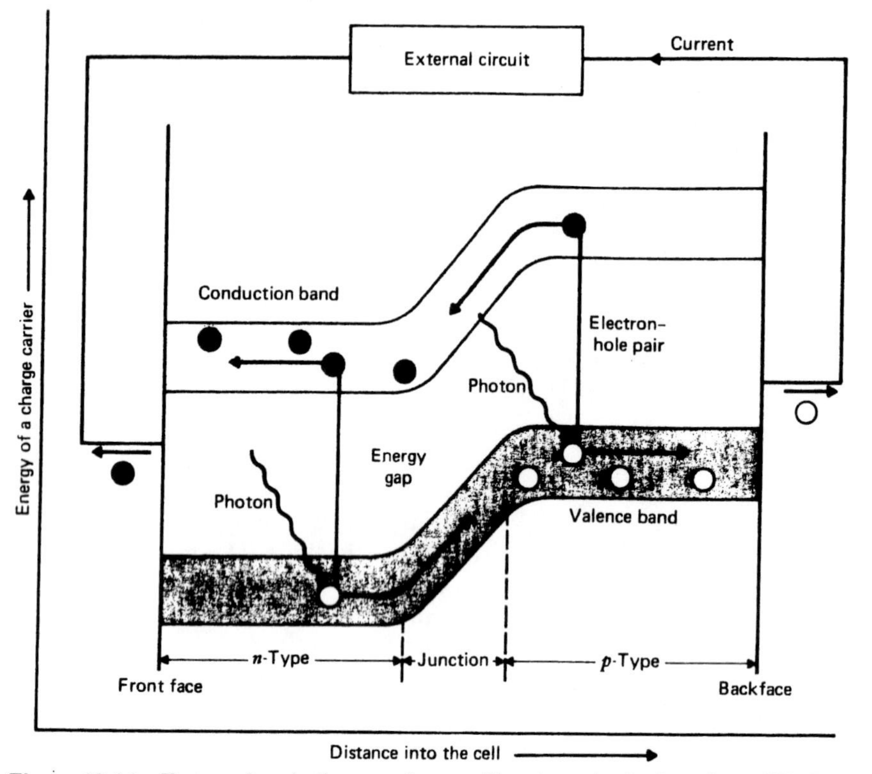

Figure 12.14 Energy band diagram for an illuminated *n/p* junction. (W. D. Johnston, Jr., *American Scientist*, Nov.–Dec. 1977.[10] Reprinted by permission of *American Scientist*, The Scientific Research Society.)

fected by the negative free electrons. The *p*-type silicon is formed by the addition of an element, such as boron or gallium, that contains one less electron than silicon. The absence of electrons in the *p*-type material creates positive charges or holes that also cause the silicon to become conductive. The boundary between the two regions, *n* type and *p* type, in the single silicon crystal, establishes the *n/p* junction. See Fig. 12.14.

In a semiconductor material, the electrical charge carriers, that is, the electrons, occupy one of two energy bands. The lower-energy valence band is filled, or nearly filled, with electrons, while the higher-energy conduction band is empty or only slightly occupied. The two bands are separated by an energy gap in which there are no electrons.

When the semiconductor wafer is exposed to sunlight, some of the solar photons are absorbed in the vicinity of the *n/p* junction; and in the process, electrons in the valence band are excited. The excited electrons move from the valence band to higher-energy levels in the conduction band and, as a consequence, leave holes in the valence band. In this manner, electron–hole pairs are created. The electron in the conduction band has a negative charge, and the hole in the valence band has a positive charge. In effect, the electrons and holes are, respectively, negative and positive charge carriers. The electric field of the *n/p* junction causes the positive and negative charges to move in opposite directions. The *n*-type semiconductor material on one side of the junction is thus negatively charged, while the *p*-type material on the other side is positively charged.

A collection grid is attached to the semiconductor as a means to placing the cell in an electric circuit. If an external load is connected to the cell, the existing potential difference created between the p and n sides will cause an electric current to flow through the external circuit and the cell. The newly formed electrons and holes flow across the junction, in opposite directions, to form the electric current.

Incidentally, the absorption of ionizing radiation, that is, sunlight, generates in the cell positive and negative charges (holes and electrons) in excess of thermal equilibrium values. The electrical output from the solar cell represents the portion of the solar energy incident to the cell that is converted into electrical energy.

The single-crystal solar cell can be modeled by the equivalent electric circuit shown in Fig. 12.15. When a load is placed on the operating cell, a portion of the current generated by the photovoltaic effect is shunted through the internal cell resistance R_J. The load resistance is indicated by R_L.

The current–voltage characteristic of a p/n junction is of the form

$$I_J = I_o[\exp(eV/kT) - 1] \qquad (12.18)$$

A voltage V applied to the cell causes a current I_J to flow across the junction, where I_J is the total current owing to the flow across the junction of both electrons and holes, and I_o is the reverse-saturation or dark current. The terms e and k are defined in Section 12.12.

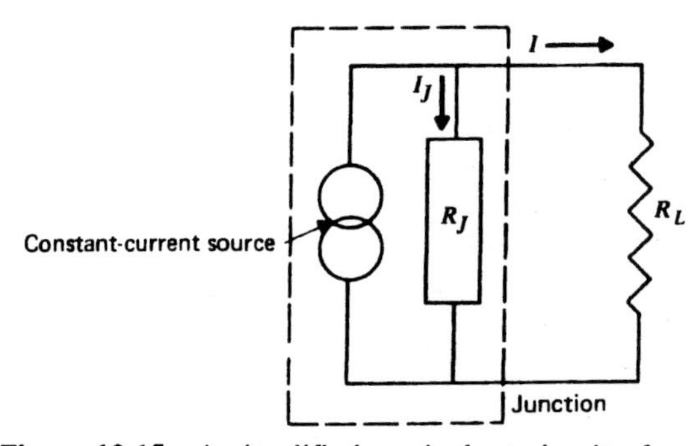

Figure 12.15 A simplified equivalent circuit of a p/n junction photovoltaic cell.

Illumination of the junction in the absence of an external load causes generation by the photons of the constant current I_{sc}, the short-circuit current. When a resistive load is connected to the cell, the current I that flows through the load is equal to the difference between the short-circuit current I_{sc} and the current I_J that flows back across the junction because of the voltage V. Then,

$$I = I_{sc} - I_o[\exp(eV/kT) - 1] \qquad (12.19)$$

The maximum voltage that can be measured on the cell will be observed under open-circuit conditions, $I = 0$. Thus,

$$V_{oc} = \frac{kT}{e} \ln\left(\frac{J_{sc}}{J_o} + 1\right) \qquad (12.20)$$

The current density J, based on the area of the exposed junction, replaces I used in earlier equations.

The power density (output) of the cell is equal to the product JV, or

$$\dot{W} = \{J_{sc} - J_o[\exp(eV/kT) - 1]\}V \qquad (12.21)$$

The voltage V_{mp} that yields the maximum power density is determined by differentiating Eq. 12.21 with respect to V and setting the result equal to zero. Hence,

$$\exp\left(\frac{eV_{mp}}{kT}\right) = \frac{(J_{sc}/J_o) + 1}{(eV_{mp}/kT) + 1} \qquad (12.22)$$

The maximum current density is determined by combining Eqs. 12.19 and 12.22. Thus,

$$J_{\max} = \frac{J_{sc} + J_o}{(kT/eV_{mp}) + 1} \qquad (12.23)$$

and the maximum power density $\dot{W}_{\max}$ is equal to the product $J_{mp}V_{mp}$, or

$$\dot{W}_{\max} = \frac{V_{mp}(J_{sc} + J_o)}{(kT/eV_{mp}) + 1} \qquad (12.24)$$

The conversion efficiency of the cell for maximum power density is evaluated by

$$\eta_{max} = \frac{\dot{W}_{max}}{\dot{W}_{in}} \tag{12.25}$$

where $\dot{W}_{in}$ is the solar power input.

Example 12.7

The prescribed power output, at design conditions, from the silicon solar cells installed in a photovoltaic electric generator is 25 W. From the characteristics of the solar radiation and the properties of the cell, the following quantities are determined: short-circuit current density 1.54×10^{-2} A/cm², reverse-saturation current density 5.35×10^{-13} A/cm², and solar radiation intensity 0.094 W/cm². The operating temperature is 22 C. For conditions of maximum power, determine the required cell area and the probable conversion efficiency.

Open-Circuit Voltage

Applying Eq. 12.20,

$$V_{oc} = \frac{kT}{e}\ln\left(\frac{J_{sc}}{J_o} + 1\right)$$

$$= \frac{295}{11605}\ln\left(\frac{1.54 \times 10^{-2}}{5.35 \times 10^{-13}} + 1\right)$$

$$= 0.612 \text{ V}$$

Maximum-Power Voltage

V_{mp} is determined from Eq. 12.22 by a trial-and-error solution. Assume $V_{mp} = 0.534$ V.

$$\exp\left(\frac{eV_{mp}}{kT}\right) = \frac{(J_{sc}/J_o) + 1}{(eV_{mp}/kT) + 1}$$

$$e/kT = 39.3390$$

$$J_{sc}/J_o = 2.8785 \times 10^{10}$$

$$\exp(39.3390 \times 0.534) - \frac{(2.8785 \times 10^{10}) + 1}{(39.3390 \times 0.534) + 1}$$

$$= 1.328 \times 10^9 - 1.308 \times 10^9 = 0.020 \times 10^9$$

The assumed value of V_{mp} is acceptable.

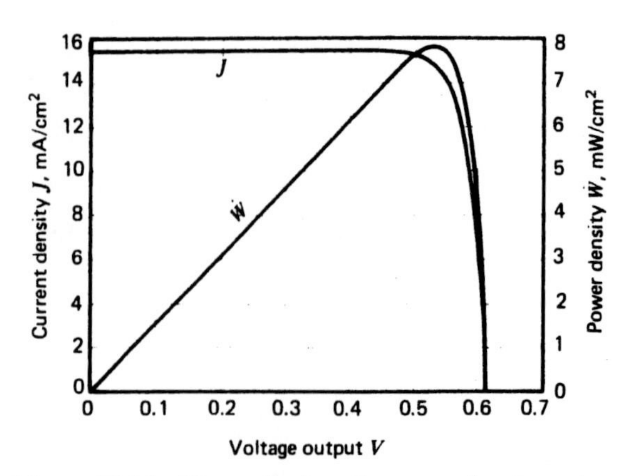

Figure 12.16 Theoretical performance for the solar cell described in Example 12.7.

Maximum-Power Density

Applying Eq. 12.24,

$$\dot{W}_{max} = \frac{V_{mp}(J_{sc} + J_o)}{(kT/eV_{mp}) + 1}$$

$$= \frac{0.534[(1.54 \times 10^{-2}) + (5.35 \times 10^{-13})]}{(0.02542/0.534) + 1}$$

$$= 7.850 \times 10^{-3} \text{ W/cm}^2$$

Solar Cell Area

$$\frac{\dot{W}}{\dot{W}/A} = \frac{25}{7.850 \times 10^{-3}} = 3185 \text{ cm}^2$$

Conversion Efficiency

$$\eta_{max} = \frac{\dot{W}_{max}}{\dot{W}_{in}} = \frac{0.007850}{0.094} = 0.0835$$

Note Fig. 12.16.

12.17 SOLAR POWER GENERATION

Solar radiation calculations are based on the solar constant, defined as the unit intensity of solar radiation on a surface normal to the sun's rays, outside the earth's atmosphere, at the mean earth–

sun distance. The average earth–sun distance is known as the "astronomical unit," A.U. The solar constant has been evaluated by several investigators. The latest and probably the most precise determinations of the solar constant were made during 1966 and 1967. The most probable value of the solar constant, as reported by Drummond and Hickey, is 136.1 mW/cm^2.[13] The Johnson solar constant, widely used during the past several decades, has a value of 140.3 mW/cm^2.

The solar arrays used for space applications can be very large, depending upon the power requirements of the system. For maximum power generation, the solar panels are oriented continuously for interception of the normal rays of the sun. Dark periods will be encountered when the space vehicle passes into the earth's shadow.

Example 12.8

An orbiting space installation is equipped with an array of solar cells. The gross area (200 m^2) of the array is 25 percent greater than the net area of the cells. The individual cells measure 2 cm × 2 cm and have a conversion efficiency of 13.5 percent. Determine the number of solar cells in the array and the power output of the array for orientation normal to the direction of the sun's rays.

$$n = \frac{0.8(200 \times 10^4)}{2 \times 2} = 400{,}000 \text{ cells}$$

$$\text{Solar input } \dot{E}_o = 1361 \text{ W/m}^2$$

$$\dot{W} = \eta_c \dot{E}_o A_{\text{net}} = 0.135 \times 1361 \times 160$$
$$= 29{,}398 \text{ W} \cong 29.4 \text{ kW}$$

The present high cost of good-quality solar cells precludes their use for commercial electric power generation. The high cost factor however is not a particularly strong deterrent in the construction of photovoltaic power plants for space and special terrestrial applications. While commercial land-based solar power plants using photovoltaic collectors are not presently economically feasible, the possibility exists that power plants of this kind will eventually be constructed, and for this reason an examination of the required collector area is in order.

On a sunny day, the solar radiation intensity varies from zero in the early morning and late afternoon to a maximum at noon when the insolation is perhaps 2 to 6 J/min per cm^2, depending upon the latitude and the season of the year.[14] Cloud cover and haze will reduce to a varying degree the solar radiation intercepted at the earth's surface. In sunny climates, a rough estimate of the terrestrial solar radiation, on a horizontal surface, is an annual average value of 2000 J/day per cm^2, or 230 W/m^2.

Example 12.9

A photovoltaic power plant has a rated generating capability of 5000 kW, based on a maximum insolation of 930 W/m^2 on the collector surface. The conversion efficiency is 0.12, based on the gross area of the collector panels. The annual average solar radiation is 212 W/m^2 on a horizontal surface and 243 W/m^2 on the inclined collector surface. Determine the area of the collector panels and the annual average power output.

$$A = \frac{\dot{W}}{\eta_c \dot{E}_i} = \frac{5000 \times 10^3}{0.12 \times 930} = 44{,}803 \text{ m}^2$$

Annual average power output:

$$\dot{W}_{\text{av}} = A \eta_c \dot{E}_{\text{av}}$$
$$= 44803 \times 0.12 \times 243 = 1.306 \times 10^6 \text{ W}$$
$$= 1306 \text{ kW}$$

The model power plant is located at latitude 35°N, and a general clear sky condition is assumed.

The output from the photovoltaic power plant varies throughout the year. For the power plant described in Example 12.9, the average power output, on or about December 21, is 2685 kW over an 8-hour period. On or about June 21, the average power output is 3490 kW over a 10-hour period.

In order to deliver a constant power output throughout the year, an isolated photovoltaic power plant would require the installation of some kind of energy storage system. For a low-capability power plant, daily storage of energy is possible, but seasonal storage of energy would not be feasible. A continuous power output could thus be achieved, but in general the daily average power output would not be as high in the winter as in the summer.

In addition to storage of energy for nighttime operation of the photovoltaic power plant, some consideration may be given to energy storage to assure operation during cloudy periods. The extent of such storage depends upon a number of factors and thus cannot be defined quantitatively.

Capital charges on the energy storage system and energy conversion losses would preclude the application of energy storage to a commercial photovoltaic power plant. The power plant could however be integrated in an electric generating system with conventional thermal and hydro machines and thus deliver electrical energy whenever solar radiation is intercepted by the collector.

Acquisition of land for the construction of a large photovoltaic collector may not prove to be difficult, particularly if the power plant is located in a desert area. The construction of a very large array of solar cells would be a good-sized undertaking, and currently capital costs are prohibitive for commercial power generation. The surface exposed to the solar rays must be clean and not liable to damage, thus maintenance of the collector could increase the operating cost. Possible replacement of solar cells because of deterioration of the conversion effectiveness should be considered.

The power generating capability of the solar cell can be enhanced by concentration of the incident solar radiation. A focusing-type collector is employed to intercept and reflect the solar radiation to the surface of the solar cells. Concentration is effected by reflecting the solar radiation from the larger collector surface to the smaller energy absorbing surface. The concentration ratio is given by A_{col}/A_{sc}, where A_{col} is the area of the effective collecting surface and A_{sc} is the area of the surface of the solar cell modules exposed to the reflected radiation.

Because of the concentration of the solar radiation, a more economical use is achieved for the solar cell. The energy input per unit area of cell surface is increased, and consequently there is a corresponding increase in the power output per unit cell area. Some reduction in the conversion efficiency can be anticipated in achieving concentration of the solar radiation because of imperfect reflection, and perhaps because of improper focusing of the collector.

The increased energy input to the solar cell tends to raise the cell operating temperature. In order to avoid deterioration of the cell structure, the energy absorbing component must be cooled, typically by a circulating fluid. The heat removed in this manner can be utilized in various kinds of processes, thus improving the overall performance of the energy conversion system.

Two rather formidable obstacles are encountered in attempting to convert solar energy into electrical energy on a commercial scale. Solar energy is intermittently available, and because of cloud cover and dust in the atmosphere, the radiation can be highly diffuse when it arrives at the earth's surface. Effective collection of solar energy extends over a period of time equivalent to about 20 percent of full time. The problems arising from intermittent and diffuse solar radiation could be avoided by placing the collector in a satellite orbit high above the surface of the earth.

This intriguing scheme has been proposed by Peter Glaser.[15] An array of solar cells would be installed on a platform hovering in synchronous orbit with the earth's rotation at an altitude of 35,600 km. The electrical energy converted from the solar energy would be transmitted through a flexible superconducting cable 3 km in length to a satellite station where the electrical energy would then be converted to microwave energy and subsequently beamed earthward, unimpeded by the atmosphere and clouds. At a station on the surface of the earth, the microwave energy would be reconverted to dc power by an antenna–rectifier array.

The satellite would be exposed to full sunlight continuously, except for a 1.2-hour interval every 25 days before and after equinox. The interruption in the energy beamed to earth during these darkened periods of 1.2-hour duration can be avoided by employing two, or perhaps more, satellites properly spaced. The more extensive collector system would assure a continuous flow of energy to the earth, and as a consequence the need for energy storage would be obviated.

Glaser's scheme has been examined in some detail and judged to be feasible, although technologically construction of the proposed satellite power system (SPS) would, in many respects, be an undertaking of great proportions. While no insurmountable technical problems have been detected, a recommendation has however been made against appropriating funds for research and development of the SPS during the next decade. Several reasons are given in support of this recommendation, mainly a very high capital cost, perhaps in excess of $10,000/kW, and the enormous effort that will be required to transport structural materials to the station placed in a geosynchronous orbit. In addition, the net energy gain, that is, the difference between the likely output of the power system during its lifetime and the total energy expended in constructing the system, could easily fail to attain an acceptable level. From both an economic and a technical point of view, the SPS is inferior to some of the advanced types of power systems currently under development.

Consideration must be given to other negative points, including the possibility of encountering, during construction of the SPS, biologically harmful bursts of strong ionizing radiation from the sun. The microwave transmission from the satellite to the earth could interfere with the operation of communications satellites that must share the same orbit. An unacceptable environmental impact is likely to develop in an attempt to erect an SPS ground station in a heavily populated area.

Fuel Cells

12.18 STATUS OF THE FUEL CELL

Development of the present-day fuel cell represents a particularly important advancement in the field of direct energy conversion. The fuel cell converts chemical energy directly into electrical energy in a reaction that eliminates combustion of the fuel. Unlike a thermal engine that operates on a thermodynamic power cycle, the performance of the fuel cell is not restricted by the second law.

The first successful fuel cell was demonstrated by Sir William Grove in 1839; but until the early years of the 1950 decade, only minor progress was achieved in fuel cell development. Then, in response to the need to develop power plants for use in space vehicles, numerous programs for fuel cell research and development were undertaken. Practical fuel cells were constructed and installed in space vehicles, and the performance of these electric power plants was judged to be satisfactory. Currently, fuel cell development is, in part, directed to commercial applications.

There are many different kinds of fuel cells and different fuels that may be used in these energy conversion devices. Some of these fuel cells operate at high temperatures, about 600 to 650 C, and employ molten salt electrolyte systems. In this section of the text, only the operation of low-temperature fuel cells supplied with typical fuels are examined. For these fuel cells, the operating temperature range extends up to the boiling point of the aqueous electrolyte. Incidentally, in the design of a fuel cell, the chemical and electrical phenomena are important and must be considered in detail along with the metallurgical requirements.

12.19 FUEL CELL OPERATING PRINCIPLES

The basic operating principle of the fuel cell is the creation of two chemical reactions. One reaction releases electrons that travel through an external circuit and return to the fuel cell. The returning electrons are in turn "absorbed" in the second reaction.

Figure 12.17 illustrates the operation of a fuel cell supplied with hydrogen and air. This device is classified as a typical hydrogen/oxygen fuel cell in which the two electrodes are separated by a liquid or a solid electrolyte. The hydrogen diffuses through one of the electrodes, the anode. On the anode surface, the hydrogen reacts with the electrolyte to produce electrons and ions. Thus,

$$2H_2 \rightarrow 4e^- + 4H^+$$

The electrons are given up to the anode and subsequently travel through the external circuit to the cathode, while the ions migrate through the electrolyte to the cathode. On the cathode surface, the ions combine with the electrons and with the oxygen that has diffused through the electrode to form water. Thus,

$$4H^+ + 4e^- + O_2 \rightarrow 2H_2O$$

Hydrogen is currently not a commercial fuel, hence for commercial applications of the fuel cell, the energy converter should be capable of operating with fuels such as methanol, methane (natural gas), propane, octane, and various hydrocarbon engine fuels. Hydrocarbon fuels have characteristically reacted incompletely at moderate temperatures and at atmospheric pressure. Earlier hydrocarbon fuel cells generally required an operating temperature in the range of 550 to 1100 C.

One method employed for circumventing the poor reactivity of hydrocarbons at low temperatures is the introduction of a preliminary re-

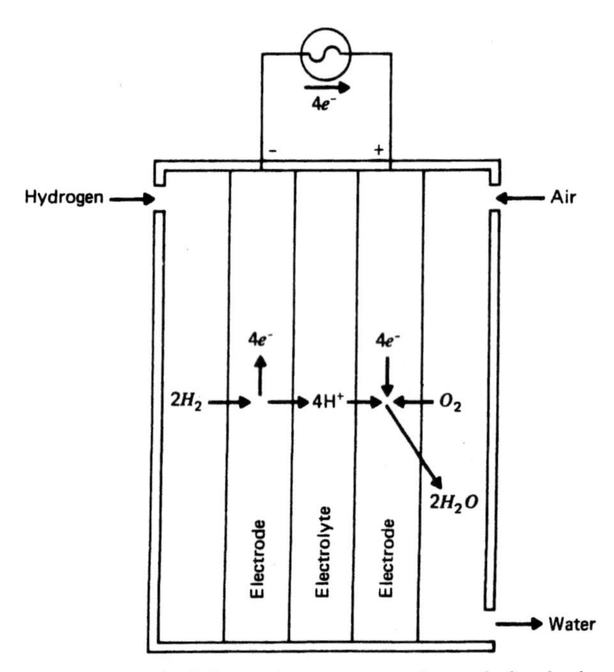

Figure 12.17 Schematic representation of the hydrogen/oxygen fuel cell.

action in which the hydrocarbon is converted to a more reactive substance. This conversion is commonly accomplished in a steam re-former, a device in which the hydrocarbon reacts with steam to form hydrogen and carbon dioxide. The hydrogen is subsequently consumed in a H_2/air fuel cell, while the carbon dioxide is discarded. Some disadvantages are observed in the operation of the re-former/fuel cell combination, namely, problems encountered in attempting to achieve rapid starting and reduced conversion efficiency in comparison with the efficiency potentially possible for a direct fuel cell.

Significant progress has been achieved in developing a fuel cell in which various hydrocarbon fuels react directly with oxygen from ambient air, at atmospheric pressure and in a temperature range of 120 to 200 C. This advanced design of the fuel cell depends for its operation upon a number of factors, including an improved electrode structure and an effective liquid electrolyte, particularly phosphoric acid.

In the fuel cell, the overall reaction process, that is, complete oxidation to carbon dioxide and water, is basically the same for each hydrocarbon, although the process differs in detail for each fuel. Methane, the principal constituent of natural gas, is used to typify the overall electrochemical reaction that produces electrical energy in the fuel cell when the cell is supplied with a hydrocarbon fuel. See Fig. 12.18.

The fuel (CH_4) is supplied to the anode side of the cell. At the anode, the methane molecule combines with water present in the electrolyte to form carbon dioxide, electrons, and ions. Thus,

$$CH_4 + 2H_2O \rightarrow CO_2 + 8e^- + 8H^+$$

The carbon dioxide is rejected from the cell, and the hydrogen ions are released to the electrolyte and travel to the cathode.

The electrons perform work as they are conducted from the cell and move through the external circuit to the cathode. Unlike the conven-

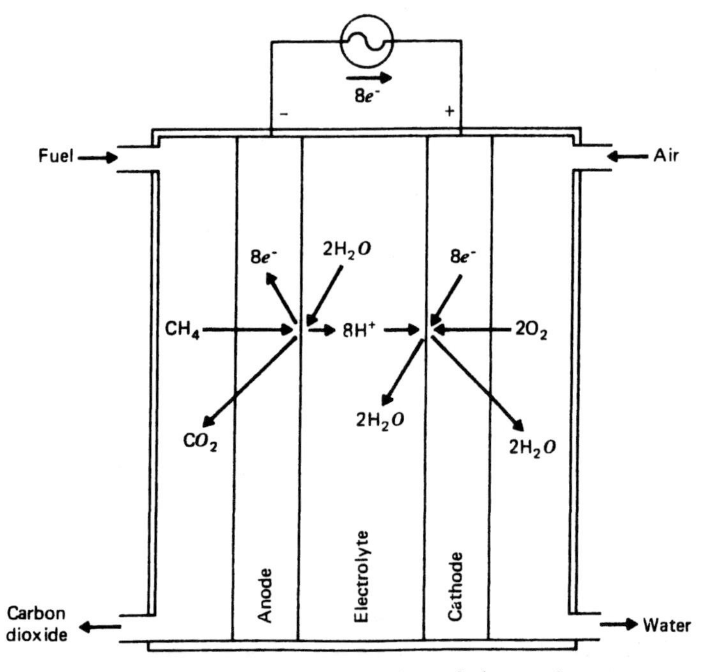

Figure 12.18 Schematic representation of the methane/oxygen fuel cell.

tional mechanical-electrical generator, the electrons released in the fuel cell make only a single circuit.

The air supplied to the cathode side of the cell provides the oxygen essential to the reaction. At the cathode, the electrons, returning from the load, combine with oxygen and the hydrogen ions to form water. Thus,

$$8e^- + 8H^+ + 2O_2 \rightarrow 4H_2O$$

The quantity of water formed at the cathode is greater than the amount of water that is removed from the electrolyte at the anode. The excess water is discharged from the cell. Incidentally, aboard a manned spaceship, the fuel cell is an important source of water.

The discharge of carbon dioxide and water from the fuel cell has no deleterious impact on the environment. Further, the high conversion efficiency of the fuel cell is an important asset. Other advantages are quiet operation and an absence of large, heavy moving parts.

12.20 FUEL CELL THEORETICAL PERFORMANCE

The fuel cell functions through a chemical reaction that proceeds in a manner that causes electrons to flow in an external circuit. Theoretically, this reaction is effected isothermally and reversibly. For a steady-state, steady-flow process that occurs in temperature equilibrium with the surroundings, the reversible or maximum work is equal to the change in the Gibbs function; $W_{rev} = G_2 - G_1$. Since $g = h - Ts$,

$$W_{rev} = \Delta G = \Delta H - T\,\Delta S \qquad (12.26)$$

The last term, $T\,\Delta S$, is the heat exchanged with the surroundings. The types of fuel cells considered for practical applications reject heat in the theoretical isothermal reaction as well as in the actual operation. Rejection of heat causes ΔG in Eq. 12.26 to be numerically smaller than ΔH.

The theoretical conversion efficiency, or thermodynamic efficiency, of the fuel cell is expressed by

$$\varepsilon_i = \frac{\Delta G}{\Delta H} = 1 - \frac{T\,\Delta S}{\Delta H} \qquad (12.27)$$

The variation in the conversion efficiency ε_i with the operating temperature of the fuel cell is illustrated by examining the hydrogen–oxygen reaction. The product, water, is assumed to be in the gaseous phase, and each gas exists at a pressure of 1 atmosphere. Thus, $H_2 + \frac{1}{2}O_2 \rightarrow (H_2O)_g$. For the prescribed reaction temperatures, ΔH, $T\,\Delta S$, and ΔG are evaluated from property values derived from the JANAF Thermochemical Tables.[16] See Table 12.2.

As the operating temperature of the fuel cell increases, the heat transferred to the surroundings increases substantially in comparison with the increase in ΔH. Thus, the theoretical conversion efficiency of the fuel cell decreases as the operating temperature increases. This trend is counter to the inherent increase in the thermal efficiency of the Carnot cycle that is observed with an increase in the temperature of heat addition.

The assumption that the fuel cell produces steam is valid because, at least in theory, the steam can be effectively utilized to produce work. On the other hand, if liquid water is produced in the fuel cell, a reduction in the theoretical conversion efficiency will be observed. For the hydrogen/oxygen fuel cell operating at 298 K and producing liquid water, the theoretical conversion efficiency is 0.830.

Example 12.10

Determine the theoretical conversion efficiency for a methane/oxygen fuel cell that operates at a temperature of 500 K.

$$CH_4 + 2O_2 \rightarrow CO_2 + 2H_2O$$

Each component gas is considered to be at a pressure of 1 atmosphere. Enthalpy and entropy

TABLE 12.2 Effect of the Reaction Temperature on the Theoretical Performance of the Hydrogen/Oxygen Fuel Cell[a]

T (K)	ΔH (kJ/kgmol)	$T\,\Delta S$ (kJ/kgmol)	ΔG (kJ/kgmol)	Fuel Cell ε_i	Carnot Cycle η_t[b]
298	−241,827	−13,221	−228,606	0.945	0
400	−242,848	−18,926	−223,922	0.922	0.255
500	−243,834	−24,755	−219,079	0.898	0.404
1000	−247,888	−55,254	−192,634	0.777	0.702
2000	−251,670	−116,107	−135,563	0.539	0.851

[a] *The evaluation of ΔH and ΔG at 500 K is determined from thermodynamic data selected from Table A.5.*

	HYDROGEN	OXYGEN	WATER
$\bar{h}_f^\circ$ (kJ/kgmol)	0	0	−241,827
$\Delta\bar{h}^\circ$ (kJ/kgmol)	5883	6088	6920
$\bar{s}^\circ$ (kJ/kgmol·K)	145.628	220.589	206.413

$$\Delta H = (\bar{h}_f^\circ + \Delta\bar{h}^\circ_{500\,K})_{H_2O} - (\bar{h}_f^\circ + \Delta\bar{h}^\circ_{500\,K})_{H_2} - {}^1/_2(\bar{h}_f^\circ + \Delta\bar{h}^\circ_{500\,K})_{O_2}$$
$$= (-241827 + 6920) - (0 + 5883) - {}^1/_2(0 + 6088)$$
$$= -243{,}834 \text{ kJ/kgmol } H_2$$

$$\Delta S = \bar{s}^\circ_{H_2O} - \bar{s}^\circ_{H_2} - {}^1/_2\bar{s}^\circ_{O_2}$$
$$= 206.413 - 145.628 - (0.5 \times 220.589)$$
$$= -49.510 \text{ kJ/K per kgmol } H_2$$

$$T\,\Delta S = 500(-49.510) = -24{,}755 \text{ kJ/kgmol } H_2$$

$$\Delta G = \Delta H - T\,\Delta S = -243834 - (-24755) = -219{,}079 \text{ kJ/kgmol } H_2$$

[b] *For the Carnot cycle, T_c = 298 K.*

values, selected at 500 K from Table A.5, are used to calculate ΔH and ΔS for the reaction. Values for the enthalpy of formation at 298 K are given in Table A.5.

	METHANE	OXYGEN	CARBON DIOXIDE	WATER
$\bar{h}_f^\circ$ (kJ/kgmol)	−74,873	0	−393,522	−241,827
$\Delta\bar{h}^\circ$ (kJ/kgmol)	8201	6088	8.314	6,920
$\bar{s}^\circ$ (kJ/kgmol·K)	206.911	220.589	234.814	206.413

$$\Delta H = (\bar{h}_f^\circ + \Delta\bar{h}^\circ_{500\,K})_{CO_2} + 2(\bar{h}_f^\circ + \Delta\bar{h}^\circ_{500\,K})_{H_2O} - (\bar{h}_f^\circ + \Delta\bar{h}^\circ_{500\,K})_{CH_4} - 2(\bar{h}_f^\circ + \Delta\bar{h}^\circ_{500\,K})_{O_2}$$
$$= (-393{,}522 + 8{,}314) + 2(-241{,}827 + 6{,}920) - (-74{,}873 + 8{,}201) - 2(0 + 6088)$$
$$= -800{,}526 \text{ kJ/kgmol methane}$$

$$\Delta S = \bar{s}^\circ_{CO_2} + 2\bar{s}^\circ_{H_2O} - \bar{s}^\circ_{CH_4} - 2\bar{s}^\circ_{O_2}$$
$$= 234.814 + (2 \times 206.413) - 206.911 - (2 \times 220.589)$$
$$= -0.449 \text{ kJ/K per kgmol methane}$$

$$\Delta G = \Delta H - T\,\Delta S$$
$$= -800526 - [500(-0.449)]$$
$$= -800{,}526 + 224$$
$$= -800{,}302 \text{ kJ/kgmol methane}$$

$$\varepsilon_i = \frac{\Delta G}{\Delta H} = \frac{-800{,}302}{-800{,}526} = 0.9997 \quad \underline{1.000}$$

The methane/oxygen fuel cell represents the limiting condition for a practical fuel cell, that is, a reaction in which the heat evolved approaches zero. It follows that the thermodynamic efficiency of the cell approaches 1 as a limit.

The maximum electrical energy, equal to W_{rev}, is obtained in the fuel cell at the highest value of E, the cell emf. The maximum value of E is E_{rev}. Then,

$$\Delta G = -E_{rev} I \tau \qquad (12.28)$$

The reaction is effected reversibly and isothermally to deliver a current I for the time τ required to consume one mole of fuel.

The relationship between the Gibbs free energy change and the theroretical electrical energy output of the fuel cell is expressed by

$$\Delta G = -z\mathcal{F}E_{rev} \qquad (12.29)$$

where

$z\mathcal{F}$ = the total charge transferred at E_{rev}
$\mathcal{F}$ = the Faraday (9.6487×10^7 coulombs/kgmol)
z = the electrochemical valence

TABLE 12.3 Standard emf (E) and Efficiency for Combustion Reactions[a]

		298 K		600 K		1000 K	
	z	E	ε	E	ε	E	ε
$C + O_2 = CO_2$	4	1.03	1.00	1.03	1.00	1.03	1.00
$2C + O_2 = 2CO$	4	0.74	1.24	0.86	1.48	1.02	1.78
$2CO + O_2 = 2CO_2$	4	1.34	0.91	1.18	0.81	1.01	0.69
$CH_4 + 2O_2 = CO_2 + 2H_2O$	8	1.04	1.00	1.04	1.00	1.04	1.00
$2H_2 + O_2 = 2H_2O$	4	1.18	0.94	1.11	0.88	1.00	0.78

Source: Weissbart, J. "Fuel Cells–Electrochemical Converters of Chemical to Electrical Energy," *Journal of Chemical Education*, **38**, May 1961.[17]

[a] *$\varepsilon = \Delta G/\Delta H$; E is expressed in volts.*

Values of z and E_{rev} are listed in Table 12.3 for several important fuel cell reactions.

Example 12.11

For the hydrogen/oxygen reaction at 298 K calculate ΔG from Eq. 12.29.

$$z = 4 \qquad E_{rev} = 1.18 \text{ V} \qquad \text{(from Table 12.3)}$$

$$\begin{aligned} \Delta G &= z\mathcal{F}E_{rev} \\ &= -4(9.6487 \times 10^7)1.18 \\ &= -455{,}420 \times 10^3 \text{ J per 2 kgmol } H_2 \end{aligned}$$

or

$$\Delta G = -227{,}710 \text{ kJ/kgmol } H_2$$

From Table 12.2, $\Delta G = -228{,}606$ kJ/kgmol H_2.

Example 12.12

For the methane/oxygen reaction at 500 K, calculate ΔG from Eq. 12.29.

$$z = 8 \qquad E_{rev} = 1.04 \qquad \text{(from Table 12.3)}$$

$$\begin{aligned} \Delta G &= -z\mathcal{F}E_{rev} \\ &= -8(9.6487 \times 10^7)1.04 \\ &= -802{,}770 \times 10^3 \text{ J/kgmol } CH_4 \end{aligned}$$

From Example 12.10, $\Delta G = -800{,}291$ kJ/kgmol CH_4.

Development of a fuel cell capable of operating on coal would be a most important accomplishment. In the overall energy conversion process, the reactions at each electrode must proceed rapidly enough to yield an acceptably high current density. The main obstacle to the use of coal in a fuel cell is the failure to meet the reactivity requirement. At least for the foreseeable future, fuel cells developed for commercial electric power production will operate on gaseous and liquid fuels.

12.21 PERFORMANCE OF REAL FUEL CELLS

Because of several irreversible effects, the actual fuel cell performs at a level below that of the comparable theoretical fuel cell. In particular, the electromotive force E_a of the actual fuel cell will drop below E_{rev}, the emf generated by a reversible reaction. Some irreversible effects are unwanted reactions occurring in the cell, impediment to reaction at the anode or cathode, and I^2R heating in the electrolyte. Although the I^2R heat need not be wasted, it is often useless and can be harmful. The cell should be designed to minimize this loss.

The first law equation for the actual fuel cell is written as

$$-E_a I\tau = \Delta H - Q \tag{12.30}$$

The product on the left-hand side of the equation represents the electrical energy delivered by the fuel cell during the time τ when one mole of fuel is consumed.

The conversion efficiency of the actual fuel cell is given by

$$\varepsilon_a = \frac{-E_a I\tau}{\Delta H} \tag{12.31}$$

This efficiency factor relates solely to the fuel cell and not to the overall system of which the fuel cell is a part. As noted earlier, ΔH includes the energy of condensation of water for cells in which liquid water is formed, but excludes this energy when the steam can be effectively utilized.

For the reversible fuel cell,

$$-E_{rev} I\tau = \Delta H - T\,\Delta S \tag{12.32}$$

The combination of Eqs. 12.30 and 12.32 yields

$$E_{rev} I\tau - E_a I\tau = T\,\Delta S - Q \tag{12.33}$$

Since $E_{rev} I\tau > E_a I\tau$, the heat transfer Q from the actual cell is numerically greater than $T\,\Delta S$, the reversible heat transfer. The difference between the two heat transfer quantities is a result of the irreversible effects that occur in the cell.

The ratio E_a/E_{rev} is a convenient index of the fuel cell irreversibility, and the magnitude of this ratio decreases with an increase in the current density. For commercial-scale installations in central power stations, values of about 0.70 for E_a/E_{rev}, at current densities above 1000 A/m^2, would be considered acceptable.

The maximum value of E_a for a near-zero current density is about 1 V for the typical fuel cell. An increase in the current density is accompanied by a reduction in E_a, as shown in Fig. 12.19. At very high current densities, the voltage developed by the cell approaches zero, and as a consequence the output power also approaches zero.

Example 12.13

A hydrogen/oxygen fuel cell operating at 600 K consumes hydrogen at the rate of 0.00835 kg/h. $E_a/E_{rev} = 0.65$; $\varepsilon_a = 0.58$. Calculate the power output and the heat transfer from the cell.

1 Enthalpy change.

$$\Delta H = -244{,}765 \text{ kJ/kgmol H}_2 \quad \text{(calculated from thermodynamic data)}$$

(Also, from Table 12.2, $\Delta H = -244{,}645$ kJ/kgmol H_2.)

2 Time required to consume one mole of hydrogen.

$$\tau = \frac{\overline{m}_{H_2}}{\dot{m}_{H_2}} = \frac{2.02}{0.00835} = 241.9h$$

3 Power.

$$E_{rev} = 1.11 \text{ V} \quad \text{(from Table 12.3)}$$

$$E_a = \frac{E_a}{E_{rev}} E_{rev}$$

$$= 0.65 \times 1.11 = 0.722 \text{ V}$$

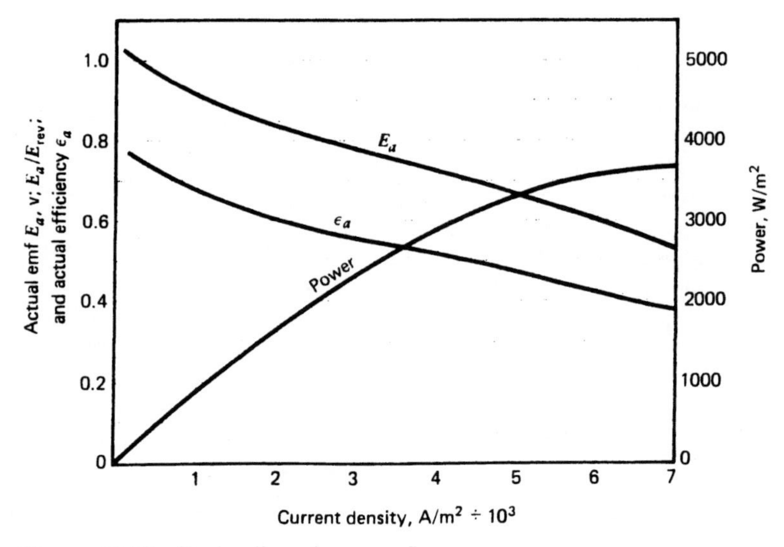

Figure 12.19 Fuel cell performance curves.

Applying Eq. 12.31,

$$\varepsilon_a = \frac{-E_a I \tau}{\Delta H}$$

$$0.58 = \frac{-0.722 I (241.9 \times 3600)}{-244765 \times 10^3}$$

$$I = 225.8 \text{ A}$$

$$\dot{W} = E_a I$$

$$= 0.722 \times 225.8 = 163.0 \text{ W}$$

4 Heat transfer.

Equations 12.30 and 12.31 are combined to yield

$$Q = (1 - \varepsilon_a)\,\Delta H$$

$$= (1 - 0.58)(-244765)$$

$$= -102{,}801 \text{ kJ/kgmol } H_2$$

$$\dot{Q} = \frac{Q}{\tau} = \frac{-102{,}801}{241.9} = -425.0 \text{ kJ/h}$$

Electrode design is an important part of fuel cell development. There are several ways of improving cell reactivity, in addition to increasing the electrode surface. High temperature and high pressure promote rapid reactions at the electrodes but cause more severe corrosion while contributing to construction difficulties, in particular with respect to cell leakage. Cell reactivity can be improved by incorporating a solid catalyst into the electrode.

A properly designed porous electrode provides for rapid movement of gas into the pores and rapid movement of the gaseous reaction products from the pores. In the fabrication of gas diffusion electrodes, it is essential to guard against appreciable gas leakage into the electrolyte and flow of electrolyte into the pores. The electrodes become inactive if the pores are filled with electrolyte or with water that is a product of the cell reaction.

12.22 FUEL CELL APPLICATIONS

In past years fuel cell research and development was directed primarily to the construction of power plants suitable for installation in space vehicles. The feasibility of extending fuel cell applications to commercial power generation is currently being explored, notably by United Technologies, Inc. The U.T.C. program, partially supported by a group of utility companies, was directed initially to the construction of a prototype fuel cell, the FCG-1. A 26-MW power plant derived from the prototype fuel cell was scheduled for commercial operation in 1978.

The FCG-1 power unit is designed to operate on a liquid or a gaseous fuel of the kind currently used by utility companies. The primary fuel supplied to the system first enters the re-former for conversion into hydrogen and carbon dioxide. The hydrogen subsequently combines with the oxygen in the air supplied to the fuel cell.

The single fuel cell, the basic element of the power plant, produces dc electric power at 1 volt and about 1000 to 2000 watts per square meter of electrode area. Fuel cells are assembled in modules to achieve a suitable voltage prior to converting the direct current to an alternating current in the inverter. A transformer subsequently raises the output voltage of the fuel cell assembly to the required line voltage.

The FCG-1 power unit, rated at 26 MW, is approximately 6 m high and can be situated on less than 0.2 hectares of land. The unit will provide sufficient electrical energy for a community of 20,000 people. Because the fuel cell operates quietly and causes no pollution, few siting problems are anticipated.

While the 26-MW fuel cell power unit is too small for base load power generation, it can serve effectively in meeting intermediate and peak power demands. Units of this kind can be located at numerous points in a power system as needed to balance the system load and at the same time reduce transmission line losses. The relatively high conversion efficiency of the fuel cell can be maintained for comparatively small units. The conversion efficiency of 0.40 for a 26-MW fuel cell power plant is approximately equal to the thermal efficiency achieved by a fossil fuel steam power plant in the 500- to 1000-MW range.

The use of fuel cells for power generation on a commercial scale could be hindered to some extent by the high cost and the uncertain supply of natural gas and petroleum fuels. There is however another point of view to be considered. Fuel cell power units would most likely be installed in a power system instead of gas turbine engine generating sets, hence in this respect the fuel problem is not so highly intensified. Further, the fuel cell would operate somewhat more efficiently than the gas turbine engine and cause no air pollution or audible disturbance.

The use of the fuel cell in transportation has been evaluated as a source of power for four vehicle types, namely, a city bus, highway bus, delivery van, and conventional automobile. The fuel cell/re-former system, motor, and batteries can be contained without difficulty in these vehicles that operate on propane or methanol. The fuel cell supplies the cruising power and the energy for recharging the batteries, while the batteries supply the transient power for acceleration and the power for operating the vehicle during fuel cell start-up. Although the fuel cell exhibited some advantages in automotive service, the fuel cell is unlikely to contribute to an early replacement of the automotive internal combustion engine.

A program has been initiated for the development of a molten carbonate fuel cell that will eventually be used to produce electrical energy on a commercial scale. The fuel cell will use a gas produced by coal conversion. The overall conversion efficiency of the coal gasifier/molten carbonate fuel cell system is expected to approach 50 percent. A mixture of hydrogen and carbon monoxide will be supplied to the fuel cell to react with oxygen from air.

PROBLEMS

12.1 The following data are applicable to a magnetohydrodynamic generator: At the inlet section, P = 5 atm, T = 2600 K, C = 720 m/s, h = −5250 kJ/kg. At the outlet section, P = 0.9 atm, T = 2295 K, C = 575 m/s, h = −6000 kJ/kg. The gas constant R = 0.205 kJ/kg·K. The gas flow is 1250 kg/s. Average values of the electrical conductivity and the magnetic induction are, respectively, 4.750 mho/m and 5.085 Wb/m^2. The heat loss factor is 0.05. The value of n, the load ratio, is taken as 0.80. Calculate the power output of the generator and the volume and the length of the generator duct.

12.2 A thermoelectric generator is fabricated from p-type and n-type materials that have similar values of the thermoelectric properties. The generator operates between 350 and 25 C. At 350 C, $Z = 1.45 \times 10^{-3}$ 1/K, $S = 160 \times 10^{-6}$ V/K, and σ = 570 mho/cm. At 25 C, $Z = 1.95 \times 10^{-3}$ 1/K, $S = 170 \times 10^{-6}$ V/K, and σ = 950 mho/cm. The cross-sectional area of the thermoelement is 1.25 cm^2, and the length is 1.10 cm. The lower heating value of the fuel is 44,000 kJ/kg. The generator is designed for an output of 20 W. Determine the required number of thermoelements and the fuel consumption, in kg/h.

12.3 A thermionic energy converter operates with a cathode temperature of 1200 K and an anode temperature of 550 K. V_c = 1.80 V, and V_a = 1.10 V. The effective converter area is 400 cm^2. Determine the maximum power that can be developed by the converter.

12.4 (a) Determine the thermal efficiency for the thermionic energy converter described in Problem 12.3. The value of the factor F_e in the radiation equation is 0.050.
(b) The energy rejected from the thermionic unit is transferred to a theoretical reheat steam power plant. The inlet steam conditions are 1.5 MPa, abs, and 270 C. The exhaust pressure is 7.5 kPa, abs. The steam is reheated to 270 C at a pressure of 0.6 MPa, abs. Calculate the improvement in the power plant thermal efficiency achieved by the addition of the thermionic topping unit to the steam power plant.

12.5 A hydrogen/oxygen fuel cell operates at 200 C and has a power output of 1 kW. The actual efficiency of the cell is 0.63, and E_a/E_{rev} = 0.70. Calculate the fuel consumption rate, in kg/h.

12.6 The following data are applicable to a char/oxygen MHD generator. At the inlet section of the generator duct, the pressure and temperature of the gas are 6 atmospheres and 2600 K, the flow area is 1.55 m^2, and the gas velocity is 710 m/s. The pressure at the duct outlet section is 1 atmosphere. The prescribed flow area ratio, outlet section to inlet section, is 6.5 to 1. The actual enthalpy drop in the gas that occurs within the duct is 78 percent of the corresponding isentropic enthalpy drop. R for the gas is 0.207 kJ/kg·K. The ratio of the generator heat loss to the generator power is 0.055 to 1. Calculate the generator electric power output.

12.7 The materials used in fabricating a thermoelectric generator have similar thermoelectric properties. At 450 C, $Z = 1.88 \times 10^{-3}$ 1/K, $\rho = 4.0 \times 10^{-3}$ ohm·cm, and $S = 238 \times 10^{-6}$ V/K. At 30 C, $Z = 1.13 \times 10^{-3}$ 1/K, $\rho = 5.0 \times 10^{-3}$ ohm·cm, and $S = 207 \times 10^{-6}$ V/K. Fifty elements are used in fabricating the thermoelectric generator. The dimensions of an element are $A = 1.35$ cm^2, $L = 1.15$ cm. Calculate the thermal efficiency and the power output for the generator when it operates between 450 and 30 C. If the generator operated continuously at the prescribed output for a period of one year, how much fuel would be consumed? The lower heating value of the fuel is 44,150 kJ/kg.

12.8 An electrical device installed in a space vehicle operates continuously and requires a power input of 0.5 kW. Power is supplied by an array of solar cells mounted on a panel that is oriented in a plane normal to the direction of the sun's rays. Determine the conversion efficiency of the solar energy collector and the solar cell area required to produce the prescribed power output.

12.9 A methane/oxygen fuel cell operates at 500 K. The conversion efficiency of the real fuel cell is 0.60. The cell has an output of 5 kW. The current density is 1000 A/m^2. $E_a/E_{rev} = 0.62$. Calculate the fuel consumption rate, in kg/h, and determine the electrode area.

12.10 A thermoelectric cooling device is designed for a hot conductor temperature of 30 C and a cold conductor temperature of -5 C. For the prescribed operating temperature range, $\bar{S}_n = -160 \times 10^{-6}$ V/K, $\bar{S}_p = 160 \times 10^{-6}$ V/K, $\bar{\rho}_n = \bar{\rho}_p = 1.10 \times 10^{-3}$ ohm·cm, $\bar{k}_n = \bar{k}_p = 0.0116$ W/cm·K. $A = 0.450$ cm^2; $L = 1.80$ cm. Calculate the cooling effect and the coefficient of performance for the prescribed design, based on an optimized value of current I.

12.11 The design prescribed in Problem 12.10 is optimized for m, with $A = 0.450$ cm^2. The current I remains unchanged. Calculate the cooling effect and the coefficient of performance.

12.12 A thermionic energy converter is constructed in the form of concentric cylindrical surfaces. The inner cylinder, the emitter, contains a radioisotope fuel. The emitter and anode temperature are, respectively, 1800 and 950 K. The corresponding surface emissivity values are 0.115 and 0.090. $V_c = 2.05$ V; $V_a = 1.20$ V. Because of closed-space construction, Eq. 12.17a is applicable to the converter design. Calculate the thermal efficiency for the device.

12.13 Determine the theoretical conversion efficiency for the carbon monoxide/oxygen fuel cell that operates at 600 K. Each gas exists at a pressure of 1 atmosphere. Calculate the value of the corresponding E_{rev}.

12.14 The theoretical fuel cell described in Problem 12.13 delivers a current of 50 A. Determine the operating time during which one mole of fuel is consumed.

12.15 A solar cell operates with an incident intensity of 0.100 W/cm^2. The short-circuit current density is 1.75×10^{-2} A/cm^2, and the reverse-saturation current density is 7.85×10^{-13} A/cm^2. The operating temperature of the cell is 25 C. The collector panel has a total cell area of 1.25 m^2. Calculate the power output and the conversion efficiency for the collector.

REFERENCES

1 Jackson, W. D., Petrick, M., and Klepeis, J. E. "Critique of MHD Power Genera-

tion," ASME Paper No. 69-WA/Pwr-12, Nov. 1969.

2 Hals, F. A. "Magnetohydrodynamic Power Generation," ASME Paper No. 67-PWR-12, Sept. 1967.

3 Postlethwaite, A. W. and Sluyter, M. M. "MHD Heat Transfer Problems—An Overview," *Mechanical Engineering*, March 1978.

4 Way, S. "Char Burning MHD Systems," *Combustion*, May 1971.

5 Angello, S. J. "Thermoelectricity," *Westinghouse Engineer*, July 1960.

6 Snyder, P. E. "Chemistry of Thermoelectric Materials," Westinghouse Electric Corp. Reprint 6045. (Reprinted from *Chemical and Engineering News*, **39**, Mar. 13, 1961.)

7 Crosby, C. R., Norwood, M. H., and West, B. R. "The Effects of Heat Transfer on Optimum Peltier Heat Pumping," ASME Paper No. 62-HT-11, Aug. 1962.

8 "Thermionic Topping Converter for a Coal-Fired Power Plant," Office of Coal Research, Department of the Interior Research and Development Report No. 52, Washington, D.C., 1970.

9 Engdahl, R. E., Cassano, A. J., and Dowdell, R. B. "Thermionics in Fossil-Fuel and Nuclear Central Power Stations," *Combustion*, March 1970.

10 Smith, A. "Status of Photovoltaic Power Technology," ASME Paper No. 68-WA/Sol-1, Dec. 1968.

11 Johnston, W. D., Jr. "The Prospects for Photovoltaic Conversion," *American Scientist*, Nov.–Dec. 1977.

12 Loferski, J. J. "Large-Scale Solar Power via the Photovoltaic Effect," *Mechanical Engineering*, Dec. 1973.

13 Yellott, J. I. "Solar Energy Progress—A World Picture," *Mechanical Engineering*, July 1970.

14 Daniels, F. "Direct Use of the Sun's Energy," *American Scientist*, Jan.–Feb. 1967.

15 Walters, S. "Power in the Year 2001—Part 3—Solar Power," *Mechanical Engineering*, Nov. 1971.

16 *JANAF Thermochemical Tables*, 2nd ed., The Dow Chemical Company, Midland, Michigan, June 1971.

17 Weissbart, J. "Fuel Cells—Electrochemical Converters of Chemical to Electrical Energy," Westinghouse Electric Corp. Reprint 6059. (Reprinted from *Journal of Chemical Education*, **38**, May 1961.)

18 Walters, S. "Commercial Fuel Cells," *Mechanical Engineering*, Feb. 1974.

BIBLIOGRAPHY

Angrist, S. W. *Direct Energy Conversion*, Allyn and Bacon, Inc., Boston, 1976.

Cambel, A. B. *Plasma Physics and Magnetofluidmechanics*, McGraw-Hill Book Co., New York, 1963.

Culp, A. W. Jr. *Principles of Energy Conversion*, McGraw-Hill Book Co., New York, 1979.

Wood, B. D. *Applications of Thermodynamics*, Addison-Wesley Publishing Co., Reading, Mass., 1982.

CHAPTER THIRTEEN

NONREACTIVE ENERGY SOURCES

A nonreactive energy source is defined at this point as energy that may be utilized directly without first being released in a chemical or nuclear reaction. Fossil and nuclear fuels occur naturally, but the energy in these fuels is made available by reactions that involve a rearrangement of the molecular or nuclear structure. Water flowing in streams and rivers has served for centuries as an important, but nonreactive, power source. Other naturally occurring and nonreactive energy sources are currently receiving considerable attention in the light of the potentialities for commercial-scale electric power generation. These sources are classified as solar, wind, geothermal, tidal, wave, ocean current, and ocean thermal energy.

The energy available from each of these sources can be used to produce mechanical and electrical power. In addition, solar and geothermal energy can be used directly for heating applications. Except for photovoltaic energy conversion, solar, geothermal, and ocean thermal energy would be utilized in a thermal power cycle for the generation of electric power.

An examination of the several nonreactive energy sources discloses that flowing water alone contributes significantly to the production of electrical energy. Other nonreactive energy sources are presently minor contributors to electric power production and to space heating. Because of dwindling supplies of natural gas and petroleum fuels, many investigations have been initiated to determine the feasibility of utilizing these nonconventional energy sources on a commercial scale. In every case, the economics of a proposed power generating system must be realistically evaluated. Some of these evaluations made to date appear to be overly optimistic and at best are subject to a fair amount of uncertainty because of a shifting cost structure.

Water Power

13.1 HYDRO POWER PLANTS

The term water power usually signifies power generated from water flowing in a river. In turn, river flow is attributed to a two-part cyclic process consisting of solar evaporation of ground level water and the subsequent fall of this water as rain. The moving water in the river possesses kinetic energy, and a portion of this energy could be extracted by some kind of mechanical system, say, a paddle wheel. A power plant of this type is however not a practical energy converter.

The customary procedure followed in a hydroelectric power development is to utilize the fall of the river by erecting a dam at a suitable point on the river and thus creating a head of water. For a typical submerged turbine runner, the head is the difference in elevation between the surface of the water in the reservoir behind the dam and the surface of the water in the tail race. In the case of the impulse turbine, the head is measured from the surface of the water in the reservoir to the center line of the wheel casing. The head represents the potential energy of a unit mass of water. If there were no losses in the conduits and in the energy converter, that is, the turbine-generator, all of this potential energy would be converted into electrical energy.

The potential energy of the stored water is high-level energy in that it is completely available. The type of water turbine installed in a particular hydro power project will be dependent upon the operating head. See Section 1.3. The ratio of the actual power output of the turbine-generator to the fluid power, measured at the unit, defines the machine efficiency. The maximum machine efficiency of a hydraulic turbine-generator is in the approximate range of 0.82 to 0.93, depending upon the type of turbine and the specific speed, a turbine performance characteristic. The machine efficiency decreases as the load on the turbine-generator is reduced. However, in some designs of propeller-type machines, adjustable blades and gates are employed to lessen the drop in efficiency.

A water power plant is in general a highly effective energy conversion system. There is no pollution of the environment, but objections are raised relative to the flooding of valuable real estate and scenic areas. Whether a particular hydroelectric installation is economically competitive with a fossil fuel or nuclear power plant will depend upon a number of factors, in particular, fuel and construction costs. In numerous instances, a hydroelectric power plant is clearly economically superior to a comparable thermal power plant. In the United States, most of the important water power sites have been developed, hence only a modest increase in the hydroelectric generating capability can be anticipated in the next two to three decades.

Recently conducted studies of water power resources disclose the existence in the United States of many low-capability hydro power sites that have potentialities for development. A typical example is the installation of a 3-MW turbine-generator at a dam that was constructed some years ago to create a water supply reservoir. A number of governmental agencies and industrial companies presently indicate an interest in developing some of these water power sites. The total potential power represented by these low-capability sites is judged to be substantial. However, the economic feasibility is likely to be the principal factor that will influence the decision to develop a particular site.

13.2 WATER TURBINE ENERGY CONVERSION

The first law is applicable to a water turbine. Then, for steady-state, steady-flow operating conditions, the energy equation is given by

$$\dot{m}[(u_1 - u_2) + v(P_1 - P_2) + g(z_1 - z_2) + (C_1^2/2 - C_2^2/2)] + \dot{Q} - \dot{W}_i = 0$$

Section 1 is taken at the surface of the water in the reservoir, and section 2 is taken at the surface of the water in the tail race or at the center line of the impulse wheel. The term z designates the elevation of the section; W_i is the turbine internal power.

At sections 1 and 2, the pressure is atmospheric, thus $v(P_1 - P_2) = 0$. The fluid velocity at section 1 is usually small, hence C_1 is assumed equal to zero. The draft tube, or turbine outlet structure, is designed to reduce C_2 to a relatively low value. For an impulse (Pelton) turbine, C_2 represents the velocity of the water leaving the wheel at pressure P_2 in the casing.

Because of the fluid losses that occur in the flow of water through the intake, turbine, and outlet structure, the energy conversion process is not frictionless, hence neither $u_1 - u_2$ nor $\dot{Q}$ will be equal to zero. These two energy quantities are difficult to measure and are replaced by a head loss term Δz_f. Then, the turbine internal power $\dot{W}_i$ is determined from

$$\dot{m}[g(z_1 - z_2) - (C_2^2/2) - g\,\Delta z_f] - \dot{W}_i = 0 \qquad (13.1)$$

The theoretical power $\dot{W}_{max}$ of the system is determined by omitting the head loss term and the residual kinetic energy term in Eq. 13.1. Thus,

$$\dot{m}g(z_1 - z_2) - \dot{W}_{max} = 0 \qquad (13.2)$$

The actual power delivered by the turbine-generator can be obtained by multiplying the theoretical power by the overall system efficiency.

Example 13.1

Water flows through a turbine at a rate of 10,000 kg/s. The operating head is 50 m. The system conversion efficiency is 0.85. The turbine mechanical efficiency is 0.99, and the generator efficiency is 0.96. The water discharge velocity is 2.5 m/s. Determine the turbine-generator power output, the turbine internal power, and the fluid head loss.

1 Theoretical power.

$$\dot{W}_{max} = \dot{m}g(z_1 - z_2)$$
$$= 10000 \times 9.807 \times 50$$
$$= 4.904 \times 10^6 \text{ W}$$

2 Turbine-generator power.

$$\dot{W}_{TG} = \eta_s \dot{W}_{max}$$
$$= 0.85 \times 4904 = 4168 \text{ kW}$$

3 Turbine internal power.

$$\dot{W}_i = \frac{\dot{W}_{TG}}{\eta_m \eta_g}$$
$$= \frac{4168}{0.99 \times 0.96} = 4386 \text{ kW}$$

4 Fluid head loss.

$$\dot{W}_i = \dot{m}[g(z_1 - z_2) - (C_2^2/2) - g\,\Delta z_f]$$
$$4386 \times 10^3 = 10{,}000[(9.807 \times 50) - (2.5)^2/2 - 9.807\,\Delta z_f]$$
$$\Delta z_f = 4.96 \text{ m}$$

13.3 OCEAN CURRENTS AND WAVES

The oceans collect a large portion of the total solar radiation that is intercepted by the earth. Absorption of the solar energy is manifested in ocean currents and waves, and numerous proposals have been made to utilize the kinetic energy of the moving ocean water.

Vast amounts of water move in the several ocean currents, notably the Gulf Stream. One proposal would extract kinetic energy from the ocean currents by means of large-diameter turbine wheels moored to the sea bottom and submerged in the flowing streams of water.

The wind blowing across the oceans and land masses of the earth is a result of the absorption of solar energy. Ocean waves, in turn, acquire their energy from wind far out at sea. A wave passing through a body of water causes the surface of the water to rise and fall. Underneath the surface of the water, molecules move in circular orbits, and only in the shallow water near shore is the undulating motion of the water transformed to a forward motion. Kinetic energy in the water, because of wave motion can be extracted by a mechanical system of some kind, such as an arrangement of floats that would rise and fall with the passage of each wave.

Several agencies and institutions in Japan are investigating a power generation system that utilizes wave action to compress air that, in turn, drives a turbine-generator.[1] No combustion process is involved in this system. Air is compressed in an open-bottomed, air piston chamber that can be either land fixed or floating offshore. Land-fixed models of very low-power capability have been placed in operation and appear to function without difficulty. The floating version of this system incorporates an air piston chamber installed in a buoy moored to the sea bottom. The economic and technical feasibility of this scheme for wave power generation will be demonstrated by constructing a full-scale power plant.

It is difficult to forecast whether production of electric power from ocean currents and waves will prove to be economically competitive with other nonconventional sources of energy, namely, solar, wind, and geothermal. Construction of various types of demonstration plants may be anticipated in an effort to reduce dependence upon fossil and nuclear fuels.

13.4 TIDAL POWER

The gravitational forces exerted on the earth by the sun and the moon cause a displacement of the ocean water of the earth. As the earth rotates, the surface of the ocean water alternately rises and falls at a particular location. During the rise of the water, the tide is designated as the flood tide, and during the fall of the water, as the ebb tide. High, or flood, tide occurs twice during each lunar day, a period of 24 hours and 51 minutes. Low tide follows high tide by 6 hours and 12.75 minutes.

The variation in the water level between high and low tide will depend upon the relative positions of the sun and the moon. At the time when the moon is new or full, the tidal range is unusually high, and when the moon is at the first or third quarter, the tidal range is unusually low. The terms spring and neap are used to describe the tides that cause, respectively, the maximum and minimum tidal ranges.

The average tidal range, the difference in elevation of the water surface between high and low tides, varies appreciably throughout the world. Along the open continental and island shores, only a modest average tidal range is observed. Of particular significance are the tides that occur in the rivers and bays connected to the oceans. Because of the topography of the area, very high tidal ranges occur in some of these connecting waterways, namely, 6 to 10 m on the average. These high tidal ranges cause strong currents in the tideway as the water advances and recedes: and, depending upon local conditions, the volume flow of water can be particularly high.

Evidentally, a great deal of energy is available in some of the tidal pools throughout the world. Some of these sites however are too far removed from an electric power market capable of absorbing the plant output. A potential site for a tidal power plant must have a sufficiently large tidal range and the capability for adequate storage of water. The plant construction cost must allow power generation costs to be competitive with other sources of energy. More specifically, a dam erected across an estuary should not be excessively long or require unduly complex construction techniques. The environmental impact on the surrounding region caused by construction of the power system must not create an adverse reaction.

13.5 TIDAL POWER PLANT OPERATION

In a tidal power plant, the potential energy of the water is converted into mechanical energy by passage through a water turbine. Although swift currents are created in the tideway, conversion of the kinetic energy in the water into mechanical energy would not be a feasible approach. In the same light, a system of floats that rise and fall with the tide would not constitute an effective energy converter.

Attention is now directed to a basic tidal power plant constructed by erecting a barrier across the tidal reach to create a basin in which water can be stored. Flood gates and water turbines are installed in the barrier. As the tide rises outside the barrier, the flood gates remain open and water flows into the basin. At high tide, the flood gates are closed, and subsequently with the ebb tide the water flows from the basin, through the turbines, to generate power. Theoretically, the outward flow of water continues until low tide occurs. The basin is filled and emptied during one tidal cycle that extends over a period of 12 hours and 25.5 minutes, that is, from low tide to high tide to low tide.

The volume of water that flows into the basin on the flood tide is equal to $A\ \Delta z$, where Δz is the tidal range and A is the corresponding mean area of the basin. Power generation on both the flood and the ebb tides is now assumed, and the volume flow of water through the turbine during one tidal cycle is equal to $2A\ \Delta z$, and the water flow rate is given by

$$\dot{m} = \frac{2\rho A\ \Delta z}{\tau}$$

where ρ is the density of the water and τ is the duration of a tidal cycle, 12.425 h.

The average head on the turbine is taken as $\Delta z/2$. The theoretical maximum average power generated in the tidal power plant is then expressed by $\dot{m}g\ \Delta z/2$, or

$$\dot{W}_{max} = \frac{\rho A g\ (\Delta z)^2}{\tau} \tag{13.3}$$

The actual power generated in the tidal power plant is a fraction of the theoretical maximum power, principally because the effective operating head on the turbine is considerably smaller than $\Delta z/2$. Incidentally, the turbines installed in a tidal power plant are capable of operating effectively under low-head conditions.

The annual average maximum power for the plant is determined by introducing into Eq. 13.3 a value of Δz equal to the annual mean value of the tidal range. The actual average power produced by the tidal power plant is equal to the product of the average maximum power and the annual plant operating factor, which may, for example, have a value of 0.11 for single-effect operation. For double-effect operation, that is, power generation on both the flood and the ebb tides, the annual plant operation factor is somewhat higher, say, 0.13.

The time schedule for power generation in a single-basin tidal power plant will be dependent upon the tidal conditions, which have a daily variation. Time control of the power generation can be achieved by utilizing a two-basin construction. The turbines are installed in the dam that separates the two basins. On the flood tide, the water enters one basin, designated as the high pool; while on the ebb tide, the water flows from the other basin, known as the low pool. The operation of the system can be regulated to generate power in accordance with a prescribed time schedule. While the two-basin system is more versatile than the single-basin system, construction costs are somewhat higher for the two-basin design.

A tidal power plant would be incorporated generally into a system comprised, in part, of other types of power plants, namely, steam and conventional hydro units. The operating schedule of the tidal power plant would be arranged for maximum utilization of the tidal energy. Consequently, the operating schedule for the entire system would be adjusted to accept the power output of the tidal plant whenever the power is available.

Example 13.2

The mean area of a tidal power plant basin is 70 km^2. The annual mean tidal range is 10 m, and the operating effectiveness of the plant is 0.12. Determine the annual average power output of the plant.

1 Theoretical maximum power.

$$\dot{W}_{max} = \frac{\rho A g (\Delta z)^2}{\tau}$$

$$= \frac{1025(70 \times 10^6)9.807(10)^2}{12.425}$$

$$= 5.6632 \times 10^{12}\ \text{J/h} \qquad 1573\ \text{MW}$$

2 Actual average power.

$$\dot{W}_a = f\dot{W}_{max}$$

$$= 0.12 \times 1573 = 188.8\ \text{MW}$$

13.6 TIDAL POWER DEVELOPMENT

The erection of water-powered grist mills was one of the first activities undertaken by the colonial settlers when establishing their communities. Grinding corn was an important function of the early water mills, but other commercial applications soon followed. Along the Atlantic seaboard, numerous rivers and streams provided many suitable sites for water power plants. Dams were constructed at these sites, and the impounded water was diverted usually to some kind of water wheel.

There was however another source of water power. Records of water power development in colonial times disclose the construction of tidal mills in sheltered locations along the coastal areas.

Mill ponds were created by constructing dams across the entrances to coves and salt marshes. A water wheel of the paddle-wheel type, placed in the mill race, turned in one direction on the flood tide and in the other direction on the ebb tide.

The tidal mills, constructed in the colonies as early as the year 1638, were low-capability power plants that ceased operation many years ago. From time to time, a few individuals have attempted to initiate tidal power development in the United States, but to date their efforts have been largely unsuccessful. On several occasions, agencies of the federal government conducted studies of the Passamaquoddy tidal energy project.

The Passamaquoddy site is located on the United States–Canadian border between Maine and New Brunswick. The average tidal range at the site is 5.5 m, and a two-basin power development has been projected, with Passamaquoddy Bay (260 km^2 area) forming the high pool and the smaller Cobscook Bay serving as the low pool. Capital costs for the project would be high, because of extensive dam construction, installation of maritime locks, and the gates required for controlling the flow of water into and from the pools. The electric generating capability of the system would, according to plan, be 1000 MW. Most reviews of the Passamaquoddy tidal power development have judged the project not to be economically feasible, but some support for the proposed installation continues.

The world's first major tidal power plant was constructed at La Rance, France, on the English channel. Construction of the 240-MW power plant was completed in 1966. The power is generated in 24 turbine-generators, each rated at 10 MW for an operating head of 5.75 m. The tidal range varies appreciably at the site, namely, from 6 to 14 m. The design of the La Rance power plant is based on a single-pool concept, but power is generated both on the flood tide and on the ebb tide. In this arrangement, the turbines are operated as the water flows into and from the basin.

The dam constructed at La Rance across the narrow estuary mouth is 700 m long. The capital cost of the plant is about $100 million, or $420/kW of installed generating capability. This cost was not particularly high, at 1966 prices, for a water power plant. The estimated capital cost, in $/kW, of the Passamaquoddy power project exceeds the capital cost of the La Rance installation by a factor of about 2, based on a comparable price structure.

Ocean Thermal Energy

13.7 OCEAN THERMAL ENERGY SOURCES

While the extraction of mechanical energy from the ocean currents is not particularly promising, utilization of ocean thermal energy (OTE) appears to be a feasible undertaking. Absorption of solar energy at the surface of the ocean creates a relatively warm layer of water that remains above the colder, more dense water in the lower depths of the ocean. Rotation of the earth causes the cold water coming from the direction of the poles to flow slowly along the ocean floor toward the tropics. In the tropical region, the cold water is heated and rises toward the surface as the density decreases. The water, warmed in this manner, flows at the surface in another current toward the polar regions. The cycle is repeated as the water cools and starts a return trip toward the tropics.

These broad currents of water transport great amounts of thermal energy and represent potential sources of electric power. The feasibility of converting ocean thermal energy into electrical energy is dependent upon the existence of two broad currents of water, one warm and the other cold, flowing in close proximity to each other. Fortunately, such juxtapositions are not uncommon. There are numerous locations within a few miles of land in and near tropical waters, such as the Caribbean Sea and the Gulf Stream, where ocean currents of vast magnitude flow within 600 to 900 m of each other. The temperature difference between the two streams is a constant 20 to 25 C, for a surface layer temperature of 27 to 29 C and a lower layer temperature of 4 to 7 C.[2]

The anticipated thermal efficiency for a power system operating within this temperature range is very low. The Carnot cycle thermal efficiency for operation between 27 and 4 C is 0.077, while the thermal efficiency of an actual Rankine cycle-type power plant would probably not exceed 0.04.

The OTE power plant would be a large and costly installation. Extremely large volumes of warm and cold water, the heat source and the heat sink, respectively, would be handled. There is of course no expenditure for fuel. The feasibility of building OTE power plants rests upon the ability to design plant components that can be constructed at a cost sufficiently low to ensure a competitive position in the electric power market.

13.8 OCEAN THERMAL ENERGY POWER PLANT DEVELOPMENT

In 1882, D'Arsonval suggested that power could be generated by utilizing the energy in the warm surface water of the ocean and rejecting heat to the colder water of a lower layer. The first attempt to utilize the ocean thermal energy was made in 1926. Georges Claude, the French scientist, constructed a 40-kW land-based OTE power plant in Matanzas Bay, Cuba.[2] A portion of the warm surface water was converted into steam in a low-pressure flash evaporator operating at a high vacuum. In the familiar flash evaporation process, sensible energy in the water is converted into latent energy. The steam produced in the evaporator was expanded in a turbine and subsequently condensed by direct contact with the cold seawater piped from the lower layer in the ocean.

Claude's attempt to utilize the ocean thermal energy for power generation failed for a number of reasons. The seawater was used for the power plant working fluid, and heat transfer problems were thus avoided; but other problems developed, including those caused by the corrosive characteristics of seawater. Water was evaporated at a low temperature and at a correspondingly low pressure. Pumping work required for removal of the noncondensable gases was consequently high. Because of the very high specific volume of the steam, a large turbine was required. A low Reynolds number was an operating characteristic of the turbine, and as a consequence the machine efficiency of the turbine was low.

Claude also encountered another problem of major significance, namely, that of conveying the cold seawater to the power plant in a relatively long (2 km) pipeline. The heat flow into the pipeline proved to be excessively high and directed attention to the inadequacy of a land-based site for an OTE power plant. Claude's experiment demonstrated that the OTE power plant will inherently require large components, have a relatively low power output, and operate at a high vacuum when the working fluid is water.

Within the past decade, improved designs for OTE power plants have been developed. The proposed power plants would be located at sea, either submerged or floating on the surface. A more appropriate working fluid than water, for example, propane or ammonia, would be used in some of these proposed installations. The possibility of constructing OTE power plants has, to some degree, been enhanced by improved designs for flash evaporators.

In 1965, J. H. Anderson and J. H. Anderson, Jr. presented in considerable detail a conceptual design for an OTE power plant encased in a submerged structure anchored in deep water of the Caribbean Sea.[3] The plant is designed to operate with a thermal differential of 20 C between the warm surface water and the colder water in a layer 600 m below the surface. Approximately 14 percent of the gross power generated by the plant would be consumed by the plant auxiliary equipment.

The power plant operates on a closed cycle with propane serving as the working fluid. The warm seawater flows through a heat exchanger, the boiler, in which the low-boiling-point propane is evaporated. Propane vapor at a pressure of about 1035 kPa, abs, expands in the turbine to a pressure of about 690 kPa, abs. Condensation of the propane vapor is effected in a surface condenser by transfer of heat to the cold seawater. The temperature increase in the cooling water is about 3 to 4 C.

Special attention must be directed to the design of the heat exchangers, because these elements are very large and all secondary temperature drops must be minimized in order to achieve the highest possible temperature drop through the turbine. The secondary temperature drops are the temperature differentials required for the transfer of heat from one fluid to another. The Anderson design employs plate-type heat exchangers to function as the boiler and as the condenser.

The operating pressures in the closed propane cycle are relatively high, hence a turbine of reasonable size and cost can be anticipated. Vacuum and deaeration problems are eliminated by operation of the thermal cycle above atmospheric pressure.

TRW and Lockheed recently completed independent comprehensive studies on the feasibility of ocean thermal energy conversion. Both studies concluded that OTEC is technically feasible. In general, the economic outlook for OTEC was reported as favorable.

The operating principle of the TRW and Lockheed systems is similar to that proposed by Anderson and Anderson. Both systems, TRW and Lockheed, use ammonia for the working fluid and employ shell-and-tube heat exchangers, with the seawater flowing inside titanium tubes. The several pieces of equipment are supported on a floating concrete structure. The TRW power system is designed to produce 100 MW(e) and the Lockheed system, 160 MW(e).[4]

Warm surface water is used for evaporation of the ammonia, while colder water from the ocean depths is employed for cooling the condenser. The TRW design for the cold-water duct is a single, 12-m-diameter tube of fiber-reinforced plastic 1200 m long. The Lockheed design for the cold-water intake line consists of five telescoping sections of posttensioned concrete pipe that has a minimum diameter of 30 m and reaches a depth of 458 m. The TRW proposed design uses a dynamic system for positioning the platform, while the Lockheed proposal specifies a single-point mooring.

Example 13.3

An OTE power plant uses propane for the working fluid. The warm water enters and leaves the evaporator at 28 and 25 C. The cold water enters and leaves the condenser at 5 and 8 C. The propane evaporating and condensing temperatures are, respectively, 22 and 11 C. The turbine machine and generator efficiencies are, respectively, 0.90 and 0.95. The power plant output is 125 MW, and 15 percent of the turbine-generator power is required for operation of the station auxiliary equipment. The heat transfer overall coefficients are 1275 $W/m^2 \cdot K$ for the evaporator and 1475 $W/m^2 \cdot K$ for the condenser. Determine the propane flow rate, the area of the heat transfer surface for the evaporator and the condenser, and the thermal efficiency of the power plant.

Propane Flow Rate

$$\dot{W}_{T\text{-}G} = \frac{\dot{W}_o}{1 - 0.15} = \frac{125}{0.85} = 147.06 \text{ MW}$$

Turbine throttle:

$T = 22$ C (295.15 K) $\quad x = 1.0$

$h_1 = 562.94$ kJ/kg $\quad s_1 = 2.0525$ kJ/kg·K

Turbine exhaust:

$T = 11$ C (284.15 K)

$h_{2,s} = 549.29$ kJ/kg

$$W_{T\text{-}G} = \eta_g \eta_T (h_1 - h_{2,s}) = 0.95 \times 0.90(562.94 - 549.29) = 11.67 \text{ kJ/kg}$$

$$\dot{m}_p = \frac{\dot{W}_{T\text{-}G}}{W_{T\text{-}G}} = \frac{147060}{11.67} = 12.602 \text{ kg/s}$$

Enthalpy and entropy values for propane are selected from *Thermodynamic Properties in SI*, by W. C. Reynolds, Stanford University, 1979.[38]

Pump Work

Entering pump:
$h_{f,2} = 194.54$ kJ/kg
$P_2 = 655.5$ kPa, abs
$v_{f,2} = 0.001961$ m³/kg

Leaving pump: $P_1 = 882.7$ kPa, abs

Assume internal pump efficiency is 0.65.

$$W_P = -\frac{v_{f,2}(P_1 - P_2)}{\eta_P} = -\frac{0.001961(882.7 - 655.5)}{0.65} = -0.69 \text{ kJ/kg}$$

Evaporator

Leaving pump:
$h_3 = h_{f,2} - W_P$
$= 194.54 - (-0.69) = 195.23$ kJ/kg
$\dot{Q} = \dot{m}_p(h_1 - h_3)$
$= 12{,}602(562.94 - 195.23)$
$= 4.634 \times 10^6$ kJ/s

Mean temperature difference:
Water $T_{in} = 28$ C $T_{out} = 25$ C
Propane: $T_{evap} = 22$ C
LMTD $= 4.33$ C

Area of heat transfer surface:

$$A_o = \frac{\dot{q}}{U_o \, \Delta T_m} = \frac{4.634 \times 10^9}{1275 \times 4.33} = 839{,}361 \text{ m}^2$$

Condenser

Entering condenser:

$$h_2 = h_1 - \eta_T(h_1 - h_{2,s}) = 562.94 - 0.90(562.94 - 549.29) = 550.66 \text{ kJ/kg}$$

Leaving condenser:
$h_{f,2} = 194.54$ kJ/kg
$\dot{Q} = \dot{m}_p(h_{f,2} - h_2)$
$= 12{,}602(194.54 - 550.66)$
$= -4.488 \times 10^6$ kJ/s

Mean temperature difference:
Water: $T_{in} = 5$ C $T_{out} = 8$ C
Propane: $T_{cond} = 11$ C
LMTD $= 4.33$ C

Area of heat transfer surface:

$$A_o = \frac{\dot{q}}{U_o \, \Delta T_m} = \frac{4.488 \times 10^9}{1475 \times 4.33} = 702{,}674 \text{ m}^2$$

Power Plant Thermal Efficiency

$\dot{Q}_s = 4.634 \times 10^6$ kJ/s $\dot{W}_o = 125 \times 10^3$ kJ/s

$$\eta_t = \frac{\dot{W}_o}{\dot{Q}_s} = \frac{125 \times 10^3}{4.634 \times 10^6} = 0.027$$

Seawater flow

Evaporator: 0.3886×10^6 kg/s 379.1 m³/s
Condenser: 0.3764×10^6 kg/s 367.2 m³/s

A design has been proposed for an OTE power plant that utilizes Claude's scheme of evaporating a small portion of the warm ocean water flow for the turbine working fluid.[5] The power plant would however be installed on a barge and draw cold water from a depth of 730 m. The principal objective of this design is the elimination of the very large heat exchangers used in the Anderson, TRW, and Lockheed designs by substituting a flash-type evaporator and a direct-contact condenser. The turbine operates through a temperature difference of 10 C and a pressure difference of 1.18 kPa. The turbine work per unit mass of vapor is small, an inherent characteristic of an OTE power plant. For this particular system, the output of the turbine-generator is 59.3 kJ/kg of vapor, and the prescribed net power plant output is 6500 kW.

Construction of an OTE power plant operating with flash evaporation can be achieved with established technology, but the economic feasibility of the system is uncertain. Corrosion by the salt ocean water may prove to be a problem or require extensive use of costly corrosion-resistant materials.

The electrical energy generated by a floating OTE power plant would be transmitted by an underwater cable to a land-based power grid. The electrical energy could also be used at the floating power plant to manufacture chemical fuels, for example, hydrogen. The product fuel however must be transported from the plant by some means, such as a barge or an underwater pipeline.

During the next two or three decades, ocean thermal energy conversion is likely to receive some attention. An effort will be made to reduce the high capital cost of the OTE power plant, in part, by substituting cheaper corrosion-resistant materials, such as aluminum, for the expensive titanium. The feasibility of commercial-scale electric power generation by ocean thermal energy conversion must first be demonstrated by constructing an OTE pilot plant of appropriate size.

Solar Energy

13.9 SOLAR ENERGY APPLICATIONS

Direct conversion of solar energy into electrical energy by means of photovoltaic converters was investigated in the preceding chapter. In this chapter, an examination will be made of the alternative use of solar energy, that is, the conversion of solar energy into thermal energy in a variety of systems.

Within the past few years, a large number of programs and projects have been initiated for examining the feasibility of utilizing solar energy for heating, air conditioning, and production of electric power. Practical applications of solar energy however have been investigated during the past 200 years. In addition to the development of various appliances, such as cookers, water heaters, and stills, a number of solar engines and furnaces were constructed.

Two solar furnaces were erected in France, the first at Mont Louis in the Western Pyrenees, and the second, a much larger unit, on a mountainside between the adjacent villages of Font Romeu and Odeillo. These furnaces, used in metallurgical processing, have thermal power ratings of 50 and 1000 kW(t), respectively.[6]

Within the past 100 years, several solar engines were constructed. In this respect, the work of John Ericsson is particularly noteworthy. He built nine solar engines, the last of which, erected in 1883, was a steam power plant of significant proportions. The reflector, 3.35 m long and 4.90 m wide, concentrated the solar energy upon a heater tube that had a diameter of 165 mm and a length of 3.35 m. Steam was generated at a pressure of 240 kPa, abs, in sufficient quantity to operate a 152-mm $\times$ 203-mm condensing engine at 120 rpm.[7]

In 1913, a solar steam power plant used to pump water was constructed on the banks of the Nile River near Cairo. Although the power plant could produce between 37 and 45 kW, it was shortly abandoned because other irrigation systems proved to be more economical to operate. From time to time, a number of investigators have built small solar engines, but the feasibility of operating solar engines on a significant commercial scale has not been demonstrated to date.

13.10 AVAILABILITY OF SOLAR ENERGY

A vast amount of solar radiation is intercepted by the earth; but, as noted earlier, this energy is diffuse, and at any particular location on the earth's surface, the energy is received intermittently. During a period of one year, the solar radiation that falls on the United States is equivalent to about 75 times as much energy as the nation consumes.[8]

The distribution of solar energy varies generally with latitude, but the distribution is not necessarily uniform. The distribution of solar energy for the United States is shown in Fig. 13.1. The lines on the chart indicate, on an annual basis, the mean quantity of solar radiation on a horizontal surface in MJ/m^2 per day. For collection of solar energy, the southwestern section of the United States appears to have the best potentiality.

The solar energy input rate at a particular location will depend upon the season of the year, the time of day, and the condition of the sky. Considerable difference is observed between the winter and the summer solar radiation values for the United States.

Figure 13.2 illustrates the variation in solar radiation with the time of day and the time of the year at latitude 43°N for a clear day. The data reported in Fig. 13.2 show the solar radiation rate, in W/m^2, during the day and the total radiation for the day, expressed in MJ/m^2. Information of this kind is important to the design of a solar energy collector.

13.11 SOLAR ENERGY COLLECTION

The solar radiation that arrives at the fringe of the earth's atmosphere is materially reduced in passing through an envelope about 150 km thick, consisting of various gases, water vapor, and dust. The amount of solar energy that reaches the earth's surface at a particular time is dependent upon a number of factors, namely, the air mass, angle, cloud and haze, and diffuse radiation factors.

The air mass factor, a measure of the length of the path through the atmosphere, is a function of the zenith distance, the angle between the zenith and the direction of the sun. The zenith angle θ_z varies throughout the day and from day to day as a function of the latitude ϕ, the solar declination δ, and the hour angle ω of the sun.

The direct radiation $\dot{E}_n$ incident upon a surface located on the earth and oriented in a direction normal to the rays of the sun is given approximately by

$$\dot{E}_n = \dot{E}_o \tau_a^m \tag{13.4}$$

where

$\dot{E}_o$ = the solar constant, 136.1 mW/cm^2
τ_a = the transmission coefficient for unit air mass
m = the relative air mass, defined as the ratio of the length of the actual path to the length of the shortest possible path

Numerically, m = secant θ_z for zenith angles up to 80°. When the sun is only 10° above the horizon, the radiation is essentially negligible and of little practical value.

The value of the transmission coefficient τ_a varies with the condition of the sky and ranges from about 0.8 on a clear day to under 0.10 on a heavily overcast day. At a particular location, the appropriate value of τ_a is predicted from meteorological data.

When the surface receiving the radiation is not normal to the sun's rays, the angle factor is applied. The direct radiation incident upon a surface not normal to the direction of the sun is determined by

$$\dot{E}_i = \dot{E}_n \cos \theta_i \tag{13.5}$$

where θ_i is the angle of incidence; θ_i is measured between the sun's direction and the normal to the surface. Inclination of the surface from the normal position evidently reduces the incident radiation on the surface.

For a horizontal surface, the angle of incidence θ_i is equal to the zenith angle θ_z. Then,

$$(\dot{E}_i)_h = \dot{E}_n(\cos \theta_i)_h = \dot{E}_n \cos \theta_z \tag{13.6}$$

With respect to the collection of solar energy, the effectiveness of a horizontal surface improves as the sun rises in the sky.

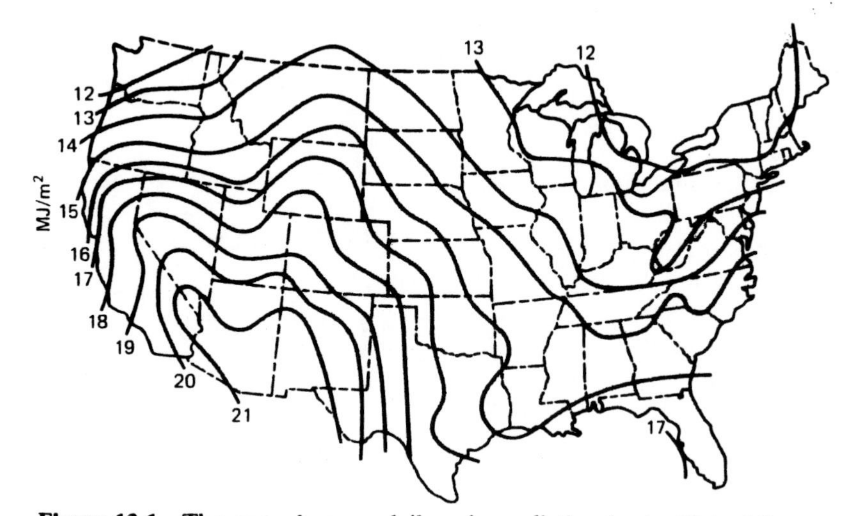

Figure 13.1 The annual mean daily solar radiation in the United States. Total radiation on a horizontal surface, in MJ/m^2 per day. (*Energy and Urban Policies/Programs*, Office of Consumer Affairs, Department of Energy, April 27, 1978.)

The total solar energy reaching a collector surface is equal to the sum of the direct radiation and the diffuse radiation that is scattered back to the earth from all parts of the sky. The direction of the direct radiation can be calculated exactly, and the amount of this radiation can be determined with reasonable precision. The diffuse radiation however varies appreciably. On a bright, sunny day, the diffuse radiation is equal to 10 percent of the total radiation that reaches a horizontal surface. In partly cloudy weather, the diffuse radiation is equal to 50 percent of the total radiation; on completely overcast days, it is equal to 100 percent.[7]

The diffuse radiation factor and the cloud-and-haze factor are dependent upon meteorological

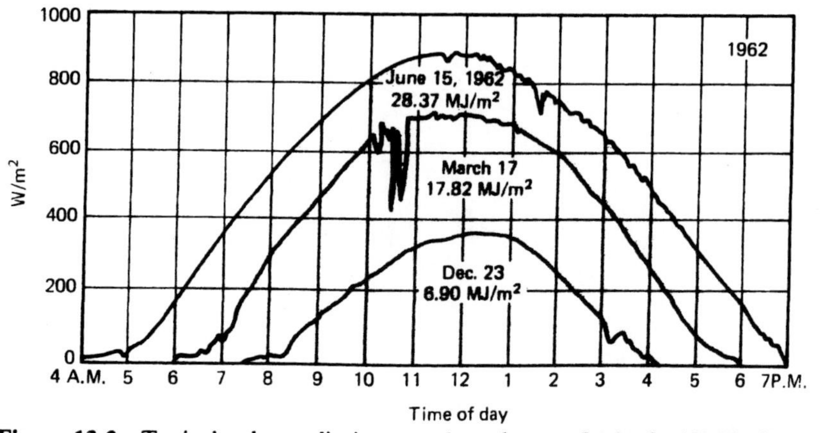

Figure 13.2 Typical solar radiation on clear days at Latitude 43° N. (F. Daniels, *American Scientist*, Jan.–Feb. 1967.[10] Reprinted by permission of *American Scientist*, The Scientific Research Society.)

conditions and are consequently statistical quantities rather than quantities capable of exact evaluation. Local Weather Bureau records are the best source of information for the probable radiation that a surface will receive during the year and should be consulted in designing a solar energy collector.

An arid or desert climate, with 4000 or more hours of sunshine expected per year, is conducive to solar energy collection. In areas such as the southwestern region of the United States, horizontal surfaces receive about 8×10^3 MJ/m^2 per year solar energy. Surfaces that track the sun can receive from 9.5×10^3 to 10×10^3 MJ/m^2 per year.

13.12 COLLECTOR GEOMETRY

Equation 13.6 indicates that the direct radiation received by a horizontal surface on the earth is equal to the product of E_n, the normal direct radiation, and the cosine of the zenith distance θ_z. A fundamental trigonometric relationship shows

$$\cos \theta_z = \sin \phi \sin \delta + \cos \phi \cos \delta \cos \omega \qquad (13.7)$$

where

θ_z = the zenith distance
ϕ = the latitude
δ = the solar declination
ω = the hour angle of the sun (15° per hour)

See Fig. 13.3.

The time of day can be expressed by the hour angle ω, which is indicative of the apparent rotation of the sun about the earth's axis. The rotation is measured with respect to the time (solar noon) when the sun crosses the south meridian of a particular location on the earth. One hour

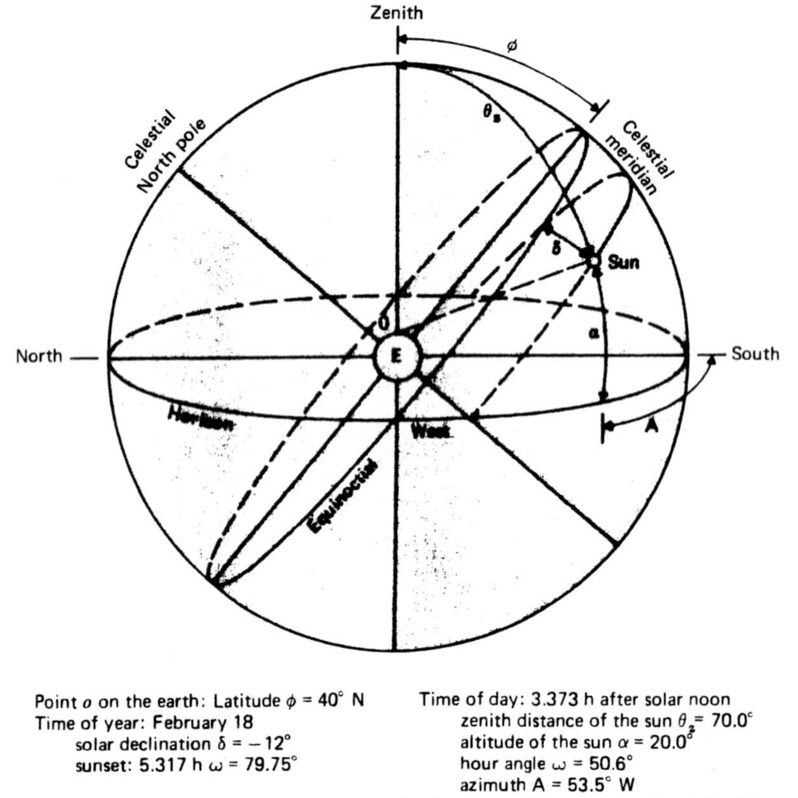

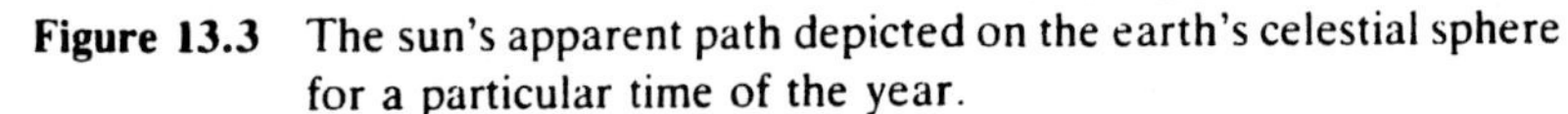

Figure 13.3 The sun's apparent path depicted on the earth's celestial sphere for a particular time of the year.

of time is equivalent to 15 degrees of rotation, and the time is expressed in hours before and after solar noon. The value of ω varies from zero at local noon to a maximum at sunrise or sunset.

A particularly significant orientation for the collection of solar energy is a surface tilted at an angle β from the horizontal plane toward the equator. The surface is aligned in an east–west direction and tilted toward the south. The incidence angle $\theta_{i,t}$ for this configuration is given by

$$\cos \theta_{i,t} = \sin (\phi - \beta) \sin \delta + \cos (\phi - \beta) \cos \delta \cos \omega \quad (13.8)$$

In Fig. 13.4, the surface s is oriented in an east–west direction and tilted from the horizontal plane at an angle β toward the south. The time of day is taken at solar noon, hence the hour angle ω is zero. The angle β is adjusted to place the surface s normal to the sun's direction, $\beta = \beta_n$. Then, $\cos \theta_{i,t} = 1$, and Eq. 13.8 is reduced to

$$1 = \sin (\phi - \beta_n) \sin \delta + \cos (\phi - \beta_n) \cos \delta$$

The right-hand side of the equation attains a value of 1, the maximum value, when $\phi - \beta_n = \delta$. The solar declination δ varies with the time of the year, and the appropriate value of δ can be obtained from a nautical almanac. The angle β_n is readily determined, since the latitude ϕ for the surface is known. It is essential to note that β_n is a unique value of the tilt angle β, established when the collector surface is normal to the sun's direction at solar noon for a particular location on the earth's surface.

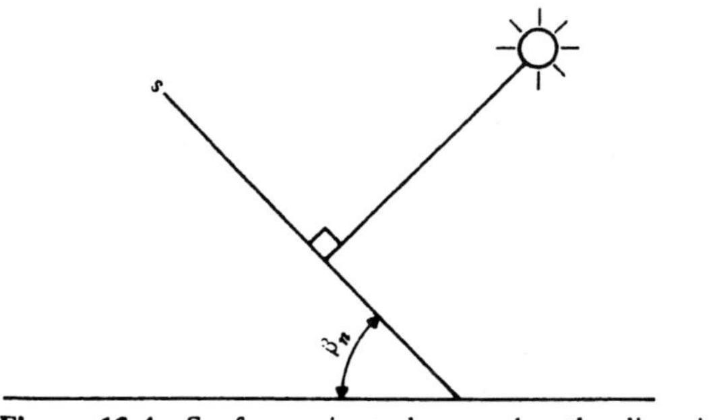

Figure 13.4 Surface oriented normal to the direction of the sun.

For any location north of the equator, both the latitude ϕ and the solar declination δ are positive values during the period extending from the vernal equinox to the autumnal equinox, about March 21 to September 21. From the autumnal equinox to the vernal equinox, the latitude is a positive value, while the solar declination is a negative value.

The altitude of the sun, α, the angular elevation of the sun above the horizon, is equal to 90° minus the zenith distance, or $\alpha = 90° - \theta_z$. See Fig. 13.3. Direct measurement of the sun's altitude can be made with a sextant by measuring the angle between the horizontal plane and the line in the direction of the sun.

The position of the sun at a particular time of the day is defined by the altitude and the azimuth of the sun. See Fig. 13.3. The azimuth is defined as the angle between the sun's direction and the south meridian measured on the horizontal plane. In other words, an arc on the horizon extending from the south meridian is the measure of the sun's azimuth.

The surface s in Fig. 13.4 can be made to track the sun continuously by rotating and tilting the surface in a manner that maintains the surface normal to the sun's direction. Further discussion of tracking will be presented at a later point.

Example 13.4

A solar collector is located at latitude 47°N. On May 15, the solar declination is +19°, and the interval extending from sunrise to sunset is 14.90 hours. Determine the magnitude of β_n and, at solar noon, the zenith angle, altitude of the sun, and the value of $\cos \theta_{i,t}$ for $\beta = \beta_n$.

1 $\beta_n = \phi - \delta = 47° - 19° = 28°$

2 $\cos \theta_z = \sin \phi \sin \delta + \cos \phi \cos \delta$
$= \sin 47° \sin 19° + \cos 47° \cos 19°$
$= 0.8830$

$\theta_z = 28.0°$ at solar noon

3 $\alpha = 90° - \theta_z = 90° - 28.0° = 62.0°$

4 $\cos \theta_{i,t} = 1$ for $\beta = \beta_n = 28.0°$

Now, determine the value of $\cos \theta_{i,t}$ at solar noon for $\beta_n - \beta = +10°$ and $-10°$.

1 For $\beta_n - \beta = +10°$,

$$\beta = 18° \qquad \phi - \beta = 47° - 18° = 29°$$

$$\cos \theta_{i,t} = \sin(\phi - \beta)\sin\delta + \cos(\phi - \beta)\cos\delta = \sin 29° \sin 19° + \cos 29° \cos 19° = 0.9848$$

2 For $\beta_n - \beta = -10°$,

$$\beta = 38° \qquad \phi - \beta = 47° - 38° = 9°$$

$$\cos \theta_{i,t} = 0.9848$$

With respect to the above example, a deviation for β of $+10°$ or $-10°$ from β_n has a minimal effect at solar noon on the magnitude of $\cos \theta_{i,t}$. Slightly larger deviations are observed, for example, at a time 4 hours before or after solar noon.

For a flat collector surface tilted to the south, an adjustment of the tilt angle β at two-week intervals is sufficient in compensating for the change in the solar declination.

Solar energy collectors used in heating applications are ordinarily installed in fixed positions. The angle β is commonly set equal to $\beta_n - 10°$, where β_n is determined at the time of the winter solstice (December 21); as a consequence, $\cos \theta_{i,t} = 0.9848$ at solar noon for $\phi = 47°$N. At the time of the summer solstice (June 21), $\cos \theta_{i,t} = 0.7986$ at solar noon for the collector in the same fixed position. A comparison of the insolation at the two different times of the year, for $\tau_a = 0.75$, is shown below.

	θ_z (Degrees)	τ_a^m	$\dot{E}_n$ (W/m²)	$\cos \theta_{i,t}$	$\dot{E}_i$ (W/m²)
December 21	70.5	0.4224	575	0.9848	566
June 21	23.5	0.7307	995	0.7986	794

Example 13.5

Determine for the surface described in Example 13.4 the magnitude of $\dot{E}_i$ at hourly intervals during the day. $\beta = \beta_n = 28°$, $\tau_a = 0.75$. The surface is fixed and faces south. The following relationships are used for obtaining the solution.

1 $\cos \theta_z = \sin\phi \sin\delta + \cos\phi \cos\delta \cos\omega = \sin 47° \sin 19° + \cos 47° \cos 19° \cos\omega = 0.2381 + 0.6448 \cos\omega$

2 $\dot{E}_n = \dot{E}_o \tau_a^m = 1361(0.75)^m$ W/m²
where $m = \text{secant}\ \theta_z$

3 $\dot{E}_i = \dot{E}_n \cos\theta_i = \dot{E}_n \cos\theta_{i,t}$

$$\cos \theta_{i,t} = \sin(\phi - \beta)\sin\delta + \cos(\phi - \beta)\cos\delta\cos\omega = \sin 19° \sin 19° + \cos 19° \cos 19° \cos\omega = 0.1060 + 0.8940 \cos\omega$$

The hour angle ω is measured from solar noon toward the time of sunrise or the time of sunset. In this example, sunrise occurs 7.45 hours before solar noon, and sunset occurs 7.45 hours after solar noon.

The calculated values for the several quantities are:

Time (hour)	$\cos \theta_z$	τ_a^m	$\dot{E}_n$ (W/m²)	$\cos \theta_{i,t}$	$\dot{E}_i$ (W/m²)
0	0.8829	0.7219	982.5	1.0000	982.5
1	0.8609	0.7159	974.4	0.9695	944.7
2	0.7965	0.6969	948.4	0.8802	834.8
3	0.6940	0.6607	899.2	0.7382	663.7
4	0.5605	0.5985	814.6	0.5530	450.5
5	0.4050	0.4915	668.9	0.3374	225.7
6	0.2381	0.2987	406.6	0.1060	43.1
7					—

A collector surface is not in every installation oriented in an east–west direction and tilted toward the south. Figure 13.5 shows a collector surface that is tilted at an angle β from the horizontal plane, but not aligned in an east–west direction. The normal to the surface establishes an angle γ, measured on the horizon, with the south meridian. The incidence angle for the general case is given by

$$\cos \theta_i = \cos\theta_z \cos\beta + \sin\theta_z \sin\beta \cos(A - \gamma) \qquad (13.9)$$

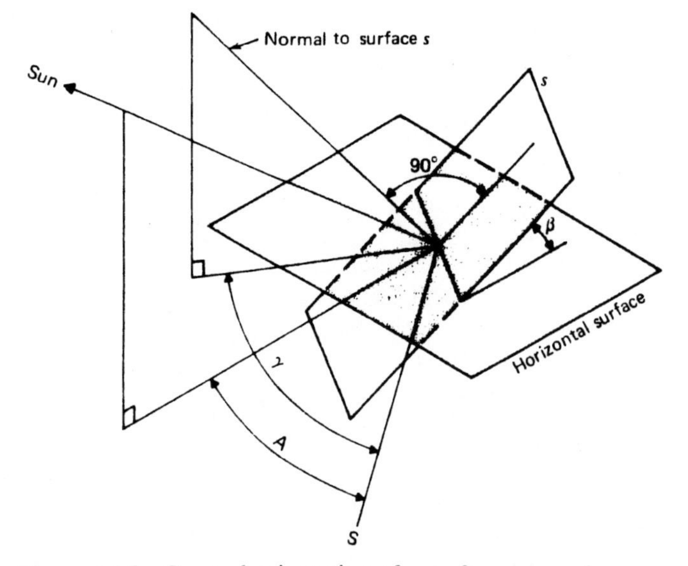

Figure 13.5 General orientation of a surface exposed to solar radiation.

where A, the azimuth of the sun, is determined from

$$A = \sin^{-1} \frac{\cos \delta \sin \omega}{\sin \theta_z} \tag{13.10}$$

As the hour angle increases, the value of $\sin A$ approaches 1 as a maximum, and subsequently continues to decrease with a further increase in ω. The decrease in the value of $\sin A$ however indicates that the azimuth angle is now in the second quadrant, or $A > 90°$. It is thus essential in the vicinity of $A = 90°$ to determine whether $A < 90°$ or $A > 90°$.

Example 13.6

A surface is exposed to the sun at 2:30 P.M. The inclination angle β is 38°. The solar declination is +15°, and the latitude is 42°N. The angle γ is 10° measured eastward from the south meridian. Calculate the incidence angle.

$$\begin{aligned} \cos \theta_z &= \sin \phi \sin \delta + \cos \phi \cos \delta \cos \omega \\ &= \sin 42° \sin 15° \\ &\quad + \cos 42° \cos 15° \cos 37.5° \\ &= 0.742671 \\ \theta_z &= 42.04° \end{aligned}$$

$$\begin{aligned} A &= \sin^{-1} \frac{\cos \delta \sin \omega}{\sin \theta_z} \\ &= \sin^{-1} \frac{\cos 15° \sin 37.5°}{\sin 42.04°} \\ &= 0.878099 \\ A &= 61.41° \\ \cos \theta_i &= \cos \theta_z \cos \beta \\ &\quad + \sin \theta_z \sin \beta \cos (A - \gamma) \\ &= \cos 42.04° \cos 38° \\ &\quad + \sin 42.04° \sin 38° \cos (61.41° + 10°) \\ &= 0.716669 \\ \theta_i &= 44.22° \end{aligned}$$

Equation 13.9 can be used to investigate the means employed to track the sun, that is, by maintaining the collector surface continuously normal to the sun's direction, or $\cos \theta_i = 1$. The surface would be rotated so that $A - \gamma$ remains equal to zero. Then, Eq. 13.9 is reduced to

$$1 = \cos (\theta_z - \beta)$$

Hence,

$$\theta_z - \beta = 0 \qquad \text{or} \qquad \beta = \theta_z$$

As the surface is rotated about a vertical axis, the tilt angle is continuously adjusted in order to maintain angle β equal to the zenith angle θ_z, while angle γ remains equal to angle A.

Example 13.7

The collector surface in Example 13.4 is arranged to track the sun continuously. Determine at hourly intervals the values of the tilt angle β and the angle γ. Values of $\beta = \theta_z$ are obtained from Example 13.5, and Eq. 13.10 is used to determine values of A and hence γ.

$\phi = 47°$ $\delta = 19°$

Time (Hour)	β *(Degrees)*	γ *(Degrees)*
0	28.0	0
1	30.6	28.7
2	37.2	51.4
3	46.0	68.3
4	55.9	81.4
5	66.1	92.6
6	76.2	103.2
7	85.9	113.7

A comparison of the values of $\dot{E}_n$ and $\dot{E}_i$ tabulated in Example 13.5 shows the improvement in solar energy collection achieved by tracking the sun. For a period of 4 hours after or before solar noon, the fixed surface would intercept about 15 percent less radiation than the tracking surface. For a period of 6 hours however, this figure is increased to about 27 percent. Somewhat different results should be anticipated for actual solar energy collectors, depending upon the type of construction.

The loss because of reflection of the solar rays from the glass cover of a collector increases very sharply as the magnitude of the incident angle increases above 60°. Thus, in Example 13.6, the values of $\dot{E}_i$ are 225.7 and 43.1 W/m^2 for the fixed surface at 5 and 6 hours respectively, would be substantially reduced, perhaps to about 160 and 15 W/m^2, respectively, principally because of reflection from a flat glass cover.

Clear sheet glass has a comparatively low absorptance factor, about 0.05, for solar radiation. Little variation is observed in the magnitude of this factor with a change in the incident angle.

13.13 DIFFUSE RADIATION

In the preceding section, the investigation of solar energy collection is limited to the interception of direct solar radiation by the collector surface. Some types of collectors are capable of collecting diffuse radiation as well as direct radiation.

On passing through the earth's atmosphere, a portion of the beam radiation incident at the outer edge of the atmosphere is changed by scattering and reflection into diffuse, or nondirectional, radiation. At a particular location, the diffuse radiation component varies quantitatively with time and is principally dependent upon the condition of the sky. The diffuse radiation is not readily evaluated generally as a component of the total radiation (direct plus diffuse) that is received by a surface.

The U.S. National Weather Service records at a number of sites the total horizontal solar radiation, along with certain meteorological data. The means to determine the direct normal radiation from the total horizontal radiation would be particularly useful in the design of solar collectors. A preliminary correlation between the direct normal radiation and the measured total horizontal radiation at several cities across the country is reported in Ref. 12.

Example 13.8

At a site in the vicinity of latitude 47°N, the mean daily total horizontal solar radiation for July is 7.10 kWh/m^2. The ratio of the direct normal radiation to the total horizontal radiation is 1.20 to 1. The mean value of the cosine of the zenith angle, from sunrise to sunset, for the month of July is 0.5311. Determine the corresponding mean daily direct normal radiation and estimate the diffuse radiation.

Direct normal radiation:

$$E_{dn} = \frac{E_{dn}}{E_{th}} E_{th}$$

$$= 1.20 \times 7.10 = 8.52 \text{ kWh/m}^2 \text{ per day}$$

Direct horizontal radiation:

$$E_{dh} = \overline{\cos \theta_z}\, E_{dn}$$

$$= 0.5311 \times 8.52 = 4.52 \text{ kWh/m}^2 \text{ per day}$$

Diffuse radiation:

$$E_{th} - E_{dh} = 7.10 - 4.52$$

$$= 2.58 \text{ kWh/m}^2 \text{ per day}$$

For this example, a uniform diffuse radiation is assumed. The total normal radiation is then equal to the sum of the direct normal radiation and the diffuse radiation, that is, 8.52 + 2.58 = 11.10 kWh/m^2 per day.

In order to intercept the direct normal radiation throughout the day, a tracking surface must be used. For a fixed, south-facing surface, inclined at an angle of 55° from the horizontal plane, the mean value of $\cos \theta_{i,t}$ is 0.5372. The direct radiation received by the surface is given by

$$E_d = \cos \theta_{i,t} E_{dn}$$

$$= 0.5372 \times 8.52 = 4.58 \text{ kWh/m}^2 \text{ per day}$$

The total radiation on the inclined surface is then equal to the direct plus the diffuse radiation, 4.58 + 2.58 = 7.16 kWh/m^2 per day.

In the preceding analysis, the diffuse radiation is assumed to be uniformly distributed over the sky. The diffuse radiation on a surface other than in a horizontal orientation is then dependent upon the portion of the sky visible from the surface. The ground and other surfaces reflect solar radiation to the inclined, or vertical, collecting surface. In the simplified case, the reflecting surfaces are assumed to have such characteristics that the diffuse radiation reflected from these surfaces is equivalent to the diffuse radiation from the sky. It follows that the collecting surface will receive the same diffuse radiation for any orientation of the surface.

The model for radiation on an inclined surface is now modified by considering three individual components, namely, the beam radiation, diffuse solar radiation, and the solar radiation diffuse-reflected from the ground. Evaluation of the second and third components requires the introduction of appropriate configuration factors that define, respectively, the portion of the sky dome visible from the collecting surface and the way in which the collecting surface sees the ground surface. Then,

$$\dot{E}_t = \dot{E}_n \cos \theta_i + F_1 \dot{E}_d + F_2 \rho (\dot{E}_n \cos \theta_z + \dot{E}_d) \quad (13.11)$$

where

$\dot{E}_t$ = total radiation received by the collecting surface
$\dot{E}_n$ = normal beam radiation on collecting and ground surfaces
$\dot{E}_d$ = diffuse radiation from the sky
θ_i = incidence angle for beam radiation on the collecting surface
F_1 = $(1 + \cos \beta)/2$, configuration factor, sky to the collecting surface (*Note:* $F_1 = 1 - F_2$)
F_2 = $(1 - \cos \beta)/2$, configuration factor, ground surface to the collecting surface (Ref.: J. A. Wiebelt, *Engineering Radiation Heat Transfer*, Holt, Rinehart and Winston, New York, 1966)[39]
ρ = ground surface diffuse reflectance, about 0.2 for most surfaces
β = inclination angle for the energy collecting surface

Example 13.9

A solar panel faces south at an angle of 50° from the horizontal surface. The intensity of the solar normal beam radiation is 760 W/m^2, and the diffuse radiation from the sky is 190 W/m^2. The zenith angle is 25°, and $\cos \theta_i = 0.8088$. The ground reflectance is 0.22. Determine the total radiation received by the panel.

Applying Eq. 13.11,

$$F_1 = \frac{1 + \cos\beta}{2}$$

$$= \frac{1 + \cos 50^\circ}{2} = 0.821$$

$$F_2 = 1 - F_1$$

$$= 1 - 0.821 = 0.179$$

$$\dot{E}_t = \dot{E}_n \cos\theta_i + F_1\dot{E}_d + F_2\rho(\dot{E}_n \cos\theta_z + \dot{E}_d)$$

$$= (760 \times 0.8088) + (0.821 \times 190) + 0.179 \times 0.22[(760 \times 0.9063) + 190]$$

$$= 614.7 + 156.0 + 34.6 = 805.3 \text{ W/m}^2$$

13.14 CONSTRUCTION OF SOLAR ENERGY COLLECTORS

Solar radiation arrives at the earth's atmosphere in the form of electromagnetic waves, although on some occasions the transfer of radiant energy by photons is the recognized mechanism. Solar radiation is characterized as short-wavelength radiation, with essentially all of the energy emitted in a band that extends from 0.25 to 3.0 microns. The band includes the invisible ultraviolet, the visible violet to red, and a portion of the infrared radiation. The function of the collector is the conversion of the incident solar radiation into thermal energy.

Two general types of collectors are used, namely, flat-plate and focusing collectors. Many different designs for both types of collectors have been developed, and numerous collectors have been constructed, tested, and placed in operation. The collection effectiveness is a measure of the ability of a solar collector to absorb and transfer the incident radiant energy.

A focusing collector employs curved reflecting surfaces to concentrate onto a small target area the solar radiation that is intercepted by the entire collector surface. The concentration of solar energy in this manner can yield very high temperatures in the energy absorbing medium, depending upon the surface reflectivity and the optical precision of the curvature of the surface. A focusing collector can intercept and concentrate only the direct rays of the sun, and hence does not perform satisfactorily when the sky is cloudy or hazy. A very small quantity of diffuse radiation will fall upon the target, but usually this source of energy is ignored.

In order to achieve satisfactory performance, a circular focusing-type collector would be designed for continuous and exact tracking of the sun. Tracking can be accomplished by using a two-axis mounting mechanism that turns the collector about a horizontal axis and about a vertical axis. The area of this type of collector is limited, and the tracking system is complex and expensive.

Figure 13.6 illustrates the construction of a focusing collector in the shape of a parabolic trough. The projected area of the reflector is equal to the product of the length of the trough and the width, or distance between the edges, of the reflecting surface. The concentration ratio is defined as the aperture area, or the projected area of the reflector, divided by the total area of the energy absorbing surface. The focal line of the parabola is indicated by the symbol *f*. The center line of the cylindrical energy absorbing element coincides with the focal line. The longitudinal fins attached to the coolant tube serve to extend the energy absorbing surface of the receiving element.

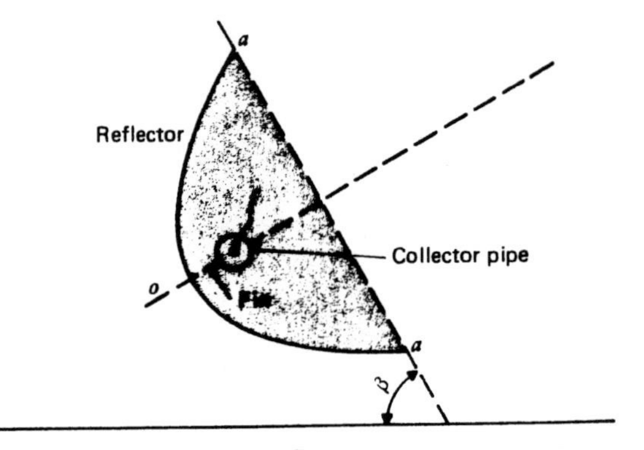

Figure 13.6 Parabolic reflector.

Because the sunlight intercepted by a focusing collector must be reflected and concentrated onto a comparatively small target, the orientation of the focusing collector is considerably more critical than the orientation of a flat-plate collector. For similar positions, each type of collector will intercept the same quantity of solar radiation per unit collector area, but the angle of incidence of the sun's rays relative to the receiving surface is particularly critical for the focusing collector.

For most effective operation of the parabolic collector, the line *o–f* should be oriented continuously in the direction of the sun. The aperture *a–a* would then move in a manner similar to that of the surface described in Example 13.6. Tracking the sun with a relatively small parabolic collector would, for general service, most likely prove to be impracticable.

In a south-facing orientation, the focal line of the parabolic reflector is aligned in an east–west direction with the aperture tilted toward the south an an angle β with the horizontal plane. When the parabolic reflector is tilted such that $\beta = \beta_n = \phi - \delta$, it will always be in focus at solar noon. The tilt angle β can be adjusted at appropriate intervals during the year in order to compensate for the change in the solar declination. Manual adjustment of the tilt angle during the day would however be impracticable.

Because exact tracking of the sun by parabolic mirrors will not in some installations be practicable, a compromise can be achieved by installing long parabolic reflectors in a north–south axial alignment and arranging the troughs to rotate daily from east to west. If the length of the parabolic collector is relatively short, say, 2.5 to 3 m, the focal line can be tilted toward the south at an angle $\beta = \beta_n$. During the day, the parabolic mirror is rotated from east to west about the inclined focal line.

Exact tracking of the sun by any parabolic reflector requires two-axis movement of the collector. Reasonably good performance can however be achieved during the midday period with movement of the reflector about a single axis. For many installations, the daily rotation of the reflector is not practicable because of the complex mounting, high capital cost, and the susceptibility to storm damage.

Tracking the sun by turning the concentrating mirror is not easily accomplished, particularly in a large-scale installation. An auxiliary mirror, called a heliostat, can be used to follow the sun's motion and reflect the sunlight onto a stationary concentrating mirror. The arrangement shown in Fig. 13.7 employs the movable mirror *m* and the intermediate fixed reflector *f*. In some installations, only a single movable mirror is used.

A flat-plate collector is generally a less complex structure than a focusing collector, but lower

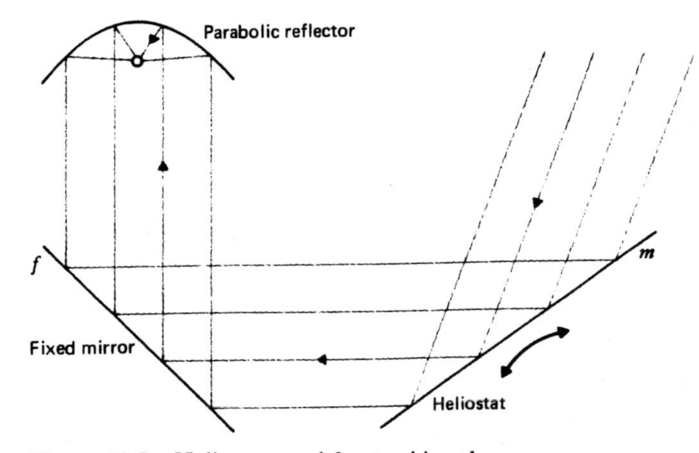

Figure 13.7 Heliostat used for tracking the sun.

temperatures are ordinarily achieved in the energy absorbing medium. The flat-plate collector is customarily installed in a fixed position, tilted toward the south with the angle β equal approximately to $\phi + 10°$. This orientation of the collector surface will provide for a relatively high value of cos θ_i at solar noon during the winter months when the zenith distance is comparatively high and τ_a^m is thus comparatively low. In the evaluation of $\beta_n = \phi - \delta$, the value of δ ranges from 0 to -23.5 to 0° during the period extending from September 21 to March 21, for an average value of about $-12°$. It is important to note that for a prescribed value of τ_a at a particular location, the noon value of τ_a^m decreases with a change in the solar declination from a maximum at the time of the summer solstice to a minimum at the time of the winter solstice.

Flat-plate, nonfocusing collectors can absorb all the sunlight, diffuse and ground-reflected, as well as the direct radiation. The diffuse and ground-reflected radiation is collected only from the hemisphere into which the collector faces. A flat-plate collector can be securely anchored and thus escape storm damage to which a focusing collector may be susceptible, depending upon the type of construction and the extent of tracking employed.

The ability of a solar collector to absorb and transfer the intercepted radiation is dependent upon the construction of the apparatus and particularly upon the collection temperature. For a constant solar energy input, the heat loss from the collector will increase with increase in the collection temperature. The limiting value of the collection temperature is attained when equilibrium is established between the solar energy input and the heat loss to the surroundings. Table 13.1 lists the equilibrium temperatures for a particular solar collector.

When equilibrium conditions exist, no useful energy is delivered by the collector, and the effectiveness of collection has a value of zero. It then becomes important in collector design to reduce the heat loss, as practicable, in order to raise the equilibrium temperature. The collector must be operated at a collection temperature somewhat below the corresponding equilibrium temperature in order to obtain a useful output.

TABLE 13.1 Equilibrium Temperatures for a Flat-Plate Collector[7]

	Incident Radiation $\dot{E}_i$ (W/m²)	
	630	*945*
One glass cover	86 C	107 C
Two glass covers	97 C	123 C
Three glass covers	109 C	139 C

Heat transfer from the collector to the surroundings is effected principally by radiation and convection. Depending upon the type of mounting, a small heat loss by conduction can occur. The convective heat loss from the frame would ordinarily be reduced by placing insulating material on the sides and on the under surface of the collector structure. The heat loss from the collector however occurs mainly from the surface that intercepts the solar radiation. Certain measures can be taken to control this heat loss.

Figure 13.8 illustrates a typical construction for a flat-plate collector. The collector energy absorbing element is a roll-bonded aluminum panel in which coolant channels are formed in the fabrication process. The glass cover plates are installed to reduce the heat loss from the upper surface of the energy absorbing element. A plastic material, such as Mylar, may be substituted for the glass.

The upper surface of the energy absorbing element is coated with a substance that has a high

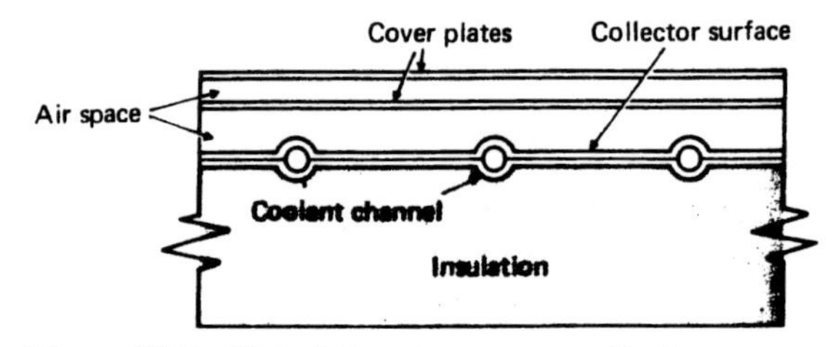

Figure 13.8 Flat-plate solar energy collector.

absorptivity, about 0.98, for the short-wavelength solar radiation. A flat, black paint is often used for this purpose. The energy reradiated from the surface of the energy absorbing element, at the relatively low steady-state temperature, is in the long-wavelength range of 5 to 10 microns. Ordinary glass is opaque to these long waves, but glass has a high transmissivity for the short-wavelength solar radiation. The glass cover plates serve effectively as a trap for the solar radiation that reaches the collector energy absorbing surface.

Flat-plate collectors are, on occasion, constructed with a selective energy absorbing surface that possesses a high absorptivity–low emissivity characteristic. The surface is highly absorptive for the short-wavelength radiation from the sun but is a very poor emitter of the long-wavelength radiation from low-temperature surfaces. The surface is prepared by application of a special coating, consisting of a bright, reflective undersurface of metal that emits very little radiation in the infrared range, covered with a very thin layer of black energy absorbing material. The solar radiation is absorbed in the black layer, while the emission of the infrared radiation is controlled by the material beneath the thin black layer.

Nickel oxide is a material that exhibits selective radiative characteristics. The absorptivity for solar radiation is about 0.9, while the emissivity for long-wavelength radiation is about 0.2.

The energy absorbing surface depicted in Fig. 13.8 is only one of the many different types of surfaces that have been developed for the nonfocusing collector. In some designs, the surface exposed to the sunlight is fabricated in the form of corrugations or troughs, in a variety of shapes, which serve to improve the capability to absorb the incident radiation.

A recently developed optical device is composed of a bank of relatively deep, mirror-lined troughs in the form of compound parabolas.[13] The aperture of the trough is several times wider than the base, and the light that enters the trough is reflected downward to achieve an energy concentration several times the normal intensity. Different light concentrations are achieved by varying the shape of the funnel. Concentration ratios of 3 and 10 were achieved in two concentrators for which the aperture width and the depth of the individual troughs are 58.4 and 305 mm, and 76.2 and 915 mm, respectively.

Because of the optical properties of the compound parabolic trough, the concentration of light is accomplished without the formation of an image. A solar collector constructed in accordance with this principle is not required to track the sun. The performance of the collector is improved however by adjusting the tilt angle periodically during the year. Commercial applications of the concentrating tubular-evacuated collector are anticipated in a variety of manufacturing processes for the production of steam and hot water in a temperature range of 70 to 320 C.

The evacuated glass tube collector, a recent innovation, is smaller in area (about 40 percent) and weighs less (about 50 percent) than a flat-plate collector of an equivalent energy absorbing capability.[14] The single evacuated tube collector has an exterior appearance of a fluorescent light tube. A cylindrical, black, energy absorbing element is contained within the concentric glass cover tube, and the space between the two tubes is evacuated. The coolant enters the unit through an inner concentric delivery tube and absorbs energy while flowing in the annular channel between the absorber and delivery tubes. The evacuated tube solar collector is constructed by installing, on a flat plate, an array of the single evacuated glass tube collectors, which incidentally function with a very small heat loss to the surroundings.

13.15 PERFORMANCE OF SOLAR ENERGY COLLECTORS

The convective heat loss from the collector is dependent upon the surface temperature of the glass cover. Installation of multiple sheets of glass will reduce the outside surface temperature of the glass cover and thus cause a decrease in the convective heat loss. The improvement in the collection effectiveness achieved by using two or three sheets of glass increases with an increase in the difference between the collection temperature and the atmospheric temperature, $T_c - T_\infty$. See Fig. 13.9. In the low range of $T_c - T_\infty$, the use of multiple sheets of glass causes a reduction in the collection effectiveness. Each additional sheet of glass increases the quantity of sunlight reflected from the glass cover. Consequently, in the low range of $T_c - T_\infty$, the increase in the energy loss because of reflection exceeds the reduction in the convective heat loss attributed to the installation of two or three sheets of glass. In general for commercial installations, good collector performance is achieved with two cover sheets of glass or plastic material. In comparison with glass, some plastic materials demonstrate superior transmission of solar radiation.

Figure 13.10 shows the energy conversion effectiveness for typical flat-plate and evacuated tubular collectors. The collection temperature is limited by the equilibrium temperature. In practice however, the collection temperature must be somewhat below the limiting temperature in order to attain an acceptable collection effectiveness.

The convective heat loss from the collector is attributed to the transfer of heat by convection currents from the energy absorbing surface to the cover glass, from one glass sheet to another, and from the outermost glass sheet to the ambient air. Test results demonstrate that, for some installations, a reduction in the convective heat loss can be achieved by partially evacuating the space between the energy absorbing surface and the glass cover and the space between the glass sheets where more than one sheet of glass is installed. Collector construction will however become more complex and costly; thus, reduction of the heat loss in this manner may not prove to be economically justified.

Under very favorable conditions, flat-plate collectors can be operated to achieve a collection temperature of about 70 C with a coolant temperature increase of, say, 10 C.

On a day when the ambient temperature is 20 C and the insolation I is 950 W/m^2, a single-glass, flat-plate collector operates with a coolant inlet temperature of 55 C and an outlet temperature of 65 C. The collector temperature is taken as 60 C. Then, $(T_c - T_a)/I = (60 - 20)/950 = 0.0421$. The collector effectiveness, obtained from

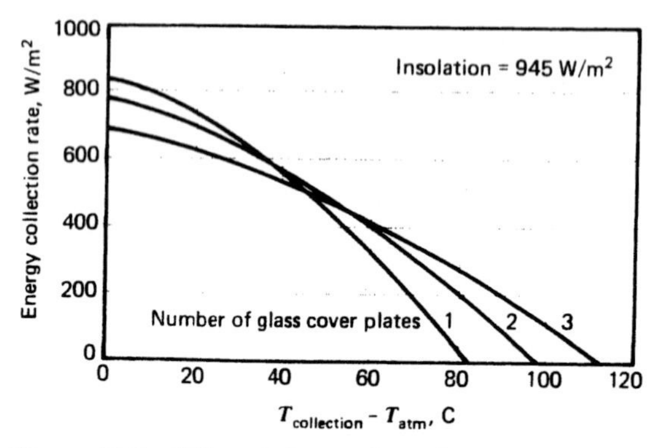

Figure 13.9 Effect of the number of glass cover plates on the performance of a solar energy collector. (J. I. Yellott, ASME Paper No. 56-F-15, 1956.[7])

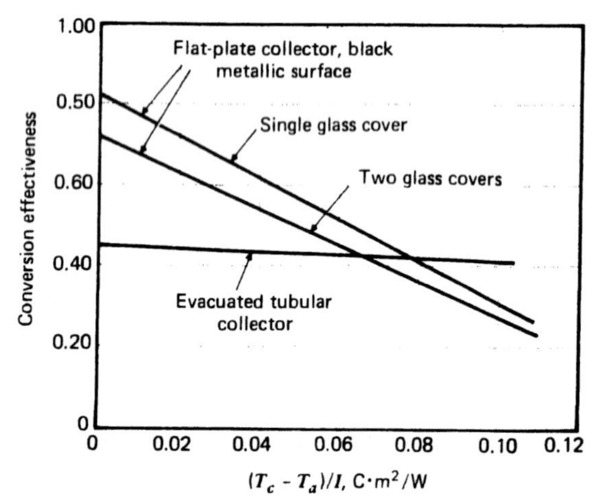

Figure 13.10 Performance of typical solar energy collectors. (T_c = collector temperature, T_a = air temperature, I = insolation.)

Fig. 13.10, is 0.61. The collector output is determined by $\eta_c I = 0.61 \times 950 = 580$ W/m^2.

At a different time of the year, the ambient temperature is -18 C, and the insolation I is 630 W/m^2. In order to operate the system effectively, the prescribed minimum collection temperature is 38 C. The collector temperature is taken as 33 C. Then, $(T_c - T_a)/I = (33 + 18)/630 = 0.0810$, and the collector effectiveness is 0.41. See Fig. 13.10. The collector output is now equal to $0.41 \times 630 = 258$ W/m^2. Evidentally, as the outdoor conditions become more severe and the building heat loss increases, perhaps substantially, the solar collector tends to provide a smaller fraction of the energy input required for heating the building.

The evacuated glass tube collector is capable of operating effectively with a collection temperature of 82 C for heating applications and a collection temperature of 115 C for cooling applications. The collector effectiveness is about 0.48 for heating and 0.43 for cooling applications over a wide range in the ambient temperature, that is, roughly from -20 to 40 C. The collector effectiveness improves from the noontime value by 3 or 4 percentage points with an hour angle change of 30° in either direction.

Improved collection temperatures can be achieved by the use of a selective energy absorbing surface, say, from about 75 C for a nonselective surface to about 125 C for a selective surface. Collection temperatures on the order of 200 to 250 C may be anticipated by evacuating the collector equipped with a selective surface. It is important to note that selective surfaces are fairly expensive, and fabrication costs for evacuated flat-plate collectors will be relatively high.

In general, a fixed focusing collector will not perform as effectively as a fixed flat-plate collector. When the diffuse radiation fraction is high, the performance of the focusing collector will be particularly poor. The performance of a parabolic collector oriented in a fixed, south-facing position is particularly time dependent. As the hour angle increases after solar noon, the reflected sunlight tends to move off the target or energy absorbing surface. This effect is especially noticeable during the early and late hours of the day, that is, outside the period of 5 or 6 hours at midday for a location in the vicinity of latitude 40°N.[15]

13.16 ARCHITECTURAL ASPECTS OF SOLAR ENERGY

Buildings are exposed to direct and diffuse solar radiation in varying amounts during the daylight hours throughout the year. During the summer months, the solar radiation can contribute significantly to the building heat input. The temperature of certain exterior surfaces exposed directly to the sun's rays will be appreciably higher than the ambient temperature and thus increase the conduction of heat into the building. The sun's rays are readily transmitted through window glass, which in turn is essentially opaque to long-wavelength radiation from the interior surfaces of the building.

Certain measures can be taken to reduce the solar heat gain in a building. Exterior surfaces can be coated with a paint that has a low absorptivity for the short-wavelength solar radiation. Building walls and roof structures are customarily insulated to reduce the conductive heat transfer. Attics and spaces between the ceiling and the roof should be well ventilated during the summer months in order to remove the warm air that would otherwise be trapped in these spaces.

The solar energy transmitted through glass usually represents an appreciable fraction of the total heat input to a building. Interior screens or shades ordinarily have a moderate effect on reducing this energy input, unless heavy insulating draperies are used. Exterior sun screens however are effective, but they obscure to some degree the view from the window. If the windows generally face south, the direct solar radiation can be effectively screened by constructing permanent, properly proportioned exterior overhangs. The sun's altitude for the summer months will indicate the required extension of the overhang from the building wall to provide the desired shielding. Window awnings are effective for screening the sun, but they are susceptible to storm damage and may be considered unsightly for some installations.

A vertical screen erected at some distance from the building wall can effectively intercept the direct solar radiation. A barrier of this kind can be particularly effective in screening a west wall when the altitude of the sun is low in the late afternoon. Vertical screens must be strongly supported to avoid wind damage, and the aesthetics of the structure can be improved by building the screen in the form of a trellis.

Large shade trees, when properly located, can intercept to a significant degree the solar radiation that would otherwise fall on the building roof and walls. The solar input through the roof of a building can be reduced by applying a reflecting coating to the exterior surface. A reduction in the energy input can also be achieved by flooding the roof with a shallow pool of water, but maintenance of the roof structure may be a problem, and sealing against leaks is most essential.

During the winter months, absorption of solar energy in the building structure will reduce fuel consumption for space heating. Some of the measures taken to reduce the solar energy input in the summer months will, during the winter months, prove to be disadvantageous. Exterior sun screens however can be removed from the windows, and deciduous trees, following the loss of their leaves, cause little shading of the building.

An expanse of glass in the south wall of the building will admit during the winter months a substantial quantity of solar radiation. If properly proportioned, an overhang installed for summer screening causes no problem in the winter. Because of the low altitude of the sun in the winter, the sun's rays will be directed under the overhang and fall on the glass wall or windows.

An interesting scheme provides for the solar radiation transmitted through the glass to be absorbed in a large mass, say, of concrete. Following sunset, the energy stored in the concrete is gradually released to the interior of the building and thus contributes supplemental heating. In a like manner, supplemental heating of the hot water supply is achieved by passing the water through pipes embedded in the concrete, before the water enters a gas or electric water heater. A solar energy heating system of this kind is known as a "passive system," and many different variations of this scheme are used for space and water heating.

Fuel conservation, an important endeavor in today's economy, is achieved, in part, by reducing the heat loss from buildings. The walls, and particularly the ceiling of the top floor of the building, should be well insulated. Installation of storm windows or thermopanes will effectively reduce the heat loss from the glass surfaces. The infiltration loss, which results primarily from loosely fitted doors and windows, can be largely eliminated by sealing all cracks.

The dependence upon artificial cooling can be reduced, perhaps substantially, by designing a building in a way to promote natural ventilation of the structure. A residential building can be cooled effectively during the nighttime hours by means of an attic fan. Schemes of this kind assist in reducing the electrical load imposed by air conditioning equipment.

13.17 SOLAR ENERGY HEATING SYSTEMS

The solar radiation admitted to a house through a glass south wall may, in the northern latitudes of the United States, provide as much as 25 percent of the total energy required for heating the building. In the southern region of the country, the percentage will be somewhat higher. However, in order to take advantage of the abundant quantity of solar energy available for space heating, some kind of exterior solar collector is normally required.

The feasibility of employing solar collectors for space and water heating has been recognized for many years. Only a few solar collectors however were installed, because fossil fuels and electrical energy have been available at low cost. Too many years were required before the capital cost of the collector could be recovered through a reduction in the consumption of the purchased energy. The increasing cost of energy has improved to some degree the economic feasibility of utilizing solar energy for certain heating applications. Fuel conservation is a further incentive to expand the utilization of solar energy.

In principle, a building could be heated entirely by solar energy, provided that the collector is sufficiently large and an adequate energy storage capability is built into the system. A self-sufficient design would in general not be economically feasible for solar heating systems installed in the United States. Current designs vary to some extent in the fraction of the total energy required for space heating and the production of hot water that will be contributed by solar energy. In the northern region of the United States, 60 percent is probably a reasonably good design figure, although for commercial buildings the figure may reach a value of 85 percent.

Local meteorological data are important to the design of a solar heating system. Such data will provide information on the quantity of solar energy that probably can be absorbed per unit area of collector surface throughout the heating season. The meteorological records will also indicate the ambient temperatures that may be expected during this period.

The collector absorbs solar energy during the daylight hours for heating the building and for charging the energy storage unit. During the nighttime hours and cloudy periods, the storage unit provides the energy required for heating the building. A flat-plate collector however will absorb during the overcast periods diffuse radiation in varying amounts, depending upon the characteristics of the cloud cover and the haze that may be present. The duration of the cloudy periods is important to prescribing the energy storage capability of the system. The energy stored in a unit of reasonable size can easily be depleted during a prolonged cloudy period. Depletion of the stored energy on occasion does not indicate an inadequate design, because such occurrences are anticipated in developing a design that will achieve the most economical operation for the system.

The main components of the building heating system are the solar collector, energy storage unit, pumps, piping or ductwork, controls, and the auxiliary heating equipment. The arrangement of the piping or ductwork will vary from one system to another and will be considered here only in a general sense. Automatic operation of the system certainly will be a requirement, and the control apparatus is designed to achieve this objective.

Flat-plate collectors are installed ordinarily in a fixed position on the roof of the building; for new construction, the collector may be built into the roof structure. In comparison with the floor area of the building, a large collector area is usually prescribed. In general, the area of the solar panel array is equal approximately to 25 percent of the building floor area.

For best performance, the collector should face south and be inclined to the horizontal plane at an appropriate value of the angle β. Selection of the tilt angle β is discussed in Section 13.14.

Various kinds of solar energy absorbing surfaces have been developed and are commercially available, usually in the form of panels that can be assembled in multiple to form the collector surface. For the flat-plate collector, Fig. 13.8 illustrates a typical construction in which the channels for conveying the coolant are fabricated by deforming the flat sheets of metal used in manufacturing the energy absorbing surface. Other designs specify a finned-tube construction or a flat sheet to which the coolant tubes are fastened. Channel spacing is about 125 to 200 mm.

A glycol–water mixture is commonly employed as the collector coolant in a solar heating system. The coolant circulates through the collector and a coil submerged in the water contained in the energy storage tank. For a typical residence, a tank of 6- to 10-m^3 capacity would be installed in the basement or beneath the basement floor. The hot coolant returning from the collector may pass through a heat exchanger for the purpose of preheating the domestic hot water supply.

If the temperature of the water in the tank is sufficiently high, the building can be heated by stored energy alone. The water from the tank is circulated through room convectors or through a central heat exchanger that is used to heat the air distributed throughout the building. Supplemental heating is employed when the temperature of the water in the tank is below the level required for adequate space heating. The supplemental heater is likely to be either gas fired or electrically operated. Incidentally, the energy storage tank is continuously charged by circulating the coolant whenever solar energy can be collected, that is, when the temperature of the coolant returning from the collector exceeds the temperature of the water in the tank by a suitable increment.

A heat pump may be incorporated into the building heating system. Warm water heated by solar energy is circulated from the energy storage tank through the heat pump evaporator. A higher coefficient of performance is achieved by the heat pump when the heat source is the warm water in the tank rather than the cold outside air. The heat rejected by the heat pump is usually transferred to the building interior by a forced-air distribution system. The same system can be used for summer cooling when the heat pump operation is reversed. In the event that the heat pump operating with the stored solar energy cannot maintain the building at the proper temperature, a supplemental heater will be activated.

In some solar energy heating systems, air is employed as the collector coolant. Whether air or water is used for this purpose, the system operating principles are essentially similar. Different types of energy storage units however are required. The warm air leaving the solar energy collector is circulated through the building and through the energy storage unit that consists primarily of a thick bed of pebbles. The air flow is divided in accordance with the building heating requirements and the quantity of solar energy collected; that is, the portion of the collected energy not used for heating is diverted to storage. Energy is stored in the rock as sensible energy in the same way that energy is stored in liquid water.

In general, energy is stored during the daylight hours. At night and during cloudy periods, energy for heating the building is supplied by the energy storage unit. Incidentally, on a cloudy day a nonfocusing collector will intercept some diffuse radiation. The supplemental heater is activated whenever the energy available from the collector and the energy storage unit is insufficient for heating the building.

Numerous experiments have been conducted on the storage of energy in systems that undergo chemical changes or physical changes such as fusion and solidification. The objective of these investigations is to develop a means to increasing the energy storage capability in relation to the mass of the substance. There are however numerous problems associated with systems that utilize a salt of some kind for storing energy. For example, in achieving energy storage by using the heat of fusion of crystals, solid crystals do not form on cooling because of a tendency to supercool. After many cycles of melting and freezing, supercooling is likely to intensify. Corrosion of the system components can cause serious maintenance problems.

In the absence of a heat pump installation, building cooling may be accomplished by the use of an air conditioning system equipped with an absorption-type refrigeration unit capable of utilizing solar energy. Sufficiently high collection temperatures for operating an absorption-type refrigeration unit can be obtained with an evacuated glass tube collector, but not with the ordinary flat-plate collector. Thus, when flat-plate collectors are used for cooling applications, supplemental heating of the generator is required.

Example 13.10

A house that has a floor area of 272 m^2 is located in a region of the country where the available solar energy on December 15 is 9.10 MJ/m^2 per day. A solar collector installed on the roof of the building has an area of 68 m^2. The effectiveness of the solar heating system is 0.40. The mean outdoor temperature for a 24-hour period is −9.5 C. For an indoor temperature of 21 C, the average heat loss from the building is 37,300 kJ/h. Determine the fraction of the heating load carried by the solar collector.

Solar energy heating:

$$\dot{Q}_s = \varepsilon \dot{E} A_c$$
$$= 0.40 \times 9100 \times 68 = 247{,}520 \text{ kJ/day}$$

Total heat loss:

$$\dot{Q}_l = 24 \times 37300 = 895{,}200 \text{ kJ/day}$$

Solar fraction of the heating load:

$$\frac{\dot{Q}_s}{\dot{Q}_l} = \frac{247520}{895200} = 0.276$$

During the early and the late periods of the heating season, the availability of the solar energy is higher and the building heat loss is lower than the quantities cited in the above example. Thus, for the entire heating season, solar energy could reasonably be expected to provide about 60 percent of the total energy supplied for heating the house.

The use of solar energy for heating buildings is expected to increase, perhaps sharply, over the next two decades. In attempting to forecast the extent to which active solar heating systems are likely to be employed, a number of factors should be examined, particularly the investment or capital cost. Depending upon the size of the house and the local conditions, a rooftop flat-plate collector system for space and water heating would cost between $12,000 and $20,000 installed under contract. A system of lower capability, adequate for heating water, would cost about $2000.

When the solar energy heating system is installed in an existing house, the conventional heating system would be retained. For new construction, a supplemental heating unit would be required; and as a safety measure, this unit should have the capability to carry the entire heating load under the most severe weather conditions.

An annual operating cost for the solar energy heating system must be anticipated, say, $200 to $250 for maintenance and power for operating the pumps. Increased insurance premiums must be expected, and higher property taxes are a possibility. The total annual cost for the solar heating system will then consist of the interest on the capital expenditure, of operating and maintenance costs, and of the increased cost for insurance, assuming no increase in the property tax. The total annual cost is partially offset by the reduced cost for fuel or electrical energy.

The economy of solar energy collection and use becomes more complex when, in addition to heating, space cooling is a function of the system. Then, the capital and operating costs for heat pumps and absorption refrigeration units are significant factors. The total cost of operating the entire system must be properly charged to both functions, namely, heating as well as cooling the building.

At the present time, solar heating for the production of hot water is economically feasible, but such is not the case generally for space heating by active systems. For space heating, the annual charges are too high in comparison with the reduction in the cost of purchased energy. It is uncertain whether this situation will in time improve. The cost of purchased energy will increase, but the rise in construction costs is also expected to continue. A solar heating system installed today may however in later years prove to be a judicious investment, depending upon future costs for energy.

The experience to date with many of the active solar heating systems has been unsatisfactory. Poor performance is attributed to inadequate and faulty design, and excessive maintenance cost has resulted from shoddy construction. All in all, the economy of these defective systems is disappointing. There is an indication that solar heating and cooling by large-scale commercial installations may not in every case prove to be an economical undertaking.

13.18 SOLAR POWER PLANTS

Through the years, the solar power plant has attracted the attention of numerous investigators who seek to tap the vast source of energy radiated by the sun. A number of small power plants operating on the Rankine and Stirling cycles have been built and tested. For the most part, these experimental engines had a power output of less than 1 kW. A solar engine operating on a vapor power cycle was developed in Italy circa 1950. The engine, rated at 1.1 kW, was subsequently produced on a commercial scale and sold for about $800.

Today, the main interest is directed to large-scale solar power generation in a system that incorporates a thermal engine. A few studies have been made to investigate the feasibility of constructing such systems, and they conclude that operating temperatures in the range of 450 to 650 C could be attained. Steam generated at a temperature within this range can be used effectively in a turbine-generator for the production of electrical energy. The anticipated temperatures however are not sufficiently high for effective operation of a gas turbine engine unless the solar energy in some manner supplements the energy supplied by a conventinal engine fuel.

Solar power plant designs usually show a boiler or heat exchanger mounted on the top of a tower about 150 m high. An array of mirrors, or heliostats, placed around the base of the tower is intended to reflect and concentrate the sun's rays onto the target or energy absorbing surface. The boiler or heat exchanger would probably be constructed in the form of a cylinder with a cavity at the bottom into which the sun's rays would be directed.

The motion of the mirrors, placed around the base of the tower, must be precisely controlled in order to maintain the point focus, which is essential to the proper operation of the power plant. Two-axis control of a heliostat can be accomplished by a hydraulically or electrically operated servomechanism that responds to a signal from a position-sensing phototube. Alternatively, the heliostats could be positioned by a computer control system programmed to track the sun continuously.

Steam generated in the elevated boiler would be piped to a turbine-generator located at the base of the tower. In an alternative arrangement, the solar energy would be absorbed by a heat transfer medium such as a liquid metal. Subsequently, energy is transferred from the carrier fluid to water in the boiler and to the energy storage medium.

High-temperature energy storage is an important requirement for solar power generation as a means to allowing the plant to operate after sunset and during cloudy periods. Storage of energy in eutectic salts is considered to be a possibility. These salts are chemicals that have high melting temperatures and the capability of storing a large quantity of energy in relation to the volume occupied by the substance. In a two-fluid system, the carrier fluid can serve as a thermal energy storage medium, in addition to transferring energy to the water in the boiler.

Generation of electrical energy in solar power plants, except for demonstration units, is not expected before the year 1990. Meanwhile, numerous investigations and studies will be conducted and conceptual designs developed. Various kinds of collector surfaces will be designed in an effort to improve the power plant conversion efficiency. Before solar power generation on a commercial scale can be realized, the technical feasibility of solar power plants must be established through successful operation of demonstration plants. In the United States, the best sites for solar power plants are located in the southwest, because the comparatively clear skies that prevail in this region are conducive to the use of focusing collectors.

The high capital cost for solar power plants is

demonstrated by comparing the electric power generating costs for different types of plants. The average cost of producing electrical energy in the United States for the year 1979 was 1.7 ¢/kWh for nuclear, 1.9 ¢/kWh for coal-fired, and 4.2 ¢/kWh for oil-fired power plants. An estimated cost for electrical energy produced in solar power plants is approximately 5 ¢/kWh. Energy storage capability added to the solar power plant would increase the cost of the generated energy by an estimated 25 percent. Provision for energy storage however can be avoided by operating the solar power plant in a system that consists mainly of hydro and fuel-burning power plants.

The first solar power plant constructed in the United States was a pilot plant, rated at 15 MW. Construction of the plant was completed in 1982. The power plant is expected to have an average electrical output of 10 MW over a period of 4000 hours per year. The general design of the pilot plant provides for a tower-mounted boiler and tilt–tilt (two-axis) mounting of the heliostats placed on the ground in a circular array around the base of the 91-m-high tower. The reflecting area of each heliostat is 40 m^2, and 1818 heliostats have been installed. Each heliostat consists of 12 slightly concave reflectors that focus the solar radiation on the receiver energy absorbing surfaces. Control of the heliostats for tracking the sun is accomplished from a central computer arranged for automatic recalibration several times during the day. The heliostats are designed to operate with a wind velocity up to about 60 km per hour. At higher wind velocities, the mirrors are stowed in an inverted position.

The single-pass boiler, placed on top of the tower, is about 15 m high. Superheated steam is generated in the boiler at 10.1 MPa, abs, and 516 C. Steam in excess of that required for an electrical output of 10 MW is diverted to a thermal energy storage unit. The turbine-generator may be operated wholly or partly by steam generated from stored energy. The power plant is equipped with a wet cooling tower that operates in conjunction with a surface condenser.

The power plant overall conversion efficiency is 0.135 when operating on steam generated in the solar boiler. The conversion efficiency is based on the radiation incident to the heliostats. When the power plant operates on steam generated from stored energy, a thermal efficiency of about 0.11 is anticipated.

The power plant is located on a site near Barstow, California. The land area occupied by the power plant is about 10 times the land area required for a coal-fired power plant, on the basis of a unit kilowatt of generating capability. Experience derived from the construction and operation of the pilot plant will contribute substantially to the design of later-day commercial-scale solar thermal-electric power plants.

A program has been initiated to develop a conceptual design for a 100-MW solar-electric power plant that utilizes an array of some 20,000 mirrors to track the sun and focus the solar radiation onto an energy receiver mounted on the top of a tower about 180 m high.[16] The mirrors, which cover an area of about 2.5 km^2, are under microcomputer control. The temperature of the receiver surface is expected to approach 650 C; and, in turn, the liquid sodium that circulates through the receiver will attain a temperature of 600 C. The power-producing equipment, located at the base of the tower, consists mainly of the sodium/water heat exchanger, the reheat steam turbine-generator, and the hot-sodium and the cold-sodium storage units. The sodium functions as the energy collection medium and as the thermal energy storage medium. Because the pressure of the sodium in the receiver is comparatively low, about 300 kPa, heavy construction is avoided for the equipment at the top of the tower. The system is designed for the production of high-temperature (538 C) superheat and reheat steam, thus ensuring a high thermal efficiency for the steam cycle.

Example 13.11

A solar steam-electric power plant is equipped with 20,000 tracking mirrors, each of which has a 40-m^2 effective collection area. The mirror reflectance is 0.85. The solar radiation is reflected to an elevated energy receiver that is cooled by circulating sodium. The sodium enters the receiver at 350 C and leaves at 600 C. The sodium flows through a heat exchanger and produces high-pressure steam at 540 C for operation of a turbine-generator. The thermal efficiency of the steam plant is 0.41. The energy concentration of the mirror–receiver system is 1000 to 1. On a particular day, the average incident radiation on the mirrors is 850 W/m^2 over a 12-hour period. Determine the electric power output for the plant.

Energy Transferred to the Receiver

$$\dot{Q} = \rho \dot{E}_n A_m$$
$$= 0.85 \times 850(20000 \times 40)$$
$$= 578.0 \times 10^6 \text{ W}$$

Radiation Heat Loss from the Receiver

$$A_m = 800{,}000 \text{ m}^2$$
$$A_m/A_r = 1000$$
$$A_r = 800 \text{ m}^2$$

The effective mean temperature of the receiver surface is taken as 535 C. The ambient temperature is 22 C. The emissivity of the receiver surface is 0.98 for long-wavelength radiation.

$$\dot{q}_r = e_s A_r \sigma (T_s^4 - T_\infty^4)$$
$$= 0.98 \times 800 \times 5.6697$$
$$\times 10^{-8}[(808)^4 - (295)^4]$$
$$= 18.610 \times 10^6 \text{ W}$$

Energy Transferred to the Water in the Heat Exchanger

Energy transferred to the sodium:

$$\dot{E}_s = \dot{Q} - \dot{q}_r$$
$$= 578.0 - 18.61 = 559.39 \text{ MW}$$

Assume 2 percent heat loss in the sodium loop.

$$\dot{E}_w = 0.98 \times 559.39 = 548.20 \text{ MW}$$

Steam Plant Output

$$\dot{W}_{TG} = \eta_t \dot{E}_w$$
$$= 0.41 \times 548.20 = 224.76 \text{ MW}$$

In order to achieve continuous, 24-hour operation of the power plant, a portion of the energy transferred to the sodium is diverted during the daylight hours to the sodium energy storage unit. A 5 percent heat loss is assumed for the energy storage operation. Then,

For the 12-hour daytime period:
To the steam plant: 267.07 MW
To the storage unit: 281.13 MW
For the 12-hour nighttime period:
Heat loss = 0.05 × 281.13 = 14.06 MW
From storage: $\dot{E}_{\text{out}} = \dot{E}_{\text{in}} - \dot{E}_{\text{loss}} = 281.13 - 14.06 = 267.07$ MW

Continuous, 24-hour operation of the steam plant:

$$\dot{W}_{T\text{-}G} = \eta_t \dot{E}_w$$
$$= 0.41 \times 267.07 = 109.50 \text{ MW}$$

Continuous operation of the solar steam-electric power plant requires installation of an energy storage unit. Further, some heat loss from the energy storage system must be anticipated. Continuous, 24-hour operation results however in a substantial reduction in the size of the steam plant, in particular the sodium/water heat exchanger and the turbine-generator.

Small-scale solar power plants that employ salt ponds for the collection of solar energy are presently operating in Israel. Characteristically, the salinity of the water in a salt pond increases with the depth, and as a consequence solar radiation is transmitted through the upper layer of comparatively fresh water and absorbed primarily in the bottom layer of water that has a high salt concentration. Because of the density gradient, convective currents are not created, and the temperature of the water at the bottom of the pond can approach the boiling point.

In the operation of the power plant, the hot water from the pond is circulated through a boiler in which an organic working fluid is evaporated. Following expansion in the turbine, the working fluid enters a surface condenser that is cooled by cold water circulated from the upper layer of the pond. If the pond is sufficiently large, the turbine-generator can be operated continuously during the daytime and the nighttime hours.

For a hot-water temperature of 80 C and a cold-water temperature of 20 C, the power plant Carnot cycle thermal efficiency is 0.17. Various thermal and mechanical losses would reduce the actual power plant thermal efficiency to about 0.07.

Pond operation can be improved by placing a layer of transparent material above the surface of the water in the pond. Evaporation of water is prevented and the salt concentration gradient is not altered; otherwise, convective currents would be created. The protective layer also prevents the formation of wind-induced currents in the water body. Currents of any kind would cause mixing of the hot and cold water and thus impair the collection effectiveness of the pond.

Wind Power

13.19 WIND POWER DEVELOPMENT

Wind machines have been used for centuries to provide power for various tasks, such as pumping water and milling grain. The windmills of Holland represent perhaps the most publicized application of wind power. At one time in the United States, a windmill erected on a farm was a familiar sight. These windmills, used principally for pumping water and producing small amounts of electric power, largely disappeared with the creation of the rural electrification program.

The largest wind power machine constructed prior to 1979 was erected in 1941 on Grandpa's Knob in the Green Mountains about 10 miles west of Rutland, Vermont. Electrical energy produced by the turbine-generator was fed into the transmission lines of the local utility company. The machine, a test unit installation of the Smith-Putnam wind turbine, was designed to drive a 1250-kW electric generator. The site was selected, following extensive meteorological investigations, because the strong prevailing wind ensured an availability factor greater than that for stored water as used in New England and at most other hydroelectric sites. An analysis of the operation indicated that the rated generating capability of the machine would be achieved with a wind velocity of 48 km/h and that this velocity would prevail at Grandpa's Knob between 4000 and 5000 hours during the year.[17]

The entire machine, including the electric generator, shafts, and blades, was mounted on the top of a structural steel tower 37 m high. The windmill rotor, 53.3 m from tip to tip, consisted of the hub structure and two blades each 19.8 m long and 3.66 m wide. Because of the large diameter of the rotor, rotation was restricted to about 29 rpm or to a tip speed of 290 km/h. The rated speed of the machine was attained at a minimum wind velocity of about 24 km/h; thus, the operating tip speed-to-wind velocity ratio was within the range of 6 to 1 to 12 to 1 for a corresponding wind velocity range of 48 to 24 km/h.

When the wind velocity increased to 24 km/h, the switchgear automatically synchronized the unit to the electrical system to which it was connected and caused the main breaker to close. Any increase in the wind velocity above 24 km/h increased the power output of the machine until the generator became fully loaded at a wind velocity of 48 km/h. The operating speed of the generator was 600 rpm.

Four principal motions are significant to the operation of the wind turbine, namely, rotation, pitching, yawing, and conning. Under normal operating conditions, the turbine blades rotated at essentially constant speed under the control of the speed-sensitive governor. Constant speed under varying wind velocities was achieved by altering the pitch of both blades in a like manner. In changing the pitch, the blade is turned about its longitudinal axis of rotation by an actuating device that is part of the speed control system.

The turbine axis was maintained closely parallel to the wind direction by rotating the entire machine about a vertical pintle shaft mounted in the tower cap. The blades were kept downwind of the tower, and yawing about the vertical axis was accomplished by a hydraulic motor. The flow of oil to this motor was controlled by a weather vane.

Each blade of the turbine was free to pivot independently on pins attached to the rotor hub. The conning motion was normal to the plane of rotation, that is, upwind and downwind. This motion was provided in order to avoid the extremely high stresses that would otherwise develop because of severe gusts of wind. The position of the blades with respect to conning is influenced by the gravitational, aerodynamic, and centrifugal forces.

The Smith-Putnam wind turbine was operated intermittently from 1941 to 1945 and was then removed from service because of failure (metal fatigue) of one of the blades. The blade could most likely have been repaired and operation of the machine continued, but the project was judged to be uneconomical in comparison with the generation of power from other energy sources. Principally because of this reason, operation of the wind turbine-generator was terminated.

During the 1950 and 1960 decades, wind power technology was further investigated, with the activity now shifted to Europe. Particularly noteworthy was the construction of a 100-kW wind turbine-generator in Germany and a 200-kW WTG in Denmark. The latter machine operated for a decade on the national power grid. The performance of most European wind turbine-generators demonstrated that, with respect to power generation, a WTG is economically inferior to a fossil fuel-fired power plant.

From an engineering point of view, the design and construction of the Smith-Putnam wind turbine was successfully completed. Today, horizontal-axis wind turbines will be constructed in much the same fashion as the forerunner that operated on Grandpa's Knob in the early years of the 1940 decade. Rotor diameters will, in some installations, be greater, and blades will be lighter as a result of employing aircraft fabrication technology.

13.20 WIND ENERGY

The air that approaches a horizontal-axis wind turbine is considered to flow in a cylinder of radius r, where r is the radius of the disk or rotor. The power associated with the moving stream of air is expressed by

$$\dot{W}_i = \tfrac{1}{2}\dot{m}C^2 = \tfrac{1}{2}(\rho A_r C)C^2$$

or

$$\dot{W}_i = \tfrac{1}{2}\pi r^2 \rho C^3 \tag{13.12}$$

where ρ is the mean density of the air and C is the free-stream air velocity. $\dot{W}_i$ thus represents the power that was originally in the undisturbed air, that is, without the windmill in place.

Because of aerodynamic factors, the maximum theoretical power that can be developed by the windmill is less than $\dot{W}_i$. Figure 13.11 shows the control volume for the air that flows through the windmill disk. At section e, upstream from the disk, the air is moving at the

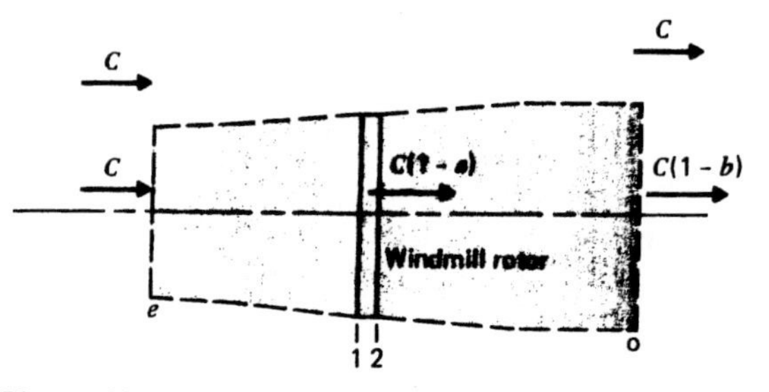

Figure 13.11 Air flow through windmill rotor.

undisturbed wind velocity C. The pressure P_e at this section is equal to the atmospheric pressure P_a, and the flow area A_e is less than πr^2, where r is the radius of the windmill disk.

At the disk, the air velocity decreases to $C(1 - a)$, where a is an interference factor that has a value less than 1. Pressure P_1, higher than pressure P_e, is determined from the energy equation written for incompressible fluid flow. Thus,

$$\frac{P_e - P_1}{\rho} = \frac{C_1^2 - C_e^2}{2} = \frac{C^2(1 - a)^2 - C^2}{2}$$

The pressure of the air drops as the air flows through the windmill rotor, hence $P_2 < P_1$. Downstream from the disk, the air velocity continues to decrease until the air pressure, which is simultaneously increasing, reaches the atmospheric pressure at section o, hence $P_o = P_a$. At section o, $C_o = C(1 - b)$, where $b > a$. Again from the energy equation,

$$\frac{P_2 - P_o}{\rho} = \frac{C_o^2 - C_2^2}{2} = \frac{C^2(1 - b)^2 - C^2(1 - a)^2}{2}$$

The above two equations are combined with each other and with $P_e = P_o = P_a$ to yield

$$P_1 - P_2 = \frac{\rho}{2} C^2[1 - (1 - b)^2]$$

and the axial thrust on the rotor is given by

$$F_x = (P_1 - P_2)A_r = \frac{\rho A_r C^2}{2}[1 - (1 - b)^2] \quad (13.13)$$

The axial thrust on the rotor can also be determined from the momentum equation. Then,

$$F_x = \dot{m}(C_e - C_o)$$

where $\dot{m} = \rho A_r C(1 - a)$, $C_e = C$, and $C_o = C(1 - b)$. Hence,

$$F_x = \rho A_r C(1 - a)[C - C(1 - b)]$$

or

$$F_x = \rho A_r C^2(1 - a)b \quad (13.14)$$

Equations 13.13 and 13.14 are combined and show that

$$b = 2a$$

The change in the transverse area of the fluid stream is particularly significant in relation to the air that flows through the rotor. At section e, where the air velocity is C, the transverse area of the stream is less than πr^2, the area of the disk. At the downstream section o, where the air velocity is $C(1 - 2a)$, the transverse area of the stream is greater than πr^2. Thus, as the velocity of the air decreases from C to $C(1 - a)$ to $C(1 - 2a)$, the boundary of the fluid stream is described by an expanding envelope.

The mass flow rate of the air is evaluated at the disk for a velocity $C(1 - a)$. Then,

$$\dot{m} = \rho\pi r^2 C(1 - a)$$

The overall rate of change of the kinetic energy of the air is determined with respect to a change in the velocity of the air from C to $C(1 - 2a)$. Then,

$$\dot{E}_k = \tfrac{1}{2}\rho\pi r^2 C(1 - a) \cdot [C^2(1 - 2a)^2 - C^2]$$

The theoretical power developed by the windmill is equivalent to the rate of change of the kinetic energy of the air. Thus,

$$\dot{W}_{th} = -\dot{E}_k$$

$$\dot{W}_{th} = 2\pi r^2 \rho C^3 a(1 - a)^2 \quad (13.15)$$

In the development of the above expression for the theoretical power of the windmill, the rotational and drag losses are neglected.

The maximum value of $\dot{W}_{th}$ is particularly significant. When the quantity $a(1 - a)^2$ is differentiated and the result set equal to zero, $a = 1/3$. Then, the maximum theoretical power of the windmill is given by

$$\dot{W}_{max} = \tfrac{8}{27}\pi r^2 \rho C^3 \tag{13.16}$$

A comparison of Eqs. 13.12 and 13.16 shows that the maximum theoretical power developed by the windmill is equal to 59.3 percent of the power originally in the air. The power coefficient 0.593, designated as the Betz limit, cannot be exceeded under steady-state steady-flow conditions of operation.

Because of fluid losses, the actual power delivered by the windmill blades to the shaft is less than $\dot{W}_{max}$. The conversion efficiency is dependent upon the type of rotor construction and the speed ratio, that is, the blade tip speed to the undisturbed wind velocity. For a two-blade rotor of good aerodynamic design, operating with a speed ratio of about 6, a conversion efficiency of approximately 0.75 may be anticipated. Thus, $\dot{W}_a \cong 0.75\dot{W}_{max}$. Above a speed ratio of 6, the conversion efficiency tends to decrease.

Addition of a third blade to the rotor would achieve a slight increase in the conversion efficiency, but the added capital cost could not be economically justified. The conversion efficiencies for the American multiblade-type rotor and for the Dutch four-arm-type rotor are comparatively low. These machines operate with characteristically low speed ratios but develop a comparatively high torque.

There is an advantage to increasing the length of the rotor blades, since the power output of the windmill is proportional to the square of the rotor diameter. Because of structural limitations, rotor diameters may not exceed a value of about 100 m.

13.21 WIND TURBINE DESIGN

The operating principles of the horizontal-axis wind turbine are basically similar to those of the steam or gas turbine. Fluid velocities and rotating speeds of the windmill are however much lower, and as a consequence blade profiles are substantially different for the two types of turbines. Steam and gas turbine blades are fabricated to produce comparatively high fluid deflection angles. The blade profiles for the wind turbine are slightly cambered airfoil sections, designed for optimum camber over the operating range of the machine. Unlike the rotating blades for the steam and gas turbines, the chord angle of the windmill blades can be changed in response to changes in the wind velocity. The horizontal-axis windmill is essentially an axial-flow turbine equipped with two, or possibly three, long, relatively slender propeller-type blades.

The air moving through the windmill rotor, at a velocity C, creates a dynamic pressure on the surface of the blades. The resultant force F_r acts on the blade section, at the center of pressure, in a direction normal to the chord line. See Fig. 13.12. There are two components of this force, namely, the lift force F_L and the drag force F_D,

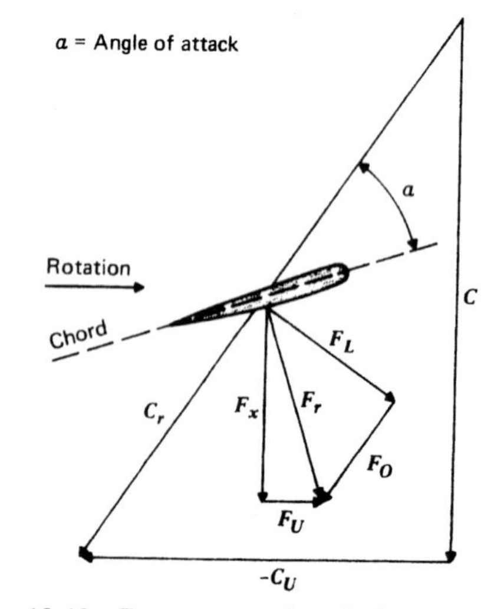

Figure 13.12 Force system for windmill rotor blade.

respectively, normal to and parallel to the direction of the relative wind velocity vector C_r. F_r is also resolved into the axial and rotational force components, F_x and F_U, respectively.

The force F_U produces the torque that turns the rotor against an external resistance. The force F_x produces the axial thrust that tends to overturn the tower.

The wind velocity is the principal parameter that influences the design of a wind turbine-generator (WTG), in part because the power developed by the turbine is proportional to the cube of the wind velocity. The turbine is designed to operate within a wind velocity range that is selected in relation to the local velocity–duration curve for the wind. Although strong winds are conductive to the generation of significant amounts of power, a high average wind velocity, over a moderate range, is a more important characteristic of a good site for a WTG. Shifting and gusting winds are in general undesirable.

Figure 13.13 is a typical local wind velocity–duration curve. For this particular site, the wind velocity will exceed 40 km/h for about 1000 hours per year and exceed 16 km/h for 6000 hours per year. A WTG operating at this site is designed for a rated output of 1000 kW at a wind velocity of 40 km/h. Because the rated output cannot be exceeded, the machine will generate 1000 kW for 1000 hours per year. See Fig. 13.14. During the remainder of the time for the year, the output from the machine will be less than 1000 kW. The generator will cut out when the wind velocity falls below 16 km/h, hence the total operating time for the WTG is 6000 hours per year.

The area under the power–duration curve, Fig. 13.14, extending from 0 to 6000 hours is equivalent to 3.55×10^6 kWh/yr. The rated output of 1000 kW can be developed by a 61-m-diameter WTG that operates with a conversion efficiency of 0.75.[19] The corresponding annual plant factor is about 0.40. The annual plant factor is defined as the ratio of the energy produced during the year to the maximum possible energy output, which for this particular WTG is equal to $8760 \times 1000 = 8.76 \times 10^6$ kWh/yr.

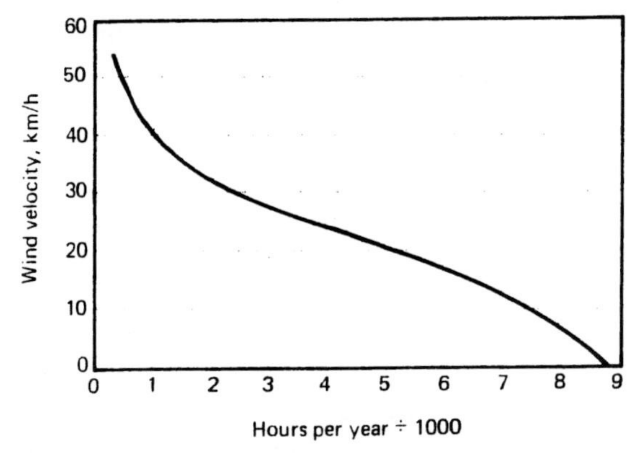

Figure 13.13 Velocity–duration curve for the wind at Amarillo, Texas. (C. C. Johnson, R. T. Smith, and R. K. Swanson, "Wind Power Development and Applications," *Power Engineering*, October 1974. Copyright 1974 Technical Publishing Co.)

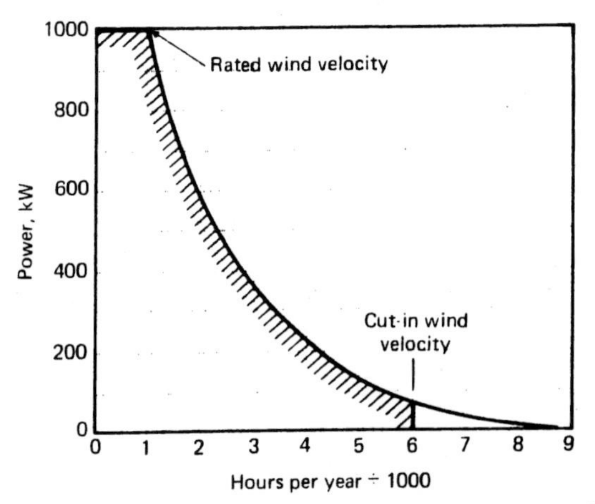

Figure 13.14 Power–duration curve for conceptual windmill configuration at Amarillo, Texas. (C. C. Johnson, R. T. Smith, and R. K. Swanson, "Wind Power Development and Applications," *Power Engineering*, October 1974. Copyright 1974 Technical Publishing Co.)

Example 13.12

A WTG has a rotor diameter of 60 m and operates in a 40-km/h wind. The conversion efficiency is 0.75, and the combined mechanical and electrical efficiency is 0.90. The air density is 1.185 kg/m^3. Determine the power output of the WTG, the power coefficient, and the energy output, in kWh/yr, for an annual plant factor of 0.45.

1 Power output.

$$\dot{W}_{max} = \frac{8}{27}\pi r^2 \rho C^3$$
$$= \frac{8}{27}\pi(30)^2 \times 1.185(11.11)^3$$
$$= 1362 \times 10^3 \text{ W}$$

$$\dot{W}_a = \eta_{m\text{-}e}\eta_c \dot{W}_{max}$$
$$= 0.90 \times 0.75 \times 1362$$
$$= 919 \text{ kW}$$

2 Power coefficient.

$$\dot{W}_i = 0.5\pi r^2 \rho C^3$$
$$= 0.5\pi(30)^2 \times 1.185(11.11)^3$$
$$= 2297 \times 10^3 \text{ W}$$

$$\text{PC} = \frac{\dot{W}_a}{\dot{W}_i} = \frac{919}{2297}$$
$$= 0.400 \quad \text{(for the turbine-generator)}$$

Rotor power coefficient:

$$\dot{W}_r = \eta_c \dot{W}_{max} = 0.75 \times 1362$$
$$= 1022 \text{ kW} \quad \text{(rotor power)}$$

$$\text{PC} = \frac{\dot{W}_r}{\dot{W}_i} = \frac{1022}{2297}$$
$$= 0.445 \quad \text{(for the rotor)}$$

3 Energy output.

Maximum power output = 919 kW (assumed rated output)

Maximum possible energy output = 8760 × 919 = 8.050 × 10^6 kWh/yr

Actual energy output = 0.45(8.050 × 10^6) = 3.623 × 10^6 kWh/yr

Two-bladed rotors of good aerodynamic design, operating at a speed ratio of about 6, are capable of achieving rotor power coefficients as high as 0.47. Maximum power coefficients for the Dutch four-blade and the American multiblade rotors are, respectively, 0.17 and 0.15.

13.22 CURRENT WIND TURBINE DEVELOPMENT

Interest in wind power has recently been revived in the United States because of the need to explore alternative sources of energy. An important current development is the erection of the experimental 100-kW wind turbine-generator at NASA's Plum Brook Station at Sandusky, Ohio. Synchronous operation in phase with the local utility was achieved by the WTG in October 1976.

The turbine-generator at the Plum Brook Station is mounted on the top of a 30-m, open-truss steel tower. See Fig. 13.15. The rotor diameter

Figure 13.15 Experimental 100-kW horizontal-axis wind turbine generator. (Courtesy NASA Lewis Research Center.)

is 38 m, and each of the two propeller blades has a mass of 910 kg. The mass of the assembly, including the rotor, installed on the top of the tower is 18 metric tons. The Plum Brook installation is the first of a number of wind turbine-generators scheduled for construction in the United States. The proposed machines are rated at 200 to 2500 kW, and the rotor diameters range up to about 90 m.

The synchronous alternator of the Plum Brook WTG operates at a constant speed of 1800 rpm. The corresponding rotor speed of 40 rpm is controlled by altering the pitch of the rotor blades in response to changes in the wind velocity. Power is generated whenever the wind velocity exceeds 13 km/h. An output of 50 kW is achieved with a wind velocity of 22.5 km/h, and the rated output of 100 kW is attained when the wind reaches a velocity of 29 km/h. At higher wind velocities, the blade pitch is adjusted to spill the excess energy. When the wind velocity exceeds 96 km/h, the blades are fully feathered, and no power is generated. The power coefficient for the rotor blades is 0.375, and the power train efficiency is 0.75.

The wind turbine-generator located at Boone, North Carolina, began preliminary operation in 1979. The diameter of the propeller-type rotor is 61 m, and the machine is installed, on a tower, 40 m above ground level. The rotor turns at 35 rpm whenever the wind velocity is in the range of 18 to 56 km/h. When the average wind velocity is outside this range, the blades are feathered, and the machine is shut down. The rated output of 2000 kW is developed when the wind attains a velocity of 40 km/h.

Three propeller-type wind turbine-generators were erected circa 1981 at a site along the Columbia River Gorge in southern Washington. Rotor diameters are 91 m, and the rated electrical power output of each WTG is 2.5 MW. Power generation starts with a wind velocity of 16 km/h, and full power output is achieved with a wind velocity in the range of 32 to 56 km/h. The average annual wind velocity at the site is 24 km/h, measured at a height of 9 m. Field tests will be conducted on the wind turbine-generators for a period of about two years in order to evaluate the potentialities of wind power as a future energy source. These machines are considered to be prototypes of the first large wind turbine-generators intended for commercial power generation, and if manufactured in quantity, the cost of a single machine is expected to be about $2 million.

The three wind turbine-generators erected near Goldendale, Washington, are placed in a triangular pattern from 460 to 910 m apart. The cluster arrangement was adopted to observe the effects of air turbulence that each machine creates on the other two machines. The rotors of these machines operate in an upwind position to avoid the interference caused by the tower in the air flow to a downwind rotor. Rotors operate at constant speed with the speed control achieved by varying the pitch of the outer section of each of the two blades. The length of this variable-pitch section is 30 percent of the length of the blade.

The motions of the rotor and the blades are similar to those of the Smith-Putnam machine. Control is exercised over the pitch and rotation of the blades, and over the yaw, or rotation of the machine around the vertical axis installed on the top of the tower. The blades have an additional degree of freedom, that is, limited motion normal to the plane of rotation as a means to minimizing the effect of induced blade forces caused by nonuniform air flow.

The horizontal-axis wind turbine-generator erected near Palm Springs, California, represents some interesting developments. The entire structure, including the tower, turns on a circular track to place the rotor in an upwind position. The three-bladed rotor operates at variable rotational speeds in order to maintain a constant relationship between the speed of the blade tips and the velocity of the wind. The rotor diameter is 58 m, and the machine is rated at 3 MW in a 64-km/h wind.

In addition to the horizontal-axis wind turbine, a large number of other types of wind power machines have been designed and constructed, notably the vertical-axis wind turbines (VAWT). The simplest vertical-axis machines are designated as Darrieus rotors.

A VAWTG, equipped with a Darrieus rotor is under test at the Sandia Laboratories. See Fig. 13.16. The rotor has a height of 19 m and a diameter of 17 m. The turbine-generator has a power output of 60 kW when operating in a 45-km/h wind and of 30 kW in a 35-km/h wind. The electrical system produces 60-Hz ac, and when synchronized with a utility network, the network

Figure 13.16 Vertical-axis wind turbine—Darrieus rotor. A cross section of the blade is shown in the foreground. (Courtesy Sandia Laboratories, Albuquerque, New Mexico.)

frequency controls the operating speed of the turbine-generator. The power output is then dependent upon the wind velocity.

The Darrieus rotor requires no orientation with respect to the direction of the wind. The rotor however is not practically self-starting. Regardless of the wind velocity, the rotor will not accelerate unassisted to the normal operating speed from a self-starting speed of about 15 rpm. The Darrieus rotor can however be made self-starting by installing starter buckets at the top and at the bottom of the vertical shaft. See Fig. 13.16.

As the Darrieus rotor turns, the angle of attack on the symmetrical airfoil sections of the blades is continuously changing. As a consequence of the change in the angle of attack, a marked change is observed in the magnitude of the lift and drag forces with each turn of the rotor. Because the lift forces are predominant, the rotor can turn against an external resistance and function with a reasonably good conversion efficiency. The maximum power coefficient for the Darrieus rotor is approximately 0.35, achieved at a speed ratio of about 6.

Small horizontal- and vertical-axis windmills have been constructed for the purpose of exploring the feasibility of installing low-capability wind machines on dwellings. To be successful, the wind turbine-generator must be installed at low cost and be free of operating difficulties. Because energy storage is a major problem, the small WTG would most likely be used to supplement the electrical energy supplied by the local utility. In an installation of this kind, a switchgear would be required to cut the generator "in and out" automatically with changes in the wind velocity. One relatively inexpensive installation would connect the WTG to an electric water heater equipped with a thermal cutout to prevent overheating.

Some difficulty is encountered with the typical location of a small WTG. Erratic winds near the ground level often cause unstable or sporadic operation of the windmill, and as a consequence the conversion efficiency of the machine is sharply impaired.

13.23 WIND POWER ON A COMMERCIAL SCALE

In common with that of solar power, the development of wind power is currently receiving support from several quarters, principally the Department of Energy. Both solar and wind energy are available intermittently and in varying amounts, depending upon the geographic location and upon local conditions. At a particular site, local meteorologic records can be consulted to predict wind velocities and other wind data. While solar energy can be collected only during a limited period of time during the day, the wind may blow at any time, and in some locations the highest wind velocities are observed in the winter and spring months of the year.

Because of the cost and limited facilities for energy storage, wind turbine-generators will most likely be integrated with existing hydro and steam power plants. Electrical energy generated by the wind machines will then be fed into the power grid whenever the wind blows at a sufficiently high velocity. A WTG can operate automatically, unattended except for the periodic visits required for inspection and maintenance of the machinery.

On the basis of current technology, the horizontal-axis wind turbine-generator has a higher power capability than a vertical-axis machine of the same diameter, and also a somewhat higher conversion efficiency. Large-scale wind turbines are likely to be horizontal-axis machines, because of superior structural and aerodynamic characteristics.

The wind turbine-generator in any installation must be designed for protection against storm damage resulting from icing and very high wind. Incidentally, the Plum Brook wind turbine-generator is designed to withstand wind velocities up to 240 km/h. Heavy ice formation on the blades could unbalance the rotor and cause severe vibration in the structure.

Of immediate interest is the number of sites in the United States suitable for the erection of wind turbine-generators. A study recently conducted by the Lockheed-California Company places this number at 54,000. The report envisions 54,000 windmills placed in remote, out-of-sight locations, mainly in the western region of the United States.

The most suitable sites for wind turbine power plants for the United States are in areas where the annual average wind velocities exceed 29 km/h, at an elevation of 46 m above ground level. The largest area is essentially the high Great Plains region, while four smaller areas are centered in New England, West Virginia, eastern Washington, and southeastern California. See Fig. 13.17.

An average wind velocity of about 26 km/h is essential to establishing a good power–duration curve. In certain windy regions, WTG plant factors range as high as 0.60 to 0.70. A high plant factor is important because the return on the investment for the plant declines when a turbine-generator is shut down or operates below the rated output. Little useful power is produced when the wind velocity drops below 25 km/h, say, for a WTG rated at a 50-km/h wind velocity.

Wind power technology has reached a practical stage, and the capability exists to build the main structure and fabricate all of the system components for a WTG. A decision to undertake the construction of a large number of wind turbine-generators will be influenced by the cost to produce the electrical energy generated by the wind machines. Bus-bar costs of 20 to 30 mills/kWh are predicted for power plants located at sites where reasonably high average wind velocities are known to occur. These costs are somewhat high in comparison with the cost of producing electrical energy in nuclear or coal-fired power plants but are below the predicted cost for electrical energy produced in solar-thermal-electric power plants. See Section 13.18.

Because of the environmental impact, opposition to the construction of thousands of windmills may be anticipated. While there are varied reasons for the opposition, the aesthetic aspect is certainly a primary factor. The potential for causing electromagnetic radiation interference

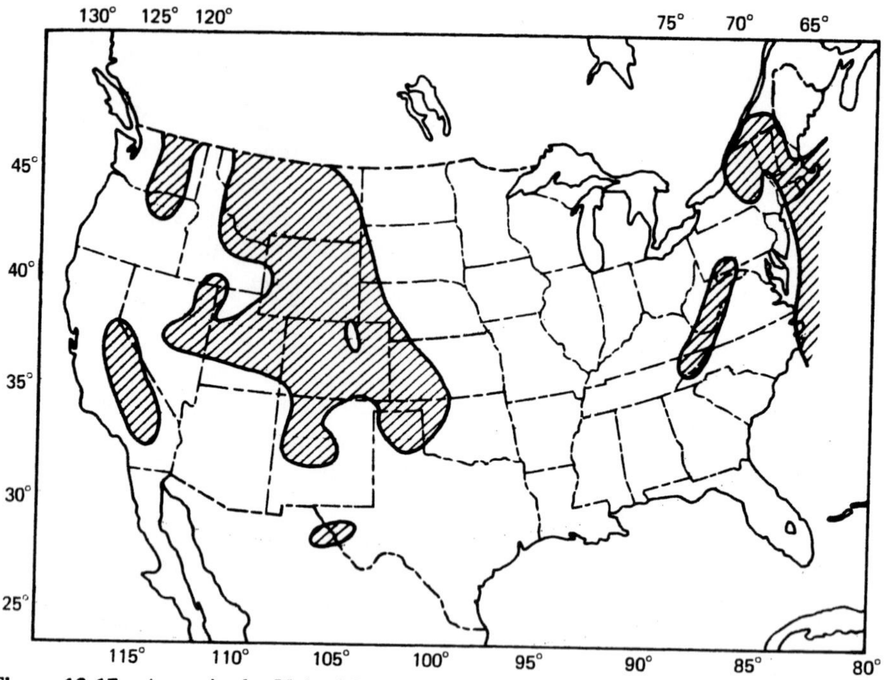

Figure 13.17 Areas in the United States where the annual average wind speeds exceed 29 km/h at 46 m elevation above ground level. (Report prepared for National Science Foundation, Grant No. AER-75-12937, by Frank R. Eldridge, The Mitre Corporation.)

places an additional constraint upon the selection of a site for a WTG. The main problem arises in the interference with the signal from a TV transmitter. Under the worst conditions, the potential for TV interference may extend to a distance of about 3 km from the WTG.

Another disturbance arising in the operation of a wind turbine-generator has been observed during the tests conducted on the WTG at Boone, N.C. Under certain conditions, the turbine will emit very-low-frequency noise, below the range of hearing. The generation of the noise, or low-frequency vibration, is influenced by the prevailing weather conditions, the terrain, the direction the turbine faces, and the level of the power output. During the investigation to determine the cause of the noise emission, the operation of the machine has been limited to generating power only on weekdays between 8 A.M. and 5 P.M. The curtailed operation alleviates to some degree the disturbance experienced in the neighboring communities.

A WTG is inherently a low-capability power plant. A single 1300-MW nuclear power plant would generate, during a period of one year, a quantity of electrical energy equal to that produced by about 1100 wind turbine-generators, each rated at 2.5 MW. The assumed annual plant factors are 0.70 and 0.33 for the nuclear power plant and the wind turbine-generators, respectively. For power generation on a commercial scale, large wind turbine-generators would normally be erected, because the unit capital cost, in $/kW, of the WTG decreases with an increase in the size of the machine.

Geothermal Power

13.24 GEOTHERMAL ENERGY

Geothermal energy is usually defined as thermal energy stored in underground deposits of steam, hot water, and hot dry rock. The inner core of the earth is highly radioactive, and as a consequence a natural flow of heat occurs from the core to the surface of the earth. Virtually all of this energy is transferred by conduction; but in a few places, limited quantities of steam or hot water flow unaided to the surface through fissures, faults, or pervious strata. Such occurrences are rare, and utilization of geothermal energy typically requires the drilling of wells.

The normal temperature gradient in the rock structure of the earth is 29 C/km. A gradient of this magnitude is not sufficiently high to support practical utilization of the earth's internal energy. Substantially higher temperature gradients exist throughout numerous regions that have a total area equivalent to about 10 percent of the earth's surface. Where a temperature gradient of at least 110 C/km is observed in a particular region, consideration can be given to the possibility of effectively utilizing the geothermal energy.[22]

Geothermal energy is normally found in two basic forms, namely, in subterranean hot water or hot dry rock. In some locations, the vapor phase of the hot water is predominant, hence the geothermal energy source is described as steam. Where the hot water is entirely in the liquid phase, the term geohydrothermal, or geopressurized, is applied.

The geologic aspects of geothermal energy sources have been examined in considerable detail. Briefly, the hot water or steam discharged at the surface originates from local cold surface water that penetrates the rocks from the upper level through fractures and porous formations. The circulation of water through a geothermal system is believed to be extremely slow. On contact with the hot rock, the surface water is heated, and density differences cause the creation of a convection cycle.[23]

Steam and hot water tend to rise to the surface along routes of highest permeability, while escape at the surface is dependent upon the nature of the cap rocks. In the United States, the geothermal activity in Yellowstone National Park is particularly striking. There are in this area many geysers, hot-water pools, boiling pools, and bubbling pots.

In considering the installation of a geothermal power plant, the most important factor is the steam supply. Geologic exploration will determine whether a sufficiently high temperature gradient exists in the field and, if so, the heat supplied by the hot rock is likely to be adequate for a period of time beyond that deemed necessary from an economic point of view. There is then the question whether surface water will penetrate to the hot rock in sufficient quantity to maintain an undiminished flow of steam from the well.

After an initial period of time, subsequent to drilling the well, a steady-state inflow–outflow condition is established with respect to hot water levels, temperatures, and saturated steam pressures. Tests conducted on actual producing wells indicate that, over a period of several decades, the steam flow remains unchanged. The possibility exists however that the flow from a well may be reduced by a self-sealing activity.

Geothermal resources in the United States are concentrated in the 11 westernmost contiguous states and in Alaska and Hawaii. Table 13.2 shows the subdivision of the geothermal resource base.

A small fraction of the total geothermal resource is identified as steam. The temperatures in the steam reservoirs typically range from 220 to 250 C.[25] In the geohydrothermal system, the pore space in the reservoir is filled with water under the hydrostatic pressure at a temperature as high as 360 C. Geopressurized resources occur in regions where natural gas (methane) coexists with the geopressurized water at temperatures typically in the range of 150 to 200 C. Hot dry rock temperatures are in the range of 150 to 400 C and may possibly approach 650 C in some

TABLE 13.2 United States Geothermal Resource Base

Type	*Energy (q Units)*[a]
Geohydrothermal	12,000
Geopressurized	190,000
Magma	200,000
Hot dry rock	13,000,000

Source: From Ref. 24.

[a] *At T > 150 C.*

instances. Magma temperatures are related to the melting point of rock, about 650 to 1200 C. Extraction of energy from the magma resource will pose formidable drilling problems, in part because of the high temperatures that will be encountered.

A number of high-temperature geothermal areas exist throughout the world. Most of these areas are located in regions of current, or recent, volcanism or tectonic activity. The regions around the rim of the Pacific Ocean, in this respect, are major areas of geothermal energy sources. Another major concentration of geothermal areas is clustered roughly about a north–south axis that extends through eastern Africa and eastern Europe.

The earth is believed to consist of a core about 6900 km in diameter, a rock mantle about 2900 km thick, and an outer crust that ranges in thickness from 15 to 50 km. The core, composed principally of iron, is subdivided into an outer core and a highly radioactive inner core. The crust is assumed to "float" on the dense rock mantle that exists in a semiplastic state. Geologic observations indicate the division of the crust into a number of major pieces, perhaps ten, called tectonic plates.

The cracks in the crust at the boundaries of the tectonic plates are known as rifts, and in the vicinity of these rifts are located the major earthquake zones and the areas of volcanic activity. Of particular importance is the flow of molten rock through the rifts to within a comparatively short distance below the surface of the earth. As a consequence of this upward flow of hot material, high-temperature gradients are established in the rock structure near the surface, that is, at depths of 1.5 to 3 km.

13.25 UTILIZATION OF GEOTHERMAL ENERGY

Today, in the light of the necessity to examine nonconventional energy sources, the utilization of geothermal energy is being investigated through a number of research and development programs. Like solar and wind energy, geothermal energy is quantitatively significant, but the extraction of this energy from the ground and subsequent conversion to electrical energy is not cost free and not without certain operating problems. Whether a particular geothermal power plant can be economically competitive with other sources of power will depend generally upon the capital and fuel costs of the different systems and upon local factors, most of which are of a subterranean nature. Geothermal power generation however is not subject to interruptions that are inherent in solar and wind power generation. The geothermal power plant is capable of continuous operation, provided that the generating capability properly matches the energy supply.

There are many locations throughout the world where geothermal steam and hot water are utilized for heating buildings. In addition to heating single houses or small groups of houses, a geothermal energy source may be adequate for district heating. In Iceland, widespread use is made of naturally heated water for heating both public and private buildings and for industrial processing. Reykjavik, the capital city of Iceland, has the largest geothermal installation in the country. Hot water, extracted from 16 boreholes that tap aquifiers at various levels beneath the city, supplies nearly 75 percent of the city's energy requirements for heating buildings and producing hot water. The temperature of the water flowing from these wells ranges from 86 to 127 C.[26] In Iceland, the cost of heating buildings and water by geothermal energy is substantially below the cost of heating by fuel oil that must be imported.

In the United States, geothermal energy is currently utilized for heating buildings only on a small scale in scattered locations. In Klamath Falls, Oregon, a number of homes in the city and buildings on the campus of the Oregon Institute of Technology are heated by geothermal hot water. Well drilling is presently under way in Idaho for tapping what is believed to be large sources of geothermal energy. Similar explorations in other parts of the country can be anticipated.

13.26 GEOTHERMAL POWER PLANTS

The world's first geothermal power plant to operate on a commercial scale was constructed at Larderello in Tuscany, Italy. The plant initially had a generating capability of 250 kW and began production in 1913. The present generating capability at Larderello is 406 MW. The average and maximum depths of the drillholes are 600 and 1600 m, respectively. At the Larderello field, steam is produced at an average temperature of 200 C, with the maximum steam temperature reaching 260 C.[23]

The sole geothermal power development in the United States is located in Sonoma and Lake Counties, about 145 km north of San Francisco, in an area called The Geysers. In 1847, William Bell Elliott, while searching for grizzly bears, came upon a narrow canyon that appeared to be "the gates of Hell." Crowded between the banks of this canyon, for a distance of half a kilometer, were numerous hot springs and fumaroles or steam vents, but no geysers. The area, given the picturesque name The Geysers, is probably the richest source of geothermal steam in the world.

During the period 1922–1925, eight shallow steam wells were drilled at The Geysers. At that time however, geothermal power generation at The Geysers was not considered to be economically feasible, because of the abundant, low-cost hydroelectric resources and the relatively cheap natural gas available for steam-electric power generation. Certain operating problems, resulting from corrosion and abrasion of the piping and turbines, were anticipated.

Well drilling was resumed at The Geysers in 1957, and subsequently a 12.5-MW steam-electric power plant was constructed at the site. This initial unit began commercial operation in 1960. Since that time, additional steam wells have been drilled and additional turbine-generators installed. The present generating capability at The Geysers is 632 MW, with a planned addition of 305 MW.

Geothermal steam at The Geysers has a constant enthalpy of 2789 to 2801 kJ/kg, and the steam wells have a shutoff pressure of 3.1 to 3.3 MPa, gage. Steam in this pressure range with an enthalpy value of 2800 kJ/kg is essentially dry, saturated. When the steam flows from the well, the pressure at the wellhead is less than the shutoff pressure; and, in addition, there is a pressure loss in the line conveying the steam to the turbine. A steam temperature of about 180 C is observed for a turbine inlet pressure of 690 kPa, gage (170 C saturation temperature).[26] The slight amount of superheat (10 C) is advantageous with respect to improving the turbine machine efficiency.

The drillholes at The Geysers have an average depth of 1500 m and a maximum depth of 2900 m. The average temperature of the fluid in the wells is 250 C, while the maximum temperature is 285 C.[23]

The steam supply at The Geysers is apparently extensive, since the pressure and steam flow show only minor effects from the continued operation of the field, although the wells are spaced relatively close to each other. The well pattern established at The Geysers is not expected to lead in the foreseeable future to a diminished steam flow. Fumaroles in the area are known to have been in existence for more than 100 years, and the activity of these vents, according to hearsay, has remained essentially constant.

The third largest geothermal power installation in the world, with an installed generating capability of 203 MW, is located at Wairakei in New Zealand. Commercial operation of the power plant began in 1958. Well temperatures at Wairakei are about 20 C below the well temperatures at The Geysers, while the average depth of the drillholes at Wairakei is 800 m. In the New Zealand fields, the steam is formed from water, under high pressure and temperature, that partially flashes into steam as the pressure is reduced. The discharge from the well is consequently a mixture of steam and water.

TABLE 13.3 Generating Capability of Geothermal Power Plants (as of December 1979)

	Generating Capability (MW)	
Country	*Installed*	*Under Construction*[a]
United States	773	641
Italy	421	—
New Zealand	203	150
Japan	166	100
Mexico	150	30
El Salvador	60	35
Iceland[b]	33	30
U.S.S.R.[b]	5.7	58
Philippines	59.2	605
Turkey	0.5	—
Total	1871.4	1649

Source: Potter, R. W., II "Geothermal Energy: An Assessment," *Mechanical Engineering*, May 1981.[40]

[a] *Scheduled completions are prior to 1985.*

[b] *Some of these countries employ geothermal resources for space and process heating, notably Iceland, 475 MW(t), the U.S.S.R., 300 MW(t), and Hungary (not included in the tabulation), 363 MW(t), for a total of 1281 MW(t).*

Table 13.3 provides a summary of the generating capability of geothermal power plants in operation and under construction.

13.27 GEOTHERMAL POWER PLANT OPERATION

The geothermal power plant operates on a comparatively simple, low-pressure steam power cycle. At The Geysers, the turbines operate with inlet steam pressures of 450 and 690 kPa, gage. A boiler and fuel handling equipment are not required. Because there is no need to conserve the condensate, a direct-contact condenser can be used. In comparison with a surface condenser, a barometric or jet condenser can be installed at a lower cost, but the condensing pressure is somewhat higher.

The geothermal steam discharged from a well contains a quantity of noncondensable gases that can cause operating difficulties, including corrosion in the condensing system. These gases are removed from the steam in the condenser by the vacuum pump, usually a steam jet ejector, and expelled into the atmosphere. The principal problem encountered in the discharge of these gases arises from objectionable odors.

At The Geysers, the total amount of the noncondensable gases is significant with respect to the pumping cost, while the hydrogen sulfide and ammonia contents are important with respect to the environmental impact. The inhabitants of the neighboring communities were understandably displeased with the offensive odor attributed to hydrogen sulfide discharged into the atmosphere. A substantial reduction in the discharge of hydrogen sulfide has been achieved by the installation of pollution control equipment.

Most of the wells drilled for geothermal power production discharge a mixture of steam and water. If the hydrostatic pressure is sufficiently high at the bottom of the well, the water will flow unaided to the surface. Hot water rising in the well and subjected to reduced pressure partially flashes into vapor. At the wellhead, the water is mechanically removed from the mixture in cyclone separators, and the relatively dry steam is transported to the power station. In an alternative arrangement, the two-phase mixture is conveyed to the turbine installation where the water is removed and some of the energy in the water is recovered.

The steam-and-water mixture flowing from the geothermal well contains dissolved solids that are particularly troublesome. For the several geothermal sites in the world, the amount of dissolved solids ranges from 1 to 20 grams per kilogram of water. The solid material in geothermal water consists of a relatively large number of different substances in quantities that vary widely from one geothermal site to another. In addition to containing dissolved solids, the water may be acidic.

In general, the solids dissolved in geothermal water cause scaling and corrosion. Scale formation can be particularly severe in the outflow pipeline in which the discarded water is carried away from the separator. Scaling and corrosion can be caused, to some degree, in the steam lines and in the turbines by the dissolved solids in the water droplets carried over in the steam discharged from the separator.

Vast amounts of fluids are extracted from the ground in geothermal power production, and as a consequence various geologic changes can occur in the area of the power plant site. In a steam–water well, large quantities of water removed from the mixture must be discarded. Disposal of this water in a natural waterway would most likely not meet with approval, because of the deleterious effects of the dissolved solids and the relatively high temperature of the water.

An alternative scheme to surface disposal of the geothermal effluent provides for reinjection of the water into the field. The discarded water, in this arrangement, would be pumped into wells or perhaps into natural fissures at the periphery of the field. The water reinjected into the ground must not cause interference with the flow from the producing wells or cause cooling of the hot rock. Certain long-term difficulties may occur which limit this method for disposal of the geothermal effluent; for example, scale formation in the rock structure could restrict the flow of the water.

Geothermal power production is evidently not pollution free. Air and water pollution is inherent in the operation of the geothermal power plant. Ear-splitting noise caused by escaping steam and the escape of radioactive radon gas are other objectionable characteristics that have been observed.

13.28 GEOTHERMAL POWER DEVELOPMENT

Geothermal exploration in the United States, and most likely in other regions of the world, will in the next several years remain fairly active. There is however presently no basis for predicting how rapidly the construction of geothermal power plants will proceed in the United States. A higher priority may be given to the construction of solar and wind power plants than to the development of geothermal power. Production costs are competitive for electrical energy generated at The Geysers power plant, but there is need to determine probable capital and operating costs for new geothermal power plants.

In 1975, a production-size exploratory well was drilled in the Raft River Valley of Idaho. Spontaneous flow from the well occurred when a depth of 1415 m was reached. The flow increased until a rate of 5700 liters/min was established and a bottom-hole temperature of 138 C was recorded. A well that yields 5700 liters/min water at 150 C could produce in a binary system a net power output of about 3400 kW.[28]

An abnormally high heat flow from the earth has been detected in a 25- to 30-km^2 area near Marysville, Montana. There are in the area no surface manifestations such as geysers, steam vents, or hot springs. The rock at this site is believed to be hot, with temperatures on the order of 375 to 425 C at a depth of 1800 m. Drilling is expected to reveal the condition of the rock, that is, whether the rock is wet or dry and, if dry, whether the rock is fractured.

Wet rock will yield steam and hot water that may be utilized for power generation. The energy in the hot dry rock can be extracted by introducing surface water that is heated on contact with the hot rock. The effectiveness of energy extraction in this scheme will be dependent upon the contact area that exists between the water and the rock. An extensively fractured rock structure is conducive to heat transfer. If the rock is dry and unfractured, some procedure would be employed to make the rock permeable to the injected water. Various methods for fracturing rock are likely to be investigated in the exploration of geothermal energy resources.

An investigation of extraction of energy from hot dry rock is under way in an area near the Valles Caldera in the Jemez Mountains of New Mexico.[29] Two holes, about 75 m apart at the surface, were drilled to a depth of about 3 km. Water under high pressure was used to fracture the rock and create a system of cracks in the hot granite bedrock, thus exposing a large heat exchange surface of rock at a temperature of about 200 C. Cold water, under a pressure of 6.2 to 6.9 MPa, was pumped down the hole. Subsequent to circulating through the crack system, the water flowed from the second hole at a temperature of 130 C. The water in the primary loop is pressurized in order to prevent the formation of vapor that could cause mineral deposits to precipitate into the system and clog the underground cracks and the piping. Increased electric generating capability can be achieved by drilling into hot granite (260 C) at a depth of about 4000 m.

Geopressurized zones exist along the Texas and Louisiana Gulf Coast, extending perhaps for a distance of 150 km inland and 150 km offshore. Over an area of some 400,000 km^2, porous rock, 3 to 15 km thick, is saturated with hot, pressurized salt water. At these depths, pressures and temperatures on the order of 85 MPa and 120 C are reached. The conditions of high pressure and temperature have forced natural gas into solution with the brine. The quantity of natural gas existing in geopressurized Gulf Coast zones is estimated as high as 115 trillion m^3, an amount that is roughly equivalent to 200 times the current annual production of natural gas in the United States.

Extraction of energy from the geopressurized resource will be difficult and costly and is likely to remain uneconomical unless the price of natural gas produced from conventional wells is increased substantially. Great depth, heat, and pressure cause severe drilling conditions. Disposal of large quantities of effluent brine would be a difficult undertaking. Reinjecting the salt water into the geopressurized zone would require an excessive consumption of energy. Other possibilities are offshore disposal of the effluent water or injecting the waste brine into the earth at shallow depths. Problems could result from possible surface subsidence caused by removal of large quantities of water from the subterranean formation.

At a depth of 5000 m, brine temperatures are expected to be 125 to 180 C. Thermal energy extracted from the brine would be used for electric generation in a binary-fluid power plant. Extraction of the natural gas from the brine would be a primary objective in any scheme for the utilization of the geopressurized resource. Figure 13.18 shows a possible mechanical arrangement that achieves the generation of power and pro-

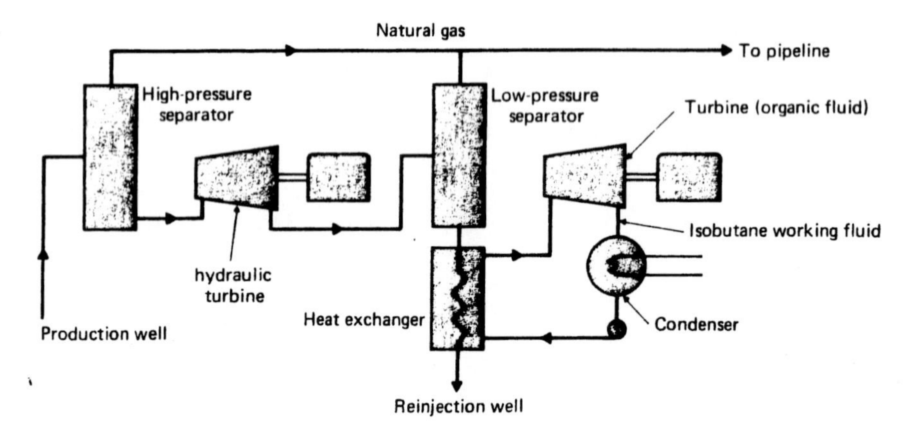

Figure 13.18 Flow diagram for a power plant supplied by a geopressurized resource. (Adapted from "Geopressured Geothermal Resources," Division of Geothermal Energy, United States Department of Energy, and from *Sun Magazine*, Spring 1980.)

duction of natural gas through the use of hot, geopressurized brine. Because of the high pressure under which the brine exists, a hydraulic turbine is used to extract the potential energy in the source fluid. Among other problems, corrosion of the turbine parts by the hot brine would certainly be cause for concern.

Extraction of energy from hot dry rock is likely to remain in the investigative stage for at least a decade. Meanwhile, most of the new wells will be drilled to tap water deposits that are at temperatures in the range of 150 to more than 300 C. The locations and extent of these deposits are as yet not exactly defined, but they occur generally in the western states at depths that range approximately from 900 to 3000 m. Almost all of these wells will require pumping in order to lift the water to the surface, because the natural well pressure will not be sufficient for this purpose. Further, these wells are expected to produce hot water that has a high content of dissolved solids.

When well pumping is employed, the energy is extracted most effectively from the hot rock in a binary-fluid geothermal power system. The water is pumped from the well through a heat exchanger and into a collecting basin. Subsequently, the water is taken from the basin and injected into the field by a second pump. The turbine working fluid, an organic substance, for example, isobutane, is evaporated in the heat exchanger, which is part of the closed power system operating on the Rankine cycle. The actual pumping of the hot water is accomplished by a downwell pump located near the bottom of the well.

The binary-fluid geothermal power system is also used effectively in extracting energy from hot rock. In this case, the water is pumped from the surface into the rock and then through the heat exchanger into the collecting basin.

In comparison with a system in which energy is supplied by steam–water wells, certain advantages are derived through hot brine pumping. The hot brine is maintained under a pressure sufficiently high to avoid flashing. As a result, the water reaches the wellhead at a temperature above that of a flashed system, hence power generation can be achieved at a higher thermal efficiency. Further, pumping generally increases the yield of the geothermal energy source.

The environmental impact is minimized through hot water pumping. Noxious gases are not released to the atmosphere, and there is no need for surface disposal of warm water containing dissolved solids. Reinjection of the geothermal water establishes a continuous circulating system and largely prevents ground subsidence. The operating life of the geothermal energy source is thus prolonged.

The dissolved solids content of geothermal hot water varies widely from less than 0.1 to over 25 percent. Precipitation of the solids is inhibited by maintaining the pressure on the hot brine, thereby substantially reducing corrosion and scale formation. Severe problems are encountered in pumping high-temperature brine, and equipment reliability must be assured.

Steam for power generation can be produced by flashing the hot geothermal water. A single-flash system has however a comparatively low energy conversion efficiency; that is, only about 10 percent of the energy in the fluid at the wellhead is converted into electrical energy. A very high fraction of the energy is carried away in the discarded brine. Multistage flashing will reduce the thermal losses, but a more complex mechanical system is required. In comparison with the flashed system, the binary-fluid geothermal power system shows a somewhat improved energy conversion efficiency.

Example 13.13

A geothermal power plant is supplied with hot water at a wellbottom temperature of 225 C (P = 8 MPa, abs). The geothermal fluid flows to a flash separator that operates at a pressure of 40.0 kPa, abs. The condenser pressure is 11.0 kPa, abs. The overall efficiency of the turbine-generator is 0.83. Neglect the heat loss from the steam and hot water pipelines. Calculate the hot water flow rate for a turbine-generator output of 25 MW.

Properties of water are selected from *Thermodynamic Properties in SI*, by W. C. Reynolds, Stanford University, 1979.[38]

Separator

$$h_o = 968.16 \text{ kJ/kg} \quad \text{at} \quad T = 225 \text{ C}, \quad P = 8 \text{ MPa, abs}$$

$$h_{f.1} = 318.1 \text{ kJ/kg} \quad \text{and}$$

$$h_{fg.1} = 2318.1 \text{ kJ/kg} \quad \text{at } 40.0 \text{ kPa, abs}$$

$$h_o = h_{f.1} + xh_{fg.1}$$

$$968.2 = 318.1 + 2318.1x \qquad x = 0.280$$

Turbine

Inlet:
$P_1 = 40.0$ kPa, abs assume $x = 1.0$
$h_1 = 2636.1$ kJ/kg $s_1 = 7.6704$ kJ/kg·K

Exhaust:
$P_2 = 11.0$ kPa, abs $s_2 = 7.6704$ kJ/kg·K
From isentropic expansion, $h_{2.s} = 2443.9$ kJ/kg

$$W_{T\text{-}G} = \eta_o(h_1 - h_{2.s}) = 0.83(2636.1 - 2443.9) = 159.5 \text{ kJ/kg}$$

Hot Water Flow

Steam flow:

$$\dot{m}_s = \frac{\dot{W}_{T\text{-}G}}{W_{T\text{-}G}} = \frac{25000}{159.5} = 156.74 \text{ kg/s}$$

Water flow:

$$\dot{m}_w = \frac{\dot{m}_s}{m_s/m_w} = \frac{156.74}{0.280}$$

$$= 559.8 \text{ kg/s} \qquad 2.015 \times 10^6 \text{ kg/h}$$

The power plant output is less than 25 MW because of the energy consumed in pumping the condenser cooling water, operating the cooling tower fans, and operating the condenser vacuum pumps. The geothermal water may contain a substantial amount of noncondensable gases. Power may also be required for pumping the hot water.

The heat supplied to the system may be arbitrarily taken equal to the enthalpy of the hot water at the bottom of the well minus the enthalpy of the water at the ambient temperature, say, 20 C. Thus,

$$Q_s = h_o - h_{f.20\text{ C}} = 968.2 - 82.9 = 885.3 \text{ kJ/kg}$$

The power plant thermal efficiency is

$$\eta_t = \frac{W_{T\text{-}G}}{Q_s} = \frac{159.5}{885.3} = 0.180$$

The net thermal efficiency of the system will be below 0.180 because of the power consumed by the auxiliary equipment. For a single-flash system, the thermal efficiency is typically about 0.10 to 0.11.

Because single-stage flashing of hot water in a geothermal power system does not yield a sufficiently high energy conversion efficiency, alternative schemes are being explored. The total flow concept is proposed as a means to achieving the direct expansion of the total wellhead product. In this scheme, the hot water from the wellhead is conveyed directly from the main pump to an impulse turbine of special design. Expansion of the fluid in convergent–divergent nozzles produces a high-velocity, low-temperature fluid stream that, in turn, is directed into the single-row impulse wheel to effect conversion of the kinetic energy of the jet into mechanical energy. Whether the low-quality, two-phase fluid can be effectively expanded in a turbine of practical proportions remains to be demonstrated. A turbine employed in this manner would necessarily be fabricated from materials resistant to both erosion and corrosion.

A different kind of total flow expander will be tested in a 1-MW demonstration plant intended for the production of electric power from geothermal water. The device is described as a helical screw expander.

An analysis of comparable geothermal power systems reports the following values for the overall energy conversion efficiency: flashed steam and binary-fluid power systems, about 0.11; total flow system, 0.177.[31] The reservoir temperature in this analysis is assumed to be 300 C.

13.29 RAFT RIVER GEOTHERMAL POWER PLANT

Construction of the Raft River 5-MW(e) geothermal pilot power plant is an important step in the program initiated by the Department of Energy for development of commercial geothermal power generation in the United States. The pilot power plant, which was preceded at the site by a small prototype geothermal power plant, began operation in 1980. The objective of the project is to demonstrate the technical feasibility and environmental acceptability of generating electric power from a moderate-temperature (143 C) geothermal resource.

The hot (143 C) geothermal water, the primary fluid, is supplied to the power plant at the rate of 0.142 m^3/s. A portion, 0.028 m^3/s, of the effluent geothermal water is used for the wet cooling tower make-up. The temperature of the water is 60 C at injection to the field. The power plant fluid supply and injection system uses three production wells, two injection wells, and a reserve well of each kind, for a total of seven wells. The depth of the production wells is 1525 m, and the effluent water is injected into the field at a depth of 1200 m. Adequate flow of hot water from the wells is achieved by drilling into a fissure. Pumping is accomplished by downwell, submersible pumps and by booster pumps that provide adequate pressure for causing the flow of hot water through the four heat exchangers. The pressure is sufficiently high to preclude boiling of the water during passage through the system.

The flow diagram for the binary-fluid power system is shown in Fig. 13.19. The hot geothermal water flows successively, in order of decreasing temperature, through the high-pressure boiler, high-pressure preheater, low-pressure boiler, and low-temperature preheater. The total flow of the working fluid, isobutane, subsequent to passage through the low-temperature preheater is divided by the flow-split valve into two streams, one of which enters the low-pressure boiler. The other stream of isobutane, about 66 percent of the total flow, passes through the high-pressure preheater and then through the high-pressure boiler. The inlet conditions for the high-pressure turbine and the low-pressure turbine are, respectively, 2.62 MPa, abs (T_{sat} = 115 C) and 1.39 MPa, abs (T_{sat} = 81 C).

The anticipated thermal efficiency for the pilot power plant is in the range of 11 to 13 percent, excluding the power consumed in pumping the geothermal water. In order to achieve economical operation on a commercial scale, a minimum generating capability of about 50 MW(e) is a probable requirement for this particular power system. It is essential to observe that the Raft River installation utilizes a moderate-temperature geothermal resource.

Example 13.14

A two-fluid (water/isobutane) geothermal power plant operates in accordance with the flow diagram shown in Fig. 13.19. The operating data are

Geothermal water:
- T_{in} = 148 C
- T_{out} = 70 C

Isobutane:
- Condenser pressure 0.40 MPa, abs
- High-pressure turbine
  - Flow rate 100 kg/s
  - Inlet condition 120 C, dry saturated
- Low-pressure turbine
  - Flow rate 50 kg/s
  - Inlet condition 85 C, dry saturated
- Low-temperature heater outlet 80 C
- High-pressure heater outlet 115 C
- Pump internal work −9.55 kJ/kg

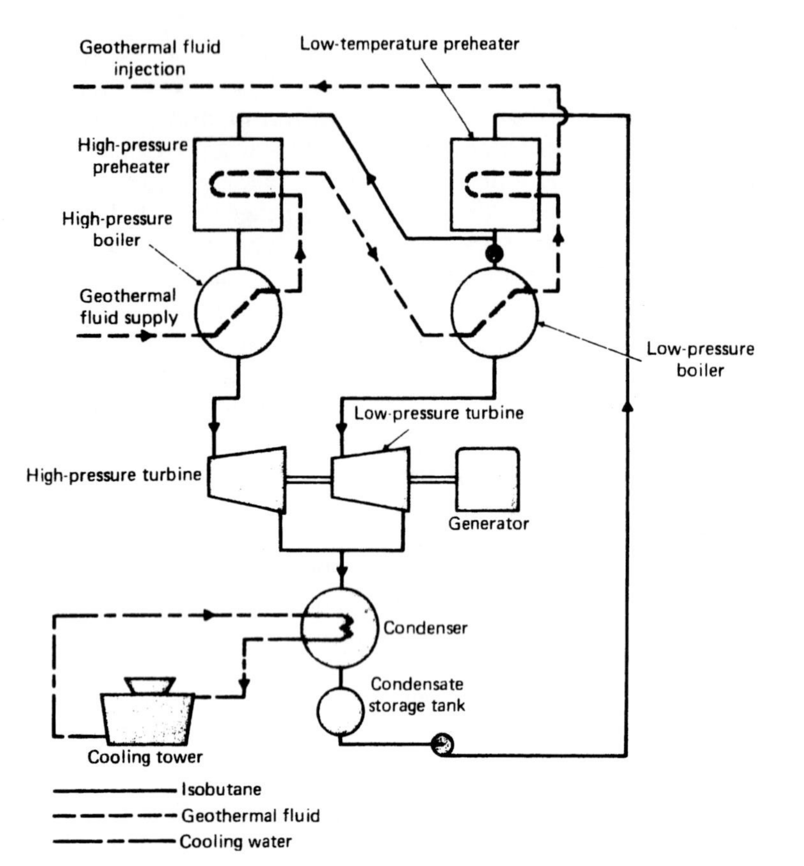

Figure 13.19 Flow diagram for the Raft River 5-MW(e) geothermal pilot power plant. (United States Department of Energy.)

Efficiency values:

Turbine internal	0.90
Mechanical	0.98
Generator	0.95
Pump motor	0.94

Calculate the geothermal water flow rate, electric power output, and the area of the heat transfer surface for each heat exchanger. Property values for isobutane are selected from *Thermodynamic Properties in SI*, by W. C. Reynolds, Stanford University, 1979.[38]

1 Geothermal water flow rate.

A control volume is prescribed so as to contain the four heat exchangers.

The isobutane leaves the condenser saturated at a pressure of 0.40 MPa, abs; $h_f = 228.85$ kJ/kg. At the pump outlet,

$$h = h_f - W_P = 228.85 - (-9.55)$$
$$= 238.40 \text{ kJ/kg} \qquad (33.0 \text{ C})$$

The flow rate at the pump outlet is 150 kg/s.

The isobutane leaves the high-pressure boiler dry, saturated, at 120 C ($h = 655.01$ kJ/kg) and leaves the low-pressure boiler at 85 C ($h = 627.47$ kJ/kg). The flow rates are, respectively, 100 and 50 kg/s from the high- and low-pressure boilers.

The energy transferred to the isobutane is determined from

$$\dot{Q} = (100 \times 655.01) + (50 \times 627.47) - (150 \times 238.40) = 61115 \text{ kJ/s}$$

The geothermal water enters the power plant at 148 C (h = 623.22 kJ/kg) and leaves at 70 C (h = 292.82 kJ/kg), or Δh = −330.40 kJ/kg. The water flow rate is given by

$$\dot{m}_w = \frac{\dot{Q}}{\Delta h} = \frac{-61115}{-330.4} = 184.97 \text{ kg/s}$$

2 Power plant electrical output.

High-pressure turbine:
Inlet 120 C, x = 1.0, h_g = 655.01 kJ/kg, s_g = 2.0750 kJ/kg·K
Outlet 0.400 MPa, abs, $h_{2,s}$ = 579.88 kJ/kg

$$W_T = \eta_m \eta_T (h_g - h_{2,s}) = 0.98 \times 0.90(655.01 - 579.88) = 66.26 \text{ kJ/kg} \quad \text{(shaft output)}$$

Low-pressure turbine:
Inlet 85 C, x = 1.0, h_g = 627.47 kJ/kg, s_g = 2.0607 kJ/kg·K
Outlet 0.400 MPa, abs, $h_{2,s}$ = 575.28 kJ/kg

$$W_T = \eta_m \eta_T (h_g - h_{2,s}) = 0.98 \times 0.90(627.47 - 575.28) = 46.03 \text{ kJ/kg} \quad \text{(shaft output)}$$

Generator output:

$$\dot{W}_g = \eta_g[(\dot{m}W_T)_{h\text{-}p} + (\dot{m}W_T)_{l\text{-}p}] = 0.95(100 \times 66.26) + (50 \times 46.03) = 8481 \text{ kW}$$

Feed pump motor input:

$$\dot{W}_m = \dot{m}\frac{\Delta h}{\eta_m \eta_e} = 150\frac{-9.55}{0.98 \times 0.94} = -1555 \text{ kW}$$

Net power plant:

$$\dot{W}_{net} = \dot{W}_g + \dot{W}_m = 8481 - 1555 = 6926 \text{ kW}$$

Power plant thermal efficiency:

$$\eta_t = \frac{\dot{W}_{net}}{\dot{Q}} = \frac{6926}{61,115} = 0.113$$

A substantial amount of the net power is commonly required for pumping the geothermal water, thus the overall thermal efficiency for the entire geothermal power system is, in this case, less than 0.113.

3 Heat exchangers.

The temperature of the water leaving a heat exchanger is determined from an energy balance on the unit. See the following tabulation.

	Isobutane Flow Rate (kg/s)	*Isobutane*		*Water*	
		h_{out} *(kJ/kg)*	h_{in} *(kJ/kg)*	h_{out} *(kJ/kg)*	h_{in} *(kJ/kg)*
Low-temperature preheater	150	365.30 (80 C)	238.40 (33.0 C)	292.82 (70 C)	395.73 (94.5 C)
Low-pressure boiler	50	627.47 (85 C)	365.30 (80 C)	395.73 (94.5 C)	466.60 (111.3 C)
High-pressure preheater	100	487.46 (115 C)	365.30 (80 C)	466.60 (111.3 C)	532.64 (126.9 C)
High-pressure boiler	100	655.01 (120 C)	487.46 (115 C)	532.64 (126.9 C)	623.22 (148 C)

The area of the heat transfer surface is determined from

$$A_o = \frac{\dot{q}}{U_o F_c \, \Delta T_m} = \frac{\dot{m}_w(h_{in} - h_{out})_w}{U_o F_c \, \Delta T_m}$$

where

$\dot{m}_w$ = geothermal water flow rate, 184.97 kg/s
U_o = overall coefficient of heat transfer
ΔT_m = log mean temperature difference
F_c = correction factor

For the two boilers, ΔT_m is evaluated from the inlet and outlet water temperatures and the evaporating temperature of the isobutane, 120 and 85 C for the high-pressure and the low-pressure boilers, respectively.

The correction factor F_c applicable to the two preheaters is selected from charts developed by Bowman, Mueller, and Nagle, *Trans. ASME*, 1940.[41] The geothermal water flows through the preheaters in four tube-side passes, and the isobutane, in two shell-side passes. Pertinent data are given in the following tabulation.

	ΔT_m (C)	F_c	U_o (W/m²·K)	$\dot{q}$ (kJ/s)	A_o (m²)
Low-temperature preheater	24.0	0.90	2000	19035	441
Low-pressure boiler	16.5	1	3500	13109	227
High-pressure preheater	20.1	0.94	2000	12215	323
High-pressure boiler	15.1	1	3500	16755	317

Energy Storage

13.30 ENERGY STORAGE DEVICES

Because of the intermittent operation of solar and wind power generating units, a continuous output from a self-contained system can only be achieved by providing some kind of energy storage facility. A storage system or device must be economically and technically acceptable. Capital and operating costs are expected to be reasonable, and the energy conversion should be achieved with low losses and free of severe operating problems. Some other important characteristics are high energy density, long operating life, capability of receiving and delivering energy at high rates, and a capability of a large number of charge–discharge cycles, or indefinite cycling.

In addition to possible applications to solar power and wind power plants, energy storage systems have other functions, to be described in the following several sections. The methods available for storing energy can be divided generally into three classes, namely, electrochemical, mechanical, and thermal. Energy storage, in the sense used here, does not include the storage of fuel.

13.31 ELECTROCHEMICAL ENERGY STORAGE

The storage battery has been in use for many years in a variety of applications. The operating principle of the storage battery is a reversible chemical reaction. Transfer of electrical energy to the cell causes the reaction to proceed in one direction, while the reversed reaction delivers electrical energy to the external load.

Many different kinds of storage batteries have been constructed and evaluated. In the class of batteries that operate at the ambient temperature, the lead–acid battery has the widest use. Current research is directed to improving the material utilization and to increasing the cycle life of the lead–acid battery, in addition to developing other types of batteries to the point where they can be considered for commercial-scale applications.

Two other batteries that operate at ambient temperatures and have potential power plant use are the zinc–nickel oxide battery and the zinc–air battery. The energy storage capability of the zinc–nickel oxide battery is more than two times that of the lead–acid battery of the same mass; and the cycle capability is, because of recent technical improvements, essentially similar to the cycle life of the lead–acid battery in vehicle service. The zinc–air battery has a theoretical energy storage capability nearly eight times that of the lead–acid battery, but the operation of the zinc–air battery requires complex auxiliary equipment.

The open-circuit voltage of the lead–acid cell is 2.1 V. Representative values of the practical energy density and the practical power density are, respectively, 5 to 35 Wh/kg and 175 to 35 W/kg.[33] The lead–acid battery is evidently a relatively heavy energy storage device, but this battery provides reasonably good service with an acceptable operating economy and a minor loss of water from the electrolyte. The lead–acid battery may be recharged some 300 times before losing 20 percent of the capability to hold a full charge.

The discharge efficiency of the lead–acid battery may be taken as 0.75 to 0.85. For a relatively slow charging rate, a charge efficiency of 0.75 can be assumed, hence the overall conversion efficiency for the lead–acid battery is about 0.60. The conversion efficiency as applied in this manner relates only to the battery and is a ratio of the output energy to the input energy. Incidentally, the discharge efficiency varies with the discharge rate and decreases with an increase in the discharge rate.

A recently developed zinc–chlorine battery consists basically of a power section equipped with graphite plates, a storage section, a pump, and a refrigerator. See Fig. 13.20. The pump circulates a zinc chloride solution through the two sections of the battery. During the charging operation, zinc is deposited on the negative graphite plates, and chlorine gas is released at the positive plates. The chlorine gas is carried by the zinc chloride solution into the storage section where it is chilled to 9 C and forms a soft chlorine hydrate ($Cl_2 \cdot 8H_2O$) deposit.

On discharge, the zinc chloride solution is pumped into the storage section and causes some of the chlorine hydrate to liquefy and release chlorine gas. The chlorine gas, carried in the zinc chloride solution, is pumped back to the power section of the battery where it reacts electrochemically with the zinc deposited on the negative plates. The zinc atom, on loosing two electrons, becomes a positively charged ion in the

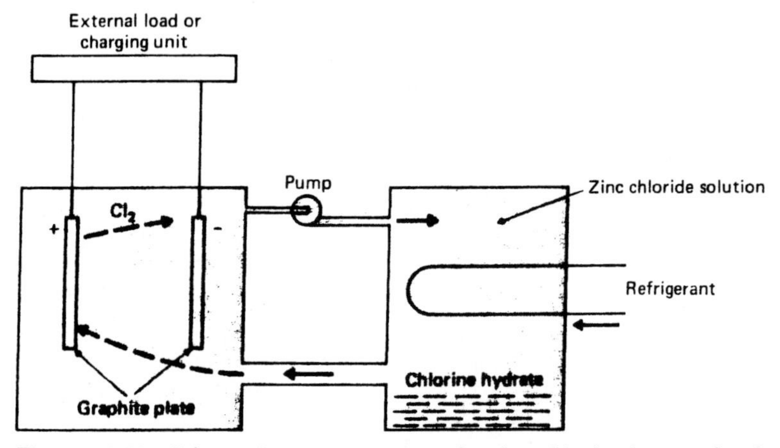

Figure 13.20 Schematic representation of a zinc–chlorine battery developed by Gulf & Western Industries.

zinc chloride solution. The electrons flow from the negative plate through the external circuit to the positive plate and thus create an electric current. The chlorine atoms in contact with the positive plate acquire the electrons and become negatively charged ions in the zinc chloride solution. The reaction is given by

$$Zn + Cl_2 \rightleftharpoons Zn^{2+} + 2Cl^-$$
$$\leftarrow \text{Charge}$$
$$\text{Discharge} \rightarrow$$

The zinc–chlorine battery is a relatively complex device, and corrosion, because of the presence of chlorine, could be a problem. The graphite plates however are strongly resistant to deterioration. A possible application of the zinc–chlorine battery is to automotive propulsion.

Batteries in the high-temperature, molten-salt electrolyte class are regarded as having good possibilities for developing batteries that characteristically have a high specific energy (200 Wh/kg) and a high specific power (50 to 200 W/kg). The open-circuit voltages are generally higher for molten-salt batteries than for aqueous batteries. Most molten salts have relatively high electrical conductivities; hence, in comparison with aqueous batteries, the internal resistance losses are lower for batteries using a molten-salt electrolyte. The high temperature at which a molten-salt battery operates improves the electrode kinetics of the cell. These batteries must necessarily operate at a temperature above the melting point of the electrolyte.

The molten-salt batteries are complex devices subject to severe corrosion. High operating temperatures aggravate the corrosion problems and require some means for start-up, that is, raising the cell temperature from the ambient to the operating level. Provision must be made to compensate for heat losses, possibly by utilizing a portion of the energy output from the cell for this purpose. The economic, and perhaps the technical, feasibility of these energy storage systems needs to be demonstrated.

In the advanced, high-temperature class of batteries, the lithium–chlorine system was the first molten-salt battery to be investigated. The reactants for this cell are lithium and chlorine gas (Li and Cl_2). A cell consists of a Li electrode, or a Li–Al alloy anode, a Cl_2 electrode, and the LiCl electrolyte. Another high-temperature battery under development utilizes a lithium–sulfur reaction.

In comparison with the lead–acid cell, the lithium–chlorine cell is a high-performance energy storage unit. The open-circuit voltage is 3.5 V, and the projected values of the energy density and the power density are, respectively, 350 to 450 Wh/kg and 200 to 400 W/kg.[33]

Some effort is currently directed to expanding the use of automotive vehicles powered by electric storage batteries. The principal incentive for this move is the conservation of petroleum fuels achieved by shifting the energy load to coal-fired and nuclear power plants. Presently, vehicles powered by lead–acid batteries travel at comparatively low speeds with a limited range of operation. Improved batteries, for example, a lithium–sulfur battery, are needed to extend the driving range of an automobile to about 240 km between battery recharges when operating at normal highway speeds.

The overall thermal efficiency for an automobile operated by a lead–acid battery is perhaps about 0.2. This figure allows for all of the losses in the energy conversion train, starting with the fuel burned in the electric generating station. A value of 0.2 for the thermal efficiency of a battery-powered automobile is somewhat below a representative value of the thermal efficiency achieved by a conventional automobile equipped with an internal combustion engine and used in a similar class of service. An economic comparison between the two different power systems should however be made on the basis of capital, energy, and maintenance costs.

The hydrogen–oxygen reaction is the basis for constructing an energy storage facility that could conceivably be used in conjunction with a solar or wind power plant. Excess electrical energy generated in one of these power plants is used to produce hydrogen by the electrolysis of water. At a later time, the hydrogen would be utilized in some kind of energy conversion device. If used to fuel a thermal engine, the conversion efficiency, from chemical to mechanical energy, would be limited to approximately 0.40.

The most effective utilization of hydrogen is likely to be achieved by generating electrical energy in a fuel cell. Application of the fuel cell to a solar power plant, for example, would represent a particularly effective combination of these two power systems, that is, from a technical point of view. In effect, an electrochemical energy storage system is created in the conversion of electrical energy to chemical energy in the hydrogen and then from the chemical energy to electrical energy. The actual system however is not reversible because of the conversion losses.

Self-containment achieved by energy storage may be a practical arrangement for certain isolated, low-capability power plants that normally operate intermittently. In general for commercial-scale power generation, wind, solar, and tidal power plants can be operated most effectively when integrated with steam and hydro turbine-generators.

13.32 MECHANICAL ENERGY STORAGE

A rotating mass, in particular a flywheel, is a reasonably good energy storage device. The flywheel energy storage system is currently being examined for application to automotive vehicles, particularly to buses. The system operation is comparatively simple. Charging is accomplished through an increase in the rotating speed of the flywheel by means of a driving motor. The vehicle is subsequently operated by utilizing the kinetic energy of the flywheel.

In order to store a significant quantity of kinetic energy, the flywheel must be capable of turning at a high speed. Stress considerations are thus important factors in flywheel design. Other critical areas in a mechanical drive system relate to the design of the bearings and the transmission that consists of a hydraulic coupling and some kind of variable-speed drive. Energy conversion losses include the electrical losses in the system, the mechanical losses in the bearings and the variable-speed drive, and the fluid losses in the hydraulic coupling. When an electric drive system is used, the system losses are primarily electrical.

A flywheel has an inherent high energy density. Flywheels constructed of high-strength steel can achieve an energy density that approaches three times the energy density of a lead–acid battery. Flywheel energy densities are presently in the range of 25 to 110 J/g. Current research with fiber composite materials is directed to improving the energy density to the vicinity of 250 J/g. A comparative value for the lead–acid battery is 65 J/g, while gas compression would achieve an energy density of about 30 J/g.[34]

The flywheel drive for a highway vehicle is not a new concept. Experimental applications to urban buses have achieved some success, notably the Oerlikon Electrogyro buses that were operated in European and African transit service for about two decades up to 1969. At certain stops along the bus route, contact was made with an electric power line in order to energize the flywheel driving motor, a generator/motor unit, and thus periodically build up the store of kinetic energy in the flywheel. When the flywheel reached a maximum speed of nearly 3000 rpm, charging was discontinued and the vehicle was free to run over any desired route up to an average distance of 1.2 km before recharging was required.

The marginal revenue operation of the Electrogyro buses contributed largely to their eventual removal from service. Some difficulty was experienced by the bus operator in controlling the electrical propulsion system, and the passengers were required to sit idly during the frequent and relatively lengthy charging periods.

When used for propelling automobiles or light commercial vehicles, the flywheel drive system should have a capability of storing a sufficient amount of kinetic energy to provide a reasonable operating range. The charging schedule is then adjusted in accordance with the prescribed availability of the vehicle. For the typical automobile however, most effective propulsion would not be achieved by a flywheel drive system because of the inability to meet the customary range requirement.

The results of various studies indicate that the hybrid flywheel/thermal engine configuration is a more appropriate power plant for the passenger automobile than the flywheel drive system. A small thermal engine of low pollution and low fuel consumption characteristics would provide an average level of energy, while a relatively small flywheel provides the peak power required for vehicle acceleration. Incidentally, because of the comparatively low energy density of a storage battery, a hybrid storage battery/thermal engine automotive power plant is believed to have greater potential than the storage battery electric drive.

The flywheel drive system must be contrived to avoid parasitic losses while the vehicle is operating or temporarily removed from service. The fluid drag that would dissipate the mechanical energy in the rotor and cause overheating of the machine is avoided by evacuating the space in which the rotor turns. A continuously running vacuum pump is used to maintain the flywheel housing at a small fraction of the atmospheric pressure.

The overall conversion efficiency of the flywheel energy storage system is likely to be somewhat below the conversion efficiency achieved by a Diesel engine-powered transit bus, but the flywheel drive provides pollution-free operation on city streets. Further, the original source of energy for the flywheel drive system can be derived from a fuel other than oil, for example, a coal or nuclear fuel. Incidentally, the flywheel drive system is conducive to utilizing automotive regenerative braking.

The flywheel energy storage system has the potential for service in utility power plants. During the time when the system load is low, the base load machines would provide energy to spin up the flywheel. Subsequently, the stored kinetic energy in the flywheel would be used for peak power generation. In an application of this kind, the flywheel energy storage unit would replace other types of peak load machines such as gas turbine engine generator sets.

13.33 THERMAL ENERGY STORAGE

A comparatively simple energy storage system can be devised by utilizing changes in the sensible energy of a substance. The temperature change in a system of this kind is not likely to be high, hence a relatively large amount of material must be supplied for charging the system. Thermal energy storage in water or rock is commonly employed in solar heating installations. The systems are easily constructed, and few operating problems are experienced.

Thermal energy storage can be accomplished by utilizing the fusion energy of a substance. The fusion energy varies widely with different substances, but individual values are substantially higher than the corresponding specific heat values. A wide variation is observed in the fusion temperature for different substances. Some fusion temperatures are very high, but certain salts melt at temperatures below 40 C. A low-melting-point salt can be used, perhaps advantageously, for thermal energy storage in a solar heating system.

Some control of the fusion temperature can be exercised by the use of eutectic salts. A eutectic salt is a combination of two, or possibly more, salts in a unique proportion that yields a melting temperature that is lower than the melting temperatures of the individual components. There may however be more than a single eutectic mixture and hence more than one eutectic temperature for a combination of certain selected salts.

Thermal energy can also be stored by means of a liquid–vapor phase change, for example, in a steam accumulator. The system, essentially a tank containing water, is charged by an inflow of steam. The water in the tank remains at a temperature below that of the incoming steam, hence the steam condenses on contact with the cooler water. The tank pressure is equal to the steam pressure; thus, the water is subcooled, although the temperature of the water increases during the charging operation.

Discharge of steam from the accumulator is effected by reducing the pressure in the outflow line to cause flashing of some of the water in the tank into steam. Continued discharge of steam from the tank requires a progressive reduction in the tank pressure. As a consequence of the pressure reduction, the temperature of the water in the tank continues to decrease. Steam accumulators are used primarily for industrial applications, but not in central power station operation.

13.34 PUMPED WATER STORAGE

The only feasible method for large-scale storage of energy achieved to date has been accomplished through pumped water storage. Basically, this method provides for conversion of electrical energy into gravitational potential energy in water and the subsequent reconversion of this potential energy into electrical energy. A single machine is used for this purpose, namely, a reversible pump/turbine coupled to a generator/motor.

A substantial portion of the electrical energy produced in the country is generated in large nuclear power plants or in large high-pressure, high-temperature fossil fuel-fired power plants. These power plants are operated most effectively under continuous high-load conditions. Unless other types of units are available for carrying the peak loads and some of the midrange loads, these large power plants would necessarily be shut down or placed on standby at night and over the weekend. The alternate periods of cooling following shutdown and warming prior to loading the unit can cause severe stress conditions in the boiler structure that lead to increased outage for maintenance of the equipment.

Peak electrical loads are usually carried by older and smaller steam units, gas turbine engine generator sets, or hydro turbine-generators. A system consisting of steam and hydro units has an inherent operating flexibility with respect to daily, weekly, and seasonal load schedules. Hydro turbine-generators are readily started and shut down without standby losses and incur no temperature-induced stresses.

Many of the electrical utility systems in the United States lack adequate sites for development of conventional hydroelectric generating capability. It may be expedient for these utility companies to consider the feasibility of resorting to pumped water storage as a means to carrying the peak loads.

A pumped water storage plant is a hydroelectric power plant that operates between two reservoirs of water maintained at different elevations. During the off-peak hours, electrical energy generated in a steam station is used to pump water from the lower body of water, say, a river or lake, to the elevated basin. Subsequently, the same machine is used to generate electrical energy during periods of peak loading on the system by the return flow of water from the upper basin to the lower body of water.

Development of the reversible pump/turbine-generator/motor was a highly important step in advancing the application of pumped water storage to the generation of electric power. The vertical-shaft machine is equipped with an impeller that resembles a conventional Francis runner. For a relatively low operating head, inclined-shaft tube-type machines capable of reversible operation may be installed. The maximum efficiency, for pumping or power generation, for a reversible pump/turbine is relatively high, on the order of 0.87. A high machine efficiency is important because losses are incurred during pumping and subsequently when generating electrical energy. In general, because of the losses in the pumping/electric power generating cycle, 3 kWh of pumping energy is required to produce 2 kWh of electrical energy.

An efficiency of 0.67 for the conversion of

electrical energy to potential energy to electrical energy is not particularly high. For the electrical energy delivered by the pumped hydro power plant, the overall thermal efficiency would be about 0.27, based on the energy of the fuel fired in the primary steam-electric power plant. The use of pumped water storage however can prove to be economical because of the improved operating schedule of the steam-electric generating units. Installation of a pumped hydro power plant in an electric power system would only be undertaken when economic justification is demonstrated through a comprehensive study of the system operating characteristics.

Installation of a pumped hydro power plant requires construction of the upper basin at a sufficiently high elevation. The topography of the area must allow the storage of an adequate quantity of water and accommodate a fairly wide fluctuation in the water level. Weekend pumping is highly important. The system load is comparatively low during this time, and a substantial quantity of water can thus be pumped to the upper basin to augment the water pumped during the nighttime hours. Pumping is consequently scheduled on a weekly cycle rather than on a daily cycle.

Prior to 1961, only four pumped hydro power projects, all relatively small, had been constructed in the United States. The largest of the four projects was the TVA Hiwassee development, which included a 59.5-MW reversible unit. During the past 18 years, there has occurred in the United States a significant expansion in the pumped hydro electric power generating capability. Not only has there been an increase in the number of units installed, but the generating capability of the individual projects is considerably greater than that of the Hiwassee machine.

A large pumped hydro power project was completed circa 1974 at Ludington, Michigan. The power plant, located on the shore of Lake Michigan, has a generating capability of 1872 MW. The 341-hectare reservoir excavated on the bluff above the lake is capable of storing more than 100 million m^3 of water. Under full-load operation, 66.2×10^6 m^3 of water can be transferred from the upper basin to the lake at a rate exceeding 125,000 m^3/min. The power equipment consists of six reversible pump/turbine-generator/motor units, each connected to the upper basin by a penstock 8.5 m in diameter and 400 m long.

Pumped water storage can on occasion be combined advantageously with the operation of conventional hydroelectric generating units. In the event that solar, wind, or tidal power plants are constructed on a commercial scale, pumped water storage may be employed as a means to compensating for the interruptable operating characteristics of these presently nonconventional power sources.

13.35 COMPRESSED AIR STORAGE

Various proposals have been made to utilize air compression as a means to storing energy. Compressed air storage would be employed on a large scale, similarly to pumped water storage, for peak electric power generation. Underground storage of compressed air is anticipated. Fabricated containers are likely to be too costly for aboveground storage of large amounts of air.

Several different types of subterranean cavities are suitable for compressed air storage, notably depleted gas and oil fields, abandoned mines, dissolved-out salt caverns, aquifers, and mined rock caverns. In this group, the salt and rock caverns are constructed, respectively, by solution mining and hard-rock mining, and such caverns can be located in the vicinity of the load center, provided that the geologic formation is suitable for construction of the cavity. The use of depleted gas and oil fields, abandoned mines, and aquifers will be influenced to some degree by their proximity to the power grid.

A mined rock cavern can be water compensated by an arrangement that connects the cavern with a surface reservoir. The water in the cavern is displaced by the inflow of compressed air. If the surface reservoir has a sufficiently large capacity, the water will maintain an essentially constant pressure on the air. A noncompensated cavern functions as a constant-volume storage vessel, hence the cavern pressure will cycle with the alternate flow of air into and from the cavern.

An aquifer is used virtually in the same manner as a mined rock cavern. The compressed air entering the aquifer displaces the water that normally fills the pore spaces. When air is discharged from the aquifer, the water migrates through the pore structure toward the well bore. Because the flow of water is impeded to some degree, depending upon the air withdrawal rate, the aquifer does not function exactly as a water-compensated mined rock cavern. Thus, under normal operating conditions, a small amount of air expansion will occur in the aquifer.

A gas turbine engine is used in conjunction with compressed air storage. The basic system consists of a combustion chamber and a generator/motor connected at one end of the machine to a compressor and at the other end to a turbine. Clutches are used to disengage the compressor and the turbine individually from the generator/motor. If desired, the gas turbine engine can be operated as a conventional power plant independently of the compressed air storage reservoir.

During the off-peak hours, the generator/motor, energized by the steam-electric units, drives the compressor, and air is delivered to the storage reservoir. Peak power is produced by using the turbine to drive the generator/motor with the compressor disengaged. Compressed air is withdrawn from the storage reservoir while the turbine is operating. Because the turbine does not drive the compressor, the full output of the turbine can be used to generate electric power.

An unfired compressor-turbine and air storage system would operate in the same manner as a pumped water storage system. Both of these systems would generate about 0.70 kWh of peak energy for each kilowatt hour of off-peak energy diverted to storage. Because of the fuel burned in the gas turbine engine, the compressed air energy storage system produces, depending upon the turbine inlet temperature, 1.3 to 1.7 kWh of peak energy for each kilowatt hour of energy that is transferred to the motor-driven compressor during off-peak periods. An energy ratio of 1.5 is achieved with a turbine inlet temperature of 950 C. Air storage is accomplished in a water-compensated reservoir.

Compressed air storage may prove to be superior to pumped water storage for some installations. Mined rock caverns and naturally occurring aquifers are particularly suitable for compressed air storage, with the mined rock cavern providing a wide choice in the geographic location. Water compensation is important in order to prevent a decrease in the turbine inlet pressure with the discharge of air from the reservoir. A reduction in the turbine inlet pressure causes a decrease in the operating pressure ratio of the machine and a consequent loss of power.

Example 13.15

A compressed-air energy storage system consists primarily of a gas turbine engine and a water-compensated air reservoir. The constant air pressure in the reservoir is 500 kPa, abs. The pressure loss in the pipeline, that runs between the engine and the reservoir, is 20 kPa. The turbine inlet temperature is 1095 C. The fuel is liquid dodecane; the lower heating value is 44,085 kJ/kg, and $\overline{m}$ = 170.34. The combustion efficiency is 0.97. Efficiency values are: compressor η_C = 0.85, turbine η_T = 0.90, generator/motor $\eta_{G/M}$ = 0.95, and mechanical η_m = 0.98. The turbine exhaust pressure is 122 kPa, abs. The temperature of the air in the reservoir is 15 C. The atmospheric conditions are 100 kPa, abs and 25 C. Determine the work ratio for the energy storage system.

Note: Thermodynamic properties of air and the combustion products are selected from Tables A 1 and A.2.

Compressor

P_1 = 100 kPa, abs $\qquad T_1$ = 298 K

P_2 = 520 kPa, abs

k_{av} = 1.393

$$T_{2,s} = T_1\left(\frac{P_2}{P_1}\right)^{\frac{k-1}{k}}$$

$$= 298\left(\frac{520}{100}\right)^{0.2821} = 474 \text{ K}$$

h_1 = 298.56 kJ/kg $\qquad h_{2,s}$ = 476.77 kJ/kg

Motor input:

$$W_M = \frac{h_1 - h_{2,s}}{\eta_M \eta_m \eta_C} = \frac{298.56 - 476.77}{0.95 \times 0.98 \times 0.85} = -225.20 \text{ kJ/kg air}$$

Combustion Chamber

Assume theoretical combustion of the fuel occurs with 230 percent of theoretical air. Then for the combustion gas

$$\bar{h}_p \text{ at } 1368 \text{ K} = 44489.8 \text{ kJ/kgmol}$$
$$\bar{h}'_p \text{ at } 25 \text{ C} = 8769.0 \text{ kJ/kgmol}$$

For air:

$$\bar{h}_a \text{ at } 288 \text{ K} = 8353.8 \text{ kJ/kgmol}$$
$$\bar{h}'_a \text{ at } 25 \text{ C} = 8648.5 \text{ kJ/kgmol}$$

For fuel:

$$\bar{h}_f - \bar{h}'_f = 0 \qquad (T_f = T_{atm} = 25 \text{ C})$$
$$N_p - N_a = 6.5 \qquad \text{(See Example 2.4.)}$$

Applying Eq. 7.7 for theoretical combustion

$$-\bar{h}^\circ_{RP} = N_p(\bar{h}_p - \bar{h}'_p) - N_a(\bar{h}_a - \bar{h}'_a)$$
$$-[170.34(-44085)] = (N_a + 6.5)(44489.8 - 8769.0) - N_a(8353.8 - 8648.5)$$
$$N_a = 202.1$$

For combustion with the stoichiometric air-fuel ratio, $N_a = 88.06$. Combustion thus occurs with 229.5 percent of theoretical air, and the selected values for $\bar{h}_p$ are acceptable. Then

$$(f/a)_{theo} = \frac{\bar{m}_f}{N_a \bar{m}_a} = \frac{170.34}{202.1 \times 28.97} = 0.02909$$

Actual fuel/air ratio:

$$(f/a)_{actual} = \frac{(f/a)_{theo}}{\eta_B} = \frac{0.02909}{0.97} = 0.02999$$

Turbine

$$P_3 = 480 \text{ kPa, abs} \qquad T_3 = 1368 \text{ K}$$
$$P_4 = 122 \text{ kPa, abs}$$
$$k_{av} = 1.304$$
$$T_{4,s} = T_3\left(\frac{P_4}{P_3}\right)^{(k-1)/k} = 1368\left(\frac{122}{480}\right)^{0.2331} = 994 \text{ K}$$
$$h_3 = 1539.0 \text{ kJ/kg} \qquad h_{4,s} = 1076.5 \text{ kJ/kg}$$

Generator output:

$$W_G = \eta_G \eta_m \eta_T (h_3 - h_{4,s}) = 0.95 \times 0.98 \times 0.90(1539.0 - 1076.5) = 387.53 \text{ kJ/kg of gas}$$
$$W'_G = (1 + f/a)W_G = (1 + 0.02999)387.53 = 399.15 \text{ kJ/kg air}$$

System Work Ratio—Generator Output/Motor Input

$$\frac{W'_G}{W_M} = \frac{399.15}{225.20} = 1.772$$

PROBLEMS

13.1 During a 30-day period, water flows into the reservoir of a hydroelectric power plant at an average rate of 1450 m^3/s. Because of "drawing down" the reservoir, the head on the water turbine is reduced during this period from 70 m to 60 m. The mean area of the basin for this 10-m decrement is 75 km^2. The turbine-generator machine efficiency is 0.91. The fluid head loss in the conduits is 3 m, and the discharge velocity of the water from the turbine is 2.75 m/s. Calculate the turbine-generator power output (MW) for constant-load operation.

13.2 An OTE power plant operates with an evaporator temperature of 20 C and a condenser temperature of 10 C. Warm ocean water is supplied to the flash-type evaporator at 25 C. The evaporator produces dry, saturated vapor at the rate of 10^6 kg/h. The turbine machine efficiency is 0.77, and the generator efficiency is 0.96. Determine the rate at which the warm ocean water flows to the evaporator and calculate the generator power output.

13.3 A solar energy collector is located at latitude 42°N with the energy absorbing surface inclined at an angle of 50° from the horizontal plane. The solar declination is +15°. The angle γ is 10°, measured eastward from the south meridian. Determine the value of $\cos \theta_i$ at 2.5 hours after solar noon.

13.4 A solar energy collector is located at latitude 40°N, and the energy absorbing surface is tilted toward the south at an angle of 50° from the horizontal plane. The collector area is 1 m^2. Estimate the rate at which solar radiation is intercepted by the collector at noon on a clear day at the time of (a) the winter solstice and (b) the summer solstice.

13.5 A wind turbine-generator is designed to produce 3.5×10^6 kWh per year with a plant factor of 0.40. The design wind velocity is 45 km/h. The conversion efficiency is 0.73, and the mechanical and generator efficiencies are, respectively, 0.90 and 0.96. The ambient conditions are P = 100 kPa, abs, and T = 25 C. Calculate the required rotor diameter and the power coefficient.

13.6 A proposed site for a tidal power plant has an average tidal range of 8.5 m. The prescribed annual average power output of the plant is 250 MW. The power plant operating effectiveness is 0.10. Estimate the mean area of the tidal basin required to produce the prescribed power output.

13.7 An OTE power plant operates with warm surface water at 28 C and cold water at 5 C supplied to the condenser. The working fluid for the Rankine cycle power system is ammonia. The evaporating temperature of the ammonia is 5 C below the temperature of the warm water, and the condensing temperature is 5 C above the temperature of the cold water. The turbine machine efficiency is 0.80, and the generator efficiency is 0.95. Ten percent of the generator output is used for pumping the warm and cold water. The plant output is 50 MW. Calculate the ammonia circulation rate, in kg/h, the rate at which heat is supplied, in kJ/h, and the overall power plant thermal efficiency.

13.8 A solar power plant located in the southwestern region of the United States has a conversion efficiency of 0.70 and a steam plant thermal efficiency of 0.30. The average power plant output is 5 MW(e). Estimate the required area of the solar collector.

13.9 A solar energy collector installed on a building located at latitude 44°N has an area of 70 m^2. The collector is tilted toward the south at an angle of 54° with the horizontal plane. On February 15, the sky is clear and the sun sets at 5:13 P.M. The solar declination is −13°. The average heat loss from the building for a 24-hour period is 30,000 kJ/h. The conversion factor for the solar heating system is taken as 0.45. Calculate for the specified day the fraction of the heating load carried by the solar energy collector.

13.10 Assume that the solar collector in Problem 13.9 is arranged to track the sun continuously. Determine the probable improvement in the fraction of the heating load carried by the solar energy collector. Discuss the feasibility of tracking the sun in an installation of this kind.

13.11 A wind turbine-generator is designed to attain the full-load output with a wind velocity of 48 km/h. The rotor diameter is 50 m. The combined mechanical and generator efficiency is 0.85. The ambient air conditions are 100 kPa, abs, and 22 C. Calculate the power plant rated load.

13.12 A geothermal power plant has an output of 50 MW(e). Hot water is pumped to the surface and enters a flash chamber at 240 C. The flashed steam enters the turbine at 800 kPa, abs, and with a liquid content of 0.5 percent. The turbine exhaust pressure is 15 kPa, abs. The turbine machine efficiency is 0.88, and the generator efficiency is 0.95. Calculate the rate, in kg/h, at which the hot water flows from the well.

13.13 A suggested location for a flat-plate solar collector is in a vertical position on the south-facing building wall. Assume that the collector for the building described in Problem 13.9 is installed on a south-facing wall. Determine the required area of the collector that will have the same solar energy input per day as the roof-mounted collector.

13.14 A municipal transit bus operating on a level roadway encounters a resisting force of 4000 N. A flywheel drive system is used for propelling the bus. The efficiency of the electric drive is 0.90. The flywheel has a diameter of 1.60 m and a mass of 1500 kg. The moment of inertia of the flywheel about the axis of rotation equals $0.65mr^2$. Charging is terminated when the flywheel speed reaches 3000 rpm. A reduction of 10 percent of the maximum speed is prescribed before the flywheel is recharged. How far can the bus run between recharging stops?

13.15 An electric vehicle has a mass of 955 kg, including 455 kg for the lead–acid battery. The energy density of the battery is 27 Wh/kg. When operating at a speed of 65 km/h, the road and air resistance is 420 N. The overall operating efficiency of the vehicle is 0.77. How far can the vehicle travel at the indicated speed on a single battery charge?

13.16 A 500-MW steam turbine-generator operates continuously at the rated output. The load factor on the unit because of the utility demand is 0.80 Monday to Friday and 0.45 over the weekend. The excess energy is utilized in pumped storage. The pumped storage energy is used to produce peaking power for a 4-hour period during each day, Monday to Friday. Determine the peaking power.

13.17 A geothermal power plant is constructed for total flow expansion of the hot water. At the turbine inlet, the pressure and temperature of the water are 3.0 MPa, abs, and 227 C (500 K). The turbine exhaust pressure is 108.2 kPa, abs. The turbine efficiencies are: nozzle 0.85, wheel 0.71, and mechanical 0.985. Calculate the water flow for a turbine output of 1000 kW.

13.18 A binary-fluid power system is supplied with geothermal water at 575 K (302 C). Isobutane is evaporated at 390 K in the heat exchanger. Subsequent to expansion in the turbine, the isobutane condenses at 400 kPa, abs. The water leaves the heat exchanger at 400 K. The machine efficiencies are: turbine 0.91, pump 0.65. The mechanical efficiency is 0.985, and the motor and generator efficiencies are 0.95. The hot water flow rate is 50 kg/s. The overall heat transfer coefficient for the evaporator is 3250 W/m^2·K. Calculate the thermal efficiency for the isobutane power plant, the power plant output, and the heat exchanger area.

13.19 A compressed air energy storage system operates at a constant air pressure of 600 kPa, abs, in the water-compensated reservoir. A pressure drop of 25 kPa is incurred in the pipeline that connects the gas turbine engine to the reservoir. The turbine inlet temperature is 1000 C, and the turbine exhaust pressure is 125 kPa, abs. The fuel is liquid $C_{11}H_{24}$. The temperature of the air in the reservoir is 17 C. The atmospheric conditions are 98 kPa, abs, and 25 C. Efficiency values can be taken from Example 13.14. Calculate the ratio of the generator output to the motor input for the energy storage system.

13.20 A south-facing solar panel is inclined at an angle of 55° from the horizontal plane. The latitude is 45°N, and the solar declination is −10°. At 2 P.M. solar time, the normal beam radiation is 550 W/m^2, and the diffuse radiation from the sky is 130 W/m^2. The ground reflectance is 0.2. Calculate the total radiation received by the panel.

REFERENCES

1 *Mechanical Engineering*, Aug. 1976, page 51.

2 Walters, S. "Power in the Year 2001—Part 2—Thermal Sea Power," *Mechanical Engineering*, Oct. 1971.

3 Anderson, J. H. and Anderson, J. H., Jr. "Thermal Power from Seawater," *Mechanical Engineering*, April 1966.

4 "Ocean Energy Resources," *Mechanical Engineering*, July 1975, page 73.

5 Ting, H.-T. "A Possible Breakthrough of Exploiting Thermal Power from Sea," *Combustion*, August 1970.

6 Yellott, J. I. "Solar Energy Progress—A World Picture," *Mechanical Engineering*, July 1970.

7 Yellott, J. I. "Power from Solar Energy—Some Fundamental Factors," ASME Paper No. 56-F-15, New York, Sept. 1956.

8 "Is the Sun our Lucky Star?" *Exxon USA*, First Quarter, 1976.

9 Yellott, J. I. "Energy from the Sun," *Power Engineering*, Feb. 1957.

10 Daniels, F. "Direct Use of the Sun's Energy," *American Scientist*, Jan.–Feb. 1967.

11 Kreith, F. *Principles of Heat Transfer*, 3rd ed., Intext Educational Publishers, New York, 1973.

12 Boes, E. C., Hall, I. J., Prairie, R. R., Stromberg, R. P., and Anderson, H. E. "Distribution of Direct and Total Solar Radiation Availabilities for the USA," Proceedings of the Meeting of the International Solar Energy Society held in Winnipeg, Manitoba, Canada, Aug. 15–20, 1976.

13 "Solar Age Nudged Forward," *Mechanical Engineering*, June 1976, page 52.

14 Walters, S. *Mechanical Engineering*, Feb. 1976, page 41.

15 Tester, J. W., Mayer, R. M., and Fraas, A. P. "Comparative Performance Characteristics of Cylindrical Parabolic and Flat Plate Solar Energy Collectors," ASME Paper No. 74-WA/Ener-3, Nov. 1974.

16 *Mechanical Engineering*, March 1979, page 50.

17 "Smith-Putnam Wind Turbine," *Mechanical Engineering*, June 1941, pages 473–474.

18 Black, T. W. "Megawatts from the Wind," *Power Engineering*, March 1976.

19 Johnson, C. C., Smith, R. T., and Swanson, R. K. "Wind Power Development and Applications," *Power Engineering*, Oct. 1974.

20 Hirschfeld, F. "Wind Power—Pipe Dream or Reality?" *Mechanical Engineering*, Sept. 1977.

21 Vann, H. E. "Energy Options," *Combustion*, Aug. 1976.

22 "Practical Geothermal Pumping System," *Combustion*, May 1975.

23 Ellis, A. J. "Geothermal Systems and Power Development," *American Scientist*, Sept.–Oct. 1975.

24 Walters, S. *Mechanical Engineering*, Sept. 1978, page 47.

25 Roberts, V. W. "Geothermal Energy," *Mechanical Engineering*, Nov. 1977.

26 Schuster, R. "Turning Turbines with Geothermal Steam," *Power Engineering*, March 1972.

27 Bruce, A. W. and Albritton, B. C. "Generation of Power from Geothermal Steam at The Geysers Power Plant," Presented at the ASCE Spring Convention, Los Angeles, Calif., Feb. 1959.

28 "Geothermal Dig," *Mechanical Engineering*, June 1975.

29 *Mechanical Engineering*, Nov. 1977, page 60; Jan. 1978, page 55.

30 *Mechanical Engineering*, April 1979, page 48.

31 Austin, A. L. and Lundberg, A. W. "A Comparison of the Methods for Electric Power Generation from Geothermal Hot Water Deposits," ASME Paper No. 74-WA/Ener-10, Nov. 1974.

32 Aronson, D. *Mechanical Engineering*, May 1976, page 39.

33 Hietbrink, E. H. and Tricklebank, S. B. "Electric Storage Batteries for Vehicle Propulsion," ASME Paper No. 70-WA/Ener-7, Nov.–Dec. 1970.

34 Lawson, J. L. "Kinetic Energy Storage for Mass Transportation," *Mechanical Engineering*, Sept. 1974.

35 Chang, G. C. and Hirschfeld, F. "For the Latest in Energy Storage, Try the Flywheel!" *Mechanical Engineering*, Feb. 1978.

36 Ayers, D. L. and Strong, R. E. "Compressed Air Storage—Another Answer to the Peaking Problem," *Power Engineering*, Aug. 1975.

37 Olsson, E. K. A. "Air Storage Power Plant," *Mechanical Engineering*, Nov. 1970.

38 Reynolds, W. C. *Thermodynamic Properties in SI*, Stanford University, 1979.

39 Wiebelt, J. A. *Engineering Radiation Heat Transfer*, Holt, Rinehart and Winston, New York, 1966.

40 Potter, R. W., II "Geothermal Energy: An Assessment," *Mechanical Engineering*, May 1981.

41 Bowman, R. A., Mueller, A. C., and Nagel, W. M. "Mean Temperature Difference in Design," *Trans. ASME*, 1940.

BIBLIOGRAPHY

Blair, P. D., Cassel, T. A. V., and Edelstein, R. H. *Geothermal Energy: Investment Decisions and Commercial Development*, Wiley–Interscience, New York, 1982.

Duffie, J. A. and Beckman, W. A. *Solar Engineering and Thermal Processes*, John Wiley & Sons, New York, 1980.

Eldridge, F. R. *Wind Machines*, U.S. Government Printing Office, Washington, D.C., 1976.

Kreider, J. F. and Kreith, F. *Solar Heating and Cooling: Active and Passive Design*, Hemisphere Publishing Corporation, Washington, D.C., 1982.

Lunde, P. J. *Solar Thermal Engineering*, John Wiley & Sons, New York, 1980.

Mazria, E. *The Passive Solar Energy Book*, Rodale Press, Emmaus, Pa., 1979.

Milora, S. L. and Tester, J. W. *Geothermal Energy as a Source of Electric Power*, M.I.T. Press, Cambridge, Mass., 1976.

Schmidt, F. W. and Willmott, A. J. *Thermal Energy Storage and Regeneration*, Hemisphere Publishing Corporation, Washington, D.C., 1981.

Schubert, R. C. and Ryan, L. D. *Fundamentals of Solar Heating*, Prentice-Hall, Inc., Englewood Cliffs, N.J., 1981.

Williams, J. R. *Solar Energy: Technology and Applications*, Ann Arbor Science Publishers, Ann Arbor, Mich., 1974.

APPENDIX

FIGURES

TABLES

CHART

NOMENCLATURE

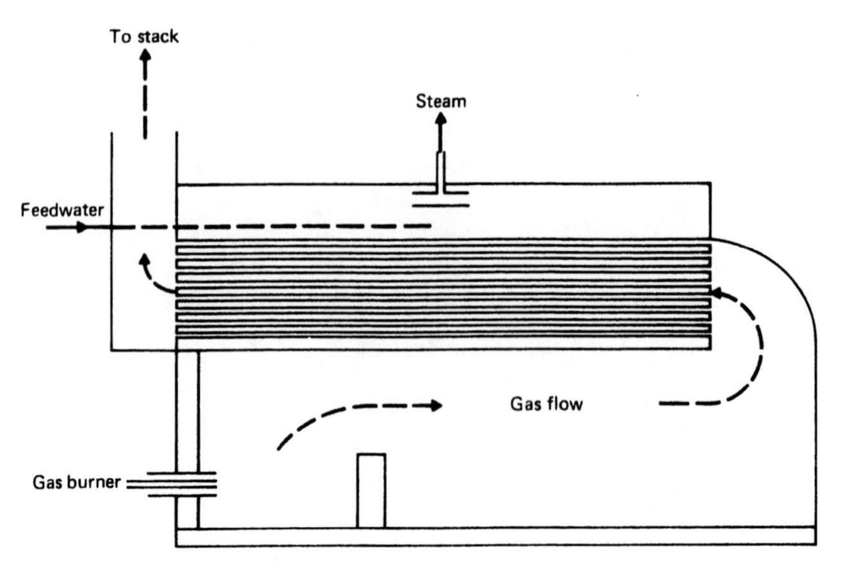

Figure A.1 Schematic illustration of a horizontal, return fire tube boiler.

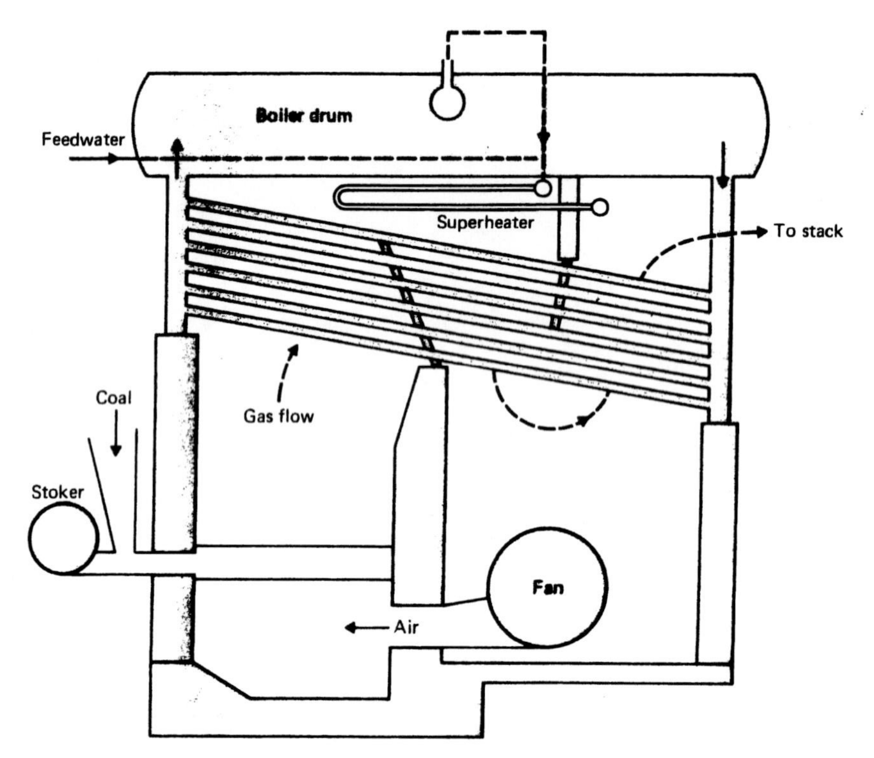

Figure A.2 Schematic illustration of a longitudinal-drum, water tube boiler.

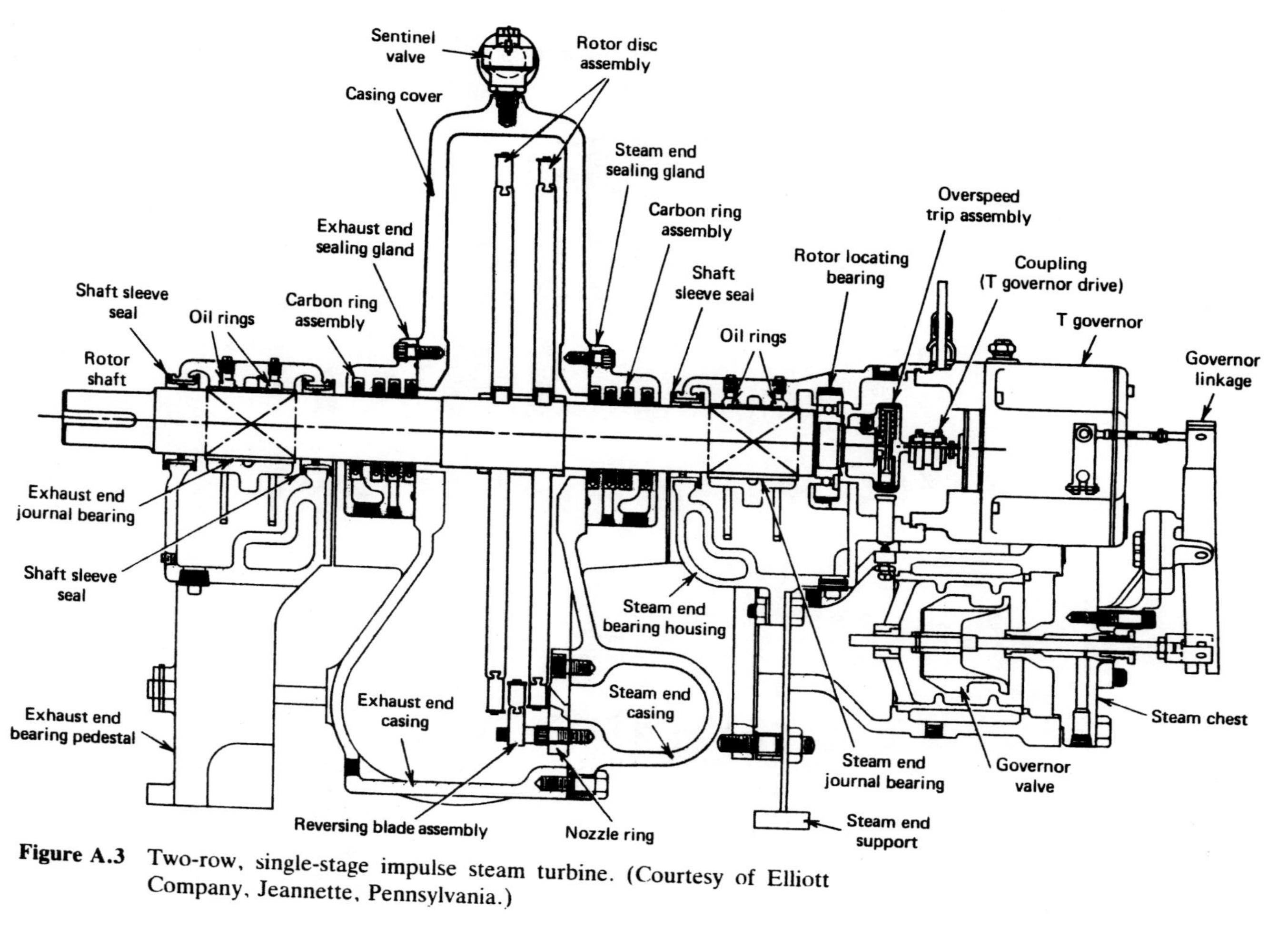

Figure A.3 Two-row, single-stage impulse steam turbine. (Courtesy of Elliott Company, Jeannette, Pennsylvania.)

TABLE A.1 Specific Heat Ratio *k* for Air and Combustion Gases at Low Pressures

		Products of Combustion—Fuel $(CH_2)_n$		
T (*K*)	*Air*	*100 Percent Air*[a]	*200 Percent Air*[a]	*400 Percent Air*[a]
200	1.401	1.383	1.392	1.397
300	1.400	1.372	1.384	1.392
400	1.395	1.360	1.376	1.385
500	1.387	1.348	1.366	1.376
600	1.376	1.335	1.354	1.364
700	1.364	1.324	1.342	1.353
800	1.354	1.313	1.332	1.342
900	1.344	1.303	1.322	1.332
1000	1.336	1.295	1.313	1.324
1100	1.330	1.288	1.307	1.318
1200	1.324	1.281	1.300	1.311
1300	1.319	1.276	1.295	1.306
1400	1.315	1.272	1.291	1.302
1500	1.311	1.268	1.287	1.298
1600	1.308	1.265	1.283	1.295
1700	1.305	1.261	1.281	1.292
1800	1.303	1.259	1.278	1.290
1900	1.301	1.257	1.276	1.288
2000	1.298	1.254	1.274	1.285

[a] *Percent theoretical air.*

Source: Tables 2, 4, 6, and 8 in *Gas Tables*, by Joseph H. Keenan, Jing Chao, and Joseph Kaye. Copyright 1980 by Estate of Joseph H. Keenan and Jing Chao. Reprinted by permission of John Wiley & Sons, Inc., New York, 1980.

TABLE A.2 Enthalpy Values for Air and Combustion Gases at Low Pressures (kJ/kgmol)

		Products of Combustion[b]		
T (*K*)	*Air*	*100 Percent Air*[a]	*200 Percent Air*[a]	*400 Percent Air*[a]
200	5797.1	5921.8	5861.5	5829.9
298.15	8648.5	8899.7	8778.6	8714.8
300	8702.4	8956.5	8833.9	8769.3
400	11623.2	12061.0	11849.3	11738.2
500	14579.0	15240.6	14921.1	14753.1
600	17590.8	18505.0	18063.3	17830.9
700	20667.9	21859.5	21284.1	20981.5
800	23815.1	25305.3	24585.4	24206.9
900	27029.0	28838.7	27964.3	27504.7
1000	30304.2	32453.7	31415.2	30869.0
1100	33633.9	36142.8	34930.7	34293.4
1200	37011.5	39899.3	38504.4	37771.0
1300	40432.3	43715.8	42129.7	41295.6
1400	43890.1	47585.3	45800.1	44861.6
1500	47380.9	51502.3	49511.2	48464.3
1600	50900.0	55461.6	53258.0	52099.4
1700	54445.3	59458.8		55763.5
1800	58014.3	63489.1		59453.5
1900	61604.2	67549.4		63166.9
2000	65212.9	71636.8		
$\overline{m}$ (kg/kgmol)	28.9669	28.9072	28.9360	28.9512
R (kJ/kg·K)	0.28703	0.28762	0.28734	0.28719

[a] *Percent theoretical air.*

[b] *Products of combustion are for a hydrocarbon fuel of composition* $(CH_2)_n$.

Source: Adapted from Tables 1, 3, 5, and 7 in *Gas Tables*, by Joseph H. Keenan, Jing Chao, and Joseph Kaye. Copyright 1980 by Estate of Joseph H. Keenan and Jing Chao. Reprinted by permission of John Wiley & Sons, Inc., New York, 1980.

TABLE A.3 Enthalpy of Combustion At 1 Atmosphere Pressure, 25 C

Compound	Formula	$\bar{m}$	Phase	H_2O Liquid, CO_2 Gas (kJ/kg)	H_2O Gas, CO_2 Gas (kJ/kg)
Hydrogen	H_2	2.02	Gas	−141,693	−119,874
Carbon	C	12.01	Solid (graphite)		−32,744
Carbon monoxide	CO	28.01	Gas		−10,097
Methane	CH_4	16.04	Gas	−55,464	−49,980
Ethane	C_2H_6	30.07	Gas	−51,844	−47,456
Propane	C_3H_8	44.10	Gas	−50,315	−46,324
Propane	C_3H_8	44.10	Liquid	−49,952	−45,964
n-Butane	C_4H_{10}	58.12	Gas	−49,471	−45,687
n-Butane	C_4H_{10}	58.12	Liquid	−49,102	−45,317
n-Pentane	C_5H_{12}	72.15	Gas	−48,981	−45,324
n-Pentane	C_5H_{12}	72.15	Liquid	−48,613	−44,955
n-Hexane	C_6H_{14}	86.18	Gas	−48,646	−45,073
n-Hexane	C_6H_{14}	86.18	Liquid	−48,281	−44,706
n-Heptane	C_7H_{16}	100.21	Gas	−48,407	−44,894
n-Heptane	C_7H_{16}	100.21	Liquid	−48,042	−44,529
n-Octane	C_8H_{18}	114.23	Gas	−48,225	−44,760
n-Octane	C_8H_{18}	114.23	Liquid	−47,863	−44,397
2,2,4-Trimethylpentane	C_8H_{18}	114.23	Gas	−48,088	−44,622
2,2,4-Trimethylpentane	C_8H_{18}	114.23	Liquid	−47,781	−44,316
n-Decane	$C_{10}H_{22}$	142.29	Gas	−47,972	−44,571
n-Decane	$C_{10}H_{22}$	142.29	Liquid	−47,612	−44,211
n-Dodecane	$C_{12}H_{26}$	170.34	Gas	−47,800	−44,443
n-Dodecane	$C_{12}H_{26}$	170.34	Liquid	−47,442	−44,085
Ethene (ethylene)	C_2H_4	28.05	Gas	−50,266	−47,130
Propene (propylene)	C_3H_6	42.08	Gas	−48,888	−45,752
Ethyne (acetylene)	C_2H_2	26.04	Gas	−49,883	−48,195
Benzene	C_6H_6	78.12	Gas	−42,240	−40,552
Benzene	C_6H_6	78.12	Liquid	−41,807	−40,118
Methanol	CH_3OH	32.04	Gas	−23,867	−21,122
Methanol	CH_3OH	32.04	Liquid	−22,698	−19,953
Ethanol	C_2H_5OH	46.07	Gas	−30,626	−27,762
Ethanol	C_2H_5OH	46.07	Liquid	−29,706	−26,842
Propanol	C_3H_7OH	60.10	Gas	−34,360	−31,431
Propanol	C_3H_7OH	60.10	Liquid	−33,588	−30,659
Butanol	C_4H_9OH	74.12	Gas	−36,580	−33,612
Butanol	C_4H_9OH	74.12	Liquid	−35,913	−32,945

Source: for hydrocarbons: "Selected Values of Properties of Hydrocarbons," National Bureau of Standards Circular No. C461, Washington, D.C., 1947.

TABLE A.4 Typical Higher Heating Values for Some Commercial Fuels[a]

Coal	In kJ/kg, as Fired
Anthracite	30,450
Semianthracite	31,600
Low-volatile bituminous	32,870
Medium-volatile bituminous	33,240
High-volatile A bituminous	30,900
High-volatile B bituminous	28,600
High-volatile C bituminous	26,300
Subbituminous A	24,900
Subbituminous B	22,300
Subbituminous C	20,000
Lignite	17,000
Petroleum fuels	*In kJ/kg*
Gasoline	47,900
Kerosene	45,560
Diesel oil	45,330
Fuel oil	43,700
Gaseous fuels	*In kJ/m³, at 1 Atm, 25 C*
Natural gas	39,500
Blast-furnace gas	3,600
Refinery gas	53,900
Coke-oven gas	19,800

[a] *Heating values listed in this table should be taken as selected but not as average values.*

TABLE A.5 Ideal Gas Enthalpy and Absolute Entropy at 101.325 kPa, abs, Pressure

	Carbon Dioxide (CO_2) $\bar{m}$ = 44.011 $(\bar{h}_f^\circ)_{298}$ = −393,522 kJ/kgmol		Carbon Monoxide (CO) $\bar{m}$ = 28.011 $(\bar{h}_f^\circ)_{298}$ = −110,529 kJ/kgmol	
Temp. (K)	$\bar{h}^\circ - \bar{h}_{298}^\circ$ (kJ/kgmol)	$\bar{s}^\circ$ (kJ/kgmol·K)	$\bar{h}^\circ - \bar{h}_{298}^\circ$ (kJ/kgmol)	$\bar{s}^\circ$ (kJ/kgmol·K)
0	−9,364	0	−8,669	0
100	−6,456	178.899	−5,770	165.741
200	−3,414	199.865	−2,858	185.916
298	0	213.685	0	197.543
300	67	213.915	54	197.723
400	4,008	225.225	2,975	206.125
500	8,314	234.814	5,929	212.719
600	12,916	243.199	8,941	218.204
700	17,761	250.663	12,021	222.953
800	22,815	257.408	15,175	227.162
900	28,041	263.559	18,397	230.957
1000	33,405	269.215	21,686	234.421

TABLE A.5 Continued

	Carbon Dioxide (CO_2) $\bar{m}$ = 44.011 $(\bar{h}_f^\circ)_{298}$ = −393,522 kJ/kgmol		Carbon Monoxide (CO) $\bar{m}$ = 28.011 $(\bar{h}_f^\circ)_{298}$ = −110,529 kJ/kgmol	
Temp. (K)	$\bar{h}^\circ - \bar{h}_{298}^\circ$ (kJ/kgmol)	$\bar{s}^\circ$ (kJ/kgmol·K)	$\bar{h}^\circ - \bar{h}_{298}^\circ$ (kJ/kgmol)	$\bar{s}^\circ$ (kJ/kgmol·K)
1100	38,894	274.445	25,033	237.609
1200	44,484	279.307	28,426	240.563
1300	50,158	283.847	31,865	243.316
1400	55,907	288.106	35,338	245.889
1500	61,714	292.114	38,848	248.312
1600	67,580	295.901	42,384	250.592
1700	73,492	299.482	45,940	252.751
1800	79,442	302.884	49,522	254.797
1900	85,429	306.122	53,124	256.743
2000	91,450	309.210	56,739	258.600
2100	97,500	312.160	60,375	260.370
2200	103,575	314.988	64,019	262.065
2300	109,671	317.695	67,676	263.692
2400	115,788	320.302	71,346	265.253
2500	121,926	322.808	75,023	266.755
2600	128,085	325.222	78,714	268.203
2700	134,256	327.549	82,408	269.596
2800	140,444	329.800	86,115	270.943
2900	146,645	331.975	89,826	272.249
3000	152,862	334.084	93,542	273.508
3200	165,331	338.109	100,998	275.914
3400	177,849	341.904	108,479	278.182
3600	190,405	345.490	115,976	280.324
3800	202,999	348.895	123,495	282.357
4000	215,635	352.134	131,026	284.286
4200	228,304	355.226	138,578	286.131
4400	241,003	358.180	146,147	287.888
4600	253,734	361.012	153,724	289.575
4800	266,500	363.728	161,322	291.190
5000	279,295	366.338	168,929	292.742
5200	292,123	368.853	176,548	294.240
5400	304,984	371.280	184,184	295.679
5600	317,884	373.627	191,832	297.068
5800	330,821	375.895	199,489	298.411
6000	343,791	378.096	207,162	299.712

TABLE A.5 Continued

	Oxygen (O_2) $\overline{m}$ = 31.999 $(\bar{h}_f^\circ)_{298}$ = 0 kJ/kgmol		Nitrogen (N_2) $\overline{m}$ = 28.013 $(\bar{h}_f^\circ)_{298}$ = 0 kJ/kgmol	
Temp. (K)	$\bar{h}^\circ - \bar{h}_{298}^\circ$ (kJ/kgmol)	$\bar{s}^\circ$ (kJ/kgmol·K)	$\bar{h}^\circ - \bar{h}_{298}^\circ$ (kJ/kgmol)	$\bar{s}^\circ$ (kJ/kgmol·K)
0	−8,682	0	−8,669	0
100	−5,778	173.197	−5,770	159.703
200	−2,866	193.376	−2,858	179.879
298	0	205.033	0	191.502
300	54	205.213	54	191.682
400	3,029	213.765	2,971	200.071
500	6,088	220.589	5,912	206.631
600	9,247	226.346	8,891	212.066
700	12,502	231.363	11,937	216.756
800	15,841	235.814	15,046	220.907
900	19,246	239.827	18,221	224.647
1000	22,707	243.475	21,460	228.057
1100	26,217	246.818	24,757	231.199
1200	29,765	249.906	28,108	234.116
1300	33,351	252.776	31,501	236.831
1400	36,966	255.454	34,936	239.375
1500	40,610	257.969	38,405	241.768
1600	44,279	260.337	41,903	244.028
1700	47,970	262.575	45,430	246.166
1800	51,689	264.701	48,982	248.195
1900	55,434	266.726	52,551	250.128
2000	59,199	268.655	56,141	251.969
2100	62,986	270.504	59,748	253.726
2200	66,802	272.278	63,371	255.412
2300	70,634	273.981	67,007	257.027
2400	74,492	275.625	70,651	258.580
2500	78,375	277.207	74,312	260.073
2600	82,274	278.738	77,981	261.513
2700	86,199	280.219	81,659	262.902
2800	90,144	281.654	85,345	264.241
2900	94,111	283.048	89,036	265.538
3000	98,098	284.399	92,738	266.793
3200	106,127	286.988	100,161	269.186
3400	114,232	289.445	107,608	271.445
3600	122,399	291.780	115,081	273.579
3800	130,629	294.005	122,570	275.604
4000	138,913	296.127	130,076	277.529
4200	147,248	298.160	137,603	279.366
4400	155,628	300.110	145,143	281.119
4600	164,046	301.984	152,699	282.801
4800	172,502	303.784	160,272	284.412
5000	180,987	305.511	167,858	285.960
5200	189,502	307.181	175,456	287.449

TABLE A.5 Continued

Temp. (K)	Oxygen (O_2) $\overline{m}$ = 31.999 $(\overline{h}_f^\circ)_{298}$ = 0 kJ/kgmol $\overline{h}^\circ - \overline{h}^\circ_{298}$ (kJ/kgmol)	$\overline{s}^\circ$ (kJ/kgmol·K)	Nitrogen (N_2) $\overline{m}$ = 28.013 $(\overline{h}_f^\circ)_{298}$ = 0 kJ/kgmol $\overline{h}^\circ - \overline{h}^\circ_{298}$ (kJ/kgmol)	$\overline{s}^\circ$ (kJ/kgmol·K)
5400	198,037	308.792	183,071	288.884
5600	206,593	310.348	190,703	290.273
5800	215,166	311.854	198,347	291.616
6000	223,756	313.310	206,008	292.913

Temp. (K)	Hydrogen (H_2) $\overline{m}$ = 2.016 $(\overline{h}_f^\circ)_{298}$ = 0 kJ/kgmol $\overline{h}^\circ - \overline{h}^\circ_{298}$ (kJ/kgmol)	$\overline{s}^\circ$ (kJ/kgmol·K)	Hydroxyl (OH) $\overline{m}$ = 17.007 $(\overline{h}_f^\circ)_{298}$ = 39,463 kJ/kgmol $\overline{h}^\circ - \overline{h}_{298}$ (kJ/kgmol)	$\overline{s}^\circ$ (kJ/kgmol·K)
0	−8,468	0	−9,171	0
100	−5,293	102.035	−6,138	149.478
200	−2,770	119.328	−2,975	171.481
298	0	130.574	0	183.594
300	54	130.754	54	183.782
400	2,958	139.105	3,033	192.355
500	5,883	145.628	5,991	198.953
600	8,812	150.967	8,941	204.334
700	11,749	155.498	11,903	208.895
800	14,703	159.440	14,878	212.869
900	17,682	162.950	17,887	216.413
1000	20,686	166.113	20,933	219.622
1100	23,723	169.008	24,025	222.568
1200	26,794	171.682	27,158	225.296
1300	29,907	174.172	30,342	227.840
1400	33,062	176.510	33,568	230.233
1500	36,267	178.724	36,840	232.488
1600	39,522	180.820	40,150	234.626
1700	42,815	182.802	43,501	236.660
1800	46,150	184.724	46,890	238.593
1900	49,522	186.548	50,308	240.442
2000	52,932	188.297	53,760	242.216
2100	56,379	189.979	57,241	243.910
2200	59,860	191.598	60,752	245.542
2300	63,371	193.159	64,283	247.115
2400	66,915	194.669	67,839	248.630
2500	70,492	196.125	71,417	250.090
2600	74,090	197.539	75,015	251.500
2700	77,718	198.907	78,634	252.864
2800	81,370	200.234	82,266	254.186
2900	85,044	201.527	85,918	255.467

TABLE A.5 Continued

Temp. (K)	Hydrogen (H_2) $\bar{m}$ = 2.016 $(\bar{h}_f^\circ)_{298}$ = 0 kJ/kgmol		Hydroxyl (OH) $\bar{m}$ = 17.007 $(\bar{h}_f^\circ)_{298}$ = 39,463 kJ/kgmol	
	$\bar{h}^\circ - \bar{h}_{298}^\circ$ (kJ/kgmol)	$\bar{s}^\circ$ (kJ/kgmol·K)	$\bar{h}^\circ - \bar{h}_{298}$ (kJ/kgmol)	$\bar{s}^\circ$ (kJ/kgmol·K)
3000	88,743	202.778	89,584	256.709
3200	96,199	205.183	96,960	259.090
3400	103,738	207.468	104,387	261.341
3600	111,361	209.648	111,859	263.479
3800	119,064	211.731	119,378	265.508
4000	126,845	213.727	126,934	267.450
4200	134,700	215.643	134,528	269.299
4400	142,624	217.484	142,156	271.073
4600	150,620	219.263	149,816	272.776
4800	158,682	220.978	157,502	274.412
5000	166,808	222.635	165,222	275.989
5200	174,996	224.241	172,967	277.508
5400	183,247	225.798	180,736	278.972
5600	191,556	227.308	188,531	280.391
5800	199,924	228.777	196,351	281.763
6000	208,346	230.204	204,192	283.094

Temp. (K)	Water (H_2O) $\bar{m}$ = 18.015 $(\bar{h}_f^\circ)_{298}$ = −241,827 kJ/kgmol		Methane (CH_4) $\bar{m}$ = 16.043 $(\bar{h}_f^\circ)_{298}$ = −74,873 kJ/kgmol	
	$\bar{h}^\circ - \bar{h}_{298}^\circ$ (kJ/kgmol)	$\bar{s}^\circ$ (kJ/kgmol·K)	$\bar{h}^\circ - \bar{h}_{298}^\circ$ (kJ/kgmol)	$\bar{s}^\circ$ (kJ/kgmol·K)
0	−9,904	0	−10,025	0
100	−6,615	152.281	−6,699	149.394
200	−3,280	175.377	−3,368	172.473
298	0	188.724	0	186.146
300	63	188.929	67	186.368
400	3,452	198.673	3,862	197.250
500	6,920	206.413	8,201	206.911
600	10,498	212.928	13,129	215.882
700	14,184	218.610	18,636	224.354
800	17,991	223.693	24,673	232.413
900	21,924	228.321	31,204	240.099
1000	25,978	232.597	38,179	247.446
1100	30,167	236.584	45,551	254.467
1200	34,476	240.333	53,271	261.182
1300	38,903	243.877	61,304	267.609
1400	43,447	247.241	69,609	273.763
1500	48,095	250.450	78,153	279.659
1600	52,844	253.513	86,910	285.311

TABLE A.5 Continued

	Water (H_2O) $\bar{m}$ = 18.015 $(\bar{h}_f^\circ)_{298}$ = −241,827 kJ/kgmol		Methane (CH_4) $\bar{m}$ = 16.043 $(\bar{h}_f^\circ)_{298}$ = −74,873 kJ/kgmol	
Temp. (K)	$\bar{h}^\circ - \bar{h}_{298}^\circ$ (kJ/kgmol)	$\bar{s}^\circ$ (kJ/kgmol·K)	$\bar{h}^\circ - \bar{h}_{298}^\circ$ (kJ/kgmol)	$\bar{s}^\circ$ (kJ/kgmol·K)
1700	57,685	256.450	95,855	290.729
1800	62,609	259.262	104,960	295.934
1900	67,613	261.969	114,215	300.938
2000	72,689	264.571	123,595	305.750
2100	77,831	267.081	133,089	310.382
2200	83,036	269.500	142,687	314.846
2300	88,295	271.839	152,373	319.151
2400	93,604	274.098	162,143	323.310
2500	98,964	276.286	171,988	327.327
2600	104,370	278.408	181,895	331.214
2700	109,813	280.462	191,866	334.979
2800	115,294	282.453	201,891	338.624
2900	120,813	284.391	211,961	342.159
3000	126,361	286.273	222,083	345.586
3200	137,553	289.884	242,438	352.155
3400	148,854	293.307	262,931	358.368
3600	160,247	296.566	283,541	364.259
3800	171,724	299.666	304,256	369.857
4000	183,280	302.633	325,055	375.192
4200	194,903	305.465	345,925	380.284
4400	206,585	308.185	366,866	385.154
4600	218,325	310.792	387,861	389.819
4800	230,120	313.302	408,902	394.300
5000	241,957	315.720	429,994	398.601
5200	253,839	318.051	451,123	402.748
5400	265,768	320.298	472,286	406.739
5600	277,738	322.478	493,482	410.593
5800	289,746	324.582	514,707	414.316
6000	301,796	326.624	535,958	417.919

[a] *The thermochemical data in this table were derived from the JANAF Thermochemical Tables, Thermal Research Laboratory, The Dow Chemical Company, Midland, Michigan.*

$\bar{h}_f^\circ$ = enthalpy of formation at 25 C (298.15 K).

TABLE A.6 Thermodynamic Properties of Ideal Gases[a]

Gas	Chemical Formula	Molecular Weight $\bar{m}$	R (kJ/kg·K)	k, c_p/c_v	c_p (kJ/kg·K)
Acetylene	C_2H_2	26.039	0.3193	1.26	1.603
Air	—	28.970	0.2870	1.400	1.004
Ammonia	NH_3	17.032	0.4882	1.310	2.188
Carbon dioxide	CO_2	44.011	0.1889	1.294	0.850
Carbon monoxide	CO	28.011	0.2968	1.401	1.041
Ethane	C_2H_6	30.072	0.2765	1.186	1.766
Helium	He	4.003	2.0770	1.667	5.230
Hydrogen	H_2	2.016	4.1242	1.405	14.293
Methane	CH_4	16.043	0.5183	1.299	2.253
Nitric oxide	NO	30.006	0.2771	1.400	0.975
Nitrogen	N_2	28.013	0.2968	1.401	1.041
Oxygen	O_2	31.999	0.2598	1.396	0.919
Sulfur dioxide	SO_2	64.063	0.1298	1.29	0.636
Water vapor	H_2O	18.015	0.4615	1.327	1.872

[a] $R = \bar{R}/\bar{m} = 8.3143/\bar{m}$ kJ/kg·K; c_p and k are at 1 atm, 25 C.

Source: Data for Table A.6 are taken from several sources, including National Bureau of Standards Circular 564, 1955.

TABLE A.7 Physical Properties

TABLE A.7.1 Air (P = 1 atm)[a]

Temp. K	Specific Heat c_p kJ/kg·K	Viscosity $\mu \times 10^5$ N·s/m²	Thermal Conductivity k W/m·K	Prandtl Number Pr
250	1.0046	1.599	0.02227	0.722
300	1.0055	1.846	0.02624	0.708
350	1.0084	2.075	0.03003	0.697
400	1.0134	2.286	0.03365	0.689
450	1.0204	2.485	0.03710	0.684
500	1.0291	2.670	0.04041	0.680
550	1.0393	2.848	0.04357	0.680
600	1.0505	3.017	0.04661	0.680
650	1.0622	3.177	0.04954	0.682
700	1.0743	3.332	0.05236	0.684
750	1.0862	3.482	0.05509	0.686
800	1.098	3.624	0.05774	0.689
850	1.110	3.763	0.06030	0.692
900	1.120	3.897	0.06276	0.695
950	1.131	4.026	0.06520	0.699
1000	1.141	4.153	0.06754	0.702

[a] *Adapted from "Tables of the Thermal Properties of Gases," National Bureau of Standards Circular 564, November 1955.*

TABLE A.7.2 Water (P = 7 MPa, abs)[b]

Temp. °C	Specific Heat c_p kJ/kg·K	Viscosity $\mu \times 10^4$ N·s/m²	Thermal Conductivity k W/m·K	Prandtl Number Pr
0	4.183	17.49	0.5744	12.73
25	4.162	8.830	0.6169	5.95
50	4.164	5.518	0.6489	3.54
100	4.202	2.806	0.6855	1.720
150	4.290	1.824	0.6913	1.132
200	4.466	1.352	0.6693	0.902
250	4.832	1.075	0.6207	0.837

[b] *Adapted from "1967 ASME Steam Tables," The American Society of Mechanical Engineers.*

TABLE A.7.3 Steam (P = 1 MPa, abs)[c]

Temp. °C	Specific Heat c_p kJ/kg·K	Viscosity $\mu \times 10^5$ N·s/m²	Thermal Conductivity k W/m·K	Prandtl Number Pr
200	2.447	1.586	0.03532	1.099
250	2.236	1.801	0.03931	1.024
300	2.142	2.025	0.04442	0.976
350	2.118	2.233	0.04988	0.948
400	2.123	2.445	0.05576	0.931
450	2.138	2.661	0.06189	0.919
500	2.164	2.832	0.06818	0.899

[c] *Adapted from "1967 ASME Steam Tables," The American Society of Mechanical Engineers.*

TABLE A.7.4 Steam (P = 10 MPa, abs)[d]

Temp. °C	Specific Heat c_p kJ/kg·K	Viscosity $\mu \times 10^5$ N·s/m²	Thermal Conductivity k W/m·K	Prandtl Number Pr
350	4.093	2.351	0.07267	1.324
400	3.095	2.589	0.06978	1.148
450	2.717	2.762	0.07292	1.029
500	2.548	2.968	0.07789	0.971
550	2.474	3.152	0.08369	0.932
600	2.447	3.353	0.08989	0.913
650	2.437	3.548	0.09648	0.896

[d] *Adapted from "1967 ASME Steam Tables," The American Society of Mechanical Engineers.*

TABLE A.8 Thermodynamic Properties of Water and Steam*

TABLE A.8.1 Saturated Steam and Saturated Water (Temperature)

Temp. °C T	Press. kPa P	Volume, m³/kg Water v_f	Steam v_g	Enthalpy, kJ/kg Water h_f	Evap. h_{fg}	Steam h_g	Entropy, kJ/kg·K Water s_f	Evap. s_{fg}	Steam s_g
0	0.6108	0.0010002	206.31	−0.04	2501.6	2501.6	−0.0002	9.1579	9.1577
0.01	0.6112	0.0010002	206.16	0.00	2501.6	2501.6	0.0000	9.1575	9.1575
5	0.8718	0.0010000	147.16	21.01	2489.7	2510.7	0.0762	8.9507	9.0269
10	1.2270	0.0010003	106.43	41.99	2477.9	2519.9	0.1510	8.7510	8.9020
15	1.7040	0.0010008	77.98	62.94	2466.1	2529.0	0.2243	8.5582	8.7825
20	2.377	0.0010017	57.84	83.86	2454.3	2538.2	0.2963	8.3721	8.6684
25	3.166	0.0010029	43.40	104.77	2442.5	2547.3	0.3670	8.1922	8.5592
30	4.241	0.0010043	32.93	125.66	2430.7	2556.4	0.4365	8.0181	8.4546
35	5.622	0.0010060	25.25	146.56	2418.8	2565.4	0.5049	7.8494	8.3543
40	7.375	0.0010078	19.546	167.45	2406.9	2574.4	0.5721	7.6861	8.2583
45	9.582	0.0010098	15.276	188.35	2394.9	2583.3	0.6383	7.5277	8.1661
50	12.335	0.0010121	12.046	209.26	2382.9	2592.2	0.7035	7.3741	8.0776
55	15.741	0.0010145	9.579	230.17	2370.8	2601.0	0.7677	7.2248	7.9925
60	19.920	0.0010171	7.679	251.09	2358.6	2609.7	0.8310	7.0798	7.9108
65	25.010	0.0010199	6.202	272.03	2346.3	2618.3	0.8933	6.9388	7.8322
70	31.16	0.0010228	5.046	292.97	2334.0	2626.9	0.9548	6.8017	7.7565
75	38.55	0.0010259	4.134	313.93	2321.5	2635.4	1.0154	6.6681	7.6835
80	47.36	0.0010292	3.409	334.92	2308.8	2643.8	1.0753	6.5380	7.6132
85	57.80	0.0010326	2.829	355.91	2296.1	2652.0	1.1343	6.4111	7.5454
90	70.11	0.0010361	2.3613	376.94	2283.2	2660.1	1.1925	6.2873	7.4799
95	84.53	0.0010398	1.9822	397.99	2270.2	2668.2	1.2501	6.1665	7.4166
100	101.33	0.0010437	1.6730	419.06	2256.9	2676.0	1.3069	6.0485	7.3554
105	120.80	0.0010477	1.4193	440.17	2243.6	2683.7	1.3630	5.9331	7.2962
110	143.27	0.0010519	1.2099	461.32	2230.0	2691.3	1.4185	5.8203	7.2388
115	169.06	0.0010562	1.0363	482.50	2216.2	2698.7	1.4733	5.7099	7.1832
120	198.54	0.0010606	0.8915	503.72	2202.2	2706.0	1.5276	5.6017	7.1293
125	232.1	0.0010652	0.7702	524.99	2188.0	2713.0	1.5813	5.4957	7.0769
130	270.1	0.0010700	0.6681	546.31	2173.6	2719.9	1.6344	5.3917	7.0261
135	313.1	0.0010750	0.5818	567.68	2158.9	2726.6	1.6869	5.2897	6.9766
140	361.4	0.0010801	0.5085	589.10	2144.0	2733.1	1.7390	5.1894	6.9284
145	415.5	0.0010853	0.4460	610.59	2128.7	2739.3	1.7906	5.0910	6.8815
150	476.0	0.0010908	0.3924	632.15	2113.2	2745.4	1.8416	4.9941	6.8358
155	543.3	0.0010964	0.3464	653.77	2097.4	2751.2	1.8923	4.8989	6.7911
160	618.1	0.0011022	0.3068	675.47	2081.3	2756.7	1.9425	4.8050	6.7475
165	700.8	0.0011082	0.2724	697.25	2064.8	2762.0	1.9923	4.7126	6.7048
170	792.0	0.0011145	0.2426	719.12	2047.9	2767.1	2.0416	4.6214	6.6630
175	892.4	0.0011209	0.21654	741.07	2030.7	2771.8	2.0906	4.5314	6.6221
180	1002.7	0.0011275	0.19380	763.12	2013.2	2776.3	2.1393	4.4426	6.5819
185	1123.3	0.0011344	0.17386	785.26	1995.2	2780.4	2.1876	4.3548	6.5424
190	1255.1	0.0011415	0.15632	807.52	1976.7	2784.3	2.2356	4.2680	6.5036
195	1398.7	0.0011489	0.14084	829.88	1957.9	2787.8	2.2833	4.1821	6.4654
200	1554.9	0.0011565	0.12716	852.37	1938.6	2790.9	2.3307	4.0971	6.4278
210	1907.7	0.0011726	0.10424	897.73	1898.5	2796.2	2.4247	3.9293	6.3539

TABLE A.8.1 Continued

Temp. °C T	Press. kPa P	Volume, m³/kg Water v_f	Volume, m³/kg Steam v_g	Enthalpy, kJ/kg Water h_f	Enthalpy, kJ/kg Evap. h_{fg}	Enthalpy, kJ/kg Steam h_g	Entropy, kJ/kg·K Water s_f	Entropy, kJ/kg·K Evap. s_{fg}	Entropy, kJ/kg·K Steam s_g
220	2319.8	0.0011900	0.08604	943.67	1856.2	2799.9	2.5178	3.7639	6.2817
230	2798.	0.0012087	0.07145	990.27	1811.7	2802.0	2.6102	3.6006	6.2107
240	3348.	0.0012291	0.05965	1037.60	1764.6	2802.2	2.7020	3.4386	6.1406
250	3978.	0.0012513	0.05004	1085.78	1714.7	2800.4	2.7935	3.2773	6.0708
260	4694.	0.0012756	0.04213	1134.94	1661.5	2796.4	2.8848	3.1161	6.0010
270	5506.	0.0013025	0.03559	1185.23	1604.6	2789.9	2.9763	2.9541	5.9304
280	6420.	0.0013324	0.03013	1236.84	1543.6	2780.4	3.0683	2.7903	5.8586
290	7446.	0.0013659	0.02554	1290.01	1477.6	2767.6	3.1611	2.6237	5.7848
300	8593.	0.0014041	0.021649	1345.05	1406.0	2751.0	3.2552	2.4529	5.7081
310	9870.	0.0014480	0.018334	1402.39	1327.6	2730.0	3.3512	2.2766	5.6278
320	11289.	0.0014995	0.015480	1462.60	1241.1	2703.7	3.4500	2.0923	5.5423
330	12863.	0.0015615	0.012989	1526.52	1143.6	2670.2	3.5528	1.8962	5.4490
340	14605.	0.0016387	0.010780	1595.47	1030.7	2626.2	3.6616	1.6811	5.3427
350	16535.	0.0017411	0.008799	1671.94	895.7	2567.7	3.7800	1.4376	5.2177
360	18675.	0.0018959	0.006940	1764.17	721.3	2485.4	3.9210	1.1390	5.0600
370	21054.	0.0022136	0.004973	1890.21	452.6	2342.8	4.1108	0.7036	4.8144
374.15	22120.	0.00317	0.00317	2107.37	0.0	2107.4	4.4429	0.0	4.4429

[a] *Abstracted from "ASME Steam Tables in SI (Metric) Units," The American Society of Mechanical Engineers. Copyright 1977.*

TABLE A.8.2 Saturated Steam and Saturated Water (Pressure)

Press. kPa P	Temp. °C T	Volume, m³/kg Water v_f	Volume, m³/kg Steam v_g	Enthalpy, kJ/kg Water h_f	Enthalpy, kJ/kg Evap. h_{fg}	Enthalpy, kJ/kg Steam h_g	Entropy, kJ/kg·K Water s_f	Entropy, kJ/kg·K Evap. s_{fg}	Entropy, kJ/kg·K Steam s_g	Energy, kJ/kg Water u_f	Energy, kJ/kg Steam u_g
1.0	6.983	0.0010001	129.21	29.34	2485.0	2514.4	0.1060	8.8706	8.9767	29.33	2385.2
1.5	13.036	0.0010006	87.98	54.71	2470.7	2525.5	0.1957	8.6332	8.8288	54.71	2393.5
2.0	17.513	0.0010012	67.01	73.46	2460.2	2533.6	0.2607	8.4639	8.7246	73.46	2399.6
3.0	24.100	0.0010027	45.67	101.00	2444.6	2545.6	0.3544	8.2241	8.5785	101.00	2408.6
4.0	28.983	0.0010040	34.80	121.41	2433.1	2554.5	0.4225	8.0530	8.4755	121.41	2415.3
5.0	32.898	0.0010052	28.19	137.77	2423.8	2561.6	0.4763	7.9197	8.3960	137.77	2420.6
7.5	40.316	0.0010079	19.239	168.77	2406.2	2574.9	0.5763	7.6760	8.2523	168.76	2430.6
10.0	45.833	0.0010102	14.675	191.83	2392.9	2584.8	0.6493	7.5018	8.1511	191.82	2438.0
15.0	53.997	0.0010140	10.023	225.97	2373.2	2599.2	0.7549	7.2544	8.0093	225.96	2448.9
20.0	60.086	0.0010172	7.650	251.45	2358.4	2609.9	0.8321	7.0774	7.9094	251.43	2456.9
30.0	69.124	0.0010223	5.229	289.30	2336.1	2625.4	0.9441	6.8254	7.7695	289.27	2468.6
40.0	75.886	0.0010265	3.993	317.65	2319.2	2636.9	1.0261	6.6448	7.6709	317.61	2477.1
50.0	81.345	0.0010301	3.240	340.56	2305.4	2646.0	1.0912	6.5035	7.5947	340.51	2484.0
75.0	91.785	0.0010375	2.2169	384.45	2278.6	2663.0	1.2131	6.2439	7.4570	384.37	2496.7
100.0	99.632	0.0010434	1.6937	417.51	2257.9	2675.4	1.3027	6.0571	7.3598	417.41	2506.1
150.0	111.37	0.0010530	1.1590	467.13	2226.2	2693.4	1.4336	5.7898	7.2234	466.97	2519.5
200.0	120.23	0.0010608	0.8854	504.70	2201.6	2706.3	1.5301	5.5967	7.1268	504.49	2529.2
300.0	133.54	0.0010735	0.6056	561.4	2163.2	2724.7	1.6716	5.3193	6.9909	561.11	2543.0
400.0	143.62	0.0010839	0.4622	604.7	2133.0	2737.6	1.7764	5.1179	6.8943	604.24	2552.7
500.0	151.84	0.0010928	0.3747	640.1	2107.4	2747.5	1.8604	4.9588	6.8192	639.57	2560.2
600.0	158.84	0.0011009	0.3155	670.4	2085.0	2755.5	1.9308	4.8267	6.7575	669.76	2566.2

TABLE A.8.2 Continued

Press. kPa P	Temp. °C T	Volume, m³/kg Water v_f	Volume, m³/kg Steam v_g	Enthalpy, kJ/kg Water h_f	Enthalpy, kJ/kg Evap. h_{fg}	Enthalpy, kJ/kg Steam h_g	Entropy, kJ/kg·K Water s_f	Entropy, kJ/kg·K Evap. s_{fg}	Entropy, kJ/kg·K Steam s_g	Energy, kJ/kg Water u_f	Energy, kJ/kg Steam u_g
700.0	164.96	0.0011082	0.27268	697.1	2064.9	2762.0	1.9918	4.7134	6.7052	696.29	2571.1
800.0	170.41	0.0011150	0.24026	720.9	2046.5	2767.5	2.0457	4.6139	6.6596	720.04	2575.3
900.0	175.36	0.0011213	0.21481	742.6	2029.5	2772.1	2.0941	4.5250	6.6192	741.63	2578.8
1000.0	179.88	0.0011274	0.19429	762.6	2013.6	2776.2	2.1382	4.4446	6.5828	761.48	2581.9
1500.0	198.29	0.0011539	0.13166	844.7	1945.2	2789.9	2.3145	4.1261	6.4406	842.93	2592.4
2000.0	212.37	0.0011766	0.09954	908.6	1888.6	2797.2	2.4469	3.8898	6.3367	906.24	2598.2
3000.0	233.84	0.0012163	0.06663	1008.4	1793.9	2802.3	2.6455	3.5382	6.1837	1004.70	2602.4
4000.0	250.33	0.0012521	0.04975	1087.4	1712.9	2800.3	2.7965	3.2720	6.0685	1082.4	2601.3
5000.0	263.91	0.0012858	0.03943	1154.4	1639.7	2794.2	2.9206	3.0529	5.9735	1148.0	2597.0
7500.0	290.50	0.0013677	0.025327	1292.7	1474.2	2766.9	3.1657	2.6153	5.7811	1282.4	2577.0
10000.0	310.96	0.0014526	0.018041	1408.0	1319.7	2727.7	3.3605	2.2593	5.6198	1393.5	2547.3
15000.0	342.13	0.0016579	0.010340	1611.0	1004.0	2615.0	3.6859	1.6320	5.3178	1586.1	2459.9
20000.0	365.70	0.0020370	0.005877	1826.5	591.9	2418.4	4.0149	0.9263	4.9412	1785.7	2300.8
22120.0	374.15	0.00317	0.00317	2107.4	0.0	2107.4	4.4429	0.0	4.4429	2037.3	2037.3

TABLE A.8.3 Superheated Steam

v (specific volume), m³/kg; *h* (enthalpy), kJ/kg; *s* (entropy), kJ/kg·K

Pressure kPa (T_{sat}) °C		Temperature, °C 50	100	150	200	300	400	500	600	700	800
1.0	*v*	149.09	172.19	195.28	218.35	264.51	310.66	356.81	402.97	449.12	495.27
(6.983)	*h*	2594.6	2688.6	2783.7	2880.1	3076.8	3279.7	3489.2	3705.6	3928.9	4158.7
	s	9.2430	9.5136	9.7527	9.9679	10.3450	10.6711	10.9612	11.2243	11.4663	11.6911
5.0	*v*	29.783	34.417	39.042	43.661	52.897	62.129	71.360	80.592	89.822	99.053
(32.90)	*h*	2593.7	2688.1	2783.4	2879.9	3076.7	3279.7	3489.2	3705.6	3928.8	4158.7
	s	8.4981	8.7698	9.0094	9.2248	9.6021	9.9283	10.2184	10.4815	10.7235	10.9483
10	*v*	14.871	17.195	19.513	21.825	26.445	31.062	35.679	40.295	44.910	49.526
(45.83)	*h*	2592.7	2687.5	2783.1	2879.6	3076.6	3279.6	3489.1	3705.5	3928.8	4158.7
	s	8.1758	8.4486	8.6889	8.9045	9.2820	9.6083	9.8984	10.1616	10.4036	10.6284
50	*v*		3.4181	3.8893	4.3560	5.2839	6.2091	7.1335	8.0574	8.9810	9.9044
(81.35)	*h*		2682.6	2780.1	2877.7	3075.7	3279.0	3488.7	3705.2	3928.6	4158.5
	s		7.6953	7.9406	8.1587	8.5380	8.8649	9.1552	9.4185	9.6606	9.8855
100	*v*		1.6955	1.9363	2.1723	2.6387	3.1025	3.5653	4.0277	4.4898	4.9517
(99.63)	*h*		2676.2	2776.3	2875.4	3074.5	3278.2	3488.1	3704.8	3928.2	4158.3
	s		7.3618	7.6138	7.8349	8.2166	8.5442	8.8348	9.0982	9.3405	9.5654
200	*v*			0.9596	1.0804	1.3162	1.5492	1.7812	2.0129	2.2442	2.4754
(120.2)	*h*			2768.5	2870.5	3072.1	3276.7	3487.0	3704.0	3927.6	4157.8
	s			7.2794	7.5072	7.8937	8.2226	8.5139	8.7776	9.0201	9.2452
300	*v*			0.6338	0.7164	0.8753	1.0314	1.1865	1.3412	1.4957	1.6499
(133.5)	*h*			2760.4	2865.5	3069.7	3275.2	3486.0	3703.2	3927.0	4157.3
	s			7.0772	7.3119	7.7034	8.0338	8.3257	8.5898	8.8325	9.0577

TABLE A.8.3 Continued

Pressure kPa (T_{sat}) °C		v (specific volume), m³/kg	h (enthalpy), kJ/kg	s (entropy), kJ/kg·K							
		Temperature, °C									
		50	100	150	200	300	400	500	600	700	800
400	v			0.4706	0.5343	0.6549	0.7725	0.8892	1.0054	1.1214	1.2372
(143.6)	h			2752.0	2860.4	3067.2	3273.6	3484.9	3702.3	3926.4	4156.9
	s			6.9286	7.1708	7.5675	7.8994	8.1919	8.4563	8.6992	8.9246
500	v				0.4250	0.5226	0.6172	0.7108	0.8039	0.8968	0.9896
(151.8)	h				2855.1	3064.8	3272.1	3483.8	3701.5	3925.8	4156.4
	s				7.0592	7.4614	7.7948	8.0879	8.3526	8.5957	8.8213
600	v				0.3520	0.4344	0.5136	0.5918	0.6696	0.7471	0.8245
(158.8)	h				2849.7	3062.3	3270.6	3482.7	3700.7	3925.1	4155.9
	s				6.9662	7.3740	7.7090	8.0027	8.2678	8.5111	8.7368
800	v				0.2608	0.3241	0.3842	0.4432	0.5017	0.5600	0.6181
(170.4)	h				2838.6	3057.3	3267.5	3480.5	3699.1	3923.9	4155.0
	s				6.8148	7.2348	7.5729	7.8678	8.1336	8.3773	8.6033
1000	v				0.2059	0.2580	0.3065	0.3540	0.4010	0.4477	0.4943
(179.9)	h				2826.8	3052.1	3264.4	3478.3	3697.4	3922.7	4154.1
	s				6.6922	7.1251	7.4665	7.7627	8.0292	8.2734	8.4997
1500	v				0.1324	0.1697	0.2029	0.2350	0.2667	0.2980	0.3292
(198.3)	h				2794.7	3038.9	3256.6	3472.8	3693.3	3919.6	4151.7
	s				6.4508	6.9207	7.2709	7.5703	7.8385	8.0838	8.3108

Pressure kPa (T_{sat}) °C		300	400	500	600	700	800
2000	v	0.1255	0.1511	0.1756	0.1995	0.2232	0.2467
(212.4)	h	3025.0	3248.7	3467.3	3689.2	3916.5	4149.4
	s	6.7696	7.1296	7.4323	7.7022	7.9485	8.1763
3000	v	0.08116	0.09931	0.1161	0.1323	0.1483	0.1641
(233.8)	h	2995.1	3232.5	3456.2	3681.0	3910.3	4144.7
	s	6.5422	6.9246	7.2345	7.5079	7.7564	7.9857
4000	v	0.05883	0.07338	0.08634	0.09876	0.1109	0.1229
(250.3)	h	2962.0	3215.7	3445.0	3672.8	3904.1	4140.0
	s	6.3642	6.7733	7.0909	7.3680	7.6187	7.8495
5000	v	0.04530	0.05779	0.06849	0.07862	0.08845	0.09809
(263.9)	h	2925.5	3198.3	3433.7	3664.5	3897.9	4135.3
	s	6.2105	6.6508	6.9770	7.2578	7.5108	7.7431
10000	v		0.02641	0.03276	0.03832	0.04355	0.04858
(311.0)	h		3099.9	3374.6	3622.7	3866.8	4112.0
	s		6.2182	6.5994	6.9013	7.1660	7.4058
15000	v		0.01566	0.02080	0.02488	0.02859	0.03209
(342.1)	h		2979.1	3310.6	3579.8	3835.4	4088.6
	s		5.8876	6.3487	6.6764	6.9536	7.2013

TABLE A.8.3 Continued

Pressure kPa (T_{sat}) °C		300	400	500	600	700	800
		v (specific volume), m³/kg *h (enthalpy), kJ/kg* *s (entropy), kJ/kg·K* — *Temperature, °C*					
20000	v		0.009947	0.01477	0.01816	0.02111	0.02385
(365.7)	h		2820.5	3241.1	3535.5	3803.8	4065.3
	s		5.5585	6.1456	6.5043	6.7953	7.0511
30000	v		0.002831	0.008681	0.01144	0.01365	0.01562
	h		2161.8	3085.0	3443.0	3739.7	4018.5
	s		4.4896	5.7972	6.2340	6.5560	6.8288

TABLE A.8.4 Compressed Water

Pressure kPa (T_{sat}) °C		50	100	150	200	300
		v (specific volume), m³/kg *h (enthalpy), kJ/kg* *s (entropy), kJ/kg·K* — *Temperature, °C*				
1000	v	0.0010117	0.0010432	0.0010903		
(179.9)	h	210.1	419.7	632.4		
	s	0.7031	1.3062	1.8410		
1500	v	0.0010114	0.0010430	0.0010901		
(198.3)	h	210.5	420.1	632.8		
	s	0.7028	1.3058	1.8405		
2000	v	0.0010112	0.0010427	0.0010897	0.0011560	
(212.4)	h	210.9	420.5	633.1	852.6	
	s	0.7025	1.3054	1.8399	2.3300	
5000	v	0.0010099	0.0010412	0.0010877	0.0011530	
(263.9)	h	213.6	422.7	635.0	853.8	
	s	0.7012	1.3030	1.8366	2.3253	
10000	v	0.0010077	0.0010386	0.0010843	0.0011480	0.0013979
(311.0)	h	217.8	426.5	638.1	855.9	1343.4
	s	0.6989	1.2992	1.8312	2.3176	3.2488
15000	v	0.0010056	0.0010361	0.0010811	0.0011433	0.0013779
(342.1)	h	222.1	430.3	641.2	858.1	1338.3
	s	0.6966	1.2954	1.8259	2.3102	3.2278
20000	v	0.0010034	0.0010337	0.0010779	0.0011387	0.0013606
(365.7)	h	226.4	434.0	644.4	860.4	1334.3
	s	0.6942	1.2916	1.8207	2.3030	3.2089
30000	v	0.0009993	0.0010289	0.0010718	0.0011301	0.0013316
	h	234.9	441.6	650.9	865.2	1328.7
	s	0.6897	1.2843	1.8106	2.2891	3.1757

TABLE A.9 Thermodynamic Properties of Ammonia[a]

TABLE A.9.1 Saturated Ammonia

Temp. °C	Abs. Pressure kPa P	Specific Volume m^3/kg Sat. Liquid v_f	Sat. Vapor v_g	Enthalpy kJ/kg Sat. Liquid h_f	Evap. h_{fg}	Sat. Vapor h_g	Entropy kJ/kg·K Sat. Liquid s_f	Evap. s_{fg}	Sat. Vapor s_g
−50	40.89	0.001425	2.6251	−44.4	1416.4	1372.0	−0.1941	6.3485	6.1544
−45	54.54	0.001436	2.0052	−22.3	1402.7	1380.4	−0.0962	6.1494	6.0532
−40	71.77	0.001450	1.5520	0.0	1388.8	1388.8	0.0	5.9574	5.9574
−35	93.22	0.001463	1.2161	22.3	1374.3	1396.6	0.0945	5.7721	5.8666
−30	119.56	0.001476	0.9633	44.6	1359.7	1404.3	0.1874	5.5926	5.7800
−25	151.62	0.001490	0.7716	67.2	1344.6	1411.8	0.2786	5.4190	5.6976
−20	190.23	0.001504	0.6237	89.7	1329.0	1418.7	0.3681	5.2509	5.6190
−15	236.28	0.001518	0.5088	112.2	1313.0	1425.2	0.4568	5.0869	5.5437
−10	290.82	0.001534	0.4185	135.2	1296.5	1431.7	0.5438	4.9280	5.4718
−5	354.87	0.001549	0.3468	158.0	1279.5	1437.5	0.6295	4.7728	5.4023
0	429.47	0.001566	0.2895	181.0	1262.1	1443.1	0.7145	4.6209	5.3354
5	515.73	0.001583	0.2433	204.3	1244.0	1448.3	0.7981	4.4733	5.2714
10	614.94	0.001601	0.2056	227.5	1225.4	1452.9	0.8805	4.3286	5.2091
15	728.09	0.001619	0.1748	251.2	1206.1	1457.3	0.9625	4.1868	5.1493
20	857.02	0.001639	0.1494	274.9	1186.4	1461.3	1.0432	4.0475	5.0907
25	1002.50	0.001659	0.1283	298.6	1165.9	1464.5	1.1231	3.9111	5.0342
30	1166.59	0.001680	0.1106	322.8	1144.7	1467.5	1.2026	3.7768	4.9794
35	1349.99	0.001702	0.0958	347.2	1122.9	1470.1	1.2808	3.6447	4.9255
40	1554.08	0.001726	0.0833	371.6	1100.3	1471.9	1.3586	3.5142	4.8728
45	1781.61	0.001750	0.0726	396.5	1076.6	1473.1	1.4360	3.3845	4.8205
50	2032.57	0.001777	0.0635	421.5	1051.8	1473.3	1.5134	3.2552	4.7686

[a] *Adapted from "Tables of Thermodynamic Properties of Ammonia," National Bureau of Standards Circular No. 142.*

TABLE A.9.2 Superheated Ammonia

Abs. Press. kPa (Sat. Temp.) °C		v (specific volume), m^3/kg; h (enthalpy), kJ/kg; s (entropy), kJ/kg·K — Temperature, °C: −20	−10	0	10	20	30	40	50	60	70	80	90	100
50	v	2.4467	2.5482	2.6481	2.7480	2.8473	2.9468	3.0455	3.1443	3.2430	3.3412	3.4400		
(−46.54)	h	1435.4	1456.7	1477.9	1498.8	1520.2	1541.3	1562.6	1584.2	1605.7	1627.5	1649.3		
	s	6.3242	6.4061	6.4850	6.5610	6.6342	6.7056	6.7749	6.8426	6.9083	6.9724	7.0356		
75	v	1.6233	1.6915	1.7589	1.8265	1.8930	1.9600	2.0261	2.0927	2.1585	2.2244	2.2903		
(−39.18)	h	1432.6	1454.3	1475.8	1497.1	1518.4	1540.0	1561.4	1582.9	1604.7	1626.4	1648.6		
	s	6.1174	6.2014	6.2813	6.3583	6.4322	6.5041	6.5739	6.6418	6.7079	6.7724	6.8356		
100	v	1.2108	1.2630	1.3144	1.3655	1.4161	1.4657	1.5166	1.5664	1.6164	1.6657	1.7157	1.7647	
(−33.61)	h	1429.8	1451.8	1473.7	1495.4	1516.9	1538.5	1560.2	1581.8	1603.6	1625.6	1647.7	1669.9	
	s	5.9681	6.0539	6.1348	6.2130	6.2878	6.3604	6.4305	6.4986	6.5652	6.6301	6.6933	6.7553	

TABLE A.9.2 Superheated Ammonia

v (specific volume), m³/kg; h (enthalpy), kJ/kg; s (entropy), kJ/kg·K

Abs. Press. kPa (Sat. Temp.) °C		Temperature, °C												
		−20	−10	0	10	20	30	40	50	60	70	80	90	100
125	v	0.9635	1.0059	1.0476	1.0889	1.1297	1.1705	1.2107	1.2508	1.2910	1.3306	1.3708	1.4102	
(−29.08)	h	1426.8	1449.4	1471.6	1493.5	1515.2	1537.3	1559.0	1580.8	1602.7	1624.7	1646.9	1669.0	
	s	5.8499	5.9374	6.0204	6.0993	6.1747	6.2478	6.3185	6.3872	6.4541	6.5190	6.5825	6.6448	
150	v	0.7983	0.8345	0.8697	0.9042	0.9389	0.9730	1.0068	1.0405	1.0740	1.1075	1.1409	1.1740	1.2070
(−25.23)	h	1423.7	1447.0	1469.5	1491.8	1513.7	1535.8	1557.6	1579.5	1601.7	1623.6	1645.9	1668.2	1690.7
	s	5.7514	5.8410	5.9251	6.0050	6.0816	6.1553	6.2263	6.2957	6.3625	6.4281	6.4916	6.5540	6.6151
200	v		0.6199	0.6471	0.6740	0.6998	0.7264	0.7520	0.7774	0.8026	0.8279	0.8532	0.8787	0.9033
(−18.87)	h		1441.7	1465.2	1488.0	1510.6	1532.8	1555.2	1577.4	1599.5	1621.9	1644.3	1666.6	1689.1
	s		5.6848	5.7722	5.8544	5.9329	6.0076	6.0798	6.1498	6.2175	6.2835	6.3476	6.4102	6.4716
250	v		0.4910	0.5135	0.5354	0.5568	0.5780	0.5989	0.6196	0.6402	0.6603	0.6810	0.7011	0.7212
(−13.67)	h		1436.4	1460.5	1484.1	1507.1	1530.0	1552.6	1575.0	1597.3	1620.0	1642.6	1665.1	1687.7
	s		5.5596	5.6503	5.7351	5.8153	5.8914	5.9646	6.0355	6.1039	6.1703	6.2349	6.2982	6.3598
300	v			0.4243	0.4430	0.4613	0.4792	0.4968	0.5143	0.5316	0.5488	0.5658	0.5828	0.5997
(−9.23)	h			1456.0	1480.1	1503.7	1527.0	1549.9	1572.7	1595.3	1618.0	1640.6	1663.5	1686.3
	s			5.5481	5.6353	5.7170	5.7951	5.8691	5.9410	6.0100	6.0771	6.1422	6.2058	6.2676
350	v			0.3605	0.3769	0.3929	0.4086	0.4240	0.4391	0.4541	0.4689	0.4837	0.4983	0.5129
(−5.35)	h			1451.1	1476.1	1500.3	1524.1	1547.2	1570.2	1593.2	1616.1	1638.9	1661.9	1684.8
	s			5.4587	5.5490	5.6330	5.7121	5.7877	5.8599	5.9298	5.9975	6.0633	6.1272	6.1893
400	v			0.3126	0.3273	0.3416	0.3556	0.3692	0.3826	0.3959	0.4090	0.4220	0.4349	0.4477
(−1.89)	h			1446.1	1471.9	1496.9	1521.1	1544.5	1567.9	1591.2	1614.2	1637.2	1660.2	1683.2
	s			5.3792	5.4720	5.5582	5.6391	5.7159	5.7892	5.8598	5.9281	5.9943	6.0585	6.1214
450	v				0.2887	0.3017	0.3143	0.3267	0.3387	0.3506	0.3624	0.3740	0.3856	0.3971
(1.26)	h				1467.7	1493.2	1517.8	1541.9	1565.4	1588.8	1612.3	1635.4	1658.6	1681.8
	s				5.4027	5.4913	5.5737	5.6517	5.7260	5.7977	5.8663	5.9331	5.9978	6.0609

Abs. Press. kPa (Sat. Temp.) °C		10	20	30	40	50	60	70	80	100	120	140	160	180
500	v	0.2578	0.2698	0.2813	0.2925	0.3036	0.3144	0.3251	0.3357	0.3565	0.3771	0.3975		
(4.14)	h	1463.5	1489.4	1514.6	1539.1	1563.0	1586.6	1610.2	1633.6	1680.5	1727.0	1774.4		
	s	5.3393	5.4301	5.5148	5.5934	5.6689	5.7412	5.8104	5.8778	6.0063	6.1290	6.2452		
600	v	0.2113	0.2217	0.2317	0.2414	0.2508	0.2600	0.2691	0.2780	0.2957	0.3130	0.3302		
(9.29)	h	1454.8	1482.0	1508.2	1533.5	1558.1	1582.4	1606.3	1630.0	1677.3	1724.5	1772.1		
	s	5.2248	5.3209	5.4088	5.4910	5.5683	5.6421	5.7129	5.7811	5.9115	6.0347	6.1528		
700	v		0.1873	0.1963	0.2048	0.2131	0.2212	0.2291	0.2369	0.2522	0.2672	0.2821		
(13.81)	h		1474.1	1501.6	1527.7	1553.0	1577.9	1602.2	1626.4	1674.1	1721.9	1769.8		
	s		5.2247	5.3165	5.4017	5.4813	5.5569	5.6288	5.6983	5.8301	5.9549	6.0736		
800	v		0.1615	0.1696	0.1773	0.1848	0.1920	0.1990	0.2060	0.2196	0.2328	0.2459		
(17.86)	h		1465.9	1494.6	1521.8	1547.8	1573.4	1598.1	1622.7	1671.2	1719.4	1767.6		
	s		5.1380	5.2338	5.3217	5.4039	5.4813	5.5550	5.6256	5.7590	5.8849	6.0042		
900	v			0.1488	0.1559	0.1627	0.1694	0.1757	0.1819	0.1942	0.2061	0.2178	0.2294	
(21.54)	h			1487.6	1515.8	1542.7	1568.7	1594.0	1619.0	1668.1	1716.7	1765.2	1813.9	
	s			5.1582	5.2496	5.3341	5.4133	5.4885	5.5601	5.6956	5.8224	5.9428	6.0578	

TABLE A.9.2 Superheated Ammonia

Abs. Press. kPa (Sat. Temp.) °C		*v* (specific volume), m³/kg	*h* (enthalpy), kJ/kg	*s* (entropy), kJ/kg·K										
		Temperature, °C												
		10	20	30	40	50	60	70	80	100	120	140	160	180
1000	*v*			0.1321	0.1387	0.1450	0.1511	0.1570	0.1627	0.1739	0.1847	0.1953	0.2058	
(24.91)	*h*			1480.3	1509.6	1537.3	1564.0	1589.9	1615.3	1665.0	1714.2	1763.0	1811.9	
	s			5.0878	5.1828	5.2701	5.3511	5.4278	5.5006	5.6379	5.7659	5.8873	6.0033	
1200	*v*				0.1129	0.1185	0.1238	0.1289	0.1338	0.1434	0.1526	0.1616	0.1705	
(30.96)	*h*				1496.6	1526.3	1554.3	1581.5	1607.6	1658.8	1708.8	1758.4	1808.1	
	s				5.0617	5.1548	5.2404	5.3203	5.3957	5.5362	5.6673	5.7904	5.9078	
1400	*v*				0.0944	0.0994	0.1042	0.1088	0.1132	0.1216	0.1297	0.1376	0.1453	
(36.28)	*h*				1482.9	1514.8	1544.3	1572.5	1599.7	1652.4	1703.5	1753.8	1804.1	
	s				4.9522	5.0519	5.1422	5.2259	5.3041	5.4490	5.5821	5.7072	5.8259	
1600	*v*					0.0851	0.0895	0.0937	0.0977	0.1053	0.1125	0.1195	0.1264	0.1330
(41.05)	*h*					1502.6	1533.9	1563.6	1591.9	1645.9	1698.1	1749.4	1800.0	1850.7
	s					4.9574	5.0532	5.1405	5.2218	5.3710	5.5071	5.6342	5.7539	5.8684
1800	*v*					0.0739	0.0781	0.0820	0.0857	0.0926	0.0991	0.1055	0.1116	0.1177
(45.39)	*h*					1489.6	1523.1	1554.2	1583.8	1639.5	1692.6	1744.7	1796.0	1847.3
	s					4.8682	4.9702	5.0623	5.1470	5.3004	5.4394	5.5687	5.6900	5.8055
2000	*v*					0.0647	0.0688	0.0725	0.0760	0.0824	0.0885	0.0943	0.0999	0.1054
(49.38)	*h*					1476.1	1511.7	1544.4	1575.2	1632.8	1687.2	1740.1	1791.9	1843.6
	s					4.7818	4.8918	4.9891	5.0773	5.2357	5.3779	5.5092	5.6318	5.7485

TABLE A.10 Thermodynamic Properties of Freon-12 (Dichlorodifluoromethane)*

TABLE A.10.1 Saturated Freon-12

		Specific volume m³/kg			Enthalpy kJ/kg			Entropy kJ/kg·K		
Temp. °C	Abs. Press. MPa P	Sat. Liquid v_f	Evap. v_{fg}	Sat. Vapor v_g	Sat. Liquid h_f	Evap. h_{fg}	Sat. Vapor h_g	Sat. Liquid s_f	Evap. s_{fg}	Sat. Vapor s_g
−90	0.0028	0.000 608	4.414 937	4.415 545	−43.243	189.618	146.375	−0.2084	1.0352	0.8268
−85	0.0042	0.000 612	3.036 704	3.037 316	−38.968	187.608	148.640	−0.1854	0.9970	0.8116
−80	0.0062	0.000 617	2.137 728	2.138 345	−34.688	185.612	150.924	−0.1630	0.9609	0.7979
−75	0.0088	0.000 622	1.537 030	1.537 651	−30.401	183.625	153.224	−0.1411	0.9266	0.7855
−70	0.0123	0.000 627	1.126 654	1.127 280	−26.103	181.640	155.536	−0.1197	0.8940	0.7744
−65	0.0168	0.000 632	0.840 534	0.841 166	−21.793	179.651	157.857	−0.0987	0.8630	0.7643
−60	0.0226	0.000 637	0.637 274	0.637 910	−17.469	177.653	160.184	−0.0782	0.8334	0.7552
−55	0.0300	0.000 642	0.490 358	0.491 000	−13.129	175.641	162.512	−0.0581	0.8051	0.7470
−50	0.0391	0.000 648	0.382 457	0.383 105	−8.772	173.611	164.840	−0.0384	0.7779	0.7396
−45	0.0504	0.000 654	0.302 029	0.302 682	−4.396	171.558	167.163	−0.0190	0.7519	0.7329
−40	0.0642	0.000 659	0.241 251	0.241 910	0.000	169.479	169.479	0.0000	0.7269	0.7269
−35	0.0807	0.000 666	0.194 732	0.195 398	4.416	167.368	171.784	0.0187	0.7027	0.7214
−30	0.1004	0.000 672	0.158 703	0.159 375	8.854	165.222	174.076	0.0371	0.6795	0.7165
−25	0.1237	0.000 679	0.130 487	0.131 166	13.315	163.037	176.352	0.0552	0.6570	0.7121
−20	0.1509	0.000 685	0.108 162	0.108 847	17.800	160.810	178.610	0.0730	0.6352	0.7082

TABLE A.10.1 Continued

		Specific volume m^3/kg			*Enthalpy* kJ/kg			*Entropy* $kJ/kg{\cdot}K$		
Temp. °C	*Abs. Press. MPa* P	*Sat. Liquid* v_f	*Evap.* v_{fg}	*Sat. Vapor* v_g	*Sat. Liquid* h_f	*Evap.* h_{fg}	*Sat. Vapor* h_g	*Sat. Liquid* s_f	*Evap.* s_{fg}	*Sat. Vapor* s_g
−15	0.1826	0.000 693	0.090 326	0.091 018	22.312	158.534	180.846	0.0906	0.6141	0.7046
−10	0.2191	0.000 700	0.075 946	0.076 646	26.851	156.207	183.058	0.1079	0.5936	0.7014
−5	0.2610	0.000 708	0.064 255	0.064 963	31.420	153.823	185.243	0.1250	0.5736	0.6986
0	0.3086	0.000 716	0.054 673	0.055 389	36.022	151.376	187.397	0.1418	0.5542	0.6960
5	0.3626	0.000 724	0.046 761	0.047 485	40.659	148.859	189.518	0.1585	0.5351	0.6937

		Specific volume m^3/kg			*Enthalpy* kJ/kg			*Entropy* $kJ/kg{\cdot}K$		
Temp. °C	*Abs. Press. MPa* P	*Sat. Liquid* v_f	*Evap.* v_{fg}	*Sat. Vapor* v_g	*Sat. Liquid* h_f	*Evap.* h_{fg}	*Sat. Vapor* h_g	*Sat. Liquid* s_f	*Evap.* s_{fg}	*Sat. Vapor* s_g
10	0.4233	0.000 733	0.040 180	0.040 914	45.337	146.265	191.602	0.1750	0.5165	0.6916
15	0.4914	0.000 743	0.034 671	0.035 413	50.058	143.586	193.644	0.1914	0.4983	0.6897
20	0.5673	0.000 752	0.030 028	0.030 780	54.828	140.812	195.641	0.2076	0.4803	0.6879
25	0.6516	0.000 763	0.026 091	0.026 854	59.653	137.933	197.586	0.2237	0.4626	0.6863
30	0.7449	0.000 774	0.022 734	0.023 508	64.539	134.936	199.475	0.2397	0.4451	0.6848
35	0.8477	0.000 786	0.019 855	0.020 641	69.494	131.805	201.299	0.2557	0.4277	0.6834
40	0.9607	0.000 798	0.017 373	0.018 171	74.527	128.525	203.051	0.2716	0.4104	0.6820
45	1.0843	0.000 811	0.015 220	0.016 032	79.647	125.074	204.722	0.2875	0.3931	0.6806
50	1.2193	0.000 826	0.013 344	0.014 170	84.868	121.430	206.298	0.3034	0.3758	0.6792
55	1.3663	0.000 841	0.011 701	0.012 542	90.201	117.565	207.766	0.3194	0.3582	0.6777
60	1.5259	0.000 858	0.010 253	0.011 111	95.665	113.443	209.109	0.3355	0.3405	0.6760
65	1.6988	0.000 877	0.008 971	0.009 847	101.279	109.024	210.303	0.3518	0.3224	0.6742
70	1.8858	0.000 897	0.007 828	0.008 725	107.067	104.255	211.321	0.3683	0.3038	0.6721
75	2.0874	0.000 920	0.006 802	0.007 723	113.058	99.068	212.126	0.3851	0.2845	0.6697
80	2.3046	0.000 946	0.005 875	0.006 821	119.291	93.373	212.665	0.4023	0.2644	0.6667
85	2.5380	0.000 976	0.005 029	0.006 005	125.818	87.047	212.865	0.4201	0.2430	0.6631
90	2.7885	0.001 012	0.004 246	0.005 258	132.708	79.907	212.614	0.4385	0.2200	0.6585
95	3.0569	0.001 056	0.003 508	0.004 563	140.068	71.658	211.726	0.4579	0.1946	0.6526
100	3.3440	0.001 113	0.002 790	0.003 903	148.076	61.768	209.843	0.4788	0.1655	0.6444
105	3.6509	0.001 197	0.002 045	0.003 242	157.085	49.014	206.099	0.5023	0.1296	0.6319
110	3.9784	0.001 364	0.001 098	0.002 462	168.059	28.425	196.484	0.5322	0.0742	0.6064
112	4.1155	0.001 792	0.000 005	0.001 797	174.920	0.151	175.071	0.5651	0.0004	0.5655

TABLE A.10.2 Superheated Freon-12

Temp. °C	v m³/kg	h kJ/kg	s kJ/kg·K	v m³/kg	h kJ/kg	s kJ/kg·K	v vm³/kg	h kJ/kg	s kJ/kg·K
	0.05 MPa			0.10 MPa			0.15 MPa		
-20.0	0.341 857	181.042	0.7912	0.167 701	179.861	0.7401			
-10.0	0.356 227	186.757	0.8133	0.175 222	185.707	0.7628	0.114 716	184.619	0.7318
0.0	0.370 508	192.567	0.8350	0.182 647	191.628	0.7849	0.119 866	190.660	0.7543
10.0	0.384 716	198.471	0.8562	0.189 994	197.628	0.8064	0.124 932	196.762	0.7763
20.0	0.398 863	204.469	0.8770	0.197 277	203.707	0.8275	0.129 930	202.927	0.7977
30.0	0.412 959	210.557	0.8974	0.204 506	209.866	0.8482	0.134 873	209.160	0.8186
40.0	0.427 012	216.733	0.9175	0.211 691	216.104	0.8684	0.139 768	215.463	0.8390
50.0	0.441 030	222.997	0.9372	0.218 839	222.421	0.8883	0.144 625	221.835	0.8591
60.0	0.455 017	229.344	0.9565	0.225 955	228.815	0.9078	0.149 450	228.277	0.8787
70.0	0.468 978	235.774	0.9755	0.233 044	235.285	0.9269	0.154 247	234.789	0.8980
80.0	0.482 917	242.282	0.9942	0.240 111	241.829	0.9457	0.159 020	241.371	0.9169
90.0	0.496 838	248.868	1.0126	0.247 159	248.446	0.9642	0.163 774	248.020	0.9354
	0.20 MPa			0.25 MPa			0.30 MPa		
0.0	0.088 608	189.669	0.7320	0.069 752	188.644	0.7139	0.057 150	187.583	0.6984
10.0	0.092 550	195.878	0.7543	0.073 024	194.969	0.7366	0.059 984	194.034	0.7216
20.0	0.096 418	202.135	0.7760	0.076 218	201.322	0.7587	0.062 734	200.490	0.7440
30.0	0.100 228	208.446	0.7972	0.079 350	207.715	0.7801	0.065 418	206.969	0.7658
40.0	0.103 989	214.814	0.8178	0.082 431	214.153	0.8010	0.068 049	213.480	0.7869
50.0	0.107 710	221.243	0.8381	0.085 470	220.642	0.8214	0.070 635	220.030	0.8075
60.0	0.111 397	227.735	0.8578	0.088 474	227.185	0.8413	0.073 185	226.627	0.8276
70.0	0.115 055	234.291	0.8772	0.091 449	233.785	0.8608	0.075 705	233.273	0.8473
80.0	0.118 690	240.910	0.8962	0.094 398	240.443	0.8800	0.078 200	239.971	0.8665
90.0	0.122 304	247.593	0.9149	0.097 327	247.160	0.8987	0.080 673	246.723	0.8853
100.0	0.125 901	254.339	0.9332	0.100 238	253.936	0.9171	0.083 127	253.530	0.9038
110.0	0.129 483	261.147	0.9512	0.103 134	260.770	0.9352	0.085 566	260.391	0.9220
	0.40 MPa			0.50 MPa			0.60 MPa		
20.0	0.045 836	198.762	0.7199	0.035 646	196.935	0.6999			
30.0	0.047 971	205.428	0.7423	0.037 464	203.814	0.7230	0.030 422	202.116	0.7063
40.0	0.050 046	212.095	0.7639	0.039 214	210.656	0.7452	0.031 966	209.154	0.7291
50.0	0.052 072	218.779	0.7849	0.040 911	217.484	0.7667	0.033 450	216.141	0.7511
60.0	0.054 059	225.488	0.8054	0.042 565	224.315	0.7875	0.034 887	223.104	0.7723
70.0	0.056 014	232.230	0.8253	0.044 184	231.161	0.8077	0.036 285	230.062	0.7929
80.0	0.057 941	239.012	0.8448	0.045 774	238.031	0.8275	0.037 653	237.027	0.8129
90.0	0.059 846	245.837	0.8638	0.047 340	244.932	0.8467	0.038 995	244.009	0.8324
100.0	0.061 731	252.707	0.8825	0.048 886	251.869	0.8656	0.040 316	251.016	0.8514
110.0	0.063 600	259.624	0.9008	0.050 415	258.845	0.8840	0.041 619	258.053	0.8700
120.0	0.065 455	266.590	0.9187	0.051 929	265.862	0.9021	0.042 907	265.124	0.8882
130.0	0.067 298	273.605	0.9364	0.053 430	272.923	0.9198	0.044 181	272.231	0.9061

TABLE A.10.2 Continued

Temp. °C	v m³/kg	h kJ/kg	s kJ/kg·K	v m³/kg	h kJ/kg	s kJ/kg·K	v m³/kg	h kJ/kg	s kJ/kg·K
	0.70MPa			*0.80 MPa*			*0.90 MPa*		
40.0	0.026 761	207.580	0.7148	0.022 830	205.924	0.7016	0.019 744	204.170	0.6982
50.0	0.028 100	214.745	0.7373	0.024 068	213.290	0.7248	0.020 912	211.765	0.7131
60.0	0.029 387	221.854	0.7590	0.025 247	220.558	0.7469	0.022 012	219.212	0.7358
70.0	0.030 632	228.931	0.7799	0.026 380	227.766	0.7682	0.023 062	226.564	0.7575
80.0	0.031 843	235.997	0.8002	0.027 477	234.941	0.7888	0.024 072	233.856	0.7785
90.0	0.033 027	243.066	0.8199	0.028 545	242.101	0.8088	0.025 051	241.113	0.7987
100.0	0.034 189	250.146	0.8392	0.029 588	249.260	0.8283	0.026 005	248.355	0.8184
110.0	0.035 332	257.247	0.8579	0.030 612	256.428	0.8472	0.026 937	255.593	0.8376
120.0	0.036 458	264.374	0.8763	0.031 619	263.613	0.8657	0.027 851	262.839	0.8562
130.0	0.037 572	271.531	0.8943	0.032 612	270.820	0.8838	0.028 751	270.100	0.8745
140.0	0.038 673	278.720	0.9119	0.033 592	278.055	0.9016	0.029 639	277.381	0.8923
150.0	0.039 764	285.946	0.9292	0.034 563	285.320	0.9189	0.030 515	284.687	0.9098
	1.00 MPa			*1.20 MPa*			*1.40 MPa*		
50.0	0.018 366	210.162	0.7021	0.014 483	206.661	0.6812			
60.0	0.019 410	217.810	0.7254	0.015 463	214.805	0.7060	0.012 579	211.457	0.6876
70.0	0.020 397	225.319	0.7476	0.016 368	222.687	0.7293	0.013 448	219.822	0.7123
80.0	0.021 341	232.739	0.7689	0.017 221	230.398	0.7514	0.014 247	227.891	0.7355
90.0	0.022 251	240.101	0.7895	0.018 032	237.995	0.7727	0.014 997	235.766	0.7575
100.0	0.023 133	247.430	0.8094	0.018 812	245.518	0.7931	0.015 710	243.512	0.7785
110.0	0.023 993	254.743	0.8287	0.019 567	252.993	0.8129	0.016 393	251.170	0.7988
120.0	0.024 835	262.053	0.8475	0.020 301	260.441	0.8320	0.017 053	258.770	0.8183
130.0	0.025 661	269.369	0.8659	0.021 018	267.875	0.8507	0.017 695	266.334	0.8373
140.0	0.026 474	276.699	0.8839	0.021 721	275.307	0.8689	0.018 321	273.877	0.8558
150.0	0.027 275	284.047	0.9015	0.022 412	282.745	0.8867	0.018 934	281.411	0.8738
160.0	0.028 068	291.419	0.9187	0.023 093	290.195	0.9041	0.019 535	288.946	0.8914
	1.60 MPa			*1.80 MPa*			*2.00 MPa*		
70.0	0.011 208	216.650	0.6959	0.009 406	213.049	0.6794			
80.0	0.011 984	225.177	0.7204	0.010 187	222.198	0.7057	0.008 704	218.859	0.6909
90.0	0.012 698	233.390	0.7433	0.010 884	230.835	0.7298	0.009 406	228.056	0.7166
100.0	0.013 366	241.397	0.7651	0.011 526	239.155	0.7524	0.010 035	236.760	0.7402
110.0	0.014 000	249.264	0.7859	0.012 126	247.264	0.7739	0.010 615	245.154	0.7624
120.0	0.014 608	257.035	0.8059	0.012 697	255.228	0.7944	0.011 159	253.341	0.7835
130.0	0.015 195	264.742	0.8253	0.013 244	263.094	0.8141	0.011 676	261.384	0.8037
140.0	0.015 765	272.406	0.8440	0.013 772	270.891	0.8332	0.012 172	269.327	0.8232
150.0	0.016 320	280.044	0.8623	0.014 284	278.642	0.8518	0.012 651	277.201	0.8420
160.0	0.016 864	287.669	0.8801	0.014 784	286.364	0.8698	0.013 116	285.027	0.8603
170.0	0.017 398	295.290	0.8975	0.015 272	294.069	0.8874	0.013 570	292.822	0.8781
180.0	0.017 923	302.914	0.9145	0.015 752	301.767	0.9046	0.014 013	300.598	0.8955

TABLE A.10.2 Continued

Temp. °C	v m³/kg	h kJ/kg	s kJ/kg·K	v m³/kg	h kJ/kg	s kJ/kg·K	v m³/kg	h kJ/kg	s kJ/kg·K
	2.50 MPa			3.00 MPa			3.50 MPa		
90.0	0.006 595	219.562	0.6823						
100.0	0.007 264	229.852	0.7103	0.005 231	220.529	0.6770			
110.0	0.007 837	239.271	0.7352	0.005 886	232.068	0.7075	0.004 324	222.121	0.6750
120.0	0.008 351	248.192	0.7582	0.006 419	242.208	0.7336	0.004 959	234.875	0.7078
130.0	0.008 827	256.794	0.7798	0.006 887	251.632	0.7573	0.005 456	245.661	0.7349
140.0	0.009 273	265.180	0.8003	0.007 313	260.620	0.7793	0.005 884	255.524	0.7591
150.0	0.009 697	273.414	0.8200	0.007 709	269.319	0.8001	0.006 270	264.846	0.7814
160.0	0.010 104	281.540	0.8390	0.008 083	277.817	0.8200	0.006 626	273.817	0.8023
170.0	0.010 497	289.589	0.8574	0.008 439	286.171	0.8391	0.006 961	282.545	0.8222
180.0	0.010 879	297.583	0.8752	0.008 782	294.422	0.8575	0.007 279	291.100	0.8413
190.0	0.011 250	305.540	0.8926	0.009 114	302.597	0.8753	0.007 584	299.528	0.8597
200.0	0.011 614	313.472	0.9095	0.009 436	310.718	0.8927	0.007 878	307.864	0.8775
	4.00 MPa								
120.0	0.003 736	224.863	0.6771						
130.0	0.004 325	238.443	0.7111						
140.0	0.004 781	249.703	0.7386						
150.0	0.005 172	259.904	0.7630						
160.0	0.005 522	269.492	0.7854						
170.0	0.005 845	278.684	0.8063						
180.0	0.006 147	287.602	0.8262						
190.0	0.006 434	296.326	0.8453						
200.0	0.006 708	304.906	0.8636						
210.0	0.006 972	313.380	0.8813						
220.0	0.007 228	321.774	0.8985						
230.0	0.007 477	330.108	0.9152						

[a] *Copyright 1955 and 1956, E. I. du Pont de Nemours & Company, Inc. Reprinted by permission. Adapted from English units, and reprinted by permission from "Fundamentals of Classical Thermodynamics," by G. J. Van Wylen & R. E. Sonntag, New York, John Wiley & Sons, Inc., 1976.*

TABLE A.11 Thermodynamic Properties of Saturated Mercury[a]

Temp. °C	Absolute Pressure kPa	Specific Volume m^3/kg v_g	Enthalpy, kJ/kg Sat. Liquid h_f	Evap. h_{fg}	Sat. Vapor h_g	Entropy, kJ/kg·K Sat. Liquid s_f	Evap. s_{fg}	Sat. Vapor s_g
110	0.0625	250.0	15.25	297.17	312.42	0.04685	0.77558	0.82243
125	0.1281	126.5	17.31	296.89	314.20	0.05218	0.74568	0.79786
150	0.3773	45.76	20.73	296.42	317.15	0.06049	0.70048	0.76097
175	0.9824	18.61	24.14	295.95	320.09	0.06829	0.66039	0.72868
200	2.308	8.369	27.53	295.49	323.02	0.07561	0.62449	0.70010
225	4.968	4.098	30.92	295.01	325.93	0.08258	0.59220	0.67478
250	9.922	2.157	34.29	294.54	328.83	0.08914	0.56302	0.65216
275	18.580	1.205	37.65	294.07	331.72	0.09534	0.53648	0.63182
300	32.893	0.7158	40.99	293.60	334.59	0.10126	0.51225	0.61351
325	55.457	0.4428	44.32	293.13	337.45	0.10691	0.49004	0.59695
350	89.562	0.2856	47.64	292.66	340.30	0.11229	0.46964	0.58193
375	139.25	0.1910	50.95	292.19	343.14	0.11740	0.45077	0.56817
400	209.33	0.1320	54.25	291.72	345.97	0.12232	0.43335	0.55567
450	433.60	0.06855	60.80	290.78	351.58	0.13154	0.40208	0.53362
500	815.03	0.03905	67.31	289.84	357.15	0.14006	0.37489	0.51495
550	1415.32	0.02399	73.77	288.89	362.66	0.14797	0.35097	0.49894
600	2301.96	0.01567	80.17	287.95	368.12	0.15529	0.32979	0.48508
650	3544.66	0.01077	86.53	287.01	373.54	0.16211	0.31089	0.47300
700	5212.03	0.00773	92.83	286.07	378.90	0.16849	0.29395	0.46244
750	7366.76	0.00574	99.08	285.13	384.21	0.17447	0.27866	0.45313

[a] *Adapted from "Thermodynamic Properties of Mercury Vapor," by Lucian A. Sheldon, ASME Paper No. 49–A–30, 1949.*

TABLE A.12 Conversion Factors

(J = joule N = newton W = watt K = Kelvin)
(TC = thermochemical IST = International Steam Table)

	To Convert From	To	Multiply by
Density	lb_m/ft^3	kg/m^3	16.018463
Energy	Btu (IST)	J	1055.04
	Btu (TC)	J	1054.350264
	ft-lb_f	J	1.355818
	kWh	J	3.60×10^6 [a]
Force	lb_f	N	4.448222[a]
Length	ft	m	0.3048[a]
Mass	lb_m	kg	0.45359237[a]
	ton (metric)	kg	1000.0[a]
	ton (2000 lb_m)	kg	907.18474[a]
Miscellaneous	Btu (IST)/lb_m	J/kg	2325.965
	Btu (TC)/lb_m	J/kg	2324.4444
	Btu (IST)/lb_m R	J/kg·K	4186.737
	Btu (TC)/lb_m R	J/kg·K	4184.0[a]

TABLE A.12 Conversion Factors

(J = joule N = newton W = watt K = Kelvin)
(TC = thermochemical IST = International Steam Table)

	To Convert		
	From	*To*	*Multiply by*
	ft-lb_f/lb_m R	J/kg·K	5.380321
	Btu (TC)/h ft R	W/m·K	1.729577
	Btu (TC)/h ft^2 R	W/m^2·K	5.674466
	Btu (TC)/h ft^2	W/m^2	3.152481
Power	Btu (TC)/sec	W	1054.350264
	Btu (TC)/min	W	17.572504
	ft-lb_f/min	W	0.0225970
	horsepower (550 ft-lb_f/sec)	W	745.69987
	horsepower (electric)	W	746.0[a]
Pressure	atmosphere	N/m^2	101325[a]
	bar	N/m^2	1×10^5[a]
	cm of mercury (0 C)	N/m^2	1333.22
	cm of mercury (25 C)	N/m^2	1332.653
	in. of mercury (60 F)	N/m^2	3376.85
	in. of water (60 F)	N/m^2	248.84
	lb_f/ft^2	N/m^2	47.880258
	lb_f/in^2 (psi)	N/m^2	6894.7572
Viscosity	lb_m/sec ft	N·s/m^2	1.488164
	lb_f sec/ft^2	N·s/m^2	47.880258
	poise	N·s/m^2	0.100[a]

[a] *Exact relationship by definition.*

Source: NASA SP-7012

TABLE A.13 Constants

Constants	*Symbol*	*Value*	
Physical constants[a]			
Boltzmann constant	k	1.38054×10^{-23}	J/K
Electron-volt	eV	1.60210×10^{-19}	J/eV
Elementary charge	e	1.60210×10^{-19}	C
Faraday constant	$\mathcal{F}$	9.64870×10^7	C/kgmol
Ideal gas constant	$\overline{R}$	8.3143	kJ/kgmol·K
Stefan-Boltzmann constant	σ	5.6697×10^{-8}	$W/m^2 \cdot K^4$
Velocity of light in vacuum	c	2.997925×10^8	m/s
Miscellaneous constants			
Solar constant	$\dot{E}_o$	136.1	mW/cm^2
Standard acceleration owing to gravity	g_0	9.8066	m/s^2
Standard atmospheric pressure		101.325	kPa

[a] *National Bureau of Standards Technical News Bulletin, October 1963.*

Note: C (coulomb) = A·s (ampere-second).

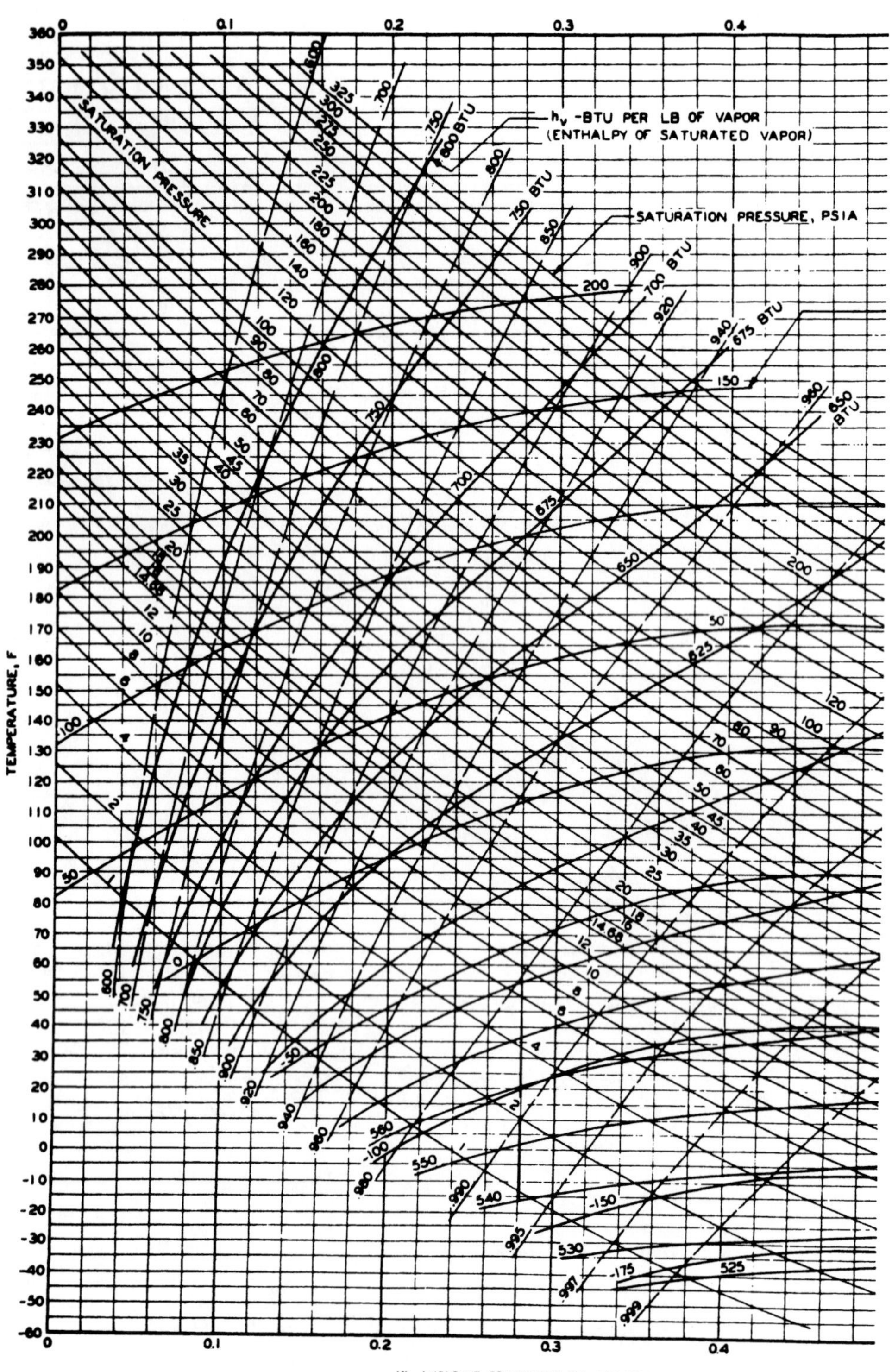
hv -BTU PER LB OF VAPOR
(ENTHALPY OF SATURATED VAPOR)
SATURATION PRESSURE, PSIA
SATURATION PRESSURE
TEMPERATURE, F
X'—WEIGHT FRACTION OF AMMONIA IN SATURATED
LIQUID—LB NH3 PER LB OF LIQUID

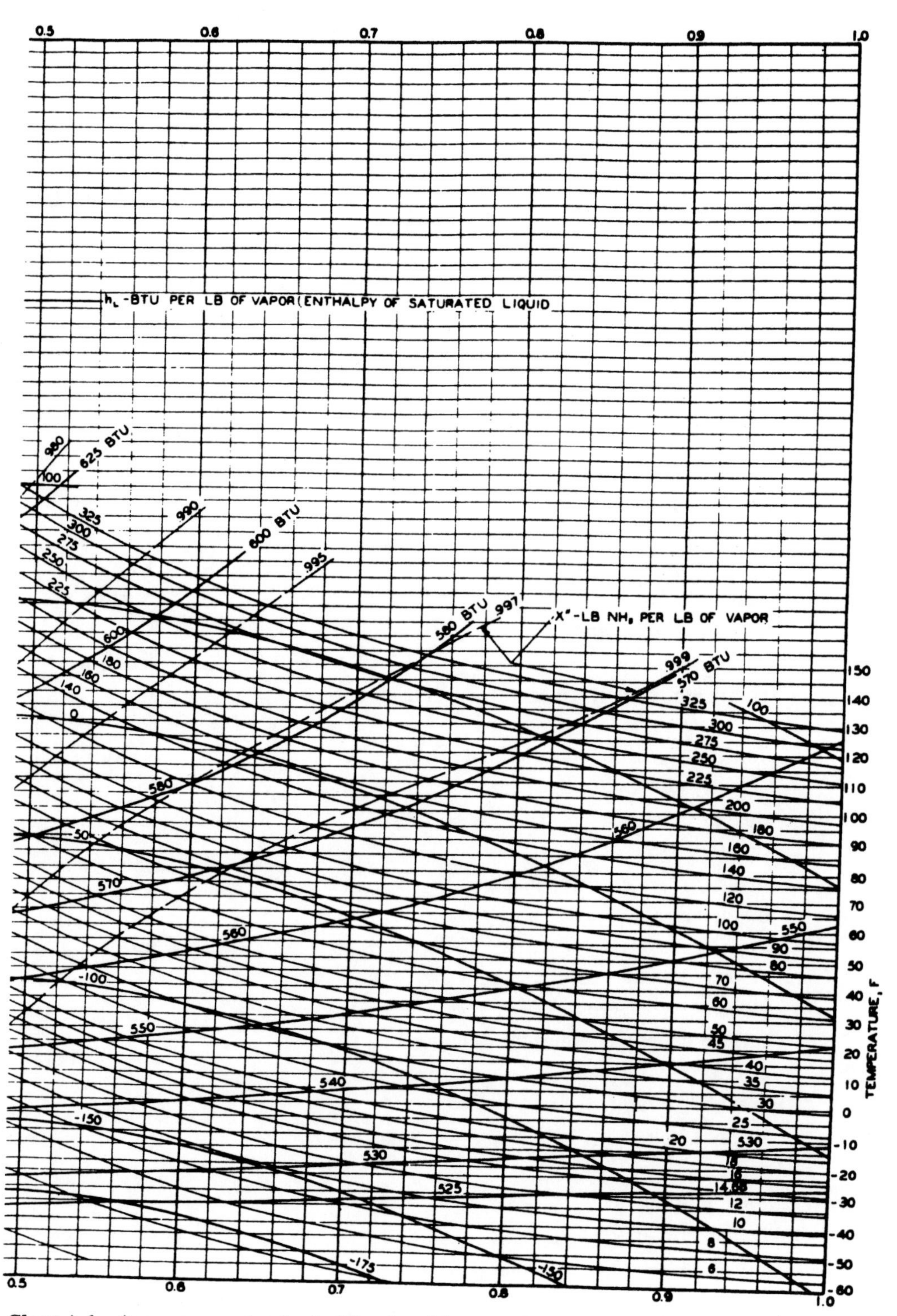

Chart A.1 Aqua–ammonia chart. (Reprinted by permission from *Refrigerating Engineering*, October, 1950.)

NOMENCLATURE

SYMBOL	QUANTITY	SI UNITS
a/f	air/fuel ratio	
a	area	m^2
atm	atmosphere	kPa
A	ampere	A
A	annual contribution to a sinking fund	
A	area	m^2
A	azimuth of the sun	rad
bfr	brake fuel rate	kg/kWh
bmep	brake mean effective pressure	kPa
BAER	furnace burning area energy release rate	$kJ/h \cdot m^2$
B	magnetic flux density	$V \cdot s/m^2$
Btu	British thermal unit	
c	specific heat	$kJ/kg \cdot K$
cop	coefficient of performance for a refrigeration system	
c_p	constant-pressure specific heat	$kJ/kg \cdot K$
C	velocity	m/s
C	degree Celsius	C
C_D	drag coefficient	
C_{fn}	nuclear reactor power factor	
d	diameter	m
D	diameter	m
D	velocity (relative)	m/s
e	elementary charge	$A \cdot s$
e	surface emissivity (thermal radiation)	
e^-	electron	
e_{av}	availability ratio	
eV	electron volt	J
E	emf	V
E	voltage gradient	V/m
E_{av}	available energy	kJ
$\dot{E}_d$	diffuse radiation from the sky	W/m^2

SYMBOL	QUANTITY	SI UNITS
E_{dh}	direct-horizontal insolation	kWh/m^2 per day
E_{dn}	direct-normal insolation	kWh/m^2 per day
E_f	energy per reaction (nuclear)	J
$\dot{E}_i$	insolation for general orientation of a surface	W/m^2
$\dot{E}_n$	insolation for a surface normal to the sun's rays	W/m^2
$\dot{E}_o$	solar constant	W/m^2
E_r	furnace energy release	kJ
E_{th}	total-horizontal insolation	kWh/m^2 per day
E_{un}	unavailable energy	kJ
f	friction factor (Eq. 8.7)	
f	mass fraction of ^{235}U in the uranium-oxide fuel	
f/a	fuel/air ratio	
f_{ab}	furnace energy absorption factor	
F	degrees Fahrenheit	
F	force	N
F_c	heat transfer temperature correction factor	
F_1	configuration factor, diffuse solar radiation to collecting surface	
F_2	configuration factor, solar radiation reflected from ground surface to collecting surface	
$F_{\Delta h}$	enthalpy rise hot channel factor (nuclear reactor)	
$\mathcal{F}$	Faraday constant	A·s/kgmol
g	acceleration owing to gravity (local)	m/s^2
g_0	acceleration owing to gravity (standard, sea-level)	m/s^2
G	Gibbs function	kJ
G	mass velocity	kg/s·m^2
h	altitude	m
h	film heat transfer coefficient	W/m^2·K
h	hour	h
h	specific enthalpy	kJ/kg
$\Delta\bar{h}^\circ$	enthalpy measured above 25 C	kJ/kgmol
$(\bar{h}_f^\circ)_{298}$	enthalpy of formation at 25 C	kJ/kgmol
$\bar{h}_{RP}^\circ$	enthalpy of combustion at 25 C	kJ/kgmol
H	enthalpy	kJ

SYMBOL	QUANTITY	SI UNITS
H	head (fluid)	m
HHV	higher heating value	kJ/kg
H°_{RP}	enthalpy of combustion at 25 C	kJ
ΔH°_R	enthalpy of reactants measured above 25 C	kJ
ΔH°_P	enthalpy of products measured above 25 C	kJ
i	compressor, or turbine, blade incidence angle	rad
i	interest rate	
imep	indicated mean effective pressure	kPa
I	conventional electric current	A
I	insolation	W/m^2
I_{sp}	rocket engine specific impulse	N per kg/s
J	electric current density	A/m^2
J	joule	J
J_{max}	saturation current (thermionic generator)	A/m^2
k	Boltzmann constant	J/K
k	specific heat ratio	
k	thermal conductivity	W/m·K
k_b	impulse turbine blade velocity coefficient	
k_{eff}	effective neutron multiplication factor	
kg	kilogram	kg
kW	kilowatt	kW
kW(e)	kilowatt (electrical)	kW
kW(t)	kilowatt (thermal)	kW
K	degree Kelvin	K
K	nuclear reactor power characteristic	
l	length	m
L	length	m
LHV	lower heating value	kJ/kg
m	mass	kg
m	meter	m
m	relative air mass	
$\overline{m}$	molecular weight	
$\overline{m}$	molal mass	kg/mol
mep	mean effective pressure	kPa
min	minute	min
M	Mach number	
MeV	million electron volts	J

SYMBOL	QUANTITY	SI UNITS
MON	motor octane number	
n	polytropic exponent	
n	revolutions per second	1/s
${}_0^1n$	neutron	
n/p	solar cell junction	
N	newton	N
N	number	
N	number of moles	
N_{ff}	density of fissionable fuel	$1/m^3$
N_s	pump specific speed	
${}_1^1p$	proton	
Pa	pascal	Pa
P_a	atmospheric pressure	kPa
P_b	barometric pressure	kPa
Pe	Peclet number	
pH	a number that indicates acidity or alkalinity	
Pr	Prandtl number	
PH	compressor preheat factor	
q	10^{15} Btu	
$\dot{q}$	heat transfer rate	kJ/s
$\dot{q}_{l,\text{av}}$	average heat generation rate per unit length of fuel rod for a nuclear reactor	kJ/s·m
$\dot{q}_t$	nuclear reactor thermal power	W
$\dot{q}_v$	specific energy release rate	W/m^3
$\dot{q}_{v,c}$	specific energy release rate at the center point of a fuel rod	W/m^3
$\dot{q}_{v,\text{co}}$	specific energy release rate at the geometrical center of a nuclear reactor core	W/m^3
Q	quantity of heat	kJ
Q	10^{18} Btu	
r	radius	m
r	compression ratio	
r	expansion ratio	
R	gas constant	kJ/kg·K
R	electrical resistance	V/A
R	thermal resistance	m^2·K/W
$\overline{R}$	ideal gas constant	kJ/kgmol·K
RCR	Rankine cycle ratio or machine efficiency	

SYMBOL	QUANTITY	SI UNITS
Re	Reynolds number	
RH	turbine reheat factor	
RON	research octane number	
s	second	s
s	specific entropy	kJ/kg·K
S	Seebeck coefficient	V/K
S	entropy	kJ/K
S	sinking fund	
T	temperature	K
T	torque	N·m
TEL	tetraethyl lead	
u	specific internal energy	kJ/kg
U	blade linear velocity	m/s
U	internal energy	kJ
U	overall heat transfer coefficient	W/m^2·K
v	specific volume	m^3/kg
V	volume	m^3
V	voltage	V
W	quantity of work	kJ
W_{rev}	reversible work	kJ
$\dot{W}$	power	W
z	compressor power input factor	
z	electrochemical valence	
z	elevation	m
z_c	nozzle velocity coefficient	
Z	figure of merit for thermoelectric material	1/K
α	altitude of the sun	rad
α	fluid or blade angle	rad
${}^4_2\alpha$	alpha particle	
β	beta particle	
β	fluid or blade angle	rad
β	inclination angle measured from the horizontal plane	rad
β_n	inclination angle for a surface oriented normal to the sun's rays	rad
γ	gamma rays	

SYMBOL	QUANTITY	SI UNITS
γ	surface orientation, measured on the horizon from the south meridian	rad
δ	solar declination	rad
ε	conversion effectiveness	
η	efficiency or effectiveness	
η_{ab}	effectiveness of energy absorption	
η_b	turbine blade efficiency	
η_e	electrical efficiency	
η_g	electric generator efficiency	
η_m	mechanical efficiency	
η_o	gas turbine engine overall efficiency	
η_p	aircraft gas turbine engine propulsive efficiency	
η_p	turbine or compressor polytropic efficiency	
η_r	machine or relative efficiency	
η_{re}	regenerator effectiveness	
η_s	propeller efficiency	
η_{st}	stage efficiency (compressor or turbine)	
η_t	thermal efficiency	
η_v	volumetric efficiency	
η_B	combustion efficiency	
η_C	compressor internal efficiency	
η_D	diffuser efficiency	
η_F	fan static efficiency	
η_I	effectiveness of gas turbine engine inlet	
η_N	nozzle efficiency	
η_P	pump efficiency	
η_R	effectiveness of aircraft gas turbine engine inlet	
η_T	turbine internal efficiency	
θ	fluid deflection angle	rad
θ^*	compressor blade camber angle	rad
θ_i	incidence angle	rad
$\theta_{i,t}$	incidence angle for a surface in a south-facing position	rad
θ_z	zenith distance, or angle	rad
λ	nozzle divergence correction factor	
μ	dynamic viscosity	$\mathrm{N \cdot s/m^2}$

SYMBOL	QUANTITY	SI UNITS
ν	neutrino	
ρ	density	kg/m^3
ρ	electrical resistivity	$V \cdot m/I$
ρ	ground-surface diffuse reflectance for solar radiation	
ρ	neutron reactivity	
ρ	turbine speed ratio	
σ	electrical conductivity	$I/V \cdot m$
σ	microscopic nuclear cross section	m^2
σ	slip factor (centrifugal compressor)	
σ	Stefan-Boltzmann constant	$W/m^2 \cdot K^4$
σ_f	effective fission microscopic cross section	m^2
Σ_f	effective fission macroscopic cross section	$1/m$
τ	time	s
τ^*	neutron lifetime	s
τ_a	solar transmission coefficient for unit air mass	
ϕ	availability	kJ/kg
ϕ	latitude	rad
ϕ	neutron flux	$1/s \cdot m^2$
ϕ	relative humidity	
ϕ_a	anode work function (thermionic generator)	eV
ϕ_c	cathode work function (thermionic generator)	eV
ϕ_E	reaction turbine expansion energy coefficient	
ϕ_i	reaction turbine incidence coefficient	
ϕ_v	reaction turbine kinetic energy coefficient	
Φ	equivalence ratio, defined as the actual fuel/air ratio divided by the stoichiometric fuel/air ratio, also corresponds to the reciprocal of the percent of theoretical air	
ψ	degree of reaction (turbomachines)	
ω	angular velocity	rad/s
ω	hour angle of the sun	rad
ω	load ratio (Diesel cycle)	
ω	specific humidity	

T